Bioactive Sphingolipids in Cancer Biology and Therapy

Yusuf A. Hannun • Chiara Luberto
Cungui Mao • Lina Marie Obeid
Editors

Bioactive Sphingolipids in Cancer Biology and Therapy

Springer

Editors
Yusuf A. Hannun
Stony Brook Cancer Center
Stony Brook University
Stony Brook, NY, USA

Chiara Luberto
Department of Physiology & Biophysics
Stony Brook University
Stony Brook, NY, USA

Cungui Mao
Department of Medicine
Stony Brook University
Stony Brook, NY, USA

Lina Marie Obeid
Department of Medicine
Stony Brook University
Stony Brook, NY, USA

ISBN 978-3-319-20749-0 ISBN 978-3-319-20750-6 (eBook)
DOI 10.1007/978-3-319-20750-6

Library of Congress Control Number: 2015947501

Springer Cham Heidelberg New York Dordrecht London

Printed on acid-free paper

Springer International Publishing AG Switzerland is part of Springer Science+Business Media (www.springer.com)

We dedicate this volume to the memory of our esteemed colleague Bob Bittman whose commitment and scientific advancement of sphingolipid research is undisputed and can also be appreciated first hand by his own contribution to this book. His research will be a legacy for years to come.

Yusuf, Lina, Chiara, and Cungui

Preface

This volume presents and discusses the evidence regarding the roles and functions of various sphingolipids (SLs), including ceramides and sphingoid bases and their phosphorylated derivatives, and more complex SLs in the development, progression, and treatment of various cancers. (Gangliosides, a class of complex SLs comprising hundreds of different molecules, and of undisputed relevance in cancer biology, are not discussed in depth in this volume.)

The modern revolution of molecular biology and biochemistry has brought us the era of genomics and proteomics, and the rediscovered importance of cellular metabolism and energetics in various diseases, including cancer. In turn, these have led to the integrated study of metabolic pathways and the novel metabolomic approach. More recently, the understanding of the bioactive functions of lipids in pathophysiology has propelled the study of this class of molecules, historically somewhat unrecognized, into the limelight. The appreciation of the metabolic interconnections among different lipid pathways has led us to the current era of lipidomics. Indeed, lipidomics has become the new frontier in the effort to uncover the global imprint of each specific patho/physiological condition. Among the various classes of lipids, SLs represent the youngest, but quite prolific, research domain.

The delay in the recognition of the functional significance of lipids, in general, and of sphingolipids, in particular, is largely due to the conceptual and technical challenges that face researchers involved in SLs, challenges brought by the intrinsic nature of these molecules, and of their metabolic pathways.

At the conceptual level, first, it is clear that bioactive SLs form a network of interconvertible metabolites. Thus, unlike "dedicated" signaling pathways such as the paradigmatic cAMP pathway, bioactive SLs are metabolically interconnected so that one lipid (e.g., ceramide) can be further metabolized to another lipid (e.g., S1P) which often exerts opposing effects. Second, pathways of lipid metabolism are compartmentally, and often topologically, constrained. Therefore, understanding how they function requires a careful determination of where they function. Third, the belated molecular identification of enzymes of SL metabolism has disclosed the existence of a multiplicity of pathways involving these lipids (e.g., ceramide) that may show distinct regulation and distinct functions. Fourth, the chemico-physical properties of lipids impose several barriers to their study. They are poorly soluble, which hinders in vitro and cell studies. This also coerces their existence to specific subcellular membranes. Moreover, accurately analyzing their levels requires sophisticated technologies, such as tandem mass spectrometry. Even their synthetic chemistry tends to be an order of magnitude more complicated than other "soluble" molecules and, as such, requires specific expertise. This is important not only for creating standards, but also for generating cellular probes and lead enzyme inhibitors.

Research in the SL field has overcome these challenges by adopting ever-evolving experimental approaches and analytical methods that have proven invaluable to uncover a critical role for many of these lipids in physiology and pathophysiology. Indeed, different SLs have been linked to the development of several diseases such as diabetes, neurodegenerative diseases, emphysema, and infections, to mention a few. Notably, different bioactive SLs have also been linked to initiation and/or progression of various cancers.

Indeed, a large body of work has suggested that certain bioactive SLs, such as ceramide and sphingosine, constitute *tumor-suppressor lipids*, whereas others, like S1P, function as *tumor-promoting lipids*. This general hypothesis is buttressed by increasing evidence that enzymes that attenuate ceramide and sphingosine or increase S1P are increased in cancers, whereas those that increase ceramide or attenuate S1P are decreased in cancers. Many emerging studies show that interfering with these pathways can exert profound effects on cancer cells (such as induction of apoptosis, differentiation, tumor senescence, and regulation of angiogenesis).

In addition to an introductory overview that illustrates the key concepts of sphingolipid metabolism and SL-mediated signaling, and that serves as reference for all other chapters, the book is divided into three main parts. The first part addresses how the SLs mentioned above are involved in the development of specific cancers, each of which is treated in a stand-alone chapter. Each chapter in this part covers basic as well as clinical aspects of SLs or SL-metabolizing enzymes in the context of each different cancer. The discussion that centers on the basic experimental evidence systematically addresses the specific involvement of the aforementioned SLs in key cellular functions known to be altered in the complex processes that lead to cancer (e.g., proliferation, the various forms of cell death, differentiation, senescence, adhesion/migration, and multidrug resistance).

The second part focuses on the most innovative techniques/approaches for quantitative analysis and imaging of SLs and SL-metabolizing enzymes, including the use of novel chemical probes.

The third and final part discusses the potential diagnostic and/or prognostic value of SL and/or SL-metabolizing enzymes and the potential benefit of targeting SL metabolism to develop novel cancer therapeutics for de novo treatment or to overcome resistance to already-established regiments.

Stony Brook, NY, USA

Yusuf A. Hannun
Chiara Luberto
Cungui Mao
Lina Marie Obeid

Contents

Basics of Sphingolipid Metabolism and Signalling

Céline Colacios, Frédérique Saboursdy, Nathalie Andrieu-Abadie, Bruno Ségui, and Thierry Levade

Abstract The term sphingolipid was coined by J.L.W. Thudichum before the turn of the nineteenth century, referring to the enigmatic (related to the Sphinx myth) nature of this class of molecules. One hundred thirty years later, the enigma is not yet completely solved. Nevertheless, much progress has been made, shedding light on the numerous roles these lipids play in eukaryotes. How sphingolipids are synthesized, transformed and degraded in mammalian cells, and how some of them transduce signals and regulate biological functions is reviewed in this chapter. Special attention is given to those sphingolipid species which regulate key aspects of the development of malignancies in humans, and therefore represent potential targets for therapy.

Keywords Ceramide • Sphingosine 1-phosphate • Sphingomyelin • Receptor • Oncometabolite • Membrane • Glycolipid

Abbreviations

Cer	Ceramide
Cer1P	Ceramide 1-phosphate
DAG	Diacylglycerol

C. Colacios • N. Andrieu-Abadie • B. Ségui
Institut National de la Santé et de la Recherche Médicale (INSERM) UMR1037, Centre de Recherches en Cancérologie de Toulouse (CRCT), Oncopole de Toulouse, 2 Avenue Hubert Curien, CS 53717, 31037 Toulouse, Cedex 1, France

Equipe Labellisée Ligue Nationale Contre le Cancer 2013, Centre de Recherches en Cancérologie de Toulouse (CRCT), Université Paul Sabatier Toulouse III, Toulouse, France
e-mail: thierry.levade@inserm.fr

F. Sabourdy • T. Levade, M.D., Ph.D. (✉)
Institut National de la Santé et de la Recherche Médicale (INSERM) UMR1037, Centre de Recherches en Cancérologie de Toulouse (CRCT), Oncopole de Toulouse, 2 Avenue Hubert Curien, CS 53717, 31037 Toulouse, Cedex 1, France

Equipe Labellisée Ligue Nationale Contre le Cancer 2013, Centre de Recherches en Cancérologie de Toulouse (CRCT), Université Paul Sabatier Toulouse III, Toulouse, France

Laboratoire de Biochimie Métabolique, Institut Fédératif de Biologie, CHU Purpan, Toulouse, France

Y.A. Hannun et al. (eds.), *Bioactive Sphingolipids in Cancer Biology and Therapy*, DOI 10.1007/978-3-319-20750-6_1

ER	Endoplasmic reticulum
GalCer	Galactosylceramide
GlcCer	Glucosylceramide
GSL	Glycosphingolipid
S1P	Sphingosine 1-phosphate
SL	Sphingolipid
SM	Sphingomyelin
SPC	Sphingosylphosphocholine

1 Introduction: Sphingolipids Are Bioactive Metabolites

Sphingolipids (SLs) are found in all eukaryotes and represent a class of lipids with considerable structural diversity. In mammals, there are likely tens of thousands of SL molecular species [1]. Because of their amphiphilic nature, most of them are membrane components; some are present in biological fluids, being constituents of circulating lipoproteins or transported by other proteins.

The smallest molecule common to all SLs is an aliphatic amino-alcohol, also known as long chain base, the most frequent one being 4-sphingenine (also termed sphingosine). Its condensation to a fatty acid, through an amide bond rather unusual in the lipid world, forms the core of most SLs, ceramide (Cer). This structural analogue of diacylglycerol (DAG) constitutes the hydrophobic anchor of all complex SLs, determining the membrane location of both, glycosphingolipids (GSLs) and sphingomyelin (SM), the two major classes of complex SLs. The analogy between DAG and Cer extends to their common ability to convey signals; these signals, however, often are of diverging nature (see Sect. 3). With regard to the signalling properties of SLs, the best documented roles are undoubtedly attributed to sphingosine 1-phosphate (S1P), the phosphorylated derivative of sphingosine, which is abundant in plasma while one of the least represented SLs within the cells. Nevertheless, Cer and S1P are not the only SLs that behave as SL signalling intermediates or second messengers. Glycosylated or *N*-methylated forms of sphingoid bases, sphingosylphosphocholine (SPC), phosphorylated Cer, as well as some GSLs can transduce signals and/or modulate biological functions.

The last 2–3 decades have witnessed an enormous improvement in our knowledge of the structural diversity, metabolism, cell biology and pathology of SLs. Such a progress has been facilitated thanks to (1) the development of new technologies for analyzing the sphingolipidome (see Part II of this book), (2) the cloning of most enzymes of SL metabolism, (3) the synthesis and use of SL analogues for studying their metabolism, location, targets (or interacting partners) and mechanism of action, (4) the generation of animal models harboring alterations of SL metabolism [2, 3], and (5) the identification of new human genetic disorders characterized by disturbed homeostasis of SLs [4]. Moreover, the involvement of SLs in the pathophysiology of numerous conditions, including multiple types of solid cancers and hematological malignancies, has been explored. The intent of this overview is not to cover in a comprehensive manner the many effects and modes of actions of SLs, but to highlight some important aspects of SL metabolism and signalling in mammalian organisms. The reader is also referred to recent excellent reviews [5–9]. In the field of cancer biology and treatment, SLs are currently viewed as multifaceted mediators and, as a consequence, potentially new therapeutic targets. The biological and physiological effects modulated by SLs include cancer (and cancer stem) cell death, survival, differentiation, cell cycle arrest, cell motility, autophagy, epithelial–mesenchymal transition, but also effects on the tumor microenvironment through angiogenesis, recruitment of inflammatory and immune cells, or phenotypic changes of adjacent fibroblasts. This astounding variety is SL-dependent, sometimes SL species-dependent (underscoring the importance of the metabolic source of SLs) [10], and cellular context-dependent. Figure 1 illustrates some of these diverse effects.

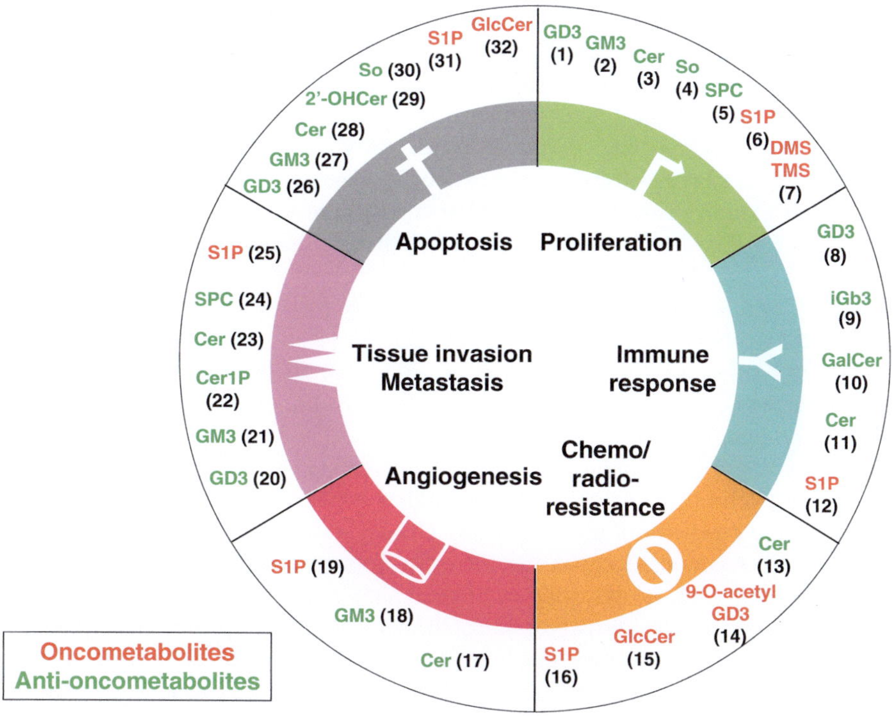

Fig. 1 Sphingolipid-mediated biological effects in cancer. The figure summarizes how SLs affect hallmarks of cancer (adapted from [95]). Note that not all bioactive SLs are indicated in this scheme. SLs written in red promote tumor development whereas those written in green have anti-tumor effects. Numbers in brackets indicate some key references. *Cer* ceramide, *Cer1P* ceramide 1-phosphate, 2′-OHCer, 2′-hydroxyceramide, *DMS N,N*-dimethyl-sphingosine, *GalCer* galactosylceramide, *GlcCer* glucosylceramide, *S1P* sphingosine 1-phosphate, *So* sphingosine, *SPC* sphingosylphosphocholine, *TMS N,N,N*-trimethyl-sphingosine. References cited: (1) [96], (2) [97], (3) [8], (4) [98], (5) [99], (6) [54], (7) [100], (8) [101], (9) [102], (10) [103], (11) [104], (12) [56], (13) [105], (14) [106], (15) [107], (16) [108], (17) [109], (18) [110], (19) [111], (20) [112], (21) [113], (22) [114], (23) [115], (24) [116], (25) [117], (26) [118], (27) [119], (28) [120], (29) [121], (30) [122], (31) [52], (32) [123]

2 How Are Sphingolipids Formed, Transformed and Degraded?

The astonishing variety of biological effects of SLs is not only related to their structural diversity but also their tissue and subcellular distribution [thanks to the newly developed imaging mass spectrometry methods, subtle differences in tissue distribution of particular molecular species of SLs begin to be appreciated; see Part II of this book]. Understanding which, where (in the cell or in the body) and how a given SL can exert a biological effect requires prior knowledge of SL metabolism and transport. Although there are still a few gaps, the picture of SL biosynthesis and catabolism is almost complete. Figures 2 and 3 depict the pathways of SL synthesis and turnover in humans (for details, the reader is referred to www.sphingomap.org).

SL biosynthesis starts in the endoplasmic reticulum (ER) with the condensation of L-serine and a fatty acyl-CoA, usually palmitoyl-CoA, by serine-palmitoyl transferase (note that instead of L-serine, L-alanine or L-glycine can be used under some pathological conditions, leading to the formation of 1-deoxysphingolipids [11, 12]). The resultant 3-keto-sphinganine is then reduced to sphinganine (also named dihydrosphingosine) before *N*-acylation by Cer synthases. In the *de novo* pathway, these enzymes form dihydroceramides. They are encoded by six different genes that exhibit distinct tissue expression and produce enzymes with different acyl-CoA (or 2-hydroxyacyl-CoA) chain length specificities, leading to numerous non-hydroxy and 2-hydroxy-dihydroceramide species with possibly distinct properties [5]. The *de novo* synthesis of Cer is completed by a desaturase that introduces a *trans*-$\Delta 4$ double bond in the sphingoid backbone.

Newly synthesized Cer can be transformed to five different lipids: (1) ceramide 1-phosphate (Cer1P), by Cer kinase; (2) sphingomyelin, one of the most abundant SL, by sphingomyelin synthases; (3) ceramide phosphoethanolamine, by sphingomyelin synthase 2 or sphingomyelin synthase-related protein; (4) galactosylceramide, a myelin lipid, by galactosylceramide synthase; and (5) glucosylceramide (GlcCer), the precursor of most GSLs, by glucosylceramide synthase. As shown in Fig. 2, these reactions occur in the ER or in discrete compartments of the Golgi apparatus, either on the cytosolic or luminal side. Selection of one of these pathways may be dictated by the expression of the corresponding enzymes. Nevertheless, it implies some sophisticated regulation of Cer channelling to these distinct locations in order to serve specific metabolic needs. Although this issue is imperfectly solved, recent progress has been made regarding SL transport. Cer can be transported from the ER to the Golgi either by a vesicular process [13] or by a CERT-mediated non-vesicular mechanism [14]. Once synthesized, GlcCer might be retrogradely transported to the ER by FAPP2 [15], and Cer1P carried from the trans-Golgi to other membrane destinations by CPTP [16]. Besides Cer-metabolizing enzymes, these transporters represent additional targets to manipulate SL metabolism and biological effects, as illustrated by the role of CERT in modulating the sensitivity of cancer cells to chemotherapeutic drugs [17].

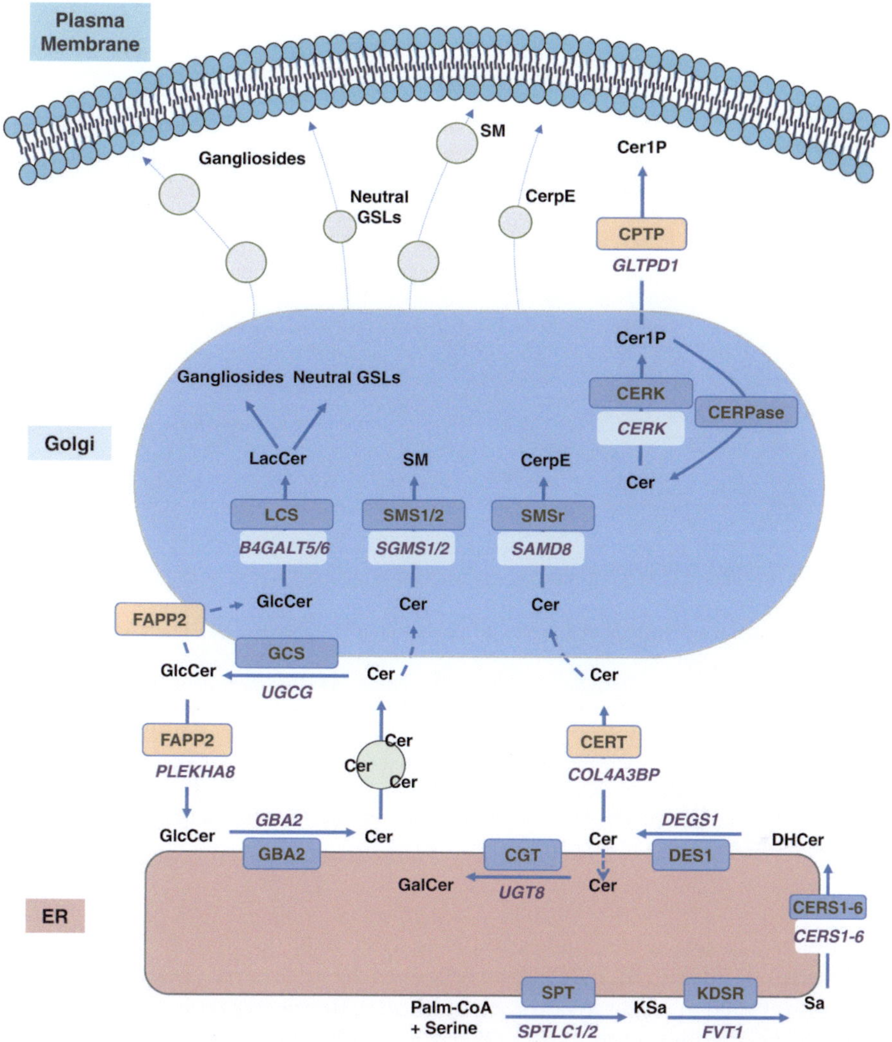

Fig. 2 Subcellular compartmentalization of sphingolipid biosynthesis. For each enzymatic reaction or transport event, the names of the corresponding protein (in *boxes*) and gene are indicated. *Cer* ceramide, *Cer1P* ceramide 1-phosphate, *CERK* ceramide kinase, *CERPase* ceramide 1-phosphate phosphatase, *CerpE* ceramide phosphoethanolamine, *CERS* ceramide synthase, *CERT* ceramide transport protein, *CGT* UDP-galactose: ceramide galactosyltransferase, *CPTP* ceramide 1-phosphate transfer protein, *DES* dihydroceramide desaturase, *DHCer* dihydroceramide, *FAPP* four-phosphate-adaptor protein, *GalCer* galactosylceramide, *GBA* beta-glucosidase, *GCS* glucosylceramide synthase, *GlcCer* glucosylceramide, *GSLs* glycosphingolipids, *KSa* 3-ketosphinganine, *KDSR* ketosphinganine reductase, *LacCer* lactosylceramide, *LCS* lactosylceramide synthase, *Palm-CoA* palmitoyl-CoA, *Sa* sphinganine, *SM* sphingomyelin, *SMS* sphingomyelin synthase, *SMSr* sphingomyelin synthase-related protein, *SPT* serine palmitoyltransferase

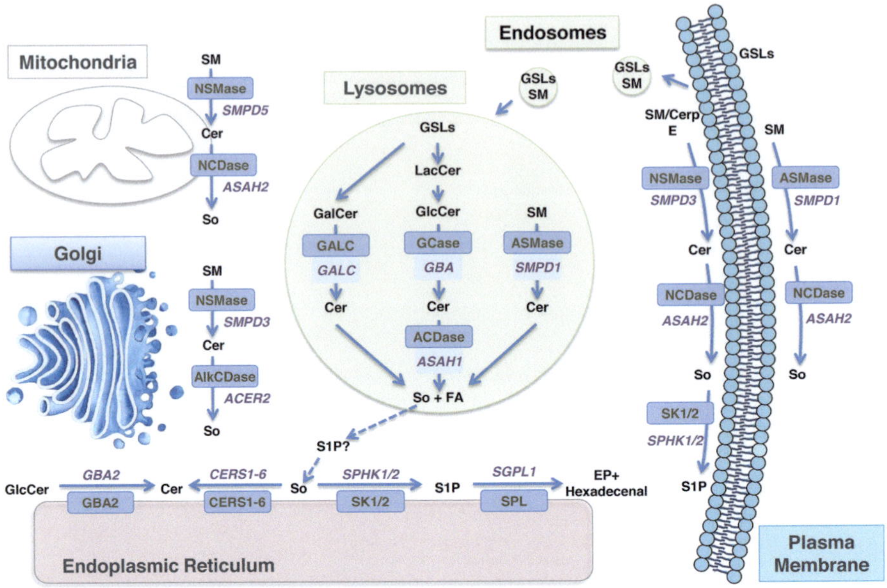

Fig. 3 Subcellular compartmentalization of sphingolipid catabolism and interconversion. For each enzymatic reaction, the names of the corresponding protein (in *boxes*) and gene are indicated. *ACDase* acid ceramidase, *AlkCDase* alkaline ceramidase, *ASMase* acid sphingomyelinase, *Cer* ceramide, *CerpE* ceramide phosphoethanolamine, *CERS* ceramide synthase, *EP* ethanolamine 1-phosphate, *FA* fatty acid, *GALC* galactosylceramidase, *GalCer* galactosylceramide, *GBA* beta-glucosidase, *GCase* glucosylceramidase, *GlcCer* glucosylceramide, *GSLs* glycosphingolipids, *LacCer* lactosylceramide, *NCDase* neutral ceramidase, *NSMase* neutral sphingomyelinase, *SM* sphingomyelin, *So* sphingosine, *S1P* sphingosine 1-phosphate, *SK* sphingosine kinase, *SPL* sphingosine 1-phosphate lyase

Most of the GSLs are synthesized from GlcCer at the luminal face of Golgi vesicles (as this part of SL metabolism is beyond the topic of the present chapter, the reader is referred to recent reviews, e.g., [18, 19]). Because of their luminal orientation and vesicular transport to the cell surface, GSLs and SM become constituents of the extracellular leaflet of the plasma membrane, and concentrate, along with cholesterol, within the so-called microdomains. A fraction of these complex SLs is incorporated in the circulating lipoproteins.

Constitutive turnover of plasma membrane elements, endocytosis of extracellular lipoproteins as well as (macro)autophagy result in the entry of SLs into the endosomal/lysosomal compartment where the bulk of SL catabolism takes place (Fig. 3). Stepwise degradation of GSLs occurs through a unique sequence of reactions catalyzed by exoglycosidases that ends by the release of Cer [20]. Formation of Cer in endosomes/lysosomes also results from the action of acid sphingomyelinase. Eventually, Cer is cleaved by acid ceramidase, which releases a fatty acid and sphingosine. How these end-products exit the lysosome is still unclear.

Whether protein-mediated or not, the efflux out of the acidic compartments of the sphingoid base initiates the SL 'salvage' pathway that allows re-use of the sphingoid moiety for SL synthesis. Unless sphingosine gets access to the ER because of close vicinity with lysosomal membranes, a prerequisite for this recycling pathway might be the phosphorylation of sphingosine by sphingosine kinase(s). The transient formation of S1P, a much less hydrophobic molecule than sphingosine, would represent a way to transport the sphingoid base to the ER, where S1P is dephosphorylated prior to *N*-acylation by Cer synthases. An alternative destination for S1P is its irreversible degradation by a single enzyme, S1P lyase, which produces phosphoethanolamine and a fatty aldehyde, two molecules that connect SL and glycerophospholipid metabolisms [21]. What directs S1P either to be recycled (after dephosphorylation) or broken down is still unknown; clarifying this issue is, however, critical for understanding the bioactive properties of S1P and manipulating its levels for therapeutic purposes.

Of note, turnover of GSLs and SM is not restricted to the acidic compartments, and it can occur at the plasma membrane level, both—at least for SM—on the extracellular side and the cytosolic side [22]. This location has long been envisioned as the starting point for the generation, or modulation, of bioactive signals. As a matter of fact, cleavage of SM and Cer, followed by phosphorylation of sphingosine can occur at the level of the plasma membrane. These reactions can be catalyzed by plasma membrane-located enzymes but also by secreted or translocated enzymes such as acid sphingomyelinase [23]. In addition, secreted hydrolases or ectoenzymes ensure SL degradation in the intestinal lumen to digest dietary lipids [24], providing not only nutrients but also potential bioactive lipids to prevent colon carcinogenesis. Moreover, although information remains scarce, GSL, SM and Cer turnover likely exists at the level of mitochondria, the nucleus, Golgi and ER membranes that are closely associated with mitochondria [25–27]. For instance, neutral and alkaline ceramidases localize to mitochondria, ER and Golgi [28].

Finally, the pathway that leads to the generation of some minor SLs, including sphingosylphosphocholine (SPC), psychosine or glucosylsphingosine, remains to be unambiguously identified. These lysosphingolipids are normally present in trace amounts but they mediate biological effects and underlie the molecular pathogenesis of some inherited disorders [29].

In summary, SL metabolism starts from structurally simple molecules, i.e., an amino acid and an acyl-CoA, produces thousands of distinct lipids, and ends by the release of small metabolites that are recycled into glycerophospholipids. Complex SLs are abundant membrane constituents that act as reservoirs for the production of simple, bioactive lipids. Transformations occur in multiple subcellular compartments, serving different functions and possibly also spatially restricting the biological effects of these lipids. Such a sophisticated metabolism is undoubtedly regulated to comply with the cell's demand and to control the levels of highly bioactive metabolites. However, how this metabolism is regulated and how the many pathways aforementioned are coordinated is still poorly understood [30]. These are key issues to appreciate the importance of the dysregulated expression of SL-metabolizing enzymes seen in tumor cells [8].

3 How Do Sphingolipids Signal and Mediate Biological Effects?

SLs are no longer considered just as inert membrane components. Some of them behave as intracellular second messengers to transduce extracellular signals; some elicit cellular outcomes through binding to cell surface receptors; some others alter the cell's responses by changing the physicochemical properties of membranes. Here, we will not describe the numerous physiological effects that can be modulated by SLs (see Fig. 1) but will emphasize their mode of action at the cellular level through some selected examples that are particularly relevant to cancer biology or therapy.

3.1 History of Signalling Sphingolipids: Where Are We One Generation Later?

Perhaps, the first SL recognized to mediate pathological changes is psychosine (i.e., galactosylsphingosine). While almost undetectable in normal brain, this lysolipid was found to accumulate in the cerebral white matter of patients affected with Krabbe disease (globoid cell leukodystrophy) [31, 32], a lysosomal disorder due to deficient activity of beta-galactosylceramidase. Psychosine was already known to be a cytotoxic substance; later studies suggested that this lysolipid, like sphingosine, inhibits protein kinase C (PKC) [29]. It also disturbs cytokinesis (thus explaining the formation of globoid, multinucleated cells) through binding to the G-protein coupled receptor TDAG8 [33]. However, whether psychosine regulates cytokinesis via TDAG8 [34] or antagonizes the proton-sensing properties of TDAG8 [35] is still debated.

Further indication that SLs could act as signal transducers was provided 25 years ago by the demonstration that the effects of vitamin D3 on leukemic cell differentiation were preceded by an increased activity of a neutral sphingomyelinase, transient hydrolysis of SM and concomitant production of Cer, and recapitulated by treatment with an exogenous sphingomyelinase [36]. This report immediately followed the observation that the phenothiazine trifluoroperazine stimulated SM breakdown and Cer generation in pituitary cells [37]. These data were the first of a series documenting the "sphingomyelin cycle", that is the transient formation of Cer from SM via activation of a sphingomyelinase by a wide variety of physiological or pathological agents [38, 39]. This pathway was viewed as a means to serve the production of bioactive SLs mediating the biological outcome triggered by the applied stimulus. However, it soon appeared that SM turnover is not the only source of signalling Cer. Activation of the Cer biosynthetic (*de novo*) pathway was reported as an alternative mechanism, especially in mediating cell death [40, 41]. Other mechanisms, such as the inhibition of SM synthesis [42, 43] or of ceramidase activity [44], likely contribute to stress-induced Cer generation. In addition, more recently production of bioactive Cer from glycolipids has been reported [45, 46].

Of note, whatever its metabolic source, Cer may not be necessarily the "signalling" SL that directly transduces the observed effect. Other SLs display biological roles, including in cancer (see Fig. 1). For instance, bioactive S1P, some effects of which were already described in 1990–1991 [47, 48], can originate from the metabolic cascade starting with SM breakdown [49]. Even products of S1P catabolism appear to be biologically active [50, 51].

Studies reported in the beginning of the 90s disclosed opposite effects of some SLs. Such a dichotomy is typically illustrated by the antagonistic actions of Cer and S1P on cell growth/death and survival, which led to the "rheostat" concept [52]. This concept initially proposed that the balance between Cer and S1P levels controls the cell fate via distinct effects on members of the MAPK family. However, such an equilibrium is presumably not maintained by the overall intracellular concentrations of these two SLs, which differ by about two orders of magnitude. One can imagine that the subcellular location (and thus, the accessibility to the protein target) and/or the nature of the molecular species are key determinants in the action of these lipid mediators. Accordingly, even a minute amount of a given Cer species may counteract the action of S1P if appropriately located.

To substantiate such hypotheses, considerable technological progress has been made during the last two decades that allowed a "higher-resolution view" of SL metabolism and signalling (see Part II of this book). In particular, mass spectrometry-based techniques for analysis of individual SL molecular species have permitted to reveal the specific role of particular SL molecules. Examples of the diversity/divergence of biological outcomes include differences due to the acyl chain-length, unsaturation or hydroxylation of Cer [10, 53]. Thus, current questions in the SL signalling field include which precise SL molecule is bioactive, in which subcellular compartment or even in which cell membrane domain it is formed or transported to, and which protein it targets. The answer to these questions is linked to the identification of the pathway that generates the candidate SL molecule. However, one should not forget that SL metabolism is a very dynamic process involving multiple reactions occurring in different compartments (and likely regulated by mechanisms that we have yet to discover). This implies that changes in the local concentration of one SL bioactive molecule may be accompanied by changes in the content of another bioactive SL, in the same or another subcellular location, indicating the need for a full spatiotemporal picture of SLs in order to accurately appreciate any variation of potential biological significance.

3.2 Sphingosine 1-Phosphate: A Paradigmatic Signalling Molecule

S1P meets the criteria that define a lipid mediator, coupling cell stimulation and functional activation, with regulation of cell proliferation being one of the first examples of its role as second messenger [54]. Not only is this small lipid, characterized by very low intracellular levels, able to activate specific cell surface

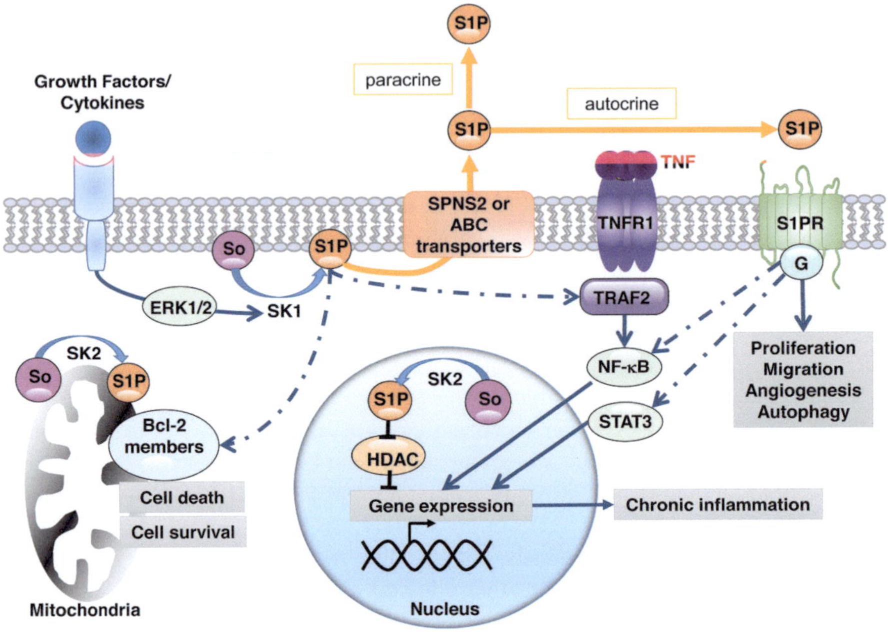

Fig. 4 Signalling cascades regulated by sphingosine 1-phosphate. Upon stimulation by growth factors and cytokines, sphingosine kinase 1 (SK1) gets activated by ERK1/2 and phosphorylates sphingosine (So) to sphingosine 1 phosphate (S1P). S1P controls cell fate by regulation of Bcl-2 members [124–127]. In the nucleus S1P, produced by sphingosine kinase 2 (SK2), binds and inhibits histone deacetylases (HDACs) that regulates gene transcription [60]. S1P can function as an intracellular messenger or is secreted out of the cell by SPNS2 or ATP-binding cassette (ABC) transporters to signal through G-protein-coupled receptors (S1PR) in an autocrine and/or paracrine manner to regulate proliferation, migration, angiogenesis or autophagy in cancer cells and tumor microenvironment. Activation of S1P1 receptor leads to the activation of signal transducer and activator of transcription 3 (STAT3) [128]. STAT3 is a transcription factor for S1P1, which then reciprocally activates STAT3, resulting in its persistent activation and interleukin-6 production, a pro-inflammatory cytokine. S1P binds to and activates TNF receptor-associated factor 2 (TRAF2) that is implicated in activation of NF-κB in response to TNF. Activation of NF-κB and STAT3 induce chronic inflammation that promotes cancer cell progression

receptors but also some intracellular targets (see Figs. 4 and 5). For detailed information regarding S1P signalling, the reader is referred to some recent reviews [6, 7, 9, 55, 56].

Upon activation of sphingosine kinase(s), for instance in response to cytokines or growth factors, S1P can be transported out of the cell (possibly by SPNS2) to activate its cognate G-protein coupled receptors. These receptors belong to a family of proteins, initially described as products of endothelial differentiation genes, some of which can bind the S1P-related lipid lysophosphatidic acid [57]. Engagement of S1P receptors modulates cell growth, migration, angiogenesis, lymphangiogenesis and the immune response [58]. Distinct outcomes arise from differential coupling of these five receptors to heterotrimeric G proteins and their downstream effectors (Fig. 5). Whereas no *bona fide* intracellular receptor of S1P has been reported, S1P

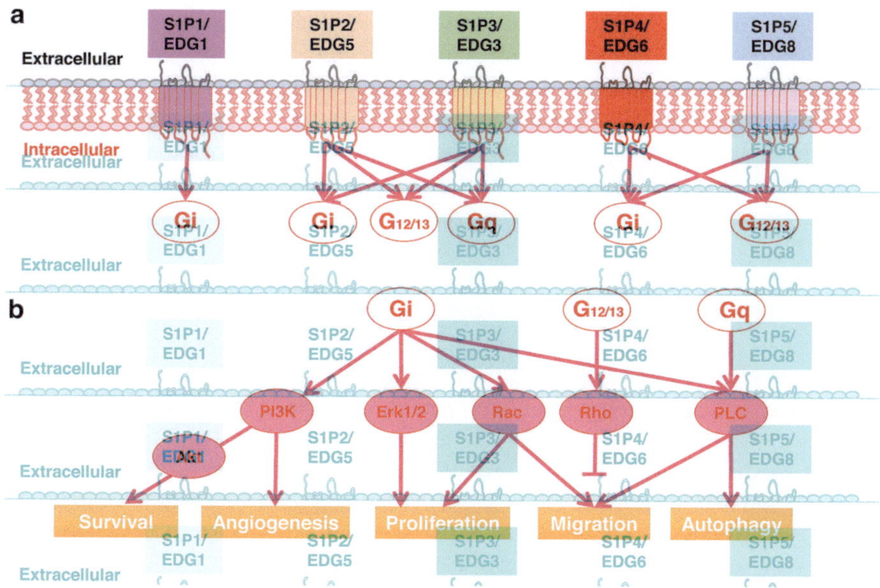

Fig. 5 Receptor-mediated effects of sphingosine 1-phosphate. S1P receptors couple to various heterotrimeric G proteins (**a**), activate partially overlapping downstream signalling pathways, and regulate several functions such as angiogenesis, proliferation, migration and autophagy (**b**)

can physically interact with and regulate some intracellular proteins. Indeed, S1P can function as a cofactor for the E3 ubiquitin ligase activity of the adaptor protein tumor necrosis factor receptor-associated factor 2 (TRAF2), regulating NF-κB activation [59] and thus explaining, among other processes, the cytoprotective function of S1P. In the nucleus, S1P binds to and inhibits HDAC1 and HDAC2, leading to increased levels of histone acetylation and gene transcription [60]. At the level of mitochondria, S1P is able to bind prohibitin-2 [61] and BAK [51], providing links with mitochondrial respiration and apoptosis, respectively. Finally, interaction of S1P with Cer synthase 2 provides a potential regulatory mechanism of the functional crosstalk between Cer and S1P [62].

3.3 How Are Other Sphingolipids Bioactive?

Early reports demonstrated that free sphingoid bases can inhibit PKC by preventing its interaction with DAG/phorbol esters [63–65]. Subsequent studies indicated that sphingosine regulates the function of many additional proteins, including members of the MAPK pathway, the sphingosine-dependent protein kinase 1, Akt and caspases (for a review see [66]). Modulation of these effectors by long-chain bases or sphingoid base analogues (i.e., *N,N*-dimethylsphingosine or *N,N,N-trimethylsphingosine*) has been classically associated with induction of cell death. Interestingly, natural unusual sphingoid bases called sphingadienes (because of the presence of an additional double bond at the C6 or C8 position) were found to inhibit colon tumorigenesis in animal models [67]. These sphingadienes, which derive from the catabolism of SLs present in food sources, inhibit the phosphoinositide 3-kinase/Akt and Wnt signalling pathways and promote apoptosis [68, 69]. They may also suppress intestinal tumorigenesis by inducing S1P lyase expression and reducing S1P colonic levels [70].

The way Cer mediates biological effects has attracted much attention. However, unlike DAG—a structurally similar lipid—the answer is not univocal and remains unclear. The search for potential Cer-binding proteins has faced technical difficulties [71] and has not yet provided unambiguous targets that could transduce all Cer actions. Nevertheless, diverse cellular proteins have been identified that relay the anti-mitogenic effects of Cer (Fig. 6; see also reviews by [8, 72–74]). Indeed, through modulation of cell cycle effectors, caspase-dependent and independent pathways and macroautophagy, Cer-activated cascades converge to antiproliferative signals. Some of them appear to be initiated by direct interaction of Cer with kinases such as PKC isoforms [75, 76], KSR1 [77], or the inhibitor 2 of protein phosphatase 2A [78]. Because of its rather hydrophobic nature, this membrane-localized SL may interact with proteins in distinct subcellular compartments, including the plasma membrane, ER, mitochondria or mitochondria-associated membranes. Of note, however, Cer may also modulate signalling pathways by changing the properties of membranes (i.e., membrane fluidity and curvature) or microdomains. For instance, ceramide synthase 6 expression is down-regulated during the epithelial-to-mesenchymal transition process, increasing plasma membrane fluidity as a consequence of reduced C16-Cer levels, and enhancing breast cancer cell motility [79]. Even modest variations (a few percent) in overall Cer levels largely exceed total S1P levels, suggesting that some actions of Cer might be related to membrane alterations that in turn influence the recruitment of key protein effectors or the activation of membrane receptors or channels [10, 80–82].

What about glycolipids? Although this topic is not discussed in this book, it is important to note that, because they localize in the extracellular leaflet of the plasma membrane, these SLs not only can affect membrane processes but also actively interact with the environment [83–85]. Hence, gangliosides, whose expression is altered in many tumors, can modulate the proliferation of malignant cells as well as cell–cell and cell–matrix adhesion. The biological effects of GSLs are likely mediated by interactions of their carbohydrate moieties, either on the same cell surface or between two different cells [86]. Perturbations of the membrane properties, the subsequent binding of ligands to membrane receptors, as well as the activity of multidrug resistance transporters could also be implicated [87]. In addition, shedding of tumor gangliosides provides both immunosuppressive and angiogenic signals [88, 89].

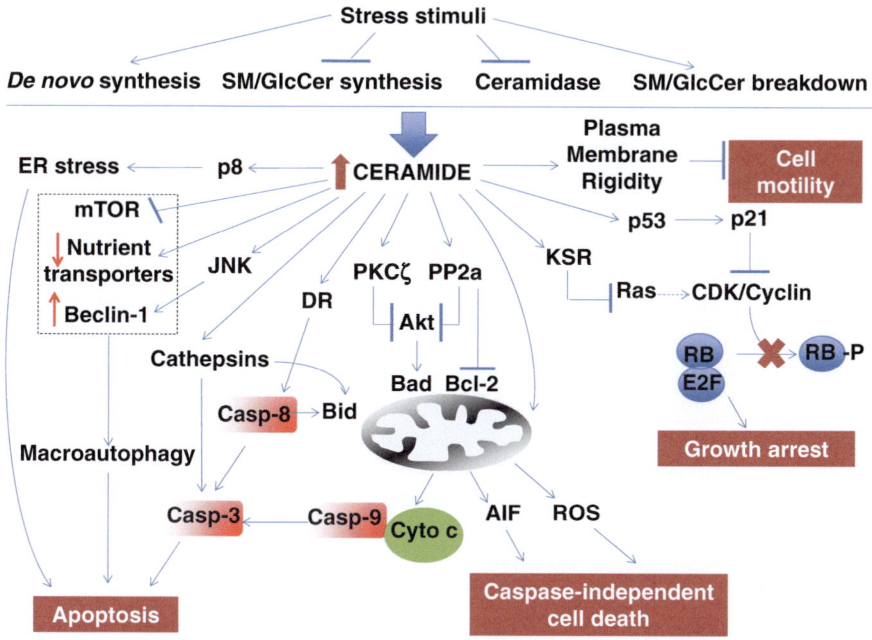

Fig. 6 Signalling cascades regulated by ceramides. Various stress stimuli, such as chemotherapy, radiotherapy and death receptor (DR) ligands, trigger the accumulation of Cer as a consequence of (1) increased *de novo* Cer synthesis, (2) inhibition of Cer conversion to complex SLs, (3) hydrolysis of SM or GlcCer and/or (4) inhibition of ceramidase. Cer behaves as an anti-oncometabolite, inhibiting cancer cell motility and proliferation or inducing cancer cell death. For the sake of clarity, not all signalling pathways modulated by Cer are depicted. The anti-proliferative signalling pathway triggered by Cer involves the activation of both Kinase Suppressor of Ras/Cer-Activated Protein Kinase (KSR/CAPK) and p53, inhibiting the cdk-dependent phosphorylation of Rb, which sequesters E2F. Cer can activate the extrinsic apoptotic signalling pathway by facilitating DR oligomerization and subsequent caspase cascade activation. Alternatively, Cer stimulates the mitochondrial (i.e., intrinsic) apoptotic signalling pathway as a consequence of (1) PKCζ and PP2a-dependent Akt inhibition and subsequent Bad activation, (2) PP2a-dependent Bcl-2 inhibition, (3) cathepsins and/or caspase-8-dependent Bid cleavage. All these events lead to the mitochondrial outer membrane permeabilization followed by cytochrome c (Cyto c) release and caspase-9 activation. Both initiator caspases-8 and -9 cleave and activate effector caspases, such as caspase-3. AIF release from the mitochondria and ROS production are involved in Cer-induced caspase-independent cell death. Cer-triggered cell death can be amplified by ER stress and macroautophagy. Whereas p8 is involved in Cer-induced ER stress, the inhibition of mTOR and nutrient transporters, as well as the JNK-dependent Beclin-1 up-regulation and activation, facilitate the macroautophagy process

4 Current and Future Challenges

A renewed interest in SLs manifested 25 years ago with the discovery of specific biological functions mediated by members of this class. Since then, not only has our knowledge of metabolic pathways, enzymes and transport proteins greatly improved but also the way some SL molecules exert their regulatory roles has been molecularly characterized. Their abilities to modulate critical events in cancer development and progression (Fig. 1) suggest that certain SLs (e.g., S1P) behave as "oncometabolites" while some others (e.g., Cer) as "anti-oncometabolites". Even though mutations in the genes encoding SL-metabolizing enzymes that would confer neomorphic activity (as mutations in the *IDH1* gene do) have not—yet—been identified in tumors or leukemic cells, some SL molecules share properties with known oncometabolites such as 2-hydroxyglutarate [90]. For instance, levels of S1P-forming enzymes (mRNA and/or protein) are higher in tumor vs. normal tissues, and S1P acts as a pro-oncogenic signal, influencing the epigenome [60], transcriptional programs (e.g., [91]), hypoxia-inducible factor biology [92], tumor development [93] (see also numerous examples of reduced tumor growth upon treatment with inhibitors of the sphingosine kinase/S1P/S1P receptors axis) and sensitivity to anticancer regimens (for a review see [6]).

That said, despite considerable advances in our understanding, the world of SLs-and-cancer still faces a number of challenges. First, the metabolism of some SLs needs to be unequivocally elucidated. This includes the way: (1) some minor sphingoid bases (i.e., sphingadiene), omega-esterified Cer species and other unusual SLs are produced; (2) some lysosphingolipids, e.g., SPC, glucosylsphingosine and psychosine, are synthesized; (3) free sphingoid bases are transported into the cell, (4) S1P gets out of the cell; (5) Cer is oriented to spatially distinct compartments for the biosynthesis of SM, GlcCer, GalCer or Cer1P. Enzymes and transport proteins/transporters for such pathways are to be identified. Equally important is the need to fully understand how SL metabolism is regulated (little is known about transcriptional regulation of the genes encoding SL-metabolizing enzymes, regulation by microRNAs, or possible sensors of SL levels).

Second, the mode of action of bioactive SLs has to be further deciphered, implying the identification of direct protein targets and receptors [71]. In this regard, one should consider the possibility that changes in SL metabolism and composition in a given cell (e.g., the cancer cell) impact surrounding cells (i.e., the tumor microenvironment) through exosomes [94]. Third, efforts to characterize alterations in the sphingolipidome of tumor cells (or even plasma from patients with cancer) as well as (epi)genomic and transcriptomic changes in the genes of SL metabolism will hopefully define novel markers. They could serve as diagnostic, prognostic and/or predictive biomarkers, useful for the stratification of cancers or the tumor response to therapy. It is anticipated that advances in the above directions will help understand the roles of SLs in cancer and develop targeted therapies based on the manipulation of SL metabolism.

Acknowledgement Support to TL's team by INSERM, Université Paul Sabatier, ANR (SphingoDR program), RITC, Ligue Nationale Contre le Cancer (Equipe Labellisée 2013), and VML is gratefully acknowledged.

References

1. Merrill AH Jr, Stokes TH, Momin A et al (2009) Sphingolipidomics: a valuable tool for understanding the roles of sphingolipids in biology and disease. J Lipid Res 50(Suppl):S97–102
2. Sabourdy F, Kedjouar B, Sorli SC et al (2008) Functions of sphingolipid metabolism in mammals—lessons from genetic defects. Biochim Biophys Acta 1781:145–183
3. Jennemann R, Grone HJ (2013) Cell-specific in vivo functions of glycosphingolipids: lessons from genetic deletions of enzymes involved in glycosphingolipid synthesis. Prog Lipid Res 52:231–248
4. Albinet V, Bats ML, Bedia C et al (2013) Genetic disorders of simple sphingolipid metabolism. Handb Exp Pharmacol 7:127–152. doi:10.1007/978-3-7091-1368-4
5. Hannun YA, Obeid LM (2008) Principles of bioactive lipid signalling: lessons from sphingolipids. Nat Rev Mol Cell Biol 9:139–150
6. Pyne NJ, Pyne S (2010) Sphingosine 1-phosphate and cancer. Nat Rev Cancer 10:489–503
7. Maceyka M, Harikumar KB, Milstien S, Spiegel S (2012) Sphingosine-1-phosphate signaling and its role in disease. Trends Cell Biol 22:50–60
8. Morad SA, Cabot MC (2013) Ceramide-orchestrated signalling in cancer cells. Nat Rev Cancer 13:51–65
9. Kunkel GT, Maceyka M, Milstien S, Spiegel S (2013) Targeting the sphingosine-1-phosphate axis in cancer, inflammation and beyond. Nat Rev Drug Discov 12:688–702
10. Grosch S, Schiffmann S, Geisslinger G (2012) Chain length-specific properties of ceramides. Prog Lipid Res 51:50–62
11. Penno A, Reilly MM, Houlden H et al (2010) Hereditary sensory neuropathy type 1 is caused by the accumulation of two neurotoxic sphingolipids. J Biol Chem 285:11178–11187
12. Bertea M, Rutti MF, Othman A et al (2010) Deoxysphingoid bases as plasma markers in diabetes mellitus. Lipids Health Dis 9:84
13. Funato K, Riezman H (2001) Vesicular and nonvesicular transport of ceramide from ER to the Golgi apparatus in yeast. J Cell Biol 155:949–959
14. Hanada K, Kumagai K, Yasuda S et al (2003) Molecular machinery for non-vesicular trafficking of ceramide. Nature 426:803–809
15. Halter D, Neumann S, van Dijk SM et al (2007) Pre- and post-Golgi translocation of glucosylceramide in glycosphingolipid synthesis. J Cell Biol 179:101–115
16. Simanshu DK, Kamlekar RK, Wijesinghe DS et al (2013) Non-vesicular trafficking by a ceramide-1-phosphate transfer protein regulates eicosanoids. Nature 500:463–467
17. Swanton C, Marani M, Pardo O et al (2007) Regulators of mitotic arrest and ceramide metabolism are determinants of sensitivity to paclitaxel and other chemotherapeutic drugs. Cancer Cell 11:498–512
18. Wennekes T, van den Berg RJ, Boot RG, van der Marel GA, Overkleeft HS, Aerts JM (2009) Glycosphingolipids—nature, function, and pharmacological modulation. Angew Chem Int Ed Engl 48:8848–8869
19. D'Angelo G, Capasso S, Sticco L, Russo D (2013) Glycosphingolipids: synthesis and functions. FEBS J 280:6338–6353
20. Kolter T, Sandhoff K (2010) Lysosomal degradation of membrane lipids. FEBS Lett 584:1700–1712
21. Kihara A (2014) Sphingosine 1-phosphate is a key metabolite linking sphingolipids to glycerophospholipids. Biochim Biophys Acta 1841:766–772
22. Tani M, Ito M, Igarashi Y (2007) Ceramide/sphingosine/sphingosine 1-phosphate metabolism on the cell surface and in the extracellular space. Cell Signal 19:229–237
23. Jenkins RW, Canals D, Hannun YA (2009) Roles and regulation of secretory and lysosomal acid sphingomyelinase. Cell Signal 21:836–846
24. Duan RD (2011) Physiological functions and clinical implications of sphingolipids in the gut. J Dig Dis 12:60–70

25. Birbes H, El Bawab S, Obeid LM, Hannun YA (2002) Mitochondria and ceramide: intertwined roles in regulation of apoptosis. Adv Enzyme Regul 42:113–129
26. Morales A, Colell A, Mari M, Garcia-Ruiz C, Fernandez-Checa JC (2004) Glycosphingolipids and mitochondria: role in apoptosis and disease. Glycoconj J 20:579–588
27. Lucki NC, Sewer MB (2012) Nuclear sphingolipid metabolism. Annu Rev Physiol 74: 131–151
28. Mao C, Obeid LM (2008) Ceramidases: regulators of cellular responses mediated by ceramide, sphingosine, and sphingosine-1-phosphate. Biochim Biophys Acta 1781:424–434
29. Hannun YA, Bell RM (1987) Lysosphingolipids inhibit protein kinase C: implications for the sphingolipidoses. Science 235:670–674
30. Breslow DK, Weissman JS (2010) Membranes in balance: mechanisms of sphingolipid homeostasis. Mol Cell 40:267–279
31. Miyatake T, Suzuki K (1972) Globoid cell leukodystrophy: additional deficiency of psychosine galactosidase. Biochem Biophys Res Commun 48:539–543
32. Vanier MT, Svennerholm L (1975) Chemical pathology of Krabbe's disease. III. Ceramide-hexosides and gangliosides of brain. Acta Paediatr Scand 64:641–648
33. Im DS, Heise CE, Nguyen T, O'Dowd BF, Lynch KR (2001) Identification of a molecular target of psychosine and its role in globoid cell formation. J Cell Biol 153:429–434
34. Radu CG, Cheng D, Nijagal A et al (2006) Normal immune development and glucocorticoid-induced thymocyte apoptosis in mice deficient for the T-cell death-associated gene 8 receptor. Mol Cell Biol 26:668–677
35. Wang JQ, Kon J, Mogi C et al (2004) TDAG8 is a proton-sensing and psychosine-sensitive G-protein-coupled receptor. J Biol Chem 279:45626–45633
36. Okazaki T, Bell RM, Hannun YA (1989) Sphingomyelin turnover induced by vitamin D3 in HL-60 cells. Role in cell differentiation. J Biol Chem 264:19076–19080
37. Kolesnick RN (1989) Sphingomyelinase action inhibits phorbol ester-induced differentiation of human promyelocytic leukemic (HL-60) cells. J Biol Chem 264:7617–7623
38. Hannun YA (1994) The sphingomyelin cycle and the second messenger function of ceramide. J Biol Chem 269:3125–3128
39. Levade T, Jaffrezou JP (1999) Signalling sphingomyelinases: which, where, how and why? Biochim Biophys Acta 1438:1–17
40. Bose R, Verheij M, Haimovitz-Friedman A, Scotto K, Fuks Z, Kolesnick R (1995) Ceramide synthase mediates daunorubicin-induced apoptosis: an alternative mechanism for generating death signals. Cell 82:405–414
41. Park JW, Park WJ, Futerman AH (2014) Ceramide synthases as potential targets for therapeutic intervention in human diseases. Biochim Biophys Acta 1841:671–681
42. Watanabe M, Kitano T, Kondo T et al (2004) Increase of nuclear ceramide through caspase-3-dependent regulation of the "sphingomyelin cycle" in Fas-induced apoptosis. Cancer Res 64:1000–1007
43. Lafont E, Milhas D, Carpentier S et al (2010) Caspase-mediated inhibition of sphingomyelin synthesis is involved in FasL-triggered cell death. Cell Death Differ 17:642–654
44. Bedia C, Casas J, Andrieu-Abadie N, Fabrias G, Levade T (2011) Acid ceramidase expression modulates the sensitivity of A375 melanoma cells to dacarbazine. J Biol Chem 286: 28200–28209
45. Valaperta R, Chigorno V, Basso L et al (2006) Plasma membrane production of ceramide from ganglioside GM3 in human fibroblasts. FASEB J 20:1227–1229
46. Sorli SC, Colie S, Albinet V et al (2013) The nonlysosomal beta-glucosidase GBA2 promotes endoplasmic reticulum stress and impairs tumorigenicity of human melanoma cells. FASEB J 27:489–498
47. Ghosh TK, Bian J, Gill DL (1990) Intracellular calcium release mediated by sphingosine derivatives generated in cells. Science 248:1653–1656
48. Zhang H, Desai NN, Olivera A, Seki T, Brooker G, Spiegel S (1991) Sphingosine-1-phosphate, a novel lipid, involved in cellular proliferation. J Cell Biol 114:155–167

49. Auge N, Nikolova-Karakashian M, Carpentier S et al (1999) Role of sphingosine 1-phosphate in the mitogenesis induced by oxidized low density lipoprotein in smooth muscle cells via activation of sphingomyelinase, ceramidase, and sphingosine kinase. J Biol Chem 274: 21533–21538
50. Kumar A, Byun HS, Bittman R, Saba JD (2011) The sphingolipid degradation product trans-2-hexadecenal induces cytoskeletal reorganization and apoptosis in a JNK-dependent manner. Cell Signal 23:1144–1152
51. Chipuk JE, McStay GP, Bharti A et al (2012) Sphingolipid metabolism cooperates with BAK and BAX to promote the mitochondrial pathway of apoptosis. Cell 148:988–1000
52. Cuvillier O, Pirianov G, Kleuser B et al (1996) Suppression of ceramide-mediated programmed cell death by sphingosine-1-phosphate. Nature 381:800–803
53. Mesicek J, Lee H, Feldman T et al (2010) Ceramide synthases 2, 5, and 6 confer distinct roles in radiation-induced apoptosis in HeLa cells. Cell Signal 22:1300–1307
54. Olivera A, Spiegel S (1993) Sphingosine-1-phosphate as second messenger in cell proliferation induced by PDGF and FCS mitogens. Nature 365:557–560
55. Strub GM, Maceyka M, Hait NC, Milstien S, Spiegel S (2010) Extracellular and intracellular actions of sphingosine-1-phosphate. Adv Exp Med Biol 688:141–155
56. Spiegel S, Milstien S (2011) The outs and the ins of sphingosine-1-phosphate in immunity. Nat Rev Immunol 11:403–415
57. Lee MJ, Van Brocklyn JR, Thangada S et al (1998) Sphingosine-1-phosphate as a ligand for the G protein-coupled receptor EDG-1. Science 279:1552–1555
58. Blaho VA, Hla T (2014) An update on the biology of sphingosine 1-phosphate receptors. J Lipid Res 55:1596–1608
59. Alvarez SE, Harikumar KB, Hait NC et al (2010) Sphingosine-1-phosphate is a missing cofactor for the E3 ubiquitin ligase TRAF2. Nature 465:1084–1088
60. Hait NC, Allegood J, Maceyka M et al (2009) Regulation of histone acetylation in the nucleus by sphingosine-1-phosphate. Science 325:1254–1257
61. Strub GM, Paillard M, Liang J et al (2011) Sphingosine-1-phosphate produced by sphingosine kinase 2 in mitochondria interacts with prohibitin 2 to regulate complex IV assembly and respiration. FASEB J 25:600–612
62. Laviad EL, Albee L, Pankova-Kholmyansky I et al (2008) Characterization of ceramide synthase 2: tissue distribution, substrate specificity, and inhibition by sphingosine 1-phosphate. J Biol Chem 283:5677–5684
63. Hannun YA, Loomis CR, Merrill AH Jr, Bell RM (1986) Sphingosine inhibition of protein kinase C activity and of phorbol dibutyrate binding in vitro and in human platelets. J Biol Chem 261:12604–12609
64. Merrill AH Jr, Sereni AM, Stevens VL, Hannun YA, Bell RM, Kinkade JM Jr (1986) Inhibition of phorbol ester-dependent differentiation of human promyelocytic leukemic (HL-60) cells by sphinganine and other long-chain bases. J Biol Chem 261:12610–12615
65. Wilson E, Olcott MC, Bell RM, Merrill AH Jr, Lambeth JD (1986) Inhibition of the oxidative burst in human neutrophils by sphingoid long-chain bases. Role of protein kinase C in activation of the burst. J Biol Chem 261:12616–12623
66. Cuvillier O (2002) Sphingosine in apoptosis signaling. Biochim Biophys Acta 1585: 153–162
67. Symolon H, Schmelz EM, Dillehay DL, Merrill AH Jr (2004) Dietary soy sphingolipids suppress tumorigenesis and gene expression in 1,2-dimethylhydrazine-treated CF1 mice and ApcMin/+ mice. J Nutr 134:1157–1161
68. Fyrst H, Oskouian B, Bandhuvula P et al (2009) Natural sphingadienes inhibit Akt-dependent signaling and prevent intestinal tumorigenesis. Cancer Res 69:9457–9464
69. Kumar A, Pandurangan AK, Lu F et al (2012) Chemopreventive sphingadienes downregulate Wnt signaling via a PP2A/Akt/GSK3beta pathway in colon cancer. Carcinogenesis 33: 1726–1735

70. Degagne E, Pandurangan A, Bandhuvula P et al (2014) Sphingosine-1-phosphate lyase downregulation promotes colon carcinogenesis through STAT3-activated microRNAs. J Clin Invest 124:5368–5384
71. Snook CF, Jones JA, Hannun YA (2006) Sphingolipid-binding proteins. Biochim Biophys Acta 1761:927–946
72. Saddoughi SA, Ogretmen B (2013) Diverse functions of ceramide in cancer cell death and proliferation. Adv Cancer Res 117:37–58
73. Young MM, Kester M, Wang HG (2013) Sphingolipids: regulators of crosstalk between apoptosis and autophagy. J Lipid Res 54:5–19
74. Garcia-Barros M, Coant N, Truman JP, Snider AJ, Hannun YA (2014) Sphingolipids in colon cancer. Biochim Biophys Acta 1841:773–782
75. Huwiler A, Fabbro D, Pfeilschifter J (1998) Selective ceramide binding to protein kinase C-alpha and -delta isoenzymes in renal mesangial cells. Biochemistry 37:14556–14562
76. Wang G, Silva J, Krishnamurthy K, Tran E, Condie BG, Bieberich E (2005) Direct binding to ceramide activates protein kinase Czeta before the formation of a pro-apoptotic complex with PAR-4 in differentiating stem cells. J Biol Chem 280:26415–26424
77. Yin X, Zafrullah M, Lee H, Haimovitz-Friedman A, Fuks Z, Kolesnick R (2009) A ceramide-binding C1 domain mediates kinase suppressor of ras membrane translocation. Cell Physiol Biochem 24:219–230
78. Mukhopadhyay A, Saddoughi SA, Song P et al (2009) Direct interaction between the inhibitor 2 and ceramide via sphingolipid-protein binding is involved in the regulation of protein phosphatase 2A activity and signaling. FASEB J 23:751–763
79. Edmond V, Dufour F, Poiroux G et al (2015) Downregulation of ceramide synthase-6 during epithelial-to-mesenchymal transition reduces plasma membrane fluidity and cancer cell motility. Oncogene 34:996–1005. doi:10.1038/onc.2014.55
80. van Blitterswijk WJ, van der Luit AH, Veldman RJ, Verheij M, Borst J (2003) Ceramide: second messenger or modulator of membrane structure and dynamics? Biochem J 369: 199–211
81. Goni FM, Alonso A (2006) Biophysics of sphingolipids I. Membrane properties of sphingosine, ceramides and other simple sphingolipids. Biochim Biophys Acta 1758:1902–1921
82. Zhang Y, Li X, Becker KA, Gulbins E (2009) Ceramide-enriched membrane domains—structure and function. Biochim Biophys Acta 1788:178–183
83. Hakomori S (2002) Glycosylation defining cancer malignancy: new wine in an old bottle. Proc Natl Acad Sci U S A 99:10231–10233
84. Fredman P, Hedberg K, Brezicka T (2003) Gangliosides as therapeutic targets for cancer. BioDrugs 17:155–167
85. Furukawa K, Hamamura K, Aixinjueluo W, Furukawa K (2006) Biosignals modulated by tumor-associated carbohydrate antigens: novel targets for cancer therapy. Ann N Y Acad Sci 1086:185–198
86. Handa K, Hakomori SI (2012) Carbohydrate to carbohydrate interaction in development process and cancer progression. Glycoconj J 29:627–637
87. Giussani P, Tringali C, Riboni L, Viani P, Venerando B (2014) Sphingolipids: key regulators of apoptosis and pivotal players in cancer drug resistance. Int J Mol Sci 15:4356–4392
88. Hossain DM, Mohanty S, Ray P, Das T, Sa G (2012) Tumor gangliosides and T cells: a deadly encounter. Front Biosci (Schol Ed) 4:502–519
89. Liu Y, Wondimu A, Yan S, Bobb D, Ladisch S (2013) Tumor gangliosides accelerate murine tumor angiogenesis. Angiogenesis 17:563–571. doi:10.1007/s10456-013-9403-4
90. Ward PS, Patel J, Wise DR et al (2010) The common feature of leukemia-associated IDH1 and IDH2 mutations is a neomorphic enzyme activity converting alpha-ketoglutarate to 2-hydroxyglutarate. Cancer Cell 17:225–234
91. Avery K, Avery S, Shepherd J, Heath PR, Moore H (2008) Sphingosine-1-phosphate mediates transcriptional regulation of key targets associated with survival, proliferation, and pluripotency in human embryonic stem cells. Stem Cells Dev 17:1195–1205

92. Ader I, Malavaud B, Cuvillier O (2009) When the sphingosine kinase 1/sphingosine 1-phosphate pathway meets hypoxia signaling: new targets for cancer therapy. Cancer Res 69:3723–3726
93. Xia P, Gamble JR, Wang L et al (2000) An oncogenic role of sphingosine kinase. Curr Biol 10:1527–1530
94. Trajkovic K, Hsu C, Chiantia S et al (2008) Ceramide triggers budding of exosome vesicles into multivesicular endosomes. Science 319:1244–1247
95. Hanahan D, Weinberg RA (2000) The hallmarks of cancer. Cell 100:57–70
96. Moon SK, Kim HM, Lee YC, Kim CH (2004) Disialoganglioside (GD3) synthase gene expression suppresses vascular smooth muscle cell responses via the inhibition of ERK1/2 phosphorylation, cell cycle progression, and matrix metalloproteinase-9 expression. J Biol Chem 279:33063–33070
97. Wang H, Isaji T, Satoh M, Li D, Arai Y, Gu J (2013) Antitumor effects of exogenous ganglioside GM3 on bladder cancer in an orthotopic cancer model. Urology 81(210):e211–215
98. Shirahama T, Sweeney EA, Sakakura C et al (1997) In vitro and in vivo induction of apoptosis by sphingosine and N, N-dimethylsphingosine in human epidermoid carcinoma KB-3-1 and its multidrug-resistant cells. Clin Cancer Res 3:257–264
99. Desai NN, Carlson RO, Mattie ME et al (1993) Signaling pathways for sphingosylphosphorylcholine-mediated mitogenesis in Swiss 3T3 fibroblasts. J Cell Biol 121:1385–1395
100. Endo K, Igarashi Y, Nisar M, Zhou QH, Hakomori S (1991) Cell membrane signaling as target in cancer therapy: inhibitory effect of N,N-dimethyl and N,N,N-trimethyl sphingosine derivatives on in vitro and in vivo growth of human tumor cells in nude mice. Cancer Res 51:1613–1618
101. Webb TJ, Li X, Giuntoli RL 2nd et al (2012) Molecular identification of GD3 as a suppressor of the innate immune response in ovarian cancer. Cancer Res 72:3744–3752
102. Zhou D (2006) The immunological function of iGb3. Curr Protein Pept Sci 7:325–333
103. Shimizu K, Kurosawa Y, Taniguchi M, Steinman RM, Fujii S (2007) Cross-presentation of glycolipid from tumor cells loaded with alpha-galactosylceramide leads to potent and long-lived T cell mediated immunity via dendritic cells. J Exp Med 204:2641–2653
104. Maru M, Haraguchi M, Higashi H et al (1993) Anti-tumor activity of ceramides and glycosphingolipids in a murine tumor system. Int J Cancer 53:645–650
105. van Vlerken LE, Duan Z, Little SR, Seiden MV, Amiji MM (2010) Augmentation of therapeutic efficacy in drug-resistant tumor models using ceramide coadministration in temporal-controlled polymer-blend nanoparticle delivery systems. AAPS J 12:171–180
106. Parameswaran R, Lim M, Arutyunyan A et al (2013) O-acetylated N-acetylneuraminic acid as a novel target for therapy in human pre-B acute lymphoblastic leukemia. J Exp Med 210:805–819
107. Lavie Y, Cao H, Bursten SL, Giuliano AE, Cabot MC (1996) Accumulation of glucosylceramides in multidrug-resistant cancer cells. J Biol Chem 271:19530–19536
108. Ponnusamy S, Meyers-Needham M, Senkal CE et al (2010) Sphingolipids and cancer: ceramide and sphingosine-1-phosphate in the regulation of cell death and drug resistance. Future Oncol 6:1603–1624
109. Bocci G, Fioravanti A, Orlandi P et al (2012) Metronomic ceramide analogs inhibit angiogenesis in pancreatic cancer through up-regulation of caveolin-1 and thrombospondin-1 and down-regulation of cyclin D1. Neoplasia 14:833–845
110. Seyfried TN, Mukherjee P (2010) Ganglioside GM3 is antiangiogenic in malignant brain cancer. J Oncol 2010:961243
111. Visentin B, Vekich JA, Sibbald BJ et al (2006) Validation of an anti-sphingosine-1-phosphate antibody as a potential therapeutic in reducing growth, invasion, and angiogenesis in multiple tumor lineages. Cancer Cell 9:225–238
112. Cazet A, Groux-Degroote S, Teylaert B et al (2009) GD3 synthase overexpression enhances proliferation and migration of MDA-MB-231 breast cancer cells. Biol Chem 390:601–609

113. Yan Q, Bach DQ, Gatla N et al (2013) Deacetylated GM3 promotes uPAR-associated membrane molecular complex to activate p38 MAPK in metastatic melanoma. Mol Cancer Res 11:665–675
114. Ratajczak MZ, Suszynska M, Borkowska S, Ratajczak J, Schneider G (2014) The role of sphingosine-1 phosphate and ceramide-1 phosphate in trafficking of normal stem cells and cancer cells. Expert Opin Ther Targets 18:95–107
115. Takenaga M, Igarashi R, Matsumoto K et al (1999) Lipid microsphere preparation of a lipophilic ceramide derivative suppresses colony formation in a murine experimental pulmonary metastasis model. J Drug Target 7:187–195
116. Beil M, Micoulet A, von Wichert G et al (2003) Sphingosylphosphorylcholine regulates keratin network architecture and visco-elastic properties of human cancer cells. Nat Cell Biol 5:803–811
117. Sadahira Y, Ruan F, Hakomori S, Igarashi Y (1992) Sphingosine 1-phosphate, a specific endogenous signaling molecule controlling cell motility and tumor cell invasiveness. Proc Natl Acad Sci U S A 89:9686–9690
118. Birks SM, Danquah JO, King L, Vlasak R, Gorecki DC, Pilkington GJ (2011) Targeting the GD3 acetylation pathway selectively induces apoptosis in glioblastoma. Neuro Oncol 13:950–960
119. Bennaceur K, Popa I, Chapman JA et al (2009) Different mechanisms are involved in apoptosis induced by melanoma gangliosides on human monocyte-derived dendritic cells. Glycobiology 19:576–582
120. Obeid LM, Linardic CM, Karolak LA, Hannun YA (1993) Programmed cell death induced by ceramide. Science 259:1769–1771
121. Kota V, Dhople VM, Fullbright G et al (2013) 2′-Hydroxy C16-ceramide induces apoptosis-associated proteomic changes in C6 glioma cells. J Proteome Res 12:4366–4375
122. Sakakura C, Sweeney EA, Shirahama T, Hakomori S, Igarashi Y (1996) Suppression of bcl-2 gene expression by sphingosine in the apoptosis of human leukemic HL-60 cells during phorbol ester-induced terminal differentiation. FEBS Lett 379:177–180
123. Bleicher RJ, Cabot MC (2002) Glucosylceramide synthase and apoptosis. Biochim Biophys Acta 1585:172–178
124. Bektas M, Jolly PS, Muller C, Eberle J, Spiegel S, Geilen CC (2005) Sphingosine kinase activity counteracts ceramide-mediated cell death in human melanoma cells: role of Bcl-2 expression. Oncogene 24:178–187
125. Sauer B, Gonska H, Manggau M et al (2005) Sphingosine 1-phosphate is involved in cytoprotective actions of calcitriol in human fibroblasts and enhances the intracellular Bcl-2/Bax rheostat. Pharmazie 60:298–304
126. Li QF, Wu CT, Guo Q, Wang H, Wang LS (2008) Sphingosine 1-phosphate induces Mcl-1 upregulation and protects multiple myeloma cells against apoptosis. Biochem Biophys Res Commun 371:159–162
127. Colie S, Van Veldhoven PP, Kedjouar B et al (2009) Disruption of sphingosine 1-phosphate lyase confers resistance to chemotherapy and promotes oncogenesis through Bcl-2/Bcl-xL upregulation. Cancer Res 69:9346–9353
128. Lee H, Deng J, Kujawski M et al (2010) STAT3-induced S1PR1 expression is crucial for persistent STAT3 activation in tumors. Nat Med 16:1421–1428

Part I
Roles and Functions of Sphingolipids in the Development of Different Types of Cancers

Role of Sphingolipids in Hematological Malignancies: Lymphoproliferative Disorders

Hirofumi Sawai, Makoto Taniguchi, and Toshiro Okazaki

Abstract Among sphingolipids, ceramide was first reported to induce cell differentiation and death in human leukemia cells. The localization of signaling ceramide, the putative different functions of ceramide species, and their metabolic regulation have been investigated in regard to induction of cell death of many types of cancers. Recently not only ceramide but also sphingosine-1-phosphate (S1P), sphingomyelin (SM) and ceramide-1-phosphate are being appreciated to act as biological lipid regulators of many cell functions. It is critical to understand the role of sphingolipids in the regulation of the balance between proliferation and cell death for developing novel therapies for hematological malignant diseases. Programmed cell death is mainly classified into four categories: apoptosis, regulated necrosis/necroptosis, pyroptosis and autophagic cell death. In this chapter, we summarize the role of ceramide and other sphingolipids in programmed cell death, proliferation/survival, differentiation, migration and secretion of lymphoid lineage cells to discuss potential applications of sphingolipid-based treatment for lymphoproliferative disorders such as acute and chronic lymphoid leukemia, malignant lymphoma and multiple myeloma.

Keywords Lymphoproliferative disorders • Acute lymphoid leukemia • Chronic lymphoid leukemia • Malignant lymphoma • Multiple myeloma • Ceramide • Sphingosine 1-phosphate • Sphingomyelin • Ceramide-1-phosphate

H. Sawai
Department of Internal Medicine, Osaka Dental University, Osaka 573-1121, Japan

M. Taniguchi
Medical Research Institute, Kanazawa Medical University, Ishikawa 920-0293, Japan

T. Okazaki (✉)
Division of Hematology/Immunology, Department of Medicine,
Kanazawa Medical University, Ishikawa 920-0293, Japan
e-mail: toshiroo@kanazawa-med.ac.jp

Y.A. Hannun et al. (eds.), *Bioactive Sphingolipids in Cancer Biology and Therapy*, DOI 10.1007/978-3-319-20750-6_2

1 Introduction

Sphingolipids have emerged as lipid mediators to regulate many cell functions including cell death, proliferation/survival, migration and immunity. Among sphingolipids, ceramide was first reported to induce cell differentiation and death in human leukemia cells [1]. The localization of signaling ceramide, the putative different functions of ceramide species, and their metabolic regulation have been investigated in regard to induction of cell death of many types of cancers. Recently not only ceramide but also sphingosine-1-phosphate (S1P), sphingomyelin (SM) and ceramide-1-phosphate are being appreciated to act as biological lipid regulators of many cell functions. It is critical to understand the role of sphingolipids in the regulation of the balance between proliferation and cell death for developing novel therapies for hematological malignant diseases. Programmed cell death is mainly classified into four categories: apoptosis, regulated necrosis/necroptosis, pyroptosis and autophagic cell death. In this chapter, we summarize the role of ceramide and other sphingolipids in programmed cell death, proliferation/survival, differentiation, migration and secretion of lymphoid lineage cells to discuss potential applications of sphingolipid-based treatment for lymphoproliferative disorders such as acute and chronic lymphoid leukemia, malignant lymphoma and multiple myeloma (Fig. 1).

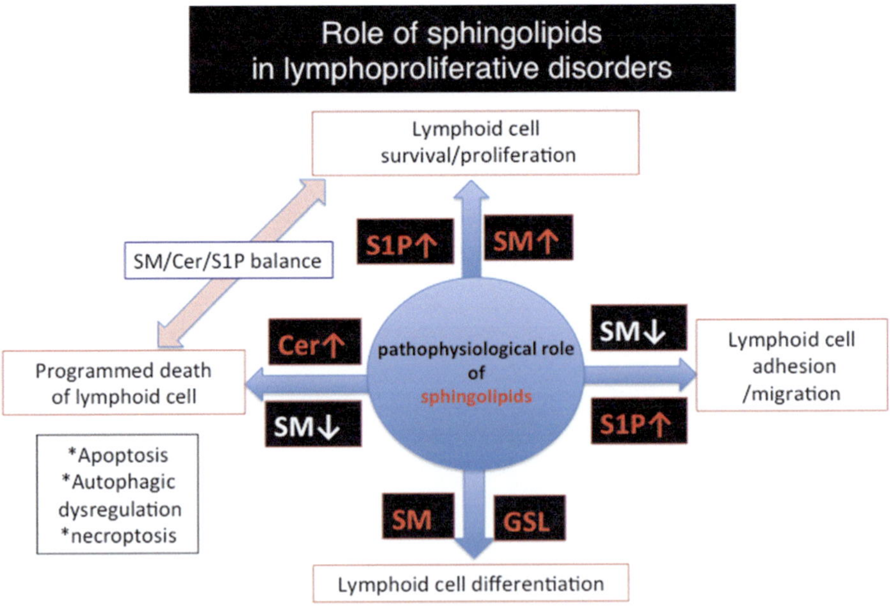

Fig. 1 Role of bioactive sphingolipids such as sphingomyelin (SM), ceramide (Cer) and sphingosine-1-phosphate (S1P) in lymphoproliferative disorders. The balance between SM, Cer and S1P plays an important role in the maintenance of homeostasis in cell function. The decrease of SM and the increase of Cer in the cells induce cell death whereas the increase of S1P and SM enhances cell survival and proliferation. In terms of cell migration and adhesion S1P has a strong chemo-attractant function, and the decrease of SM in the plasma membrane of lymphoid cells shows higher extent of migration than the control cells

1.1 Sphingolipids in Lymphoid Cell Functions

1.1.1 Apoptosis and the Mechanisms by Which Ceramide Induces Apoptosis

The involvement of ceramide in apoptosis was first reported in 1993. Obeid et al. reported that DNA fragmentation, a hallmark of apoptosis, was induced in response to N-acetyl sphingosine (C_2-ceramide) in various cells including the human T cell line CEM as well as primary mouse thymocytes [2]. Thereafter, the role for ceramide in apoptosis has been intensively investigated.

C2-ceramide induced apoptosis in WEHI 231 murine B-cell lymphoma cells. Furthermore, intracellular accumulation of ceramide was detected in WEHI 231 cells exposed to apoptosis inducers, including irradiation and dexamethazone [3]. Ceramide was shown to accumulate following cross-linking with anti-IgM on parental WEHI-231 cells, whereas only a minor increase of ceramide was detected in WEHI-231 sublines resistant to anti-IgM-induced apoptosis [4]. Exposure of SKW6.4 B-lymphoblastoid cells to anti-Fas antibody induced apoptosis and a 200–300 % elevation in endogenous levels of ceramide. In contrast, similar treatment of Fas-resistant MUTU-BL Burkitt lymphoma cells caused insignificant changes in ceramide levels. Exposure of the resistant cells to exogenous ceramide induced apoptosis. These results provided evidence for ceramide as a necessary and sufficient lipid mediator of Fas-mediated apoptosis [5]. A subline of WEHI-231 lymphoma cells deficient in ceramide production was found to be resistant to radiation-induced apoptosis compared with the parental cells, suggesting that loss of ceramide production represented an extranuclear mechanism for the development of resistance [6].

Several downstream effectors and other mechanisms have been implicated in ceramide-induced apoptosis and in the following sections we will discuss them in detail.

1.1.2 Rb

In Molt-4 cells, growth arrest induced by serum withdrawal was associated with dephosphorylation of Rb and elevation of endogenous ceramide. The addition of short-chain ceramide resulted in dephosphorylation of Rb and growth arrest, indicating Rb as a downstream target of ceramide [7].

1.1.3 Ras and Raf-1

Ligation of Fas induced SM consumption and concomitant ceramide production in Jurkat T leukemia cells. Exogenous ceramide stimulated the activation of Ras and MAP kinases, as did Fas ligation. These results implicated that ceramide acted as a second messenger in Fas signaling via Ras [8]. In murine pre-B 70Z/3 cells, TNF-induced activation of Raf-1 kinase was not inhibited by tricyclodecan-9-yl-xanthogenate (D609), an inhibitor of ASMase signaling pathway, implicating the involvement of neutral sphingomyelinase (NSMase) rather than ASMase in Raf-1 activation [9]. However, these conclusions may not be completely convincing since the specific target of D609 is unknown, as D609 has been shown to inhibit phosphatidylcholine-specific phospholipase C (PC-PLC), sphingomyelin synthase (SMS) as well as acid sphingomyelinase (ASMase) [10].

1.1.4 Protein Kinase C

In Molt-4 acute T-lymphoblastic leukemia cells, PMA induced Rb phosphorylation, and ceramide inhibited this effect. Although ceramide did not directly inhibit PKCα in vitro, exposure of Molt-4 cells to ceramide inhibited PKCα activity. Furthermore, ceramide inhibited basal and PMA-induced phosphorylation of PKCα, and okadaic acid blocked these effects suggesting that ceramide action may be mediated by a protein phosphatase [11]. In human T-lymphoblastic leukemia HPB-ALL cells, treatment with short-chain ceramide induced translocation of PKC-δ and -ε from the membrane to the cytosol within 5 min. Similarly, anti-Fas antibody induced cytosolic translocation of PKC-δ and -ε in HPB-ALL cells but not in the Fas-resistant subline [12].

1.1.5 Bcl-2 and Bcl-x_L

WEHI 231 murine B-cell lymphoma cells transfected with Bcl-2 or Bcl-x_L were resistant to C_2-ceramide-induced apoptosis [13]. Treatment of Molt-4 cells with exogenous ceramide induced proteolytic cleavage of PARP, which was blocked by caspases inhibitors or overexpression of *bcl-2* [14]. In ALL-697 human pre-B leukemia cells and Molt-4 cells, vincristine induced apoptosis in association with a prolonged and sustained accumulation of ceramide. Although overexpression of *bcl-2* resulted in complete protection of apoptosis in response to vincristine, ceramide formation was not affected by *bcl-2* overexpression. On the other hand, ceramide-induced apoptosis was prevented by overexpression of *bcl-2*, suggesting that Bcl-2 may act downstream of ceramide in vincristine-induced apoptosis. In contrast, Bcl-2 did not interfere with the ability of ceramide to activate Rb by dephosphorylation or to induce cell cycle arrest. These results suggest that the effect of ceramide on cell cycle arrest can be dissociated from that on Bcl-2-dependent inhibition of apoptosis [15].

1.1.6 ROS

In WEHI 231 cells, treatment with C_2-ceramide elevated intracellular peroxide production, which was not prevented by overexpression of *bcl-x_L*. Since C_2-ceramide-induced apoptosis was blocked by overexpression of *bcl-x_L*, these results suggested that Bcl-x_L may function downstream of ROS production [13].

Short-chain ceramide-induced apoptosis in Jurkat T cells was preceded by a rise in mitochondrial peroxide production and loss of cellular glutathione, suggesting the involvement of reactive oxygen species in ceramide-induced apoptosis [16].

1.1.7 Caspase

Treatment of Molt-4 cells with exogenous ceramide induced proteolytic cleavage of PARP, suggesting that ceramide induced activation of caspase [14]. In Jurkat T cells, DEVD-CHO, a specific inhibitor of caspase-3, blocked anti-Fas-induced apoptosis and ceramide accumulation, suggesting the requirement of caspase-3 activation for ceramide generation in Fas-induced apoptosis [17]. In another report, C_2-ceramide induced cleavage and activation of caspase-3 in Jurkat T cells. C_2-ceramide-induced apoptosis was blocked by a caspase-3-specific inhibitor DEVD-CHO. Furthermore, Jurkat cells resistant to anti-Fas-induced apoptosis were also resistant to C_2-ceramide. These results suggested that ceramide functioned upstream of caspase-3 activation [18]. In the human acute lymphoblastic T-cell line CEM-C7H2, short-chain ceramide-induced apoptosis was prevented by Bcl-2, but not by CrmA, suggesting that CrmA-inhibitable proteases such as Interleukin-1(IL-1) beta converting enzyme and related proteases were involved in ceramide generation by TNF-α but not in ceramide-induced apoptosis [19].

In Jurkat T cells, ceramide formation induced by CD95, etoposide, or irradiation was abrogated by Z-VAD-fmk, but only partially inhibited by DEVD-CHO. CD95-induced ceramide accumulation was completely blocked in $FLIP_L$-transfectant, whereas ceramide production in response to etoposide or irradiation was unaffected by $FLIP_L$. CD95, but not etoposide or irradiation, required caspase-8 to signal to mitochondrial apoptotic pathway. Although *bcl-2* overexpression abrogated ceramide accumulation in response to etoposide or irradiation, it only partially inhibited anti-Fas-induced ceramide formation. These results suggested that ceramide response to DNA damage fully depended on mitochondrial caspase-dependent pathway, whereas the response to CD95 only partially relied on these caspases [20]. Moreover, in wild-type Jurkat cells and their clones mutated in caspase-8 or FADD, both daunorubicin (DNR) and Ara-C similarly induced early ceramide generation (within 5–15 min) and interphasic apoptosis, while cleavage of caspase-8 was observed only in DNR-treated cells and after several hours in wild-type or FADD-mutant. These results suggested that caspase-8 activation was not a requisite for the interphasic apoptotic pathway mediated by drug-induced early generation of ceramide [21].

1.1.8 p53

Treatment of Molt-4 acute T-lymphoblastic leukemia cells with actinomycin D or gamma-irradiation induced p53-dependent apoptosis. Activation of p53 was followed by an increase in endogenous ceramide levels which was not observed in cells lacking functional p53. In contrast, serum deprivation accumulated ceramide irrespective of p53 status. Exogenous ceramide induced growth suppression without an increase in p53 levels. Furthermore, exogenous ceramide induced dephosphorylation of Rb irrespective of p53 status. These results suggest that p53 functions upstream of ceramide [22].

1.1.9 Lipid Rafts

In Jurkat T-cell leukemia cells, ceramide elevation was detected as early as 15–30 s after anti-Fas antibody treatment and peaked at 1 min. Capping of Fas was also detected 30 s after Fas ligation and peaked at 2 min. Pretreatment with natural *N*-palmitoyl sphingosine (C16-ceramide) enabled soluble Fas ligand to cap and kill without forced cross-linking of FasL. The presence of sphingolipid-enriched membrane domains may be essential for Fas capping since their disruption with cholesterol-depleting agents abrogated capping and prevented apoptosis [23]. Application of C16-ceramide at the cell surface triggered clustering of CD95 in JY B lymphocytes. Neutralization of surface ceramide by GST-CRDs almost completely inhibited CD95 clustering and apoptosis, and CD95 clustering was not observed in B lymphocytes from a patient with Niemann–Pick disease. Exogenous ASMase or transfection with a plasmid for ASMase expression restored CD95 clustering and apoptosis in ASMase-deficient cells. These results suggested that surface ceramide generated by ASMase may mediate CD95 clustering and initiation of apoptosis [24]. CD95 triggered translocation of ASMase to the outer surface of the plasma membrane, leading to the generation of extracellularly orientated ceramide, which enabled CD95 clustering in sphingolipid-rich membrane rafts and induction of apoptosis. ASMase deficiency, destruction of lipid rafts, or neutralization of surface ceramide by anti-ceramide antibody prevented CD95 clustering and apoptosis, whereas natural ceramide restored it in ASMase-deficient cells [23]. Comparing differences between ASMase-deficient versus wild type B lymphocytes, it was concluded that CD95 ligation induced a first circumscribed activation of caspase 8 which in turn was sufficient to trigger translocation of ASMase onto the outer leaflet of the plasma membrane. The ASMase-induced ceramide accumulation at the plasma membrane then mediated CD95 clustering and amplified the CD95 signaling inducing apoptosis [25].

In comparison between SMS-defective WR19L mouse lymphoid cells and those that was functionally restored by transfection with *SMS1*, expression of membrane SM enhanced Fas-mediated apoptosis through ceramide generation within lipid rafts, translocation of Fas into lipid rafts, increased DISC formation, and caspase activation [26].

1.1.10 Autophagy

Autophagy was first described as a “self-eating” system to protect the cells from starvation-induced death by recycling proteins and organelles, such as mitochondria. It is divided into three classes: macroautophagy, chaperon-dependent autophagy and microautophagy [27]. In macroautophagy, cytosolic components including mitochondria are separated by the double-membrane wall and sequestered into autophagosomes, which finally fuse with lysosomes to digest the contents. In microautophagy, vesicles are transferred into lysosomes with the intact plasma membrane. In mammalian cells, the role of macroautophagy (referred to as autophagy) in cancer cells has been investigated vigorously because cancer cells showed longer survival in starved conditions. Autophagy plays a role in various lymphocytic functions such as T cell activation and homeostasis, generation of T cell repertoire and B cell development [28]. NK cell-mediated autophagy was reported to induce autophagy of T cells to promote cancer cell survival [29], while, as a cause of immunodeficiency, autophagy induced by HIV-1 envelope glycoproteins (Env) was reported to cause apoptosis of CD4+ cells [30].

However, since a large body of literature suggests the role of autophagy in cell survival and differentiation of T cells and plasma cells, the role of autophagy in cell death of lymphocytes is controversial. We recently reported that ceramide induced autophagy in amino acids deprivation through mTOR suppression while S1P inhibited autophagy through the activation of the Akt pathway [31]. Dysregulation of autophagy may be involved in the induction of apoptosis, but the precise mechanism by which ceramide is involved in autophagy and how it is induced by a certain stress for cell survival or apoptosis remains unclear [32, 33]. Cell death by autophagy induced by amino acids deprivation was enhanced in SMS1 and 2 defective-T lymphocytes (WR19L/Fas-SM(−)). These cells are characterized by little SM on the cell surface as measured by LC-MS/MS and SM-binding lysenin staining, as compared to wild type WR19l/Fas-SM(+) cells and SMS1-overexpressing WR19l/Fas-SM(−) cells (WR19L/Fas-SMS1) (Fig. 2).

When the levels of sphingolipids were measured by LC-MS/MS after amino acids deprivation for 1 h, ceramide content in the cellular compartments, including autophagosomes, was significantly increased in WR19L/Fas-SM(−) cells (data not shown). These findings may point to a role for ceramide in autophagosomes in inducing apoptosis after the merge of the autophagosomes with lysosomes.

1.1.11 Regulated Necrosis/Necroptosis

TNF/zVAD induced caspase-independent cell death in Jurkat T, and this cytotoxic effect was significantly blocked by loss of Rip-1, suggesting the role of ceramide in necroptosis [34]. Smac mimetic BV6, which inhibits the function of the inhibitor of apoptosis (IAP), enhanced AraC-induced cell death through Rip-1 or the MLKL-dependent necroptosis pathway in drug-resistant AML cells [35]. U-937 monocytic leukemia cells showed that TNF, which is a well-known inducer of ceramide generation, caused cell death in a caspase-independent manner, suggesting the induction of necroptosis by ceramide pathway [36]. Previously, ceramide was reported to increase degradation of XIAP by the proteasome and TNF induced an increase of ceramide [37], strongly suggesting the relationship between ceramide and the necroptosis pathway even in lymphoid malignancies. Since the inhibition of

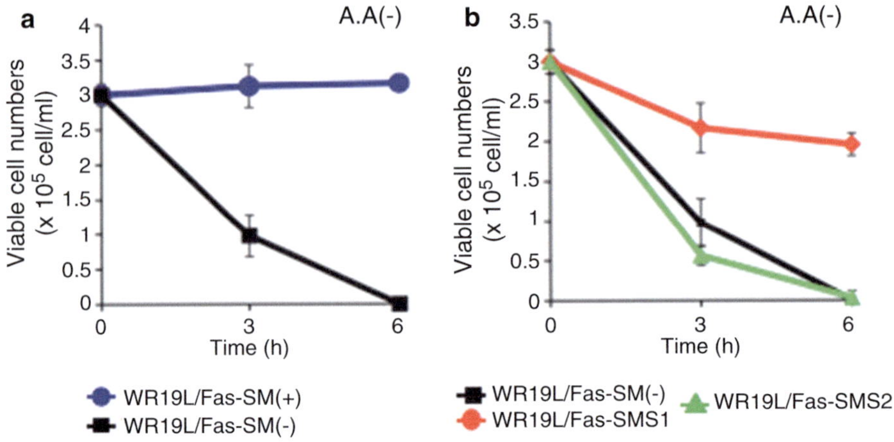

Fig. 2 SMS1 expression rescued WR19L/Fas-SM(−) cells from autophagic cell death. (**a**, **b**) Cell sensitivity to amino acid deprivation (AA(−)) on WR19L/Fas-SM(+) and WR19L/Fas-SM(−) cells (**a**) and WR19L/Fas-SMS1, WR19L/Fas-hSMS2 cells (**b**). Cells were incubated for amino acid free RPMI1640 supplemented with 5 μg/mL linsulin and 5 μg/mL transferrin for indicated time. Cell viability was quantified by MTS assay. Values reported are the mean ± SD of three independent experiments

necroptosis in CD4+ cells in the double KO mice of caspase8 and RIP-1 showed the increase of B220 and CD3 positive cells with a similar phenotype to Fas-defective lpr mice, necroptosis may play an important role in lymphoproliferative disorders [38]. In terms of the relationship between necroptosis and sphingolipids in lymphoid cells, it was reported that the selective inhibition of the glycolytic pathway in chronic lymphoid leukemia (CLL) cells induced caspase-independent necrotic cell death following treatment with nanoliposomal *N*-hexanoyl sphingosine (C6-ceramide) [39]. In addition to apoptosis and autophagic cell death, regulated necrosis/necroptosis through sphingolipid regulation may become a novel mechanism of cell death that could be exploited in apoptosis-resistant lymphoproliferative disorders.

1.1.12 Migration

We recently showed that SMS1-defective T cells and SMS-double KO mouse-derived embryonic fibroblasts (MEFs), which contain little SM in the plasma membrane, showed significant increase of migration following the CXCL12 concentration gradient [40]. A similar increase of migration to fetal bovine serum was detected in T lymphoid WR19L/Fas-SM(−) cells (Fig. 3).

Since CXCR4, a CXCL12 receptor, was dimerized and concentrated in microdomain of plasma membrane in WR19l/Fas-SM(−) cells as compared to WR19L/Fas-SM(+) cells, the regulation of SM in the microdomain by SMS is critical to migration induced by chemokine such as CXCL12 [41] (Fig. 4).

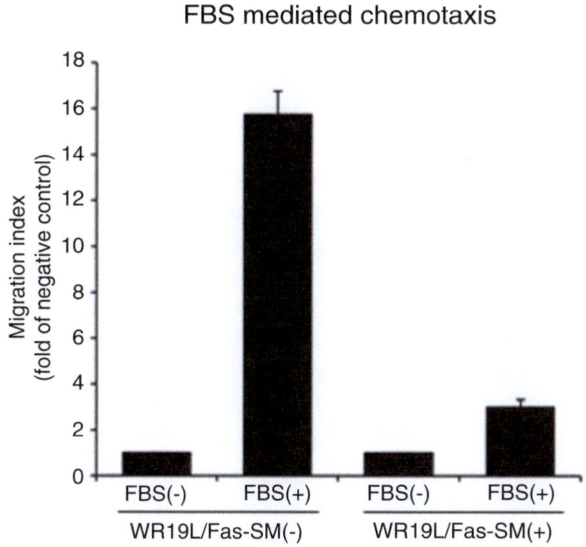

Fig. 3 SMS deficiency promotes FBS-induced cell migration. WR cells migration was measured by transwell migration assay. DMEM containing 10 % FBS was added to lower chamber of the vessel used to measure cell migration. Cells were added to upper chamber and incubated for 6 h at 34 °C. The migrant cells moved to the far side of the upper chamber were stained with DAPI, and counted. The number of cells migrating in response to the agonist relative to the basal level of cell, divided by that of SM(+)and FBS(−) cells and represented as migration index. Values are mean ± SD (n = 3). **, P < 0.01

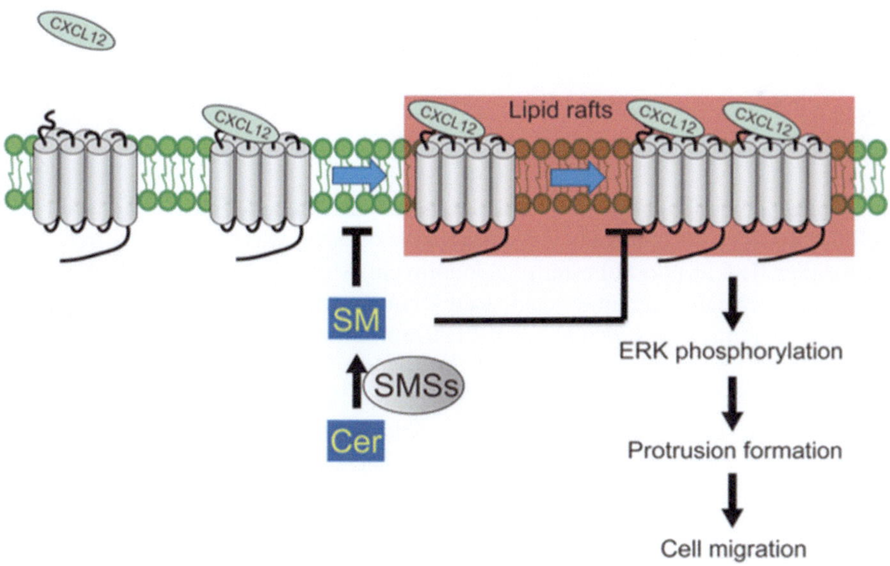

Fig. 4 The regulation of cell migration through CXCL12/CXCR4 axis by SMS-generated SM. Generally, CXCL12 binds to its receptor CXCR4 on non-raft region of plasma membrane, and then move to lipid rafts and form the dimer to induce migration through ERK phosphorylation and protrusion formation. SMS-generated SM controls migration via regulation of CXCL12/CXCR4 transfer to lipid raft and dimerization

In macrophages and monocytes, C1P induced migration through the release of the monocyte chemoattractant protein-1 (MCP-1), although this was not the case for migration of lymphocytes [42].

S1P has been studied as a strong inducer of T and B cell egress from lymph nodes and bone marrow under normal physiological conditions [43]. S1P binds to S1P receptors ($S1P_1$–$S1P_5$) transducing its signal via small G proteins. In terms of cell trafficking by S1P, $S1P_1$ showed a chemotactic action while $S1P_2$ showed a negative effect on chemotaxis [44]. In fact, overexpression of $S1P_2$ induced B cell confinement in germinal center niche, suggesting the inhibition of egress of B cells from germinal center of lymph nodes by $S1P_2$ [45]. NK cells express $S1P_5$, and $S1P_5$ deficient mice showed abnormality in NK cell distribution, suggesting each lymphocyte utilizes different trafficking systems involving specific S1P receptors [46]. It is still unclear whether the regulation of migration, attachment and egress by S1P plays a role in the proliferation and infiltration of lymphoid malignant cells.

1.1.13 Differentiation

Recently, the differentiation of mouse ES cells was guided towards neuronal and glial lineage by the balance of ceramide and S1P [47]. In the lymphoid cells, the combination of short-chain ceramide, including C6- and *N*-octanoyl sphingosine (C8-ceramide), with TGF-β induced a long-term expression of FoxP3 in regulatory T cells (Treg) as compared to TGF-β alone. Surprisingly, other ceramides, such as C2-, C16- and C24-ceramide did not show the same effect [48]. These results suggested the role of ceramide in physiological differentiation of lymphoid cells.

1.1.14 Proliferation

In human T-lymphocytes and Jurkat cells, an inverse relationship between the cellular levels of ceramide and the proliferative capacity was observed by treatment with either IL-2 or phorbol ester plus ionomycin. Moreover, treatment with exogenous cell-permeable ceramide decreased DNA synthesis. These results suggested the role of ceramide in the regulation of T cell proliferation [49]. In the IL-2-dependent mouse T cell line CTLL-2, addition of short-chain ceramide blocked IL-2-induced cell cycle entry, as well as apoptosis triggered by IL-2 deprivation. The protective effect of short-chain ceramide towards apoptosis was achieved only at the early stage of IL-2 deprivation [50]. We showed that SM in the microdomain of the plasma membrane played a critical role in Transferrin (Tf)-induced T cell proliferation in a clathrin-dependent manner as SMS-defective T cells proliferated much slower than wild type. This was because of the deterioration of Tf trafficking into the recycle pathway [51].

Thus, it is suggested that SM, which was previously mainly considered solely as a source for ceramide production, it now begins to be considered as a regulator of cell function through the modulation of the integrity of microdomains.

1.1.15 Secretion

In splenocytes of ASMase deficient mice, stimulation with a combination of anti-CD3 and anti-CD28 antibodies induced IL-2 expression. However, secretion of IL-2 was significantly reduced and the intracellular IL-2 levels were elevated. Proliferation of ASMase-deficient splenocytes stimulated with anti-CD3/anti-CD28 was reduced by 50 % in comparison to cells with intact ASMase. These results suggested that ceramide generated by ASMase was not involved in CD28 signal transduction per se but rather in the secretion of cytokines [52]. Moreover $CD8^+$ T cells from ASMase-deficient mice were defective in exocytosis of cytolytic effector molecules and showed attenuated cytotoxic activity, and a delayed elimination of lymphocytic choriomeningitis virus was observed in ASMase-KO mice. Secretory granules underwent shrinkage after fusion with the plasma membrane in wild-type $CD8^+$ T cells, whereas the contraction of secretory granules was markedly impaired in ASMase-KO $CD8^+$ T cells. It was postulated that an increase in the tension of the lipid bilayer of the vesicle following cleavage of SM by ASMase would add contractile force facilitating the extrusion of the contents from cytotoxic granules [53].

S1P generated by sphingosine kinase 1 (SphK1) increased IL12p70 through $S1P_1$ in splenocytes [54]. On the other hand, S1P was reported to decrease the secretion of IL-12 and -23 and to increase that of IL-27 through $S1P_1$ in LPS-stimulated dendritic cells, suggesting the importance of further investigation of the involvement of S1P/S1P receptors in cytokine secretion [55].

1.2 Mechanisms of Ceramide Generation

1.2.1 Acid Sphingomyelinase (ASMase)

TNF treatment of Jurkat cells resulted in rapid hydrolysis of SM with a concomitant production of ceramide [56]. Exogenous addition of SMase or ceramide mediated NF-κB activation in permeabilized Jurkat cells. D609 pretreatment of Jurkat cells resulted in a dose-dependent inhibition of TNF-induced NF-κB activation. DAGs stimulated SM hydrolysis at pH 5.0 in a cell-free system. These results suggested that TNF-responsive PC-PLC activated ASMase via DAG, resulting in ceramide generation, which triggered NF-κB activation. In murine 70Z/3 pre-B cells, TNF treatment of a clone overexpressing wild-type TNF-R55 simultaneously elicited NSMase and ASMase activity, whereas TNF-R55 mutants truncated in the C-terminus were defective in ASMase activation but retained NSMase activation. These results suggested that both NSMase and ASMase were independently activated by distinct cytoplasmic domains of TNF-R55 [57]. In human T cell lymphoma HuT78 cells, CD95 cross-linking resulted in SM hydrolysis and ceramide generation by the action of both NSMase and ASMase. On the other hand, CD95 cross-linking failed to activate ASMase in HuT78 mutants expressing death domain-defective CD95, while NSMase, ERK-2 and PLA2 activities were induced. These results suggested that ASMase-mediated pathway contributed to CD95-induced apoptotic signal [58]. Inhibition of caspases by Ac-YVAD-cmk or transient CrmA transfection prevented stimulation of ASMase, release of ceramide and activation of JNK and p38 MAPK upon Fas-receptor crosslinking [59].

Lymphoblasts from Niemann–Pick patients, which have an inherited deficiency of ASMase activity, failed to accumulate ceramide generation and induce apoptosis in response to ionizing radiations. These abnormalities were reversible upon restoration of ASMase activity by overexpression of human ASMase cDNA. ASMase knockout mice expressed defects in radiation-induced ceramide generation and apoptosis in vivo. These results directly addressed the role of ceramide generation via ASMase in radiation-induced apoptosis [60]. Intravenous injection of anti-CD4 antibodies induced apoptosis of peripheral blood lymphocytes from normal mice via CD95/CD95 ligand, whereas ASMase knock-out mice failed to immunodeplete T cells. Furthermore, ASMase-deficient mice or lymphocytes ex vivo manifested resistance to anti-CD95 treatment. However, if a strong stimulus was applied, ASMase was dispensable for CD95-induced apoptosis [61]. In another report, $ASMase^{-/-}$ mice were completely resistant to the lethal effect of intravenous injection of low dose anti-Fas antibody, whereas both $ASMase^{+/+}$ and $ASMase^{-/-}$ mice were similarly susceptible to high dose anti-Fas. These results indicated a role for ASMase in some but not all forms of Fas-induced apoptosis [62]. The CD28 costimulatory signaling triggered the activation of ASMase (but not NSMase) resulting in the generation of ceramide in resting as well as in activated primary T cells or Jurkat cells. In contrast, ligation of CD3 or CD2 did not induce ASMase activation [63].

1.2.2 Neutral Sphingomyelinase (NSMase)

Exogenously applied bacterial NSMase in Molt-4 cells was unable to induce apoptosis, whereas transfection and induction of bacterial NSMase gene induced PARP cleavage and cell death, supporting the existence of a signal transducing pool of SM distinct from the pool accessible to exogenous SMase [64]. Treatment of Molt-4 cells with BSO, an inhibitor of GSH synthesis, resulted in a time-dependent depletion of GSH, accompanied by increased SM hydrolysis and ceramide generation, consistent with the notion that GSH inhibited NSMase [65]. In apoptosis induced by TCR cross-linking in the T cell hybridoma 3DO, NSMase activation and production of ceramide were elicited. Pharmacological inhibition of ceramide production by fumonisin B1 impaired TCR-induced IL-2 production and apoptosis. Moreover, specific inactivation of NSMase by antisense cDNA reduced IL-2 production and ERK activation after TCR cross-linking in Jurkat cells [66]. Treatment of WEHI-231 murine B-cell lymphoma cells with *N*-oleoylethanolamine (OE) activated NSMase, induced ceramide production, and apoptosis. OE-resistant cells were also resistant to ionizing radiation-induced apoptosis [67].

Although Jurkat cells overexpressing NSMase1 showed dramatically elevated NSMase activity, steady-state levels of SM and ceramide were not altered.

Moreover, Fas-induced ceramide response and apoptosis in NSMase1-transfectant was identical to vector-transfectant, suggesting that NSMase1 was not involved in CD95-induced ceramide generation and apoptosis [68]. Accordingly, it was demonstrated that NSMase1 does not act as a NSMase in cells but it is rather a lyso-platelet activating factor-phospholipase C [69].

1.2.3 De Novo Ceramide Synthesis

The anticancer drug DNR stimulated ceramide elevation and apoptosis in P388 murine lymphocytic leukemia cells. DNR-induced ceramide elevation did not result from SM hydrolysis, but rather from activation of ceramide synthase, since fumonisin B1, a ceramide synthase inhibitor, blocked DNR-induced ceramide elevation and apoptosis [70]. In Molt-4 cells, accumulation of ceramide mass by the chemotherapeutic agent etoposide was found to occur primarily due to activation of the de novo pathway through activation of serine palmitoyltransferase. Ceramide generated from de novo synthesis was not involved in caspase-induced PARP cleavage but rather played a role in disrupting membrane integrity [71]. Treatment with either anti-Fas IgM (CH-11) or exogenous ceramide induced the dephosphorylation of SR (serine/arginine-rich) proteins in Jurkat acute leukemia T-cells. Caliculin A, an inhibitor of both PP1 and PP2A, but not okadaic acid, a PP2A-specific inhibitor, inhibited dephosphorylation of SR proteins, indicating that PP1 was responsible for it. Pretreatment with myriocin or fumonisin B1 blocked Fas-induced SR protein dephosphorylation. These results suggested that ceramide generated from de novo synthesis activated PP1, leading to the dephosphorylation of SR proteins in Fas-induced apoptosis [72]. An increase in ceramide mass was also observed in response to heat-shock in Molt-4 acute leukemic lymphocytes. The de novo sphingolipid biosynthesis pathway was rapidly induced upon heat-shock and it also induced dephosphorylation of SR proteins. Similarly to Fas treatment, heat-shock-induced dephosphorylation of SR proteins was dependent upon de novo-generated ceramide [73].

In Ramos B-cells, induction of apoptosis by B-cell receptor cross-linking was associated with an early rise of C16-ceramide derived from the de novo ceramide synthesis. Fumonisin B1 but not Z-VAD-fmk inhibited BcR-induced loss of mitochondrial transmembrane potential, implicating de novo-generated C_{16}-ceramide in mitochondrial damage upstream of caspase activation [74]. Furthermore, in BcR-induced apoptosis, an early increase in long-chain (predominantly C16-) ceramide was caspase-independent, whereas very long-chain (mainly C24) ceramide was generated later in a caspase-dependent manner. BcR-induced formation of long-chain ceramide species resulted in proteasomal activation, leading to degradation of XIAP and subsequent activation of effector caspases. These results demonstrated that de novo-generated long-chain ceramide species were involved in the activation of effector caspases and subsequent formation of very long-chain ceramide species [37].

1.2.4 Sphingomyelin Synthase (SMS)

WR19L/Fas-SM(−) mouse lymphoid cells showed a defect of SM at the plasma membrane due to the lack of SM synthase activity. WR19L/Fas-SM(−) cells expressing SMS1 restored the accumulation of SM at the plasma membrane. The growth of WR19L/Fas-SM(−) cells was severely inhibited under transferrin and insulin-containing, serum-free conditions, whereas WR19L/Fas-SMS1 cells restored growth [75]. In Jurkat T cells overexpressing SMS1, the increase of ceramides following photodynamic therapy (PDT) was lower than in vector-transfected cells. PDT-induced DEVDase activation was substantially reduced in SMS-overexpressors. Furthermore, in SMS1 siRNA-transfected cells the increase of ceramides was higher than in control siRNA-transfectants after PDT, and PDT-induced DEVDase activation was enhanced in SMS1 siRNA-transfectants. These results suggested that SMS1 played a pivotal role in ceramide response and apoptosis induced by photodamage [76, 77].

In Jurkat cells, Fas ligand treatment inhibited SMS activity before nuclear fragmentation. SMS1 was cleaved by active caspase-8, and the subsequently accumulated ceramide enhanced apoptotic cell death. Conversely, overexpression of SMS1 protected cells from FasL-induced ceramide generation and cell death [78]. Fas ligand was also shown to regulate caspase-9 dependent apoptosis through modulation of ceramide via SMS1, because Fas-induced apoptosis was enhanced by knock down of SMS1 [79].

1.2.5 Glucosylceramide Synthase (GCS)

In Jurkat cells overexpressing GCS, ceramide induced by CD95 cross-linking, etoposide, or irradiation was not glycosylated. In contrast, de novo synthesized ceramide as well as exogenous short-chain ceramide were efficiently glycosylated. These results suggested that GCS may convert de novo synthesized but not SM-derived ceramide [80].

IL-2 rescued KHYG-1 human NK cell line from apoptosis along with a reduction of ceramide accumulation. Conversely, IL-2 deprivation induced an increase in ceramide due to activation of ASMase and inhibition of GCS and SMS. LY294002, a PI-3 kinase inhibitor, inhibited IL-2-rescued survival concomitantly with the increase in ceramide due to activation of ASMase and inhibition of GCS and SMS. These results suggested the involvement of PI-3 kinase in the regulation of ceramide [81].

GalCer-enriched lymphoblasts from patients with Krabbe's disease were significantly more resistant to DNR- and cytosine arabinoside-induced apoptosis than normal lymphoblasts, whereas GluCer-enriched cells from patients with Gaucher's disease were more sensitive, suggesting that GalCer may be a "protector" of apoptosis [82].

1.3 Sphingolipids Other than Ceramide in Lymphoid Cell Functions

1.3.1 Sphingosine (Sph)

Besides its inhibitory effect on PKC [83], several other roles for Sph have been reported. Sph was found to cause prominent phosphorylation of a number of cytosolic proteins in Jurkat T cells. It was suggested that Sph activated more than one protein kinase. Activation of kinases showed a very specific requirement for D-erythro-sphingoid bases. Furthermore, other related molecules including C2-ceramide and SM were not active [84]. Sph potently induced early dephosphorylation of the retinoblastoma gene product (Rb), hypophosphorylated forms of which were observed in the G_0/G_1 phase of the cell cycle. Induction of Rb dephosphorylation by Sph preceded inhibition of growth and a specific arrest in the G_0/G_1 phase of the cell cycle [85]. Cotreatment of Molt-4 cells with Sph and fumonisin B1 demonstrated that the growth inhibitory effects of Sph were potentiated by fumonisin B1. Fumonisin B1 also potentiated Sph-induced dephosphorylation of Rb protein. These results suggested that Sph itself (without conversion into ceramide) induced dephosphorylation of Rb and growth arrest [86]. Sph potently inhibited growth of Molt-4 lymphoblastic leukemia cells through the activation of Rb through its dephosphorylation, the loss of the interaction of E2F and its specific DNA sequence element, and subsequent c-myc down-regulation. These results suggested that activation of Rb by Sph induced sequestration of E2F with the resultant loss of its DNA-binding and gene-transcribing abilities [87].

ISP-1 (or myriocin), an inhibitor of serine palmitoyltransferase, induced apoptosis in an IL-2-dependent cytotoxic T cell line, CTLL-2. The addition of a low dose of Sph inhibited ISP-1-induced apoptosis, while a high dose of Sph itself induced apoptosis in CTLL-2 cells, suggesting dual roles of Sph in pro- and anti-apoptotic signaling [88].

1.3.2 Sphingosine 1-Phosphate (S1P)

The role for S1P in cell proliferation and survival has been also proposed. In Jurkat cells, cotreatment with S1P attenuated apoptosis induced by agonistic anti-Fas antibody [89], and Fas ligation or exogenous short-chain ceramide induced caspase activation followed by PARP cleavage, which was suppressed by S1P [90]. In Jurkat T cells, overexpression of SK1 suppressed serum deprivation and ceramide-induced apoptosis, and, to a lesser extent, Fas-induced apoptosis [91]. S1P prevented apoptosis by inhibiting the translocation of cytochrome c and Smac/DIABLO from mitochondria to the cytosol induced by anti-Fas, TNF, serum deprivation, and short-chain ceramide [92].

DNA damaging agents including actinomycin D, doxorubicin, etoposide, and γ-irradiation caused a reduction in the protein levels of SK1 with a concomitant decrease in SK1 activity in Molt-4 cells. Z-VAD-fmk and CA-074-Me inhibited this down-regulation of SK1, implicating the involvement of cysteine proteases. When p53 up-regulation was inhibited, SK1 down-regulation was rescued, indicating p53 dependence of SK1 modulation. Exogenous S1P partially attenuated Act D-induced cell death. These results suggested the significance of SK1 and S1P in maintaining cell growth [93].

1.3.3 Sphingomyelin (SM)

While ceramide is well known as a lipid mediator for many cell functions, sphingomyelin (SM) has been mainly regarded as a lipid with structural functions for the membranes. However, recently the roles of SM in cell proliferation and migration were reported. In fact SM regulates the organization of microdomains in the plasma membrane, which act as platforms for the membrane receptors [41]. Recently SM production in membrane microdomains was reported to play a role in the resistance to Fas ligand-induced cell death by regulating Fas expression on the plasma membrane [94]. It is proposed that SMS1 engages in the regulation of SM in the plasma membrane after its production in the Golgi apparatus while SMS2 in the plasma membrane may intervene in SM depletion when it is not recovered by the compensation of SMS1 [41]. Dysfunction of membrane microdomains caused by deficiency of SM in SMS1-KO conditions impaired the secretion of cytokines such as IL-2, IL-6 and IFN-gamma, resulting in the dysfunction of CD4+ lymphocytes in vivo and in vitro [95]. SM deficiency in the clathrin-dependent pathway due to loss of SMS1 also caused the inhibition of T cell proliferation by impairing the recycle of transferrin/transferrin receptor in the vesicular trafficking (Fig. 5) [51]. SM regulated by both SMS1 and 2 has been shown to be an important lipid to keep the integrity and homeostasis of membrane microdomains, which affects the function of pro- and anti-apoptotic signaling through trans-membrane receptors in hematological malignancies [96].

1.3.4 GD3/GM3

Cross-linking of CD95 in human cutaneous T cell lymphoma HuT78 cells triggered transient accumulation of GD3, which was dependent on the death domain of CD95 protein and caspase activation. GD3 disrupted mitochondrial transmembrane potential and induced apoptosis in a caspase-independent manner. Transient overexpression of GD3 synthase directly induced apoptosis, while either pharmacological inhibition of GD3 synthesis by PDMP or exposure to GD3 synthase antisense oligodeoxynucleotides prevented CD95-induced apoptosis [97]. Furthermore, NPD lymphoblasts failed to accumulate GD3 and displayed substantially inefficient apoptosis after CD95 cross-linking. Apoptosis could be efficiently triggered by either exogenous short-chain ceramide or mannose-receptor-mediated transfer of ASMase into NPD lymphoblasts [98].

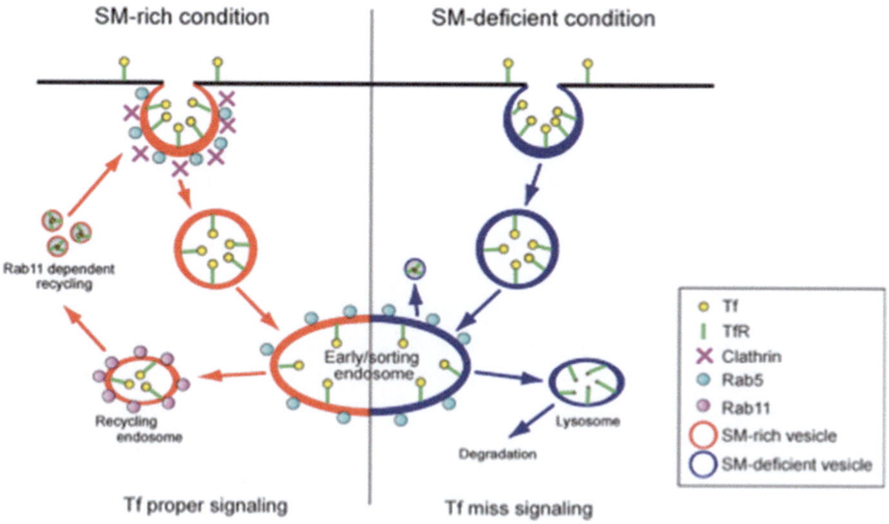

Fig. 5 The trafficking pathway of internalized Tf/TfR in SM-rich or SM-deficient conditions. In SM-rich condition, Tf/TfR goes to recycling endosome through early/sorting endosome and returns to plasma membrane. On the other hand, Tf/TfR proceeds to lysosome via early/sorting endosome and degradation pathway in SM-deficient condition induced by SMS1 deficiency

The *N*-glycosylated variant of GM3 ganglioside (NGcGM3) induced anti-inflammatory cytokines such as IL-4 and IL-10 in CD4+CD25− cells, resulting in tumor growth. Reduction of NGcGM3 in myeloma cells showed the deterioration of in vivo tumor growth [99]. Gangliosides seem to play a role in hematological malignancies by affecting not only the immune surveillance system against the tumor progression but also proliferation of tumor cells.

1.4 Clinical Aspects of Sphingolipids

1.4.1 Leukemia [Acute Lymphoid Leukemia (ALL) and Chronic Lymphoid Leukemia (CLL)]

The levels of ceramide were lower in chemo-resistant leukemia (including both myeloid and lymphoid) patients than in chemo-sensitive patients, and the activities of GCS and SMS were higher in chemo-resistant leukemia cells than in chemo-sensitive cells. The results suggested that the mechanism by which ceramide is metabolized to SM and GC to reduce the apoptotic effects is enhanced for survival of leukemia cells [100]. In another report, mutations in NSMase gene *SMPD3* were found in 5 % of acute myeloid leukemias and 6 % of acute lymphoid leukemias but not in other tumor types, suggesting the existence in a small fraction of these leukemias of a block of ceramide generation [101]. In addition to the involvement of SMS, GCS and NSMase, an increase of de novo synthesis of ceramide was reported in Arsenic trioxide (ATO)-treated adult T cell leukemia (ATL) cells [102].

In Jurkat T leukemia cells, gamma-ray exposure induced progressive accumulation of ceramide with time by activation of different ceramide metabolizing pathways while, in Molt-4T acute leukemia cells, 7.5 Gy X-ray irradiation did not induce ceramide generation even though apoptosis occurred [103, 104]. However, in general, an increase of ceramide levels after irradiation correlated to the radiation-induced cell death in many cell types including prostate, pancreatic, colorectal, and breast [105].

In B CLL cells, WSU and JVM-2 anti-cancer agent fludarabine induced the activation of caspases and the increase of ceramide through ceramide synthase, as fumonisin B1 inhibited ceramide generation [106]. The SM surface content in B lymphocytes from 36 patients with CLL and 13 patients with non-Hodgkin's lymphoma was lower in CD-20 antibody-resistant group as compared to the sensitive group, suggesting the involvement of SM in the plasma membrane in the diversity of B cell immunotherapy-sensitivity through the regulation of CD-20 function in microdomain [107]. C6-ceramide nanoliposomes induced a decrease of GAPDH expression and protein in CLL cell line in an caspase-independent manner, suggesting that ceramide may be involved in necrosis by targeting the Warburg effect. In fact, C6-ceramide liposomes were effective to inhibit tumor growth of CLL cells through energy depletion by the suppression of glycolysis in a mouse xenograft model [39]. Expression of S1P receptors such as $S1P_1$ and $S1P_2$ were reduced in CLL B cells. It was suggested that $S1P_1$ expression was controlled by p66Shc through its pro-oxidant activity [108].

In leukemia cells isolated from a mouse model of p185 BCR/ABL-expressing acute lymphoid Leukemia (ALL), loss of SK2 reduced MYC expression and hindered their proliferation. A xenograft model of human ALL cells in NOD/SCID$\gamma c^{-/-}$ mice showed prolonged survival upon treatment with an SK2 inhibitor, suggesting a role for S1P inhibition or ceramide increase in the suppression of human ALL progression [109]. Treatment of primary ALL leukemia cells with withanolide D, a pure herbal compound isolated from *Withania somnifera*, induced an increase of ceramide through activation of NSMase2 and it induced apoptosis through the activation of JNK and p38MAPK [110].

FTY720 is a synthetic analogue of ISP-1 and an agonist of S1P receptors; its phosphorylation by SK2 results in the inhibition of the immune system through the modulation of T cell trafficking. In primary B cells from CLL patients, FTY720 induced caspase-independent cell death probably through the activation of PP2A [111, 112]. In human ALL xenografts in NOD/SCIDγc(−/−) mice, FTY720 was effective towards Ph(+) ALL but not Ph(−) ALL, suggesting activation of PP2A by FTY720 to induce cell death [113]. In the syngeneic Fischer rat model of NK-cell leukemia, FTY720 increased caspase-dependent apoptosis with an increase of Sph and a decrease of Mcl-1 due to cathepsin B-induced degradation [114]. FTY720 also induced cell death in Jurkat T cells and multiple myeloma cells by inhibiting proproliferative signaling, such as Akt and NF-κB, but the mechanism by which FTY720 induces cell death is unclear in terms of the involvement of S1P receptors and S1P-induced survival signal [115, 116].

When the gene expression signature of peripheral blood mononuclear cells was investigated in 30 LGL leukemia patients, acid ceramidase was overexpressed and $S1P_5$ was increased, suggesting that alteration of the balance between ceramide and S1P signal is a possible strategy for the treatment of LGL leukemia [117]. Nanoliposomal C6-ceramide induced caspase-dependent apoptosis through the suppression of survival in NK-LGL leukemia cells. Similar effects by C_6-ceramide were observed in Fischer F344 rat xenograft model of aggressive NK-LGL leukemia [118]. Co-administration of C6-ceramide nanoliposomes and an inhibitor of GCS, PPMP elicited an increase in endogenous long-chain ceramide species, which led to cellular apoptosis in a synergistic manner via the mitochondrial intrinsic cell death pathway in leukemic NK cells [119].

1.4.2 Malignant Lymphoma

The treatment of malignant B cell lymphoma has been improved after the development of the anti-CD20 antibody, rituximab. Treatment of Daudi and RLB-lymphoma cells with rituximab induced a rapid and transient increase in ASMase activity and concomitant ceramide generation in raft microdomains, resulting in accumulation in G1 phase and growth inhibition (but no apoptosis) [120]. It was recently reported that responsiveness of rituximab is dependent on the microdomain structure regulated by sphingolipids, such as GM1. Marginal zone lymphoma showed much higher levels of GM1 than CLL and effectiveness of rituximab seemed to correlate with the level of GM1 in microdomains [121]. The relationship between cell death induced by rituximab with sphingolipids is still ambiguous, and the cytotoxic mechanism of rituximab remains to be clarified.

In clinical tissue samples from patients with non-Hodgkin lymphomas, SK1 was overexpressed and increased with progressing clinical grade [122]. In mantle cell lymphoma, cannabinoid induced cell death with the generation of ceramide through de novo ceramide synthesis with involvement of ceramide synthase 3 and 6 [123, 124]. Like leukemia cells, irradiation caused cell death of Burkitt's lymphoma cells with an increase of ceramide, and abrogation of ceramide generation with 12-*O*-tetradecanoylphorbol 13-acetate (TPA) treatment conferred resistance to radiation-induced apoptosis [125]. Malignant cells from primary effusion lymphoma (PEL), which is a very deadly tumor, etiologically linked to the Kaposi's sarcoma-associated herpes virus (KSHV), were treated with the SK2 inhibitor ABC294640, and they subsided to cell death with accumulation of ceramide and decrease of S1P. Furthermore, its effectiveness in inhibiting tumor growth was evident in a PEL xenograft mouse (NOD/SCID) model [126].

Hodgkin lymphoma (HL) cells showed the expression of $S1P_1$ in cell lines as well as patient samples. Immunohistochemical assessment of the tissues from classical Hodgkin lymphoma patients revealed that a subset of cases (12 %) showed strong, membranous staining for $S1P_1$ in Hodgkin-Reed-Sternberg cells. Antagonist of $S1P_1$ inhibited S1P-induced migration of HL cells, suggesting $S1P_1$ as a novel target for HL treatment [127].

The usefulness of $S1P_1$ expression for the diagnosis of lymphoma was also investigated. Strong expression of $S1P_1$ was observed in all classical mantle cell lymphoma cases, but in no cases of follicular lymphoma, marginal zone lymphoma, B lymphoblastic leukemia/lymphoma, or Burkitt's lymphoma. Also small fractions of chronic lymphocytic leukemia (CLL) in bone marrow, lymphoplasmacytic lymphomas in the lymph node, and diffuse large B-cell lymphomas exhibited positive staining. In the case of mantle cell lymphoma expressing low level of cyclin D1, the examination for $S1P_1$ may be valuable for the assistance of precise diagnosis [128]. Interestingly, $S1P_2$-KO mice developed germinal center (GC)-derived diffuse large B-cell lymphoma (DLBCL) by 1.5–2 years and in humans, 28 of 106 (26 %) DLBCL samples showed to harbor multiple somatic mutations in the 5′ sequences of the $S1P_2$ gene. These results strongly suggest the importance of $S1P_2$ to protect the patients from DLBCL progression [129].

In murine Gaucher's disease, sporadic fatal B cell lymphomas developed in 11 of 21 mice (6–24 months) but only two of eight control animals developed tumors by age 24 months. Unexpectedly, most mice with overt lymphoma had no or few Gaucher cells, but local inflammatory macrophages were present. Eleven of 39 of Gaucher mice developed monoclonal gammopathy, but in the control group only 1 animal of 25 had clonal immunoglobulin abnormalities. Although the positive relation of Gaucher disease with progression of lymphoma is well known, the implications of sphingolipids such as ceramide and glycolipids in B cell malignancies remains to be elucidated in the future [130].

1.4.3 Multiple Myeloma

There are limited data to show a relationship between sphingolipids and multiple myeloma (MM), so far. Treatment of MM cells with prokinetucin1 caused a time- and dose-dependent phosphorylation of MAPK, AKT and STAT3 and induced cell survival with up-regulation of SK1 expression and cellular activity [131]. In addition, it was shown in vitro and in vivo that S1P up-regulated myeloma cell adhesion mediated by $\alpha 4\beta 1$ and transendothelial migration stimulated by CXCL12, suggesting that the cooperation of S1P and CXCL12 plays a role in MM cell progression [132]. IL-6 increased SK, and IL-6 or overexpression of SK induced Mcl-1. Treatment with S1P inhibited dexamethasone-induced MM cell death by up-regulating Mcl-1. MM cells express SIP_1, $S1P_2$ and $S1P_3$, but it remains unclear which S1P receptor is most involved in Mcl-1-induced anti-apoptotic effect in MM cells [133, 134]. FTY720 induced potent cytotoxicity against drug-sensitive and drug-resistant multiple myeloma cell lines as well as freshly isolated tumor cells from multiple myeloma patients, who do not respond to conventional agents. FTY720 may overcome drug resistance in multiple myeloma cells and provide the rationale for its clinical evaluation to improve patient outcome in multiple myeloma [116].

2 Concluding Remarks

Firstly, it was reported that Sph inhibited PKC [83] and subsequently ceramide was shown to play a role in cell differentiation and programmed cell death of hematopoietic cells [1, 2]. Since then sphingolipids have been described as bioactive lipids able to regulate a diversity of cell functions including cell death, proliferation, differentiation, migration, senescence and secretion. Based on the great research progress it is well accepted now that ceramide acts as a lipid mediator, and SM works as a platform for trans-membrane signals not only in in vitro systems but also in in vivo models. In this manuscript the recent progress of the role of sphingolipids such as ceramide, S1P, Sph and SM and their metabolizing enzymes was summarized in terms of the functions of normal and malignant lymphoid cells. We provided key piece of evidence to understand the involvement of sphingolipids in the etiology of lymphoproliferative disorders such as lymphoid leukemia, malignant lymphoma and multiple myeloma and the mechanism by which sphingolipids mainly regulate cell death and proliferation. It must be acknowledged that our present knowledge seems to be far from the research goal to develop novel medication for lymphoproliferative disorders through the regulation of active sphingolipids. The following research challenges should be resolved in the future: (1) whether ceramides carrying different fatty acids exert different and specific actions in diverse cell functions, (2) which cell compartment or tissue is regulated by different species of ceramide and S1P for execution of cell death and other cell functions, and (3) what biochemical manipulations through sphingolipid metabolizing enzymes will be efficient to achieve the targeted balance among ceramide, SM, Sph and S1P. Thus, to develop a novel medication for lymphoproliferative disorders, the investigation of the implication for cell death and proliferation of the different sphingolipid molecular species and of the intracellular localization as well as the precise way of regulation of sphingolipid metabolism is required in not only in vitro but also in vivo pathophysiological conditions.

Acknowledgments This study was supported by the Takeda Science Foundation (2012–2016), and performed with the joint research project with Shalome Co. (2012–2014), Ltd., and with ONO Pharmaceutical Co. Ltd. (2011–2014). This work was also supported in part by a grant from Strategic Research Foundation Grant-aided Project for Private Universities from the Ministry of Education, Culture, Sports, Science and Technology, Japan (MEXT) (No. S1201004, 2012–2016) and from Kanazawa Medical University (2012–2013), Mizutani Foundation for Glycoscience (2013–2014), the Collaborative Research from Kanazawa Medical University (C2012-4, C2013-1. 2012–2013). We greatly appreciate Drs. Asano S. and Shakor A.B. kindly granting us the use of Figs. 4 and 5.

References

1. Okazaki T, Bell RM, Hannun YA (1989) Sphingomyelin turnover induced by vitamin D3 in HL-60 cells. Role in cell differentiation. J Biol Chem 264:19076–19080
2. Obeid LM, Linardic CM, Karolak LA, Hannun YA (1993) Programmed cell death induced by ceramide. Science 259:1769–1771

3. Quintans J, Kilkus J, McShan CL, Gottschalk AR, Dawson G (1994) Ceramide mediates the apoptotic response of WEHI 231 cells to anti-immunoglobulin, corticosteroids and irradiation. Biochem Biophys Res Commun 202:710–714
4. Gottschalk AR, McShan CL, Kilkus J, Dawson G, Quintans J (1995) Resistance to anti-IgM-induced apoptosis in a WEHI-231 subline is due to insufficient production of ceramide. Eur J Immunol 25:1032–1038
5. Tepper CG, Jayadev S, Liu B, Bielawska A, Wolff R, Yonehara S, Hannun YA, Seldin MF (1995) Role for ceramide as an endogenous mediator of Fas-induced cytotoxicity. Proc Natl Acad Sci U S A 92:8443–8447
6. Chmura SJ, Nodzenski E, Beckett MA, Kufe DW, Quintans J, Weichselbaum RR (1997) Loss of ceramide production confers resistance to radiation-induced apoptosis. Cancer Res 57:1270–1275
7. Dbaibo GS, Pushkareva MY, Jayadev S, Schwarz JK, Horowitz JM, Obeid LM, Hannun YA (1995) Retinoblastoma gene product as a downstream target for a ceramide-dependent pathway of growth arrest. Proc Natl Acad Sci U S A 92:1347–1351
8. Gulbins E, Bissonnette R, Mahboubi A, Martin S, Nishioka W, Brunner T, Baier G, Baier-Bitterlich G, Byrd C, Lang F et al (1995) FAS-induced apoptosis is mediated via a ceramide-initiated RAS signaling pathway. Immunity 2:341–351
9. Belka C, Wiegmann K, Adam D, Holland R, Neuloh M, Herrmann F, Kronke M, Brach MA (1995) Tumor necrosis factor (TNF)-alpha activates c-raf-1 kinase via the p55 TNF receptor engaging neutral sphingomyelinase. EMBO J 14:1156–1165
10. Adibhatla RM, Hatcher JF, Gusain A (2011) Tricyclodecan-9-yl-xanthogenate (D609) mechanism of actions: a mini-review of literature. Neurochem Res 37:671–679
11. Lee JY, Hannun YA, Obeid LM (1996) Ceramide inactivates cellular protein kinase Calpha. J Biol Chem 271:13169–13174
12. Sawai H, Okazaki T, Takeda Y, Tashima M, Sawada H, Okuma M, Kishi S, Umehara H, Domae N (1997) Ceramide-induced translocation of protein kinase C-delta and -epsilon to the cytosol. Implications in apoptosis. J Biol Chem 272:2452–2458
13. Fang W, Rivard JJ, Ganser JA, LeBien TW, Nath KA, Mueller DL, Behrens TW (1995) Bcl-xL rescues WEHI 231 B lymphocytes from oxidant-mediated death following diverse apoptotic stimuli. J Immunol 155:66–75
14. Smyth MJ, Perry DK, Zhang J, Poirier GG, Hannun YA, Obeid LM (1996) prICE: a downstream target for ceramide-induced apoptosis and for the inhibitory action of Bcl-2. Biochem J 316(Pt 1):25–28
15. Zhang J, Alter N, Reed JC, Borner C, Obeid LM, Hannun YA (1996) Bcl-2 interrupts the ceramide-mediated pathway of cell death. Proc Natl Acad Sci U S A 93:5325–5328
16. Phillips DC, Allen K, Griffiths HR (2002) Synthetic ceramides induce growth arrest or apoptosis by altering cellular redox status. Arch Biochem Biophys 407:15–24
17. Gamen S, Marzo I, Anel A, Pineiro A, Naval J (1996) CPP32 inhibition prevents Fas-induced ceramide generation and apoptosis in human cells. FEBS Lett 390:232–237
18. Mizushima N, Koike R, Kohsaka H, Kushi Y, Handa S, Yagita H, Miyasaka N (1996) Ceramide induces apoptosis via CPP32 activation. FEBS Lett 395:267–271
19. Geley S, Hartmann BL, Kofler R (1997) Ceramides induce a form of apoptosis in human acute lymphoblastic leukemia cells that is inhibited by Bcl-2, but not by CrmA. FEBS Lett 400:15–18
20. Tepper AD, de Vries E, van Blitterswijk WJ, Borst J (1999) Ordering of ceramide formation, caspase activation, and mitochondrial changes during CD95- and DNA damage-induced apoptosis. J Clin Invest 103:971–978
21. Terrisse AD, Bezombes C, Lerouge S, Laurent G, Jaffrezou JP (2002) Daunorubicin- and Ara-C-induced interphasic apoptosis of human type II leukemia cells is caspase-8-independent. Biochim Biophys Acta 1584:99–103
22. Dbaibo GS, Pushkareva MY, Rachid RA, Alter N, Smyth MJ, Obeid LM, Hannun YA (1998) p53-dependent ceramide response to genotoxic stress. J Clin Invest 102:329–339

23. Cremesti A, Paris F, Grassme H, Holler N, Tschopp J, Fuks Z, Gulbins E, Kolesnick R (2001) Ceramide enables fas to cap and kill. J Biol Chem 276:23954–23961
24. Grassme H, Schwarz H, Gulbins E (2001) Molecular mechanisms of ceramide-mediated CD95 clustering. Biochem Biophys Res Commun 284:1016–1030
25. Grassme H, Cremesti A, Kolesnick R, Gulbins E (2003) Ceramide-mediated clustering is required for CD95-DISC formation. Oncogene 22:5457–5470
26. Miyaji M, Jin ZX, Yamaoka S, Amakawa R, Fukuhara S, Sato SB, Kobayashi T, Domae N, Mimori T, Bloom ET, Okazaki T, Umehara H (2005) Role of membrane sphingomyelin and ceramide in platform formation for Fas-mediated apoptosis. J Exp Med 202:249–259
27. Santambrogio L, Cuervo AM (2011) Chasing the elusive mammalian microautophagy. Autophagy 7:652–654
28. Gianchecchi E, Delfino DV, Fierabracci A (2014) Recent insights on the putative role of autophagy in autoimmune diseases. Autoimmun Rev 13:231–241
29. Buchser WJ, Laskow TC, Pavlik PJ, Lin HM, Lotze MT (2012) Cell-mediated autophagy promotes cancer cell survival. Cancer Res 72:2970–2979
30. Espert L, Denizot M, Grimaldi M, Robert-Hebmann V, Gay B, Varbanov M, Codogno P, Biard-Piechaczyk M (2006) Autophagy is involved in T cell death after binding of HIV-1 envelope proteins to CXCR4. J Clin Invest 116:2161–2172
31. Taniguchi M, Kitatani K, Kondo T, Hashimoto-Nishimura M, Asano S, Hayashi A, Mitsutake S, Igarashi Y, Umehara H, Takeya H, Kigawa J, Okazaki T (2012) Regulation of autophagy and its associated cell death by "sphingolipid rheostat": reciprocal role of ceramide and sphingosine 1-phosphate in the mammalian target of rapamycin pathway. J Biol Chem 287:39898–39910
32. Jiang W, Ogretmen B (2014) Autophagy paradox and ceramide. Biochim Biophys Acta 1841:783–792
33. Young MM, Kester M, Wang HG (2013) Sphingolipids: regulators of crosstalk between apoptosis and autophagy. J Lipid Res 54:5–19
34. Thon L, Mohlig H, Mathieu S, Lange A, Bulanova E, Winoto-Morbach S, Schutze S, Bulfone-Paus S, Adam D (2005) Ceramide mediates caspase-independent programmed cell death. FASEB J 19:1945–1956
35. Chromik J, Safferthal C, Serve H, Fulda S (2014) Smac mimetic primes apoptosis-resistant acute myeloid leukaemia cells for cytarabine-induced cell death by triggering necroptosis. Cancer Lett 344:101–109
36. Sawai H (2014) Characterization of TNF-induced caspase-independent necroptosis. Leuk Res 38:706–713
37. Kroesen BJ, Jacobs S, Pettus BJ, Sietsma H, Kok JW, Hannun YA, de Leij LF (2003) BcR-induced apoptosis involves differential regulation of C16 and C24-ceramide formation and sphingolipid-dependent activation of the proteasome. J Biol Chem 278:14723–14731
38. Ch'en IL, Tsau JS, Molkentin JD, Komatsu M, Hedrick SM (2011) Mechanisms of necroptosis in T cells. J Exp Med 208:633–641
39. Ryland LK, Doshi UA, Shanmugavelandy SS, Fox TE, Aliaga C, Broeg K, Baab KT, Young M, Khan O, Haakenson JK, Jarbadan NR, Liao J, Wang HG, Feith DJ, Loughran TP Jr, Liu X, Kester M (2013) C6-ceramide nanoliposomes target the Warburg effect in chronic lymphocytic leukemia. PLoS One 8, e84648
40. Asano S, Kitatani K, Taniguchi M, Hashimoto M, Zama K, Mitsutake S, Igarashi Y, Takeya H, Kigawa J, Hayashi A, Umehara H, Okazaki T (2012) Regulation of cell migration by sphingomyelin synthases: sphingomyelin in lipid rafts decreases responsiveness to signaling by the CXCL12/CXCR4 pathway. Mol Cell Biol 32:3242–3252
41. Taniguchi M, Okazaki T (2014) The role of sphingomyelin and sphingomyelin synthases in cell death, proliferation and migration-from cell and animal models to human disorders. Biochim Biophys Acta 1841:692–703
42. Arana L, Ordonez M, Ouro A, Rivera IG, Gangoiti P, Trueba M, Gomez-Munoz A (2013) Ceramide 1-phosphate induces macrophage chemoattractant protein-1 release: involvement

in ceramide 1-phosphate-stimulated cell migration. Am J Physiol Endocrinol Metab 304:E1213–E1226
43. Cyster JG, Schwab SR (2012) Sphingosine-1-phosphate and lymphocyte egress from lymphoid organs. Annu Rev Immunol 30:69–94
44. Matloubian M, Lo CG, Cinamon G, Lesneski MJ, Xu Y, Brinkmann V, Allende ML, Proia RL, Cyster JG (2004) Lymphocyte egress from thymus and peripheral lymphoid organs is dependent on S1P receptor 1. Nature 427:355–360
45. Green JA, Suzuki K, Cho B, Willison LD, Palmer D, Allen CD, Schmidt TH, Xu Y, Proia RL, Coughlin SR, Cyster JG (2011) The sphingosine 1-phosphate receptor S1P(2) maintains the homeostasis of germinal center B cells and promotes niche confinement. Nat Immunol 12:672–680
46. Walzer T, Chiossone L, Chaix J, Calver A, Carozzo C, Garrigue-Antar L, Jacques Y, Baratin M, Tomasello E, Vivier E (2007) Natural killer cell trafficking in vivo requires a dedicated sphingosine 1-phosphate receptor. Nat Immunol 8:1337–1344
47. Bieberich E (2012) Ceramide and sphingosine-1-phosphate signaling in embryonic stem cell differentiation. Methods Mol Biol 874:177–192
48. Kue CS, Lim HX, Jung MY, Hong HJ, Cho D, Kim TS (2013) C6-ceramide in combination with transforming growth factor-beta enhances Treg cell differentiation and stable FoxP3 expression in vitro and in vivo. Immunobiology 218:952–959
49. Borchardt RA, Lee WT, Kalen A, Buckley RH, Peters C, Schiff S, Bell RM (1994) Growth-dependent regulation of cellular ceramides in human T-cells. Biochim Biophys Acta 1212:327–336
50. Flores I, Martinez AC, Hannun YA, Merida I (1998) Dual role of ceramide in the control of apoptosis following IL-2 withdrawal. J Immunol 160:3528–3533
51. Shakor AB, Taniguchi M, Kitatani K, Hashimoto M, Asano S, Hayashi A, Nomura K, Bielawski J, Bielawska A, Watanabe K, Kobayashi T, Igarashi Y, Umehara H, Takeya H, Okazaki T (2011) Sphingomyelin synthase 1-generated sphingomyelin plays an important role in transferrin trafficking and cell proliferation. J Biol Chem 286:36053–36062
52. Stoffel B, Bauer P, Nix M, Deres K, Stoffel W (1998) Ceramide-independent CD28 and TCR signaling but reduced IL-2 secretion in T cells of acid sphingomyelinase-deficient mice. Eur J Immunol 28:874–880
53. Herz J, Pardo J, Kashkar H, Schramm M, Kuzmenkina E, Bos E, Wiegmann K, Wallich R, Peters PJ, Herzig S, Schmelzer E, Kronke M, Simon MM, Utermohlen O (2009) Acid sphingomyelinase is a key regulator of cytotoxic granule secretion by primary T lymphocytes. Nat Immunol 10:761–768
54. Schroder M, Richter C, Juan MH, Maltusch K, Giegold O, Quintini G, Pfeilschifter JM, Huwiler A, Radeke HH (2011) The sphingosine kinase 1 and S1P1 axis specifically counteracts LPS-induced IL-12p70 production in immune cells of the spleen. Mol Immunol 48:1139–1148
55. Schaper K, Kietzmann M, Baumer W (2014) Sphingosine-1-phosphate differently regulates the cytokine production of IL-12, IL-23 and IL-27 in activated murine bone marrow derived dendritic cells. Mol Immunol 59:10–18
56. Schutze S, Potthoff K, Machleidt T, Berkovic D, Wiegmann K, Kronke M (1992) TNF activates NF-kappa B by phosphatidylcholine-specific phospholipase C-induced "acidic" sphingomyelin breakdown. Cell 71:765–776
57. Wiegmann K, Schutze S, Machleidt T, Witte D, Kronke M (1994) Functional dichotomy of neutral and acidic sphingomyelinases in tumor necrosis factor signaling. Cell 78:1005–1015
58. Cifone MG, Roncaioli P, De Maria R, Camarda G, Santoni A, Ruberti G, Testi R (1995) Multiple pathways originate at the Fas/APO-1 (CD95) receptor: sequential involvement of phosphatidylcholine-specific phospholipase C and acidic sphingomyelinase in the propagation of the apoptotic signal. EMBO J 14:5859–5868
59. Brenner B, Ferlinz K, Grassme H, Weller M, Koppenhoefer U, Dichgans J, Sandhoff K, Lang F, Gulbins E (1998) Fas/CD95/Apo-I activates the acidic sphingomyelinase via caspases. Cell Death Differ 5:29–37

60. Santana P, Pena LA, Haimovitz-Friedman A, Martin S, Green D, McLoughlin M, Cordon-Cardo C, Schuchman EH, Fuks Z, Kolesnick R (1996) Acid sphingomyelinase-deficient human lymphoblasts and mice are defective in radiation-induced apoptosis. Cell 86:189–199
61. Kirschnek S, Paris F, Weller M, Grassme H, Ferlinz K, Riehle A, Fuks Z, Kolesnick R, Gulbins E (2000) CD95-mediated apoptosis in vivo involves acid sphingomyelinase. J Biol Chem 275:27316–27323
62. Lin T, Genestier L, Pinkoski MJ, Castro A, Nicholas S, Mogil R, Paris F, Fuks Z, Schuchman EH, Kolesnick RN, Green DR (2000) Role of acidic sphingomyelinase in Fas/CD95-mediated cell death. J Biol Chem 275:8657–8663
63. Boucher LM, Wiegmann K, Futterer A, Pfeffer K, Machleidt T, Schutze S, Mak TW, Kronke M (1995) CD28 signals through acidic sphingomyelinase. J Exp Med 181:2059–2068
64. Zhang P, Liu B, Jenkins GM, Hannun YA, Obeid LM (1997) Expression of neutral sphingomyelinase identifies a distinct pool of sphingomyelin involved in apoptosis. J Biol Chem 272:9609–9612
65. Liu B, Hannun YA (1997) Inhibition of the neutral magnesium-dependent sphingomyelinase by glutathione. J Biol Chem 272:16281–16287
66. Tonnetti L, Veri MC, Bonvini E, D'Adamio L (1999) A role for neutral sphingomyelinase-mediated ceramide production in T cell receptor-induced apoptosis and mitogen-activated protein kinase-mediated signal transduction. J Exp Med 189:1581–1589
67. Chmura SJ, Nodzenski E, Kharbanda S, Pandey P, Quintans J, Kufe DW, Weichselbaum RR (2000) Down-regulation of ceramide production abrogates ionizing radiation-induced cytochrome c release and apoptosis. Mol Pharmacol 57:792–796
68. Tepper AD, Ruurs P, Borst J, van Blitterswijk WJ (2001) Effect of overexpression of a neutral sphingomyelinase on CD95-induced ceramide production and apoptosis. Biochem Biophys Res Commun 280:634–639
69. Sawai H, Domae N, Nagan N, Hannun YA (1999) Function of the cloned putative neutral sphingomyelinase as lyso-platelet activating factor-phospholipase C. J Biol Chem 274:38131–38139
70. Bose R, Verheij M, Haimovitz-Friedman A, Scotto K, Fuks Z, Kolesnick R (1995) Ceramide synthase mediates daunorubicin-induced apoptosis: an alternative mechanism for generating death signals. Cell 82:405–414
71. Perry DK, Carton J, Shah AK, Meredith F, Uhlinger DJ, Hannun YA (2000) Serine palmitoyltransferase regulates de novo ceramide generation during etoposide-induced apoptosis. J Biol Chem 275:9078–9084
72. Chalfant CE, Ogretmen B, Galadari S, Kroesen BJ, Pettus BJ, Hannun YA (2001) FAS activation induces dephosphorylation of SR proteins; dependence on the de novo generation of ceramide and activation of protein phosphatase 1. J Biol Chem 276:44848–44855
73. Jenkins GM, Cowart LA, Signorelli P, Pettus BJ, Chalfant CE, Hannun YA (2002) Acute activation of de novo sphingolipid biosynthesis upon heat shock causes an accumulation of ceramide and subsequent dephosphorylation of SR proteins. J Biol Chem 277:42572–42578
74. Kroesen BJ, Pettus B, Luberto C, Busman M, Sietsma H, de Leij L, Hannun YA (2001) Induction of apoptosis through B-cell receptor cross-linking occurs via de novo generated C16-ceramide and involves mitochondria. J Biol Chem 276:13606–13614
75. Yamaoka S, Miyaji M, Kitano T, Umehara H, Okazaki T (2004) Expression cloning of a human cDNA restoring sphingomyelin synthesis and cell growth in sphingomyelin synthase-defective lymphoid cells. J Biol Chem 279:18688–18693
76. Separovic D, Hanada K, Maitah MY, Nagy B, Hang I, Tainsky MA, Kraniak JM, Bielawski J (2007) Sphingomyelin synthase 1 suppresses ceramide production and apoptosis post-photodamage. Biochem Biophys Res Commun 358:196–202
77. Separovic D, Semaan L, Tarca AL, Awad Maitah MY, Hanada K, Bielawski J, Villani M, Luberto C (2008) Suppression of sphingomyelin synthase 1 by small interference RNA is associated with enhanced ceramide production and apoptosis after photodamage. Exp Cell Res 314:1860–1868

78. Lafont E, Milhas D, Carpentier S, Garcia V, Jin ZX, Umehara H, Okazaki T, Schulze-Osthoff K, Levade T, Benoist H, Segui B (2010) Caspase-mediated inhibition of sphingomyelin synthesis is involved in FasL-triggered cell death. Cell Death Differ 17:642–654
79. Lafont E, Dupont R, Andrieu-Abadie N, Okazaki T, Schulze-Osthoff K, Levade T, Benoist H, Segui B (2012) Ordering of ceramide formation and caspase-9 activation in CD95L-induced Jurkat leukemia T cell apoptosis. Biochim Biophys Acta 1821:684–693
80. Tepper AD, Diks SH, van Blitterswijk WJ, Borst J (2000) Glucosylceramide synthase does not attenuate the ceramide pool accumulating during apoptosis induced by CD95 or anti-cancer regimens. J Biol Chem 275:34810–34817
81. Taguchi Y, Kondo T, Watanabe M, Miyaji M, Umehara H, Kozutsumi Y, Okazaki T (2004) Interleukin-2-induced survival of natural killer (NK) cells involving phosphatidylinositol-3 kinase-dependent reduction of ceramide through acid sphingomyelinase, sphingomyelin synthase, and glucosylceramide synthase. Blood 104:3285–3293
82. Grazide S, Terrisse AD, Lerouge S, Laurent G, Jaffrezou JP (2004) Cytoprotective effect of glucosylceramide synthase inhibition against daunorubicin-induced apoptosis in human leukemic cell lines. J Biol Chem 279:18256–18261
83. Hannun YA, Loomis CR, Merrill AH Jr, Bell RM (1986) Sphingosine inhibition of protein kinase C activity and of phorbol dibutyrate binding in vitro and in human platelets. J Biol Chem 261:12604–12609
84. Pushkareva M, Khan WA, Alessenko AV, Sahyoun N, Hannun YA (1992) Sphingosine activation of protein kinases in Jurkat T cells. In vitro phosphorylation of endogenous protein substrates and specificity of action. J Biol Chem 267:15246–15251
85. Chao R, Khan W, Hannun YA (1992) Retinoblastoma protein dephosphorylation induced by D-erythro-sphingosine. J Biol Chem 267:23459–23462
86. Pushkareva M, Chao R, Bielawska A, Merrill AH Jr, Crane HM, Lagu B, Liotta D, Hannun YA (1995) Stereoselectivity of induction of the retinoblastoma gene product (pRb) dephosphorylation by D-erythro-sphingosine supports a role for pRb in growth suppression by sphingosine. Biochemistry 34:1885–1892
87. Dbaibo GS, Wolff RA, Obeid LM, Hannun YA (1995) Activation of a retinoblastoma-protein-dependent pathway by sphingosine. Biochem J 310(Pt 2):453–459
88. Nakamura S, Kozutsumi Y, Sun Y, Miyake Y, Fujita T, Kawasaki T (1996) Dual roles of sphingolipids in signaling of the escape from and onset of apoptosis in a mouse cytotoxic T-cell line, CTLL-2. J Biol Chem 271:1255–1257
89. Cuvillier O, Pirianov G, Kleuser B, Vanek PG, Coso OA, Gutkind S, Spiegel S (1996) Suppression of ceramide-mediated programmed cell death by sphingosine-1-phosphate. Nature 381:800–803
90. Cuvillier O, Rosenthal DS, Smulson ME, Spiegel S (1998) Sphingosine 1-phosphate inhibits activation of caspases that cleave poly(ADP-ribose) polymerase and lamins during Fas- and ceramide-mediated apoptosis in Jurkat T lymphocytes. J Biol Chem 273:2910–2916
91. Olivera A, Kohama T, Edsall L, Nava V, Cuvillier O, Poulton S, Spiegel S (1999) Sphingosine kinase expression increases intracellular sphingosine-1-phosphate and promotes cell growth and survival. J Cell Biol 147:545–558
92. Cuvillier O, Levade T (2001) Sphingosine 1-phosphate antagonizes apoptosis of human leukemia cells by inhibiting release of cytochrome c and Smac/DIABLO from mitochondria. Blood 98:2828–2836
93. Taha TA, Osta W, Kozhaya L, Bielawski J, Johnson KR, Gillanders WE, Dbaibo GS, Hannun YA, Obeid LM (2004) Down-regulation of sphingosine kinase-1 by DNA damage: dependence on proteases and p53. J Biol Chem 279:20546–20554
94. van Blitterswijk WJ, Klarenbeek JB, van der Luit AH, Alderliesten MC, van Lummel M, Verheij M (2009) Fas/CD95 down-regulation in lymphoma cells through acquired alkyllysophospholipid resistance: partial role of associated sphingomyelin deficiency. Biochem J 425:225–234
95. Dong L, Watanabe K, Itoh M, Huan CR, Tong XP, Nakamura T, Miki M, Iwao H, Nakajima A, Sakai T, Kawanami T, Sawaki T, Masaki Y, Fukushima T, Fujita Y, Tanaka M, Yano M,

Okazaki T, Umehara H (2012) CD4+ T-cell dysfunctions through the impaired lipid rafts ameliorate concanavalin A-induced hepatitis in sphingomyelin synthase 1-knockout mice. Int Immunol 24:327–337

96. Lafont E, Kitatani K, Okazaki T, Segui B (2011) Regulation of death and growth signals at the plasma membrane by sphingomyelin synthesis: implications for hematological malignancies. Recent Pat Anticancer Drug Discov 6:324–333
97. De Maria R, Lenti L, Malisan F, d'Agostino F, Tomassini B, Zeuner A, Rippo MR, Testi R (1997) Requirement for GD3 ganglioside in CD95- and ceramide-induced apoptosis. Science 277:1652–1655
98. De Maria R, Rippo MR, Schuchman EH, Testi R (1998) Acidic sphingomyelinase (ASM) is necessary for fas-induced GD3 ganglioside accumulation and efficient apoptosis of lymphoid cells. J Exp Med 187:897–902
99. de Leon J, Fernandez A, Clavell M, Labrada M, Bebelagua Y, Mesa C, Fernandez LE (2008) Differential influence of the tumour-specific non-human sialic acid containing GM3 ganglioside on CD4+CD25- effector and naturally occurring CD4+CD25+ regulatory T cells function. Int Immunol 20:591–600
100. Itoh M, Kitano T, Watanabe M, Kondo T, Yabu T, Taguchi Y, Iwai K, Tashima M, Uchiyama T, Okazaki T (2003) Possible role of ceramide as an indicator of chemoresistance: decrease of the ceramide content via activation of glucosylceramide synthase and sphingomyelin synthase in chemoresistant leukemia. Clin Cancer Res 9:415–423
101. Kim WJ, Okimoto RA, Purton LE, Goodwin M, Haserlat SM, Dayyani F, Sweetser DA, McClatchey AI, Bernard OA, Look AT, Bell DW, Scadden DT, Haber DA (2008) Mutations in the neutral sphingomyelinase gene SMPD3 implicate the ceramide pathway in human leukemias. Blood 111:4716–4722
102. Dbaibo GS, Kfoury Y, Darwiche N, Panjarian S, Kozhaya L, Nasr R, Abdallah M, Hermine O, El-Sabban M, de The H, Bazarbachi A (2007) Arsenic trioxide induces accumulation of cytotoxic levels of ceramide in acute promyelocytic leukemia and adult T-cell leukemia/lymphoma cells through de novo ceramide synthesis and inhibition of glucosylceramide synthase activity. Haematologica 92:753–762
103. Ardail D, Maalouf M, Boivin A, Chapet O, Bodennec J, Rousson R, Rodriguez-Lafrasse C (2009) Diversity and complexity of ceramide generation after exposure of jurkat leukemia cells to irradiation. Int J Radiat Oncol Biol Phys 73:1211–1218
104. Takahashi E, Inanami O, Asanuma T, Kuwabara M (2006) Effects of ceramide inhibition on radiation-induced apoptosis in human leukemia MOLT-4 cells. J Radiat Res 47:19–25
105. Morad SA, Levin JC, Tan SF, Fox TE, Feith DJ, Cabot MC (2013) Novel off-target effect of tamoxifen–inhibition of acid ceramidase activity in cancer cells. Biochim Biophys Acta 1831:1657–1664
106. Biswal SS, Datta K, Acquaah-Mensah GK, Kehrer JP (2000) Changes in ceramide and sphingomyelin following fludarabine treatment of human chronic B-cell leukemia cells. Toxicology 154:45–53
107. Hammadi M, Youinou P, Tempescul A, Tobon G, Berthou C, Bordron A, Pers JO (2011) Membrane microdomain sphingolipids are required for anti-CD20-induced death of chronic lymphocytic leukemia B cells. Haematologica 97:288–296
108. Capitani N, Patrussi L, Trentin L, Lucherini OM, Cannizzaro E, Migliaccio E, Frezzato F, Gattazzo C, Forconi F, Pelicci P, Semenzato G, Baldari CT (2012) S1P1 expression is controlled by the pro-oxidant activity of p66Shc and is impaired in B-CLL patients with unfavorable prognosis. Blood 120:4391–4399
109. Wallington-Beddoe CT, Powell JA, Tong D, Pitson SM, Bradstock KF, Bendall LJ (2014) Sphingosine kinase 2 promotes acute lymphoblastic leukemia by enhancing MYC expression. Cancer Res 74:2803–2815
110. Mondal S, Mandal C, Sangwan R, Chandra S, Mandal C (2010) Withanolide D induces apoptosis in leukemia by targeting the activation of neutral sphingomyelinase-ceramide cascade mediated by synergistic activation of c-Jun N-terminal kinase and p38 mitogen-activated protein kinase. Mol Cancer 9:239

111. Wallington-Beddoe CT, Hewson J, Bradstock KF, Bendall LJ (2011) FTY720 produces caspase-independent cell death of acute lymphoblastic leukemia cells. Autophagy 7:707–715
112. Liu Q, Zhao X, Frissora F, Ma Y, Santhanam R, Jarjoura D, Lehman A, Perrotti D, Chen CS, Dalton JT, Muthusamy N, Byrd JC (2008) FTY720 demonstrates promising preclinical activity for chronic lymphocytic leukemia and lymphoblastic leukemia/lymphoma. Blood 111:275–284
113. Wallington-Beddoe CT, Don AS, Hewson J, Qiao Q, Papa RA, Lock RB, Bradstock KF, Bendall LJ (2012) Disparate in vivo efficacy of FTY720 in xenograft models of Philadelphia positive and negative B-lineage acute lymphoblastic leukemia. PLoS One 7, e36429
114. Liao A, Broeg K, Fox T, Tan SF, Watters R, Shah MV, Zhang LQ, Li Y, Ryland L, Yang J, Aliaga C, Dewey A, Rogers A, Loughran K, Hirsch L, Jarbadan NR, Baab KT, Liao J, Wang HG, Kester M, Desai D, Amin S, Loughran TP Jr, Liu X (2011) Therapeutic efficacy of FTY720 in a rat model of NK-cell leukemia. Blood 118:2793–2800
115. Matsuoka Y, Nagahara Y, Ikekita M, Shinomiya T (2003) A novel immunosuppressive agent FTY720 induced Akt dephosphorylation in leukemia cells. Br J Pharmacol 138: 1303–1312
116. Yasui H, Hideshima T, Raje N, Roccaro AM, Shiraishi N, Kumar S, Hamasaki M, Ishitsuka K, Tai YT, Podar K, Catley L, Mitsiades CS, Richardson PG, Albert R, Brinkmann V, Chauhan D, Anderson KC (2005) FTY720 induces apoptosis in multiple myeloma cells and overcomes drug resistance. Cancer Res 65:7478–7484
117. Shah MV, Zhang R, Irby R, Kothapalli R, Liu X, Arrington T, Frank B, Lee NH, Loughran TP Jr (2008) Molecular profiling of LGL leukemia reveals role of sphingolipid signaling in survival of cytotoxic lymphocytes. Blood 112:770–781
118. Liu X, Ryland L, Yang J, Liao A, Aliaga C, Watts R, Tan SF, Kaiser J, Shanmugavelandy SS, Rogers A, Loughran K, Petersen B, Yuen J, Meng F, Baab KT, Jarbadan NR, Broeg K, Zhang R, Liao J, Sayers TJ, Kester M, Loughran TP Jr (2010) Targeting of survivin by nanoliposomal ceramide induces complete remission in a rat model of NK-LGL leukemia. Blood 116:4192–4201
119. Watters RJ, Fox TE, Tan SF, Shanmugavelandy S, Choby JE, Broeg K, Liao J, Kester M, Cabot MC, Loughran TP, Liu X (2013) Targeting glucosylceramide synthase synergizes with C6-ceramide nanoliposomes to induce apoptosis in natural killer cell leukemia. Leuk Lymphoma 54:1288–1296
120. Bezombes C, Grazide S, Garret C, Fabre C, Quillet-Mary A, Muller S, Jaffrezou JP, Laurent G (2004) Rituximab antiproliferative effect in B-lymphoma cells is associated with acid-sphingomyelinase activation in raft microdomains. Blood 104:1166–1173
121. Meyer zum Buschenfelde C, Feuerstacke Y, Gotze KS, Scholze K, Peschel C (2008) GM1 expression of non-Hodgkin's lymphoma determines susceptibility to rituximab treatment. Cancer Res 68:5414–5422
122. Bayerl MG, Bruggeman RD, Conroy EJ, Hengst JA, King TS, Jimenez M, Claxton DF, Yun JK (2008) Sphingosine kinase 1 protein and mRNA are overexpressed in non-Hodgkin lymphomas and are attractive targets for novel pharmacological interventions. Leuk Lymphoma 49:948–954
123. Gustafsson K, Sander B, Bielawski J, Hannun YA, Flygare J (2009) Potentiation of cannabinoid-induced cytotoxicity in mantle cell lymphoma through modulation of ceramide metabolism. Mol Cancer Res 7:1086–1098
124. Gustafsson K, Christensson B, Sander B, Flygare J (2006) Cannabinoid receptor-mediated apoptosis induced by R(+)-methanandamide and Win, 55,212-2 is associated with ceramide accumulation and p38 activation in mantle cell lymphoma. Mol Pharmacol 70:1612–1620
125. Michael JM, Lavin MF, Watters DJ (1997) Resistance to radiation-induced apoptosis in Burkitt's lymphoma cells is associated with defective ceramide signaling. Cancer Res 57:3600–3605
126. Qin Z, Dai L, Trillo-Tinoco J, Senkal C, Wang W, Reske T, Bonstaff K, Del Valle L, Rodriguez P, Flemington E, Voelkel-Johnson C, Smith CD, Ogretmen B, Parsons C (2014) Targeting

sphingosine kinase induces apoptosis and tumor regression for KSHV-associated primary effusion lymphoma. Mol Cancer Ther 13:154–164
127. Kluk MJ, Ryan KP, Wang B, Zhang G, Rodig SJ, Sanchez T (2013) Sphingosine-1-phosphate receptor 1 in classical Hodgkin lymphoma: assessment of expression and role in cell migration. Lab Invest 93:462–471
128. Nishimura H, Akiyama T, Monobe Y, Matsubara K, Igarashi Y, Abe M, Sugihara T, Sadahira Y (2010) Expression of sphingosine-1-phosphate receptor 1 in mantle cell lymphoma. Mod Pathol 23:439–449
129. Cattoretti G, Mandelbaum J, Lee N, Chaves AH, Mahler AM, Chadburn A, Dalla-Favera R, Pasqualucci L, MacLennan AJ (2009) Targeted disruption of the S1P2 sphingosine 1-phosphate receptor gene leads to diffuse large B-cell lymphoma formation. Cancer Res 69:8686–8692
130. Pavlova EV, Wang SZ, Archer J, Dekker N, Aerts JM, Karlsson S, Cox TM (2013) B cell lymphoma and myeloma in murine Gaucher's disease. J Pathol 231:88–97
131. Li QF, Zhu HY, Yang YF, Liu J, Xiao FJ, Zhang QW, Wu CT, Wang H, Wang LS (2010) Prokineticin-1/endocrine gland-derived vascular endothelial growth factor is a survival factor for human multiple myeloma cells. Leuk Lymphoma 51:1902–1912
132. Garcia-Bernal D, Redondo-Munoz J, Dios-Esponera A, Chevre R, Bailon E, Garayoa M, Arellano-Sanchez N, Gutierrez NC, Hidalgo A, Garcia-Pardo A, Teixido J (2013) Sphingosine-1-phosphate activates chemokine-promoted myeloma cell adhesion and migration involving alpha4beta1 integrin function. J Pathol 229:36–48
133. Li QF, Wu CT, Duan HF, Sun HY, Wang H, Lu ZZ, Zhang QW, Liu HJ, Wang LS (2007) Activation of sphingosine kinase mediates suppressive effect of interleukin-6 on human multiple myeloma cell apoptosis. Br J Haematol 138:632–639
134. Li QF, Wu CT, Guo Q, Wang H, Wang LS (2008) Sphingosine 1-phosphate induces Mcl-1 upregulation and protects multiple myeloma cells against apoptosis. Biochem Biophys Res Commun 371:159–162

Role of Sphingolipids in Hematological Malignancies: Myeloproliferative Disorders

Sitapriya Moorthi and Chiara Luberto

Abstract From an historical point of view, hematological malignancies have been instrumental for the discovery of the first link between sphingolipids and their bioactive role in apoptosis and differentiation. Ever since, hematological malignancies have represented a powerful model to discover and dissect sphingolipid-related functions in the context of cell proliferation, differentiation, autophagy, immunological responses and neoplastic transformation. Moreover, more recent studies have also translated the discoveries carried out in cell culture to clinically relevant models of patient samples and/or *in vivo* animal models.

In the present chapter, we provide a comprehensive discussion of the roles and functions of sphingolipids and sphingolipid-metabolizing enzymes in leukemias of myeloid origins. The discussion will cover the molecular mechanisms at the basis of sphingolipid actions and the therapeutic applications of such discoveries.

Keywords Hematological malignancies • Myeloid neoplasms • Chronic myelogenous leukemia • Acute myelogenous leukemia • Myelodysplastic syndrome • Ceramide • Sphingosine • Sphingosine-1-phosphate • Differentiation • Apoptosis

1 Introduction

Hematological malignancies have been a powerful model to discover and dissect sphingolipid-related functions in the context of cell proliferation, differentiation, autophagy, immunological responses and neoplastic transformation. Moreover, more recent studies have also translated the discoveries carried out in cell culture to clinically relevant models of patient samples and/or *in vivo* animal models.

S. Moorthi
Department of Molecular Genetics and Microbiology, Stony Brook University, 100 Nicolls Road, Stony Brook, NY 11794-8155, USA

C. Luberto (✉)
Department of Physiology and Biophysics, Stony Brook University, 100 Nicolls Road, Stony Brook, NY 11794-8155, USA
e-mail: chiara.luberto@stonybrook.edu

Y.A. Hannun et al. (eds.), *Bioactive Sphingolipids in Cancer Biology and Therapy*, DOI 10.1007/978-3-319-20750-6_3

From an historical point of view, hematological malignancies have been instrumental in the discovery of the first link between sphingolipids and their bioactive role in apoptosis and differentiation. Molt-4 and HL-60 cell lines were in fact the experimental models used in those first seminal reports [1, 2].

In the present chapter, we will discuss the role of sphingolipids in hematological malignancies of myeloid origins, following the World Health Organization (WHO) classification of myeloid neoplasms and acute leukemias revised in 2008 [3]. As of today, most of the published studies addressing the role of sphingolipids in hematological malignancies of myeloid origins have been performed in cell culture using cell lines derived from patients with either chronic myelogenous leukemia or acute myelogenous leukemia, thus the majority of the observations discussed in this chapter will address these conditions.

1.1 Myeloproliferative Neoplasms (MPNs)

1.1.1 Chronic Myelogenous Leukemia, *BCR-ABL1* Positive (CML)

Chronic myelogenous (or Myelocytic/Myeloid/Granulocytic) Leukemia (CML) is a myeloproliferative disorder of the hematopoietic stem cells (HSC) and accounts for 11 % of all leukemia's affecting adults [4]. It is characterized by a consistent chromosomal abnormality, which involves the reciprocal translocation between chromosome 9 and 22 [t(9;22)(q34;q11)] resulting in a shortened chromosome 22, known as the Philadelphia Chromosome [5, 6]. This translocation results in the fusion gene, *BCR-ABL* that is translated into an abnormal protein kinase, which is constitutively active and is responsible for CML pathology [7, 8]. CML progresses through three clinical stages, each characterized by specific hematological and molecular features. Several reviews have extensively discussed the details of these different clinical stages [9–12] and briefly these are: (1) the Chronic Phase (CP) characterized by the expansion of the granulocytic cell lineage with a median duration of 3-4 years, if left untreated; (2) the Accelerated Phase (AP) which involves the acquisition of additional cytogenetic aberrations and is characterized by the presence of 10–20 % blast cells in blood and bone marrow; and (3) the Blast Phase (BP) characterized by a block in differentiation of HSC and increased proliferation of immature myeloid or lymphoid (blast) cells in the bone marrow and peripheral blood which account for more than 30 % of cells. Apart from distinguished clinical characteristics, CML presents certain molecular features unique to this BCR-ABL-driven disorder. The BCR-ABL onco-protein prolongs cell survival and inhibits cell death rather than solely promoting cellular proliferation [13–16]. In fact it has been shown that CML cells, precursors and progenitors display proliferation characteristics (mitotic indices, rate and maturation) similar to their normal counterparts, yet there is an accumulation of cells at disease initiation (CP) [17–19]. Additionally, the prolonged cell survival of CML cells fosters the acquisition of secondary genetic aberrations that promote the progression from the CP to the BP of the disease [20, 21]. This mitotic "immortality" is characteristic of cancer cells and it is elicited by alterations in molecular pathways that govern normal cellular turnover. These include (not limited to) expression of growth factor ligands and/or of their receptors, hyper-responsiveness to limiting proliferative signals and stimulation of stromal cells to supplement available growth factor repertoire, in other words cancer cells become 'growth factor-independent'. In CML, BCR-ABL constitutively activates signaling pathways that normally mediate growth factor-dependent hematopoietic signaling and these pathways are key drivers of neoplastic transformation in CML [22]. Studies have shown that CML cells from CP are still partially dependent on extracellular supplemented factors whereas cells from BP acquire total independence from such factors through the acquisition of secondary genetic changes [23].

The role of sphingolipids in the regulation of BCR-ABL-mediated cellular processes has been investigated mainly in regard to the ability to resist cell death and to sustain cell survival.

Role of Sphingolipids in CML

1. To sustain cell proliferation.
   The role of the sphingolipid pathway in sustaining proliferation of CML cells has only been recently explored, and the link is *via* the sphingomyelin synthase enzyme (SMS) [24]. SMS catalyzes the transfer of phosphocholine from phosphatidylcholine (PC) to ceramide to generate diacylglycerol (DAG) [25] and sphingomyelin (SM) [26, 27]. Thus SMS is at a critical juncture in the regulation of the levels of two bioactive lipids, ceramide and DAG, with opposing effects on cell growth (Reviewed here: [28]). Additionally, this suggests that the deregulation of SMS may potentially play a role in pro-proliferative pathologies. Burns et al. have shown that BCR-ABL specifically up-regulated the *SMS1* isoform in K562 cells (a human cell line derived from a patient in BC). Pharmacological inhibition or silencing of *SMS1* caused a clear block in cell growth, with minimal effect on cell death, highlighting SMS1 specific role in sustaining cellular proliferation of BCR-ABL positive CML cells. Importantly, inhibition of SMS1 in these cells caused an accumulation of ceramide (substrate) and a drastic reduction of DAG (product), suggesting that the effect of SMS1 on cell proliferation may be elicited through the regulation of either or both these bioactive lipids [24]. A possible mechanism by which ceramide may exert an anti-proliferative effect is *via* stimulation of the tumor suppressor protein phosphatase PP2A, an established ceramide target with a well documented role in regulating cell cycle and apoptosis also in CML cells [29, 30]. In fact, it has been reported that BCR-ABL up-regulates SET, a physiological inhibitor of PP2A, resulting in the repression of PP2A, and that reactivation of PP2A has detrimental effects on CML cells [31, 32]. All together these observations suggest that BCR-ABL may employ at least two different mechanisms to keep the activity of PP2A under-check, and the up-regulation of *SMS1* by quenching the

"PP2A-activating pool of ceramides" could potentially be one such contributing arm. Additionally, given the profound effects that inhibition of SMS1 exerts on DAG levels in K562 cells and the recent work on the role of DAG-responsive classical and novel PKCs and PKDs in cell proliferation (including work in BCR-ABL positive cells), a DAG-mediated mechanism by which *SMS1* may contribute to sustaining proliferation of CML cells can also potentially take place [33–45]. All together, these results and their potential implications in the regulation of important signaling pathways warrant further investigations.

2. To resist cell death.
   The sphingolipid pathway is known to regulate cell death in different cell types [46]. In the context of CML, the major breadth of research focuses on understanding the role of ceramides and its metabolizing enzymes in: 1) mediating apoptosis and 2) mechanisms involved to keep ceramide levels in check and promote survival. Ceramides are known to mediate apoptosis [47–50] but we are just beginning to understand their role and regulation in the context of CML. Initial studies tested the effects of exogenous synthetic ceramide treatments. However, reports on the sensitivity of CML cells to treatment with such exogenous analogues (C2-, C6- and C8-ceramide) have been inconsistent. For instance, Maguer-Satta et al. showed that primary CML progenitor cells and BCR-ABL transformed cell lines (BAF3-BCR-ABL+) were more sensitive than their 'normal' or non-transformed counterparts to apoptosis by exogenous treatment with C2-ceramide. In fact such a treatment resulted in a transient increase of BCR-ABL phosphorylation, which appeared to surprisingly accelerate the rate of apoptosis in these cells [51, 52]. However in contrast to this study, Amarante-Mendes et al., showed that exogenous C2-ceramide treatment of CML cell lines (K562 and BCR-ABL transformed HL60 cells) did not result in apoptosis [53]. This treatment in these cell lines failed to induce release of cytochrome C from the mitochondria (possibly due to the high levels of anti-apoptotic Bcl-xl in these cells), and to activate caspase 3, which instead occurred in control cells (HL60). Moreover, Nica and colleagues showed that, in BCR-ABL positive K562 and KBM5 cells, C6-ceramide induced apoptosis in a caspase 8-dependent manner. This was shown to occur via activation of the JNK/SAPK cascade, which in turn led to phosphorylation and inhibition of the anti-apoptotic protein, Mcl-1 (a member of the Bcl-2 family of proteins) [54, 55]. Additionally, the authors describe that ceramide treatment did not affect Bcl-xl levels in K562 cells nor did it activate PP2A, thus eliminating two other possible mechanisms of induction of apoptosis [56]. In contrast to these investigations, McGahon and co-workers showed that K562 cells were protected from apoptosis induced by C6-ceramide [57].

   Thus studies based on treatment with exogenous analogues do not provide clear insights into the role of ceramides in cell death of CML cells. However studies looking at endogenous ceramides have shed light upon the possible pro-apoptotic roles of specific ceramide species and re-directed our understanding on how endogenous ceramide levels are regulated in CML cells in order to resist cell death. Baran et al., for instance, have shown that, in K562 cells, upon induction of apoptosis by pharmacological inhibition of the kinase activity of BCR-ABL with Imatinib [tyrosine kinase inhibitor (TKI)] [58], there was an increase in endogenous C18-ceramide generated via the activation of the ceramide synthase 1 enzyme (CerS1). This increase in C18-ceramide preceded apoptosis and it was not observed in imatinib-resistant cells [59]. Accumulation of ceramide following Imatinib treatment was also observed in the other CML-derived cell line, LAMA84 [60].

While there seems to be a consensus on the accumulation of ceramide in response to inhibition of the kinase activity of BCR-ABL [24, 59, 60], the role of Sphingosine-1-Phosphate (S1P) in Imatinib-induced cell death is still not clear. In fact, while in LAMA84 cells Imatinib treatment caused a decrease of S1P possibly due to inhibition of Sphingosine kinase 1 (SK1) [60], in K562 cells the drug caused an increase of S1P; thus the contribution of Imatinib-induced S1P changes to the overall functional effects of Imatinib needs further investigation.

Changes in sphingolipid-metabolizing enzymes consistent with elevation of ceramide levels were also reported upon treatment of K562 cells with dasatinib and nilotinib, second generation TKIs.[61, 62] In fact both TKIs caused an up-regulation of the expression of ceramide synthase genes, mainly CerS1 and an inhibition of *SK1*. Therefore, all together these results describe that ceramide generation precedes and signals TKI-mediated programmed cell death in CML cells.

In addition to these observations, Beverly et al. showed that, in K562 cells, pharmacological inhibition of anti-apoptotic Bcl-2 by ABT-263 activated pro-apoptotic BAK and BAX, causing an increase in the activity of CerS and resulted in the accumulation of C16-ceramide, which possibly drove forward the pro-apoptotic response [63]. This study implicates the activation of pro-apoptotic molecules along with ceramide generation in the induction of apoptosis thus perhaps providing an explanation for why exogenous ceramide alone could not consistently induce apoptosis in CML cells.

Thus, all together these observations point to an important role for ceramides as effectors of apoptotic cell death of CML cells, and the studies that employed TKIs suggest that BCR-ABL activity contributes to keep ceramide levels in check in these cells.

These conclusions are further corroborated by studies investigating the role of other sphingolipid metabolic pathways in CML. For instance, Hu et al. have shown that Acid Ceramidase (A-CDase), the enzyme that cleaves ceramide and produces sphingosine, is overexpressed in CML cells through a mechanism involving regulation of its transcription [64]. In particular, while A-CDase is normally transcriptionally repressed by IFN regulatory factor 8 (IRF8), activation of STAT5 by BCR-ABL leads to IRF-8 down-regulation and consequent lift of the A-CDase transcriptional repression [65]. In line with the functional significance of these observations, IRF8 is frequently lost in myeloid leukemias [66]. Accordingly, restoration of IRF8 expression in CML cells caused repression of A-CDase expression, accumulation of C16-ceramide and restored susceptibility of myelogenous leukemia cells to Fas L-induced apoptosis [64]. Thus all together these results suggest that CML cells have in place a fine mechanism to keep ceramide levels in check via elevation of A-CDase expression.

Another enzyme whose regulation is also affected by BCR-ABL is the SK, which catalyzes the phosphorylation of sphingosine to S1P and it is known to promote cell proliferation and to inhibit apoptosis [67]. Li et al. have shown that BCR-ABL is capable of up-regulating SK1 expression and its cellular activity [68]. Moreover, it was also shown that SK1 and S1P up-regulated the antiapoptotic protein and pro-survival factor of early hematopoietic progenitors, Mcl-1 [69, 70] implicating an anti-apoptotic function of SK1/S1P in CML. This conclusion is also supported by the fact that pharmacological inhibition of SK with dimethylsphingosine (DMS) in CML cell lines and blast cells derived from patients was shown to be cytotoxic [71].

An additional mechanism employed by CML cells to regulate ceramide levels may be represented by inactivation of the first step of sphingolipid *de novo* synthesis regulated by the serine palmitoyltransferase (SPT). In Taouji et al., data are provided to show that BCR-ABL inactivates SPT through phosphorylation of the SPTLC-1 subunit at Tyr164 in K562 and LAMA84 cells [72]. Additionally the authors showed that the inhibition of BCR-ABL by Imatinib reversed the phosphorylation on SPTLC-1, activated the enzyme, and resulted in an increase in ceramide levels preceding apoptosis. All these events were also accompanied by the unexpected re-localization of SPTLC-1 from the ER to the Golgi.

Importantly, all together these studies illustrate how expression of BCR-ABL causes the re-programming of the sphingolipid pathway with the net effect of keeping ceramide levels in check.

3. TKI resistance and therapeutic strategies.

The completion of phase III clinical trial of Imatinib, has established this drug as the first line of therapy for CP, with the best cytological response of 82 % (as determined by IRIS - International Randomized Study of Interferon and STI571) [73–75]. While the patient outcome for CP patients on Imatinib is very successful, its inability to kill BCR-ABL positive stem cells necessitates lifelong treatment [76, 77]. A problem associated with this long-term usage of Imatinib has been the insurgence of resistance to the drug, which, if not bypassed with alternative therapies, can lead to the clinical progression from CP to the resilient BP [78–83]. The development of second and third generation TKIs has tackled in part this problem, however resistance to TKIs remains an obstacle in CML therapy [84]. Thus the understanding of the molecular mechanisms that lead to TKI resistance has become one of the most studied aspect of CML research [85, 86]. As discussed, the sphingolipid pathway is significantly altered in CML cells and evidence has suggested that regulation of ceramide is central to its pathogenesis (Fig. 1). Thus targeting ceramide metabolism poses as an attractive clinical strategy and several approaches for inducing ceramide accumulation in CML cells have been explored as potential novel therapeutic means, mainly to overcome TKI resistance. For instance, it has been reported that, in Imatinib-resistant K562 CML cells, there is an increased expression of glucosylceramide synthase (GCS) and the over-expression of GCS in Imatinib-sensitive K562

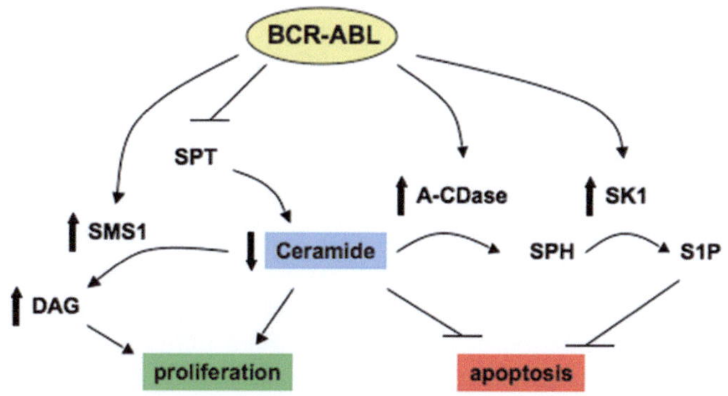

Fig. 1 Reprogramming of sphingolipid-metabolizing enzymes in CML cells. Expression of BCR-ABL induces the reprogramming of some sphingolipid metabolizing enzymes ultimately favoring a proliferative and antiapoptotic phenotype in CML cells. The reprogramming results in the overall control of ceramide production through reduced SPT activity and increased SMS1 and A-CDase expression, and it feeds into the formation of S1P through the combination of the increased A-CDase and SK1 activities

cells conferred resistance to Imatinib [87]. While the levels of ceramide don't dramatically differ between the resistant and sensitive sublines, inhibition of GCS with PDMP (a GCS inhibitor) in the resistant cells dramatically elevated ceramide accumulation and enhanced apoptosis. All together these results suggest that GCS may have a role to play in mediating imatinib resistance by preventing the accumulation of C18-ceramide and that GCS could represent a viable target to re-sensitize TKI resistant cells.

Although the precise mechanism by which ceramide exerts its cytotoxic effects in CML cells is not entirely clear, one possible mechanism may be through reactivation of the glycogen synthase kinase 3 (GSK3) [88]. GSK3 is a well-described tumor suppressor with a wide array of targets [89–91] and, in CML cells, GSK-3B is inactivated by BCR-ABL [92]. On the other hand, GSK3 is potently activated by ceramide [93]. Interestingly, Huang et al. have shown that Imatinib resistant patient samples harboring the BCR-ABL mutation T315I or cells transformed with mutant BCR-ABL could be re-sensitized to TKIs by forced accumulation of ceramide following inhibition of GCS, and they demonstrated that this combined cytotoxic effect occurred through reactivation of GSK3 [88].

In the context of re-sensitization to TKI, the natural phenol resveratrol has also been shown to cause an increase in the intracellular ceramide level in K562 cells and to promote apoptosis, and along with inhibitors of GCS has been proposed as a potential combinatorial option for CML [94]. Thus all together these observations support the idea that increasing intracellular levels of ceramide represents an intervention that could sensitize TKI resistant CML cells.

SK1 up-regulation has also been implicated in TKI resistance in CML. In imatinib resistant-K562 cells, Baran et al. have observed that there is an increased expression of SK1 and a concomitant shift of the ceramide to S1P ratio, in favor of S1P [59]. Additionally, upon SK1 silencing in resistant K562 cells, sensitivity to Imatinib was significantly increased. This was also the case in Imatinib-resistant LAMA84 cells [60]. These data point to a role of SK1 in contributing to TKI resistance in CML cells. Moreover, overexpression of SK1 and production of S1P in parent K562 or LAMA84 cells was sufficient to induce resistance suggesting an active role of SK1 in this process, independently from the regulation of ceramide levels [59, 60]. In line with these observations, Salas et al. have shown that engagement of the S1P2 receptor *via* S1P promoted the stabilization of BCR-ABL through inhibition of PP2A. In fact, PP2A causes the dephosphorylation and subsequent proteosomal degradation of BCR-ABL, thus SK1 signaling, by blocking PP2A, promotes TKI-resistance [95]. All together these observations strongly support a link between resistance to TKI and expression of SK1.

Recently, resistance to nilotinib, a second generation TKI, has been associated with increased GCS and SK1, and decreased CerS1. Down-regulation of GCS and SK-1 restored nilotinib sensitivity [96] reinforcing the concept that targeting the sphingolipid pathway might represent a viable strategy to overcome TKI resistance.

In terms of viable drugs, FTY720 is interesting since it can reactivate the tumor suppressor PP2A [31, 97–99] and, once converted into its phosphorylated form, FTY720-P, it can also downregulate S1P receptors (except S1PR2), factually inhibiting S1P-receptor mediated signaling. Unexpectedly, FTY720 has also been reported to induce ceramide accumulation in cells [100]. Thus these multifactorial functions of FTY720, all converging towards a proapoptotic phenotype, make it an attractive candidate for an alternative CML therapeutic strategy.

1.1.2 Sphingolipids and Other MPNs

Involvement of SK1/S1P has been also shown to play a role in the oncogenic function of mutant Jak2 (V617F) [101]. In fact, mutant Jak2 (V617F) has been shown to utilize also an SK1-mediated pathway to inhibit the activity of the tumor suppressor PP2A. Since constitutively active Jak2 (V617F) is a characteristic feature of almost all polycythemia vera and approximately 60 % of essential thrombocythemia (ET) and primary myelofibrosis, a protumorigenic function for SK1 is likely to exist in these myeloproliferative neoplasms. Additionally, in serum of patients with ET, the levels of S1P and dihydro-S1P were significantly elevated and they strictly correlated with the number of platelets (and not red blood cells), suggesting that these sphingoid bases may derive from the platelet fraction and their increased levels may have significant implications in atherogenic complications linked to this condition [102].

1.2 Myelodysplastic Syndrome (MDS)

Little is known about dysregulation of sphingolipids in MDS and the only available observation in this regard is the overall increased *SK1* expression in patients with MDS and the fact that *SK1* expression seems to increase with the severity of the disease [103]. A number of other sphingolipid-metabolizing enzymes were also analyzed in the same MDS samples and the only other one that showed some changes was neutral sphingomyelinase 2 (*NSMase2*) which was decreased in samples collected from patients in the early phases of the disease. None of the other sphingolipid genes showed significant changes. Since no follow up was carried out to study the potential functional role of SK1 and NSMase2 in MDS, the functional significance of these observations remains unknown.

1.3 Acute Myeloid Leukemia (AML) and Other Related Neoplasms

AML is the most common acute leukemia affecting adults; nevertheless, it is a rare disease and in the United States, it accounts for 1.2 % of all cancer-related deaths. AML arises from the combination of a block of differentiation and additional genetic alterations that promote unchecked proliferation of immature cells. Differently from CML, AML is characterized by great heterogeneity because genetic alterations can occur in myeloid cells at different stages of differentiation. These alterations often consist of recurrent chromosomal translocations leading to the generation of fusion proteins with oncogenic functions, and specific genetic alterations are often associated with different prognosis. Progression of the disease occurs rapidly and, depending on the AML subtype, the 5 years survival rates ranges from 15 to 70 %. The most current classification of AML has been revised by the WHO and takes into consideration differences in morphology, cytochemistry, immunophenotype, genetic and clinical features to group disease entities. On the other hand, few older manuscripts refer to the French-American-British (FAB) classification of AML which instead consists of 8 subtypes (M0 to M7), organized by the type of cell from which the leukemia developed and its level of maturity.

Chemotherapy is the treatment of choice for AML, and most commonly it consists in the combination of cytarabine and anthracycline. This first intervention (induction phase of therapy) aims to bring the patients in complete remission (no detectable leukemic cells) and it is followed by the consolidation therapy that aims at the eradication of any residual leukemic cell. The consolidation therapy varies depending on the specific prognostic factors of the patient, and can range from high dose of cytarabine to allogeneic hematopoietic stem cell transplantation for patients at high risk of relapse, with a 5-year overall survival rate of 40–45 % in younger patients [104]. Generally, the relapse rate varies from 33 to 78 % depending on the subtype of AML, and once AML relapses or does not initially respond to induction phase therapy, treatment options are almost non existent. Thus alternative therapies targeting resilient AML or relapsed disease are warranted.

The sphingolipid metabolic pathway has been implicated in the regulation of differentiation, cell cycle progression, apoptosis, and autophagy of AML cells and SPLs have been linked to the response and/or resistance to treatment of AML. Hence regulation of SPL metabolism may represent a way to trigger therapeutic response of AML.

1.3.1 Sphingolipids and Regulation of Differentiation of AML Cells and Other Cell Models of Related Neoplasms

Most of the studies addressing the roles of SPLs in differentiation of AML cells were performed in cell culture models mainly employing HL-60 or U937 cells. HL-60 cells are a valuable model to study the molecular mechanisms leading to maturation since they retain the ability to differentiate either towards the monocytic or the granulocytic lineages. Retinoic acid (RA) induces differentiation of HL-60 cells towards the granulocytic lineage and during the differentiation process accumulation of ceramide has been reported. Indeed treatment with all trans RA (ATRA) in HL-60 cells induced accumulation of ceramide in the nucleus whereas treatment of promyelocytic NB4 cells induced elevation of ceramide levels potentially via increased expression of acid sphingomyelinase [105, 106]. Since, at the time of these reports, very limited molecular tools were available to dissect and study the SPL metabolism, the causative link between these observations and the differentiation pathway was not established.

Moreover it has been also shown that the addition of sphinganine enhanced the differentiating activity of RA, possibly through inhibition of Protein kinase C (PKC) and it blocked proliferation [104, 107]. Similarly, sphinganine enhanced the granulocytic differentiation of HL-60 cells also induced by Adriamycin, Daunomycin and DMSO [108, 109]. Thus treatments that would increase sphinganine levels in combination with inducers of granulocytic differentiation might have clinical benefits. On the other hand, C2-ceramide was shown to inhibit the granulocytic differentiation induced by ATRA [106], highlighting the potential counteracting effects of the interconversion of sphinganine into ceramide and raising the question whether the accumulation of ceramide in response to ATRA is the actual effector in induction of differentiation or it is a step for the formation of other active intermediate(s), like sphinganine.

Contrary to granulocytic differentiation, low concentrations of sphinganine and sphingosine or treatment with exogenously added sphingomyelinase (which also caused an increase of sphingoid bases) were found to inhibit the monocytic differentiation of HL-60 cells induced by the phorbol ester, PMA and thus activation of PKC [110, 111]. On the other hand, glucosylceramide and GM3 were shown to specifically favor adherence of HL-60 cells undergoing monocytic differentiation in response to PMA as the GCS inhibitor, PDMP hindered attachment of these cells to the plates without affecting other morphological aspects characteristic of monocytic differentiation or growth inhibition [112, 113]. Interestingly, the effect of sphinganine on monocytic differentiation depends specifically on the inducer. In fact sphinganine was unable to block monocytic differentiation of HL-60 cells induced by 1 alpha-25-dihydroxyvitamin D3 (Vit D3). These results suggest that, in the case of PMA treatment, it is the inhibitory effect of sphinganine on PKC responsible for the effects and that PKC may not play a role in differentiation to Vit D3 [114].

Furthermore, sphingolipids have been implicated in the regulation of the monocytic differentiation process itself. Indeed a role for ceramide generated from hydrolysis of sphingomyelin (SM) was proposed to be critical for monocytic differentiation of HL-60 cells induced by Vit D3, TNFα and IFNγ [2, 115]. These were among the first reports on the existence of a SM cycle that, in the interval of few hours (2–4 h), caused the generation of ceramide from SM and SM resynthesis thereafter. Interestingly, the activation of the SM cycle seemed to be inducer-specific since it did not occur during granulocytic differentiation or upon monocytic differentiation promoted by PMA. The activation of a neutral sphingomyelinase enzyme, responsible for SM hydrolysis, was described and the involvement of its activation in the differentiation process was strengthened by the addition of exogenous bacterial sphingomyelinase which enhanced the effects of subthreshold concentrations of Vit D3. Following studies supported a signaling role for ceramide generated through the SM cycle in Vit D3 and IFNγ-induced differentiation, as the kinetics of ceramide generation matched those of SM hydrolysis, treatment with C2-ceramide enhanced the effect of subthreshold concentrations of Vit D3 and higher concentrations of C2-ceramide were sufficient to mimic Vit D3 effects [116, 117]. In the case of IFNγ, production of arachidonic acid through PLA2 seemed to be responsible for the activation of the SM cycle whereas the PI3-K/PKC/JNK/ERK pathways were later indicated as responsible for the differentiation-enhancing activity of ceramide on Vit D3 [118]. Of interest, it is that mutations in the neutral sphingomyelinase 2 gene (*SMPD3*) were found in 5 % of acute myeloid leukemias and 6 % of acute lymphoid leukemias but not in other tumor types, suggesting, in a small fraction of these leukemias, the existence of a block of ceramide generation [119].

All together these observations suggest that the SPL metabolism has different outcomes depending on the specific molecular pathway invoked during differentiation and that increasing sphinganine/ceramide levels might represent a beneficial strategy against AML when used in combination with differentiating agents.

Involvement of sphingosine phosphorylation has been instead implicated in PMA-induced differentiation of human erythroleukemia (HEL) and megakaryoblastic cells [120, 121]. In fact, in both cases, prolonged treatment with PMA induced an increase of SK activity, which was inhibited by staurosporine and calphostin C, implicating the upstream activation of PKC. The PKC-mediated activation of SK1 was not through direct phosphorylation but rather through regulation of SK1 expression. Interestingly though, other differentiating agents did not exert the same activating effects on SK, implicating the existence of a specific PKC-SK pathway.

1.3.2 Sphingolipids and Regulation of Apoptosis in AML Cells

1. Ceramide treatment and apoptosis.
   Several studies have shown that treatment of AML cell lines with exogenous ceramide causes cell death. Indeed treatment with low micromolar concentrations of short chain ceramide (C2-and C8-ceramide) or addition of bacterial sphingomyelinase was reported to induce DNA fragmentation in both U937 and HL-60 cells [1, 122–124]. Interestingly, it was also noted that activation of PKC, either via treatment with DAG analogues or phorbol ester, counteracted the cytotoxic effects of ceramide. Ceramide-induced DNA fragmentation was shown to occur through stimulation of p46-JNK1/p54-JNK2 activity, increased expression of c-jun and ultimately the activation of the transcription factor AP-1; indeed AP-1 inhibition blocked ceramide-induced DNA damage [125–127]. Importantly, while activation of AP-1 by ceramide was shown to regulate apoptosis, this pathway was not involved in ceramide-induced differentiation [128]. Together with the activation of AP-1, downregulation of the antiapoptotic protein Bcl2 was also observed upon treatment with C2-ceramide in both HL-60 and U937 cells [129]. Involvement of Bcl2 in ceramide-induced cell death was strengthened by the observation that stable overexpression of this protein in HL-60 cells protected from C2-ceramide-induced caspase 3 activation and cell death [130]. It has also been reported that treatment with C2-ceramide in HL-60 cells exerts a G1 cell cycle arrest via the induction of the cdk inhibitor p27(kip1), and that this growth arrest may be necessary for the onset of apoptosis [123, 131, 132]; furthermore, ceramide-induced expression of p27(kip1) was also significantly decreased by overexpression of Bcl2 [131].

   Another proposed indirect downstream target of ceramide in HL-60 cells is Bax. In fact ceramide has been shown to increase Bax expression and to induce its translocation to the mitochondria, while antisense Bax inhibited ceramide-induced caspase activation and apoptosis [133]. Following studies have proposed a regulatory role for p38 MAPK in the induction of ceramide-dependent translocation of Bax to mitochondria, as both pharmacological inhibition and dominant negative p38 attenuated this process, as well as DNA fragmentation and caspase 3 activation [134].

   An additional downstream target for C2-ceramide in HL-60 cells is the oncogene c-Myc [135]. In fact low concentrations of ceramide downregulated c-Myc mRNA and this process seemed to be mediated by activation of an okadaic acid sensitive protein phosphatase, most likely PP2A.

   Treatment with short chain ceramide analogs has been linked also to oxidative damage in HL-60 cells, and it has been proposed that this may be due to the inhibitory effect of ceramide-induced caspase-3 activation on catalase [136, 137]. On the other hand, accumulation of ceramide (and dihydrosphingosine) is also downstream of H_2O_2 treatment suggesting that potentially the cytotoxic effect of H_2O_2 is amplified by the inhibitory effect of ceramide on catalase [138].

Complementary observations seem to indicate that conditions in which ceramide metabolism is stimulated (i.e., conversion into glucosylceramide promoted by the expression of p-glycoproteins) confer resistance to ceramide in AML cells [139].

2. (Chemo)therapeutics and ceramide-mediated apoptosis.
Ceramide accumulation has been observed in response to different chemotherapeutic agents. In particular, daunorubicin treatment of HL-60 and U937 cells caused an early increase in ceramide (10 min post treatment) through activation of a neutral SMase [140]. Daunorubicin-induced activation of neutral SMase, generation of ceramide and induction of apoptosis were all blocked by inhibition of serine proteases or by activation of PKC [141, 142]. Interestingly though, a report suggested that ceramide accumulation per se is not the apoptotic signal rather its metabolic conversion to produce the ganglioside GD3 [143].

In addition to daunorubicin, ceramide accumulation was observed following adriamycin treatment of HL-60 cells in both microsomal fraction and nucleus, although the role of these pools of ceramide in the cytotoxic effect of the drug was not established [144]. Moreover, it has been reported that ceramide increased also in response to 1-beta-D-Arabinofuranosylcytosine (AraC) in HL-60 cells [145] but, in this case, its contribution to drug-induced apoptosis seems marginal since it was later shown that inhibition of neutral SMase activity and ceramide accumulation did not affect AraC-mediated cell death [146]. On the other hand, co-treatment with AraC and sphingoid bases enhanced AraC cytotoxicity possibly via inhibition of PKC [146].

Neutral SMase, ceramide synthase activation and ceramide accumulation were all reported following treatment of HL-60 cells with irradiated riboflavin. These events were followed by inhibition of MAPK and upregulation of Bax and they seemed to be specific to leukemic cells since normal cells were not sensitive to irradiated riboflavin, highlighting the potential beneficial therapeutic effect of this treatment [147].

More recently, involvement of acid and neutral SMase activation and consequent ceramide generation in mitochondrial-mediated cell death induced by Stichoposide C was reported, and the antitumor activity of Stichoposide C, accompanied by ceramide accumulation, was also confirmed in an *in vivo* HL-60 xenograft model [148]. Other studies pointed to a role for ceramide in mediating in part the cytotoxic effect of arsenic trioxide in acute promyelocitic leukemia (APL) cells, NB4 [149]. In this model, ceramide accumulation was shown to derive from both *de novo* biosynthesis and reduced conversion to glucosylceramide, and the study suggested that ceramide accumulation could represent an important step in the therapeutic benefit exerted by arsenic trioxide in combination with ATRA as standard treatment for APL or for treatment of relapsed APL. Ceramide accumulation through ceramide synthase was also involved in fenretinide-mediated apoptosis of HL-60 cells since the ceramide synthase inhibitor, fumonisin B1 blocked fenretinide-mediated ceramide accumulation and its cytotoxic effects [150]; in this model, ceramide generation seemed to be triggered by production of reactive oxygen species.

Another clinically relevant chemotherapeutic agent found to induce ceramide accumulation is DT(388)-GM-CSF, a fusion toxin which conjugates the human granulocyte-macrophage colony-stimulating factor (GM-CSF) with the catalytic and translocation domains of diphteria toxin. In cell culture using either parental or vincristine resistant HL-60 cells, DT(388)-GM-CSF caused a substantial accumulation of ceramide (potentially generated through SM hydrolysis) and subsequent caspase activation and inhibition of protein synthesis, resulting in cell death [151]. A phase I clinical trial with DT(388)-GM-CSF for patients with highly chemotherapy refractory AML produced some clinical remissions but it also showed dose-limiting toxicity for liver injury [152].

A potentially beneficial drug treatment for AML, found to induce ceramide production, is the combination of histone deacetylase inhibitors (HDACIs) and the alkyl-lysophospholipid, perifosine [153]. This drug combination in fact synergistically induced apoptosis in HL-60 and U937 cells by triggering mitochondrial dysfunction and caspase activation. Several events concurred to this highly cytotoxic phenotype, such as inhibition of antiapototic proteins ERK and Akt and induction of proapoptotic factors such as ceramide accumulation, oxidative damage, translocation of Bax to mitochondria and elevated Bak. Beside the fact that inhibition of ERK and Akt was shown to play an active role in the regulation of ceramide accumulation, the apoptotic chain of events was not ordered.

Finally, FTY720, a sphingosine analogue approved for treatment of multiple sclerosis, has been shown to induce apoptosis of AML-M2 cells by increasing intracellular ceramide levels [154]. Indeed FTY720 has shown antitumorigenic activity against kasumi-1 cells, xenograft mouse models and blast cells isolated from AML-M2 patients with the t(8;21) translocation. In Kasumi-1 cells, FTY720 induced ceramide accumulation possibly through activation of both neutral SMase2 and ceramide synthase, and it was proposed that the accumulated ceramide exerts its cytotoxic effects through activation of the tumor suppressor protein phosphatase 2A by binding and displacing its inhibitor, IPP2A.

Early on, lack of ceramide accumulation has been also proposed to play a role in conferring resistance of human erythromyeloblastic cells to apoptosis induced by ionizing radiations (IR) [155]. In fact, two subclones of the human TF-1 cells (one CD34 positive and one CD34 negative) were characterized by different sensitivity to IR which was not due to different DNA repair mechanisms; on the other hand, sensitivity to IR of the CD34 negative cells corresponded to activation of a neutral SMase and consequent increase of ceramide which did not occur in CD34 positive cells.

3. Sphingoid Bases and Apoptosis in AML Cell Culture Models.

Sphingoid bases have also been shown to be an integral part of the apoptosis pathway in AML cells.

Treatment with exogenously added sphingosine or sphinganine induced apoptosis in HL-60 and U937 cells, causing DNA fragmentation, nuclear condensation, production of apoptotic bodies and reduced clonogenic capability, and similarly to ceramide, exogenously added sphingosine also down-regulated c-Myc mRNA [156–159]. Also, addition of sub-lethal concentrations of sphingoid bases (below 750nM) synergized with the ability of short chain ceramides or bacterial sphingomyelinase to induce apoptosis. This synergistic effect was also mimicked by selective pharmacological inhibitors of PKC, suggesting that the cytotoxic actions of sphingoid bases might be due to their inhibitory effect on PKC [157]. Interestingly, a study of the effect of different caspase inhibitors on cell death induced by sphingoid bases versus ceramide revealed that sphingoid bases and ceramide operate independently on the apoptotic pathway with sphingoid bases acting upstream of caspase 8 and ceramides activating late caspases [160].

Endogenous sphingosine accumulation was observed upon monocytic differentiation of HL-60 cells induced by PMA and the elevation of sphingosine paralleled the amount of apoptotic cells during differentiation. Interestingly, during this process, the conversion of ceramide into sphingosine was greatly stimulated suggesting the activation of a ceramidase [156].

Finally, the optimization of the use of sphingoid bases as anti-leukemic agents was investigated by studying a liposomal encapsulation of safingol (L-THREO-SPHINGANINE) with minimal hemolytic activity in a panel of AML cell lines and patients samples. It was found that this preparation exhibited effective cytotoxicity in cells and it also prolonged the median survival time of SCID mice inoculated with U937 cells [161]. Moreover, a synergistic effect of a 1:1 mixture of safingol and C2-ceramide was also reported in AML cell culture and in *in vivo* xenografts of U937 cells [162].

4. S1P and Apoptosis in AML Cell Culture and *In Vivo* Models.
A cytoprotective role for SK and S1P has been observed in AML cell lines in response to different stimuli.

For instance, long-term treatment with Vit D3 in HL-60 cells was shown to induce a late activation of SK, and activation of SK achieved through prolonged Vit D3 treatments provided protection from ceramide-induced apoptosis [163]. Exogenous treatment with S1P also protected from ceramide-induced apoptosis and dimethylsphingosine (DMS), a competitive inhibitor of SK, eliminated the protective effect of long-term Vit D3 treatment. Furthermore, addition of S1P bypassed the cytotoxic effects of inhibition of SK by DMS, suggesting that S1P mediates the cytoprotective effects of long-term treatment with Vit D3.

S1P was found to prevent apoptosis of HL-60 and U937 cells induced by Fas ligation, TNF, ceramide and serum deprivation by preventing the translocation of cytochrome c and Smac/DIABLO from mitochondria to the cytosol [164]. In line with these observations, inhibition of SK with DMS enhanced the cytotoxic effects of these treatments and accelerated cytochrome c and Smac/DIABLO release from mitochondria, while addition of S1P reverted these effects. More recent studies using an SK1 specific inhibitor (SK1-I), revealed that the activity of this SK isoform is critical for survival of U937 cells and blasts isolated from AML patients, and implicated SK1/S1P in the positive regulation of phosphorylation of ERK1/2 and Akt. Importantly SK1-I also significantly reduced growth of AML xenografts [165]. The cytoprotective effect of SK1 in AML cells was confirmed by the use of SKI-178 [166]. Importantly, multidrug resistant AML cells were similarly sensitive to the inhibitor highlighting the potential broad-spectrum therapeutic application of this drug.

In line with an antiapoptotic and proproliferative function of SK1 in AML, an oncogenic role for this protein was reported in erythroleukemic progression [167]. Indeed, taking advantage of a transgenic multistage erythroleukemic mouse model in which sp-1/PU.1 is overexpressed in hematopoietic cells, it was first shown that expression of *SK1* was significantly higher in tumorigenic pro-erythroblasts (HS2) as compared to that of non tumorigenic proerythroblasts (HS1). Furthermore, overexpression of SK1 in HS1 cells conferred increased proliferation, clonogenicity, resistance to apoptosis and tumorigenicity when engrafted *in vivo*. Complementary to these observations, expression of dominant negative SK1 in HS2 cells reduced cell growth and resistance to apoptosis, suggesting that SK1 might represent a viable drug target for AML.

5. Sphingolipid Metabolism and Chemoresistance in AML.
As discussed, several chemotherapeutic agents exert their cytotoxic effects by elevating intracellular levels of ceramide. Importantly, it has been observed that some chemoresistant AML cell lines (i.e., HL-60/ADR) are unable to accumulate ceramide in response to chemotherapeutic drugs (i.e., doxorubicine) even though the uptake of the drug is comparable to chemosensitive cells [168]. In these cells, a lower level of endogenous ceramide was reported and this was accompanied by an augmented basal activity of enzymes that convert ceramide into more complex sphingolipids, like glucosylceramide and SM synthases. Importantly, these observations were also confirmed in chemoresistant leukemic patients samples. Since the levels of Bcl2 were also increased in the chemoresistant samples, it is possible that the lower level of ceramide might have consequences on Bcl2 expression, thus bolstering the antiapoptotic capacity of these cells. Interestingly, ceramide resistant HL-60 cells were also found to be resistant to anticancer drugs (daunorubicin, etoposide, ara-C) and differentiation-inducing agents (vit D3, RA, PMA) [169, 170]. In these cells, the levels of Bax and PKCδ were significantly decreased while the levels of antiapoptotic PKCs, such as PKCα and βI, were increased.

All together these results indicate that ceramide action is able to modulate both proapoptotic (PKCδ delta, Bax) and antiapoptotic effectors (Bcl2, PKCα and βI) and they suggest that the block of the metabolic conversion of ceramide into complex sphingolipids may be a valuable approach to sensitize resistant AML cells.

Finally, sphingolipid metabolism has been found to influence p-glycoprotein (Pgp) function, and, in particular, inhibition of glucosylceramide synthase resulted in increased drug retention and cytotoxicity [171]. In immature AML KG1a cells, it was observed that inhibition of glucosylceramide synthase activity with PDMP induced accumulation of ceramide and concomitant reduction of the levels of glucosylceramide and other glycosylated derivatives, including gangliosides GD3 and GM3 [172]. On the other hand, only the addition of gangliosides and not of ceramide or other glycosylated sphingolipids stimulated P-gp function, indicating a specific role for gangliosides in the regulation of P-gp.

In addition to the glycosylation pathway, SK activity was also linked to multidrug resistance. In fact, it was observed that while chemosensitive HL-60 cells responded to doxorubicine and etoposide treatment by accumulating ceramide and inhibiting SK1 activity, chemorestistant HL-60 cells overexpressing MRP1 and MDR1 retained SK1 activity and failed to increase intracellular ceramide [173]. Incubation of chemoresistant cells with short chain ceramide caused inhibition of SK1 and apoptosis, which was also triggered by pharmacological inhibition of SK1 with F-12509a. Thus altogether, these observations suggest that inhibition of SK1 may be an effective way to overcome multi drug resistance.

2 Conclusions

It is clear from the above discussion that investigations of the basic molecular mechanisms contributing to myelocytic transformation are essential to the development of targeted interventions. As for the role of sphingolipids in the homeostasis of the myelocytic arm of hematopoiesis, they intervene in critical aspect of physiological processes such as differentiation and their dysregulation contributes to the development of myelogenous neoplasms. As such, one could envision targeting of sphingolipid metabolism to prevent progression of these disorders.

It is also increasingly evident that the alteration of the sphingolipid metabolism plays a role in drug resistance. Thus, combination therapy is being actively investigated in order to maximize the effectiveness of each intervention and decrease the toxic side effects of chemotherapy.

These represent the challenges and, at the same time, the inspiration for sphingolipid research applied to cancer biology and therapy.

Acknowledgement This work is partly supported by NCI grant P01 CA097132 to CL and by the "Scholars in Biomedical Sciences" fellowship from Stony Brook University, School of Medicine to SM.

References

1. Obeid LM, Linardic CM, Karolak LA, Hannun YA (1993) Programmed cell death induced by ceramide. Science 259:1769–1771
2. Okazaki T, Bell RM, Hannun YA (1989) Sphingomyelin turnover induced by vitamin D3 in HL-60 cells. Role in cell differentiation. J Biol Chem 264:19076–19080
3. Vardiman JW, Thiele J, Arber DA, Brunning RD, Borowitz MJ, Porwit A, Harris NL, Le Beau MM, Hellstrom-Lindberg E, Tefferi A, Bloomfield CD (2009) The 2008 revision of the World Health Organization (WHO) classification of myeloid neoplasms and acute leukemia: rationale and important changes. Blood 114:937–951

4. American Cancer Society (2014) Cancer facts & figures 2014. American Cancer Society, Atlanta, p 2014
5. Rowley JD (1973) Letter: A new consistent chromosomal abnormality in chronic myelogenous leukaemia identified by quinacrine fluorescence and Giemsa staining. Nature 243: 290–293
6. Nowell P, Hungerford D (1960) A minute chromosome in human chronic granulocytic leukemia. Science 132:1497
7. Bartram CR, Kleihauer E, de Klein A, Grosveld G, Teyssier JR, Heisterkamp N, Groffen J (1985) C-abl and bcr are rearranged in a Ph1-negative CML patient. EMBO J 4:683–686
8. Lugo TG, Pendergast AM, Muller AJ, Witte ON (1990) Tyrosine kinase activity and transformation potency of bcr-abl oncogene products. Science 247:1079–1082
9. Di Bacco A, Keeshan K, McKenna SL, Cotter TG (2000) Molecular abnormalities in chronic myeloid leukemia: deregulation of cell growth and apoptosis. Oncologist 5:405–415
10. Kantarjian HM, Deisseroth A, Kurzrock R, Estrov Z, Talpaz M (1993) Chronic myelogenous leukemia: a concise update. Blood 82:691–703
11. Ohmine K, Ota J, Ueda M, Ueno S, Yoshida K, Yamashita Y, Kirito K, Imagawa S, Nakamura Y, Saito K, Akutsu M, Mitani K, Kano Y, Komatsu N, Ozawa K, Mano H (2001) Characterization of stage progression in chronic myeloid leukemia by DNA microarray with purified hematopoietic stem cells. Oncogene 20:8249–8257
12. Savage DG, Szydlo RM, Goldman JM (1997) Clinical features at diagnosis in 430 patients with chronic myeloid leukaemia seen at a referral centre over a 16-year period. Br J Haematol 96:111–116
13. Bedi A, Zehnbauer BA, Barber JP, Sharkis SJ, Jones RJ (1994) Inhibition of apoptosis by BCR-ABL in chronic myeloid leukemia. Blood 83:2038–2044
14. Cortez D, Kadlec L, Pendergast AM (1995) Structural and signaling requirements for BCR-ABL-mediated transformation and inhibition of apoptosis. Mol Cell Biol 15:5531–5541
15. Cotter TG (1995) BCR-ABL: an anti-apoptosis gene in chronic myelogenous leukemia. Leuk Lymphoma 18:231–236
16. Evans CA, Owen-Lynch PJ, Whetton AD, Dive C (1993) Activation of the Abelson tyrosine kinase activity is associated with suppression of apoptosis in hemopoietic cells. Cancer Res 53:1735–1738
17. Alenzi FQ (2012) Is there a link between apoptosis and chronic leukemia? J Cell Sci Ther 3
18. McGahon A, Bissonnette R, Schmitt M, Cotter KM, Green DR, Cotter TG (1994) BCR-ABL maintains resistance of chronic myelogenous leukemia cells to apoptotic cell death. Blood 83:1179–1187
19. Amos TA, Lewis JL, Grand FH, Gooding RP, Goldman JM, Gordon MY (1995) Apoptosis in chronic myeloid leukaemia: normal responses by progenitor cells to growth factor deprivation. X-irradiation and glucocorticoids. Br J Haematol 91:387–393
20. Handa H, Hegde UP, Kotelnikov VM, Mundle SD, Dong LM, Burke P, Rose S, Gaskin F, Raza A, Preisler HD (1997) Bcl-2 and c-myc expression, cell cycle kinetics and apoptosis during the progression of chronic myelogenous leukemia from diagnosis to blastic phase. Leuk Res 21:479–489
21. Ren R (2005) Mechanisms of BCR-ABL in the pathogenesis of chronic myelogenous leukaemia. Nat Rev Cancer 5:172–183
22. Sattler M, Salgia R (1997) Activation of hematopoietic growth factor signal transduction pathways by the human oncogene BCR/ABL. Cytokine Growth Factor Rev 8:63–79
23. Kabarowski JH, Allen PB, Wiedemann LM (1994) A temperature sensitive p210 BCR-ABL mutant defines the primary consequences of BCR-ABL tyrosine kinase expression in growth factor dependent cells. EMBO J 13:5887–5895
24. Burns TA, Subathra M, Signorelli P, Choi Y, Yang X, Wang Y, Villani M, Bhalla K, Zhou D, Luberto C (2013) Sphingomyelin synthase 1 activity is regulated by the BCR-ABL oncogene[S]. J Lipid Res 54:794–805

25. Gu L, Vogiatzi P, Puhr M, Dagvadorj A, Lutz J, Ryder A, Addya S, Fortina P, Cooper C, Leiby B, Dasgupta A, Hyslop T, Bubendorf L, Alanen K, Mirtti T, Nevalainen MT (2010) Stat5 promotes metastatic behavior of human prostate cancer cells in vitro and in vivo. Endocr Relat Cancer 17:481–493
26. Ullman MD, Radin NS (1974) The enzymatic formation of sphingomyelin from ceramide and lecithin in mouse liver. J Biol Chem 249:1506–1512
27. Tafesse FG, Huitema K, Hermansson M, van der Poel S, van den Dikkenberg J, Uphoff A, Somerharju P, Holthuis JC (2007) Both sphingomyelin synthases SMS1 and SMS2 are required for sphingomyelin homeostasis and growth in human HeLa cells. J Biol Chem 282:17537–17547
28. Bocker MT, Hellwig I, Breiling A, Eckstein V, Ho AD, Lyko F (2011) Genome-wide promoter DNA methylation dynamics of human hematopoietic progenitor cells during differentiation and aging. Blood 117:e182–e189
29. Perrotti D, Neviani P (2008) Protein phosphatase 2A (PP2A), a drugable tumor suppressor in Ph1(+) leukemias. Cancer Metastasis Rev 27:159–168
30. Mukhopadhyay A, Saddoughi SA, Song P, Sultan I, Ponnusamy S, Senkal CE, Snook CF, Arnold HK, Sears RC, Hannun YA, Ogretmen B (2009) Direct interaction between the inhibitor 2 and ceramide via sphingolipid-protein binding is involved in the regulation of protein phosphatase 2A activity and signaling. FASEB J 23:751–763
31. Neviani P, Santhanam R, Trotta R, Notari M, Blaser BW, Liu S, Mao H, Chang JS, Galietta A, Uttam A, Roy DC, Valtieri M, Bruner-Klisovic R, Caligiuri MA, Bloomfield CD, Marcucci G, Perrotti D (2005) The tumor suppressor PP2A is functionally inactivated in blast crisis CML through the inhibitory activity of the BCR/ABL-regulated SET protein. Cancer Cell 8:355–368
32. Perrotti D, Neviani P (2006) ReSETting PP2A tumour suppressor activity in blast crisis and imatinib-resistant chronic myelogenous leukaemia. Br J Cancer 95:775–781
33. Azoitei N, Kleger A, Schoo N, Thal DR, Brunner C, Pusapati GV, Filatova A, Genze F, Moller P, Acker T, Kuefer R, Van Lint J, Baust H, Adler G, Seufferlein T (2011) Protein kinase D2 is a novel regulator of glioblastoma growth and tumor formation. Neuro Oncol 13:710–724
34. Cerbon J, del Carmen Lopez-Sanchez R (2003) Diacylglycerol generated during sphingomyelin synthesis is involved in protein kinase C activation and cell proliferation in Madin-Darby canine kidney cells. Biochem J 373:917–924
35. Ellwanger, K. and Hausser, A. (2013) Physiological functions of protein kinase D in vivo. IUBMB Life
36. Fu Y, Rubin CS (2011) Protein kinase D: coupling extracellular stimuli to the regulation of cell physiology. EMBO Rep 12:785–796
37. Karam M, Legay C, Auclair C, Ricort JM (2012) Protein kinase D1 stimulates proliferation and enhances tumorigenesis of MCF-7 human breast cancer cells through a MEK/ERK-dependent signaling pathway. Exp Cell Res 318:558–569
38. LaValle CR, Zhang L, Xu S, Eiseman JL, Wang QJ (2012) Inducible silencing of protein kinase D3 inhibits secretion of tumor-promoting factors in prostate cancer. Mol Cancer Ther 11:1389–1399
39. Markova B, Albers C, Breitenbuecher F, Melo JV, Brummendorf TH, Heidel F, Lipka D, Duyster J, Huber C, Fischer T (2010) Novel pathway in Bcr-Abl signal transduction involves Akt-independent, PLC-gamma1-driven activation of mTOR/p70S6-kinase pathway. Oncogene 29:739–751
40. Meng A, Luberto C, Meier P, Bai A, Yang X, Hannun YA, Zhou D (2004) Sphingomyelin synthase as a potential target for D609-induced apoptosis in U937 human monocytic leukemia cells. Exp Cell Res 292:385–392
41. Onishi Y, Kawamoto T, Kishimoto K, Hara H, Fukase N, Toda M, Harada R, Kurosaka M, Akisue T (2012) PKD1 negatively regulates cell invasion, migration and proliferation ability of human osteosarcoma. Int J Oncol 40:1839–1848

42. Shabelnik MY, Kovalevska LM, Yurchenko MY, Shlapatska LM, Rzepetsky Y, Sidorenko SP (2011) Differential expression of PKD1 and PKD2 in gastric cancer and analysis of PKD1 and PKD2 function in the model system. Exp Oncol 33:206–211
43. Subathra M, Qureshi A, Luberto C (2011) Sphingomyelin synthases regulate protein trafficking and secretion. PLoS One 6, e23644
44. Sundram V, Chauhan SC, Jaggi M (2011) Emerging roles of protein kinase D1 in cancer. Mol Cancer Res 9:985–996
45. Valverde AM, Sinnett-Smith J, Van Lint J, Rozengurt E (1994) Molecular cloning and characterization of protein kinase D: a target for diacylglycerol and phorbol esters with a distinctive catalytic domain. Proc Natl Acad Sci U S A 91:8572–8576
46. Morales A, Lee H, Goni FM, Kolesnick R, Fernandez-Checa JC (2007) Sphingolipids and cell death. Apoptosis 12:923–939
47. Kolesnick R, Hannun YA (1999) Ceramide and apoptosis. Trends Biochem Sci 24:224–225
48. Hannun YA, Obeid LM (2002) The ceramide-centric universe of lipid-mediated cell regulation: stress encounters of the lipid kind. J Biol Chem 277:25847–25850
49. Pettus BJ, Chalfant CE, Hannun YA (2002) Ceramide in apoptosis: an overview and current perspectives. Biochim Biophys Acta 1585:114–125
50. Ruvolo PP (2001) Ceramide regulates cellular homeostasis via diverse stress signaling pathways. Leukemia 15:1153–1160
51. Maguer-Satta V, Burl S, Liu L, Damen J, Chahine H, Krystal G, Eaves A, Eaves C (1998) BCR-ABL accelerates C2-ceramide-induced apoptosis. Oncogene 16(2):237–248
52. Maguer-Satta V (1998) CML and apoptosis: the ceramide pathway. Hematol Cell Ther 40:233–236
53. Amarante-Mendes GP, Naekyung Kim C, Liu L, Huang Y, Perkins CL, Green DR, Bhalla K (1998) Bcr-Abl exerts its antiapoptotic effect against diverse apoptotic stimuli through blockage of mitochondrial release of cytochrome C and activation of caspase-3. Blood 91: 1700–1705
54. Domina AM, Vrana JA, Gregory MA, Hann SR, Craig RW (2004) MCL1 is phosphorylated in the PEST region and stabilized upon ERK activation in viable cells, and at additional sites with cytotoxic okadaic acid or taxol. Oncogene 23:5301–5315
55. Maurer U, Charvet C, Wagman AS, Dejardin E, Green DR (2006) Glycogen synthase kinase-3 regulates mitochondrial outer membrane permeabilization and apoptosis by destabilization of MCL-1. Mol Cell 21:749–760
56. Nica AF, Tsao CC, Watt JC, Jiffar T, Kurinna S, Jurasz P, Konopleva M, Andreeff M, Radomski MW, Ruvolo PP (2008) Ceramide promotes apoptosis in chronic myelogenous leukemia-derived K562 cells by a mechanism involving caspase-8 and JNK. Cell Cycle 7:3362–3370
57. McGahon AJ, Nishioka WK, Martin SJ, Mahboubi A, Cotter TG, Green DR (1995) Regulation of the Fas apoptotic cell death pathway by Abl. J Biol Chem 270:22625–22631
58. Schindler T, Bornmann W, Pellicena P, Miller WT, Clarkson B, Kuriyan J (2000) Structural mechanism for STI-571 inhibition of Abelson tyrosine kinase. Science 289:1938–1942
59. Baran Y, Salas A, Senkal CE, Gunduz U, Bielawski J, Obeid LM, Ogretmen B (2007) Alterations of ceramide/sphingosine 1-phosphate rheostat involved in the regulation of resistance to imatinib-induced apoptosis in K562 human chronic myeloid leukemia cells. J Biol Chem 282:10922–10934
60. Bonhoure E, Lauret A, Barnes DJ, Martin C, Malavaud B, Kohama T, Melo JV, Cuvillier O (2008) Sphingosine kinase-1 is a downstream regulator of imatinib-induced apoptosis in chronic myeloid leukemia cells. Leukemia 22:971–979
61. Gencer EB, Ural AU, Avcu F, Baran Y (2011) A novel mechanism of dasatinib-induced apoptosis in chronic myeloid leukemia; ceramide synthase and ceramide clearance genes. Ann Hematol 90:1265–1275
62. Camgoz A, Gencer EB, Ural AU, Avcu F, Baran Y (2011) Roles of ceramide synthase and ceramide clearance genes in nilotinib-induced cell death in chronic myeloid leukemia cells. Leuk Lymphoma 52:1574–1584

63. Beverly LJ, Howell LA, Hernandez-Corbacho M, Casson L, Chipuk JE, Siskind LJ (2013) BAK activation is necessary and sufficient to drive ceramide synthase-dependent ceramide accumulation following inhibition of BCL2-like proteins. Biochem J 452:111–119
64. Hu X, Yang D, Zimmerman M, Liu F, Yang J, Kannan S, Burchert A, Szulc Z, Bielawska A, Ozato K, Bhalla K, Liu K (2011) IRF8 regulates acid ceramidase expression to mediate apoptosis and suppresses myelogeneous leukemia. Cancer Res 71:2882–2891
65. Waight JD, Banik D, Griffiths EA, Nemeth MJ, Abrams SI (2014) Regulation of the interferon regulatory factor-8 (IRF-8) tumor suppressor gene by the signal transducer and activator of transcription 5 (STAT5) transcription factor in chronic myeloid leukemia. J Biol Chem 289:15642–15652
66. Schmidt M, Nagel S, Proba J, Thiede C, Ritter M, Waring JF, Rosenbauer F, Huhn D, Wittig B, Horak I, Neubauer A (1998) Lack of interferon consensus sequence binding protein (ICSBP) transcripts in human myeloid leukemias. Blood 91:22–29
67. Maceyka M, Payne SG, Milstien S, Spiegel S (2002) Sphingosine kinase, sphingosine-1-phosphate, and apoptosis. Biochim Biophys Acta 1585:193–201
68. Li QF, Huang WR, Duan HF, Wang H, Wu CT, Wang LS (2007) Sphingosine kinase-1 mediates BCR/ABL-induced upregulation of Mcl-1 in chronic myeloid leukemia cells. Oncogene 26:7904–7908
69. Zhou P, Qian L, Kozopas KM, Craig RW (1997) Mcl-1, a Bcl-2 family member, delays the death of hematopoietic cells under a variety of apoptosis-inducing conditions. Blood 89:630–643
70. Opferman JT, Iwasaki H, Ong CC, Suh H, Mizuno S, Akashi K, Korsmeyer SJ (2005) Obligate role of anti-apoptotic MCL-1 in the survival of hematopoietic stem cells. Science 307:1101–1104
71. Jendiroba DB, Klostergaard J, Keyhani A, Pagliaro L, Freireich EJ (2002) Effective cytotoxicity against human leukemias and chemotherapy-resistant leukemia cell lines by N-N-dimethylsphingosine. Leuk Res 26:301–310
72. Taouji S, Higa A, Delom F, Palcy S, Mahon FX, Pasquet JM, Bosse R, Segui B, Chevet E (2013) Phosphorylation of serine palmitoyltransferase long chain-1 (SPTLC1) on tyrosine 164 inhibits its activity and promotes cell survival. J Biol Chem 288:17190–17201
73. Baccarani M, Cortes J, Pane F, Niederwieser D, Saglio G, Apperley J, Cervantes F, Deininger M, Gratwohl A, Guilhot F, Hochhaus A, Horowitz M, Hughes T, Kantarjian H, Larson R, Radich J, Simonsson B, Silver RT, Goldman J, Hehlmann R (2009) Chronic myeloid leukemia: an update of concepts and management recommendations of European LeukemiaNet. J Clin Oncol 27:6041–6051
74. O'Brien S, Berman E, Borghaei H, Deangelo DJ, Devetten MP, Devine S, Erba HP, Gotlib J, Jagasia M, Moore JO, Mughal T, Pinilla-Ibarz J, Radich JP, Shah Md NP, Shami PJ, Smith BD, Snyder DS, Tallman MS, Talpaz M, Wetzler M (2009) NCCN clinical practice guidelines in oncology: chronic myelogenous leukemia. J Natl Compr Canc Netw 7:984–1023
75. O'Brien SG, Guilhot F, Goldman JM et al (2008) International randomized study of interferon versus STI571 (IRIS) 7-year follow-up: sustained survival, low rate of transformation and increased rate of major molecular response (MMR) in patients (pts) with newly diagnosed chronic myeloid leukemia in chronic phase (CML-CP) treated with Imatinib (IM). Blood 112(11):76
76. Corbin AS, Agarwal A, Loriaux M, Cortes J, Deininger MW, Druker BJ (2011) Human chronic myeloid leukemia stem cells are insensitive to imatinib despite inhibition of BCR-ABL activity. J Clin Invest 121:396–409
77. Barnes DJ, Melo JV (2006) Primitive, quiescent and difficult to kill: the role of non-proliferating stem cells in chronic myeloid leukemia. Cell Cycle 5:2862–2866
78. Krystal GW (2001) Mechanisms of resistance to imatinib (STI571) and prospects for combination with conventional chemotherapeutic agents. Drug Resist Updat 4:16–21
79. Perrotti D, Jamieson C, Goldman J, Skorski T (2010) Chronic myeloid leukemia: mechanisms of blastic transformation. J Clin Invest 120:2254–2264

80. Skorski T (2012) Genetic mechanisms of chronic myeloid leukemia blastic transformation. Curr Hematol Malig Rep 7:87–93
81. Walz C, Sattler M (2006) Novel targeted therapies to overcome imatinib mesylate resistance in chronic myeloid leukemia (CML). Crit Rev Oncol Hematol 57:145–164
82. Druker BJ, Guilhot F, O'Brien SG, Gathmann I, Kantarjian H, Gattermann N, Deininger MW, Silver RT, Goldman JM, Stone RM, Cervantes F, Hochhaus A, Powell BL, Gabrilove JL, Rousselot P, Reiffers J, Cornelissen JJ, Hughes T, Agis H, Fischer T, Verhoef G, Shepherd J, Saglio G, Gratwohl A, Nielsen JL, Radich JP, Simonsson B, Taylor K, Baccarani M, So C, Letvak L, Larson RA (2006) Five-year follow-up of patients receiving imatinib for chronic myeloid leukemia. N Engl J Med 355:2408–2417
83. Hochhaus A, O'Brien SG, Guilhot F, Druker BJ, Branford S, Foroni L, Goldman JM, Muller MC, Radich JP, Rudoltz M, Mone M, Gathmann I, Hughes TP, Larson RA (2009) Six-year follow-up of patients receiving imatinib for the first-line treatment of chronic myeloid leukemia. Leukemia 23:1054–1061
84. Jain P, Kantarjian H, Cortes J (2013) Chronic myeloid leukemia: overview of new agents and comparative analysis. Curr Treat Options Oncol 14:127–143
85. Ottmann OG, Druker BJ, Sawyers CL, Goldman JM, Reiffers J, Silver RT, Tura S, Fischer T, Deininger MW, Schiffer CA, Baccarani M, Gratwohl A, Hochhaus A, Hoelzer D, Fernandes-Reese S, Gathmann I, Capdeville R, O'Brien SG (2002) A phase 2 study of imatinib in patients with relapsed or refractory Philadelphia chromosome-positive acute lymphoid leukemias. Blood 100:1965–1971
86. Sawyers CL, Hochhaus A, Feldman E, Goldman JM, Miller CB, Ottmann OG, Schiffer CA, Talpaz M, Guilhot F, Deininger MW, Fischer T, O'Brien SG, Stone RM, Gambacorti-Passerini CB, Russell NH, Reiffers JJ, Shea TC, Chapuis B, Coutre S, Tura S, Morra E, Larson RA, Saven A, Peschel C, Gratwohl A, Mandelli F, Ben-Am M, Gathmann I, Capdeville R, Paquette RL, Druker BJ (2002) Imatinib induces hematologic and cytogenetic responses in patients with chronic myelogenous leukemia in myeloid blast crisis: results of a phase II study. Blood 99:3530–3539
87. Baran Y, Bielawski J, Gunduz U, Ogretmen B (2011) Targeting glucosylceramide synthase sensitizes imatinib-resistant chronic myeloid leukemia cells via endogenous ceramide accumulation. J Cancer Res Clin Oncol 137:1535–1544
88. Huang WC, Tsai CC, Chen CL, Chen TY, Chen YP, Lin YS, Lu PJ, Lin CM, Wang SH, Tsao CW, Wang CY, Cheng YL, Hsieh CY, Tseng PC, Lin CF (2011) Glucosylceramide synthase inhibitor PDMP sensitizes chronic myeloid leukemia T315I mutant to Bcr-Abl inhibitor and cooperatively induces glycogen synthase kinase-3-regulated apoptosis. FASEB J 25: 3661–3673
89. Ding Q, He X, Xia W, Hsu JM, Chen CT, Li LY, Lee DF, Yang JY, Xie X, Liu JC, Hung MC (2007) Myeloid cell leukemia-1 inversely correlates with glycogen synthase kinase-3beta activity and associates with poor prognosis in human breast cancer. Cancer Res 67: 4564–4571
90. Luo J (2009) Glycogen synthase kinase 3beta (GSK3beta) in tumorigenesis and cancer chemotherapy. Cancer Lett 273:194–200
91. Rayasam GV, Tulasi VK, Sodhi R, Davis JA, Ray A (2009) Glycogen synthase kinase 3: more than a namesake. Br J Pharmacol 156:885–898
92. Huang WC, Lin YS, Chen CL, Wang CY, Chiu WH, Lin CF (2009) Glycogen synthase kinase-3beta mediates endoplasmic reticulum stress-induced lysosomal apoptosis in leukemia. J Pharmacol Exp Ther 329:524–531
93. Lin CF, Chen CL, Chiang CW, Jan MS, Huang WC, Lin YS (2007) GSK-3beta acts downstream of PP2A and the PI 3-kinase-Akt pathway, and upstream of caspase-2 in ceramide-induced mitochondrial apoptosis. J Cell Sci 120:2935–2943
94. Kartal M, Saydam G, Sahin F, Baran Y (2011) Resveratrol triggers apoptosis through regulating ceramide metabolizing genes in human K562 chronic myeloid leukemia cells. Nutr Cancer 63:637–644

95. Salas A, Ponnusamy S, Senkal CE, Meyers-Needham M, Selvam SP, Saddoughi SA, Apohan E, Sentelle RD, Smith C, Gault CR, Obeid LM, El-Shewy HM, Oaks J, Santhanam R, Marcucci G, Baran Y, Mahajan S, Fernandes D, Stuart R, Perrotti D, Ogretmen B (2011) Sphingosine kinase-1 and sphingosine 1-phosphate receptor 2 mediate Bcr-Abl1 stability and drug resistance by modulation of protein phosphatase 2A. Blood 117:5941–5952
96. Camgoz A, Gencer EB, Ural AU, Baran Y (2013) Mechanisms responsible for nilotinib resistance in human chronic myeloid leukemia cells and reversal of resistance. Leuk Lymphoma 54:1279–1287
97. Matsuoka Y, Nagahara Y, Ikekita M, Shinomiya T (2003) A novel immunosuppressive agent FTY720 induced Akt dephosphorylation in leukemia cells. Br J Pharmacol 138:1303–1312
98. Neviani P, Santhanam R, Oaks JJ, Eiring AM, Notari M, Blaser BW, Liu S, Trotta R, Muthusamy N, Gambacorti-Passerini C, Druker BJ, Cortes J, Marcucci G, Chen CS, Verrills NM, Roy DC, Caligiuri MA, Bloomfield CD, Byrd JC, Perrotti D (2007) FTY720, a new alternative for treating blast crisis chronic myelogenous leukemia and Philadelphia chromosome-positive acute lymphocytic leukemia. J Clin Invest 117:2408–2421
99. Perrotti D, Neviani P (2013) Protein phosphatase 2A: a target for anticancer therapy. Lancet Oncol 14:e229–e238
100. Lahiri S, Park H, Laviad EL, Lu X, Bittman R, Futerman AH (2009) Ceramide synthesis is modulated by the sphingosine analog FTY720 via a mixture of uncompetitive and noncompetitive inhibition in an Acyl-CoA chain length-dependent manner. J Biol Chem 284(24): 16090–16098
101. Oaks JJ, Santhanam R, Walker CJ, Roof S, Harb JG, Ferenchak G, Eisfeld AK, Van Brocklyn JR, Briesewitz R, Saddoughi SA, Nagata K, Bittman R, Caligiuri MA, Abdel-Wahab O, Levine R, Arlinghaus RB, Quintas-Cardama A, Goldman JM, Apperley J, Reid A, Milojkovic D, Ziolo MT, Marcucci G, Ogretmen B, Neviani P, Perrotti D (2013) Antagonistic activities of the immunomodulator and PP2A-activating drug FTY720 (Fingolimod, Gilenya) in Jak2-driven hematologic malignancies. Blood 122:1923–1934
102. Ono Y, Kurano M, Ohkawa R, Yokota H, Igarashi K, Aoki J, Tozuka M, Yatomi Y (2013) Sphingosine 1-phosphate release from platelets during clot formation: close correlation between platelet count and serum sphingosine 1-phosphate concentration. Lipids Health Dis 12:20
103. Sobue S, Iwasaki T, Sugisaki C, Nagata K, Kikuchi R, Murakami M, Takagi A, Kojima T, Banno Y, Akao Y, Nozawa Y, Kannagi R, Suzuki M, Abe A, Naoe T, Murate T (2006) Quantitative RT-PCR analysis of sphingolipid metabolic enzymes in acute leukemia and myelodysplastic syndromes. Leukemia 20:2042–2046
104. Schlenk RF, Dohner H (2013) Genomic applications in the clinic: use in treatment paradigm of acute myeloid leukemia. Hematology Am Soc Hematol Educ Program 2013:324–330
105. Murate T, Suzuki M, Hattori M, Takagi A, Kojima T, Tanizawa T, Asano H, Hotta T, Saito H, Yoshida S, Tamiya-Koizumi K (2002) Up-regulation of acid sphingomyelinase during retinoic acid-induced myeloid differentiation of NB4, a human acute promyelocytic leukemia cell line. J Biol Chem 277:9936–9943
106. Bertolaso L, Gibellini D, Secchiero P, Previati M, Falgione D, Visani G, Rizzoli R, Capitani S, Zauli G (1998) Accumulation of catalytically active PKC-zeta into the nucleus of HL-60 cell line plays a key role in the induction of granulocytic differentiation mediated by all-trans retinoic acid. Br J Haematol 100:541–549
107. Yung BY, Hui EK (1993) Differential regulation of protein kinase C isoenzymes during sphinganine potentiation of retinoic acid-induced granulocytic differentiation in human leukemia HL-60 cells. Biochem Biophys Res Commun 196:1390–1400
108. Yung BY, Luo KJ, Hui EK (1992) Interaction of antileukemia agents adriamycin and daunomycin with sphinganine on the differentiation of human leukemia cell line HL-60. Cancer Res 52:3593–3597
109. Yung BY, Hsiao TF, Wei LL, Hui EK (1994) Sphinganine potentiation of dimethyl sulfoxide-induced granulocyte differentiation, increase of alkaline phosphatase activity and

decrease of protein kinase C activity in a human leukemia cell line (HL-60). Biochem Biophys Res Commun 199:888–896
110. Merrill AH Jr, Sereni AM, Stevens VL, Hannun YA, Bell RM, Kinkade JM Jr (1986) Inhibition of phorbol ester-dependent differentiation of human promyelocytic leukemic (HL-60) cells by sphinganine and other long-chain bases. J Biol Chem 261:12610–12615
111. Kolesnick RN (1989) Sphingomyelinase action inhibits phorbol ester-induced differentiation of human promyelocytic leukemic (HL-60) cells. J Biol Chem 264:7617–7623
112. Kan CC, Kolesnick RN (1992) A synthetic ceramide analog, D-threo-1-phenyl-2-decanoylamino-3-morpholino-1-propanol, selectively inhibits adherence during macrophage differentiation of human leukemia cells. J Biol Chem 267:9663–9667
113. Nojiri H, Takaku F, Terui Y, Miura Y, Saito M (1986) Ganglioside GM3: an acidic membrane component that increases during macrophage-like cell differentiation can induce monocytic differentiation of human myeloid and monocytoid leukemic cell lines HL-60 and U937. Proc Natl Acad Sci U S A 83:782–786
114. Stevens VL, Winton EF, Smith EE, Owens NE, Kinkade JM Jr, Merrill AH Jr (1989) Differential effects of long-chain (sphingoid) bases on the monocytic differentiation of human leukemia (HL-60) cells induced by phorbol esters, 1 alpha, 25-dihydroxyvitamin D3, or ganglioside GM3. Cancer Res 49:3229–3234
115. Kim MY, Linardic C, Obeid L, Hannun Y (1991) Identification of sphingomyelin turnover as an effector mechanism for the action of tumor necrosis factor alpha and gamma-interferon. Specific role in cell differentiation. J Biol Chem 266:484–489
116. Okazaki T, Bielawska A, Bell RM, Hannun YA (1990) Role of ceramide as a lipid mediator of 1 alpha,25-dihydroxyvitamin D3-induced HL-60 cell differentiation. J Biol Chem 265:15823–15831
117. Visnjic D, Batinic D, Banfic H (1997) Arachidonic acid mediates interferon-gamma-induced sphingomyelin hydrolysis and monocytic marker expression in HL-60 cell line. Blood 89:81–91
118. Kim DS, Kim SH, Song JH, Chang YT, Hwang SY, Kim TS (2007) Enhancing effects of ceramide derivatives on 1,25-dihydroxyvitamin D(3)-induced differentiation of human HL-60 leukemia cells. Life Sci 81:1638–1644
119. Kim WJ, Okimoto RA, Purton LE, Goodwin M, Haserlat SM, Dayyani F, Sweetser DA, McClatchey AI, Bernard OA, Look AT, Bell DW, Scadden DT, Haber DA (2008) Mutations in the neutral sphingomyelinase gene SMPD3 implicate the ceramide pathway in human leukemias. Blood 111:4716–4722
120. Buehrer BM, Bardes ES, Bell RM (1996) Protein kinase C-dependent regulation of human erythroleukemia (HEL) cell sphingosine kinase activity. Biochim Biophys Acta 1303:233–242
121. Nakade Y, Banno Y, T-Koizumi K, Hagiwara K, Sobue S, Koda M, Suzuki M, Kojima T, Takagi A, Asano H, Nozawa Y, Murate T (2003) Regulation of sphingosine kinase 1 gene expression by protein kinase C in a human leukemia cell line, MEG-O1. Biochim Biophys Acta 1635:104–116
122. Jarvis WD, Fornari FA Jr, Browning JL, Gewirtz DA, Kolesnick RN, Grant S (1994) Attenuation of ceramide-induced apoptosis by diglyceride in human myeloid leukemia cells. J Biol Chem 269:31685–31692
123. Bartova E, Spanova A, Janakova L, Bobkova M, Rittich B (1997) Apoptotic damage of DNA in human leukaemic HL-60 cells treated with C2-ceramide was detected after G1 blockade of the cell cycle. Physiol Res 46:155–160
124. Jarvis WD, Kolesnick RN, Fornari FA, Traylor RS, Gewirtz DA, Grant S (1994) Induction of apoptotic DNA damage and cell death by activation of the sphingomyelin pathway. Proc Natl Acad Sci U S A 91:73–77
125. Sawai H, Okazaki T, Yamamoto H, Okano H, Takeda Y, Tashima M, Sawada H, Okuma M, Ishikura H, Umehara H et al (1995) Requirement of AP-1 for ceramide-induced apoptosis in human leukemia HL-60 cells. J Biol Chem 270:27326–27331
126. Jarvis WD, Fornari FA Jr, Auer KL, Freemerman AJ, Szabo E, Birrer MJ, Johnson CR, Barbour SE, Dent P, Grant S (1997) Coordinate regulation of stress- and mitogen-activated

protein kinases in the apoptotic actions of ceramide and sphingosine. Mol Pharmacol 52: 935–947
127. Westwick JK, Bielawska AE, Dbaibo G, Hannun YA, Brenner DA (1995) Ceramide activates the stress-activated protein kinases. J Biol Chem 270:22689–22692
128. Yamamoto H (1995) Interrelation of differentiation, proliferation and apoptosis in cancer cells. J Osaka Dent Univ 29:51–60
129. Chen M, Quintans J, Fuks Z, Thompson C, Kufe DW, Weichselbaum RR (1995) Suppression of Bcl-2 messenger RNA production may mediate apoptosis after ionizing radiation, tumor necrosis factor alpha, and ceramide. Cancer Res 55:991–994
130. Zhang QH, Sheng HP, Loh TT (1999) Redistribution of cytochrome c is not an essential requirement in C2-ceramide induced apoptosis in HL-60 cells. Life Sci 65:1715–1723
131. Kim WH, Ghil KC, Lee JH, Yeo SH, Chun YJ, Choi KH, Kim DK, Kim MY (2000) Involvement of p27(kip1) in ceramide-mediated apoptosis in HL-60 cells. Cancer Lett 151:39–48
132. Kim HJ, Ghil KC, Kim MS, Yeo SH, Chun YJ, Kim MY (2005) Potentiation of ceramide-induced apoptosis by p27kip1 overexpression. Arch Pharm Res 28:87–92
133. Kim HJ, Mun JY, Chun YJ, Choi KH, Kim MY (2001) Bax-dependent apoptosis induced by ceramide in HL-60 cells. FEBS Lett 505:264–268
134. Kim HJ, Oh JE, Kim SW, Chun YJ, Kim MY (2008) Ceramide induces p38 MAPK-dependent apoptosis and Bax translocation via inhibition of Akt in HL-60 cells. Cancer Lett 260:88–95
135. Wolff RA, Dobrowsky RT, Bielawska A, Obeid LM, Hannun YA (1994) Role of ceramide-activated protein phosphatase in ceramide-mediated signal transduction. J Biol Chem 269:19605–19609
136. Kondo T, Kitano T, Iwai K, Watanabe M, Taguchi Y, Yabu T, Umehara H, Domae N, Uchiyama T, Okazaki T (2002) Control of ceramide-induced apoptosis by IGF-1: involvement of PI-3 kinase, caspase-3 and catalase. Cell Death Differ 9:682–692
137. Kondo T, Suzuki Y, Kitano T, Iwai K, Watanabe M, Umehara H, Daido N, Domae N, Tashima M, Uchiyama T, Okazaki T (2002) Vesnarinone causes oxidative damage by inhibiting catalase function through ceramide action in myeloid cell apoptosis. Mol Pharmacol 61:620–627
138. Son JH, Yoo HH, Kim DH (2007) Activation of de novo synthetic pathway of ceramides is responsible for the initiation of hydrogen peroxide-induced apoptosis in HL-60 cells. J Toxicol Environ Health A 70:1310–1318
139. Turzanski J, Grundy M, Shang S, Russell N, Pallis M (2005) P-glycoprotein is implicated in the inhibition of ceramide-induced apoptosis in TF-1 acute myeloid leukemia cells by modulation of the glucosylceramide synthase pathway. Exp Hematol 33:62–72
140. Jaffrezou JP, Levade T, Bettaieb A, Andrieu N, Bezombes C, Maestre N, Vermeersch S, Rousse A, Laurent G (1996) Daunorubicin-induced apoptosis: triggering of ceramide generation through sphingomyelin hydrolysis. EMBO J 15:2417–2424
141. Mansat V, Bettaieb A, Levade T, Laurent G, Jaffrezou JP (1997) Serine protease inhibitors block neutral sphingomyelinase activation, ceramide generation, and apoptosis triggered by daunorubicin. FASEB J 11:695–702
142. Mansat V, Laurent G, Levade T, Bettaieb A, Jaffrezou JP (1997) The protein kinase C activators phorbol esters and phosphatidylserine inhibit neutral sphingomyelinase activation, ceramide generation, and apoptosis triggered by daunorubicin. Cancer Res 57:5300–5304
143. Grazide S, Terrisse AD, Lerouge S, Laurent G, Jaffrezou JP (2004) Cytoprotective effect of glucosylceramide synthase inhibition against daunorubicin-induced apoptosis in human leukemic cell lines. J Biol Chem 279:18256–18261
144. Kawase M, Watanabe M, Kondo T, Yabu T, Taguchi Y, Umehara H, Uchiyama T, Mizuno K, Okazaki T (2002) Increase of ceramide in adriamycin-induced HL-60 cell apoptosis: detection by a novel anti-ceramide antibody. Biochim Biophys Acta 1584:104–114
145. Strum JC, Small GW, Pauig SB, Daniel LW (1994) 1-Beta-D-Arabinofuranosylcytosine stimulates ceramide and diglyceride formation in HL-60 cells. J Biol Chem 269:15493–15497

146. Jarvis WD, Fornari FA Jr, Tombes RM, Erukulla RK, Bittman R, Schwartz GK, Dent P, Grant S (1998) Evidence for involvement of mitogen-activated protein kinase, rather than stress-activated protein kinase, in potentiation of 1-beta-D-arabinofuranosylcytosine-induced apoptosis by interruption of protein kinase C signaling. Mol Pharmacol 54:844–856
147. de Souza AC, Kodach L, Gadelha FR, Bos CL, Cavagis AD, Aoyama H, Peppelenbosch MP, Ferreira CV (2006) A promising action of riboflavin as a mediator of leukaemia cell death. Apoptosis 11:1761–1771
148. Yun SH, Park ES, Shin SW, Na YW, Han JY, Jeong JS, Shastina VV, Stonik VA, Park JI, Kwak JY (2012) Stichoposide C induces apoptosis through the generation of ceramide in leukemia and colorectal cancer cells and shows in vivo antitumor activity. Clin Cancer Res 18:5934–5948
149. Dbaibo GS, Kfoury Y, Darwiche N, Panjarian S, Kozhaya L, Nasr R, Abdallah M, Hermine O, El-Sabban M, de The H, Bazarbachi A (2007) Arsenic trioxide induces accumulation of cytotoxic levels of ceramide in acute promyelocytic leukemia and adult T-cell leukemia/lymphoma cells through de novo ceramide synthesis and inhibition of glucosylceramide synthase activity. Haematologica 92:753–762
150. Jiang L, Pan X, Chen Y, Wang K, Du Y, Zhang J (2011) Preferential involvement of both ROS and ceramide in fenretinide-induced apoptosis of HL60 rather than NB4 and U937 cells. Biochem Biophys Res Commun 405:314–318
151. Senchenkov A, Han TY, Wang H, Frankel AE, Kottke TJ, Kaufmann SH, Cabot MC (2001) Enhanced ceramide generation and induction of apoptosis in human leukemia cells exposed to DT(388)-granulocyte-macrophage colony-stimulating factor (GM-CSF), a truncated diphtheria toxin fused to human GM-CSF. Blood 98:1927–1934
152. Frankel AE, Powell BL, Hall PD, Case LD, Kreitman RJ (2002) Phase I trial of a novel diphtheria toxin/granulocyte macrophage colony-stimulating factor fusion protein (DT388GMCSF) for refractory or relapsed acute myeloid leukemia. Clin Cancer Res 8:1004–1013
153. Rahmani M, Reese E, Dai Y, Bauer C, Payne SG, Dent P, Spiegel S, Grant S (2005) Coadministration of histone deacetylase inhibitors and perifosine synergistically induces apoptosis in human leukemia cells through Akt and ERK1/2 inactivation and the generation of ceramide and reactive oxygen species. Cancer Res 65:2422–2432
154. Chen L, Luo LF, Lu J, Li L, Liu YF, Wang J, Liu H, Song H, Jiang H, Chen SJ, Luo C, Li KK (2014) FTY720 induces apoptosis of M2 subtype acute myeloid leukemia cells by targeting sphingolipid metabolism and increasing endogenous ceramide levels. PLoS One 9, e103033
155. Bruno AP, Laurent G, Averbeck D, Demur C, Bonnet J, Bettaieb A, Levade T, Jaffrezou JP (1998) Lack of ceramide generation in TF-1 human myeloid leukemic cells resistant to ionizing radiation. Cell Death Differ 5:172–182
156. Ohta H, Sweeney EA, Masamune A, Yatomi Y, Hakomori S, Igarashi Y (1995) Induction of apoptosis by sphingosine in human leukemic HL-60 cells: a possible endogenous modulator of apoptotic DNA fragmentation occurring during phorbol ester-induced differentiation. Cancer Res 55:691–697
157. Jarvis WD, Fornari FA, Traylor RS, Martin HA, Kramer LB, Erukulla RK, Bittman R, Grant S (1996) Induction of apoptosis and potentiation of ceramide-mediated cytotoxicity by sphingoid bases in human myeloid leukemia cells. J Biol Chem 271:8275–8284
158. Sweeney EA, Sakakura C, Shirahama T, Masamune A, Ohta H, Hakomori S, Igarashi Y (1996) Sphingosine and its methylated derivative N, N-dimethylsphingosine (DMS) induce apoptosis in a variety of human cancer cell lines. Int J Cancer 66:358–366
159. Suzuki E, Handa K, Toledo MS, Hakomori S (2004) Sphingosine-dependent apoptosis: a unified concept based on multiple mechanisms operating in concert. Proc Natl Acad Sci U S A 101:14788–14793
160. Sweeney EA, Inokuchi J, Igarashi Y (1998) Inhibition of sphingolipid induced apoptosis by caspase inhibitors indicates that sphingosine acts in an earlier part of the apoptotic pathway than ceramide. FEBS Lett 425:61–65

161. Tan KB, Ling LU, Bunte RM, Chng WJ, Chiu GN (2012) In vivo efficacy of a novel liposomal formulation of safingol in the treatment of acute myeloid leukemia. J Control Release 160:290–298
162. Tan KB, Ling LU, Bunte RM, Chng WJ, Chiu GN (2013) Liposomal codelivery of a synergistic combination of bioactive lipids in the treatment of acute myeloid leukemia. Nanomedicine (Lond) 9:1665–1679
163. Kleuser B, Cuvillier O, Spiegel S (1998) 1Alpha,25-dihydroxyvitamin D3 inhibits programmed cell death in HL-60 cells by activation of sphingosine kinase. Cancer Res 58: 1817–1824
164. Cuvillier O, Levade T (2001) Sphingosine 1-phosphate antagonizes apoptosis of human leukemia cells by inhibiting release of cytochrome c and Smac/DIABLO from mitochondria. Blood 98:2828–2836
165. Paugh SW, Paugh BS, Rahmani M, Kapitonov D, Almenara JA, Kordula T, Milstien S, Adams JK, Zipkin RE, Grant S, Spiegel S (2008) A selective sphingosine kinase 1 inhibitor integrates multiple molecular therapeutic targets in human leukemia. Blood 112:1382–1391
166. Dick TE, Hengst JA, Fox TE, Colledge AL, Kale VP, Sung SS, Sharma A, Amin S, Loughran TP Jr, Kester M, Wang HG, Yun JK (2015) The apoptotic mechanism of action of the sphingosine kinase 1 selective inhibitor SKI-178 in human acute myeloid leukemia cell lines. J Pharmacol Exp Ther 352:494–508
167. Le Scolan E, Pchejetski D, Banno Y, Denis N, Mayeux P, Vainchenker W, Levade T, Moreau-Gachelin F (2005) Overexpression of sphingosine kinase 1 is an oncogenic event in erythroleukemic progression. Blood 106:1808–1816
168. Itoh M, Kitano T, Watanabe M, Kondo T, Yabu T, Taguchi Y, Iwai K, Tashima M, Uchiyama T, Okazaki T (2003) Possible role of ceramide as an indicator of chemoresistance: decrease of the ceramide content via activation of glucosylceramide synthase and sphingomyelin synthase in chemoresistant leukemia. Clin Cancer Res 9:415–423
169. Yakushiji K, Sawai H, Kawai S, Kambara M, Domae N (2003) Characterization of C2-ceramide-resistant HL-60 subline (HL-CR): involvement of PKC delta in C2-ceramide resistance. Exp Cell Res 286:396–402
170. Sawai H, Kawai S, Domae N (2004) Reduced expression of Bax in ceramide-resistant HL-60 subline. Biochem Biophys Res Commun 319:46–49
171. Sietsma H, Veldman RJ, Kolk D, Ausema B, Nijhof W, Kamps W, Vellenga E, Kok JW (2000) 1-Phenyl-2-decanoylamino-3-morpholino-1-propanol chemosensitizes neuroblastoma cells for taxol and vincristine. Clin Cancer Res 6:942–948
172. Plo I, Lehne G, Beckstrom KJ, Maestre N, Bettaieb A, Laurent G, Lautier D (2002) Influence of ceramide metabolism on P-glycoprotein function in immature acute myeloid leukemia KG1a cells. Mol Pharmacol 62:304–312
173. Bonhoure E, Pchejetski D, Aouali N, Morjani H, Levade T, Kohama T, Cuvillier O (2006) Overcoming MDR-associated chemoresistance in HL-60 acute myeloid leukemia cells by targeting sphingosine kinase-1. Leukemia 20:95–102

Sphingolipids as Mediators of Breast Cancer Progression, Metastasis, Response and Resistance to Chemotherapy

Benjamin Newcomb and Yusuf A. Hannun

Abstract Approximately 250,000 people are diagnosed with breast cancer and 40,000 people die from the disease annually in the United States. Significant effort has been put into developing a thorough understanding of the molecular and genetic events that contribute to tumor progression due to mounting evidence that breast cancers are driven by predictable and identifiable genetic mutations, Sphingolipids modulate many of the growth, apoptosis, inflammatory, and angiogenic pathways that breast carcinomas rely on. Recent work in the field of sphingolipidomics has identified many key roles for the sphingolipids in breast tumor progression and identified signaling roles in each of the molecular subtypes of breast cancer. For example, in luminal type tumors, increased expression of ceramide producing enzymes and high levels of complex sphingolipids are associated with poor outcomes and multi-drug resistance. In basal type breast tumors CERT and GCS play a role in resistance to taxanes suggesting that targeting these pathways may be an effective route for the treatment of aggressive triple negative tumors. The role of sphingolipids in Her2-like tumors has not been thoroughly studied. However, CERK has been identified as potential driver of Her2-like tumors. Clearly, sphingolipid signaling in tumor modulation is both complex and vital. By integrating sphingolipid signaling into the molecular characterization of breast tumors, new opportunities for targeted intervention are identified. It is increasingly apparent that modulating sphingolipid levels in tumors will be an effective and powerful method for the treatment of breast cancer.

Keywords Breast cancer • Sphingolipids • Ceramide • Luminal • Basal • Her2 • Triple negative • Metastasis

B. Newcomb
Department of Medicine, Stony Brook University,
178 Health Sciences Center, Stony Brook, NY 11794, USA

Stony Brook University Cancer Center,
178 Health Sciences Center, Stony Brook, NY 11794, USA

Y.A. Hannun (✉)
Stony Brook Cancer Center, Stony Brook University,
Stony Brook, NY, USA
e-mail: yusuf.hannun@stonybrookmedicine.edu

Y.A. Hannun et al. (eds.), *Bioactive Sphingolipids in Cancer Biology and Therapy*, DOI 10.1007/978-3-319-20750-6_4

1 Introduction to Breast Cancer

Breast cancer results from abnormal growth of the breast tissue. The most common form is ductal carcinoma resulting from hyperplasia and neoplastic change in epithelial cells lining the lactiferous ducts. Breast tumors are a heterogeneous mix of tumor, immune, and mesenchymal cells, and vary widely in their histopathologic make up, aggressiveness, and response to therapy. Annually in the US, roughly 250,000 people are diagnosed with breast cancer and 40,000 die [1].

Breast cancer is one of the earliest malignancies to be identified, and has caused significant morbidity and mortality throughout history. The earliest written accounts of cancer originate in an ancient Egyptian medical treatise known as the Edwin Smith Papyrus. In the text, dated around 2500 BC, the authors describe treatment of 48 medical cases using a scientific rather than mythological approach [2]. Included therein is a description of a palpable breast tumor and the conclusion that it is incurable. Early attempts at treatment were made using a number of remedies such as arsenic paste and cautery [3]. Other civilizations of the same era used various concoctions and salves made from heavy metals or sulfur to treat breast cancer. For the most part, these early treatments were used, with only minimal modification, for several thousand years.

In the eighteenth and nineteenth centuries, the clinicopathalogic signature of breast cancer was becoming clearer. Theodor Schwann (1810–1882) demonstrated that tissues were made up of living cells instead of vacant cell structures as Hook had proposed 175 years earlier. Schwann, together with Johannes Muller (1801–1858), described cancers as new cells aberrantly growing in tissues and causing dysfunction of the organs [4]. Their work was one of the first clinicopathologic correlations of cancer, and subsequent studies by Walter Walshe (1812–1880) suggested that cancer was caused by chronic inflammation, trauma, and smoking [5]. As the study of breast cancer entered the twentieth century, research greatly accelerated as a result of accumulated pathologic knowledge as well as new techniques such as transcutaneous fine needle aspiration [6]. In 1904, Hugo Ribbert (1855–1920) established the theory that breast cancer resulted in the abnormal growth of epithelial cells as a result of chronic inflammation, and scientists began probing the root causes of breast cancer [7].

Despite the growing body of knowledge of the origins and progression of breast cancer, the standard of care had not changed appreciably since the sixteenth century. William Halsted (1852–1922) attempted to standardize mastectomy techniques, and he demonstrated that his technique of deep excision of the breast and pectoral muscle was highly efficacious at preventing local recurrence of breast cancer [7, 8]. In the late nineteenth and early twentieth centuries, aggressive surgical techniques and early diagnosis significantly reduced the number of deaths from breast cancer, and the 3 year survival rate was over 50 % [9].

In the early twentieth century, focus began to shift to finding cures for inoperable breast cancers. In the late 1800s, George Beatson (1848–1933) observed that oophorectomy (removal of the ovaries) reduced the growth and spread of breast tumors in laboratory animals. He attempted oophorectomy combined with treatment with thyroid extracts, and saw reduction in tumor growth in some patients [10]. Oophorectomy came into regular clinical practice, but when Stanley Boyd conducted a national survey of cases of breast cancer with oophorectomy, he found that oophorectomy had limited efficacy in a small number of patients [11]. Nevertheless, this hinted at hormonal regulation of breast tumors. Then, in 1910, one of the most important events in cancer biology occurred. Peyton Rous (1879–1970) published his work demonstrating that sarcomas in hens could be retrovirally induced [12]. His work demonstrated that tumors could be induced by expression of specific genes. Rous' work further revolutionized cancer research when it was recognized that his virus could immortalize cells and allow them to be grown *ex vivo*. Tissue culture techniques progressed rapidly, and new models of breast cancer developed rapidly. Explanted tumors were grown in rodents, and the modern era of chemotherapeutic testing began.

These efforts consolidated much of the knowledge of breast cancer, and breast cancer researchers began taking a multidisciplinary approach to understanding the disease. The work of James Ewing (1866–1943) and Albert Broders (1885–1964) produced a histologic grading system for tumors that is still in use today [9]. Their systems were based on cellular atypia, tumor vascularity, and necrosis, as well as size and level of invasion. Based on these parameters, tumor outcome and response to treatment could be reasonably predicted.

Today, treatment of breast cancer is highly complex, multimodal, and interdisciplinary. By utilizing knowledge gained from decades of research, current breast cancer therapies include targeted therapies, radiation treatment, systemic administration of drugs, and surgery to ablate primary early stage tumors. Our current understanding of and approach to treat breast cancer have increased the overall 10 year survival of breast cancer patients to 83 % [1]. However, locally advanced or metastatic tumors have poor outcomes. For example, patients presenting with stage III disease, have a roughly 50 % 5-year survival [1]. Therefore, it is easy to understand the value of early detection, as well as the need to improve therapy for aggressive tumors.

Indeed, breast cancer therapy matured throughout the twentieth century, and several classes of chemotherapeutics were discovered, beginning with the nitrogen mustards and antimetabolites [13]. In the early 1950s Heidelberger and colleagues found that rat hepatomas had increased uptake of uracil. Therefore, they developed drugs targeting uracil metabolism and achieved a durable reduction in tumor size with 5-florouracil. 5-Florouracil was one of the first antimetabolites, followed by its pro-drug capecitabine in the early 2000s [14, 15]. Antimetabolites interfere with metabolic processes, often causing defects in the production of nucleic acids and replication of DNA, leading to replication stress and apoptosis in rapidly dividing or checkpoint deficient cells.

Following the introduction of antimetabolites, were the antitumor antibiotics. This class of chemotherapeutics functions primarily through cytotoxic mechanisms including disruption of DNA structure. The earliest antitumor antibiotic was bleomycin, discovered in the mid 1960s [16]. Shortly after the discovery of bleomycin, adriamycin was introduced [17]. Although the antitumor antibiotics were highly effective, their low therapeutic index limited their use. For example, early Phase I/II trials with Adriamycin highlighted the significant hematologic cytotoxicity and endothelial toxicity of high dose Adriamycin treatment protocols. At doses where tissue and binding factors are saturated (i.e., in patients with high plasma concentrations of Adriamycin), patients experienced catastrophic granulocytopenia, and secondary infections, as well as premature ventricular contractions. Since Adriamycin is rapidly metabolized, achieving therapeutic plasma concentrations requires high dosage schedules that are associated with poor patient outcomes. Despite the fact that significant tumor response was observed in the patients on high dose schedules of Adriamycin, 9 of the 20 patients on high dose schedules died from side effects of the drug [18]. Following the early clinical trials with anti tumor antibiotics, investigators realized the need for targeted therapies and dose limiting combinational therapy to maximize tumor response at lower doses of chemotherapeutic drugs [19].

Drugs targeting estrogen synthesis were developed soon after the discovery of the antimetabolites. Building upon Beatson's work, estrogen ablative therapy was pursued for many patients throughout the 50s and 60s with varied success. In the early 1970s, Elwood Jensen and colleagues demonstrated that only certain breast cancers expressed estrogen receptors, and that these receptors contributed to tumor growth and predicted response to antiendocrine therapy [20]. Jensen and colleagues went on to develop robust antibodies to the estrogen receptor, and demonstrated a clinically relevant link between ER expression in small tumors and eventual response to antiendocrine therapy [21]. ICI Pharmaceuticals developed Tamoxifen (an antagonist of the estrogen receptor) in the 1960s, but it was not used for breast cancer until the 1970s [22].

In the 1990s and early 2000s, naturally derived taxane based chemotherapeutics were developed. Taxanes hinder cytoskeletal remodeling in dividing cells, and interfere with mitotic spindle formation [23, 24]. The production and distribution of taxanes was a highly politicized and controversial issue in the early 1990s, but the effectiveness of taxanes in the treatment of metastatic disease led to their rapid adoption and widespread use [25, 26]. Therapeutic employment of taxanes (docetaxel or paclitaxel) is currently recommended for treatment of a variety of locally advanced Her2 positive as well as Her2 negative breast tumors.

The discovery that tumors are driven by specific and predictable growth and development pathways led to the advancement of targeted therapies. One of the first rationally developed and targeted therapies to come to market, imatinib mesylate (Gleevec), was developed by Ciba-Geigy (Novartis) using *in silico* computational methods to design inhibitors of bcr-abl, the aberrant tyrosine kinase in chronic myelogenous leukemia [27]. The success of imatinib validated the use of pathway analysis and bioinformatics as a guide for research and development of novel therapeutics, and heralded in a new era in chemotherapy. In a subset of breast tumors, the membrane bound receptor tyrosine kinase Her2 (ERBB2) is overexpressed, and efforts to develop therapies targeting the Her2 receptor were also met with rapid success [28, 29]. One of the best characterized targeted therapies for breast cancer is trastuzumab, a monoclonal antibody that results in blockage of Her2 signaling as well as opsonization of the tumor cells [30]. Opsonization of the tumor cells results in immune mediated clearance of these cells, and future therapeutics may rely more heavily on targeted manipulation of adaptive immunity [31].

2 Molecular Genetic Portraits of Breast Cancer

Traditionally, the choice of treatment protocol has been based on anatomic and histopathologic characteristics of the tumor. The American Joint Committee on Cancer has set forth staging criteria for anatomical classification of tumors based on tumor size (T), node status (N), and presence or absence of distant metastasis (M) [32]. T refers to the size of the primary tumor and local invasion, N denotes the presence or absence of tumor cells in regional lymph nodes, and M indicates distant metastasis of the tumor. These TNM criteria are used to establish the stage of the tumor, and by extension, the likely outcomes of the disease. Generally speaking, early stage tumors include tumors up to T2N1M0. Locally advanced tumors include tumors up to T4N3M0, and metastatic tumors are any T or N status with identifiable metastasis to other organs.

However, it is difficult to predict the biomolecular make up of a tumor vis-à-vis receptor status from the clinical designation (e.g., TNM or grade). Thus, histopathologic characterization of tumors is also used to establish receptor status for the purposes of choosing targeted therapies and predicting the response to endocrine therapy. Tumors generally fall into one of three therapeutically relevant categories: ER+, Her2+, or triple negative. Based on the stage and receptor status, surgical recommendations are made, and chemotherapeutic regimens are chosen. Thus, the presence or absence of progesterone receptors (PR), estrogen receptors (ER), and the human epidermal growth factor receptor (HER2) determines the type and duration of targeted treatment [33].

The last decade has witnessed increasing intensity in molecular profiling and sub-typing of breast cancer based on gene expression profiles and presence of specific mutations in the hope of developing more accurate diagnosis of breast cancer sub-types. Perou et al. performed hierarchical clustering of gene expression data from a large microarray database and established several gene expression patterns [34]. These expression patterns are known as Luminal type/ER+, Basal type (triple negative), and Her2 like. Thus, these and other results have validated the use of receptor status to stratify tumors into molecular subtypes [35]. Subsequent analysis has demonstrated two distinct subsets of Luminal-like tumors, referred to as Luminal A and Luminal B, both ER positive but relying on different growth pathways for survival and proliferation [36]. Outcome analysis further established functional significance of these molecular subtypes with respect to response to treatment and overall survival of patients [37, 38].

In 2012, the Cancer Genome Atlas (TCGA) published a thorough and multiplatform proteomic and genetic characterization of breast carcinomas and described the main signaling pathways that are up- and down-regulated in each tumor subtype. Luminal type tumors were found to have frequent mutations in GATA3, PIK3CA and MAP3K1 genes, and Luminal B tumors in particular have increased incidence of mutations in DNA damage pathway genes, as opposed to Luminal A tumors. Basal type tumors were found to have high expression of proliferation associated genes as well as HIF1-α/ARNT. Her2 like tumors had high expression of Her2 amplicon genes,

Table 1 Molecular subtypes of breast cancer

	Luminal A	Luminal B	Basal	Her2-like
Description	High expression of hormone receptors and ER regulated genes	High expression of hormone receptors and ER regulated genes	High expression of epithelial and cytokeratin genes	High expression of growth factor receptors
Most common receptor status	ER^+ or PR^+, $Her2^-$, Low Ki67	ER^+ or PR^+, $Her2^-$ or $Her2^+$, High Ki67	ER^-, PR^-, $Her2^-$	ER^-, PR^-, $Her2^+$
General treatment protocol	Endocrine therapy	Endocrine therapy plus chemotherapy	Platin containing regiment	Anti-Her2 and anthracycline
Outcomes	Good (92 %[a])	Moderate (90 %[a])	Variable, but mostly poor (86 %[a])	Poor (79 %[a])
Approximate fraction of newly diagnosed cases	47 %	35 %	7 %	7 %

[a]10 Year local relapse free survival in patients that presented with non-metastatic disease and received breast conserving surgery
Adapted from sources [34, 38, 39, 57]

as well as active PIK3CA signaling and over expression of Cyclin D1 [39]. These signaling pathways, as they relate to sphingolipids, will be discussed thoroughly in the following sections. The general characteristics of each breast cancer subtype, most common histopathologic findings, and outcomes are summarized in Table 1.

3 Sphingolipids in Breast Cancer

It is now recognized that many convergent and divergent pathways drive breast tumor growth and spread. Modulation of these pathways by intracellular signaling is an extremely complex topic. In the 1980s it was discovered that lipids behave as potent signaling molecules in addition to being structural components of the cell. Bioactive lipids include the phosphoinositides, diacylglycerols, eicosanoids, fatty acids, and sphingolipids. Lipids represent a unique class of signaling molecules because they mostly belong in biological membranes and, as a class of biologic molecules, lipids have an extremely wide range of physicochemical properties. The structural and physiochemical diversity of lipids requires a wide range of metabolic enzymes and binding molecules. The resulting multiplicity leads to an extremely diverse set of possible combinations of signaling partners. We are just beginning to understand the diverse and interconnected nature of lipid second messengers, and the sphingolipids represent the forefront of this research.

Sphingolipids are a diverse set of signaling and homeostatic molecules. Head group composition, acyl chain length, and saturation all have significant effects on their properties. Additionally, spatiotemporal distribution of sphingolipid synthesis is highly regulated, and subcellular compartmentalization has drastic effects on the function of individual sphingolipids (thoroughly reviewed in [40] and discussed in the overview on sphingolipid metabolism and signaling of this book). Sphingolipids were shown to be powerful signaling molecules capable of modulating many of the mitogenic pathways that tumors rely on for growth. Glycosphingolipid studies in the 1970s and earlier focused heavily on the gangliosides and their antigenic function in different cancers [41–43]. Work by Hakomori and others demonstrated that the gangliosides were important membrane components that affected cell growth, differentiation, and transformation and also served as tumor antigens [44–46]. Work in the mid 1980s demonstrated a powerful role for sphingolipids in subcellular signaling cascades. The finding that sphingosine is a competitive inhibitor of PKC,[47] a known regulator of several oncogenes [48, 49], sparked the interest in the role of sphingolipids in tumors. Subsequent work in the 1990s identified the tumor suppressors Protein Phosphatases 1 and 2A as direct targets of ceramide [50]. Ever since, the metabolism and signaling functions of sphingolipids, and of ceramide in particular have been the subject of significant research.

The main sources of ceramide are *de novo* synthesis or catabolism of sphingomyelin and complex sphingolipids (for details, please refer to the overview on sphingolipid metabolism at the beginning of this book). The *de novo* biosynthetic pathway resides in the ER/Golgi and requires the Serine Palmitoyltransferase complex (SPTLC1-3) and the Ceramide Synthases (CerS1-6). Ceramide derived from the *de novo* pathway is then shuttled to the Golgi either by vesicular transport or by non-vesicular transport through the action of Ceramide Transfer Protein (CERT) [51]. Ceramide delivered to the Golgi by CERT appears to be more specifically targeted for the synthesis of sphingomyelin through the action of Sphingomyelin Synthases (SGMS1 and 2) [52, 53]. Once ceramide has been delivered to the Golgi, it can also be utilized by several other enzymes, including CERK (ceramide kinase), and Ceramide Glucosyltransferase (UCGC). Thus, ceramide-1-phosphate, sphingomyelin, and glycosphingolipids are the major lipid species exported from the Golgi.

In contrast to the *de novo* pathway, the hydrolytic (catabolic) pathways occur at the plasma membrane, in the lysosome/endolysosomal compartments, in mitochondria, and in the Golgi, resulting in the formation of ceramide. These pathways involve one or more of the Sphingomyelinases (SMPD1-5), the Ceramidases (ASAH1-2, and ACER1-3), the Glucocerebrosidases (GBA1-3), Sphingosine and Ceramide kinase (SPHK1-2 and CERK) (reviewed in [54]). Catabolism of sphingosine-1-phosphate and ceramide 1-phosphate is carried out by a S1P lyase and the lipid phosphatases. The salvage or recycling pathway involves hydrolysis of complex sphingolipids in the endolysosomal system, resulting in the formation of sphingosine that can then be salvaged for the re-synthesis of ceramide and other sphingolipids.

It is now increasingly appreciated that any perturbations in the strictly regulated sphingolipid metabolic pathway have significant effects on signaling and growth of breast tumors. The following sections will address the changes seen in sphingolipid metabolism, and the functional consequences thereof, in Luminal, Basal, and Her2-like breast cancer subtypes. A number of cell lines are available that recapitulate gene expression patterns and phenotypic characteristics of the molecular subtypes of breast cancers [55, 56]. A few example cell lines are described in Table 2.

Table 2 Commonly used cell lines in breast cancer research

	Luminal like	Basal like	Her2 like	Normal like
Cell line	MCF7	MDAMB231	SKBR3	MCF10A
	T47D	BT20	BT474	

Adapted from sources [55, 56]

4 Sphingolipids in Luminal Type Breast Cancers

Luminal A Type breast tumors have favorable outcomes and are the most heterogeneous group of breast tumors in terms of signaling characteristics [37]. Luminal Type A tumors represent about 50 % of breast tumors, and are responsive to endocrine therapy [57] as they have high expression of hormone receptors, and downstream estrogen receptor (ESR1) signaling partners such as FOXA1 and RUNX1 [58]. One of the defining characteristics of luminal type A tumors, that separates them from the other molecular subtypes, is intact p53 and RB1 signaling [39]. Additionally, luminal A type tumors frequently have phosphatidylinositol-3-kinase (PI3K) mutations, especially in the PIK3 catalytic subunit (PIK3CA), even though these mutations are not associated with enhanced activation of the PI3K pathway as evidenced by phospho-AKT status, a key downstream target of this pathway [59]. This may be due to high expression of S6K and resultant feedback inhibition. Luminal type tumors also have high frequencies of inactivating mutations of mitogen activated protein kinases (MAPK). For example, loss-of-function mutations in MAP3K1 and MAP2K4 were found with high frequency in these tumors, suggesting loss of function in the p38-JNK1 signaling cascade [58, 60].

Compared to luminal A breast cancer, Luminal B breast cancer has significantly higher risk of rapid relapse following endocrine therapy [61]. Luminal B tumors are classified as ER+, but have lower expression of ER related genes than luminal A tumors and higher expression of proliferative genes [36, 62]. Luminal B tumors often have increased HER2 expression, but since HER2 status is used to distinguish a very specific clinical subset of tumors, HER2 status is not relevant to classification of luminal type tumors. Nevertheless, luminal B tumors have increased growth factor signaling as compared to luminal A tumors [63]. In contrast to luminal A tumors, the proliferative gene signatures of luminal B tumors seem to stem from increased genomic signaling events secondary to genetic duplication of proliferative genes [39, 64]. This finding is in accordance with a high frequency of p53 inactivating mutations, loss of ATM, and deregulation of MDM2 in luminal B tumors [39, 65]. Therefore, the defining characteristics of luminal B tumors are loss of function of the TP53 signaling pathway and amplification of proliferative genes, such as cyclin D1. For the most part, luminal type breast cancers can be considered ER+, HER2−, and the distinction between luminal A and luminal B is mainly a molecular genetic definition that is not regularly made in a clinical setting [66]. Luminal tumors rely heavily on cyclin D1 and on growth factor signaling for growth. Therefore, the following section will focus on the interplay between sphingolipids and growth factor receptors, and sphingolipids and cell cycle progression.

A number of interesting correlations between estrogen receptor signaling and sphingolipid metabolic enzymes have been published recently. Many of the studies on sphingolipids in breast cancer have been conducted using breast cancer cell lines, which limits their impact, nevertheless several studies with patient samples have also been carried out. Among the studies focusing on ER+ cell lines and tumors, UGCG, CerS, sphingomyelinases, ASAH1, and Sphingosine Kinase 1 and 2 (SK1 and SK2) play prominent roles [67–72].

4.1 Glucosylceramide Transferase

Glucosylceramide transferase (GCS) produces glucocerebrosides from ceramide and glucose. Subsequent action by other glycosyltransferases leads to a wide variety of glycolipids. In a study of 200 human breast cancer samples, the authors found high *GCS* (*UGCG*, gene name) expression in luminal A type tumors, and high *GCS* expression was also associated with poor outcomes in ER+ tumors [67]. Since ceramide is known to induce apoptosis and to act as a stress signal, conversion to other lipids, such as the glycosphingolipids, can reduce cell stress. Indeed, Liu et al. have reported that breast tissue generally has low expression of *GCS*, as compared to other organs, and that high *GCS* expression is associated with multi-drug resistance and metastatic disease in ER+, HER2+ tumors [73]. In agreement with this study, several groups have reported on the ability of GCS inhibitors to reverse drug resistance in breast cancer cell lines with high *GCS* expression [74, 75]. A recent report has even suggested that GCS activity is directly inhibited by tamoxifen, and that high GCS activity can decrease sensitivity to the drug [76]. Besides a potential role in receptor function and endocytosis, several groups have also implicated GCS in resistance to chemotherapy [77, 78]. In fact, GCS may limit the efficacy of several chemotherapeutic drugs by titrating away apoptotic ceramide that accumulates following treatment with drugs such as taxol and Adriamycin, among others [79, 80].

4.2 The Ceramide Synthases

The Ceramide Synthase (CerS) enzymes produce dihydroceramide from dihydrosphingosine, and the rapid desaturation of dihydroceramide produces ceramide. In a large-scale lipidomics study of breast cancers, Schiffmann et al. demonstrated a statistically significant increase in C18:0 and C20:0 ceramide in ER+ tumor samples [81]. The authors also demonstrated increased expression of CerS2,

CerS4, and CerS6 in malignant tumor samples. Erez-Roman et al. have also reported elevations in CerS2 and CerS6 in malignant breast tissue samples as compared to normal solid tissue [82]. However, neither group could conclude whether the elevation in ceramide content or CerS expression was a consequence of or a driver of malignancy in breast carcinomas. Continuation of the CerS studies demonstrated opposite roles for long chain and very long chain ceramides. The authors found that CerS2, and by extension the very long chain C24 and C26 ceramides, increased colony formation in MCF7 breast cancer cells whereas CerS4 and CerS6, which produce long chain ceramides (mostly C14 and C16 ceramides) decreased colony formation [83]. These studies concluded that the long chain ceramides induce mitochondrial damage and inhibit cell proliferation whereas the very long chain ceramides decrease contact inhibition in MCF7 cells. Data mining has confirmed robust up-regulation of CerS2, CerS4, and CerS6 in Luminal type tumors (Fig. 1).

In line with the above observations, different studies have proposed that long chain ceramides and the enzymes responsible for their production CerS6 and CerS1 have tumor suppressive properties [84, 85]. For instance, CerS6 has been implicated in the response to folate stress, suggesting CerS6 may also play a role in response to antimetabolites [86]. Antimetabolites are employed in the treatment of some early stage breast cancers, and cell culture studies suggest that combined therapy with the antimetabolite pemetrexed and the kinase inhibitor sorafenib can induce increases in the dihydroceramides. The authors showed that the production of dihydroceramide was dependent on the activity of CerS6 and led to apoptosis of luminal type breast cancer cell lines [87]. Other researchers have reported repression of hTERT activity by CerS1 in various cancer cell lines [88, 89], suggesting an inhibitory role for the long chain ceramides in tumor growth.

4.3 The Sphingomyelinases

The Sphingomyelinases hydrolyze sphingomyelin to ceramide and phosphocholine. There are several isoforms of sphingomyelinase (SMPD1-5), and both SMPD1 (Acid Sphingomyelinase, ASMase) and SMPD3 (Neutral Sphingomyelinase 2, NSMase 2) have been implicated in breast cancer progression. As seen in Fig. 1, *SMPD3* is down regulated in all 4 subtypes of breast carcinomas. Loss of heterozygosity and epigenetic regulation of this region has been described, and may account for the down regulation of *SMPD3*. A comparative epigenetic study reported that there is aberrant CpG methylation around the *SMPD3* locus in human breast cancer cell lines as well as several primary tumor samples [90]. Additionally, another group reported that there is frequent loss of heterozygosity events at 16q22.1, a chromosomal area that includes the *SMPD3* locus [91]. These results suggest NSmase 2 (*SMPD3*) may play an anti-tumor role, however this hypothesis has never been directly tested in breast cancer.

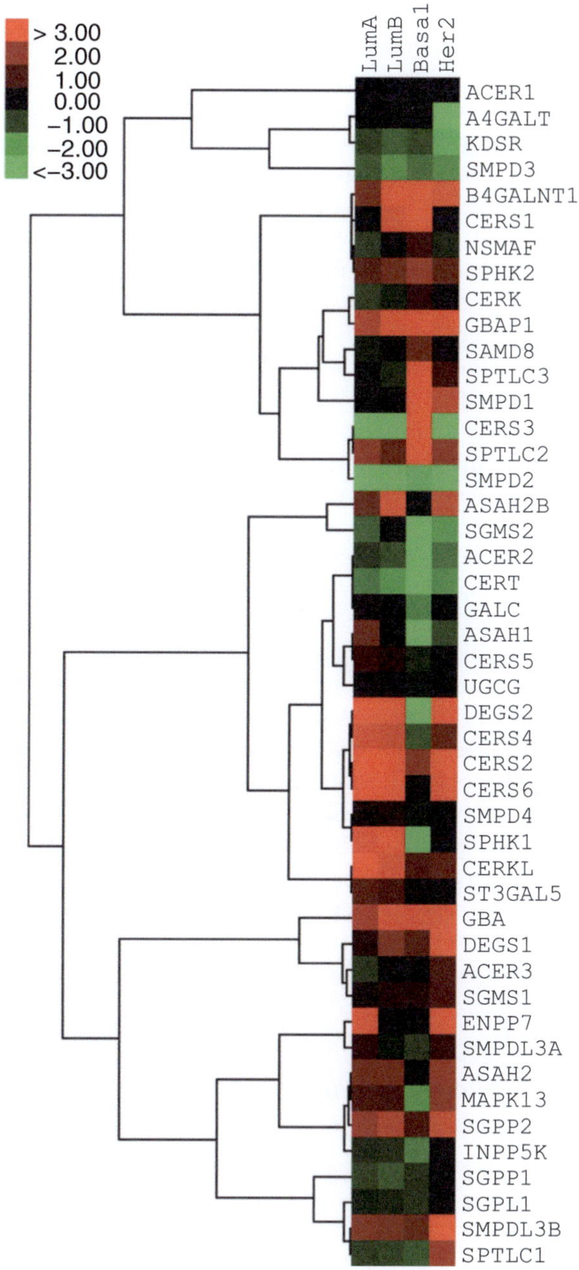

Fig. 1 Sphingolipid gene expression in molecular subtypes of breast cancer. Publicly available TCGA expression data was used to assess the expression levels of sphingolipid genes in molecular subtypes of breast cancer. The expression levels of the indicated genes were compared to expression in adjacent normal solid tissue, and the fold change was calculated. Hierarchical clustering was carried out as shown below

The role of NSMase 2 in drug response has been characterized in breast cancer. Adriamycin resistant MCF7 (MCF7-ADR) cells were found to have higher SM and lower NSMase levels as compared to MCF7 cells. Furthermore, expression and activity of NSMase was significantly up regulated in MCF7-ADR cells following treatment with a demethylating agent that decreases methylation-dependent silencing of genes. This response was not seen in MCF7 cells, suggesting MCF7-ADR cells acquire epigenetic alterations that contribute to the ADR phenotype [92]. In the MCF7-ADR cells, demethylation of DNA was correlated with the subsequent decrease in plasma membrane sphingomyelin, and increased membrane fluidity, attributable to increased activity of NSMase 2. Up-regulated NSMase 2 activity resulted in increased doxorubicin uptake and apoptosis in MCF7-ADR cells, suggesting that silencing of the *SMPD3* locus may be one mechanism by which MCF7 cells acquire resistance to Adriamycin [92]. In line with the work by Vijayaraghavalu et al. is the finding by Ito and colleagues that apoptosis of Adriamycin sensitive MCF7 cells is mediated by NSMase 2 following daunorubicin treatment [93]. Taken together, these observations suggest that re-activation of *SMPD3* expression may be a powerful tool for restoration of sensitivity to anthracyclines in resistant tumors.

In addition to NSMase 2, ASMase plays an important role in tumor biology. ASMase has been found to have significant impact on lysosomal stability and in tumor motility. Traditionally, ASMase is associated with Niemann–Pick Disease, and loss of ASMase enzymatic activity results in lysosomal dysfunction. It should come as no surprise that inhibition of ASMase causes significant lysosomal dysfunction in tumor cells as well. In MCF7-Bcl-2 cells, inhibition of ASMase with the receptor tyrosine kinase inhibitor Sunitinib, leads to lysosomal instability and apoptosis [94]. In contrast to lysosomal stability, governance of cell motility seems to be a tumor specific function of ASMase. ASMase was shown to mediate some of the cytotoxic effects of cisplatin on cytoskeletal remodeling in MCF7 cells [95]. The researchers showed that induction of ASMase and the subsequent generation of ceramide reduces ERM phosphorylation and retards cell motility [95, 96]. Besides its role in cell motility and lysosomal stability, ASMase is known to modulate inflammatory signaling in breast carcinoma cells. Studies on ASMase in MCF7 cells showed ASMase was necessary for p38-dependent expression of IL-6 following PKC activation by the cell permeable phorbol ester phorbol 12-myristate 13-acetate (PMA) [97]. However, the role of ASMase-induced IL-6 in these cells remains unclear.

4.4 *The Ceramidases*

Catabolism of ceramide to other sphingolipids is another potential route by which luminal type breast cancers are able to escape the pro-apoptotic effects of ceramide. The ceramidases rapidly convert ceramide to sphingosine, and then sphingosine is subsequently converted into S1P. Down stream products of the ceramidases, especially sphingosine and S1P, have been shown to be potent signaling molecules. Therefore, conversion of a small amount of ceramide is likely to have profound effects on the cell. Several studies have linked acid ceramidase with growth of ER+ cell lines. Indeed, inhibition of acid ceramidase causes lysosomal disruption that results in rapid cell death in ER+ cell lines [98, 99]. However, the cytotoxic effect of acid ceramidase inhibition is not only linked to its effects on lysosomal permeability. Inhibition of acid ceramidase with non-lysomatrophic compounds still results in cell death, suggesting that the functional consequence of inhibition extends beyond lysosomal homeostasis [70]. The downstream signaling function of ASAH1/Sphingosine was recently delineated by Lucki et al. The authors found that induction of *ASAH1* by phytoestrogens in MCF7 cells resulted in cyclin B2 activation and increased cell proliferation [69], and they further showed that induction of ASAH1 relied on c-SRC and ERK activation and subsequent ERα recruitment to the *ASAH1* promoter. In accordance with these findings, tumor samples with increased estrogen receptor signaling, i.e., luminal tumors, have high levels of *ASAH1* mRNA (Fig. 1).

Paradoxically, high *ASAH1* expression has also been correlated with better outcomes following treatment of ER+ breast tumors [71]. A potential reason cell culture studies may not align with *in vivo* studies is because acid ceramidase has been shown to modulate the tumor microenvironment through regulation of inflammatory cytokines and prostanoid signaling [97, 100]. These inflammatory pathways may play a role in baseline tumor growth, but may make the tumor more prone to apoptosis or immune surveillance following chemotherapy. Therefore, cell culture studies may underestimate the anti-tumor effects of ASAH1 that are mediated through the immune response to tumors following chemotherapy.

Several groups have investigated the interplay between immune modulators and sphingolipid signaling. Signaling events intrinsic to ER+ tumor cells seem to mediate both autocrine and paracrine signaling within the tumor microenvironment. Stress induced cytokines, such as Tumor Necrosis Factor (TNF-α) and Interleukin-1 (IL-1β), activate sphingolipid catabolic pathways [101–103], which in turn have been shown to regulate production of inflammatory mediators in an ASAH1 dependent manner. In breast cancer cell lines, several investigators have demonstrated a requirement for ASAH1 in the production of prostaglandin E2 (PGE2), Chemokine Ligand 5 (CCL5), and Interleukin-6 (IL-6) [97, 100, 104]. Each of these inflammatory mediators can induce tumor progression by creating a permissive tumor microenvironment. Tumor derived PGE2 induces p38/JNK activation in adjacent fibroblasts. The activated p38/JNK signaling cascade resulted in transcription of aromatase and consequent production of estrogen by fibroblasts that promotes tumor growth in estrogen responsive tumors [105]. CCL5 promotes growth and metastasis of ER+ tumors by recruiting macrophages, inducing angiogenesis, and stimulating proliferation of CD44+/CD24− cancer stem like cells [106–108]. The role of IL-6 in breast cancer is multifaceted. In clinical studies on the relationship between IL-6 and breast cancer, high IL-6 is associated with early locally advanced tumors, and good prognosis [109]. Conversely, in late stage tumors, IL-6 was associated with worse prognosis [110]. Cell line studies have demonstrated that ER+ cell lines become growth suppressed by IL-6 and undergo EMT when exposed to IL-6 for extended periods of time. Conversely, ER− cell lines are not responsive to exogenous IL-6 (reviewed in [111]). A possible reason for ER− tumors being unresponsive to IL-6 is the finding that ER− tumor cell lines have a very high level of endogenous IL-6 production, as compared to ER+ cell lines, and therefore ER− tumors already have activated IL-6 signaling pathways and are not responsive to additional IL-6 [112].

4.5 The Sphingosine Kinases

Sphingosine kinase phosphorylates sphingosine to produce sphingosine-1-phosphate (S1P). S1P acts as an intracellular signaling molecule as well as an extracellular signaling molecule that interacts with S1P receptors on the cell surface. In a thorough study of SK1 in human breast cancer samples, Watson et al. found that SK1/S1P participate in non-genomic signaling by ER and cross talk between ER and ERK-1/2. Patients with ER+ tumors displaying high ERK-1/2 and SK1 relapsed nearly 10 years earlier than patients with low SK1 and ERK-1/2. Additionally, the patients with high SK1 and ERK-1/2 were found to be tamoxifen resistant [113]. In accordance with the work by Watson et al., Zhang et al. carried out a large scale, pan-cancer meta-study demonstrating a significant correlation between SK1 expression and worse outcomes in a number of solid tumors. The authors reviewed immunohistochemical staining for SK1 in a range of breast tumor samples and calculated a hazard ratio of 1.86 for tumors with high SK1 expression [114]. Mining of microarray data has revealed luminal type tumors have high SPHK1 and SPHK2 expression (Fig. 1).

SK1 expression in ER+ tumors is modulated by ER signaling. In a study examining expression of micro RNA miR-515-5p, it was noted that in MCF7 cells this miRNA is correlated with SK1 expression. The authors reported that when the ERα receptor is activated by estradiol (E2), the ERα/E2 complex directly interacts with the promoter sequence that regulates miR-515-5p. This interaction leads to a decrease in miR-515-5p expression and de-repression of SK1 expression [115]. The authors went on to report that in a small cohort of 34 tumor samples, ER+ tumors had lower miR-515-5p expression than ER− tumors. However, they do not correlate this finding with SK1 expression in their cohort [115]. In support of the findings by Pinho et al., Takabe and colleagues found that E2 treatment resulted in increased S1P efflux from cultured MCF7 cells, and the observed increase in S1P export was dependent on the ERα receptor suggesting that active ER signaling is necessary for SK1, but not SK2, activation [116]. Taken together, active ERα signaling promotes S1P production in MCF7 cells.

In addition to Sphingosine Kinase activity, S1P receptors (S1PR1-5) seem to be modulated by ER signaling. Indeed, multiple groups have observed ERK activation by S1P/S1P receptor signaling, and correlations between S1PR expression and ERK1/2 expression have been made [117, 118]. In a study of 304 ER+ breast tumor samples, Watson et al. correlated S1P receptor 1 and 3 expression with ERK1/2 expression. The authors found that patients with S1PR1 highly localized in the membrane had shorter time to recurrence of disease, and high levels of cytoplasmic S1PR1 and 3 were associated with worse survival [113]. The authors also noted that TRANSFAC analysis (Biobase) revealed binding sites for ERα, c-Jun, and Sp1 in the $S1P_3$ promoter region [113]. This suggests that tumors with active ER signaling (i.e., tumors with high PR and ER expression [119]) also have high S1P mediated ERK-1/2 activation.

The finding that the SK1/S1P/S1PR axis is up-regulated in ER+ tumors would suggest that targeting this axis would be beneficial in the treatment of breast carcinomas. However, targeting of the SK1/S1P/S1P receptor axis in breast cancer has been met with mixed results. A group from the Department of Chemistry at Amgen utilized the crystal structure of SK1 to generate specific inhibitors of the sphingosine kinases. The inhibitors, termed 'compound A' and 'compound B,' were potent and reproduced the phenotypes of SK1−/− mice [120]. The Amgen Oncology Research group then carried out a cell line study with compound A and compound B, as well as the inhibitor SKII, and found that their compounds reduced S1P levels, but did not impact viability of tumor cell lines at concentrations that reduced S1P levels in these cells [121]. Additionally, the authors reported that ablation of SK1 or SK2 did not reduce tumor cell viability. The conclusion was that SK1 and SK2 are not high yield targets for cancer chemotherapeutics. On the other hand, as it will be discussed later on, targeting the S1P receptors is a very attractive strategy in basal type tumors.

4.6 Ceramide Kinase

In contrast to S1P, comparatively little work has been done on delineating the role of ceramide-1-phosphate (C1P). C1P is produced by CERK mediated phosphorylation of ceramide. In low grade breast cancer, it has been noted that high CERK expression correlates with poor histologic grading of tumors and ER negativity [122]. Additionally, CERK inhibition results in M-phase arrest of MCF7 cells, and sensitized cells to protein kinase inhibitors [123]. Message level of CERK is not significantly different between tumor subtypes (Fig. 1) suggesting that post-translational regulation and subcellular localization of CERK may play a significant role in its function within the tumor cell.

5 Sphingolipids in Basal Type Breast Cancer

The basal-like genetic signature was one of the first genetic portraits of disease to be described. This signature includes expression of mitogenic pathways, as well as high expression of keratins 5, 6, and 17 [34]. Basal type breast carcinomas are mostly triple negative tumors that do not express ER, PR, or HER2 receptors, but roughly 25 % of basal like tumors are not triple negative. Greater than 80 % of basal like tumors have mutant TP53 and loss of p53 signaling, as well as loss of RB1 and BRCA1 [39]. Myc activation, cyclin E1 amplification, and activation of the HIF1-α/ARNT pathways are prominent features of basal like tumors [39, 124]. Another interesting observation from TCGA is that basal-like breast carcinomas share several common features with serous ovarian carcinoma, including Myc activation, cyclin E1 amplification, and inactivation of TP53 and RB1, as well as loss of BRCA1 [39]. This finding suggests that some of the processes driving serous ovarian tumors may be active in basal type breast cancers as well. The lack of hormone receptors and the aggressive nature of basal-like tumors have created a significant need for identification of new therapies.

The approach to surgical treatment of triple negative breast cancers is similar to other breast carcinomas. The choice between adjuvant (treatment given after surgical excision of a primary tumor) and neoadjuvant (treatment given before surgical excision) therapy is made based on tumor size and local spread. Triple negative tumors often respond well to chemotherapy; however, there is a high recurrence rate of tumors [125]. Therefore, there is a significant need for more effective chemotherapeutics. In this regard, ceramide has been used to induce apoptosis of hormone insensitive breast cancer cell lines [126, 127], and some success has been achieved in murine models of breast cancer [128]. Ceramide causes clustering and activation of death receptors, and the apoptotic threshold of cells is modulated by the presence of ceramide at the plasma membrane [129]. Ceramide nanoliposomes can increase the concentration of ceramide at the plasma membrane of target cells and promote pro-apoptotic signaling events in breast cancer cell lines [130].

Beside using ceramide nanoliposomes to indiscriminately increase ceramide levels in cell membranes, ceramide nanoliposomes have also been used to improve the bioavailability and efficacy of breast cancer chemotherapeutics. Nanoliposomal preparations containing ceramide and the tyrosine kinase inhibitor sorafenib were shown to decrease AKT phosphorylation and Cyclin D1 expression, and the presence of ceramide sensitized MDA-MB-231 cells to sorafenib-induced apoptosis [131]. Sorafenib is traditionally used to treat renal carcinoma, and has powerful anti-angiogenic and anti-mitogenic properties, making it an interesting therapeutic for the treatment of basal type breast cancers [132]. Indeed recent clinical trials have demonstrated a potential role for sorafenib in Her2 negative tumors as well as in tamoxifen resistant tumors [133, 134]. Ceramide nanoliposomes have also been used to improve the cellular uptake of doxorubicin in MCF7 and SKBR3 breast cancer cell lines [135]. Ceramide nanoliposomes have been used *in vivo* and may hold promise in adjuvant or neoadjuvant treatment of breast cancers by increasing the bioavailability, stability, and efficacy of current chemotherapeutics [136].

The success of neoadjuvant treatment of triple negative breast tumors is assessed by the pathologic response of the tumor to chemotherapeutic agents. Achieving a preoperative pathologic complete response (pCR) is the goal of neoadjuvant therapy, but it is difficult to predict what patients will achieve pCR with a given chemotherapy regiment. Roughly 55 % of patients with triple negative tumors do not respond well to multi-agent neoadjuvant therapy [137]. A series of studies has recently uncovered potential roles for GCS, CERT, and the ceramide producing β-Glucosidase enzymes (GBA1 and GBA3) in resistance to taxanes in hormone insensitive breast tumors. The first study demonstrated that down regulation of CERT, or up-regulation of GBA1 or 3, sensitized tumor cells to taxanes [138]. As a follow up to their *in vitro* functional genomics study, the research group carried out a retrospective meta-study and found that up-regulation of GCS and CERT, and concomitant down-regulation of GBA1 and GBA3, was highly associated with failure of 5-fluorouracil, doxorubicin, cyclophosphamide, paclitaxel (TFAC) treatment of triple negative tumors possibly because of poor accumulation of ceramide [139]. It is important to note though that the majority of basal type tumors have low GCS and CERT expression (Fig. 1), therefore, the association between high expression of these genes and poor response to treatment may make them valuable biomarkers for a subset of triple negative patients. Altogether these results highlight the importance of regulation of ceramide levels (and of ceramide metabolizing enzymes) in triple negative breast carcinoma cells in order to sensitize tumors to adjuvant chemotherapy.

Based on the work by the TCGA and others, triple negative tumors have high expression of the hypoxia inducible genes. This suggests these tumors may be susceptible to anti-vasculogenic therapies. One sphingolipid analog with potent anti-vasculogenic properties is fingolimod (Gilenya). Fingolimod (also known as FTY720) is a sphingosine analog approved for treatment of multiple sclerosis that serves as a precursor for an agonist of the S1P receptors that often results in internalization of the receptor followed by its shut down [140, 141]. In studies with cell lines, fingolimod has been shown to disrupt ERK1/2 and HER2 cross talk in Her2-like breast cancer cells and to decrease the proliferation of luminal type breast cancer cells [142, 143]. In studies using syngeneic mouse models of breast cancer, research has shown fingolimod decreases vascular smooth muscle motility in response to tumor-derived cytokines [144]. Based on its anti-vasculogenic properties, fingolimod may also serve as a potent inhibitor of basal type breast tumor growth *in vivo*.

6 Sphingolipids in Her2-Like Breast Cancer

Genomic duplication of the Her2 gene results in overexpression of HER2 and HER2-associated genes. In these tumors, HER2 itself as well as FGFR4, EGFR, and several other receptor tyrosine kinases are up regulated. Her2-like tumors have high levels of phospho-SRC and phospho-S6 [39]. Her2-like tumors are largely treated with adjuvant chemotherapy. Effective treatment relies primarily on targeted therapies such as trastuzumab (targeting HER2) in combination with a taxane or anthracycline [145, 146]. Despite the finding that EGFR signaling is up regulated in Her2-like tumors, lapatinib, an inhibitor of the Her2 and EGFR pathways, has not been clinically successful in the adjuvant treatment of Her2-like tumors [147]. On the other hand, adjuvant treatment with trastuzumab has a fairly high disease free survival rate of 76 % [148]. Although there is a small subset of Her2-like tumors that are ER positive, the clinical relevance of combining endocrine therapy with targeted HER2 therapies has not been rigorously evaluated. Also, general clinical practice focuses on targeting the HER2 receptor, and current investigations are underway to evaluate the efficacy of targeting tyrosine kinase activity in Her2 like tumors [149].

Several groups have studied the role of sphingolipids in Her2-like tumors with a focus on CERT, CERK, and sphingomyelin homeostasis. CERT seems to play a prominent role in Her2-like tumors, and has been shown to mediate paclitaxel sensitivity in these tumors. Extending their work with functional metagenomics, the Swanton group found that CERT protein was overexpressed in $HER2^{+}$ tumors, and they demonstrated that silencing of CERT in $HER2^{+}$ tumor cell lines sensitized the cells to chemotherapeutics [139, 150]. Inhibition of CERT in these cell lines induced autophagic flux and LAMP2 expression, and increased sensitivity to paclitaxel, doxorubicin, cisplatin and trastuzumab. Therefore, CERT may be of both therapeutic and prognostic value in $HER2^+$ breast tumors that express high levels of CERT. However, mining of TCGA data has revealed that CERT mRNA is not highly expressed in Her2-Like tumors as compared to adjacent normal tissue. The reason for this discrepancy may be due to discordance between CERT protein and mRNA levels, or possibly due to the choice of controls used for comparison.

The role of CERK in $HER2^+$ tumors has recently been investigated using a mouse model of breast cancer. Utilizing a doxycycline inducible *HER2/neu* mouse model of breast cancer, Payne et al. demonstrated that CERK was required for tumor cell survival following loss of Her2 upon doxycycline withdrawal. When expression of Her2 was decreased and cells were treated with shRNA to CERK, tumor cells expressed high levels of pro-apoptotic markers [151]. This result suggests that following deprivation of Her2 signalling, tumors become reliant on CERK for survival. The authors also tested the role of CERK in BT474 (ER^+, $HER2^+$, p53 mut) and SKBR3 (ER^-, $HER2^+$, p53 mut) cells treated with the tyrosine kinase inhibitor lapatinib. As expected, based on the *in vivo* studies, Her2 cell lines were dependent on CERK expression for survival following Her2 inhibition by lapatinib. The authors extended their study to analyze microarray data from 2200 patient samples, and found significant correlation between CERK expression and aggressive tumors such as $HER2^+$ tumors [151]. The association between CERK and $HER2^+$ tumor progression makes CERK inhibition an attractive target for combination chemotherapy with Her2 inhibitors.

The role of other sphingolipids in Her2-like tumors has not been rigorously studied. However, a study investigating lipid levels in several patient samples from Her2-like tumors found elevations in the level of sphingomyelin [152]. This finding is interesting and helps validate the work by Lee et al. because overexpression of CERT is known to increase sphingomyelin levels in cells [150, 153]. Our mining of TGCA expression data sets reveals that there are a number of alterations in the expression of several sphingolipid genes that are specific to Her2-Like tumors. Members of the SPT complex, acid sphingomyelinase, DEGS1, and GBA are all up regulated in Her2-Like tumors relative to other tumor types (Fig. 1). But, the functional roles of these enzymes in Her2-like tumors remain to be elucidated.

7 Conclusions

The function of sphingolipids in breast cancer is an area of intense investigation. Sphingolipids modulate many of the growth, apoptosis, inflammatory, and perhaps even angiogenic pathways that breast carcinomas rely on. In luminal type breast cancers, conversion of ceramide to complex sphingolipids is associated with drug resistance and worse patient outcomes. The somewhat paradoxical finding by Schiffmann et al. that luminal type tumors have up-regulated synthesis of several ceramide species highlights the different functional outcomes of the ceramides with different molecular composition. Ceramide is rapidly converted to other lipid species, and several groups have demonstrated the role of sphingosine kinase 1 in tumor growth, while the role of acid ceramidase in luminal type tumors is complex. Acid ceramidase in fact promotes a number of cell-cell signaling pathways, and some of the subtleties of these signaling pathways may be lost in over simplified experimental conditions.

Basal type breast tumors are highly aggressive, and have poor response to treatment. Novel therapeutics based on ceramide, as well as novel drug delivery methods have recently shown promise in the treatment of these tumors. The studies showing potential roles for CERT and GCS suggest that targeting these pathways may be an effective route for treating triple negative tumors. Many current chemotherapeutics, such as doxorubicin and etoposide, cause cell death through generation of ceramide. Therefore, directly modulating ceramide metabolic pathways may increase the effectiveness of current chemotherapeutics and increase their therapeutic index.

The role of sphingolipids in Her2-like tumors has not been thoroughly studied. This is despite the fact that robust murine models of Her2-like tumors are readily available. In one of the only studies on sphingolipids in Her2 driven breast cancers, Payne et al. showed reliance on CERK. In addition to the role of CERK, non-vescicular transport of ceramide seems to be an important process in Her2-like tumors, and inhibition of CERT results in autophagy of $HER2^+$ cell lines.

Sphingolipid research in other systems has demonstrated the power of multimodal bioinformatics studies focused on sphingolipid metabolic pathways [154]. Application of cross platform informatics to breast cancer will reveal new opportunities for targeted therapy focused on sphingolipid metabolic enzymes. It is obvious that more studies are needed into the specific roles of individual enzymes and pathways of sphingolipid metabolism in breast cancer and, as these roles of sphingolipids become clearer, more specific rationales will emerge for targeting sphingolipid metabolism as an attractive therapeutic modality.

Acknowledgements The authors would like to acknowledge the SUNY Stony Brook Bioinformatics Facility. The results presented in Fig. 1 are based upon data generated by the TCGA Research Network: http://cancergenome.nih.gov/. This work is partly supported by NCI grant P01 CA097132. We thank Dr. Christopher J. Clarke for careful review of this manuscript.

References

1. American Cancer Society (2013) Cancer facts & figures 2013. American Cancer Society, Atlanta
2. Hajdu SI (2011) A note from history: landmarks in history of cancer, part 1. Cancer 117(5):1097–1102
3. Vargas A et al (2012) [The Edwin Smith papyrus in the history of medicine]. Rev Med Chil 140(10):1357–1362
4. Aszmann OC (2000) The life and work of Theodore Schwann. J Reconstr Microsurg 16(4):291–295
5. Triolo VA (1965) Nineteenth century foundations of cancer research advances in tumor pathology, nomenclature, and theories of oncogenesis. Cancer Res 25:75–106

6. Hajdu SI, Darvishian F (2013) A note from history: landmarks in history of cancer, part 5. Cancer 119(8):1450–1466
7. Hajdu SI (2012) A note from history: landmarks in history of cancer, part 4. Cancer 118(20):4914–4928
8. Sistrunk WE, Maccarty WC (1922) Life expectancy following radical amputation for carcinoma of the breast: a clinical and pathologic study of 218 cases. Ann Surg 75(1):61–69
9. Hajdu SI, Vadmal M (2013) A note from history: Landmarks in history of cancer, Part 6. Cancer 119(23):4058–4082
10. Beatson G (1896) On the treatment of inoperable cases of carcinoma of the mamma: suggestions for a new method of treatment with illustrative cases. Lancet 148(3803):162–165
11. Love RR, Philips J (2002) Oophorectomy for breast cancer: history revisited. J Natl Cancer Inst 94(19):1433–1434
12. Rous P (1911) A sarcoma of the fowl transmissible by an agent separable from the tumor cells. J Exp Med 13(4):397–411
13. DeVita VT, Chu E (2008) A history of cancer chemotherapy. Cancer Res 68(21):8643–8653
14. Umeda M, Heidelberger C (1968) Comparative studies of fluorinated pyrimidines with various cell lines. Cancer Res 28(12):2529–2538
15. Oshaughnessy JA et al (2001) Randomized, open-label, phase II trial of oral capecitabine (Xeloda) vs. a reference arm of intravenous CMF (cyclophosphamide, methotrexate and 5-fluorouracil) as first-line therapy for advanced/metastatic breast cancer. Ann Oncol 12(9):1247–1254
16. Kunimoto T, Hori M, Umezawa H (1967) Modes of action of phleomycin, bleomycin and formycin on HeLa S3 cells in synchronized culture. J Antibiot (Tokyo) 20(5):277–281
17. Di Marco A, Gaetani M, Scarpinato B (1969) Adriamycin (NSC-123,127): a new antibiotic with antitumor activity. Cancer Chemother Rep 53(1):33–37
18. Creasey WA et al (1976) Clinical effects and pharmacokinetics of different dosage schedules of adriamycin. Cancer Res 36(1):216–221
19. Moore FD et al (1967) Carcinoma of the breast. A decade of new results with old concepts. N Engl J Med 277(7):343–350
20. Jensen EV (1975) Estrogen receptors in hormone-dependent breast cancers. Cancer Res 35(11 Pt. 2):3362–3364
21. Block GE et al (1978) Correlation of estrophilin content of primary mammary cancer to eventual endocrine treatment. Ann Surg 188(3):372–376
22. Ward HW (1973) Anti-oestrogen therapy for breast cancer: a trial of tamoxifen at two dose levels. Br Med J 1(5844):13–14
23. Slichenmyer WJ, Von Hoff DD (1991) Taxol: a new and effective anti-cancer drug. Anticancer Drugs 2(6):519–530
24. Hanauske AR et al (1992) Effects of Taxotere and taxol on in vitro colony formation of freshly explanted human tumor cells. Anticancer Drugs 3(2):121–124
25. Seidman AD et al (2008) Randomized phase III trial of weekly compared with every-3-weeks paclitaxel for metastatic breast cancer, with trastuzumab for all HER-2 overexpressors and random assignment to trastuzumab or not in HER-2 nonoverexpressors: final results of Cancer and Leukemia Group B protocol 9840. J Clin Oncol 26(10):1642–1649
26. Sparano JA et al (2008) Weekly paclitaxel in the adjuvant treatment of breast cancer. N Engl J Med 358(16):1663–1671
27. Druker BJ et al (1996) Effects of a selective inhibitor of the Abl tyrosine kinase on the growth of Bcr-Abl positive cells. Nat Med 2(5):561–566
28. Pegram M et al (1999) Inhibitory effects of combinations of HER-2/neu antibody and chemotherapeutic agents used for treatment of human breast cancers. Oncogene 18(13):2241–2251
29. Pegram MD, Konecny G, Slamon DJ (2000) The molecular and cellular biology of HER2/neu gene amplification/overexpression and the clinical development of herceptin (trastuzumab) therapy for breast cancer. Cancer Treat Res 103:57–75
30. Baselga J et al (2001) Mechanism of action of trastuzumab and scientific update. Semin Oncol 28(5 Suppl 16):4–11

31. Hardy NM, Fowler DH, Bishop MR (2006) Immunotherapy of metastatic breast cancer: phase I trail of reduced-intensity allogeneic hematopoietic stem cell transplantation with Th2/Tc2 T-cell exchange. Clin Breast Cancer 7(1):87–89
32. Edge S, Byrd DR, Compton CC, Fritz AG, Greene FL, Trotti A (2010) AJCC cancer staging manual, 7th edn. Springer, New York, p 649
33. Schott A (2014) Systemic treatment of metastatic breast cancer in women: Chemotherapy
34. Perou CM et al (2000) Molecular portraits of human breast tumours. Nature 406(6797):747–752
35. Howlader N et al (2014) US incidence of breast cancer subtypes defined by joint hormone receptor and HER2 status. J Natl Cancer Inst 106(5):pii: dju055
36. Sørlie T et al (2001) Gene expression patterns of breast carcinomas distinguish tumor subclasses with clinical implications. Proc Natl Acad Sci 98(19):10869–10874
37. Parker JS et al (2009) Supervised risk predictor of breast cancer based on intrinsic subtypes. J Clin Oncol 27(8):1160–1167
38. Voduc KD et al (2010) Breast cancer subtypes and the risk of local and regional relapse. J Clin Oncol 28(10):1684–1691
39. Cancer Genome Atlas Network (2012) Comprehensive molecular portraits of human breast tumours. Nature 490(7418):61–70
40. Hannun YA, Obeid LM (2002) The ceramide-centric universe of lipid-mediated cell regulation: stress encounters of the lipid kind. J Biol Chem 277(29):25847–25850
41. Snyder RA, Brady RO, Kornblith PL (1970) Ganglioside patterns of cultured human glioma cells. Neurology 20(4):412
42. Keenan TW et al (1975) Exogenous glycosphingolipids suppress growth rate of transformed and untransformed 3 T3 mouse cells. Exp Cell Res 92(2):259–270
43. Manuelidis L, Yu RK, Manuelidis EE (1977) Ganglioside content and pattern in human gliomas in culture. Correlation of morphological changes with altered gangliosides. Acta Neuropathol 38(2):129–135
44. Bremer EG, Schlessinger J, Hakomori S (1986) Ganglioside-mediated modulation of cell growth. Specific effects of GM3 on tyrosine phosphorylation of the epidermal growth factor receptor. J Biol Chem 261(5):2434–2440
45. Okada Y, Matsuura H, Hakomori S (1985) Inhibition of tumor cell growth by aggregation of a tumor-associated glycolipid antigen: a close functional association between gangliotriaosylceramide and transferrin receptor in mouse lymphoma L-5178Y. Cancer Res 45(6):2793–2801
46. Kim YS et al (1986) Expression of LeY and extended LeY blood group-related antigens in human malignant, premalignant, and nonmalignant colonic tissues. Cancer Res 46(11):5985–5992
47. Hannun YA, Bell RM (1989) Regulation of protein kinase C by sphingosine and lysosphingolipids. Clin Chim Acta 185(3):333–345
48. Chen Z et al (2014) Protein kinase C-delta inactivation inhibits the proliferation and survival of cancer stem cells in culture and in vivo. BMC Cancer 14:90
49. Kim J et al (2011) Sustained inhibition of PKCalpha reduces intravasation and lung seeding during mammary tumor metastasis in an in vivo mouse model. Oncogene 30(3):323–333
50. Chalfant CE et al (1999) Long chain ceramides activate protein phosphatase-1 and protein phosphatase-2A. Activation is stereospecific and regulated by phosphatidic acid. J Biol Chem 274(29):20313–20317
51. Hanada K et al (2003) Molecular machinery for non-vesicular trafficking of ceramide. Nature 426(6968):803–809
52. Charruyer A et al (2008) Decreased ceramide transport protein (CERT) function alters sphingomyelin production following UVB irradiation. J Biol Chem 283(24):16682–16692
53. Blom T et al (2012) Tracking sphingosine metabolism and transport in sphingolipidoses: NPC1 deficiency as a test case. Traffic 13(9):1234–1243
54. Fyrst H, Saba JD (2010) An update on sphingosine-1-phosphate and other sphingolipid mediators. Nat Chem Biol 6(7):489–497

55. Bertucci F et al (2005) Gene expression profiling identifies molecular subtypes of inflammatory breast cancer. Cancer Res 65(6):2170–2178
56. Prat A et al (2013) Characterization of cell lines derived from breast cancers and normal mammary tissues for the study of the intrinsic molecular subtypes. Breast Cancer Res Treat 142(2):237–255
57. Schnitt SJ (2010) Classification and prognosis of invasive breast cancer: from morphology to molecular taxonomy. Mod Pathol 23(S2):S60–S64
58. Bange J, Zwick E, Ullrich A (2001) Molecular targets for breast cancer therapy and prevention. Nat Med 7(5):548–552
59. Stemke-Hale K et al (2008) An integrative genomic and proteomic analysis of PIK3CA, PTEN, and AKT mutations in breast cancer. Cancer Res 68(15):6084–6091
60. Wagner EF, Nebreda AR (2009) Signal integration by JNK and p38 MAPK pathways in cancer development. Nat Rev Cancer 9(8):537–549
61. Tran B, Bedard PL (2011) Luminal-B breast cancer and novel therapeutic targets. Breast Cancer Res 13(6):221
62. Wirapati P et al (2008) Meta-analysis of gene expression profiles in breast cancer: toward a unified understanding of breast cancer subtyping and prognosis signatures. Breast Cancer Res 10(4):R65
63. Sotiriou C, Pusztai L (2009) Gene-expression signatures in breast cancer. N Engl J Med 360(8):790–800
64. Troester MA et al (2006) Gene expression patterns associated with p53 status in breast cancer. BMC Cancer 6:276
65. Jiang Z et al (2010) Rb deletion in mouse mammary progenitors induces luminal-B or basal-like/EMT tumor subtypes depending on p53 status. J Clin Invest 120(9):3296–3309
66. Neve RM et al (2006) A collection of breast cancer cell lines for the study of functionally distinct cancer subtypes. Cancer Cell 10(6):515–527
67. Ruckhaberle E et al (2009) Prognostic relevance of glucosylceramide synthase (GCS) expression in breast cancer. J Cancer Res Clin Oncol 135(1):81–90
68. Zhang X et al (2012) Doxorubicin influences the expression of glucosylceramide synthase in invasive ductal breast cancer. PLoS One 7(11), e48492
69. Lucki NC, Sewer MB (2011) Genistein stimulates MCF-7 breast cancer cell growth by inducing acid ceramidase (ASAH1) gene expression. J Biol Chem 286(22):19399–19409
70. Bai A et al (2009) Synthesis and bioevaluation of omega-N-amino analogs of B13. Bioorg Med Chem 17(5):1840–1848
71. Ruckhaberle E et al (2009) Acid ceramidase 1 expression correlates with a better prognosis in ER-positive breast cancer. Climacteric 12(6):502–513
72. Ruckhaberle E et al (2008) Microarray analysis of altered sphingolipid metabolism reveals prognostic significance of sphingosine kinase 1 in breast cancer. Breast Cancer Res Treat 112(1):41–52
73. Liu YY et al (2011) Glucosylceramide synthase, a factor in modulating drug resistance, is overexpressed in metastatic breast carcinoma. Int J Oncol 39(2):425–431
74. Sun YL et al (2005) [Construction of glucosylceramide synthase-specific siRNA expression vector and its efficiency in reversal of drug resistance in breast carcinoma cells]. Zhonghua Yi Xue Za Zhi 85(8):518–521
75. Gouaze V et al (2005) Glucosylceramide synthase blockade down-regulates P-glycoprotein and resensitizes multidrug-resistant breast cancer cells to anticancer drugs. Cancer Res 65(9):3861–3867
76. Lavie Y et al (1997) Agents that reverse multidrug resistance, tamoxifen, verapamil, and cyclosporin A, block glycosphingolipid metabolism by inhibiting ceramide glycosylation in human cancer cells. J Biol Chem 272(3):1682–1687
77. Bose R et al (1995) Ceramide synthase mediates daunorubicin-induced apoptosis: an alternative mechanism for generating death signals. Cell 82(3):405–414

78. Lucci A et al (1999) Multidrug resistance modulators and doxorubicin synergize to elevate ceramide levels and elicit apoptosis in drug-resistant cancer cells. Cancer 86(2):300–311
79. Charles AG et al (2001) Taxol-induced ceramide generation and apoptosis in human breast cancer cells. Cancer Chemother Pharmacol 47(5):444–450
80. Gewirtz DA (2000) Growth arrest and cell death in the breast tumor cell in response to ionizing radiation and chemotherapeutic agents which induce DNA damage. Breast Cancer Res Treat 62(3):223–235
81. Schiffmann S et al (2009) Ceramide synthases and ceramide levels are increased in breast cancer tissue. Carcinogenesis 30(5):745–752
82. Erez-Roman R, Pienik R, Futerman AH (2010) Increased ceramide synthase 2 and 6 mRNA levels in breast cancer tissues and correlation with sphingosine kinase expression. Biochem Biophys Res Commun 391(1):219–223
83. Hartmann D et al (2012) Long chain ceramides and very long chain ceramides have opposite effects on human breast and colon cancer cell growth. Int J Biochem Cell Biol 44(4):620–628
84. Walker T et al (2009) Sorafenib and vorinostat kill colon cancer cells by CD95-dependent and -independent mechanisms. Mol Pharmacol 76(2):342–355
85. Min J et al (2007) (Dihydro)ceramide synthase 1 regulated sensitivity to cisplatin is associated with the activation of p38 mitogen-activated protein kinase and is abrogated by sphingosine kinase 1. Mol Cancer Res 5(8):801–812
86. Hoeferlin LA et al (2013) Folate stress induces apoptosis via p53-dependent de novo ceramide synthesis and up-regulation of ceramide synthase 6. J Biol Chem 288(18):12880–12890
87. Bareford MD et al (2012) Sorafenib and pemetrexed toxicity in cancer cells is mediated via SRC-ERK signaling. Cancer Biol Ther 13(9):793–803
88. Wooten-Blanks LG et al (2007) Mechanisms of ceramide-mediated repression of the human telomerase reverse transcriptase promoter via deacetylation of Sp3 by histone deacetylase 1. FASEB J 21(12):3386–3397
89. Koybasi S et al (2004) Defects in cell growth regulation by C18:0-ceramide and longevity assurance gene 1 in human head and neck squamous cell carcinomas. J Biol Chem 279(43):44311–44319
90. Demircan B et al (2009) Comparative epigenomics of human and mouse mammary tumors. Genes Chromosomes Cancer 48(1):83–97
91. Chalmers IJ et al (2001) Mapping the chromosome 16 cadherin gene cluster to a minimal deleted region in ductal breast cancer. Cancer Genet Cytogenet 126(1):39–44
92. Vijayaraghavalu S et al (2012) Epigenetic modulation of the biophysical properties of drug-resistant cell lipids to restore drug transport and endocytic functions. Mol Pharm 9(9):2730–2742
93. Ito H et al (2009) Transcriptional regulation of neutral sphingomyelinase 2 gene expression of a human breast cancer cell line, MCF-7, induced by the anti-cancer drug, daunorubicin. Biochim Biophys Acta 1789(11-12):681–690
94. Ellegaard AM et al (2013) Sunitinib and SU11652 inhibit acid sphingomyelinase, destabilize lysosomes, and inhibit multidrug resistance. Mol Cancer Ther 12(10):2018–2030
95. Zeidan YH, Jenkins RW, Hannun YA (2008) Remodeling of cellular cytoskeleton by the acid sphingomyelinase/ceramide pathway. J Cell Biol 181(2):335–350
96. Canals D et al (2010) Differential effects of ceramide and sphingosine 1-phosphate on ERM phosphorylation: probing sphingolipid signaling at the outer plasma membrane. J Biol Chem 285(42):32476–32485
97. Perry DM et al (2014) Defining a role for acid sphingomyelinase in the p38/interleukin-6 pathway. J Biol Chem 10(6):515–527
98. Ullio C et al (2012) Sphingosine mediates TNFalpha-induced lysosomal membrane permeabilization and ensuing programmed cell death in hepatoma cells. J Lipid Res 53(6):1134–1143

99. Bhabak KP et al (2013) Effective inhibition of acid and neutral ceramidases by novel B-13 and LCL-464 analogues. Bioorg Med Chem 21(4):874–882
100. Jenkins RW et al (2011) Regulation of CC ligand 5/RANTES by acid sphingomyelinase and acid ceramidase. J Biol Chem 286(15):13292–13303
101. Cain BS et al (1999) Tumor necrosis factor-alpha and interleukin-1beta synergistically depress human myocardial function. Crit Care Med 27(7):1309–1318
102. Mathias S et al (1993) Activation of the sphingomyelin signaling pathway in intact EL4 cells and in a cell-free system by IL-1 beta. Science 259(5094):519–522
103. Hannun YA (1994) The sphingomyelin cycle and the second messenger function of ceramide. J Biol Chem 269(5):3125–3128
104. Zeidan YH et al (2006) Acid ceramidase but not acid sphingomyelinase is required for tumor necrosis factor-{alpha}-induced PGE2 production. J Biol Chem 281(34):24695–24703
105. Chen D et al (2007) Prostaglandin E(2) induces breast cancer related aromatase promoters via activation of p38 and c-Jun NH(2)-terminal kinase in adipose fibroblasts. Cancer Res 67(18):8914–8922
106. Soria G, Ben-Baruch A (2008) The inflammatory chemokines CCL2 and CCL5 in breast cancer. Cancer Lett 267(2):271–285
107. Pinilla S et al (2009) Tissue resident stem cells produce CCL5 under the influence of cancer cells and thereby promote breast cancer cell invasion. Cancer Lett 284(1):80–85
108. Zhang Y et al (2009) Role of CCL5 in invasion, proliferation and proportion of CD44+/CD24- phenotype of MCF-7 cells and correlation of CCL5 and CCR5 expression with breast cancer progression. Oncol Rep 21(4):1113–1121
109. Karczewska A et al (2000) Expression of interleukin-6, interleukin-6 receptor, and glycoprotein 130 correlates with good prognoses for patients with breast carcinoma. Cancer 88(9):2061–2071
110. Salgado R et al (2003) Circulating interleukin-6 predicts survival in patients with metastatic breast cancer. Int J Cancer 103(5):642–646
111. Dethlefsen C, Hojfeldt G, Hojman P (2013) The role of intratumoral and systemic IL-6 in breast cancer. Breast Cancer Res Treat 138(3):657–664
112. Asgeirsson KS et al (1998) The effects of IL-6 on cell adhesion and e-cadherin expression in breast cancer. Cytokine 10(9):720–728
113. Watson C et al (2010) High expression of sphingosine 1-phosphate receptors, S1P1 and S1P3, sphingosine kinase 1, and extracellular signal-regulated kinase-1/2 is associated with development of tamoxifen resistance in estrogen receptor-positive breast cancer patients. Am J Pathol 177(5):2205–2215
114. Zhang Y et al (2014) Sphingosine kinase 1 and cancer: a systematic review and meta-analysis. PLoS One 9(2), e90362
115. Pinho FG et al (2013) Downregulation of microRNA-515-5p by the estrogen receptor modulates sphingosine kinase 1 and breast cancer cell proliferation. Cancer Res 73(19):5936–5948
116. Takabe K et al (2010) Estradiol induces export of sphingosine 1-phosphate from breast cancer cells via ABCC1 and ABCG2. J Biol Chem 285(14):10477–10486
117. Kim ES et al (2014) Inflammatory lipid sphingosine-1-phosphate upregulates C-reactive protein via C/EBPbeta and potentiates breast cancer progression. Oncogene 33(27):3583–3593
118. Rutherford C et al (2013) Regulation of cell survival by sphingosine-1-phosphate receptor S1P1 via reciprocal ERK-dependent suppression of Bim and PI-3-kinase/protein kinase C-mediated upregulation of Mcl-1. Cell Death Dis 4, e927
119. Hu X et al (2009) Genetic alterations and oncogenic pathways associated with breast cancer subtypes. Mol Cancer Res 7(4):511–522
120. Gustin DJ et al (2013) Structure guided design of a series of sphingosine kinase (SphK) inhibitors. Bioorg Med Chem Lett 23(16):4608–4616
121. Rex K et al (2013) Sphingosine kinase activity is not required for tumor cell viability. PLoS One 8(7), e68328

122. Ruckhaberle E et al (2009) Gene expression of ceramide kinase, galactosyl ceramide synthase and ganglioside GD3 synthase is associated with prognosis in breast cancer. J Cancer Res Clin Oncol 135(8):1005–1013
123. Pastukhov O et al (2014) The ceramide kinase inhibitor NVP-231 inhibits breast and lung cancer cell proliferation by inducing M phase arrest and subsequent cell death. Br J Pharmacol 171(24):5829–5844
124. Chandriani S et al (2009) A core MYC gene expression signature is prominent in basal-like breast cancer but only partially overlaps the core serum response. PLoS One 4(8), e6693
125. Liedtke C et al (2008) Response to neoadjuvant therapy and long-term survival in patients with triple-negative breast cancer. J Clin Oncol 26(8):1275–1281
126. Stover T, Kester M (2003) Liposomal delivery enhances short-chain ceramide-induced apoptosis of breast cancer cells. J Pharmacol Exp Ther 307(2):468–475
127. Shabbits JA, Mayer LD (2003) Intracellular delivery of ceramide lipids via liposomes enhances apoptosis in vitro. Biochim Biophys Acta 1612(1):98–106
128. Stover TC et al (2005) Systemic delivery of liposomal short-chain ceramide limits solid tumor growth in murine models of breast adenocarcinoma. Clin Cancer Res 11(9):3465–3474
129. Miyaji M et al (2005) Role of membrane sphingomyelin and ceramide in platform formation for Fas-mediated apoptosis. J Exp Med 202(2):249–259
130. Watters RJ et al (2012) Development and use of ceramide nanoliposomes in cancer. Methods Enzymol 508:89–108
131. Tran MA et al (2008) Combining nanoliposomal ceramide with sorafenib synergistically inhibits melanoma and breast cancer cell survival to decrease tumor development. Clin Cancer Res 14(11):3571–3581
132. Wilhelm SM et al (2008) Preclinical overview of sorafenib, a multikinase inhibitor that targets both Raf and VEGF and PDGF receptor tyrosine kinase signaling. Mol Cancer Ther 7(10):3129–3140
133. Tan QX et al (2014) Sorafenib-based therapy in HER2-negative advanced breast cancer: Results from a retrospective pooled analysis of randomized controlled trials. Exp Ther Med 7(5):1420–1426
134. Pedersen AM et al (2014) Sorafenib and nilotinib resensitize tamoxifen resistant breast cancer cells to tamoxifen treatment via estrogen receptor alpha. Int J Oncol 45(5):2167–2175
135. Pedrosa LR et al (2013) Improving intracellular doxorubicin delivery through nanoliposomes equipped with selective tumor cell membrane permeabilizing short-chain sphingolipids. Pharm Res 30(7):1883–1895
136. Zolnik BS et al (2008) Rapid distribution of liposomal short-chain ceramide in vitro and in vivo. Drug Metab Dispos 36(8):1709–1715
137. Sirohi B et al (2008) Platinum-based chemotherapy in triple-negative breast cancer. Ann Oncol 19(11):1847–1852
138. Swanton C et al (2007) Regulators of mitotic arrest and ceramide metabolism are determinants of sensitivity to paclitaxel and other chemotherapeutic drugs. Cancer Cell 11(6):498–512
139. Juul N et al (2010) Assessment of an RNA interference screen-derived mitotic and ceramide pathway metagene as a predictor of response to neoadjuvant paclitaxel for primary triple-negative breast cancer: a retrospective analysis of five clinical trials. Lancet Oncol 11(4):358–365
140. Hla T et al (2000) Sphingosine-1-phosphate signaling via the EDG-1 family of G-protein-coupled receptors. Ann N Y Acad Sci 905:16–24
141. Mousseau Y et al (2012) Fingolimod inhibits PDGF-B-induced migration of vascular smooth muscle cell by down-regulating the S1PR1/S1PR3 pathway. Biochimie 94(12):2523–2531
142. Long JS et al (2010) Sphingosine 1-phosphate receptor 4 uses HER2 (ERBB2) to regulate extracellular signal regulated kinase-1/2 in MDA-MB-453 breast cancer cells. J Biol Chem 285(46):35957–35966

143. Nagaoka Y et al (2008) Effects of phosphorylation of immunomodulatory agent FTY720 (fingolimod) on antiproliferative activity against breast and colon cancer cells. Biol Pharm Bull 31(6):1177–1181
144. Azuma H et al (2002) Marked prevention of tumor growth and metastasis by a novel immunosuppressive agent, FTY720, in mouse breast cancer models. Cancer Res 62(5):1410–1419
145. Slamon D et al (2011) Adjuvant trastuzumab in HER2-positive breast cancer. N Engl J Med 365(14):1273–1283
146. Joensuu H et al (2006) Adjuvant docetaxel or vinorelbine with or without trastuzumab for breast cancer. N Engl J Med 354(8):809–820
147. Goss PE et al (2013) Adjuvant lapatinib for women with early-stage HER2-positive breast cancer: a randomised, controlled, phase 3 trial. Lancet Oncol 14(1):88–96
148. Goldhirsch A et al (2013) 2 years versus 1 year of adjuvant trastuzumab for HER2-positive breast cancer (HERA): an open-label, randomised controlled trial. Lancet 382(9897):1021–1028
149. Agrawal A et al (2005) Overview of tyrosine kinase inhibitors in clinical breast cancer. Endocr Relat Cancer 12(Suppl 1):S135–S144
150. Lee AJ et al (2012) CERT depletion predicts chemotherapy benefit and mediates cytotoxic and polyploid-specific cancer cell death through autophagy induction. J Pathol 226(3):482–494
151. Payne AW et al (2014) Ceramide kinase promotes tumor cell survival and mammary tumor recurrence. Cancer Res 74(21):6352–6363
152. Kim IC et al (2013) Lipid profiles for HER2-positive breast cancer. Anticancer Res 33(6):2467–2472
153. Sano O et al (2007) Sphingomyelin-dependence of cholesterol efflux mediated by ABCG1. J Lipid Res 48(11):2377–2384
154. Montefusco DJ et al (2013) Distinct signaling roles of ceramide species in yeast revealed through systematic perturbation and systems biology analyses. Sci Signal 6(299):p. rs14

Role of Sphingolipids in Non-melanoma Skin Cancer

Chih-Li Lin and Cungui Mao

Abstract Non-melanoma skin cancer (NMSC) is the most common cancer in the world and its incidence continues to rise. NMSC includes two main types, basal cell carcinoma (BCC) and squamous cell carcinoma (SCC), both of which arise from malignant keratinocytes, the major cell type in the epidermis of the skin. Most cases of NMSC can be cured by existing new surgical techniques and chemo- or radiotherapy. However, NMSC remains a leading cause of cancer-related deaths in the United States and worldwide likely due to its extremely high incidence and recurrence rate, and limited availability of effective therapies for advanced cases. Therefore, we still need a better understanding of its pathogenesis and the development of novel and effective approaches to diagnosis, prevention, and treatment. Like many other cancers, NMSC results from dysregulation of the proliferation, differentiation, and apoptosis of cells that it originates from. Emerging evidence suggests that sphingolipids play an important role in regulating these cellular responses of keratinocytes and the homeostasis of the epidermis. In this review, we will discuss the role for sphingolipids and their metabolizing enzymes in regulating the proliferation, differentiation, and apoptosis of epidermal keratinocytes and the potential role of sphingolipids in NMSC prevention and/or therapy.

Keywords Basal cell carcinoma (BCC) • Squamous cell carcinoma (SCC) • Epidermal keratinocyte • Stem cells • Ceramide • Sphingoid base • Sphingosine-1-phosphate (S1P)

C.-L. Lin
Graduate Program of Molecular and Cellular Biology, Stony Brook University, Stony Brook, NY USA

Department of Medicine, Stony Brook University, Stony Brook, NY USA

Stony Brook University Cancer Center, T15-023, HSC, Stony Brook, NY 11794-8155, USA
e-mail: chih-li.lin@stonybrook.edu

C. Mao, Ph.D. (✉)
Graduate Program of Molecular and Cellular Biology, Stony Brook University, Stony Brook, NY USA
e-mail: cungui.mao@stonybrook.edu

Y.A. Hannun et al. (eds.), *Bioactive Sphingolipids in Cancer Biology and Therapy*, DOI 10.1007/978-3-319-20750-6_5

1 Cellular Origins of BCC and SCC

Mammalian epidermis, the outermost layer of the skin, comprises different compartments including the interfollicular epidermis (IFE), the hair follicle (HF), and the sebaceous gland (SG) (Fig. 1) [1]. The IFE is mainly composed of keratinocytes and has four distinct cell layers: the basal, spinous, and granular layers, and stratum corneum (Fig. 2) [2]. Under basal conditions, the IFE is renewed mainly from the IFE resident stem cells (SCs) in the basal layer of the epidermis whereas HFs and SGs regenerate from bulge and isthmus SCs, respectively [3]. Upon wounding, bulge and isthmus SCs move upward to repair the wounded IFE [3].

Although they both occur most frequently in the sun-exposed regions of the skin, BCC and SCC arise in different cell layers of the epidermis. BCC arises from the basal layer and SCC from the suprabasal layers. Although it was thought that BCC initiated from the bulge SCs [4], a recent study using a mouse model indicates that it actually may initiate from long-term resident progenitor cells of the IFE and the upper infundibulum [5]. Although its cellular origin remains controversial, a recent study using mice conditionally expressing a constitutively active K-Ras mutant in different epidermal lineages demonstrated that SCC can initiate from multiple

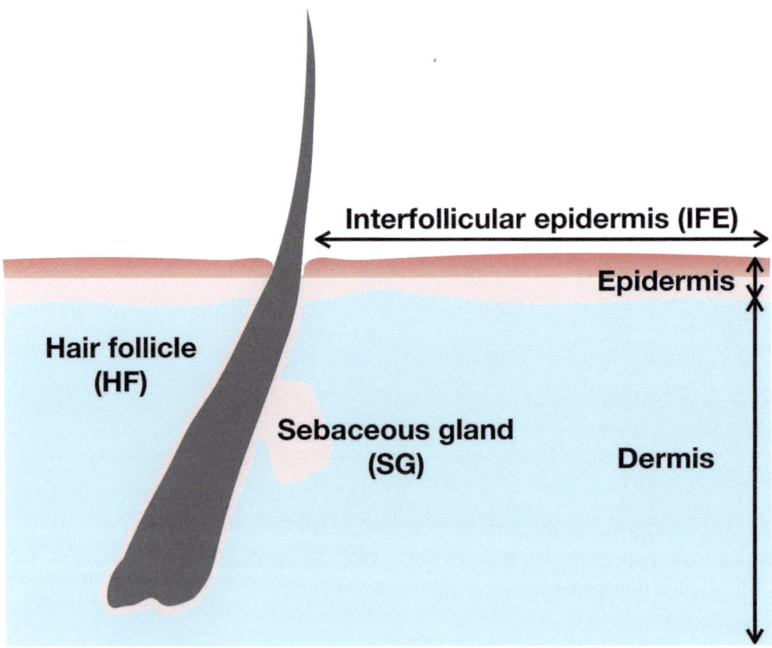

Fig. 1 The structure and compartments of epithelium. Squamous epithelium has dermal and epidermal layers. The epidermal layers include keratinocytes located in hair follicle and interfollicular epidermis (IFE)

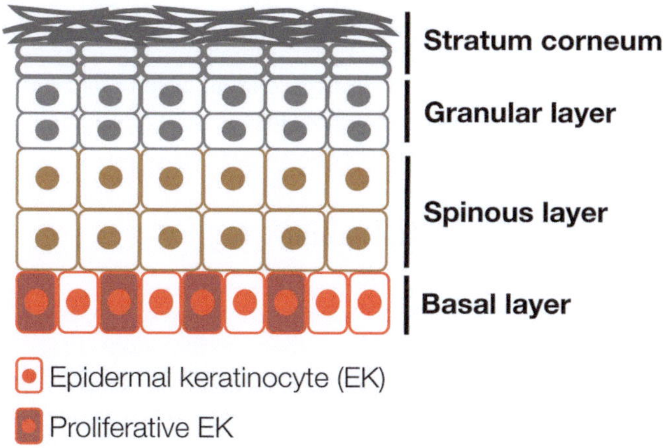

Fig. 2 Schematic detail of interfollicular epidermis (IFE). The proliferative epidermal keratinocyte (EK) is limited in the basal layer. The epidermal differentiation is unidirectional toward the upper spinous, granular, and cornified layers respectively

epidermal lineages including HF and IFE SCs but not transient amplifying cells, the immediate progeny of the IFE SC [6]. In addition, actinic keratosis (AK) is now considered a carcinoma in situ, and an early stage of invasive SCC [7].

2 BCC and SCC Incidence

BCC is the most frequent NMSC with more than two million new cases in the United States and many more worldwide being diagnosed each year [8–10]. BCC is a slow growing and locally invasive cancer and, if diagnosed and treated early, most cases can be eradicated by surgical excision [11]. If left untreated, however, BCC may cause destruction of underlying tissues [12]. SCC is the second most frequent NMSC, with approximately 700,000 in the United States and 2–3 million new cases worldwide being diagnosed [10, 13, 14], and its incidence is markedly increased in immunocompromised patients [15]. Although most SCC cases can be cured by surgical removal, about 8 % of the cases relapse and 5 % metastasize. There is only a 10–20 % survival rate over 10 years in patients with metastatic SCC [16].

3 Causal Factors of BCC and SCC

Risk factors for NMSC include chemical carcinogens and physical ultraviolet (UV) radiation [8]. Of these, UV radiation (UVR) are the most prominent and ubiquitous physical carcinogen of NMSC, especially UV-A and UV-B from sunlight [17].

Common chemical risk factors of NMSC are cigarette smoking and arsenic poisoning with evidence showing that cigarette smoking is strongly associated with SCC in a case–control study [18]. Moreover, chronic arsenic poisoning caused by arsenic contamination of drinking water may also lead to NMSC [19, 20].

4 Genetic Mutations in BCC and SCC

Solar UV radiation has been proven to provoke DNA and RNA damage by inducing the mutagenic photoproducts which lead to gene mutations [21]. Mutations in Hedgehog (HH) pathway-related genes, especially Patched 1 (PTCH1) and Smoothened (SMO), are the most frequent and significant genetic alterations in BCC [22]. Loss of heterozygosity of PTCH1 or activating mutations in SMO leads to aberrant activation of the HH pathway that may drive BCC development by transforming normal adult stem cells into cancer stem cells [22]. It has also been observed that the UV-induced p53 mutation signature is present in a significant percentage of SCC and its precursor AK [7]. The dominant active mutations of proto-oncogene Ras, the upstream activator of the Raf/Mek/Erk1/Erk2 kinase pathway also occur frequently in SCC [7]. There are 21 % of SCCs that harbor activating Ras mutations (9 % H-Ras, 7 % N-Ras, and 5 % K-Ras) [7]. EGFR is frequently overexpressed in various kinds of cancer, including SCC [23]. Recently, the gene BRM encoding the Brahma catalytic subunit of the SWI/SNF complex was found to be frequently mutated in both BCC and SCC [24]. Mutations in many other genes that may contribute to the NMSC tumorigenesis were identified by using mouse models of skin carcinogenesis induced either by the UV radiation and/or the two-stage chemical carcinogenesis (Table 1) [7, 25, 26].

5 Epidermal Homeostasis

The homeostasis of the cell layers in the IFE depends on the tight coordination of keratinocyte proliferation and differentiation [32]. The IFE resident SCs continuously generate proliferative keratinocytes that are referred to as transit amplifying

Table 1 Key genes mutated in BCC and SCC

Tumor type	Genes mutated	References
Basal cell carcinoma (BCC)	Patched 1 (PTCH1)	[27]
	Smoothened (SMO)	[28]
Squamous cell carcinoma (SCC)	TP53	[29]
	Ras	[30]
	β-Catenin	[31]

cells [33–35]. These cells cease proliferation and initiate terminal differentiation upon leaving the basal layer and migrating upwards into the superbasal layers [33]. Keratinocytes in different cell layers vary in differentiation degree [36–38]. Keratinocytes in the spinous layer cease proliferation, but remain metabolically active [39, 40]. After migrating further to the upper granular layer, keratinocytes initiate degrading processes for their intracellular contents including nuclei, but maintain an intact cytoskeletal scaffold while forming lipid-containing lamellar bodies [41, 42]. Some keratinocytes in this cell layer form a cornified envelope underneath the plasma membrane by incorporating various precursor proteins [43], and subsequently become anucleated cells and form the outermost stratum corneum [44]. These anucleated cells become permeable; and a calcium influx activates transglutaminases to irreversibly crosslink the cornified envelope proteins [43]. The lipids secreted from lamellar bodies to form the lamellar matrix to seal the cornified envelopes to create the water impermeable epidermal permeability barrier (EPB) that protects the body from pathogens, allergens, unwanted chemicals, and ultraviolet radiation while preventing the loss of water and electrolytes [37, 44, 45].

6 Extracellular Ceramides Are a Major Constituent of the Epidermal Permeability Barrier

Ceramides are composed of a sphingoid base backbone and an amide-linked acyl chain, and are a key lipid component of the EPB [46]. In mammalian cells, ceramides are synthesized de novo from serine and palmitoyl-CoA through multiple enzymatic steps including sphingoid base acylation by ceramide synthases encoded by six distinct genes LASS1-6 or CERS1-6 [47]. Once synthesized, ceramides are incorporated into complex sphingolipids, such as sphingomyelins and glucosylceramides [47]. More detailed information on the metabolism of ceramides and other sphingolipids is covered in Chap. 1 of this book. In differentiated keratinocytes, glucosylceramides are packaged into the lamellar body before being secreted to intercellular spaces between corneocytes through exocytosis [48]. Secreted glucosylceramides are converted by beta-glucocerebrosidase into ceramides, which form the lamellar matrix with cholesterol and free fatty acids [49].

There are many diverse epidermal ceramide species consisting of different sphingoid bases and amide-linked acyl chains [49]. As shown in Fig. 3, five long chain bases, including D-*erythro*(*e*)-sphingosine (SPH), D-e-dihydrosphingosine (DHS), D-*e*-6-hydroxysphingosine, D-*ribo*-phytosphingosine (PHS), and dihydroxysphinganine, were found in the epidermis. These long-chain bases differ in the number of double bonds and hydroxyl groups. In the epidermal ceramides, there are four major types of amide-linked acyl-chains that differ in hydroxylation type and extent, including non-hydroxylated acyl, alpha-hydroxyacyl, omega-hydroxtacyl, and omega-linoleyloxyacyl [37, 41]. These acyl-chains also vary in length and unsaturation degree [49]. Combinations of five sphingoid bases and four acyl-chain types

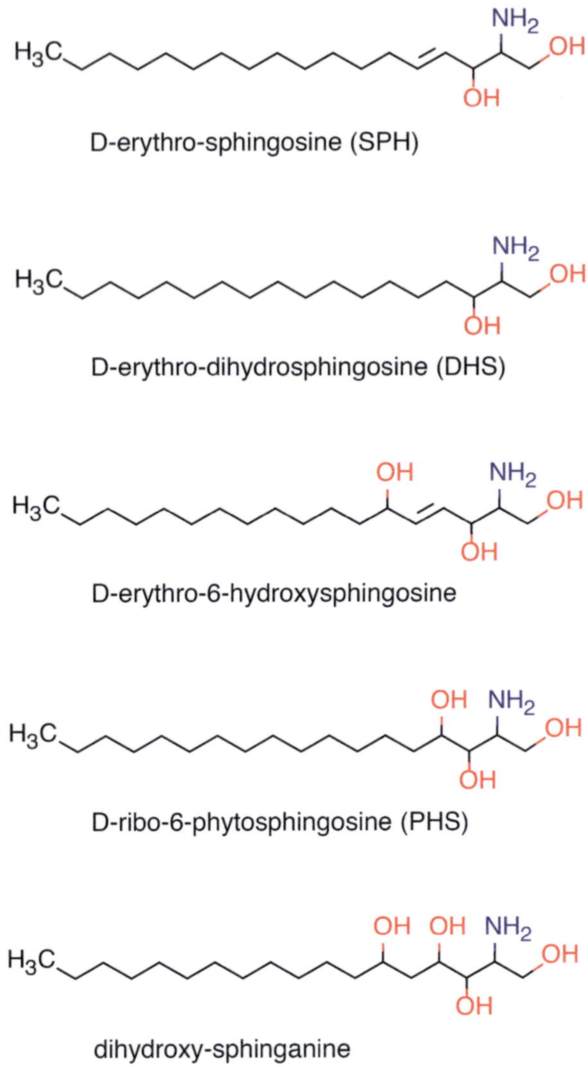

Fig. 3 Human epidermal sphingolipids: diversity of sphingosines

make numerous subclasses of skin ceramides [50, 51]. Ceramides carrying the sphingoid base D-*e*-6-hydroxysphingosine or dihydroxy-sphinganine are unique to the skin and ceramides carrying the sphingoid base phytosphingosine are much more abundant in the skin than in other tissues [49]. Although the specific role for each subclass of ceramides in the EPB remains unclear, a recent study with a genetic mouse model deficient in ceramide synthase 3 (Cers3) suggest that ceramides carrying very or ultra-long acyl-chains are essential for the formation of functional EPB [52]. Disturbances in the EPB due to a decrease in ceramide has been associated with various skin diseases including atopic dermatitis [53], psoriasis [54], harlequin ichthyosis [55], and possibly NMSC.

Because the EPB is the first line of protection for mammals from harmful environmental insults [56], ceramide in the EPB may prevent BCC and SCC by shielding human body from UVR and other causative skin carcinogens.

7 NMSC Results from an Imbalance Between Keratinocyte Proliferation and Differentiation in the Epidermis

As noted above, the homeostasis of the epidermis is controlled by the tight coordination of keratinocyte proliferation and differentiation [32] which is achieved by a balance between pro-proliferation and pro-differentiation factors. Various growth factors, mainly from the epidermal growth factor (EGF) family, such as EGF [57, 58], transforming growth factor β (TGFβ) [59], amphiregulin [60], heparin-binding EGF-like growth factor [61], and betacellulin [62], were shown to promote proliferation of keratinocytes in a paracrine or endocrine fashion [57, 59, 61]. These growth factors upregulate or activate signaling molecules that promote proliferation of keratinocytes while suppressing signaling molecules with anti-proliferative and/or pro-differentiating functions [63]. In contrast, other factors such as calcium ion (Ca^{2+}) induce growth arrest and differentiation of keratinocytes [64]. The Ca^{2+} concentration ([Ca^{2+}]) in the epidermis gradually increases from the basal layer, spinous layer, and granular layer, thus forming an epidermal [Ca^{2+}] gradient which is believed to play a key role in the epidermal differentiation [36]. A shift from a low to high [Ca^{2+}] in medium, a so-called “Ca switch”, induces the growth arrest and differentiation of keratinocytes in culture by activating or upregulating downstream effectors that mediate proliferation inhibition and/or differentiation of keratinocytes [36].

In malignant keratinocytes, proliferation and differentiation are both dysregulated [65] and they exhibit hyper-activated proliferative signaling pathways, such as the EGFR pathway [66] whereas pro-differentiation pathways are suppressed so that the Ca switch fails to induce their differentiation [67]. Dysregulation of these signaling pathways leads to an imbalance between proliferation and differentiation and therefore leads to NMSC.

8 Intracellular Ceramides and Their Metabolites Regulate Epidermal Keratinocyte Proliferation and Differentiation

As stated in other chapters of this book, ceramides and their metabolites sphingoid bases and sphingoid base phosphates are bioactive lipids that mediate various cellular responses such as cell proliferation, differentiation, and apoptosis in many mammalian cell types [68]. Increasing studies suggest that these bioactive sphingolipids may also play an important role in regulating the aforementioned biological responses in epidermal keratinocytes [69]. Indeed, it has been shown that the levels of ceramides are increased in keratinocytes during differentiation [70].

Furthermore, Takeda et al. [71] demonstrated that increasing endogenous ceramide species by treating immortalized keratinocytes with cell-permeable short-chain ceramides inhibited cell proliferation and induced apoptosis. This study supports the notion that increased ceramides may mediate growth arrest and differentiation of keratinocytes in addition to the formation of the EPB.

Similar to ceramide, sphingoid bases have also been shown to induce the growth inhibition and differentiation of keratinocytes. A study by Kim et al. demonstrated that sphingosine (sph), dihydrosphingosine and their respective phosphates are increased in keratinocytes upon differentiation induced by the combination of vitamin C and serum. SPH and other sphingoid bases have been demonstrated to display anti-proliferative and pro-apoptotic abilities in various cell types [72–77]. Similar to SPH, PHS that is expressed at higher levels in the skin than other tissues has been shown to induce the growth arrest and differentiation of human keratinocytes [78]. Kim et al. demonstrated that PHS stimulates the differentiation of human keratinocytes and inhibits 12-*O*-Tetradecanoylphorbol-13-acetate (TPA)-induced inflammatory epidermal hyperplasia in hairless mouse skin. Tolleson et al. [79] demonstrated that treatment with the mycotoxin fumonisin B1, which specifically inhibits ceramide synthase activity, increases the levels of endogenous levels of DHS, leading to the growth inhibition of keratinocytes. A recent study by Paragh et al. [80] demonstrated that DHS also promotes the differentiation of keratinocytes. Finally, Sigruener et al. [81] demonstrated that treatment with PHS and DHS increase the levels of phytoceramides and very long-chain ceramides, respectively and leads to the differentiation of keratinocytes. Collectively, this suggests that either sphingoid bases, phytoceramides, and/or ceramides mediate keratinocyte differentiation.

In contrast to ceramides and SPH, sphingosine-1-phosphate (S1P) has been demonstrated to promote the proliferation and survival of many cell types [82–85]. Interestingly, although S1P also promotes survival of epidermal keratinocytes against various insults [86–88], it induces the growth arrest and differentiation of keratinocytes by binding to its G-protein-coupled receptors [89–92]. Schaper et al. demonstrated that topic application of S1P inhibits the imiquimod-induced epidermal hyperproliferation of the ear skin in a mouse model of psoriasis [93]. Schuppel et al. demonstrated that S1P inhibits insulin-induced proliferation of keratinocytes via AKT inhibition dependent on the S1P2 receptor subtype [94]. These results suggest that ceramides, sphingoid bases, and their phosphates are important regulators of keratinocyte differentiation, proliferation, and survival.

9 Role for Sphingolipid-Metabolizing Enzymes in Regulating Keratinocyte Proliferation and Differentiation

As mentioned in various chapters of this book, ceramides can be synthesized from different pathways. Indeed, several enzymes responsible for the formation of ceramides are upregulated in keratinocytes during differentiation, resulting in an increase in the cellular levels of ceramides [69] and consistent with a potential role in this process. Supporting this, Brodesser et al. [95] demonstrated that inhibiting the formation of ceramides from dihydroceramides through inhibition of dihydroceramide desaturase attenuates the expression of differentiation markers in cultured keratinocytes. Furthermore, Kato et al. [96] demonstrated that treatment with sodium dl-a-tocopheryl-6-*O*-phosphate, a water-soluble derivative of vitamin E, increased the levels of ceramides in normal human epidermal keratinocytes by increasing the expression of serine palmitoyl-CoA transferase (SPT) that catalyzes the first and rate-limiting step for ceramide biosynthesis. Interestingly, treatment with this compound also induced the expression of various keratinocyte differentiation markers in these cells.

In many cell types, ceramides can be hydrolyzed to generate a free fatty acid and SPH through the action of ceramidases (CDases) [68, 97–100]. To date, 5 human genes encoding different CDases have been identified, including one acid ceramidase (aCDase; ASAH1), one neutral (nCDase; ASAH2), and three alkaline ceramidases (ACER1, ACER2, and ACER3) [68]. These five CDase genes have been shown to be expressed in keratinocytes [101]. We have demonstrated that ACER1 is expressed in proliferating keratinocytes at a modest level, and is markedly upregulated upon differentiation observed in response to the Ca switch [102]. We have further demonstrated that RNA interference (RNAi)-mediated inhibition of ACER1 inhibited the growth arrest of normal human keratinocytes and attenuated the expression of involucrin and keratin 1 induced in these cells by the Ca switch [102], suggesting that ACER1 plays an important role in controlling the growth and differentiation of keratinocytes. Because ACER1 upregulation causes a substantial increase in the levels of SPH and S1P while decreasing the levels of ceramides, it likely regulates the growth and differentiation of keratinocytes by controlling these bioactive lipids.

In cells, SPH is phosphorylated to generate S1P through the action of each of two SPH kinases (SPHKs), SPHK1 and SPHK2 [103, 104]. Once generated, S1P can be cleaved by S1P lyase to form hexadecenol and ethanolamine phosphate or be converted back to SPH by the action of one of two S1P phosphohydrolases, SGPP1 [105] and SGPP2 [106]. Consequently, these enzymes may regulate keratinocyte proliferation and differentiation by controlling S1P. Indeed, Hong et al. [107] demonstrated that treatment with the SPHK1 activator K6PC-5 promotes epidermal differentiation and inhibits the proliferation of keratinocytes through intracellular Ca^{2+} signaling in an S1P-dependent manner. Using a mouse model, Allende et al. [108] demonstrated that deficiency in SGPP1 aberrantly induced the expression of various genes implicated in keratinocyte differentiation likely due to an accumulation of S1P in keratinocytes, thus leading to multiple epidermal abnormalities and death of mice due to excessive body water loss.

Based on the studies discussed above, we can conclude that sphingolipid metabolizing enzymes play important roles in regulating keratinocyte proliferation, differentiation, and apoptosis by controlling the metabolism of bioactive sphingolipids.

10 Roles of Bioactive Sphingolipids in Mediating UVR-Induced Proliferation Inhibition and Apoptosis of Keratinocytes

Because of the central role that UVR plays in the pathogenesis of both BCC and SCC, methods that reduce UVR exposure and UVR-induced gene mutations in keratinocytes form the basis of the prevention of both types of malignancy [109]. In addition to gene mutations, UVR induces apoptosis of epidermal keratinocytes [110]. Because apoptosis can prevent UVB-induced gene mutations from being inherited by daughter cells, enhancing UVR-induced apoptosis may represent a novel approach to NMSC prevention.

Several studies have demonstrated that UVB induces growth arrest and apoptosis of keratinocytes in part by regulating the metabolism of bioactive sphingolipids. Magnoni et al. [111] showed that UV-B radiation rapidly induces ceramide generation in cultured normal human keratinocytes by activating neutral and acidic sphingomyelinases. Uchida et al. [112] demonstrated that treatment with UV-B at a low dose caused a temporary increase in the levels of ceramides in keratinocytes in culture by transiently reducing both aCDase and nCDase activities, leading to growth arrest but not apoptosis whereas treatment with UV-B at a high dose caused a sustained increase in the levels of ceramides by permanently reducing CDase activities, leading to both growth arrest and apoptosis in keratinocytes. Inhibiting CDase activities by either the CDase inhibitor *N*-oleoylethanolamine or small interfering RNA (siRNA) sensitized keratinocytes to apoptosis in response to treatment with the low dose of UV-B. These results suggest that increased ceramides mediate growth arrest and/or apoptosis in keratinocytes in response to UVB. Consistent with this, the same group has also shown that UVB increased the levels of ceramides by inhibiting the function of ceramide transfer protein (CERT) in delivering ceramides from the ER to the Golgi complex in HaCaT keratinocytes [113]. Furthermore, similar to UV-B stimulation, treatment with a CERT inhibitor increased the levels of ceramides while reducing those of sphingomyelins in keratinocytes and leading to apoptosis. Thus, because ceramide-mediated apoptosis may prevent UVB-induced gene mutations from being inherited by daughter cells, modulating the metabolism of ceramides may represent a novel approach for NMSC prevention.

11 Potential Roles of Sphingolipids in NMSC Therapy

Although surgical excision is the primary treatment option for both BCC and SCC cases [114], local therapy with chemotherapeutic and immune-modulating agents may be applied to some cases, especially for small and superficial BCC. Topical imiquimod or fluorouracil is approved by the US Food and Drug Administration

(FDA) for the treatment of nonfacial superficial BCCs that are less than 2 cm in diameter [65]. In addition, radiation therapy may be used as an adjuvant to surgery for BCC and SCC [115] and there are also clinical trials of targeted therapies such as hedgehog pathway inhibitors [116, 117] and EGFR inhibitors [116, 118] for metastatic BCC and SCC, respectively.

Numerous studies have demonstrated that ionizing radiation and DNA damaging chemotherapeutic drugs induce growth arrest and apoptosis of various cancer cells in part through generation of ceramide [119] or SPH [72]. Consequently, it is possible that these bioactive sphingolipids also mediate the therapeutic action of imiquimod or fluorouracil in NMSC. Supporting this notion, an increasing number of studies have found that modulating the metabolism of bioactive sphingolipids can inhibit proliferation of SCC cells or induce apoptosis of SCC cells. For example, Sugiki et al. [120] demonstrated that treatment with a cell permeable synthetic ceramide (C_2-ceramide) can potently kill HSC-1 SCC cells through apoptosis. Furthermore, studies with an animal model of skin carcinogenesis demonstrated that treatment with SPH or C_2-ceramide inhibits the conversion of papillomas to SCC [121] suggesting an inhibition of SCC progression. Based on these preliminary results, it is envisioned that modulating the metabolism of sphingolipids in cells or direct treatment with sphingolipid mimics may represent novel approaches to NMSC therapy.

12 Conclusions and Future Directions

NMSC is an increasing problem for health care services worldwide with significant morbidity owing to a lack of effective preventive and therapeutic measures for this disease. As discussed above, emerging evidence suggests that bioactive sphingolipids and their enzymes may serve as potential targets for both preventing and treating NMSC. However, to facilitate the development of sphingolipid-based approaches to NMSC prevention and treatment, it is first necessary to understand if and how the dysregulation of bioactive sphingolipid metabolism contributes to the pathogenesis of NMSC. This goal should be achievable with the advance of mass spectrometry technologies for both lipidomics and proteomics and new technologies in genome editing, such as Clustered Regularly Interspaced Short Palindromic Repeat (CRISPR). Once sphingolipid enzymes or regulators are identified as cancer drivers for NMSC, a high throughput screening can be conducted to develop small molecule inhibitors for NMSC prevention and/or therapy.

Acknowledgement This work was supported by National Institutes of Health Grants R01CA104834, R01CA163825, and P01CA097132 to C.M.

References

1. Owens DM, Watt FM (2003) Contribution of stem cells and differentiated cells to epidermal tumours. Nat Rev Cancer 3(6):444–451
2. Fuchs E (2008) Skin stem cells: rising to the surface. J Cell Biol 180(2):273–84
3. Blanpain C, Fuchs E (2009) Epidermal homeostasis: a balancing act of stem cells in the skin. Nat Rev Mol Cell Biol 10(3):207–217
4. Arora A, Attwood J (2009) Common skin cancers and their precursors. Surg Clin North Am 89(3):703–712
5. Youssef KK et al (2010) Identification of the cell lineage at the origin of basal cell carcinoma. Nat Cell Biol 12(3):299–305
6. Lapouge G et al (2011) Identifying the cellular origin of squamous skin tumors. Proc Natl Acad Sci 108(18):7431–7436
7. Ratushny V et al (2012) From keratinocyte to cancer: the pathogenesis and modeling of cutaneous squamous cell carcinoma. J Clin Invest 122(2):464–472
8. Acartürk TO, Edington H (2005) Nonmelanoma skin cancer. Clin Plast Surg 32(2):237–248
9. Eisemann N et al (2014) Non-melanoma skin cancer incidence and impact of skin cancer screening on incidence. J Invest Dermatol 134(1):43–50
10. Rogers HW et al (2010) Incidence estimate of nonmelanoma skin cancer in the United States, 2006. Arch Dermatol 146(3):283–287
11. Lallas A et al (2014) The dermatoscopic universe of basal cell carcinoma. Dermatol Pract Concept 4(3):11–24
12. Netscher DT, Spira M (2004) Basal cell carcinoma: an overview of tumor biology and treatment. Plast Reconstr Surg 113(5):74e–94e
13. Karia PS, Han J, Schmults CD (2013) Cutaneous squamous cell carcinoma: estimated incidence of disease, nodal metastasis, and deaths from disease in the United States, 2012. J Am Acad Dermatol 68(6):957–966
14. Venura Samarasinghe VM (2012) Nonmelanoma skin cancer. J Cutan Aesthet Surg 5(1):3–10
15. Martorell-Calatayud A et al (2013) Cutaneous squamous cell carcinoma: defining the high-risk variant. Actas Dermosifiliogr 104(5):367–379
16. Alam M, Ratner D (2001) Cutaneous squamous-cell carcinoma. N Engl J Med 344(13):975–983
17. de Gruijl FR (1999) Skin cancer and solar UV radiation. Eur J Cancer 35(14):2003–2009
18. Rollison DE et al (2012) Case-control study of smoking and non-melanoma skin cancer. Cancer Causes Control 23(2):245–254
19. Khan MMH et al (2003) Magnitude of arsenic toxicity in tube-well drinking water in Bangladesh and its adverse effects on human health including cancer: evidence from a review of the literature. Asian Pac J Cancer Prev 4(1):7–14
20. Knobeloch LM, Zierold KM, Anderson HA (2006) Association of arsenic-contaminated drinking-water with prevalence of skin cancer in Wisconsin's Fox River Valley. Journal of Health, Population and Nutrition, pp 206–213
21. Rastogi RP et al (2010) Molecular mechanisms of ultraviolet radiation-induced DNA damage and repair. J Nucleic Acids 2010(6551):592980–32
22. de Zwaan SE, Haass NK (2010) Genetics of basal cell carcinoma. Australas J Dermatol 51(2):81–92; quiz 93–94
23. Uribe P, Gonzalez S (2011) Epidermal growth factor receptor (EGFR) and squamous cell carcinoma of the skin: molecular bases for EGFR-targeted therapy. Pathol Res Pract 207(6):337–342
24. Moloney FJ et al (2009) Hotspot mutation of Brahma in non-melanoma skin cancer. J Investig Dermatol 129(4):1012–1015
25. Patel GK et al (2011) Identification and Characterization of Tumor-Initiating Cells in Human Primary Cutaneous Squamous Cell Carcinoma. J Investig Dermatol 132(2):401–409

26. Schwarz M, Munzel PA, Braeuning A (2013) Non-melanoma skin cancer in mouse and man. Arch Toxicol 87(5):783–98
27. Nagano T et al (1999) Overexpression of the human homologue of Drosophila patched (PTCH) in skin tumours: specificity for basal cell carcinoma. Br J Dermatol 140(2): 287–290
28. Xie J et al (1998) Activating Smoothened mutations in sporadic basal-cell carcinoma. Nature 391(6662):90–92
29. Brash DE et al (1991) A role for sunlight in skin cancer: UV-induced p53 mutations in squamous cell carcinoma. Proc Natl Acad Sci U S A 88(22):10124–10128
30. Dajee M et al (2003) NF-kappaB blockade and oncogenic Ras trigger invasive human epidermal neoplasia. Nature 421(6923):639–643
31. Malanchi I et al (2008) Cutaneous cancer stem cell maintenance is dependent on beta-catenin signalling. Nature 452(7187):650–653
32. Sotiropoulou PA, Blanpain C (2012) Development and Homeostasis of the Skin Epidermis. Cold Spring Harb Perspect Biol 4(7):a008383–a008383
33. Watt FM (1998) Epidermal stem cells: markers, patterning and the control of stem cell fate. Philos Trans R Soc Lond B Biol Sci 353(1370):831–7
34. Christiano AM (2004) Epithelial stem cells: stepping out of their niche. Cell 118(5):530–2
35. Alonso L, Fuchs E (2003) Stem cells of the skin epithelium. Proc Natl Acad Sci U S A 100(Suppl 1):11830–5, Epub 2003 Aug 11
36. Bikle DD, Xie Z, Tu C-L (2012) Calcium regulation of keratinocyte differentiation. Expert Rev Endocrinol Metab 7(4):461–472
37. Madison KC (2003) Barrier function of the skin: "la raison d'etre" of the epidermis. J Invest Dermatol 121(2):231–41
38. Gniadecki R (1998) Regulation of keratinocyte proliferation. Gen Pharmacol 30(5):619–22
39. Jetten AM, Harvat BL (1997) Epidermal differentiation and squamous metaplasia: from stem cell to cell death. J Dermatol 24(11):711–25
40. Koch PJ, Roop DR (2004) The role of keratins in epidermal development and homeostasis—going beyond the obvious. J Invest Dermatol 123(5):x–xi
41. Coderch L et al (2003) Ceramides and skin function. Am J Clin Dermatol 4(2):107–29
42. Fartasch M (2004) The epidermal lamellar body: a fascinating secretory organelle. J Invest Dermatol 122(5):XI–XII
43. Kalinin AE, Kajava AV, Steinert PM (2002) Epithelial barrier function: assembly and structural features of the cornified cell envelope. Bioessays 24(9):789–800
44. Marks R (2004) The stratum corneum barrier: the final frontier. J Nutr 134(8 Suppl): 2017S–2021S
45. Feingold KR (2007) The role of epidermal lipids in cutaneous permeability barrier homeostasis. J Lipid Res 13:13
46. Mizutani Y et al (2009) Ceramide biosynthesis in keratinocyte and its role in skin function. Biochimie 91(6):784–790
47. Levy M, Futerman AH (2010) Mammalian ceramide synthases. IUBMB Life 62(5): 347–356
48. Meckfessel MH, Brandt S (2014) The structure, function, and importance of ceramides in skin and their use as therapeutic agents in skin-care products. J Am Acad Dermatol 71(1): 177–84
49. Rabionet M, Gorgas K, Sandhoff R (2014) Ceramide synthesis in the epidermis. Biochim Biophys Acta 1841(3):422–434
50. Farwanah H et al (2007) Separation and mass spectrometric characterization of covalently bound skin ceramides using LC/APCI-MS and Nano-ESI-MS/MS. J Chromatogr B Analyt Technol Biomed Life Sci 852(1-2):562–70
51. Masukawa Y et al (2008) Characterization of overall ceramide species in human stratum corneum. J Lipid Res 49(7):1466–76
52. Jennemann R et al (2012) Loss of ceramide synthase 3 causes lethal skin barrier disruption. Hum Mol Genet 21(3):586–608

53. Imokawa G (2014) Role of ceramide in the barrier function of the stratum corneum, implications for the pathogenesis of atopic dermatitis. J Clin Exp Dermatol Res 05(01):1–12
54. Nakajima K et al (2013) Barrier abnormality due to ceramide deficiency leads to psoriasiform inflammation in a mouse model. J Investig Dermatol 133(11):2555–2565
55. Yanagi T et al (2010) Self-improvement of keratinocyte differentiation defects during skin maturation in ABCA12-deficient harlequin ichthyosis model mice. Am J Pathol 177(1): 106–118
56. Baroni A et al (2012) Structure and function of the epidermis related to barrier properties. Clin Dermatol 30(3):257–262
57. Jost M, Kari C, Rodeck U (2000) The EGF receptor—an essential regulator of multiple epidermal functions. Eur J Dermatol 10(7):505–510
58. Jost M et al (2001) Epidermal growth factor receptor-dependent control of keratinocyte survival and Bcl-xL expression through a MEK-dependent pathway. J Biol Chem 276(9):6320–6326
59. Hashimoto K (2000) Regulation of keratinocyte function by growth factors. J Dermatol Sci 24(Suppl 1):S46–50
60. Piepkorn M, Lo C, Plowman G (1994) Amphiregulin-dependent proliferation of cultured human keratinocytes: autocrine growth, the effects of exogenous recombinant cytokine, and apparent requirement for heparin-like glycosaminoglycans. J Cell Physiol 159(1):114–120
61. Piepkorn M, Pittelkow MR, Cook PW (1998) Autocrine regulation of keratinocytes: the emerging role of heparin-binding, epidermal growth factor-related growth factors. J Invest Dermatol 111(5):715–721
62. Schneider MR et al (2008) Betacellulin regulates hair follicle development and hair cycle induction and enhances angiogenesis in wounded skin. J Investig Dermatol 128(5): 1256–1265
63. Müller EJ et al (2008) Outside-in signaling through integrins and cadherins: a central mechanism to control epidermal growth and differentiation? J Investig Dermatol 128(3):501–516
64. Bikle DD (2004) Vitamin D regulated keratinocyte differentiation. J Cell Biochem 92(3):436–44
65. Dubas LE, Ingraffea A (2013) Nonmelanoma skin cancer. Facial Plast Surg Clin North Am 21:43–53
66. Degen M et al (2012) MAPK/ERK-dependent translation factor hyperactivation and dysregulated laminin γ2 expression in oral dysplasia and squamous cell carcinoma. Am J Pathol 180(6):2462–2478
67. Pillai S et al (1991) Uncoupling of the calcium-sensing mechanism and differentiation in squamous carcinoma cell lines. Exp Cell Res 192(2):567–73
68. Mao C, Obeid LM (2008) Ceramidases: regulators of cellular responses mediated by ceramide, sphingosine, and sphingosine-1-phosphate. Biochim Biophys Acta 1781(9): 424–434
69. Holleran WM, Takagi Y, Uchida Y (2006) Epidermal sphingolipids: Metabolism, function, and roles in skin disorders. FEBS Lett 580(23):5456–5466
70. Lampe MA, Williams ML, Elias PM (1983) Human epidermal lipids: characterization and modulations during differentiation. J Lipid Res 24(2):131–140
71. Takeda S et al (2006) Apoptosis occurs via the ceramide recycling pathway in human HaCaT keratinocytes. J Biochem 139(2):255–62
72. Cuvillier O et al (2001) Sphingosine generation, cytochrome c release, and activation of caspase-7 in doxorubicin-induced apoptosis of MCF7 breast adenocarcinoma cells. Cell Death Differ 8(2):162–171
73. Nava VE et al (2000) Sphingosine enhances apoptosis of radiation-resistant prostate cancer cells. Cancer Res 60(16):4468–74
74. Cuvillier O et al (1996) Suppression of ceramide-mediated programmed cell death by sphingosine-1-phosphate. Nature 381(6585):800–3
75. Nakamura H et al (1998) Survival by Mac-1-mediated adherence and anoikis in phorbol ester-treated HL-60 cells. J Biol Chem 273(25):15345–51

76. Hamada K et al (1998) Involvement of Mac-1-mediated adherence and sphingosine 1-phosphate in survival of phorbol ester-treated U937 cells. Biochem Biophys Res Commun 244(3):745–50
77. Cuvillier O (2002) Sphingosine in apoptosis signaling. Biochim Biophys Acta 1585(2-3):153–62
78. Kim S et al (2006) Phytosphingosine stimulates the differentiation of human keratinocytes and inhibits TPA-induced inflammatory epidermal hyperplasia in hairless mouse skin. Mol Med 12(1-3):17–24
79. Tolleson WH et al (1999) Fumonisin B1 induces apoptosis in cultured human keratinocytes through sphinganine accumulation and ceramide depletion. Int J Oncol 14(5):833–43
80. Paragh G et al (2008) Novel sphingolipid derivatives promote keratinocyte differentiation. Exp Dermatol 9:9
81. Sigruener A et al (2013) Effects of sphingoid bases on the sphingolipidome in early keratinocyte differentiation. Exp Dermatol 22(10):677–679
82. Lee MJ et al (1998) Sphingosine-1-phosphate as a ligand for the G protein-coupled receptor EDG-1. Science 279(5356):1552–5
83. Hla T et al (2000) Sphingosine-1-phosphate signaling via the EDG-1 family of G-protein-coupled receptors. Ann N Y Acad Sci 905:16–24
84. Hla T (2001) Sphingosine 1-phosphate receptors. Prostaglandins 64(1-4):135–142
85. Spiegel S, Milstien S (2000) Sphingosine-1-phosphate: signaling inside and out. FEBS Lett 476(1-2):55–7
86. Moriue T et al (2013) Sphingosine 1-phosphate attenuates peroxide-induced apoptosis in HaCaT cells cultured in vitro. Clin Exp Dermatol 38(6):638–45
87. Schmitz EI et al (2012) Sphingosine 1-phosphate protects primary human keratinocytes from apoptosis via nitric oxide formation through the receptor subtype S1P(3). Mol Cell Biochem 371(1-2):165–76
88. Hammer S et al (2004) Glucocorticoids mediate differential anti-apoptotic effects in human fibroblasts and keratinocytes via sphingosine-1-phosphate formation. J Cell Biochem 91(4):840–51
89. Vogler R et al (2003) Sphingosine-1-phosphate and its potentially paradoxical effects on critical parameters of cutaneous wound healing. J Invest Dermatol 120(4):693–700
90. Sauer B et al (2004) Involvement of Smad signaling in sphingosine 1-phosphate-mediated biological responses of keratinocytes. J Biol Chem 279(37):38471–9, Epub 2004 Jul 09
91. Kim DS et al (2004) Sphingosine-1-phosphate inhibits human keratinocyte proliferation via Akt/protein kinase B inactivation. Cell Signal 16(1):89–95
92. Schuppel M et al (2008) Sphingosine 1-Phosphate Restrains Insulin-Mediated Keratinocyte Proliferation via Inhibition of Akt through the S1P(2) Receptor Subtype. J Invest Dermatol 24:24
93. Schaper K et al (2013) Sphingosine-1-phosphate exhibits anti-proliferative and anti-inflammatory effects in mouse models of psoriasis. J Dermatol Sci 71(1):29–36
94. Schüppel M et al (2008) Sphingosine 1-Phosphate Restrains Insulin-Mediated Keratinocyte Proliferation via Inhibition of Akt through the S1P2 Receptor Subtype. J Investig Dermatol 128(7):1747–1756
95. Brodesser S, Kolter T (2011) Dihydroceramide desaturase inhibition by a cyclopropanated dihydroceramide analog in cultured keratinocytes. J Lipids 2011:724015
96. Kato E, Takahashi N (2012) Improvement by sodium dl-α-tocopheryl-6-O-phosphate treatment of moisture-retaining ability in stratum corneum through increased ceramide levels. Bioorg Med Chem 20(12):3837–3842
97. el Bawab S et al (2002) Ceramidases in the regulation of ceramide levels and function. Subcell Biochem 36:187–205
98. Mao C et al (2000) Cloning of an alkaline ceramidase from Saccharomyces cerevisiae. An enzyme with reverse (CoA-independent) ceramide synthase activity. J Biol Chem 275(10):6876–84

99. Mao C et al (2001) Cloning and characterization of a novel human alkaline ceramidase: a mammalian enzyme that hydrolyzes phytoceramide. J Biol Chem 16:16
100. Mao C et al (2003) Cloning and characterization of a mouse endoplasmic reticulum alkaline ceramidase: an enzyme that preferentially regulates metabolism of very long chain ceramides. J Biol Chem 278(33):31184–91
101. Houben E et al (2006) Differentiation-associated expression of ceramidase isoforms in cultured keratinocytes and epidermis. J Lipid Res 47(5):1063–70
102. Sun W et al (2007) Upregulation of the human alkaline ceramidase 1 and acid ceramidase mediates calcium-induced differentiation of epidermal keratinocytes. J Investig Dermatol 128(2):389–97
103. Kohama T et al (1998) Molecular cloning and functional characterization of murine sphingosine kinase. J Biol Chem 273(37):23722–8
104. Liu H et al (2000) Molecular cloning and functional characterization of a novel mammalian sphingosine kinase type 2 isoform. J Biol Chem 275(26):19513–20
105. Johnson KR et al (2003) Role of human sphingosine-1-phosphate phosphatase 1 in the regulation of intra- and extracellular sphingosine-1-phosphate levels and cell viability. J Biol Chem 278(36):34541–34547
106. Ogawa C et al (2003) Identification and characterization of a novel human sphingosine-1-phosphate phosphohydrolase, hSPP2. J Biol Chem 278(2):1268–1272
107. Hong JH et al (2008) K6PC-5, a direct activator of sphingosine kinase 1, promotes epidermal differentiation through intracellular Ca2+ signaling. J Investig Dermatol 128(9):2166–2178
108. Allende ML et al (2013) Sphingosine-1-phosphate phosphatase 1 regulates keratinocyte differentiation and epidermalhomeostasis. J Biol Chem 288(25):18381–91
109. Boukamp P (2005) UV-induced skin cancer: similarities–variations. J Dtsch Dermatol Ges 3(7):493–503
110. Schwarz A et al (1995) Ultraviolet-B-induced apoptosis of keratinocytes: evidence for partial involvement of tumor necrosis factor-alpha in the formation of sunburn cells. J Invest Dermatol 104(6):922–927
111. Magnoni C et al (2002) Ultraviolet B radiation induces activation of neutral and acidic sphingomyelinases and ceramide generation in cultured normal human keratinocytes. Toxicol In Vitro 16(4):349–355
112. Uchida Y et al (2010) Hydrolytic pathway protects against ceramide-induced apoptosis in keratinocytes exposed to UVB. J Investig Dermatol 130(10):2472–2480
113. Charruyer A et al (2008) Decreased ceramide transport protein (CERT) function alters sphingomyelin production following UVB irradiation. J Biol Chem 283(24):16682–16692
114. Firnhaber JM (2012) Diagnosis and treatment of Basal cell and squamous cell carcinoma. Am Fam Physician 86(2):161–168
115. Hulyalkar R, Rakkhit T, Garcia-Zuazaga J (2011) The role of radiation therapy in the management of skin cancers. Dermatol Clin 29(2):287–96, x
116. Bonner JA et al (2006) Radiotherapy plus cetuximab for squamous-cell carcinoma of the head and neck. N Engl J Med 354(6):567–578
117. De Smaele E et al (2012) Hedgehog signaling pathway and its targets for treatment in basal cell carcinoma. Journal of Experimental, Pharmacology, pp 173–13
118. Soulieres D (2003) Multicenter Phase II study of erlotinib, an oral epidermal growth factor receptor tyrosine kinase inhibitor, in patients with recurrent or metastatic squamous cell cancer of the head and neck. J Clin Oncol 22(1):77–85
119. Aureli M et al (2014) Exploring the link between ceramide and ionizing radiation. Glycoconj J 31(6-7):449–459
120. Sugiki H et al (2000) C2-ceramide induces apoptosis in a human squamous cell carcinoma cell line. Br J Dermatol 143(6):1154–1163
121. Birt DF et al (1998) Inhibition of skin carcinomas but not papillomas by sphingosine, N-methylsphingosine, and N-acetylsphingosine. Nutr Cancer 31(2):119–26

Dysregulation of Sphingolipid Metabolism in Melanoma: Roles in Pigmentation, Cell Survival and Tumor Progression

David Garandeau, Marguerite Mrad, Thierry Levade, Cristiana Perrotta, Nathalie Andrieu-Abadie, and Mona Diab-Assaf

Abstract Over the last decades, the incidence of cutaneous malignant melanoma rapidly increased in Caucasian populations. Moreover, melanoma is a leading cause of cancer death and consequently has become a significant public health problem worldwide in fair-skinned populations. The discovery of the prevalent mutation BRAFV600E, which drives melanoma cell growth, has made this oncogenic protein an ideal therapeutic target. However, the beneficial effects of BRAF-directed therapies are usually short-lived due to the appearance of resistance, which leads to disease progression. This emphasizes the need to develop new therapeutic approaches that could overcome tumor relapse. Alterations in sphingolipid metabolism are associated with melanoma progression and represent an exploitable target for the development of novel chemotherapeutics. The aim of this review is to concentrate on the critical metabolites and enzymes that contribute to this metabolic

Author contributed equally with all other contributors

D. Garandeau • M. Mrad • N. Andrieu-Abadie, Ph.D. (✉)
Inserm UMR1037, Equipe Labellisée Ligue Contre le Cancer 2013, Centre de Recherches en Cancérologie de Toulouse, Oncopole de Toulouse, 2 Avenue Hubert Curien, CS 53717, 31037 Toulouse Cedex 1, France

Université Toulouse III Paul-Sabatier, Toulouse, France
e-mail: nathalie.andrieu@inserm.fr

T. Levade
Inserm UMR1037, Equipe Labellisée Ligue Contre le Cancer 2013, Centre de Recherches en Cancérologie de Toulouse, Oncopole de Toulouse, 2 Avenue Hubert Curien, CS 53717, 31037 Toulouse Cedex 1, France

Université Toulouse III Paul-Sabatier, Toulouse, France

Laboratoire de Biochimie Métabolique, CHU, Toulouse, France

C. Perrotta
Unit of Clinical Pharmacology, National Research Council Institute of Neuroscience, Department of Biomedical and Clinical Sciences, University Hospital L. Sacco, Università di Milano, Milan, Italy

M. Diab-Assaf
Faculty of Sciences II, Chemistry Department, Lebanese University, Beirut, Lebanon

Y.A. Hannun et al. (eds.), *Bioactive Sphingolipids in Cancer Biology and Therapy*, DOI 10.1007/978-3-319-20750-6_6

dysregulation in melanoma, to discuss the emerging roles of sphingolipids on melanogenesis, tumor microenvironment and melanoma progression, and to highlight relevant therapeutic approaches applicable for melanoma treatment.

Keywords Acid sphingomyelinase • BRAFV600E • Melanoma • Metabolism • Metastasis • Microenvironment • MITF • Myofibroblasts • Resistance • Sphingosine-1-phosphate

1 Melanoma: A Dreadful Cancer

Although far less prevalent than non-melanoma skin cancers, cutaneous melanoma is the major cause of death from skin cancer. The incidence of this cancer is also one of the most rapidly growing worldwide, with a consistent increase in Caucasian populations over the past four decades. This trend is predicted to continue increasing for at least over the next two decades, with a doubling of incidence rates [1, 2]. Cutaneous melanoma is a heterogeneous disease that arises through the stepwise transformation of melanocytes within the basal epidermal layer of the skin and evolves due to a myriad of genetic and epigenetic aberrations correlated or not with the person's UV exposure behaviour. Based on the clinical and histopathological characteristics, cutaneous melanoma is classified into four categories: superficial spreading, nodular, lentigo maligna, and acral lentiginous melanoma, according to the site of origin, the UV light exposure and the duration of preinvasive growth. It can arise from a pre-existent nevus or even from healthy skin.

Patients with early stage melanomas can be treated successfully by surgery; however, one in ten will develop disseminated disease. The prognosis for patients with stage IV melanoma remains poor with a median overall survival of 9–11 months, and 1-year and 5-year survivals of about 33 % and 15 %, respectively. Highly aggressive, metastatic melanoma is notoriously refractory to standard treatments such as chemo- and radiotherapy [3]. Moreover, although very promising with meaningful effects on progression-free survival (PFS), new targeted therapies are either effective on only a few patients, e.g., in the case of ipilimumab, an immune-modulating antibody that targets CTLA-4 on T lymphocytes [4] or will lead to a relapse due to a pre-existing or acquired resistance, e.g., for vemurafenib, a BRAF inhibitor [5]. Clearly, as a multifaceted disease, skin melanoma is far from being under control. This implies that studies of the genetic and cellular events underlying melanoma development must continue to better elucidate the pathogenesis of the disease and generate more efficient and personalized approaches both for diagnosis and treatment.

2 SL Metabolism Is Deregulated in Melanoma

Ceramides play an essential role in skin homeostasis [6]. Dysregulation of ceramide metabolism has been identified in inflammatory skin diseases such as atopic dermatitis and psoriasis vulgaris [7]. Interestingly, we recently showed that ceramide formation is also strongly altered in melanoma skin cancer. In particular, a β-glucocerebrosidase, encoded by the *GBA2* gene is under-expressed in melanoma cells compared to healthy melanocytes promoting the accumulation of its substrate, glucosylceramide (GlcCer) and a decrease of ceramide, in tumor cells [8]. This might be related to the presence of tumor-associated mutations in the promoter region or coding sequence of the *GBA2* gene or to some epigenetic change. Although this hypothesis remains to be tested, some reports suggest that patients with Gaucher disease, who are molecularly characterized by a deficient activity of the lysosomal β-glucocerebrosidase GBA1 resulting in GlcCer accumulation, are at increased risk of developing specific malignancies such as melanoma [9]. In particular, the N370S mutation in the *GBA1* gene is associated with an increased risk to develop solid cancers including melanoma and hematological malignancies [9, 10]. For instance, in a study of 1525 patients, the relative risk for non-Hodgkin lymphoma was 2.54, for melanoma 3.07 and for pancreas cancer 2.37 [9]. The pathogenesis of cancers in Gaucher patients has not yet been elucidated although it has been proposed that the lipid storage that occurs primarily in monocytic cells and macrophages leads to chronic stimulation of the immune system and subsequent lymphoproliferation [10]. Interestingly, a host-related immunogenetic profile has been proposed to be of importance for susceptibility and tumorigenesis in malignant melanoma [11].

Moreover, a significant association between the presence of mutations in the *GBA1* gene and Parkinson's disease was found [12]. Intriguingly, patients with Parkinson's disease showed a sevenfold increased relative risk for melanoma [13]. In addition, it has been reported that PARKIN, encoded by the major Parkinson disease-predisposing gene *PARK2*, regulates the polyubiquitination and proteasomal degradation of misfolded mutants of GBA1 [14]. These observations suggest that some genetic determinants of SL metabolism could be associated with cutaneous melanoma predisposition. However, whether and how germline mutations of *GBA* genes lead to melanoma development in patients with Gaucher or Parkinson's diseases remain to be demonstrated.

In addition, a genome-wide association study has identified the 1q21.3 chromosomal region, containing *LASS2* gene that encodes the dihydroceramide synthase 2 (CERS2), as a locus predisposing to cutaneous melanoma [15]. CERS2, a synthase of long-chain ceramides (C22-C24), is also known as tumor metastasis suppressor gene 1 (TMSG1). Recently, it has been reported that silencing of LASS2/TMSG1 enhances prostate cancer cell invasion and metastasis [16]. However, it was not established whether cancer progression was associated with decreased ceramide levels or other modifications in SL metabolism due to the down-regulation of LASS2.

Looking at the enzymes involved in the generation of ceramide through the hydrolysis of sphingomyelinase, it has been recently reported, in human specimens of melanomas at various stages, that acid sphingomyelinase (acid SMase) expression inversely correlates with tumor malignancy, being lower in lymph node metastases with respect to benign nevi [17]. These results were confirmed in vivo in a mouse model of melanoma by two pieces of evidence: (1) after subcutaneous injection to C57BL/6 mice, the expression of acid SMase was higher in B16-F1-derived tumors than in the more metastatic B16-F10 ones; (2) the expression of acid SMase by the B16-generated tumors decreased progressively during melanoma growth [17].

Besides these observations, ceramide catabolism is also enhanced in melanoma cells. Indeed, mining of the cancer profiling database Oncomine reveal significantly elevated expression of ASAH1 in melanoma biopsies compared with benign melanocytic lesions [18, 19]. We also reported that the expression of proteins encoded by *SPHK1*, *SGPP1* and *SGPL1* genes was altered in a series of invasive or metastatic melanoma cells as compared to that of healthy melanocytes [20, 21]. These changes are expected to lead to an increased production of ceramide metabolite, sphingosine 1-phosphate (S1P), in tumor cells. This was confirmed by measuring both SPHK1 activity in vitro and expression in human melanoma specimens compared to nevi. The lipid kinase was overexpressed in the four main histologic subtypes of cutaneous melanoma. Moreover, owing to frequent BRAF or NRAS mutations in melanocytic neoplasms, which are, for the vast majority of them, single amino acid changes such as at codon 600 (BRAFV600E) or 61 (NRASQ61L), the RAS/RAF/MEK/ERK signaling pathway is activated, representing a major event in the metastatic progression of melanoma [22]. By using pharmacological and genetic approaches, we showed that SPHK1 expression/activity is increased by ERK activation due to BRAF or NRAS mutations in melanoma cells [21].

Dysregulation of SL metabolism in melanoma cells not only affects ceramide, GlcCer, and S1P but also other yet unidentified derivatives. For instance, melanoma cells contain high amounts of certain gangliosides (which derive from GlcCer), some of them being considered as melanoma-specific antigens [23]. These changes have also been associated to increased serum ganglioside levels in melanoma patients compared to melanocytic patients or healthy individuals, and could be predictive of a decrease in the overall survival [24]. Finally, as in many tumors, dysregulations in melanoma may combine many anabolic and catabolic reactions leading to the formation of multiple SL metabolites. A complete lipidomic profile associated to melanocytic tumor imaging would be highly useful tools to define new SL-based targets and assist in directing antitumoral therapy.

3 Role of SLs in Melanogenesis

Melanogenesis involves different stages, from melanocyte embryogenesis to melanosome transfer from melanocytes to keratinocytes. Melanocytes derive from neural crest melanoblasts that migrate to different destinations, including the basal layer of the epidermis and hair follicles [25]. To protect skin towards the deleterious effects of ultraviolet light, melanocytes synthesize and introduce melanin into specialized lysosome-related organelles called melanosomes, which move along arm-like structures called dendrites, so as to reach the keratinocytes.

Analysis of the biology of melanocytes and skin pigmentation is important to better understand the complex system of melanoma development and progression [26]. For instance, a dual role has been ascribed to the basic helix-loop-helix leucine zipper microphthalmia-associated transcription factor (MITF). Not only MITF activates the transcription of genes involved in melanin synthesis, melanosome biogenesis and transport, but it is also reported as an amplified oncogene in a subset of human melanomas in which it controls proliferation, migration and invasion [27]. Treatment of B16F10 murine melanoma cells with SL-enriched fraction from human placenta stimulates melanin production by inducing tyrosinase gene expression [28]. The characterization of lipids, by MALDI-TOF mass spectrometry, indicated the presence of sphingosine and S1P in the active fraction, and identified the fatty acid moiety as palmitic acid (C16:0) and oleic acid (C18:0) [29]. The human placental lipids induce the p38 stress-signalling pathway [30] leading to phosphorylation of the cyclic AMP response element binding protein (CREB), which in turn activates MITF expression and consequently increases tyrosinase expression and pigmentation [29, 31]. Interestingly, topical application of a purified SL fraction to age-onset gray-haired C57BL/6J mouse skin, that exhibit decaying MITF expression, reversed the hair greying phenotype by inducing the re-growth of thick black hair with MITF-positive differentiated melanocytes found near the dermal papille [29].

These observations strongly suggest that some sphingoid bases may have the ability to activate MITF expression when added to melanoma cells. However, addition of exogenous S1P to the spontaneously immortalized mouse cell line Mel-Ab was shown to reduce melanin synthesis by ERK activation, MITF phosphorylation at Ser73 and degradation by the proteasome. This effect was associated with subsequent down-regulation of tyrosinase and tyrosinase-related protein 1 [32] and was mediated by S1PR3 receptor activation [33]. On the other hand, modulation of acid SMase has been demonstrated to control MITF expression during melanoma growth, MITF expression being increased in B16-derived tumors at late stages of progression, i.e., when the expression of acid SMase becomes down-regulated [17]. Of importance, acid SMase induced the degradation of MITF by proteasome activation through a sustained activation of ERK. By affecting MITF expression acid SMase determined the modulation of its downstream pathway. Whenever MITF was up-regulated, because of the low expression of acid SMase, the proliferation marker cyclin-dependent kinase 2, the apoptosis modulator Bcl-2, the melanin synthesizing enzyme tyrosinase and the pro-metastatic c-Met were up-regulated, thus leading to a more aggressive melanoma cell phenotype [17].

4 Role of SLs in Melanoma–Stroma Interactions

Although genetic changes are indisputable causes of formation and progression of melanoma, they are also influenced by the adjacent microenvironment [34]. Melanoma is composed of not only the malignant cells but also normal skin cells, such as fibroblasts, endothelial, and immune cells, alongside the melanocytes, all of them embedded in the extracellular matrix (ECM). Altogether, they constitute a special microenvironment in which disturbed homeostasis may facilitate tumor development. Understanding the role of these cells or molecules that allow interactions of normal cells with melanoma cells could be useful for designing potential combinations of specific target therapies and immunotherapies that could be more efficient approaches against malignant melanoma [35].

In many tumors, after intracellular production by SPHKs, S1P is released outside cancer cells and can activate S1P receptors in the surrounding microenvironment in a paracrine manner [36]. Here, we will review the current knowledge of the roles of S1P on the remodelling of tumor microenvironment in melanoma.

4.1 Effect of S1P on Angiogenesis

Melanoma invasion can take place when cells produce critical angiogenesis-related growth factors, such as VEGF [37], bFGF/FGF-2 [38], Ang-1 and Ang-2 [39]. These factors stimulate the proliferation of endotheliocytes of the vessels of the near-connective tissue and then contribute to the formation of a new vascular system that will supply oxygen and nutrients to tumour cells [40]. Proangiogenic factors promote lymphatic and hematogenous metastasis and are associated with higher disease burdens and worse outcomes in melanoma. For instance, a positive correlation between increased serum VEGF, bFGF and IL-8 levels and a shorter disease-free survival was reported in patients with metastatic melanoma [41]. Similarly, expression of hypoxia-inducible factors (HIFs), global regulators of oxygen homeostasis, was associated with poorer prognosis in patients with metastatic melanoma [42]. The anti-angiogenic agents that have been tested so far, such as bevacizumab, axitinib or dovitinib, offer little benefit as single agents, but in combination with cytotoxic agents are promising for patients who may be ineligible for currently FDA approved first-line therapy with vemurafenib or ipilimumab or in patients for whom this therapy has failed [43].

In many tumors, hypoxia induces the release of S1P from cancer cells. Extracellularly released S1P is a robust pro-angiogenic factor that promotes angiogenesis directly as well as indirectly by stimulating VEGF secretion. Conversely, VEGF stimulation induces the transcription of S1PR1 mRNA, and thus VEGF and S1P act in a feed-forward loop promoting the development of *de novo* blood vessels [44]. In the murine melanoma B16-F10 allograft model, neutralization of extracellular S1P with a monoclonal S1P antibody was associated with decreased tumor progression likely due to inhibition of VEGF- and bFGF-stimulated endothelial cell migration and angiogenesis [45]. Additionally, treatment of mice with the S1PR modulator FTY720, which renders cells unresponsive to S1P activation by sequestering S1PRs internally, reduced melanoma progression by inhibiting tumor vascularization [46]. Emerging evidence shows that S1P exerts both positive and negative effects on angiogenesis and vascular maturation, which are mediated by S1PR1 and S1PR2, respectively [47]. Consistent with these observations, B16BL6 murine melanoma cells implanted in S1PR2-null mice displayed increased tumor growth and angiogenesis with enhanced vascular mural cell recruitment, compared to wild-type mice [48]. These effects on angiogenesis bring the targeting of S1PRs into perspective as a novel anti-melanoma therapeutic approach.

4.2 Role of S1P in Fibroblast Differentiation

One of the key cellular components in the microsystem of the reactive stroma is the myofibroblast, also known as tumor-associated fibroblast (TAF), a modulated fibroblast that has acquired during stromagenesis the ability to express α-smooth muscle actin (α-SMA), fibroblast surface protein (FSP-1) or fibroblast-activated protein (FAP). TAFs also secrete growth factors, ECM components including collagen and fibronectin, and ECM modulating factors such as matrix metalloproteinases (MMPs) [49, 50]. Altogether, these microenvironmental factors promote the conversion of otherwise incipient tumor cells into highly malignant cells which can spread to and infiltrate distant organs, due to the acquisition of invasive and metastatic phenotypes [51]. Myofibroblasts are proposed to originate from various sources including pre-existing fibroblasts, preadipocytes, smooth muscle cells, endothelial cells, epithelial cells, and bone marrow-derived progenitors [52]. Whatever their origin, under the influence of afferent signals from cancer cells, they differentiate into myofibroblasts, thereby engaging a cycle of paracrine efferent signals, sensed by cancer cells which respond by changing their motility and aggressiveness [53].

In melanoma, the dense and fibrotic nature of the tumor is thought to be a direct consequence of the presence of myofibroblasts [54]. The recruitment and infiltration of large numbers of mesenchymal cells in the microenvironment of solid tumors is often associated with the development of high-grade cancers with poor prognosis [52]. Notably, the virulence of melanoma, assessed either by Clark levels [55] or the Breslow measurement of tumor thickness [56], has been conventionally related to the invasion by melanoma cells of the dermal fibroblast-rich layer of the skin. However, whereas the role of tumor-myofibroblast interactions in melanoma progression is increasingly documented [57], how dermal fibroblasts modulate this dialogue is less known. Recently, by using a 3D-coculture model in which preformed melanoma spheroids were implanted into a collagen matrix containing dermal fibroblasts, it has been shown that melanoma cells actively recruit normal fibroblasts [58]. This event could contribute to melanoma cell growth and resistance to treatment but the molecular mechanisms underlying these processes are not completely elucidated.

Interestingly, we identified S1P and its receptor S1PR3 as a critical signaling pathway in the dialogue between melanoma cells and dermal fibroblasts to mediate tumor cell invasion [21]. Indeed, the conditioned medium from SPHK1-expressing melanoma cells (as compared to SPHK1-null cells) led to the differentiation of skin fibroblasts into myofibroblasts with increased expression of MMP-2, MMP-9 and SPHK1, and the release of both proteases and S1P. These findings were recapitulated in TAFs, isolated from human melanoma biopsies, compared with dermal fibroblasts. Therefore, dysregulation of S1P production in melanoma cells elicits a fibrotic response in the tumor microenvironment, which in turn stimulates melanoma cell migration by promoting S1PR3 expression. Several studies have shown that SPHK1/S1P mediates differentiation of fibroblasts into myofibroblasts induced by pro-fibrotic factors in fibrotic diseases. For instance, TGF-β, a key cytokine of fibroblastic differentiation, stimulated SPHK1 expression and activity in dermal

[59] and cardiac [60] fibroblasts. Down-regulation of SPHK1 by siRNA abolished TGF-β-induced increases in α-SMA, laminin, TIMP-1 [59] and fibronectin [61], suggesting that SPHK1 is a downstream mediator of TGF-β signaling. Our recent data emphasize a new role for SPHK1/S1P signaling in fibroblastic differentiation in the context of skin cancer development.

In addition, in vivo tumorigenesis experiments showed that the lack of S1P in the microenvironment (as observed in *Sphk1*$^{-/-}$ mice) prevented the growth of orthotopically injected murine melanoma cells [21]. Consistent with these observations, lung colonization of B16F10 melanoma cells, after tail vein injection, was lower in *Sphk1*$^{-/-}$ mice than in control animals [62]. Both local tumor growth and dissemination were enhanced more efficiently by co-injection of wild-type skin fibroblasts than by fibroblasts from *Sphk1*$^{-/-}$ mice [21]. Hence, these results show that melanoma invasiveness is critically dependent on S1P derived from host stromal cells. These findings also identify S1P in melanoma, derived from tumor cells, endothelial cells or TAFs and probably other cell types, as a novel therapeutic target.

5 Role of SLs on Melanoma Growth and Survival

Understanding the driving mutations, which contribute to melanomagenesis, has been crucial for the development of drugs that specifically target the underlying cellular signaling. Activating mutations of *RAS* have been detected in 15–20 % of melanomas, and almost exclusively involve the NRAS isoform. Mutations in the serine threonine kinase gene *BRAF* have been identified in 50–60 % of all metastatic melanomas, and 80–90 % of all *BRAF* mutations consist of an exchange of glutamine for valine at amino acid 600 (V600E) [63]. This genetic modification locks the kinase into a 500-fold more active conformation than wild-type BRAF and lead to constitutive activation of the MAPK pathway and ultimately to increased survival and growth. The efficacy and survival advantage of vemurafenib, which inhibits the mutant BRAF protein, over dacarbazine has been demonstrated in treatment-naïve patients with advanced BRAF V600E mutated melanoma [64]. However, in spite of a clear improvement, the PFS was still only 5.3 months (vemurafenib) compared to 1.6 months (dacarbazine). In 2013, dabrafenib, another inhibitor of mutant BRAF, and trametinib, which inhibits MEK1 and MEK2 proteins, were approved by the FDA for use as monotherapies in BRAF-mutant metastatic melanoma. Both treatments have shown benefit with a prolonged median PFS and overall survival compared with standard chemotherapy [65]. Combination of BRAF and MEK inhibitors appears to further improve PFS and overall survival [66].

Although BRAF represents one of the prime therapeutic targets in melanoma, both intrinsic and acquired resistance occurs following treatment with BRAF inhibitors; some patients do not respond to these agents while others have a short-lived response. Multiple changes within a given melanoma may contribute to this resistance [67, 68], and ongoing research tries to identify new therapeutic targets to overcome this resistance.

Among SLs, ceramide is well recognized as an important mediator of cell death in combination or not with other stresses such as radiation, chemotherapy, hypoxia, and nutrient deprivation [69]. As described above, melanoma cells develop numerous aberrations of SL metabolism that prevent the accumulation of ceramide, either by reducing its generation or enhancing its catabolism. Thanks to its central role in apoptosis, an increase of intracellular ceramide levels should result in attenuation of tumor growth and/or enhancement of drug sensitivity. This may be achieved through direct application of SLs or by the inhibition/activation of the enzymes that either produce or consume ceramide.

For instance, inhibition of SPHK1 by SKI-I leads to increased ceramide levels, decreases S1P levels and is associated with inhibition of xenografted melanoma tumor growth [70]. Moreover, small interfering RNA knockdown of SPHK1 not only inhibits anchorage-dependent growth of human melanoma cells but also enhances sensitivity of these cells to apoptosis-inducing agents such as staurosporine. These events were linked to reduced levels of pAKT and CYCLIN D1 as well as increased expression of p21 and cleaved caspase-3 [70]. Consistent with these results, addition of nanoliposomal short-chain ceramide to human melanoma cells [71] or intracellular ceramide accumulation induced by cytotoxic drugs in B16 murine melanoma cells [72, 73], was associated with a significant inhibition of the prosurvival Akt signaling leading to the activation of the mitochondrial pathway of apoptosis. High levels of SPHK1 activity, associated to high expression of BCL-2, provide a cytoprotective effect by reducing the effects of ceramide on the mitochondrial outer membrane permeabilization [74]. The inhibitory effects of ceramide on melanoma cell growth were also reproduced by the use of natural compounds. Indeed, Jaspine B, an anhydrophytosphingosine derivative isolated from the marine sponge *Jaspis* sp., was described to trigger melanoma cell death through the inhibition of sphingomyelin synthase (SMS), an enzyme that converts de novo ceramide into the membrane lipid sphingomyelin [75].

Another well-documented example is the conversion of ceramide to GlcCer, catalyzed by GlcCer synthase (GCS). Indeed, inhibition of GlcCer synthesis, by the iminosugar *N*-butyldeoxynojirimycin or OGT2378, led to delayed or reduced murine melanoma development, respectively [76, 77]. Similarly, transfection of antisense GCS in MB4 murine melanoma cells was associated with reduction of cellular gangliosides and poor tumorigenic capacity when injected into mice [78]. Conversely, elevated expression of GCS was associated with cell survival in the etoposide-resistant human melanoma cell line MeWo [79].

In addition, we showed that GBA2, an enzyme capable of degrading GlcCer, may control A375 human melanoma cell growth likely through GlcCer degradation into ceramide leading to ER-stress and subsequent apoptosis of tumor cells [8].

A fundamental role both in melanoma progression and resistance to radiotherapy has been established for the proapoptotic enzyme acid SMase [17, 80]. Analysis of human specimens as well as of murine B16 melanoma cells has shown that, during melanoma progression, acid SMase expression decreases progressively with time, resulting in the generation of very aggressive cell clones, characterized by an increased ability to proliferate and to form metastases in vitro and, more importantly, in vivo [17]. The mechanism responsible for acid SMase reduction and whether this event might be involved in the resistance to chemotherapy (especially to drugs known to involve acid SMase, e.g., cisplatin, doxorubicin, gemcitabine) needs further investigation, but these data suggest for the acid SMase pathway a clear potential in therapeutic perspective. In line with this area, data from Smith and Schuchman, showing for the first time that overexpression of acid SMase cDNA in melanoma cells, as well as the administration of the recombinant protein, can act in combination with irradiation in vitro and in vivo to induce cell death, suggests their possible use as antineoplastic agents [80].

In addition to acid SMase, acid ceramidase (acid CDase), the enzyme responsible for ceramide hydrolysis, appears to play an important role in cancer progression [81]. It played a key role in the response of human melanoma cells to the treatment with dacarbazine [82]. Indeed, using A375 human melanoma cells as a model, it was shown that dacarbazine triggers melanoma cell death, in part through the cathepsin-B mediated degradation of acid CDase and the ensuing increase of ceramide levels. These events lead to autophagy and loss of cell viability. Of importance, the modulation of the levels of acid CDase affected the sensitivity of cells to dacarbazine.

Taken together, these data suggest that the manipulation of the acid SMase/acid CDase axis, i.e., down-regulating the former and up-regulating the latter, might be considered a strategy that cancer cells utilize to reduce the levels of ceramide and thus to overcome its pro-apoptotic effects. From this point of view, it is worth speculating that genetic or therapeutic approaches aimed at restoring the appropriate levels of both enzymes may be useful in the treatment of melanoma by enhancing the efficacy of anticancer drugs.

Beside their direct effects on cancer cells, antitumoral strategies targeting SL metabolism could also impair melanoma progression by affecting the tumor microenvironment. Indeed, FTY720, which is known to interfere with S1P receptors and metabolism, leads to a decrease in CD4+Foxp3+ regulatory T cells and limits metastatic development when orally administered to syngeneic mice injected with B16F10-Nex2 melanoma cells [83]. Similarly, the peritumoral injection of a recombinant acid SMase to mice bearing B16-F10 melanoma tumors is associated to the killing of the surrounding microvascular endothelial cells leading to smaller tumors [80]. In this respect, also acid SMase expressed by the cells involved in the immune response against melanoma, i.e., dendritic cells, can be a suitable target for ameliorating the current therapeutic protocols. Indeed, pharmacological inhibition or genetic down-regulation of acid SMase in dendritic cells results in their survival within the tumor after cisplatin administration and consequently in the reduction of tumor growth [84].

Altogether, these results suggest that development of pharmacological agents targeting SL metabolism holds promise in the potentiation of anti-melanoma therapies. Table 1 summarizes aspects of SL metabolism, as well as known inhibitors of this pathway, that could be considered as potential agents and targets for improving therapeutic efficacy in the treatment of melanoma.

Table 1 In vitro and in vivo effects of SL analogues or (epi)genetic manipulation of SL metabolism on cell death and tumorigenicity of melanoma cells

Agents	Enzyme targeted	Combination therapy	Melanoma cell line	Biological effect	References
Myriocin	SPT		B16F10	Cell cycle arrest	[85, 86]
				Reduced tumor growth	
NB-DNJ	GCS		MEB4	Delayed tumor growth	[76]
OGT2378	GCS		MEB4	Reduced tumor growth	[77]
PDMP	GCS	Genistein	B16F10	Apoptosis	[73]
		Curcumin	WM-115, B16F10	Apoptosis	[87]
siRNA	GCS		MEB4	Reduced tumor growth	[78]
Jaspine B	SMS		SK-Mel28, B16F10	Apoptosis	[75, 88]
D-e-MAPP	CDase		M186, M221, A375, Mel2A	Apoptosis	[89]
D-NMAPPD					
SKI-II	SPHK1	Genistein	B16F10	Apoptosis	[73]
SKI-I	SPHK1		UACC 903	Reduced tumor growth	[70]
siRNA	SPHK1	Anti-CD95, C_6-ceramide	Mel-2a	Apoptosis	[74]
		Staurosporine	UACC 903	Apoptosis	[70]
siRNA	aSMase		B16F10	Increased tumor invasion	[17]
Recombinant protein	aSMase	Irradiation	B16F10	Reduced tumor growth	[80]
FTY720			B16F10-NEX2	Apoptosis	[83]
				Reduced tumor invasion	
C_6-ceramide		Curcumin	WM-115, B16F10	Apoptosis	[72]
		Genistein	B16F10	Apoptosis	[73]
Ceramide		Docetaxel	B16F10	Apoptosis	[90]
				Reduced tumor growth	
Nanoliposomes with C_2-ceramide		Sorafenib	UACC 903, 1205 Lu	Apoptosis	[71]
				Reduced tumor growth	
Nanovesicles with *N*-octanoyl-GlcCer		Doxorubicin	B16F10	Reduced tumor growth	[91]

CDase ceramidase, *D-e-MAPP* (1*S*,2*R*)-2-*N*-myristoylamino-1-phenyl-1-propanol, *D-NMAPPD* (1*R*,2*R*)-2-*N*-myristoylamino-1-(4-nitrophenyl)-1,3-propandiol, *GCS* glucosylceramide synthase, *NB-DNJ* *N*-butyldeoxynojirimycin, *PDMP* *N*-[2-hydroxy-1-(4-morpholinylmethyl)-2-phenylethyl]-decanamide, monohydrochloride, *aSMase* acid sphingomyelinase, *SMS* sphingomyelin synthase, *SPHK1* sphingosine kinase 1, *SPT* serine palmitoyltransferase

B16F10 and MEB4 are murine melanoma cells: A375, 1205 Lu, M186, M221, Mel-2a, SK-Mel28, UACC 903 and WM-115 are human melanoma cells

6 Concluding Remarks

With the advent of targeted therapies, great progress has been made to improve the overall survival rate of patients with metastatic melanoma. However, long-term survivors of stage IV melanoma are still rare, and among patients whose tumors respond to targeted therapies, relapse is frequent. Significant efforts are currently being made, not only to further understand chemoresistance mechanisms but also to improve treatments with new drugs and rational use of combination therapy. Thanks to the wide spectrum of interconnection and interconversion possibilities in their metabolism, SLs supply to the cell a wide repertoire of potential targets that may control fine-tuning of signal transduction pathways. In melanoma cells, SL metabolism is dysregulated and some species are able to control key molecular steps of melanoma progression such as tumor cell growth and survival. Moreover, S1P and gangliosides act on the remodeling of the stroma cells surrounding melanoma cells, which in turn, facilitate tumor invasion (Fig. 1).

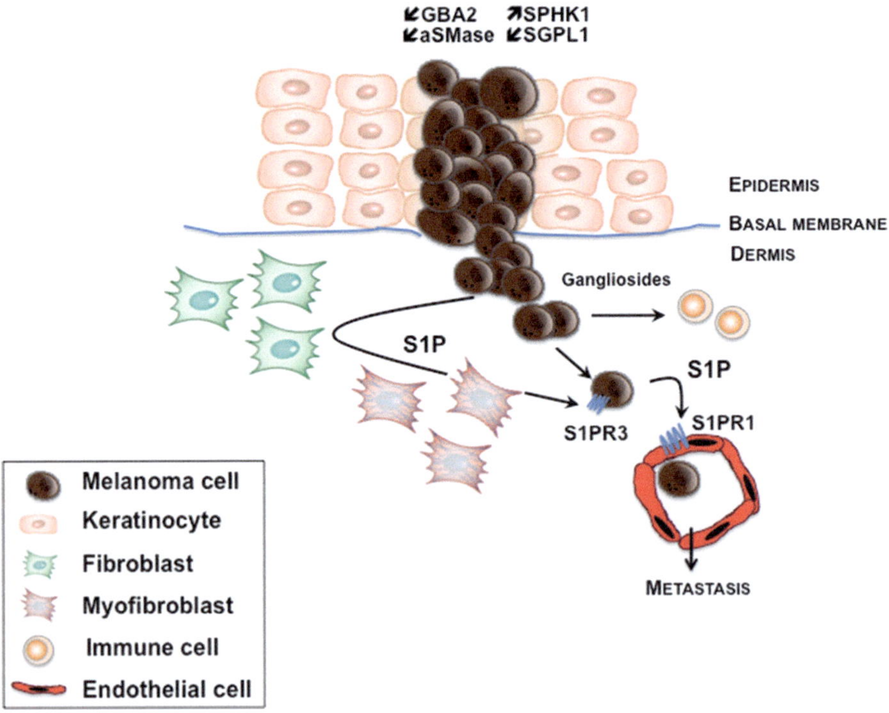

Fig. 1 Role of sphingolipids in melanoma. The figure summarizes how SL metabolism is affected in melanoma cells and how SLs could modulate melanoma progression. *aSMase* acid sphingomyelinase, *CDase* ceramidase, *GBA2* β-glucocerebrosidase 2, *SPHK1* sphingosine kinase 1, *S1P* sphingosine 1-phosphate, *S1PR* sphingosine 1-phosphate receptor

Some unmet challenges include deciphering the mechanisms that control the dysregulations of SL metabolism in melanoma cells as well as the identification and the role of the different molecular species in the multi-step process of melanomagenesis. Finally, targeting SL metabolism could be a promising area in the development of new drugs to halt melanoma progression and overcome the resistance towards recently targeted therapies.

Acknowledgements Financial support by INSERM, Paul Sabatier University, Ligue Nationale Contre le Cancer (Equipe Labellisée 2013), the Italian Association of Cancer Research (AIRC, IG11365) and the French and Lebanese ministries of foreign affairs (MAEDI) and Higher Education and Research (MENESR) (PHC CEDRE, 30750VL) is gratefully acknowledged. D.G. is a recipient of a fellowship from Ligue Nationale Contre le Cancer.

References

1. MacKie RM, Hauschild A, Eggermont AM (2009) Epidemiology of invasive cutaneous melanoma. Ann Oncol 20(Suppl 6):vi1–vi7
2. Nikolaou V, Stratigos AJ (2014) Emerging trends in the epidemiology of melanoma. Br J Dermatol 170:11–19
3. Flaherty K (2010) Advances in drug development. BRAF validation in melanoma. Clin Adv Hematol Oncol 8:31–34
4. Graziani G, Tentori L, Navarra P (2012) Ipilimumab: a novel immunostimulatory monoclonal antibody for the treatment of cancer. Pharmacol Res 65:9–22
5. Solit DB, Rosen N (2011) Resistance to BRAF inhibition in melanomas. N Engl J Med 364:772–774
6. Rabionet M, Gorgas K, Sandhoff R (2014) Ceramide synthesis in the epidermis. Biochim Biophys Acta 1841:422–434
7. Elias PM, Gruber R, Crumrine D et al (2014) Formation and functions of the corneocyte lipid envelope (CLE). Biochim Biophys Acta 1841:314–318
8. Sorli SC, Colie S, Albinet V et al (2013) The nonlysosomal beta-glucosidase GBA2 promotes endoplasmic reticulum stress and impairs tumorigenicity of human melanoma cells. FASEB J 27:489–498
9. Landgren O, Turesson I, Gridley G et al (2007) Risk of malignant disease among 1525 adult male US Veterans with Gaucher disease. Arch Intern Med 167:1189–1194
10. Taddei TH, Kacena KA, Yang M et al (2009) The underrecognized progressive nature of N370S Gaucher disease and assessment of cancer risk in 403 patients. Am J Hematol 84:208–214
11. Naumova E, Mihaylova A, Ivanova M et al (2007) Impact of KIR/HLA ligand combinations on immune responses in malignant melanoma. Cancer Immunol Immunother 56:95–100
12. Sidransky E, Nalls MA, Aasly JO et al (2009) Multicenter analysis of glucocerebrosidase mutations in Parkinson's disease. N Engl J Med 361:1651–1661
13. Pan T, Li X, Jankovic J (2011) The association between Parkinson's disease and melanoma. Int J Cancer 128:2251–2260
14. Ron I, Rapaport D, Horowitz M (2010) Interaction between parkin and mutant glucocerebrosidase variants: a possible link between Parkinson disease and Gaucher disease. Hum Mol Genet 19:3771–3781
15. Amos CI, Wang LE, Lee JE et al (2011) Genome-wide association study identifies novel loci predisposing to cutaneous melanoma. Hum Mol Genet 20:5012–5023

16. Xu X, Liu B, Zou P et al (2014) Silencing of LASS2/TMSG1 enhances invasion and metastasis capacity of prostate cancer cell. J Cell Biochem 115:731–743
17. Bizzozero L, Cazzato D, Cervia D et al (2014) Acid sphingomyelinase determines melanoma progression and metastatic behaviour via the microphtalmia-associated transcription factor signalling pathway. Cell Death Differ 21:507–520
18. Haqq C, Nosrati M, Sudilovsky D et al (2005) The gene expression signatures of melanoma progression. Proc Natl Acad Sci U S A 102:6092–6097
19. Talantov D, Mazumder A, Yu JX et al (2005) Novel genes associated with malignant melanoma but not benign melanocytic lesions. Clin Cancer Res 11:7234–7242
20. Colie S, Van Veldhoven PP, Kedjouar B et al (2009) Disruption of sphingosine 1-phosphate lyase confers resistance to chemotherapy and promotes oncogenesis through Bcl-2/Bcl-xL upregulation. Cancer Res 69:9346–9353
21. Albinet V, Bats ML, Huwiler A et al (2014) Dual role of sphingosine kinase-1 in promoting the differentiation of dermal fibroblasts and the dissemination of melanoma cells. Oncogene 33:3364–3373
22. Rebecca VW, Sondak VK, Smalley KS (2012) A brief history of melanoma: from mummies to mutations. Melanoma Res 22:114–122
23. Carr A, Mullet A, Mazorra Z et al (2000) A mouse IgG1 monoclonal antibody specific for N-glycolyl GM3 ganglioside recognized breast and melanoma tumors. Hybridoma 19:241–247
24. Nicolae I, Nicolae CD, Coman OA et al (2011) Serum total gangliosides level: clinical prognostic implication. Rom J Morphol Embryol 52:1277–1281
25. Costin GE, Hearing VJ (2007) Human skin pigmentation: melanocytes modulate skin color in response to stress. FASEB J 21:976–994
26. Gupta PB, Kuperwasser C, Brunet JP et al (2005) The melanocyte differentiation program predisposes to metastasis after neoplastic transformation. Nat Genet 37:1047–1054
27. Koludrovic D, Davidson I (2013) MITF, the Janus transcription factor of melanoma. Future Oncol 9:235–244
28. Saha B, Singh SK, Sarkar C et al (2006) Transcriptional activation of tyrosinase gene by human placental sphingolipid. Glycoconj J 23:259–268
29. Saha B, Singh SK, Mallick S et al (2009) Sphingolipid-mediated restoration of Mitf expression and repigmentation in vivo in a mouse model of hair graying. Pigment Cell Melanoma Res 22:205–218
30. Singh SK, Sarkar C, Mallick S et al (2005) Human placental lipid induces melanogenesis through p38 MAPK in B16F10 mouse melanoma. Pigment Cell Res 18:113–121
31. Mallick S, Singh SK, Sarkar C et al (2005) Human placental lipid induces melanogenesis by increasing the expression of tyrosinase and its related proteins in vitro. Pigment Cell Res 18:25–33
32. Kim DS, Hwang ES, Lee JE et al (2003) Sphingosine-1-phosphate decreases melanin synthesis via sustained ERK activation and subsequent MITF degradation. J Cell Sci 116: 1699–1706
33. Kim DS, Park SH, Jeong YM et al (2011) Sphingosine-1-phosphate decreases melanin synthesis via microphthalmia-associated transcription factor phosphorylation through the S1P3 receptor subtype. J Pharm Pharmacol 63:409–416
34. Brandner JM, Haass NK (2013) Melanoma's connections to the tumour microenvironment. Pathology 45:443–452
35. Botti G, Cerrone M, Scognamiglio G et al (2013) Microenvironment and tumor progression of melanoma: new therapeutic perspectives. J Immunotoxicol 10:235–252
36. Takabe K, Spiegel S (2014) Export of Sphingosine-1-Phosphate and Cancer Progression. J Lipid Res 55:1839–1846
37. Claffey KP, Brown LF, del Aguila LF et al (1996) Expression of vascular permeability factor/vascular endothelial growth factor by melanoma cells increases tumor growth, angiogenesis, and experimental metastasis. Cancer Res 56:172–181

38. Rodeck U, Becker D, Herlyn M (1991) Basic fibroblast growth factor in human melanoma. Cancer Cells 3:308–311
39. Pomyje J, Zivny JH, Stopka T et al (2001) Angiopoietin-1, angiopoietin-2 and Tie-2 in tumour and non-tumour tissues during growth of experimental melanoma. Melanoma Res 11:639–643
40. Hanahan D, Weinberg RA (2000) The hallmarks of cancer. Cell 100:57–70
41. Ugurel S, Rappl G, Tilgen W et al (2001) Increased serum concentration of angiogenic factors in malignant melanoma patients correlates with tumor progression and survival. J Clin Oncol 19:577–583
42. Giatromanolaki A, Sivridis E, Kouskoukis C et al (2003) Hypoxia-inducible factors 1alpha and 2alpha are related to vascular endothelial growth factor expression and a poorer prognosis in nodular malignant melanomas of the skin. Melanoma Res 13:493–501
43. Pastushenko I, Vermeulen PB, Van den Eynden GG et al (2014) Mechanisms of tumour vascularisation in cutaneous malignant melanoma: clinical implications. Br J Dermatol 171:220–233
44. Yester JW, Tizazu E, Harikumar KB et al (2011) Extracellular and intracellular sphingosine-1-phosphate in cancer. Cancer Metastasis Rev 30:577–597
45. Visentin B, Vekich JA, Sibbald BJ et al (2006) Validation of an anti-sphingosine-1-phosphate antibody as a potential therapeutic in reducing growth, invasion, and angiogenesis in multiple tumor lineages. Cancer Cell 9:225–238
46. LaMontagne K, Littlewood-Evans A, Schnell C et al (2006) Antagonism of sphingosine-1-phosphate receptors by FTY720 inhibits angiogenesis and tumor vascularization. Cancer Res 66:221–231
47. Takuwa Y, Du W, Qi X et al (2010) Roles of sphingosine-1-phosphate signaling in angiogenesis. World J Biol Chem 1:298–306
48. Du W, Takuwa N, Yoshioka K et al (2010) S1P(2), the G protein-coupled receptor for sphingosine-1-phosphate, negatively regulates tumor angiogenesis and tumor growth in vivo in mice. Cancer Res 70:772–781
49. Kalluri R, Zeisberg M (2006) Fibroblasts in cancer. Nat Rev Cancer 6:392–401
50. Cirri P, Chiarugi P (2011) Cancer-associated-fibroblasts and tumour cells: a diabolic liaison driving cancer progression. Cancer Metastasis Rev 31:195–208
51. Joyce JA, Pollard JW (2009) Microenvironmental regulation of metastasis. Nat Rev Cancer 9:239–252
52. Shimoda M, Mellody KT, Orimo A (2010) Carcinoma-associated fibroblasts are a rate-limiting determinant for tumour progression. Semin Cell Dev Biol 21:19–25
53. Kunz-Schughart LA, Knuechel R (2002) Tumor-associated fibroblasts (part I): active stromal participants in tumor development and progression? Histol Histopathol 17:599–621
54. Ruiter D, Bogenrieder T, Elder D et al (2002) Melanoma-stroma interactions: structural and functional aspects. Lancet Oncol 3:35–43
55. Clark WH Jr, Elder DE, Guerry D et al (1989) Model predicting survival in stage I melanoma based on tumor progression. J Natl Cancer Inst 81:1893–1904
56. Breslow A (1970) Thickness, cross-sectional areas and depth invasion in the prognosis of cutaneous melanoma. Ann Surg 172:902–908
57. Smalley KS, Lioni M, Herlyn M (2005) Targeting the stromal fibroblasts: a novel approach to melanoma therapy. Expert Rev Anticancer Ther 5:1069–1078
58. Flach EH, Rebecca VW, Herlyn M et al (2011) Fibroblasts contribute to melanoma tumor growth and drug resistance. Mol Pharm 8:2039–2049
59. Yamanaka M, Shegogue D, Pei H et al (2004) Sphingosine kinase 1 (SPHK1) is induced by transforming growth factor-beta and mediates TIMP-1 up-regulation. J Biol Chem 279: 53994–54001
60. Gellings Lowe N, Swaney JS, Moreno KM et al (2009) Sphingosine-1-phosphate and sphingosine kinase are critical for transforming growth factor-beta-stimulated collagen production by cardiac fibroblasts. Cardiovasc Res 82:303–312

61. Kono Y, Nishiuma T, Nishimura Y et al (2007) Sphingosine kinase 1 regulates differentiation of human and mouse lung fibroblasts mediated by TGF-beta1. Am J Respir Cell Mol Biol 37:395–404
62. Ponnusamy S, Selvam SP, Mehrotra S et al (2012) Communication between host organism and cancer cells is transduced by systemic sphingosine kinase 1/sphingosine 1-phosphate signalling to regulate tumour metastasis. EMBO Mol Med 4:761–775
63. Davies H, Bignell GR, Cox C et al (2002) Mutations of the BRAF gene in human cancer. Nature 417:949–954
64. Chapman PB, Hauschild A, Robert C et al (2011) Improved survival with vemurafenib in melanoma with BRAF V600E mutation. N Engl J Med 364:2507–2516
65. Hauschild A, Grob JJ, Demidov LV et al (2012) Dabrafenib in BRAF-mutated metastatic melanoma: a multicentre, open-label, phase 3 randomised controlled trial. Lancet 380: 358–365
66. Voskoboynik M, Arkenau HT (2014) Combination Therapies for the Treatment of Advanced Melanoma: A Review of Current Evidence. Biochem Res Int 2014:307059
67. Nazarian R, Shi H, Wang Q et al (2010) Melanomas acquire resistance to B-RAF(V600E) inhibition by RTK or N-RAS upregulation. Nature 468:973–977
68. Sun C, Wang L, Huang S et al (2014) Reversible and adaptive resistance to BRAF(V600E) inhibition in melanoma. Nature 508:118–122
69. Hannun YA, Obeid LM (2008) Principles of bioactive lipid signalling: lessons from sphingolipids. Nat Rev Mol Cell Biol 9:139–150
70. Madhunapantula SV, Hengst J, Gowda R et al (2012) Targeting sphingosine kinase-1 to inhibit melanoma. Pigment Cell Melanoma Res 25:259–274
71. Tran MA, Smith CD, Kester M et al (2008) Combining nanoliposomal ceramide with sorafenib synergistically inhibits melanoma and breast cancer cell survival to decrease tumor development. Clin Cancer Res 14:3571–3581
72. Yu T, Li J, Sun H (2010) C6 ceramide potentiates curcumin-induced cell death and apoptosis in melanoma cell lines in vitro. Cancer Chemother Pharmacol 66:999–1003
73. Ji C, Yang YL, He L et al (2012) Increasing ceramides sensitizes genistein-induced melanoma cell apoptosis and growth inhibition. Biochem Biophys Res Commun 421:462–467
74. Bektas M, Jolly PS, Muller C et al (2005) Sphingosine kinase activity counteracts ceramide-mediated cell death in human melanoma cells: role of Bcl-2 expression. Oncogene 24: 178–187
75. Salma Y, Lafont E, Therville N et al (2009) The natural marine anhydrophytosphingosine, Jaspine B, induces apoptosis in melanoma cells by interfering with ceramide metabolism. Biochem Pharmacol 78:477–485
76. Guerrera M, Ladisch S (2003) N-butyldeoxynojirimycin inhibits murine melanoma cell ganglioside metabolism and delays tumor onset. Cancer Lett 201:31–40
77. Weiss M, Hettmer S, Smith P et al (2003) Inhibition of melanoma tumor growth by a novel inhibitor of glucosylceramide synthase. Cancer Res 63:3654–3658
78. Deng W, Li R, Guerrera M et al (2002) Transfection of glucosylceramide synthase antisense inhibits mouse melanoma formation. Glycobiology 12:145–152
79. Gouaze V, Yu JY, Bleicher RJ et al (2004) Overexpression of glucosylceramide synthase and P-glycoprotein in cancer cells selected for resistance to natural product chemotherapy. Mol Cancer Ther 3:633–639
80. Smith EL, Schuchman EH (2008) Acid sphingomyelinase overexpression enhances the antineoplastic effects of irradiation in vitro and in vivo. Mol Ther 16:1565–1571
81. Truman JP, Garcia-Barros M, Obeid LM et al (2014) Evolving concepts in cancer therapy through targeting sphingolipid metabolism. Biochim Biophys Acta 1841:1174–1188
82. Bedia C, Casas J, Andrieu-Abadie N et al (2011) Acid ceramidase expression modulates the sensitivity of A375 melanoma cells to dacarbazine. J Biol Chem 286:28200–28209
83. Pereira FV, Arruda DC, Figueiredo CR et al (2013) FTY720 induces apoptosis in B16F10-NEX2 murine melanoma cells, limits metastatic development in vivo, and modulates the immune system. Clinics (Sao Paulo) 68:1018–1027

84. Perrotta C, Bizzozero L, Falcone S et al (2007) Nitric oxide boosts chemoimmunotherapy via inhibition of acid sphingomyelinase in a mouse model of melanoma. Cancer Res 67: 7559–7564
85. Lee YS, Choi KM, Lee S et al (2012) Myriocin, a serine palmitoyltransferase inhibitor, suppresses tumor growth in a murine melanoma model by inhibiting de novo sphingolipid synthesis. Cancer Biol Ther 13:92–100
86. Lee YS, Choi KM, Choi MH et al (2011) Serine palmitoyltransferase inhibitor myriocin induces growth inhibition of B16F10 melanoma cells through G(2)/M phase arrest. Cell Prolif 44:320–329
87. Yu T, Li J, Qiu Y et al (2012) 1-phenyl-2-decanoylamino-3-morpholino-1-propanol (PDMP) facilitates curcumin-induced melanoma cell apoptosis by enhancing ceramide accumulation, JNK activation, and inhibiting PI3K/AKT activation. Mol Cell Biochem 361:47–54
88. Salma Y, Ballereau S, Maaliki C et al (2010) Flexible and enantioselective access to Jaspine B and biologically active chain-modified analogues thereof. Org Biomol Chem 8:3227–3243
89. Raisova M, Goltz G, Bektas M et al (2002) Bcl-2 overexpression prevents apoptosis induced by ceramidase inhibitors in malignant melanoma and HaCaT keratinocytes. FEBS Lett 516:47–52
90. Feng LX, Li M, Liu YJ et al (2014) Synergistic enhancement of cancer therapy using a combination of ceramide and docetaxel. Int J Mol Sci 15:4201–4220
91. van Lummel M, van Blitterswijk WJ, Vink SR et al (2011) Enriching lipid nanovesicles with short-chain glucosylceramide improves doxorubicin delivery and efficacy in solid tumors. FASEB J 25:280–289

Colon Cancer: The Role of Sphingolipid Metabolic Enzymes

Hideki Furuya, Songhwa Choi, Lina M. Obeid, Toshihiko Kawamori, and Ashley J. Snider

Abstract Colorectal cancer is one of the most common tumors worldwide, with sustained incidence in developed countries and increasing incidence in developing countries. Although recent studies provide knowledge of the molecular signaling pathways that are implicated in colon carcinogenesis, treatments and outcomes still need further improvement. Bioactive sphingolipids, such as ceramide, sphingosine, and sphingosine-1-phosphate (S1P), are signaling molecules that regulate cellular events including cell proliferation, apoptosis, senescence, angiogenesis, and transformation in response to diverse stimuli. Ceramide and sphingosine mediate numerous cell-stress responses, including induction of apoptosis and cell senescence. In contrast, S1P plays pivotal roles in cell survival, migration, and inflammation. These sphingolipids with opposing roles can be quickly metabolized and catabolized within cells, suggesting that the balance between these potent bioactive lipids may dictate cell fate. In this chapter, we review the roles of these bioactive sphingolipids and their metabolic enzymes in colitis, colitis associated cancer, and colorectal cancer.

H. Furuya
Cancer Biology Program, University of Hawaii Cancer Center,
701 Ilalo Street, Honolulu, HI 96813, USA

Clinical and Translational Research Program, University of Hawaii Cancer Center,
701 Ilalo Street, Honolulu, HI 96813, USA

S. Choi
Department of Medicine, Stony Brook University, Stony Brook, NY 11794, USA

L.M. Obeid • A.J. Snider (✉)
Department of Medicine, Stony Brook University, Stony Brook, NY 11794, USA

Northport VA Medical Center, Northport, NY 11768, USA
e-mail: ashley.snider@stonybrook.edu

T. Kawamori
Cancer Biology Program, University of Hawaii Cancer Center,
701 Ilalo Street, Honolulu, HI 96813, USA

Department of Pathology, Ihicnimiya Nishi Hospital,
1 Hira, Aza, Kaimei, Ichinomiya, Aichi-Pref. 494-0001, Japan
e-mail: toshikawamori@icloud.com

Y.A. Hannun et al. (eds.), *Bioactive Sphingolipids in Cancer Biology and Therapy*, DOI 10.1007/978-3-319-20750-6_7

Keywords Colon cancer • Inflammatory bowel disease • Cyclooxygenase 2 • Signal transducer and activator of transcription 3 • Sphingosine kinase • Sphingosine-1-phosphate • Ceramide

Abbreviations

aCDase	Acid CDase
ACER	Alkaline CDase
alk-SMase	Alkaline SMase
AOM	l Azoxymethane
APC	Adenomatous polyposis coli
aSMase	Acid SMase
CAC	Colitis-associated cancer
CDases	Ceramidases
CerS	Ceramide synthases
COX2	Cyclooxygenase-2
CRC	Colorectal cancer
DSS	Dextran sodium sulfate
EGFR	Epidermal growth factor receptor
FAP	Familial adenomatous polyposis
IBD	Inflammatory bowel disease
IL-6	Interleukin-6
miRNAs	microRNAs
MMR	Mismatch repair
MSI	Microsatellite instability
NaBT	Sodium butyrate
nCDase	Neutral CDase
NF-κB	Nuclear factor kappa B
nSMase	Neutral SMase
PGE_2	Prostaglandin E_2
PI3K	Phosphatidylinositol 3-kinase
S1P	Sphingosine-1-phosphate
S1PR	S1P receptor
SM	Sphingomyelins
SMases	Sphingomyelinases
SphKs	Sphingosine kinases
SPL	Sphingosine-1-phosphate lyase
SPPs	Sphingosine-1-phosphate Phosphatases
STAT3	Signal transducer and activator of transcription 3
TIMP1	Tissue inhibitor of metalloproteinase 1
TNBS	2,4,6-Trinitrobenzenesulfonic acid
TNF-α	Tumor necrosis factor alpha
TRAIL	Tumor necrosis factor-related apoptosis-inducing ligand
WT	Wild type

1 Introduction

Colorectal cancer (CRC) is the third most common cancer worldwide with an estimated 1.36 million new cancer cases and 693,881 cancer-related deaths per year [1, 2], and is the second leading cause of all cancer-related deaths with 62,130 annual deaths in the United States [3]. Despite aggressive screening guidelines for early detection and detailed knowledge of several critical events underlying the pathogenesis of cancer, it continues to be a major health concern worldwide, especially in developed countries. Risk factors associated with sporadic CRC include obesity, alcohol consumption, history of smoking, diets rich in red meats and age [2]. Hereditary CRCs include FAP (Familial Adenomatous Polyposis), which is due to mutations in the adenomatous polyposis coli (APC) gene [4–6], and Lynch Syndrome, which is inherited in an autosomal dominant pattern and associated with numerous genetic alterations [7].

Mutations associated with hereditary and sporadic CRC have been well defined [8]. These mutations primarily affect the APC/Wnt/β-catenin pathway, where frameshift and missense mutations in APC result in the synthesis of a truncated protein. These mutations result in increased levels of Wnt signaling and translocation of β-catenin to the nucleus. Increased β-catenin signaling has also been described to be associated with direct point mutations in β-catenin as well as hypermethylation of the APC promoter. Hereditary CRC, specifically Lynch Syndrome, is associated with microsatellite instability (MSI) and alterations in mismatch repair (MMR) genes.

Several well-studied signaling pathways have been shown to play significant roles in colon cancer development and progression. Mutations in B-raf, downstream of Kras, indicate poor prognosis and survival and are associated with MSI in CRC [9, 10]. In addition to this signaling pathway, mutations in the phosphatidylinositol 3-kinase (PI3K) pathway are also known to play a role in CRC. *PI3KCA* encoding p110α is mutated in approximately 10–30 % of CRC cases [11, 12], and are thought to occur in the adenoma-carcinoma transition, because they are more prevalent in the invasive tumor than in adenomatous polyps [13]. Upstream of the MAPK and PI3K pathways, over-expression of the epidermal growth factor receptor (EGFR) has been implicated in CRC tumor progression [14, 15].

Inflammation is another risk factor for CRC, and there are well established links between inflammatory bowel disease (IBD) and colon cancer [16]. Significant literature shows that inflammation in the colon can predispose to colitis-associated cancers (CAC) [17]. In addition there are emerging roles for the microbiome in colitis and possibly in predisposing to transformation and oncogenesis. Many groups including ours have also implicated cytokines, such as tumor necrosis factor alpha (TNF-α) and interleukin-6 (IL-6), and their downstream signaling pathways cyclooxygenase-2 (COX2), prostaglandins, and signal transducer and activator of transcription 3 (STAT3) in colitis and in CAC [16, 18, 19].

Among bioactive molecules involved in signaling, sphingolipids, such as ceramide, sphingosine, and sphingosine-1-phosphate (S1P) have important roles in numerous cell events including cell proliferation, apoptosis, senescence, angiogenesis, and migration [20]. Interestingly, the key bioactive sphingolipids ceramide, sphingosine, and S1P exert distinct biologic effects in cell signaling, suggesting that the balance between the sphingolipids is essential to cell fate. Several studies have demonstrated involvement of the sphingolipid pathway in the pathogenesis of cancers [21–24]. This chapter will summarize the roles of sphingolipids and their metabolic enzymes in the regulation of IBD, CAC and CRC.

2 Modulation of Sphingomyelin and Ceramide: Sphingomyelinases

Sphingomyelinases (SMases) hydrolyze sphingomyelins (SM) into phosphocholine and ceramides, and are classified according to their pH optima of action; acid, neutral, and alkaline (aSMase, nSMase, and alk-SMase, respectively). Although they have the same enzymatic function (degradation of SM), aSMase and alk-SMase have been suggested to play opposing roles in colon carcinogenesis. This may be due at least in part to their distinct cellular localizations. ASMase exists in two forms, secretory and lysosomal [25], while alk-SMase functions primarily as an ectoenzyme.

Putative anti-colon cancer agents, ursolic acid and curcumin, inhibit aSMase activity. This has been suggested to be a mechanism of action for their anti-proliferative actions in colon cancer [26, 27], indicating that aSMase may contribute to the induction of colon cancer. In an *in vivo* model for IBD, dextran sodium sulfate (DSS) administration in drinking water, ceramide levels were increased in lipid extracts from murine intestinal epithelial cells [28], and inhibition of aSMase with SMA-7 (SMase inhibitor) significantly reduced pro-inflammatory cytokines in the colon and alleviated the severity of colonic injury induced by DSS [25, 29]. A recent study has also highlighted a potential role for aSMase in suppressing colorectal liver metastasis. ASMase deficient mice are viable, but exhibit neurologic effects over time due to lysosomal storage defects. In a liver metastasis model using SL4 colon cancer cells, tumor metastasis was increased in mice deficient in aSMase when compared to WT [30], this was shown to be attributed to reduced accumulation of macrophages in the liver and decreased expression of tissue inhibitor of metalloproteinase 1 (TIMP1) around the tumor margin. This increase in metastasis was reversed in response to hepatocyte specific overexpression of aSMase. It has been shown that lysosomal ceramide produced by lysosomal aSMase was unable to escape from this compartment and it was implicated in mediating apoptosis in response to TNF-α, interferon-γ and CD95 through Cathepsin D [25], suggesting that localization of ceramide may affect its functions. These studies suggest that aSMase may have pro-inflammatory, and anti-metastatic potential, with further studies into the direct roles of aSMase in IBD and CRC needed.

Accumulating evidence suggests that alk-SMase may exert anticancer effects in colon carcinogenesis. Alk-SMase is an ectoenzyme bound to the surface of mucosal membrane [31], where upon cleavage it is released into the lumen and its activity enhanced, thus, catalyzing the generation of ceramide in the intestinal lumen. It has been reported that alk-SMase activity in human biopsies, and murine samples, collected from CRC and FAP was lower than those from adjacent or normal tissues [32, 33]. Furthermore, ceramide levels were decreased in human colon cancer samples compared to normal colon mucosa [34]. Dietary factors and drugs that have been implicated in the inhibition of colorectal inflammation and cancer have been shown to affect expression of alk-SMase [35–37], suggesting that alk-SMase may be involved in colon carcinogenesis. In fact, chronic colitis in humans was associated with a reduction of mucosal alk-SMase activity [38]. Moreover, in an *in vivo* model for colon cancer, namely 1,2-dimethylhydrazine (DMH)-induced carcinogenesis, alk-SMase activity was decreased in colon tissues upon tumor development. In this model dietary SM reversed the DMH-induced reduction of alk-SMase activity and induction of colon cancer [39]. In another line of research, intrarectal instillation of Alk-SMase showed anti-inflammatory effects through inhibition of TNF-α expression [40]. To support the result, recent study demonstrated that alk-SMase deficiency enhanced colonic tumorigenesis in azoxymethane (AOM)/DSS-induced inflammation-related colon carcinogenesis animal model [41]. Interestingly, analysis in human specimens showed that Alk-SMase is not directly linked to APC gene mutation [42]. Together these studies suggest that alk-SMase activity may play a protective role in the intestinal epithelium, especially in sporadic CRC.

Acid and alkaline SMase thus far have been suggested to play opposing roles in colon cancer, with very little study into the role of nSMase in IBD or CRC. Further studies are necessary into the roles these enzymes play in these disease states, which may be affected by the enzymes cellular localization, tissue distribution and ceramide production.

3 Modulation of Ceramide and Sphingosine: Ceramidases and Ceramide Synthases

3.1 Ceramidases (CDases)

CDases deacylate ceramide to form free fatty acids and sphingosine, and are classified based on their pH optima: Acid (aCDase), neutral (nCDase), or alkaline (ACER) [43]. Neutral CDase is a type II integral membrane protein and is mostly highly expressed at the brush border of the intestinal epithelium [44, 45]. Furthermore, it can be cleaved from the membrane to function as an ectoenzyme. Mice deficient for nCDase, are normal and viable; however, these mice exhibit impaired degradation of dietary sphingolipids in the colon resulting in increased ceramide and decreased sphingosine in the jejunum [46]. These data, together with evidence that psyllium a water-soluble dietary fiber linked to colon cancer, decreased both nCDase activity and induction of alk-SMase [37], suggesting that nCDase may also play key roles in IBD and CRC. A recent study by our group indicated that indeed nCDase deficiency profoundly affects the local and systemic inflammatory responses in DSS-induced colitis, a mouse model of IBD [47]. This study revealed that DSS increased nCDase activity in the colon epithelium, but not in the sub-mucosa. DSS treatment in both wild type (WT) and nCDase$^{-/-}$ mice increased ceramide levels in colon epithelium, whereas S1P levels and COX2 were significantly elevated only in the epithelium of nCDase$^{-/-}$ mice. COX2 is a key player in colon inflammation and cancer [discussed in sphingosine kinases (SphKs) and S1P], and our labs and others have implicated SK1 as a critical upstream mediator of COX2 expression in colitis and CAC. nCDase deficiency also altered the systemic inflammatory response as evidenced by increased circulating white blood cells, specifically neutrophils and lymphocytes.

The roles of aCDase and ACERs in IBD and CRC have yet to be elucidated. The results obtained from nCDase deficient mice suggest that this enzyme may play a protective role in regulating inflammation. Further studies are necessary to reveal the specific roles for acid, neutral, and alkaline CDases in colon inflammation and carcinogenesis.

3.2 Ceramide Synthases (CerS)

CerSs catalyze the formation of dihydroceramide or ceramide from dihydrosphingosine or sphingosine, respectively. To date, six individual CerSs have been identified (CerS1–CerS6) in humans or mice. Importantly, these CerSs show distinct preferences for fatty acyl-CoA substrates and, therefore, generate distinct ceramide species with unique *N*-linked fatty acids [43, 48, 49]. Although roles of CerSs in colon carcinogenesis *in vivo* have not yet been reported, recent studies employing *in vitro* models have suggested that CerSs may play important roles in colon carcinogenesis. It was shown that the death receptor ligand TRAIL (tumor necrosis factor-related apoptosis-inducing ligand) induced an increase in intracellular C_{16}-ceramide in TRAIL-sensitive SW480, but not in TRAIL-resistant SW620 cells [50]. Resistance in SW620 cells was reversed by the addition of exogenous ceramide, suggesting that defective ceramide signaling contributes to TRAIL resistance. White-Gilbertson et al. found that CerS6 expression was decreased in TRAIL-resistant SW620 cells when compared with TRAIL-sensitive SW480 [51]. CerS6 down-regulation using siRNA induced TRAIL resistance in SW480 cells, while overexpression of CerS6 rescued TRAIL sensitivity in SW620 cells. These results suggest that defective ceramide signaling may contribute to TRAIL resistance in colon cancer cells. Additional studies by Schiffmann et al. demonstrated that celecoxib (COX2 selective inhibitor) selectively activated CerS6 and modulated sphingolipid synthesis in HCT116 human colon carcinoma cells [52, 53], suggesting that CerS6 upregulation may partly contribute to the anti-proliferative effects of celecoxib. Furthermore, overexpression of CerS4 and CerS6 in HCT116 cells increased

C_{16}-, C_{18}-, and C_{20}-ceramides, and led to the inhibition of cell proliferation and induction of apoptosis [54]. Although overexpression of CerS2 had no effect on ceramide production, combination of CerS2 overexpression and long chain acyl-CoA increased long chain ceramide production ($C_{24:0}$- and $C_{24:1}$-ceramides), as well as colony formation in HCT-116 cells.

These studies suggest that distinct ceramide species generated by specific CerSs may play roles in colon carcinogenesis. Several of the CerS deficient mice have been generated and further investigations employing these animal models are needed to define the role of CerS *in vivo* in colon cancer.

4 Modulation of Sphingosine and Sphingosine-1-Phosphate: Kinases, Lyase and Phosphatases, Sphingosine-1-Phosphate Receptors

The balance of sphingosine and S1P are modulated by SphKs, S1P lyase and sphingosine phosphatases [21, 43]. Briefly, sphingosine is phosphorylated to form S1P by SphKs and S1P is hydrolyzed to reversibly form sphingosine by sphingosine phosphatases. S1P lyase can irreversibly cleave S1P to generate non-sphingolipids, ethanolamine phosphate, and hexadecenal as the only exit pathway from the sphingolipid pathways.

4.1 Sphingosine Kinases (SphKs)

SphK activity is a critical regulator of cellular functions as it phosphorylates sphingosine to produce S1P a potent bioactive lipid. Two isotypes of SphK have been characterized, SphK1 and SphK2 [55]. Previous studies using mice demonstrated that SphK1 enzyme activity is detected in most tissues including colon [56]. The first report on a role for SphK1 in colon cancer was published in 2006 [57, 58]. Kawamori et al. demonstrated that SphK1 is upregulated when compared to normal mucosa in rat colon adenocarcinoma induced by AOM. In this study SphK1/S1P was suggested to regulate COX2 expression and prostaglandin E_2 (PGE_2) production [57]. Increased expression of SphK1 has also been associated with decreased survival in human patients with colon cancer [59]. In the $Apc^{Min/+}$ mouse model for intestinal tumorigenesis SphK1 deficiency drastically reduced adenoma size, but not incidence [58]. Thereafter, the direct involvement of SphK1 in acute colitis [60] and colon carcinogenesis [61] was revealed using $SphK1^{-/-}$ mice. These mice are viable, with no observable phenotype other than a 50 % reduction in circulating S1P [62]. In an acute colitis model, $SphK1^{-/-}$ mice demonstrated decreased pathology, systemic inflammation, and COX2 expression in colon tissues after DSS treatment [60]. Deficiency of SphK1 significantly inhibited the formation of precancerous lesions, aberrant crypt foci, induced by AOM, as well as CAC [61]. Mice deficient in SphK1 exhibited increased apoptosis in colon tumors, thus consolidating a role for this pathway in regulation of apoptosis *in vivo*. Moreover, these mice showed less induction of COX2 expression in colon tumors as compared to WT mice. These studies also suggest that the COX2/PGE_2 pathway implicated in colon cancer is downstream of SphK1 [63–65].

Two major pathways linking inflammation to cancer are nuclear factor kappa B (NF-κB) and STAT3 [66]. Our studies and others have implicated SphK1 in NF-κB-mediated responses, and other recent studies have implicated STAT3 activation downstream of SphK1 and S1P receptor 1 (S1PR1) [67]. The activation of the S1P/S1PR1 pathway resulted in increased STAT3 phosphorylation and CAC in the AOM/DSS model. Both the STAT3 phosphorylation and tumorigenesis were blunted by FTY720. These results suggest that the SphK1/S1P/S1PR1 pathway may play key roles in chronic inflammation and CAC by regulating NF-κB/IL-6/STAT3 and implicate this pathway.

Studies in colon cancer cell lines have also suggested critical roles for SphKs and S1P in colon cancer. SphK1 overexpression increased proliferation and invasion, via MAPK signaling and MMP2/9 in LOVO colon cancer cells [68]. Liu et al. also demonstrated that overexpression of SphK1 increased invasion via FAK and that this was inhibited by DMS (non-selective SphK inhibitor) [69]. Inhibition of SphK1 with SKI-ii or silencing with siRNA also inhibited cell growth and increased ROS formation in response to doxorubicin in HCT116 colon cancer cells [70]. Expression of SphKs has also been implicated in responses to chemotherapeutics, where RKO, LOVO, and HCT116 human colon cancer cells increased ceramide generation in response to oxaliplatin treatment. This increase in ceramide was augmented by treatment with the SphK inhibitor SKI, which also augmented apoptosis induced by oxaliplatin in these cells [71]. These studies suggest that SphK1 may play a critical role in colon cancer cell proliferation and invasion in response to chemotherapeutics.

An interesting and current focus in colon cancer, and cancer in general, is the role of cancer stem cells. These cells are thought to reside mostly in a quiescent state; however, they remain pluripotent and have a potential to induce both tumor recurrence and metastasis. In RKO and HCT116 human colon cancer cell lines SphK1 has been shown to play a role in up-regulation of CD44, a marker for chemotherapeutic resistance and cancer stem cells [72]. Up-regulation of CD44 in these cells was dependent on ERK signaling and downstream of $S1PR_2$. This study suggests that sphingolipids and their metabolic enzymes, especially, the SK1/S1P pathway, may play a role in cancer stem cells and serve as potential therapeutic targets for preventing colon cancer recurrence and metastasis. .

Several studies have indicated critical roles for SphK1 and S1P in colon inflammation and carcinogenesis. Our recent study demonstrated that SphK1 deficiency did not exacerbate angiotensin II-induced acute hypertension [73], suggesting that SphK1 inhibition may be a promising strategy for cancer treatment or chemoprevention with lack of the adverse cardiovascular side effects. Future studies should be geared at development of bioavailable specific inhibitors for SphK1.

The role of SphK2 in IBD, CAC and CRC has also been examined with conflicting results. Studies using the small molecule SphK2 selective inhibitor ABC294640 have demonstrated that inhibition of SphK2 partially reduced the disease severity by regulating cytokine induction and pathology in both acute and chronic colitis models [74]. Similarly, ABC294640 has been shown to decrease TNBS-induced colitis [75]. This compound has also been utilized in a CAC model where administration of ABC29460 resulted in a dose-dependent decrease in tumor incidence [76]. These results are in contrast to those from the studies using genetically modified mice. This discrepancy could be due to some of the off-target effects of this inhibitor as it has been shown to bind and inhibit the estrogen receptor [77]. This could also be due to SK1 up-regulation or compensation by Sphk2 knockout but not its inhibitor, as demonstrated by Liang et al. [67]. In these studies mice deficient in SphK2, which are viable and demonstrate an increase in circulating S1P [78], exhibited more severe colitis upon DSS administration and more tumor burden upon AOM/DSS treatment.

Only a few studies have examined the role of SphK2 in colon cancer cell lines. In HCT116 colon cancer cells overexpression of Sphk2 inhibited sodium butyrate (NaBT)-induced apoptosis, whereas siRNA for SphK2 sensitized these cells to this apoptotic inducer [79]. Similar results were obtained in response to doxorubicin in HCT116 [80]. The studies geared at determining the role of Sphk2 in colon cancer cells have begun to suggest that this enzyme may play a role in response and resistance to chemotherapeutics.

4.2 Sphingosine-1-Phosphate Lyase and Phosphatases (SPL and SPPs)

SPL irreversibly degrades S1P to hexadecenal and ethanolamine-phosphate and has been suggested to be a key regulator of S1P levels [81]. Oskouian et al. determined the mechanism by which SPL expression enhances cell death responses [82, 83], using chemical inhibitors and dominant negative constructs. These studies demonstrated that SPL requires the activities of both p53 and p38 MAPK to promote apoptosis in HEK293 cells. In patient samples SPL expression was significantly lower in human colon cancer tissues than in normal adjacent tissues [82]. In addition, SPL expression and activity in polyps from $Apc^{Min/+}$ mice were lower than normal intestine [82]. $SPL^{-/-}$ mice demonstrate significant increases in circulating S1P and neutrophils, and only survive up to 15 weeks post-partum [84]. Additionally, the structure of the intestinal epithelium is disrupted in these mice. A recent study demonstrated that mice deficient in SPL in intestinal epithelial cells have more severe CAC, elevated cytokines, elevated S1P, STAT3 phosphorylation and higher expression of STAT3-activated microRNAs (miRNAs) compared to WT mice [85].

Lipid phosphate phosphatases that dephosphorylate S1P, also known as SPPs, have not been widely studied in colon cancer. However, cellular studies in SW480 human colon cancer cells, demonstrated that knockdown of SPPs increased cell growth through stabilization of β-catenin [86]. These studies on S1P degrading enzymes suggest that S1P catabolism may be altered in colon cancer and that S1P may play a key role in colon cancer development and progression (Table 1).

Table 1 Sphingosine-1-phosphate in intestinal inflammation and cancer

Enzyme or receptor	Disease state	Experimental model	Results	References
SPL	Colon Cancer	(a) WT or catalytically dead mutant SPL overexpression	(a) SPL expression in HEK293 promoted apoptosis via p53 and p38-dependent pathway; catalytic domain of SPL was required for apoptosis	[82]
		(b) Gene expression in human colorectal carcinoma (CRC) tissue	(b) Gene expression of SPL was down-regulated in human CRC compared to normal adjacent tissue	
		(c) $APC^{Min/-}$ mice and human samples	(c) SPL protein expression was diminished in intestinal tumorigenesis in $APC^{Min/-}$ mice and human CRC	
SPHK1	Colon Cancer	HCT116 and RKO colon cancer cells		[72]
		(a) SPHK1 inhibitor, 5c	(a) Inhibition of SPHK1 decreased expression of CD44, a cancer stem cell marker, resulting in increased susceptibility of cell lines to oxaliplatin	
		(b) SPHK1overexpression	(b) SPHK1 overexpression increased CD44 expression	
SPHK1	Colon Cancer	(a) AOM colon cancer model in rats	(a) SPHK1 and COX2 mRNA and protein were increased in colon cancer induced by AOM	[57]
		(b) HT29 colon cancer cells	(b) COX2 induction by pro-inflammatory cytokines is abolished in SPHK1 siRNA HT29 cells	
		(c) RIE rat intestinal epithelial cells	(c) SPHK1 overexpression in RIE cells increased COX-2 expression and PGE2 production	

(continued)

Table 1 (continued)

Enzyme or receptor	Disease state	Experimental model	Results	References
SPHK1	IBD	(a) Human samples	(a) Patients with UC had increased expression of SPHK1 and COX2	[60]
		(b) DSS-induced colitis in WT and SPHK1$^{-/-}$ Mice	(b) DSS-induced colitis increased SPHK1 and COX2 expression; SPHK1$^{-/-}$ mice were partially protected from DSS-induced damage and exhibited decreased COX2 expression when compared to WT	
SPHK1	Colon Cancer	(a) Human samples	(a) SPHK1 expression correlated with stage of colon cancer and SPHK1 expression is an independent predictor for mortality in colon cancer patients	[59]
		(b) Colon cancer cell lines	(b) SPHK1 inhibition and SPHK1 siRNA increased susceptibility of colon cancer cells to 5-FU	
SPHK1	Colitis and CAC	(a) DSS-induced colitis in WT and SPHK2$^{-/-}$ mice	(a) SPHK2$^{-/-}$ mice treated with DSS demonstrated increased SPHK1 expression and more severe colitis than WT	[67]
		(b) AOM/DSS CAC model in WT and SPHK2$^{-/-}$ mice	(b) S1PR1 modulator, FTY720, suppressed CAC in SPHK2$^{-/-}$ mice	
SPHK1	Colon Cancer	LOVO colon cancer cells		[61]
		(a) SPHK1 overexpression	(a) SPHK1 overexpression promoted cell proliferation and invasion, via activation of ERK and suppression of p38 MAPK pathways	
		(b) SPHK1 shRNA	(b) SPHK1 shRNA suppressed cell proliferation and invasion	

(continued)

Table 1 (continued)

Enzyme or receptor	Disease state	Experimental model	Results	References
SPHK1	Colon cancer	(a) Human samples	(a) SPHK1 expression was increased in primary colon cancer as compared to normal colon mucosa	[69]
		(b) LOVO colon cancer cells	(b) SPHK1 overexpression increased cancer cell viability and invasion via FAK, while SK inhibition with DMS reduced cell viability and invasiveness	
SPHK1	Colon cancer	HCT116 colon cancer cells		[70]
		(a) SPHK1 shRNA	(a) SPHK1 shRNA and SKI-II inhibited cell growth	
		(b) SK inhibitor, SKI-II	(b) SPHK1 inhibition increased intracellular ROS formation and DNA damage by doxorubicin	
SPHK2	Colon cancer	AOM/DSS CAC model in WT mice	ABC294640 treatment (SPHK2 inhibitor) reduced AOM/DSS colitis-induced epithelial damage and colon cancer incidence	[76]
SPHK2	Crohn's disease	Trinitrobenzene sulfonic acid (TNBS) CD model in rats	ABC294640 treatment protected from TNBS-induced colitis	[75]
SPHK2	Colitis	(a) Rat IEC6 and human endothelial cells (b) DSS-induced colitis in WT mice	ABC294640 inhibited PGE_2 production induced by TNFα, and protected from DSS-induced pro-inflammatory cytokines in the colon	[74]
SPHK2	Colon cancer	HCT116 colon cancer cells		[79]
		(a) SPHK2 overexpression	(a) SPHK2 overexpression inhibited sodium butyrate (NaBT)-induced apoptosis.	
		(b) SPHK2 siRNA	(b) SPHK2 siRNA sensitized colon cancer cells to NaBT-induced apoptosis	
SPHK2	Colon cancer	HCT116 colon cancer cells	SPHK2 siRNA sensitized HCT116 cells to doxorubicin, resulting in increased apoptosis.	[71]

(continued)

Table 1 (continued)

Enzyme or receptor	Disease state	Experimental model	Results	References
SPHKs	Colon cancer	RKO, LOVO, HCT116 colon cancer cells treated with oxaliplatin and SK inhibitor SKI	Oxaliplatin elevated ceramide production in cancer cells. Oxaliplatin and inhibition of SK with SKI resulted in increased ceramide production and augmented apoptosis	[80]
S1PR1	Colitis	$IL10^{-/-}$ mice treated with SEW2871, S1PR1 selective agonist	SEW2871 prevented experimental colitis in $IL10^{-/-}$ mice; reducing pro-inflammatory cytokine production, $CD4^{+}CD45^{+}$ T lymphocyte involvement and phosphorylation of STAT3	[94]
S1PRs	Colitis	DSS-induced colitis and T cell transfer-induced colitis in WT mice treated with FTY720, S1PR modulator	FTY720 treatment relieved disease activity and histologic damage in both colitis disease models	[97]
S1PR1	Colitis	(a) Human samples	(a) S1PR1 was expressed in colonic vasculature in UC patient samples	[91]
		(b) $S1PR1^{-/-}$ mice with DSS-induced colitis	(b) Inducible total body deletion of S1PR1 resulted in disruption of colonic vascular integrity and increased DSS-induced bleeding tendency. This was not affected with deletion of S1PR1 in bone marrow-derived cells	
S1PR1	Colitis	DSS-induced colitis in WT mice treated with W-061, an agonist for S1PR1	W-061 reduced colonic inflammation induced by DSS, by regulating lymphocyte recruitment in peripheral blood, payer's patch, and colonic lamina propria	[93]
S1PR1	Colitis	$IL10^{-/-}$ mice treated with KRP-203, S1PR agonist	KRP-203 protected from chronic colitis in $IL10^{-/-}$ mice, by reducing lymphocyte recruitment in colonic lamina propria and pro-inflammatory cytokine production	[95]

4.3 Sphingosine-1-Phosphate Receptors (S1PRs)

S1P exerts its significant effects on cellular signaling events such as proliferation, migration, survival, angiogenesis, inflammation, and lymphocyte egress by binding to one of five different G-protein coupled receptors known as S1PR1-5 [87]. S1PRs have a distinct profile of binding to different G-protein subunits, permitting S1P to exert different biological effects due to cell type-specific distribution of S1PRs [88]. Many of the functions and roles for S1PRs, and specifically S1PR1, have emerged with the development of several S1PR modulators and the immunosuppressant, FTY720 (FDA approved to treat multiple sclerosis and marketed as Gilyena™) [89]. S1PR1 regulation of basal and inflammation-induced vascular leak *in vivo* has been demonstrated in the lung [90] and in an animal model for ulcerative colitis [91]. Expression of S1PR1 expression was increased in animal models of CAC [67] and in human samples from colitis patients [91, 92].

Chemical agonists and antagonists for S1PR1 have also been utilized in animal models of colitis and colon cancer. Oral administration of the S1PR1 agonist, W-061, decreased DSS-induced colonic inflammation and T cell recruitment to the lamina propria [93]. Similarly, SEW2871, also an agonist for S1PR1, has been shown to be effective at preventing and abrogating experimental colitis in IL10$^{-/-}$ mice, reducing cytokines, T-cell involvement and activation of STAT3 (signal transducer and activator of transcription 3) [94]. KRP-203, yet another S1PR1 agonist, has also been shown to provide protection from experimental model of colitis in the IL10$^{-/-}$ mouse model [95]. FTY720 was used in many studies that implicate S1PRs in intestinal inflammation and cancer. This immune modulator has been shown to reduce cytokines and pathological effects in oxazolone-induced colitis [96], and migration of CD4+ T cells in CD4+/CD62L+ T adaptive transfer model colitis [97]. In the TNBS (2,4,6-trinitrobenzenesulfonic acid) model, commonly used to mimic Crohn's Disease, FTY720 did not protect mice from developing the disease [98]. Decreased proliferation in colon cancer cell lines treated with FTY720 has been demonstrated [99]; and in an *in vivo* model for CAC, FTY720 decreased epithelial cell proliferation and STAT3 activation [67]. Though many of the studies defining the role of S1PRs in the colon have focused on inflammation and suggest that S1PRs may be a potential therapeutic targets for colitis, CAC and CRC, careful consideration will be needed to avoid untoward side effects (bradycardia and diarrhea) [100, 101].

5 Conclusion

Sphingolipids, specifically ceramide and S1P, are important messengers that modulate cell fate and may play crucial roles in colitis, CAC, and intestinal tumorigenesis. In the past decade numerous exciting studies employing chemical inhibitors and gene targeting approaches, such as the knockout and transgenic mice, as well as RNA interfering techniques have revealed roles for sphingolipid metabolic enzymes in colon carcinogenesis. These studies serve as foundations for proposed therapeutic potentials involving the sphingolipid pathway. As the sphingolipid pathway may be intertwined with other diverse signaling pathways, further studies examining the tumor microenvironment, such as 3D culture, co-culture, xenograft and carcinogenesis models, could identify and validate sphingolipid metabolic enzymes as targetable therapeutic modalities in IBD, CAC and CRC.

Acknowledgement This work was supported in part by National Institutes of Health grants R01CA124687 (T.K.) R01GM097741 (L.M.O.), P01CA097132 (L.M.O.) and VA MERIT Award (L.M.O.)

References

1. Ferlay J et al. GLOBOCAN 2012 v1.0, Cancer incidence and mortality worldwide: IARC CancerBase No. 11 [Internet]. 2013 [cited 2014 February 14]. http://globocan.iarc.fr
2. Stewart B, Wild C (2014) World Cancer Report 2014. International Agency for Research on Cancer. World Health Organization, Lyon, France
3. WHO Global Cancer Country Profiles. United States of America 2014 [cited 2014 February 16]. http://www.who.int/cancer/country-profiles/usa_en.pdf?ua=1
4. Miyoshi Y et al (1992) Germ-line mutations of the APC gene in 53 familial adenomatous polyposis patients. Proc Natl Acad Sci U S A 89(10):4452–4456
5. Brensinger JD et al (1998) Variable phenotype of familial adenomatous polyposis in pedigrees with 3′ mutation in the APC gene. Gut 43(4):548–552
6. Johns LE, Houlston RS (2001) A systematic review and meta-analysis of familial colorectal cancer risk. Am J Gastroenterol 96(10):2992–3003
7. Steinke V et al (2013) Hereditary nonpolyposis colorectal cancer (HNPCC)/Lynch syndrome. Dtsch Arztebl Int 110(3):32–38
8. Fearon ER, Vogelstein B (1990) A genetic model for colorectal tumorigenesis. Cell 61(5):759–767
9. Rajagopalan H et al (2002) Tumorigenesis: RAF/RAS oncogenes and mismatch-repair status. Nature 418(6901):934
10. Samowitz WS et al (2005) Poor survival associated with the BRAF V600E mutation in microsatellite-stable colon cancers. Cancer Res 65(14):6063–6069
11. Samuels Y et al (2004) High frequency of mutations of the PIK3CA gene in human cancers. Science 304(5670):554
12. Velho S et al (2005) The prevalence of PIK3CA mutations in gastric and colon cancer. Eur J Cancer 41(11):1649–1654
13. Velho S et al (2008) BRAF, KRAS and PIK3CA mutations in colorectal serrated polyps and cancer: primary or secondary genetic events in colorectal carcinogenesis? BMC Cancer 8:255
14. Jonjic N et al (1997) Epidermal growth factor-receptor expression correlates with tumor cell proliferation and prognosis in gastric cancer. Anticancer Res 17(5B):3883–3888
15. Gross ME et al (1991) Cellular growth response to epidermal growth factor in colon carcinoma cells with an amplified epidermal growth factor receptor derived from a familial adenomatous polyposis patient. Cancer Res 51(5):1452–1459
16. Janakiram NB, Rao CV (2014) The role of inflammation in colon cancer. Adv Exp Med Biol 816:25–52
17. Danese S (2008) Inflammatory bowel disease and inflammation-associated colon cancer: partners in crime. Curr Drug Targets 9(5):360

18. Setia S, Nehru B, Sanyal SN (2014) Activation of NF-kappaB: bridging the gap between inflammation and cancer in colitis-mediated colon carcinogenesis. Biomed Pharmacother 68(1):119–128
19. Pyne NJ, Pyne S (2013) Sphingosine 1-phosphate is a missing link between chronic inflammation and colon cancer. Cancer Cell 23(1):5–7
20. Hannun YA, Obeid LM (2008) Principles of bioactive lipid signalling: lessons from sphingolipids. Nat Rev Mol Cell Biol 9(2):139–150
21. Furuya H, Shimizu Y, Kawamori T (2011) Sphingolipids in cancer. Cancer Metastasis Rev 30(3-4):567–576
22. Garcia-Barros M et al (2014) Sphingolipids in colon cancer. Biochim Biophys Acta 1841(5):773–782
23. Nagahashi M et al (2014) Sphingosine-1-phosphate in chronic intestinal inflammation and cancer. Adv Biol Regul 54C:112–120
24. Heffernan-Stroud LA, Obeid LM (2013) Sphingosine kinase 1 in cancer. Adv Cancer Res 117:201–235
25. Marchesini N, Hannun YA (2004) Acid and neutral sphingomyelinases: roles and mechanisms of regulation. Biochem Cell Biol 82(1):27–44
26. Zhang P, Cheng Y, Duan RD (2013) Ursolic acid inhibits acid sphingomyelinase in intestinal cells. Phytother Res 27(2):173–178
27. Cheng Y et al (2007) Curcumin decreases acid sphingomyelinase activity in colon cancer Caco-2 cells. Planta Med 73(8):725–730
28. Bauer J et al (2009) Lipid alterations in experimental murine colitis: role of ceramide and imipramine for matrix metalloproteinase-1 expression. PLoS One 4(9), e7197
29. Sakata A et al (2007) Acid sphingomyelinase inhibition suppresses lipopolysaccharide-mediated release of inflammatory cytokines from macrophages and protects against disease pathology in dextran sulphate sodium-induced colitis in mice. Immunology 122(1):54–64
30. Osawa Y et al (2013) Liver acid sphingomyelinase inhibits growth of metastatic colon cancer. J Clin Invest 123(2):834–843
31. Duan RD et al (2003) Identification of human intestinal alkaline sphingomyelinase as a novel ecto-enzyme related to the nucleotide phosphodiesterase family. J Biol Chem 278(40): 38528–38536
32. Hertervig E et al (1997) Alkaline sphingomyelinase activity is decreased in human colorectal carcinoma. Cancer 79(3):448–453
33. Hertervig E et al (1999) Familial adenomatous polyposis is associated with a marked decrease in alkaline sphingomyelinase activity: a key factor to the unrestrained cell proliferation? Br J Cancer 81(2):232–236
34. Selzner M et al (2001) Induction of apoptotic cell death and prevention of tumor growth by ceramide analogues in metastatic human colon cancer. Cancer Res 61(3):1233–1240
35. Cheng Y et al (1999) Ursodeoxycholic acid increases the activities of alkaline sphingomyelinase and caspase-3 in the rat colon. Scand J Gastroenterol 34(9):915–920
36. Duan RD et al (1998) Effects of ursodeoxycholate and other bile salts on levels of rat intestinal alkaline sphingomyelinase: a potential implication in tumorigenesis. Dig Dis Sci 43(1): 26–32
37. Cheng Y, Ohlsson L, Duan RD (2004) Psyllium and fat in diets differentially affect the activities and expressions of colonic sphingomyelinases and caspase in mice. Br J Nutr 91(5): 715–723
38. Sjoqvist U et al (2002) Chronic colitis is associated with a reduction of mucosal alkaline sphingomyelinase activity. Inflamm Bowel Dis 8(4):258–263
39. Zhang P et al (2008) Dietary sphingomyelin inhibits colonic tumorigenesis with an up-regulation of alkaline sphingomyelinase expression in ICR mice. Anticancer Res 28(6A): 3631–3635
40. Andersson D et al (2009) Expression of alkaline sphingomyelinase in yeast cells and anti-inflammatory effects of the expressed enzyme in a rat colitis model. Dig Dis Sci 54(7): 1440–1448

41. Chen Y et al (2015) Enhanced colonic tumorigenesis in alkaline sphingomyelinase (NPP7) knockout mice. Mol Cancer Ther 14(1):259–267
42. Hertervig E et al (2003) Reduction in alkaline sphingomyelinase in colorectal tumorigenesis is not related to the APC gene mutation. Int J Colorectal Dis 18(4):309–313
43. Hannun YA, Obeid LM (2011) Many ceramides. J Biol Chem 286(32):27855–27862
44. Duan RD et al (2001) Evidence for specific ceramidase present in the intestinal contents of rats and humans. Lipids 36(8):807–812
45. Lundgren P, Nilsson A, Duan RD (2001) Distribution and properties of neutral ceramidase activity in rat intestinal tract. Dig Dis Sci 46(4):765–772
46. Kono M et al (2006) Neutral ceramidase encoded by the Asah2 gene is essential for the intestinal degradation of sphingolipids. J Biol Chem 281(11):7324–7331
47. Snider AJ et al (2012) Loss of neutral ceramidase increases inflammation in a mouse model of inflammatory bowel disease. Prostaglandins Other Lipid Mediat 99(3-4):124–130
48. Pewzner-Jung Y, Ben-Dor S, Futerman AH (2006) When do Lasses (longevity assurance genes) become CerS (ceramide synthases)?: insights into the regulation of ceramide synthesis. J Biol Chem 281(35):25001–25005
49. Mullen TD, Hannun YA, Obeid LM (2012) Ceramide synthases at the centre of sphingolipid metabolism and biology. Biochem J 441(3):789–802
50. Voelkel-Johnson C, Hannun YA, El-Zawahry A (2005) Resistance to TRAIL is associated with defects in ceramide signaling that can be overcome by exogenous C6-ceramide without requiring down-regulation of cellular FLICE inhibitory protein. Mol Cancer Ther 4(9): 1320–1327
51. White-Gilbertson S et al (2009) Ceramide synthase 6 modulates TRAIL sensitivity and nuclear translocation of active caspase-3 in colon cancer cells. Oncogene 28(8):1132–1141
52. Schiffmann S et al (2009) The selective COX-2 inhibitor celecoxib modulates sphingolipid synthesis. J Lipid Res 50(1):32–40
53. Schiffmann S et al (2010) Activation of ceramide synthase 6 by celecoxib leads to a selective induction of C16:0-ceramide. Biochem Pharmacol 80(11):1632–1640
54. Hartmann D et al (2012) Long chain ceramides and very long chain ceramides have opposite effects on human breast and colon cancer cell growth. Int J Biochem Cell Biol 44(4): 620–628
55. Liu H et al (2002) Sphingosine kinases: a novel family of lipid kinases. Prog Nucleic Acid Res Mol Biol 71:493–511
56. Fukuda Y, Kihara A, Igarashi Y (2003) Distribution of sphingosine kinase activity in mouse tissues: contribution of SPHK1. Biochem Biophys Res Commun 309(1):155–160
57. Kawamori T et al (2006) Sphingosine kinase 1 is up-regulated in colon carcinogenesis. FASEB J 20(2):386–388
58. Michaud J et al (2006) Normal acute and chronic inflammatory responses in sphingosine kinase 1 knockout mice. FEBS Lett 580(19):4607–4612
59. Tan SS et al (2014) Sphingosine kinase 1 promotes malignant progression in colon cancer and independently predicts survival of patients with colon cancer by competing risk approach in South Asian population. Clin Transl Gastroenterol 5, e51
60. Snider AJ et al (2009) A role for sphingosine kinase 1 in dextran sulfate sodium-induced colitis. FASEB J 23(1):143–152
61. Kawamori T et al (2009) Role for sphingosine kinase 1 in colon carcinogenesis. FASEB J 23(2):405–414
62. Allende ML et al (2004) Mice deficient in sphingosine kinase 1 are rendered lymphopenic by FTY720. J Biol Chem 279(50):52487–52492
63. Moreira L, Castells A (2011) Cyclooxygenase as a target for colorectal cancer chemoprevention. Curr Drug Targets 12(13):1888–1894
64. Koehne CH, Dubois RN (2004) COX-2 inhibition and colorectal cancer. Semin Oncol 31(2 Suppl 7):12–21
65. Marnett LJ, DuBois RN (2002) COX-2: a target for colon cancer prevention. Annu Rev Pharmacol Toxicol 42:55–80

66. Fan Y, Mao R, Yang J (2013) NF-kappaB and STAT3 signaling pathways collaboratively link inflammation to cancer. Protein Cell 4(3):176–185
67. Liang J et al (2013) Sphingosine-1-phosphate links persistent STAT3 activation, chronic intestinal inflammation, and development of colitis-associated cancer. Cancer Cell 23(1): 107–120
68. Liu SQ et al (2012) Sphingosine kinase 1 enhances colon cancer cell proliferation and invasion by upregulating the production of MMP-2/9 and uPA via MAPK pathways. Int J Colorectal Dis 27(12):1569–1578
69. Liu SQ et al (2013) Sphingosine kinase 1 promotes tumor progression and confers malignancy phenotypes of colon cancer by regulating the focal adhesion kinase pathway and adhesion molecules. Int J Oncol 42(2):617–626
70. Huwiler A et al (2011) Loss of sphingosine kinase-1 in carcinoma cells increases formation of reactive oxygen species and sensitivity to doxorubicin-induced DNA damage. Br J Pharmacol 162(2):532–543
71. Nemoto S et al (2009) Sphingosine kinase isoforms regulate oxaliplatin sensitivity of human colon cancer cells through ceramide accumulation and Akt activation. J Biol Chem 284(16):10422–10432
72. Kawahara S et al (2013) Sphingosine kinase 1 plays a role in the upregulation of CD44 expression through extracellular signal-regulated kinase signaling in human colon cancer cells. Anticancer Drugs 24(5):473–483
73. Furuya H et al (2013) Effect of sphingosine kinase 1 inhibition on blood pressure. FASEB J 27(2):656–664
74. Maines LW et al (2008) Suppression of ulcerative colitis in mice by orally available inhibitors of sphingosine kinase. Dig Dis Sci 53(4):997–1012
75. Maines LW et al (2013) Efficacy of a novel sphingosine kinase inhibitor in experimental Crohn's disease. Inflammopharmacology 18(2):73–85
76. Chumanevich AA et al (2010) Suppression of colitis-driven colon cancer in mice by a novel small molecule inhibitor of sphingosine kinase. Carcinogenesis 31(10):1787–1793
77. Antoon JW et al (2010) Antiestrogenic effects of the novel sphingosine kinase-2 inhibitor ABC294640. Endocrinology 151(11):5124–5135
78. Kharel Y et al (2012) Sphingosine kinase type 2 inhibition elevates circulating sphingosine 1-phosphate. Biochem J 447(1):149–157
79. Xiao M, Liu Y, Zou F (2012) Sensitization of human colon cancer cells to sodium butyrate-induced apoptosis by modulation of sphingosine kinase 2 and protein kinase D. Exp Cell Res 318(1):43–52
80. Sankala HM et al (2007) Involvement of sphingosine kinase 2 in p53-independent induction of p21 by the chemotherapeutic drug doxorubicin. Cancer Res 67(21):10466–10474
81. Reiss U et al (2004) Sphingosine-phosphate lyase enhances stress-induced ceramide generation and apoptosis. J Biol Chem 279(2):1281–1290
82. Oskouian B et al (2006) Sphingosine-1-phosphate lyase potentiates apoptosis via p53- and p38-dependent pathways and is down-regulated in colon cancer. Proc Natl Acad Sci U S A 103(46):17384–17389
83. Oskouian B, Saba J (2007) Sphingosine-1-phosphate metabolism and intestinal tumorigenesis: lipid signaling strikes again. Cell Cycle 6(5):522–527
84. Allende ML et al (2011) Sphingosine-1-phosphate lyase deficiency produces a pro-inflammatory response while impairing neutrophil trafficking. J Biol Chem 286(9): 7348–7358
85. Degagne E et al (2014) Sphingosine-1-phosphate lyase downregulation promotes colon carcinogenesis through STAT3-activated microRNAs. J Clin Invest 124(12):5368–5384
86. Chatterjee I et al (2011) Lipid phosphate phosphatase-3 regulates tumor growth via beta-catenin and CYCLIN-D1 signaling. Mol Cancer 10:51
87. Spiegel S, Milstien S (2003) Sphingosine-1-phosphate: an enigmatic signalling lipid. Nat Rev Mol Cell Biol 4(5):397–407

88. Taha TA, Argraves KM, Obeid LM (2004) Sphingosine-1-phosphate receptors: receptor specificity versus functional redundancy. Biochim Biophys Acta 1682(1-3):48–55
89. Brinkmann V (2009) FTY720 (fingolimod) in multiple sclerosis: therapeutic effects in the immune and the central nervous system. Br J Pharmacol 158(5):1173–1182
90. Camerer E et al (2009) Sphingosine-1-phosphate in the plasma compartment regulates basal and inflammation-induced vascular leak in mice. J Clin Invest 119(7):1871–1879
91. Montrose DC et al (2013) S1P(1) localizes to the colonic vasculature in ulcerative colitis and maintains blood vessel integrity. J Lipid Res 54(3):843–851
92. Nguyen AV et al (2013) STAT3 in epithelial cells regulates inflammation and tumor progression to malignant state in colon. Neoplasia 15(9):998–1008
93. Sanada Y et al (2011) Therapeutic effects of novel sphingosine-1-phosphate receptor agonist W-061 in murine DSS colitis. PLoS One 6(9), e23933
94. Dong J et al (2014) Oral treatment with SEW2871, a sphingosine-1-phosphate type 1 receptor agonist, ameliorates experimental colitis in Interleukin-10 gene deficient mice. Clin Exp Immunol 177(1):94–101
95. Song J et al (2008) A novel sphingosine 1-phosphate receptor agonist, 2-amino-2-propanediol hydrochloride (KRP-203), regulates chronic colitis in interleukin-10 gene-deficient mice. J Pharmacol Exp Ther 324(1):276–283
96. Daniel C et al (2007) FTY720 ameliorates oxazolone colitis in mice by directly affecting T helper type 2 functions. Mol Immunol 44(13):3305–3316
97. Deguchi Y et al (2006) The S1P receptor modulator FTY720 prevents the development of experimental colitis in mice. Oncol Rep 16(4):699–703
98. Radi ZA et al (2011) Pharmacologic evaluation of sulfasalazine, FTY720, and anti-IL-12/23p40 in a TNBS-induced Crohn's disease model. Dig Dis Sci 56(8):2283–2291
99. Nagaoka Y et al (2008) Effects of phosphorylation of immunomodulatory agent FTY720 (fingolimod) on antiproliferative activity against breast and colon cancer cells. Biol Pharm Bull 31(6):1177–1181
100. Budde K et al (2002) First human trial of FTY720, a novel immunomodulator, in stable renal transplant patients. J Am Soc Nephrol 13(4):1073–1083
101. Tedesco-Silva H et al (2004) FTY720, a novel immunomodulator: efficacy and safety results from the first phase 2A study in de novo renal transplantation. Transplantation 77(12): 1826–1833

Dietary Sphingolipids in Colon Cancer Prevention

Eva M. Schmelz, Hui Zhou, and Paul C. Roberts

Abstract Complex sphingolipids (sphingomyelin, cerebrosides, gangliosides) are natural components of our diet, amounting to an approximate intake of 0.3-0.4g per day. In the intestinal tract, they are digested to the bioactive metabolites ceramide and free sphingoid bases. These metabolites regulate proliferation, survival and cell death, alter gene expression levels and modulate functions such as angiogenesis, migration and invasion. They also impact local and systemic inflammation, affecting both epithelial and immune cells. Thus, they are involved in the regulation of functions that contribute to cancer cell promotion, progression and metastasis. Here we review evidence that dietary or orally administered sphingolipids are beneficial in the prevention of colon and other cancers in rodent models, and discuss the molecular mechanisms and cellular targets in the complex tumor microenvironment of the colon. We also investigate factors that contribute to a permissive tumor microenvironment such as inflammation, obesity and the intestinal microbiota as potential novel targets of dietary sphingolipids in the prevention of colon cancer.

Keywords Cancer • APC-ß-catenin • Gene expression • Inflammation • Microenvironment • Proliferation • Apoptosis • Transcription factors

---

E.M. Schmelz (✉)
Department of Human Nutrition, Foods and Exercise, Virginia Tech,
1981 Kraft Drive, ILSB 1011, Blacksburg, VA 24061, USA
e-mail: eschmelz@vt.edu

H. Zhou
Virginia Tech, Blacksburg, Virginia, USA

P.C. Roberts
Department of Biomedical Science and Pathobiology, Virginia Tech, Blacksburg, VA, USA

Y.A. Hannun et al. (eds.), *Bioactive Sphingolipids in Cancer Biology and Therapy*, DOI 10.1007/978-3-319-20750-6_8

1 Introduction

Colorectal cancer is the fourth leading cancer in the United States, with an estimated 136,830 new cases diagnosed (8.2 % of all new cancer cases), and more than 50,000 expected deaths in 2014 [1]. Similar to other types of cancer, the 5-year survival of localized disease is 90 % but is reduced to only 13 % in metastasized disease. As a result of better screening and detection techniques, prevention and treatment opportunities, colon cancer incidence and mortality has been continually decreasing (3.1 % and 2.8 % per year, respectively). However, colon cancer is prevalent among the older population with a median age at diagnosis of 74 years [1]. The most common risk factors include genetic predispositions, environmental factors such as carcinogens, and modifiable factors such as smoking, dietary patterns, obesity, and physical activity. A protective role for certain foods or specific components of the diet has been established while their therapeutic effectiveness is still controversially discussed. Here we will review what we currently understand of dietary sphingolipids in the prevention of colon cancer, and discuss potential targets of diet-derived sphingolipid metabolites in the colonic microenvironment.

2 Colon Cancer

Colon cancer is a complex, multi-stage process. The aberrant expression of genes—the result of mutations, deletions, translocations, chromosome losses etc.—leads to initiated and transformed epithelial cells. Upon clonal expansion of transformed cells, the first visible morphological changes in colon carcinogenesis are the aberrant crypt foci (ACF) [2]. While most of these putative preneoplastic lesions regress over time, those containing *Apc* (adenomatous polyposis coli) [3] or *Ctnnb1* [4] (encoding ß-catenin) mutations are more likely to progress to adenomas and carcinomas and, eventually, cells will disseminate from the primary tumor and grow secondary tumors at distant sites such as the liver [2]. The long latency of *Apc* requires more genetic alterations for progression with some involving loss of cell cycle and apoptosis control, or regulation of cell adhesion and migration. *Apc* mutations are found in all cases of familial adenomatous polyposis [5] and have also been identified as an early event in 70–80 % of sporadic cancers [6], apparent already at 30–50 years of age [7]. This provides a growth advantage for these cells and allows for clonal outgrowth and causes the dysregulation of cell turnover and differentiation. However, only *Apc* mutations (loss of function) in the intestinal stem cells located at the bottom of the colonic crypts but not differentiated colon epithelial cells may be able to drive tumorigenesis [8] and were identified as the cell(s) of origin for colonic adenomas [9]. De-differentiation of colonic epithelial cells and their acquisition of stem-like properties may be another contributor to tumorigenesis [10]. Mutations in *p53*, *Ras* or *Wnt* cannot initiate tumorigenesis by themselves but are acquired later; importantly, the sequence of their acquisition will

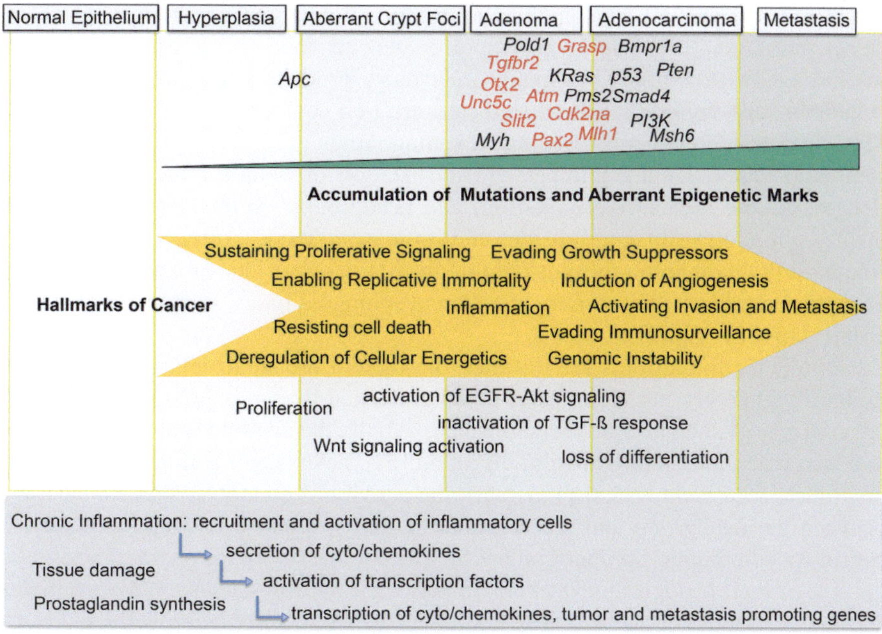

Fig. 1 Model for mutation- and inflammation-driven colon cancer development with select common mutations. Genes with changes in methylation status are in *red*

define the propensity of the tumor with early *Apc* mutations as critical determinants [8, 9]. The accumulation of more mutations provides additional advantages for the transformed cells and appears to be a constantly ongoing process; more than 70 alterations in protein-coding regions per colon tumor have been identified [7] (select genes are depicted in Fig. 1).

In addition to gene sequence-altering events, epigenetic alterations contribute to tumorigenesis. The constantly growing list of epigenetically silenced genes (hypermethylated in their promoter region) in colon cancer includes *MLH1* (mismatch repair), *Rb*, *p16*, *Rarb*, *Sfrp*, and *Cdk2na*, alterations that can either precede genetic aberrations or enhance the impact of oncogenes in tumor progression [11, 12]. In contrast to mutations, epigenetic changes are plastic and potentially reversible, and as such are targets for intervention efforts.

3 Sphingolipids in the Diet: Release of Bioactive Metabolites into the Intestinal Tract

Sphingolipids are only minor components of the human diet. The daily intake of complex sphingolipids in a Western diet has been estimated to be approximately 0.3–0.4 g/day [13]; this is similar to the Japanese diet that contains 128–292 mg cerebrosides and 45–81 mg sphingomyelin per day, mostly from cereals, vegetables, pulses and fruit [14]. The sphingolipid content of food varies greatly but milk (including low-fat milk) and many other dairy products, eggs, meat and some vegetables and legume (such as soybean) are rich sources of sphingolipids [13]. The structural components of complex sphingolipids in foods vary greatly in their sphingoid bases (length, number and localization of double bonds), fatty acids (length, double bonds, hydroxylation) and headgroups. Animal products contain mainly sphingomyelin, cerebrosides and some gangliosides, often with a sphingosine moiety. Plant sphingolipids contain mostly cerebrosides or glycosyl inositol phosphoceramides with a variety of different sphingoid bases, fatty acids and headgroups [15–18] (Fig. 2).

Complex sphingolipids are digested throughout the small intestine and colon by intestinal and bacterial enzymes to ceramide and free sphingoid bases [19–21]. All enzymes of sphingolipid metabolism, especially the alkaline sphingomyelinases that are critical for sphingomyelin digestion [22], have been identified in the intestinal tract (see recent reviews [23, 24]); the activity of endoglycosidases, cleaving between the headgroup and the ceramide moiety, may also be important for the hydrolysis of complex sphingolipids [25]. The complexity of the sphingolipids, i.e., the size of their headgroups, does not impact their hydrolysis since only ceramides and free sphingoid bases were detected after digestion of sphingomyelin and cerebrosides but not their partially hydrolyzed intermediates, suggesting either a rapid succession of cleavage events or a complete removal of the intact headgroup [25]. However, the structure of the released metabolites determined their intracellular fate.

Sources		**Structures**		
		Sphingoid bases	Fatty Acids	Headgroups
animal products	Milk and milk products meat, eggs fish	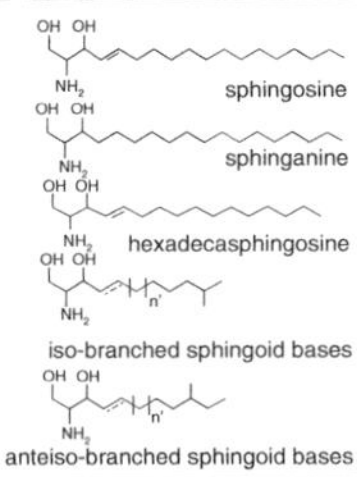 sphingosine sphinganine hexadecasphingosine iso-branched sphingoid bases anteiso-branched sphingoid bases	C<10->C26, saturated. sometimes α-hydoxylated, sometimes branched or unsaturated	glucose, galactose, mannose, sialic acid, glucosamine, galactosamine,pentoses, N-acetylated or sulfated, phosphorylcholine
plants	soy beans sweet potato pulses and nuts cereals spinach, cabbage, other vegetables and fruit (low amounts)	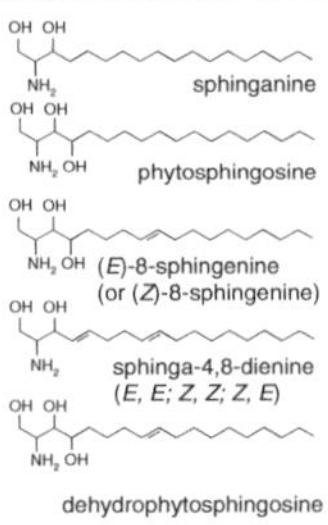 sphinganine phytosphingosine (*E*)-8-sphingenine (or (*Z*)-8-sphingenine) sphinga-4,8-dienine (*E, E; Z, Z; Z, E*) dehydrophytosphingosine	C<10->30, even or odd numbered, saturated, unsaturated, non-hydroxylated, α-hydroxylated,	glycosylceramides: glucose (most common), mannose, Glycosylinositolphosphoceramides: hexose(s), pentose(s), inositol, N-acetyl-, glucuronic acid, hexuronic acid, glucosamine, hexosamine

Fig. 2 Sources and structures of dietary sphingolipids

While there was no preference for specific ceramide species found in complex sphingolipids from milk for enzymatic hydrolysis, the ceramides resistant to further metabolism and therefore overrepresented in the intestinal extracts were ceramides with a C16:0 fatty acid while the major ceramide species found in milk sphingolipids (C22:0, C23:0, and C24:0) were mostly further metabolized [25]. Interestingly, the C16:0 ceramide has been associated with the induction of apoptosis [26], suggesting that the structure of diet-derived ceramides direct their specific metabolic fate and determines the sphingolipid profile in the intestinal cells and, thereby, directing the response of intestinal cells.

After the hydrolysis of ceramide, sphingosine is rapidly taken up by the intestinal cells and, after phosphorylation, degraded by sphingosine-1-phosphate lyase, or is recycled into complex sphingolipids [21]. Both sphingosine kinase 1 and sphingosine 1-phosphate lyase are highly expressed in the intestinal tract [27, 28]. Free sphingoid bases from maize cerebrosides, 4,8-sphingadienines (*cis* and *trans* isomers) were detected in both intestinal lumen and intestinal mucosa of rats after oral administration [29] and exhibited both a poor uptake and a low retention compared to sphingosine in Caco-2 cells. The levels of 4,8-sphingadienine in Caco-2 cells were not increased after inhibition of sphingosine kinase but increased by verapamil [29], suggesting that the small amounts absorbed are exported from the enterocytes rather than metabolized; this is supported by the observation that small amounts of *trans*-4,*cis*-8-sphingadienes appear in the lymph after oral administration of maize cerebrosides [30]. Therefore, the structure of the dietary sphingolipids affects their digestion, uptake, and intracellular fate and, thus, determines their function in the colon.

More than 80 % of dietary sphingolipids are digested in the small intestine [31]. Intact complex sphingolipids or ceramides are transported to the colon and either excreted [32] or hydrolyzed by the colonic microbiota to ceramide and sphingosine [21] which are rapidly taken up by the cells by passive diffusion [33]. The hydrolysis capacity of the intestinal tract is limited, and increased amounts of sphingolipids in the diet are not digested efficiently but excreted with the feces [32]. As a result, even high amounts of sphingolipids in the diet (1 % of the diet or 100 times more than average estimated consumption) are not releasing cytotoxic levels of bioactive metabolites into the intestinal tract [34] and thus far, deleterious side effects of dietary sphingolipids have not been reported. The generation of bioactive sphingolipid metabolites throughout the intestinal tract including the colon indicates a potential of complex sphingolipids in the diet to regulate cell processes in the colon that are involved in colon carcinogenesis.

4 Sphingolipids in the Diet: Impact on Colon Cancer

Systematic epidemiologic studies that directly correlate the sphingolipid intake to the risk of developing colon cancer are still not available due to the lack of comprehensive food analyses. While the impact of sphingolipid-rich dairy food consumption on the risk of developing breast and prostate cancer is still controversially discussed, a recent meta-analysis showed that the consumption of dairy products is associated with a reduced relative risk (RR) of developing colon cancer (RR 0.59). This was especially strong for the consumption of cheese, butter and whole milk: per two servings per day, the RR decreased by 13 % [35]. Other meta-analyses of 27 and 60 international studies also demonstrated an inverse association of total dairy consumption with a reduced risk (RR 0.83 and 0.78, respectively) [36, 37]. Here we will review rodent studies investigating the potential of specific sphingolipid classes to suppress colon cancer.

4.1 Sphingomyelin

Rodent models of chemically induced colon carcinogenesis are convenient tools to investigate early and late events in colon carcinogenesis in a time- and cost-effective manner. The carcinogens (azoxymethane, AOM, or dimethylhydrazine, DMH) are injected intraperitoneally, are metabolized by the liver and secreted into the intestinal tract via the bile to induce DNA damage, causing the formation of early stages of colon cancer—aberrant crypt foci (ACF)—that can progress to adenomas and adenocarcinomas. In the last decade, this model has been used to investigate the capacity of sphingolipids added to a semi-purified, essentially sphingolipid-free diet to reduce colon cancer in amounts that could be achieved in the human diet. Adding 0.025–0.1 % sphingomyelin from milk powder (by weight) to an AIN 76A diet *after* carcinogen treatment reduced the number and size of ACF after 4 weeks by 70 % and 30 %, respectively [38, 39]. This is not specific to the mouse model since similar results were observed in rats [40]; the differences in the magnitude of the response may be species related, or impacted by the administration of the sphingolipids as single bolus that may decrease hydrolysis and increase excretion. Synthetic sphingomyelin caused a similar reduction on ACF number and size [41], indicating that the observed effect of the natural isolate was indeed the result of the sphingomyelin and not of co-purified contaminants. A synthetic sphingomyelin with a sphinganine backbone was even more effective in suppressing ACF [41]. The metabolites generated from this sphingomyelin species are dihydroceramide and sphinganine; since dihydroceramide has been considered biologically inactive, these results suggested that the sphingoid bases may be the critical cancer-suppressing metabolites. However, both ceramides and sphingoid bases are generated in the intestinal tract [42] and can easily taken up by the intestinal cells; since more recent studies demonstrated the induction of autophagy by dihydroceramide [43], these results suggest that both ceramides and sphingoid bases may be effective in the suppression of colon cancer. The same suppression of ACF was observed when instead of the purified milk sphingomyelin the milk globule membrane was added to the diet, a complex of proteins and lipids that contains about 20 % of the polar lipids as sphingomyelin (0.11 % of the diet); this mixture also significantly reduced DMH-induced ACF in rats [44].

The reduction of early stages of colon cancer—ACF—results in the suppression of adenomas [39] and adenocarcinomas [25, 38, 45, 46] after approximately 40 weeks of feeding a diet supplemented with sphingomyelin, confirming the validity of ACF reduction as marker for sphingolipid efficacy. Sphingomyelin supplementation *after* carcinogen treatment suppressed the number of adenomas and carcinomas comparable to an administration *before* carcinogen treatment [46]. This indicates that the modulation of carcinogen metabolism or carcinogen-mediated cell transformation is not the mechanism of how dietary sphingomyelin inhibits colon cancer as has been shown for other natural compounds but is also effective when the initial damage to the colonic cells has already occurred and predisposed, transformed and pre-malignant cells are present in the colon. Therefore, the anti-cancer effect of dietary sphingomyelin is not restricted to the initiation stage. It is currently unknown if there are limits of its effectiveness determined by the disease stage, a specific geno- or phenotype of the tumor or its aggressiveness, and if in addition to a chemopreventive approach, the dietary or orally administered sphingomyelin also can exert suppressive effects in a chemotherapeutic approach against existing tumors. Nonetheless, the reduction of tumor incidence and size was independent of existing *p53* mutations [47]; if that is also true for other mutations commonly detected in colon cancer cells is not known. In none of these studies any deleterious side effects of the sphingomyelin supplements were detected, demonstrating that this is an effective and safe prevention regimen.

4.2 Ceramides, Cerebrosides and Gangliosides

Mammalian glucosylceramide, lactosylceramide, gangliosides GD3 (from milk) [25] or ganglioside GM1 (from brain) [38] added to the diet at 0.025 or 0.1 % (by weight) all have shown to significantly reduce the number of ACF in DMH-treated CF1 mice by 40–70 %. This is comparable to the effect of glucosylceramide from rice bran that contain a 4,8-sphingadiene backbone rather than the sphingosine moiety observed in the mammalian species; feeding rice bran glucosylceramide reduced the number and size of ACF in DMH-treated rats by more than 60 %. This was not limited to conventional ACF that may regress over time but was also demonstrated in ß-catenin-overexpressing ACF [48], considered to be a better prognostic marker for colon carcinogenesis. Glucosylceramide from soy also contains mostly a 4,8-sphingadienine backbone with an amid-bound α-hydroxy palmitic acid that may exert different responses in the colon cells. Adding 0.025 or 0.1 % soy glucosylceramide to the diet of CF1 mice treated with DMH reduced the resulting ACF by 38 % and 52 %, respectively [49]. These studies show that the origin of the sphingolipid class (mammalian vs. plant) and the differing structure of their sphingoid bases did not lessen their inherent anti-cancer effect. However, since more of the sphingadienines are exported from the colonic cells, it would be interesting to determine if a systemic effect or the targeting of cancers at distant sites would be more pronounced with these plant sphingolipids.

Humans are rarely exposed to DMH or its carcinogenic metabolites and the initial genetic alterations (DNA alkylation, generating N^7 and O^6-methylguanine that induce GC → AT transitions) may not accurately reflect the slow development of the human disease since ACF are already visible within five days of treatment [50]. However, the gene mutations found in carcinogen-induced rodent ACF (i.e., *Apc*, *Kras*) [51] are also found in human colon cancer. Thus, this is a quick and cost-effective model for the screening of anti-cancer agents. A rodent model that does not require chemical transformation, the C57/B6J$^{Min/+}$ mice (*m*ultiple *i*ntestinal *n*eoplasia, Min mice) mice carry an *Apc* mutation and spontaneously develop numerous adenomas in the small intestine but only few in the colon; these mice are often used as a rodent model for familial adenomatous polyposis. Min mice do not live long enough to allow for the progression of adenomas to carcinomas and a high incidence of colon tumors; this would also require treatment with a carcinogen. Feeding a mixture of milk sphingomyelin, glucosylceramide, lactosylceramide and ganglioside GD3 in a ratio that reflects their proportions in milk (65:7.5:20:7.5) at 0.1 % of the diet for four weeks significantly reduced the number of intestinal and colonic tumors in Min mice by 40 % [42]. Supplementation of the diet with 0.1 % ceramide (synthetic *N*-palmitoyl sphingosine) alone also reduced tumor formation by 40 % but adding ceramide to the sphingolipid mixture (at 40 % of the mixture, total 0.1 % of the diet) intended to increase the bioactive metabolite concentrations especially in the small intestine (where most of the tumors are located) reduced the tumor numbers significantly more to 50 % of controls [42]. Glucosylceramide from soy in amounts of 0.025–0.1 % of the diet that successfully suppressed ACF formation also reduced tumor formation in Min mice by 27 and 40 %, respectively, after 45 days of supplementation [49].

4.3 Free Sphingoid Bases

As mentioned above, the digestion of complex sphingolipids in the intestinal tract is incomplete, and increasing the amount of dietary sphingolipids results in an increased excretion of non- or only partially digested compounds. Using free sphingoid bases may bypass this loss and increase the anti-tumor effect of sphingolipids. A unique 4,6-sphingadienine from *Drosophila melanogaster* was administered per gavage to Min mice (25 mg/kg) daily for 10 days, beginning at 80 days of age. In contrast to other studies, this was a more therapeutic approach since the treatment began close to the natural end of lifespan of the Min mice when all adenomas had already been formed. This treatment reduced the number of adenomas by 35 % with a significant reduction of adenoma size seen only in small lesions [52]. Synthetic 4,6-sphingadienine exerted the same suppressive effects and, additionally, significantly increased the tumor-suppressing effect of rapamycin [53].

The hydrolysis of complex sphingolipids generates ceramides and free sphingoid bases in the intestinal tract that can modulate processes involved in colon carcinogenesis. However, colon cancer cells have shown to overexpress sphingosine kinase 1 [54] but have a low expression of sphingosine-1-phosphate lyases and phosphatases

[27] and, therefore, could generate and accumulate sphingosine-1-phosphate from the diet-derived free sphingoid bases that may exert a tumor-stimulatory rather than the desired suppressive effect. To prevent the generation of potentially pro-tumorigenic metabolites, enigmol (2S,3S,5S)-2-amino-3,5-dihydroxyoctadecane) was developed, a synthetic sphinganine analog that lacks the 1-hydroxyl group and therefore cannot be phosphorylated by sphingosine kinases; it is also a poor substrate for ceramide synthase [55]. This results in a higher toxicity towards colon cancer cell lines and significantly reduced intestinal tumors in Min mice by 37 and 52 % when added to the diet at 0.025 or 0.1 % for 45 days [56].

As shown above, the administration of complex sphingolipids via the diet in amounts that could be achieved in the human diet did not cause any toxic side effects even when added for a long period of time. In contrast, the oral administration of free sphingoid bases appears to have limited use due to their toxic side effects. Min mice fed enigmol at 0.1 % of the diet showed a lower weight gain, normal biomarkers for liver and kidney function but increased blood urea, serum protein and albumin levels [56]. Oral administration of same concentration of safingol, the L-threo stereoisomer of sphinganine, a sphingosine kinase inhibitor, caused severe liver and renal toxicity in rodents [57]; this was also observed in a human trial but appeared to be reversible after completion of the drug treatment [58]. Thus, the administration of free sphingoid bases at amounts of 0.1 % of the diet or more may provide limited value as anti-cancer agents due to their systemic toxic effects and a different route of administration may need to be considered.

5 Cancers of Distant Organs as Targets for Dietary Sphingolipids

Only small portions of the orally administered sphingolipids are absorbed and transported into the body [19–21] and do not cause an increase of circulating sphingosine levels [42]. However, a change in lipid composition or tissue function in brain [59], skin [60, 61], central nervous system [62], ear [63], and liver [64] indicates a systemic distribution and distal effects of diet-derived bioactive metabolites. Recent reports have identified cancers of different sites, e.g., liver, breast, head-neck, and prostate, as targets for orally administered sphingolipids. Dietary sphingomyelin (0.1 % of diet, from chicken egg yolk) fed for only two weeks to carcinogen-treated Sprague-Dawley rats reduced the number and size of pre-neoplastic lesions of the liver without any deleterious effect to non-transformed liver cells [65] suggesting that sphingolipids in sufficient levels reach the liver after ingestion and target specifically processes involved in the transformation of liver cells or directly affect transformed cells. Sphingomyelin from milk (0.1 % of the diet) was fed to nude mice after MCF10AT1 breast xenografts were palpable, a progression model for early stages of human breast cancer, representing a more chemotherapeutic approach. The xenografts in the sphingomyelin-fed mice were smaller (63.7 vs. 99.3 mm^3 in the controls, albeit not statistically significant) and their progression to more aggressive stages was significantly suppressed: most xenografts in the sphingomyelin-fed group were stage 2 (88 %) while the control xenografts had progressed to stage 3 (27 %), 4 (9 %) and 5 (14 %) [66]. Glucosylceramide from rice bran (300 mg/kg) was administered by daily gavage for 14 days *after* SCCKN xenografts, a model for human head-and-neck squamous carcinoma, reached a size of >7 mm (also a more therapeutic approach) and reduced the size of the lesions by 40 % without apparent side effect. Interestingly, administration of 6000 mg/kg bodyweight caused hemorrhagia and was lethal to the mice [67], suggesting that there are differences between the administration via the diet—which has been demonstrated to be safe- and administration as a bolus per gavage. Administration of enigmol per gavage (10 mg/kg bodyweight) twice weekly after lesions from the injected human prostate cancer cell line PC-3 were palpable reduced their size by 50 % without affecting the mice' health [56].

Together, these studies show that the administration of complex sphingolipids and, probably to a lesser extend due to their inherent toxicity, free sphingoid bases via the diet is a safe and effective measure to suppress cancer of the colon and several other organs. While this approach is highly effective against early stages of the disease, it is possible that this route of administration may not deliver sufficient concentrations of bioactive molecules to highly aggressive disease, or these cells may prove resistant to diet-derived metabolites. This needs to be investigated in greater detail.

6 Mechanisms of Colon Cancer Suppression by Dietary Sphingolipids

6.1 Regulation of Proliferation and Apoptosis

Hyperproliferation or unlimited proliferation is among the hallmarks of cancer that drive tumor development and progression (Fig. 1); its reversal may be critical to prevent tumor formation and metastasis. A constant increase in proliferation is already induced by the carcinogen treatment of rodents in the macroscopically normal appearing colonic mucosa and is accompanied by a decrease in the rate of apoptosis [46]. Consistently, all studies using dietary sphingolipids show a reduced proliferation of colonic crypt cells after sphingolipid treatment for short periods (4 weeks) or long-term (45 weeks) to the levels of the control group that was not treated with the carcinogen [39, 41, 42, 48, 49, 65], but not beyond [46]. A reduced rate of proliferation after sphingolipid treatment was also observed in breast [66] and head and neck cancer xenografts [67]. In addition to the sphingolipid-mediated reduction of aberrant proliferation, in most studies the rate of apoptosis was either not altered or increased by dietary sphingomyelin, but not beyond the levels of the untreated controls [46]. Only the administration of glucosylceramide per gavage that reduced tumor size by 50 % in 2 weeks caused an increase in central tumor necrosis and apoptosis [67]; however, if this treatment is comparable with long-term dietary administration regarding the concentrations of bioactive sphingolipids generated in the intestinal tract and the exposure time of the intestinal cells to these amounts remains to be determined. These studies suggest that the regulation of cell growth rather than the induction of apoptosis may be the key event in the prevention of colon cancer by orally administered sphingolipids. If this would eliminate aggressive tumors or delay their growth is not known.

6.2 APC-β-Catenin Signaling Pathways as Targets for Colon Cancer Suppression

As mentioned above, *Apc* mutations are evident in all patients with familial adenomatous polyposis and most patients with sporadic colon cancer and are considered the initiating event in colon carcinogenesis. The *Apc* gene product is a large, 310 kDa scaffolding protein with distinct binding domains for several proteins. As a crucial part of the canonical Wnt signaling pathway, APC functions as destruction complex for ß-catenin, a cell adhesion protein that connects E-cadherin to the actin cytoskeleton. APC also controls proliferation via regulation of spindle orientation and chromosome segregation, and cell migration in the colonic crypt. APC can shuttle from the cytosol to the nucleus, contributing to microtubule assembly and formation of the mitotic spindle, the association of spindle microtubule with the kinetochores of chromosomes, regulate centrosome positioning and may be involved in DNA replication and excision repair. Localization of truncated APC to the mitochondria—favored in colon cancer cells—allows for the binding of Bcl-2 and may contribute to cancer cells' resistance to apoptosis and survival (see recent review [68]).

The regulation of ß-catenin and its importance in colon carcinogenesis has been extensively studied in the last decade. ß-catenin functions as a signaling molecule in developmental systems [69, 70] and is released into the cytosol of cells responding to growth stimulation via the Wnt pathway or epidermal and hepatocyte growth factor receptor activation [71, 72]. APC binds axin to capture cytosolic ß-catenin and mediate its sequential phosphorylation by casein kinase 1 and GSK-3ß; this marks ß-catenin for ubiquitination by the SCF-ßTRCP complex and proteasomal degradation, keeping cytosolic concentrations of ß-catenin low in the absence of Wnt signaling. Binding of Wnt ligands to the frizzled and LPR5/6 receptor complex allows for the formation of the complex of dishevelled (Dvl) and the frizzled receptor, the subsequent recruitment of axin to the membrane and the dissolving of the APC/axin/GSK3ß degradation complex. Stabilized cytosolic ß-catenin accumulates in the cytosol, is translocated to the nucleus and binds to transcription factors of the LEF/TCF family, replacing the repressor protein groucho and forming a transcriptionally active complex that includes CBP/p300, Bcl9 and co-activators such as pygopus [73–75].

Apc mutations or inactivation via epigenetic silencing result in the same stabilization of ß-catenin with subsequent accumulation in the cytosol, translocation to the nucleus and activation of transcription factors to mediate transcription of genes associated with growth and altered adhesion and motility properties [76]. These include cyclin D_1, c-myc, c-jun, uPAR, fra-1, Tcf-1, MMP-7, Axin2, PPARδ, CD44, claudin, VEGF, survivin (for a more complete list of targets see http://www.stanford.edu/group/nusselab/cgi-bin/wnt/). Many of these proteins are supporting tumor growth and survival. The overexpression of conductin, a protein associated with the ß-catenin degradation complex, during mitosis is mediated by aberrant APC/ß-catenin signaling and induced chromosome instability by compromising the mitotic spindle checkpoint, causing spindle stress and a faster exit from mitosis [77]. Since the mutations in *Apc* occur early in life but colon cancer is a late event, it has been suggested that residual functions of APC such as control of mutant cells and their elimination remain to differing degrees, maintaining some adhesion and proliferation control until the function of the truncated APC is below 12 % of its capacity. This causes the reversal of the cell adhesion gradient, normally directed from the bottom of the colonic crypt to the top, effectively re-positioning highly proliferative cells towards the bottom of the colonic crypts and limiting their upward motility. The retention of these proliferating cells in the bottom of the colonic crypt drastically enhances the potential of these crypts for tumor formation. The accumulation of ß-catenin is a central event in these aberrant growth and motility patterns [78]. ß-catenin localized in the invasive front of primary tumors and metastases has been shown to play a role in the tumor cell migration and dissemination of metastases [79].

The regulation of ß-catenin metabolism and trafficking has been identified as a target of sphingolipids both *in vivo* and *in vitro*. In normal colon tissue, ß-catenin is associated with the cell membranes. Immunohistochemical analyses of the intestinal tissue from Min mice (all cells carry the *Apc* mutation) showed a high expression of ß-catenin in the cytosol of intestinal sections. However, Min mice that were fed sphingolipid supplements in their diet displayed mostly membrane-associated ß-catenin [42]. This was also observed in the macroscopically normal appearing colon of DMH-treated CF1 mice after sphingolipid treatment: the carcinogen treatment increased cytosolic ß-catenin which was reversed by the dietary sphingolipids; nuclear ß-catenin was only detected in colon carcinomas [80]. The comparison of colonic tissue from mice fed a control diet or sphingolipid supplements identified genes that respond to diet-derived bioactive lipid metabolites: there was a 50–60 % reduction in HIF1α expression, a 35–33 % reduction in the expression of its downstream target VEGF, and a 70–80 % reduction in TCF-4 expression [49], a transcription factor downstream in the Wnt/APC signaling pathway and a preferred ß-catenin binding partner in the colon to mediate ß-catenin-induced expression of genes important for proliferation, adhesion and motility. Surprisingly, sphingolipid treatment did not cause the differentially expression of a vast number of genes as often detected in Affymetrix array analyses. DMH-treated colon mucosa (not tumors) also did not exhibit changes in the mRNA levels of ß-catenin and select genes involved in colon carcinogenesis (Bax, Bad, Bcl-2, Bcl-XL, Cyclin D1, c-myc). However, the protein levels of ß-catenin were significantly downregulated, and both connexin43 and E-cadherin were significantly upregulated by dietary sphingolipids [80]. These studies suggest that sphingolipids regulate protein stability or turnover and localization rather than or in addition to transcription in intestinal tissue primed for colon cancer either by *Apc* mutations (Min mice) or carcinogen treatment in CF1 mice. However, ß-catenin is not the only sphingolipd target in cancer cells.

Detailed analyses of how dietary sphingolipids mediate ß-catenin turnover *in vivo* are not available. Treatment of colon cancer cell lines with sphingosine, enigmol, 4,8-sphingadienine from corn or natural ceramides (but not C_2-ceramide) *in vitro* for 1–48 h reduced total ß-catenin [53, 56] and nuclear ß-catenin expression levels [42, 56], transiently decreased ß-catenin-induced TCF-4 reporter activity (for 5 to >24 h dependent on the cell line) and lowered the expression levels of TCF-4 target genes *c-myc* and *cyclin D1* [53]. The effective concentrations ranged from non-toxic to toxic (2.5–30 μM) and the responsiveness of the specific cell line determined the concentration of the effective treatment and its time course. The altered ß-catenin levels were associated with a reduced AKT phosphorylation [53], activated GSK-3ß, and increased phosphorylation of ß-catenin on S33, S37 T41 (casein kinase 1 phosphorylation sites) and T41, S45 (GSK-3ß phosphorylation sites) which marks ß-catenin for ubiquitination and proteasomal degradation. Inhibition of the proteasome abrogated this degradation [56]. Since the reversal of aberrant ß-catenin expression decreases the tumorigenic potential of cancer cells [81], the regulation of ß-catenin expression and localization and its transcriptional activity may be the critical step in the prevention of colon cancer by dietary sphingolipids. This is important because *Apc* mutations and ß-catenin dysregulation are not only relevant in human colon cancer but a growing list of cancer in other organs such as esophagus, breast, liver and stomach also exhibit altered APC and ß-catenin localization and expression. However, the impact of sphingolipids on non-canonical signaling pathways of APC or ß-catenin and their importance for sphingolipid-mediated tumor suppression still needs to be evaluated. This is especially important since the targets of the bioactive sphingolipids generated in the colon may be manifold and the regulation of a single pathway may not be sufficient for tumor suppression. Targeting functional categories that are commonly altered in colon cancer and cancer of other sites may be more effective and independent of specific, individual gene alterations, and may provide a better control for sphingolipid efficacy. Recent studies have identified the modulation of cellular architecture—specifically actin cytoskeleton organization—that influenced cell biomechanical properties (softness and elasticity), motility and bioelectrical behavior [82–87] and metabolism [88] in ovarian cancer by sphingolipid metabolites; if this is also true for colon cancer has yet to be determined.

7 Microenvironment Intestinal Tract

The focus of preventive and treatment efforts has been the elimination of cancer cells. However, the intestinal tract is a complex microenvironment of epithelial, immune and stromal cells and intestinal microbiota that together provide protection from infection, inflammation, and allergic reactions and support digestion, uptake and systemic distribution of nutrients. The disturbance of the delicate balance of these players can lead to severe diseases.

The epithelial layer lining the intestinal tract provides the first line of immune defense to injury and pathogenic insult. Further support is provided by mucus layers secreted by goblet cells together with an abundance of antimicrobial peptides largely produced and secreted by Paneth cells. The intestinal epithelium or mucosa is further protected by abundant levels of secretory IgA, which is produced by plasma cells residing within the lamina propria and subsequently transported by intestinal epithelial cells (IECs) to the lumen of the GI tract. All of these factors, mucus, antimicrobial peptides and IgA help sequester microbes in the lumen and reduce epithelial overexposure to antigen and microbes and triggering of innate surface receptors such as the Toll-like receptors. This helps to maintain the normal intestinal mucosa in a state of antigenic ignorance defined as preventing antigens and microbes from gaining excessive access to immune inductive sites in the GI tract. In conjunction with antigenic ignorance, there is an active system referred to as mucosal-induced tolerance which includes both antigen specific responses as well as non-specific anti-inflammatory responses both of which help protect against food allergies as well as unwarranted responses to commensal microbes.

The gut associated lymphoid tissue (GALT) has various unique features, such as M-cells and Peyer's patches and is highly compartmentalized into inductor and effector sites, which helps to ensure that immune responses are tightly regulated and not overtly triggered in response to non-pathogenic commensal organisms or food antigens. Well-organized lymphoid aggregates, typically referred to as Peyer' Patches (PPs) and draining mesenteric lymph nodes (MLNs) are immune inductive sites, whereas the lamina propria interspersed with macrophages, fibroblasts, mesenchymal stem cells, dendritic cells, neutrophils as well as primed T and B cells makes up the extensive effector site within the intestinal tissues. Mature PPs are readily visible macroscopically and have lymph node-like architecture with respect to the organization of B and T cell follicles. The overlying follicle associated epithelium is unique to PPs and is more porous and contains M-cells (micro-fold) that serve to transport particulate matter (viruses, bacteria) to the underlying dendritic cells that subsequently sample and process antigen for presentation to B and T cells. A vast array of regulatory T and B cell subsets help micro-manage immune responses ensuring that homeostasis is maintained in a less reactive state, or to control excessive inflammatory responses that are initiated by pathogenic insults. The influx and maintenance of regulatory subsets is often controlled by the context of the chemokine and cytokine milieu maintained at these inductive sites.

The intestinal epithelial cells (IECs) themselves are to be considered as active, important players in maintenance of tissue homeostasis and immune surveillance within the gut. They can serve as weak antigen presenting cells or they can transport luminal antigens to underlying professional antigen presenting dendritic cells dispersed throughout the lamina propria, which then transport, process and present antigens to T and B cells in the draining mesenteric lymph nodes. IECs in response to various exogenous stimuli secrete a wide variety of chemokines and growth factors that can actively recruit immune cells or stimulate proliferative responses in cells residing within the lamina propria such as macrophages, fibroblasts, mesenchymal stem cells and other lymphoid cells. IECs are also responsible for transporting polymeric, secretory IgA produced by lamina propria-residing plasma cells to the intestinal lumen. Together the mucus, secretory IgA and anti-microbial peptides provide the first line of defense against pathogenic microbes and help maintain a less responsive environment to commensals.

7.1 Modulating the Inflammatory Intestinal Microenvironment

While endogenous sphingolipid metabolites (ceramide, ceramide-1-phosphate and sphingosine-1-phosphate) have been associated with acute and chronic inflammation in endothelial cells [89], adipose tissue [90], lung [91], immune cells [92], and colon [93], information of diet-derived bioactive sphingolipid metabolite involvement in the inflammation of the intestinal tract is limited. However, dietary sphingolipids have been shown to alter the composition of intestinal cells. Milk gangliosides fed to rats for 2 weeks modified the ganglioside content of the enterocytes, increasing ganglioside GD3 levels in basolateral membranes, ganglioside GM3 levels in the apical membrane. Dietary ganglioside also altered the composition of intestinal brush border microdomains and reduced the cholesterol, caveolin1, and PAF, suggesting a potentially anti-inflammatory mechanism [94]. Ganglioside GD3 fed to newborn rats reduced necrotizing enteral colitis [95]. Dietary sphingomyelin did not alter the ganglioside content but reduced the cholesterol content of enterocytes [59] which can impact membrane fluidity and signaling events. Changes in gene expression levels in the normal gut by dietary sphingolipids have not been evaluated in greater detail but dietary sphingomyelin increases the expression and activity of alkaline sphingomyelinase in the colon [45]. Alkaline sphingomyelinase is reduced in aging individuals and patients with chronic colitis which was comparable to patients with dysplasia [96], suggesting that the decreased expression levels may represent a predisposition to colon cancer that can be modulated by dietary sphingolipids. Dietary glucosylceramide from maize did not only impact colonic inflammation but also exerted systemic anti-inflammatory effects as evident by reduced levels of IFNγ and TNFα, and increased IL-10 levels in distant tissues [63]. Feeding mice milkfat globule membrane at 125 g/kg bodyweight in an AIN76A diet for 5 weeks reduced lipopolysaccharide (LPS)-induced monocyte adhesion and infiltration of intestinal tissue, gut permeability, reduced serum cytokines (IL-6, IL-10, IL-17, MCP-1, IFNγ, TNFα) and prevented LPS lethality [44]. Thus, dietary sphingolipids can modulate the inflammatory intestinal microenvironment, a role that can be contradictory to the effect of endogenously generated sphingolipid metabolites.

7.2 *Inflammation-Driven Colon Cancer*

Inflammatory bowel diseases (IBD) are chronic inflammatory disorders of the gastrointestinal tract with an unclear etiology. The most common clinical manifestations of IBD are ulcerative colitis and Crohn's disease. These patients have an increased risk of developing colorectal cancer (CRC) that is thought to be the result of prolonged inflammation within the intestinal tract rather than genetic predisposition [97]. CRC patients without chronic inflammation of the colon generally develop precancerous polyps (adenoma) that slowly progress while those who experience chronic inflammation have an early onset of multiple adenomas that constitute an increased risk for tumor development and can require removal of the tumor or potentially the entire colon/rectum to prevent subsequent colon cancer metastases [97]. In contrast to the general population (average age of diagnosis: 74 years), IBD patients are diagnosed with CRC in their 30s [98]. The inflammation is especially detrimental if the onset of IBD occurs at an early age [99]: prolonged ulcerative colitis over 10–30 years is directly associated with an increased risk of developing CRC by 2–18 % after disease onset [100]; these numbers are only slightly less for Crohn's disease. Thus, reducing chronic intestinal inflammation can significantly reduce CRC development.

The impact of dietary sphingolipids on intestinal inflammation and inflammation-driven colon cancer has not been investigated in humans. Rodent models for intestinal inflammation include the IL-10 knock-out mice that spontaneously develop enterocolitis, a model for IBD with symptoms similar to Crohn's disease [101], and the widely used dextran sodium sulfate (DSS)-induced inflammation that causes deleterious intestinal acute colonic inflammation. The combination of carcinogen treatment followed by DSS administration via the drinking water allows for the investigation of the association of inflammation and colon carcinogenesis, and the identification of targets for intervention of either process. Feeding 0.1 % milk sphingomyelin for 2 weeks before a single AOM injection and a 7-day DSS treatment reduced the magnitude of the acute inflammation at day 12, and accelerated the recovery from the inflammatory insult [102, 103]. Similar effects were observed after feeding the glycosphingolipid cerebroside D [104]. The reduction of inflammation was less pronounced in mice carrying tissue-specific deletions of peroxisome proliferator-activator receptor (PPAR)-γ in the intestinal epithelial cells, T cells and macrophages [102]. PPAR-γ is a member of a family of nuclear hormone receptors that regulate inflammation, immunity and metabolism; its activation attenuates inflammation [105], suggesting the sphingolipid-mediated reduction of inflammation is partially dependent on PPAR-γ expression or activation. In the DSS/AOM model for inflammation-driven colon cancer, the mice recover from the DSS treatment, and rapidly form multiple tumors. Feeding sphingomyelin significantly increased the survival rate of mice independent of PPAR-γ expression, and reduced the number adenocarcinomas; this was accompanied by an improved colonic tissue architecture and a reduction in infiltrating inflammatory cells in the colon and mesenteric lymph nodes [102]. In contrast, egg sphingomyelin fed via gavage and not mixed into the diet induced mild signs of inflammation even without DSS treatment and actually exacerbated DSS-induced inflammation or inflammation in IL10-/- mice [106]. This may be due to the structure of egg sphingomyelin that contains 87 % C16 fatty acids in the ceramide moiety in contrast to the predominant C22, C23, and C24 species in milk that can affect lipid absorption [107] and increase the pro-apoptotic C16 ceramide in the colon. As mentioned above, C16 ceramide remains as the major ceramide species after digestion of milk sphingolipids in the intestine [39] and therefore may accumulate above beneficial levels or shift the bioactive sphingolipid profile to elicit rather than suppress an inflammatory response. This is supported by reports demonstrating a suppression of DSS-induced inflammation by reducing ceramide and sphingosine-1-phosphate generation [108–111] or inhibiting sphingosine-1-phosphate signaling and recruitment of inflammatory cells to the intestinal tract [112–114]. Thus, dietary and oral administration per gavage are not interchangeable, and the selected sphingolipid species and their administration (in the diet vs. bolus per gavage) may be critical determinants for the inflammatory and apoptotic response (see also 6.1).

These studies show that the reduction of inflammation by the dietary sphingolipids is associated with an increased survival and reduced tumor development. The precise mechanisms of how dietary sphingolipids modulate the inflammatory response is not known. Using a PCR array with 372 genes associated with inflammatory response, autoimmunity and tissue regeneration but also with processes important in cancer such as proliferation, differentiation and angiogenesis, several genes were identified that were altered by dietary sphingomyelin alone; only a few were up-regulated (IL-11, IL-12, IL-17, Ccl12, Cxcl2, Reg3g) but many were down-regulated (many interleukins, their receptors, several chemokines) (see complete list in table 3 in [102]). During peak DSS-induced inflammation in the colon (day 12 of administration via drinking water), the drastic increase in inflammatory markers was modulated by the dietary sphingomyelin. Both pro-and anti-inflammatory genes were up-regulated, a pattern consistent with a stimulation of an inflammatory response in addition to a counter-regulatory immune response that together cause a suppression of inflammation. Interestingly, dietary sphingomyelin also lowered the expression of genes in the DSS-treated colons that are involved in tumorigenesis (ErbB2, Ltb, NfrκB) while genes with an anti-cancer effect (Gdf9) were up-regulated [102]. These results suggest that dietary sphingolipids support the balance of pro-and anti-inflammatory responses in the colon after the DSS insult but, in parallel, regulate tissue regeneration and prevent the expression of pro-tumorigenic genes. The specific mechanisms of this immune modulation by the dietary sphingolipids need further investigation. However, these studies have shown that the epithelial or cancer cells are not the only targets of dietary sphingolipids, and other cells in the colonic microenvironment are also responding to sphingolipids with changes in gene expression levels. Importantly, changes in specific gene expression levels in one cell type may not be beneficial in another. For example, ß-catenin over-expression in epithelial cells is associated with increased proliferation and tumorigenic potential while activation of ß-catenin signaling in intestinal dendritic cells is critical for the inflammatory homeostasis (inflammatory vs. regu-

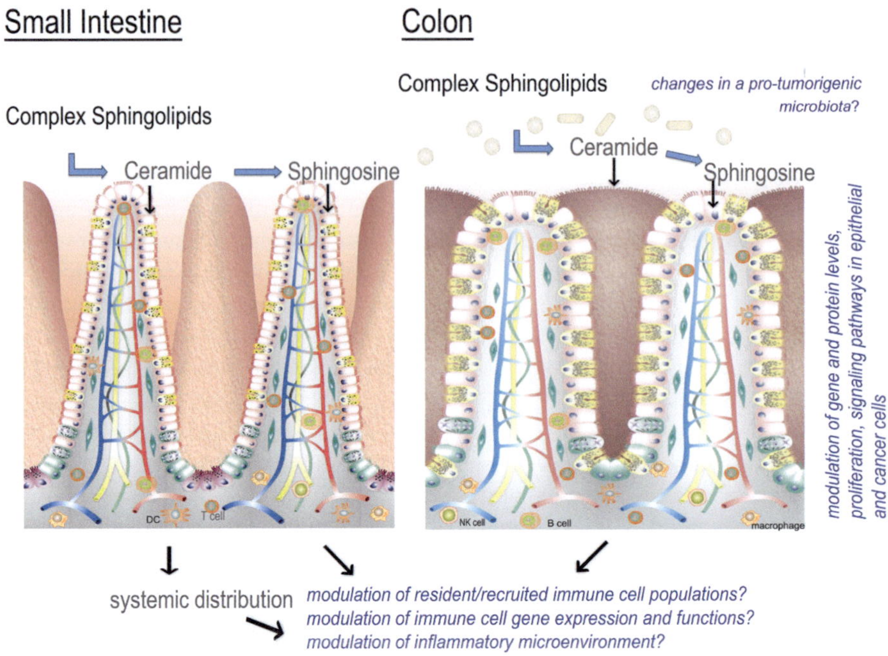

Fig. 3 Generation of sphingolipid metabolites in the intestinal tract and three groups of potential targets: colonic microbiota, colonic epithelial and cancer cells, and immune cell population. All groups may be involved and contribute to the sphingolipid-induced suppression of colon cancer

latory responses immunity vs. tolerance) [115]. In contrast, intestinal T cells and Tregs in Min mice expressed elevated levels of ß-catenin that increased during tumor formation independent of *Apc* mutations, and ß-catenin signaling was associated with sustained inflammation and tumor formation [116]. The expression of the pro-inflammatory leukotriene D4 secreted by recruited macrophages and leukocytes to the intestine causes the activation and translocation of ß-catenin to the nucleus in cancer cells, increasing their migration [117]. Thus, an indiscriminate down-regulation of ß-catenin may not be beneficial. It is not known if diet-derived sphingolipid metabolites can differentially regulate proteins in the cells that constitute the colonic microenvironment but it is an intriguing concept based on the cell-type dependent modulation of signaling events by sphingolipid metabolites in a context-dependent manner.

In addition to immune cells, progenitor and immune cells from the adipose tissue are actively recruited to the tumor site; the complex interactions of benign and cancer cells, resident and recruited immune and progenitor cells that can be differentiated into fibroblasts, endothelial cells etc. together generate a tumor microenvironment permissive of tumor cell survival, proliferation and progression. Therefore, in addition to the cancer cells themselves, other tumor-associated cells and their interactions may be targets for bioactive sphingolipid metabolites in the suppression of colon cancer (Fig. 3).

8 Emerging Target: Intestinal Microbiota

The gastrointestinal tract of vertebrates and invertebrates is inhabited by a diverse array of microorganisms comprising the gut microbiota. In humans and other mammals, the number of microbial cells is tenfold greater than cells of the host [118, 119] and the number of genes in the gut microbiota is estimated to exceed the number of host genes by a factor of 100 [120]. The gut microbiota [121] is dominated by bacterial phylogenetic types belonging to the *Firmicutes and Bacteroidetes* phyla [122] that aid the host by fermenting fiber and other indigestable dietary components and producing short chain fatty acids and small organic acids, butyrate and lactate, or stimulate the immune system and maintain colonocyte health [123–129]. Conversely, specific bacteria and viruses such as *Clostridium perfringens*, enteropathogenic *Escherichia coli*, and rotavirus are known pathogens that often result in host disease. Changes to the gut microbiota, a concept referred to as dysbiosis, is associated with disease in humans such as intestinal infections [130], inflammation [131, 132], type 2 diabetes, metabolic syndrome, and cardiovascular disease (see recent reviews [133, 134]). The reduced microbiota diversity in inflammatory bowel disease [135, 136] and an altered microbiota promotes inflammation-driven colon cancer [137, 138], lymphoma [139] and liver cancer [140]. In Min mice, microbiota-mediated local inflammation is driving polyp formation in Min mice, mediated by IL-10 producing T cells; treatment with antibiotics reduced both inflammation and adenoma growth [141]. There is only little information on dietary sphingolipid interaction with the colonic microbiota and if this would affect colon tumorigenesis. Glycosphingolipids have been shown to bind viruses and bacteria and their secreted enterotoxins (see review [108]). Intestinal bacteria are critical for the digestion of sphingolipids in the colon that can be taken up by the intestinal cells [32]; on the other hand, bacterial sphingolipids (mainly ceramides, glycosceramide, phosphoethanol ceramides) have been shown to impact iNKT cells, restricting their growth and thereby protect against colitis and autoimmune disease [142]. Adding porcine brain gangliosides for 30 days to the formula reduced the *E. coli* population while increasing bifidobacteria in infants [143]. It is not known if any of these changes affect colon tumorigenesis but it is intriguing to speculate that specific complex sphingolipids could selectively either bind pathogens that then will be excreted or prevent the enterotoxin-mediated local and systemic effects.

High-fat diets (10 % of diet as fat, fed for 8–10 weeks) have also been shown to alter the microbiota (total number, diversity, Clostridialis, Proteobacteria, Enterobacteriales [144]), resulting in inflammation of the gastrointestinal tract, higher systemic endotoxin levels and inflammation of adipose tissue, metabolic disorders and aging [145]. Since high-fat diets and obesity are associated with a higher risk of developing colon cancer [146], the intestinal dysbiosis may be a critical factor for tumor progression. Again, if dietary sphingolipids can counteract the obesity-induced dysbiosis and prevent local disease (colon inflammation, colon cancer) or systemic disease such as diabetes and metabolic syndrome is not known.

9 Summary

It is now well established that dietary sphingolipids are highly effective in suppressing colon cancer in all tested rodent models of colon cancer without any deleterious side effects. While there is clearly an effect on cancer cell growth, and the modulation of important protein mediators such as ß-catenin can contribute to the removal of the growth advantage, it also is apparent that these may not be the only targets of dietary sphingolipids. How the diet-derived metabolites target other cells in the normal colonic, inflammatory and tumor microenvironment, changing a tumor-permissive to a non-permissive state, and if a reversal of intestinal dysbiosis or at least an increase in beneficial bacteria or a reduction in pathogens is involved in the tumor suppressive outcome of feeding sphingolipids needs further investigation. This will allow for the tailoring of specific treatment regimens for individual need, e.g., select sphingolipid species based on their structure to optimize delivery to specific locations at concentrations for defined functions in specific cell types, perhaps even in combination with lower doses of conventional chemotherapeutic drugs. Importantly, the identification of novel players in colon carcinogenesis that respond to dietary sphingolipids is critical for the timely control of the efficacy of this approach in future human trials.

References

1. SEER Program N. SEER Stat fact sheets: Colon and rectum cancer (2014) http://seer.cancer.gov/statfacts/html/colorect.html
2. Fearon ER, Vogelstein B (1990) A genetic model for colorectal tumorigenesis. Cell 61:759–767
3. Femia AP, Dolara P, Giannini A, Salvadori M, Biggeri A, Caderni G (2007) Frequent mutation of Apc gene in rat colon tumors and mucin-depleted foci, preneoplastic lesions in experimental colon carcinogenesis. Cancer Res 67:445–449
4. Paulsen JE, Knutsen H, Olstorn HB, Loberg EM, Alexander J (2006) Identification of flat dysplastic aberrant crypt foci in the colon of azoxymethane-treated A/J mice. Int J Cancer 118:540–546
5. Groden J, Thliveris A, Samowitz W, Carlson M, Gelbert L, Albertsen H, Joslyn G, Stevens J, Spirio L, Robertson M et al (1991) Identification and characterization of the familial adenomatous polyposis coli gene. Cell 66:589–600
6. Bienz M, Clevers H (2000) Linking colorectal cancer to Wnt signaling. Cell 103:311–320
7. Vogelstein B, Papadopoulos N, Velculescu VE, Zhou S, Diaz LA Jr, Kinzler KW (2013) Cancer genome landscapes. Science 339:1546–1558

8. Barker N, Ridgway RA, van Es JH, van de Wetering M, Begthel H, van den Born M, Danenberg E, Clarke AR, Sansom OJ, Clevers H (2009) Crypt stem cells as the cells-of-origin of intestinal cancer. Nature 457:608–611
9. Schepers AG, Snippert HJ, Stange DE, van den Born M, van Es JH, van de Wetering M, Clevers H (2012) Lineage tracing reveals Lgr5+ stem cell activity in mouse intestinal adenomas. Science 337:730–735
10. Schwitalla S, Fingerle AA, Cammareri P, Nebelsiek T, Goktuna SI, Ziegler PK, Canli O, Heijmans J, Huels DJ, Moreaux G et al (2013) Intestinal tumorigenesis initiated by dedifferentiation and acquisition of stem-cell-like properties. Cell 152:25–38
11. Hammoud SS, Cairns BR, Jones DA (2013) Epigenetic regulation of colon cancer and intestinal stem cells. Curr Opin Cell Biol 25:177–183
12. Beggs AD, Jones A, El-Bahrawy M, Abulafi M, Hodgson SV, Tomlinson IP (2013) Whole-genome methylation analysis of benign and malignant colorectal tumours. J Pathol 229:697–704
13. Vesper H, Schmelz EM, Nikolova-Karakashian MN, Dillehay DL, Lynch DV, Merrill AH Jr (1999) Sphingolipids in food and the emerging importance of sphingolipids to nutrition. J Nutr 129:1239–1250
14. Yunoki K, Ogawa T, Ono J, Miyashita R, Aida K, Oda Y, Ohnishi M (2008) Analysis of sphingolipid classes and their contents in meals. Biosci Biotechnol Biochem 72:222–225
15. Pata MO, Hannun YA, Ng CK (2010) Plant sphingolipids: decoding the enigma of the Sphinx. New Phytol 185:611–630
16. Pruett ST, Bushnev A, Hagedorn K, Adiga M, Haynes CA, Sullards MC, Liotta DC, Merrill AH Jr (2008) Biodiversity of sphingoid bases ("sphingosines") and related amino alcohols. J Lipid Res 49:1621–1639
17. Blaas N, Humpf HU (2013) Structural profiling and quantitation of glycosyl inositol phosphoceramides in plants with Fourier transform mass spectrometry. J Agric Food Chem 61:4257–4269
18. Cacas JL, Bure C, Furt F, Maalouf JP, Badoc A, Cluzet S, Schmitter JM, Antajan E, Mongrand S (2013) Biochemical survey of the polar head of plant glycosylinositolphosphoceramides unravels broad diversity. Phytochemistry 96:191–200
19. Nilsson A (1968) Metabolism of sphingomyelin in the intestinal tract of the rat. Biochim Biophys Acta 164:575–584
20. Nilsson A (1969) The presence of spingomyelin- and ceramide-cleaving enzymes in the small intestinal tract. Biochim Biophys Acta 176:339–347
21. Schmelz EM, Crall KJ, Larocque R, Dillehay DL, Merrill AH Jr (1994) Uptake and metabolism of sphingolipids in isolated intestinal loops of mice. J Nutr 124:702–712
22. Zhang Y, Cheng Y, Hansen GH, Niels-Christiansen LL, Koentgen F, Ohlsson L, Nilsson A, Duan RD (2011) Crucial role of alkaline sphingomyelinase in sphingomyelin digestion: a study on enzyme knockout mice. J Lipid Res 52:771–781
23. Duan RD (2011) Physiological functions and clinical implications of sphingolipids in the gut. J Dig Dis 12:60–70
24. Duan RD, Nilsson A (2009) Metabolism of sphingolipids in the gut and its relation to inflammation and cancer development. Prog Lipid Res 48:62–72
25. Schmelz EM, Sullards MC, Dillehay DL, Merrill AH Jr (2000) Colonic cell proliferation and aberrant crypt foci formation are inhibited by dairy glycosphingolipids in 1, 2-dimethylhydrazine-treated CF1 mice. J Nutr 130:522–527
26. Thomas RL Jr, Matsko CM, Lotze MT, Amoscato AA (1999) Mass spectrometric identification of increased C16 ceramide levels during apoptosis. J Biol Chem 274:30580–30588
27. Oskouian B, Sooriyakumaran P, Borowsky AD, Crans A, Dillard-Telm L, Tam YY, Bandhuvula P, Saba JD (2006) Sphingosine-1-phosphate lyase potentiates apoptosis via p53- and p38-dependent pathways and is down-regulated in colon cancer. Proc Natl Acad Sci U S A 103:17384–17389
28. Fukuda Y, Kihara A, Igarashi Y (2003) Distribution of sphingosine kinase activity in mouse tissues: contribution of SPHK1. Biochem Biophys Res Commun 309:155–160

29. Sugawara T, Kinoshita M, Ohnishi M, Nagata J, Saito M (2003) Digestion of maize sphingolipids in rats and uptake of sphingadienine by Caco-2 cells. J Nutr 133:2777–2782
30. Sugawara T, Tsuduki T, Yano S, Hirose M, Duan J, Aida K, Ikeda I, Hirata T (2010) Intestinal absorption of dietary maize glucosylceramide in lymphatic duct cannulated rats. J Lipid Res 51:1761–1769
31. Ohlsson L, Hertervig E, Jonsson BA, Duan RD, Nyberg L, Svernlov R, Nilsson A (2010) Sphingolipids in human ileostomy content after meals containing milk sphingomyelin. Am J Clin Nutr 91:672–678
32. Nyberg L, Nilsson A, Lundgren P, Duan RD (1997) Localization and capacity of sphingomyelin digestion in the rat intestinal tract. J Nutr Biochem 8:112–118
33. Garmy N, Taieb N, Yahi N, Fantini J (2005) Apical uptake and transepithelial transport of sphingosine monomers through intact human intestinal epithelial cells: physicochemical and molecular modeling studies. Arch Biochem Biophys 440:91–100
34. Kobayashi T, Shimizugawa T, Osakabe T, Watanabe S, Okuyama H (1997) A long-term feeding of sphingolipids affected the levels of plasma cholesterol and hepatic triacylglycerol but not tissue phospholipids and sphingolipids. Nutr Res 17
35. Larsson SC, Bergkvist L, Wolk A (2005) High-fat dairy food and conjugated linoleic acid intakes in relation to colorectal cancer incidence in the Swedish Mammography Cohort. Am J Clin Nutr 82:894–900
36. Huncharek M, Muscat J, Kupelnick B (2009) Colorectal cancer risk and dietary intake of calcium, vitamin D, and dairy products: a meta-analysis of 26,335 cases from 60 observational studies. Nutr Cancer 61:47–69
37. Aune D, Lau R, Chan DS, Vieira R, Greenwood DC, Kampman E, Norat T (2012) Dairy products and colorectal cancer risk: a systematic review and meta-analysis of cohort studies. Ann Oncol 23:37–45
38. Dillehay DL, Webb SK, Schmelz EM, Merrill AH Jr (1994) Dietary sphingomyelin inhibits 1,2-dimethylhydrazine-induced colon cancer in CF1 mice. J Nutr 124:615–620
39. Schmelz EM, Dillehay DL, Webb SK, Reiter A, Adams J, Merrill AH Jr (1996) Sphingomyelin consumption suppresses aberrant colonic crypt foci and increases the proportion of adenomas versus adenocarcinomas in CF1 mice treated with 1,2-dimethylhydrazine: implications for dietary sphingolipids and colon carcinogenesis. Cancer Res 56:4936–4941
40. Exon JH, South EH (2003) Effects of sphingomyelin on aberrant colonic crypt foci development, colon crypt cell proliferation and immune function in an aging rat tumor model. Food Chem Toxicol 41:471–476
41. Schmelz EM, Bushnev AS, Dillehay DL, Liotta DC, Merrill AH Jr (1997) Suppression of aberrant colonic crypt foci by synthetic sphingomyelins with saturated or unsaturated sphingoid base backbones. Nutr Cancer 28:81–85
42. Schmelz EM, Roberts PC, Kustin EM, Lemonnier LA, Sullards MC, Dillehay DL, Merrill AH Jr (2001) Modulation of intracellular beta-catenin localization and intestinal tumorigenesis in vivo and in vitro by sphingolipids. Cancer Res 61:6723–6729
43. Zheng W, Kollmeyer J, Symolon H, Momin A, Munter E, Wang E, Kelly S, Allegood JC, Liu Y, Peng Q et al (1758) Ceramides and other bioactive sphingolipid backbones in health and disease: lipidomic analysis, metabolism and roles in membrane structure, dynamics, signaling and autophagy. Biochim Biophys Acta 2006:1864–1884
44. Snow DR, Ward RE, Olsen A, Jimenez-Flores R, Hintze KJ (2011) Membrane-rich milk fat diet provides protection against gastrointestinal leakiness in mice treated with lipopolysaccharide. J Dairy Sci 94:2201–2212
45. Zhang P, Li B, Gao S, Duan RD (2008) Dietary sphingomyelin inhibits colonic tumorigenesis with an up-regulation of alkaline sphingomyelinase expression in ICR mice. Anticancer Res 28:3631–3635
46. Lemonnier LA, Dillehay DL, Vespremi MJ, Abrams J, Brody E, Schmelz EM (2003) Sphingomyelin in the suppression of colon tumors: prevention versus intervention. Arch Biochem Biophys 419:129–138

47. Hu Y, Le Leu RK, Belobrajdic D, Young GP (2008) The potential of sphingomyelin as a chemopreventive agent in AOM-induced colon cancer model: wild-type and p53+/- mice. Mol Nutr Food Res 52:558–566
48. Inamine M, Suzui M, Morioka T, Kinjo T, Kaneshiro T, Sugishita T, Okada T, Yoshimi N (2005) Inhibitory effect of dietary monoglucosylceramide 1-O-beta-glucosyl-N-2′-hydroxyarachidoyl-4,8-sphingadienine on two different categories of colon preneoplastic lesions induced by 1,2-dimethylhydrazine in F344 rats. Cancer Sci 96:876–881
49. Symolon H, Schmelz EM, Dillehay DL, Merrill AH Jr (2004) Dietary soy sphingolipids suppress tumorigenesis and gene expression in 1,2-dimethylhydrazine-treated CF1 mice and ApcMin/+ mice. J Nutr 134:1157–1161
50. Wargovich MJ, Harris C, Chen CD, Palmer C, Steele VE, Kelloff GJ (1992) Growth kinetics and chemoprevention of aberrant crypts in the rat colon. J Cell Biochem Suppl 16G:51–54
51. Nagase H, Nakamura Y (1993) Mutations of the APC (adenomatous polyposis coli) gene. Hum Mutat 2:425–434
52. Fyrst H, Oskouian B, Bandhuvula P, Gong Y, Byun HS, Bittman R, Lee AR, Saba JD (2009) Natural sphingadienes inhibit Akt-dependent signaling and prevent intestinal tumorigenesis. Cancer Res 69:9457–9464
53. Kumar A, Pandurangan AK, Lu F, Fyrst H, Zhang M, Byun HS, Bittman R, Saba JD (2012) Chemopreventive sphingadienes downregulate Wnt signaling via a PP2A/Akt/GSK3beta pathway in colon cancer. Carcinogenesis 33:1726–1735
54. Kawamori T, Kaneshiro T, Okumura M, Maalouf S, Uflacker A, Bielawski J, Hannun YA, Obeid LM (2009) Role for sphingosine kinase 1 in colon carcinogenesis. FASEB J 23:405–414
55. Liotta DC, Bushnev AS, Baillie MT, Holt JJ, Menaldino DS, Merrill AH (2010) An efficient asymmetric synthesis of Enigmols (1-deoxy-5-hydroxysphingoid bases), an important class of bioactive lipid modulators. Arkivoc 8:263–277
56. Symolon H, Bushnev A, Peng Q, Ramaraju H, Mays SG, Allegood JC, Pruett ST, Sullards MC, Dillehay DL, Liotta DC, Merrill AH Jr (2011) Enigmol: a novel sphingolipid analogue with anticancer activity against cancer cell lines and in vivo models for intestinal and prostate cancer. Mol Cancer Ther 10:648–657
57. Morales PR, Dillehay DL, Moody SJ, Pallas DC, Pruett S, Allgood JC, Symolon H, Merrill AH Jr (2007) Safingol toxicology after oral administration to TRAMP mice: demonstration of safingol uptake and metabolism by N-acylation and N-methylation. Drug Chem Toxicol 30:197–216
58. Dickson MA, Carvajal RD, Merrill AH Jr, Gonen M, Cane LM, Schwartz GK (2011) A phase I clinical trial of safingol in combination with cisplatin in advanced solid tumors. Clin Cancer Res 17:2484–2492
59. Park EJ, Suh M, Ramanujam K, Steiner K, Begg D, Clandinin MT (2005) Diet-induced changes in membrane gangliosides in rat intestinal mucosa, plasma and brain. J Pediatr Gastroenterol Nutr 40:487–495
60. Hasegawa T, Shimada H, Uchiyama T, Ueda O, Nakashima M, Matsuoka Y (2011) Dietary glucosylceramide enhances cornified envelope formation via transglutaminase expression and involucrin production. Lipids 46:529–535
61. Ueda O, Hasegawa M, Kitamura S (2009) Distribution in skin of ceramide after oral administration to rats. Drug Metab Pharmacokinet 24:180–184
62. Oshida K, Shimizu T, Takase M, Tamura Y, Yamashiro Y (2003) Effects of dietary sphingomyelin on central nervous system myelination in developing rats. Pediatr Res 53:589–593
63. Duan J, Sugawara T, Sakai S, Aida K, Hirata T (2011) Oral glucosylceramide reduces 2,4-dinitrofluorobenzene induced inflammatory response in mice by reducing TNF-alpha levels and leukocyte infiltration. Lipids 46:505–512
64. Yunoki K, Renaguli M, Kinoshita M, Matsuyama H, Mawatari S, Fujino T, Kodama Y, Sugiyama M, Ohnishi M (2010) Dietary sphingolipids ameliorate disorders of lipid metabolism in Zucker fatty rats. J Agric Food Chem 58:7030–7035

65. Silins I, Nordstrand M, Hogberg J, Stenius U (2003) Sphingolipids suppress preneoplastic rat hepatocytes in vitro and in vivo. Carcinogenesis 24:1077–1083
66. Simon KW, Tait L, Miller F, Cao C, Davy KP, LeRoith T, Schmelz EM (2010) Suppression of breast xenograft growth and progression in nude mice: implications for the use of orally administered sphingolipids as chemopreventive agents against breast cancer. Food Funct 1:90–98
67. Fujiwara K, Kitatani K, Fukushima K, Yazama H, Umehara H, Kikuchi M, Igarashi Y, Kitano H, Okazaki T (2011) Inhibitory effects of dietary glucosylceramides on squamous cell carcinoma of the head and neck in NOD/SCID mice. Int J Clin Oncol 16:133–140
68. Lui C, Mills K, Brocardo MG, Sharma M, Henderson BR (2012) APC as a mobile scaffold: regulation and function at the nucleus, centrosomes, and mitochondria. IUBMB Life 64: 209–214
69. Fagotto F, Funayama N, Gluck U, Gumbiner BM (1996) Binding to cadherins antagonizes the signaling activity of beta-catenin during axis formation in Xenopus. J Cell Biol 132:1105–1114
70. Orsulic S, Peifer M (1996) An in vivo structure-function study of armadillo, the beta-catenin homologue, reveals both separate and overlapping regions of the protein required for cell adhesion and for wingless signaling. J Cell Biol 134:1283–1300
71. Hoschuetzky H, Aberle H, Kemler R (1994) Beta-catenin mediates the interaction of the cadherin-catenin complex with epidermal growth factor receptor. J Cell Biol 127:1375–1380
72. Shibamoto S, Hayakawa M, Takeuchi K, Hori T, Oku N, Miyazawa K, Kitamura N, Takeichi M, Ito F (1994) Tyrosine phosphorylation of beta-catenin and plakoglobin enhanced by hepatocyte growth factor and epidermal growth factor in human carcinoma cells. Cell Adhes Commun 1:295–305
73. Munemitsu S, Albert I, Souza B, Rubinfeld B, Polakis P (1995) Regulation of intracellular beta-catenin levels by the adenomatous polyposis coli (APC) tumor-suppressor protein. Proc Natl Acad Sci U S A 92:3046–3050
74. Rubinfeld B, Albert I, Porfiri E, Fiol C, Munemitsu S, Polakis P (1996) Binding of GSK3beta to the APC-beta-catenin complex and regulation of complex assembly. Science 272: 1023–1026
75. de Sousa EM, Vermeulen L, Richel D, Medema JP (2011) Targeting Wnt signaling in colon cancer stem cells. Clin Cancer Res 17:647–653
76. Giles RH, van Es JH, Clevers H (2003) Caught up in a Wnt storm: Wnt signaling in cancer. Biochim Biophys Acta 1653:1–24
77. Hadjihannas MV, Bruckner M, Jerchow B, Birchmeier W, Dietmaier W, Behrens J (2006) Aberrant Wnt/beta-catenin signaling can induce chromosomal instability in colon cancer. Proc Natl Acad Sci U S A 103:10747–10752
78. Song JH, Huels DJ, Ridgway RA, Sansom OJ, Kholodenko BN, Kolch W, Cho KH (2014) The APC network regulates the removal of mutated cells from colonic crypts. Cell Rep 7:94–103
79. Brabletz T, Jung A, Reu S, Porzner M, Hlubek F, Kunz-Schughart LA, Knuechel R, Kirchner T (2001) Variable beta-catenin expression in colorectal cancers indicates tumor progression driven by the tumor environment. Proc Natl Acad Sci U S A 98:10356–10361
80. Simon KW, Roberts PC, Vespremi MJ, Manchen S, Schmelz EM (2009) Regulation of beta-catenin and connexin-43 expression: targets for sphingolipids in colon cancer prevention. Mol Nutr Food Res 53:332–340
81. Kim JS, Crooks H, Foxworth A, Waldman T (2002) Proof-of-principle: oncogenic beta-catenin is a valid molecular target for the development of pharmacological inhibitors. Mol Cancer Ther 1:1355–1359
82. Creekmore AL, Silkworth WT, Cimini D, Jensen RV, Roberts PC, Schmelz EM (2011) Changes in gene expression and cellular architecture in an ovarian cancer progression model. PLoS One 6:e17676
83. Creekmore AL, Heffron CL, Brayfield BP, Roberts PC, Schmelz EM (2013) Regulation of cytoskeleton organization by sphingosine in a mouse cell model of progressive ovarian cancer. Biomolecules 3:386–407

84. Ketene AN, Roberts PC, Shea AA, Schmelz EM, Agah M (2012) Actin filaments play a primary role for structural integrity and viscoelastic response in cells. Integr Biol (Camb) 4:540–549
85. Babahosseini H, Ketene AN, Schmelz EM, Roberts PC, Agah M (2014) Biomechanical profile of cancer stem-like/tumor-initiating cells derived from a progressive ovarian cancer model. Nanomedicine 10(5):1013–9
86. Babahosseini H, Roberts PC, Schmelz EM, Agah M (2013) Bioactive sphingolipid metabolites modulate ovarian cancer cell structural mechanics. Integr Biol (Camb) 5:1385–1392
87. Salmanzadeh A, Elvington ES, Roberts PC, Schmelz EM, Davalos RV (2013) Sphingolipid metabolites modulate dielectric characteristics of cells in a mouse ovarian cancer progression model. Integr Biol (Camb) 5:843–852
88. Anderson AS, Roberts PC, Frisard MI, McMillan RP, Brown TJ, Lawless MH, Hulver MW, Schmelz EM (2013) Metabolic changes during ovarian cancer progression as targets for sphingosine treatment. Exp Cell Res 319:1431–1442
89. Lin CI, Chen CN, Lin PW, Lee H (2007) Sphingosine 1-phosphate regulates inflammation-related genes in human endothelial cells through S1P1 and S1P3. Biochem Biophys Res Commun 355:895–901
90. Wu D, Ren Z, Pae M, Guo W, Cui X, Merrill AH, Meydani SN (2007) Aging up-regulates expression of inflammatory mediators in mouse adipose tissue. J Immunol 179:4829–4839
91. Teichgraber V, Ulrich M, Endlich N, Riethmuller J, Wilker B, De Oliveira-Munding CC, van Heeckeren AM, Barr ML, von Kurthy G, Schmid KW et al (2008) Ceramide accumulation mediates inflammation, cell death and infection susceptibility in cystic fibrosis. Nat Med 14:382–391
92. Claycombe KJ, Wu D, Nikolova-Karakashian M, Palmer H, Beharka A, Paulson KE, Meydani SN (2002) Ceramide mediates age-associated increase in macrophage cyclooxygenase-2 expression. J Biol Chem 277:30784–30791
93. Nagahashi M, Hait NC, Maceyka M, Avni D, Takabe K, Milstien S, Spiegel S (2014) Sphingosine-1-phosphate in chronic intestinal inflammation and cancer. Adv Biol Regul 54:112–120
94. Park EJ, Suh M, Thomson B, Thomson AB, Ramanujam KS, Clandinin MT (2005) Dietary ganglioside decreases cholesterol content, caveolin expression and inflammatory mediators in rat intestinal microdomains. Glycobiology 15:935–942
95. Xu J, Anderson V, Schwarz SM (2013) Dietary GD3 ganglioside reduces the incidence and severity of necrotizing enterocolitis by sustaining regulatory immune responses. J Pediatr Gastroenterol Nutr 57:550–556
96. Sjoqvist U, Hertervig E, Nilsson A, Duan RD, Ost A, Tribukait B, Lofberg R (2002) Chronic colitis is associated with a reduction of mucosal alkaline sphingomyelinase activity. Inflamm Bowel Dis 8:258–263
97. Cooper HS, Everley L, Chang WC, Pfeiffer G, Lee B, Murthy S, Clapper ML (2001) The role of mutant Apc in the development of dysplasia and cancer in the mouse model of dextran sulfate sodium-induced colitis. Gastroenterology 121:1407–1416
98. Bleyer AOLM, Barr R, Ries LAG (eds) (2006) Cancer epidemiology in older adolescents and young adults 15 to 29 Years of age, including SEER incidence and survival: 1975-2000. National Cancer Institute, Bethesda, MD, NIH Pub. No. 06-5767
99. Peneau A, Savoye G, Turck D, Dauchet L, Fumery M, Salleron J, Lerebours E, Ligier K, Vasseur F, Dupas JL et al (2013) Mortality and cancer in pediatric-onset inflammatory bowel disease: a population-based study. Am J Gastroenterol 108:1647–1653
100. Lakatos PL, Lakatos L (2008) Risk for colorectal cancer in ulcerative colitis: changes, causes and management strategies. World J Gastroenterol 14:3937–3947
101. Scheinin T, Butler DM, Salway F, Scallon B, Feldmann M (2003) Validation of the interleukin-10 knockout mouse model of colitis: antitumour necrosis factor-antibodies suppress the progression of colitis. Clin Exp Immunol 133:38–43
102. Mazzei JC, Zhou H, Brayfield BP, Hontecillas R, Bassaganya-Riera J, Schmelz EM (2011) Suppression of intestinal inflammation and inflammation-driven colon cancer in mice by

dietary sphingomyelin: importance of peroxisome proliferator-activated receptor gamma expression. J Nutr Biochem 22:1160–1171
103. Furuya H, Ohkawara S, Nagashima K, Asanuma N, Hino T (2008) Dietary sphingomyelin alleviates experimental inflammatory bowel disease in mice. Int J Vitam Nutr Res 78:41–49
104. Wu XF, Wu XX, Guo WJ, Luo Q, Gu YH, Shen Y, Tan RX, Sun Y, Xu Q (2012) Cerebroside D, a glycoceramide compound, improves experimental colitis in mice with multiple targets against activated T lymphocytes. Toxicol Appl Pharmacol 263:296–302
105. Bassaganya-Riera J, Reynolds K, Martino-Catt S, Cui Y, Hennighausen L, Gonzalez F, Rohrer J, Benninghoff AU, Hontecillas R (2004) Activation of PPAR gamma and delta by conjugated linoleic acid mediates protection from experimental inflammatory bowel disease. Gastroenterology 127:777–791
106. Fischbeck A, Leucht K, Frey-Wagner I, Bentz S, Pesch T, Kellermeier S, Krebs M, Fried M, Rogler G, Hausmann M, Humpf HU (2011) Sphingomyelin induces cathepsin D-mediated apoptosis in intestinal epithelial cells and increases inflammation in DSS colitis. Gut 60:55–65
107. Noh SK, Koo SI (2004) Milk sphingomyelin is more effective than egg sphingomyelin in inhibiting intestinal absorption of cholesterol and fat in rats. J Nutr 134:2611–2616
108. Andersson D, Kotarsky K, Wu J, Agace W, Duan RD (2009) Expression of alkaline sphingomyelinase in yeast cells and anti-inflammatory effects of the expressed enzyme in a rat colitis model. Dig Dis Sci 54:1440–1448
109. Bauer J, Liebisch G, Hofmann C, Huy C, Schmitz G, Obermeier F, Bock J (2009) Lipid alterations in experimental murine colitis: role of ceramide and imipramine for matrix metalloproteinase-1 expression. PLoS One 4, e7197
110. Chumanevich AA, Poudyal D, Cui X, Davis T, Wood PA, Smith CD, Hofseth LJ (2010) Suppression of colitis-driven colon cancer in mice by a novel small molecule inhibitor of sphingosine kinase. Carcinogenesis 31:1787–1793
111. Sakata A, Ochiai T, Shimeno H, Hikishima S, Yokomatsu T, Shibuya S, Toda A, Eyanagi R, Soeda S (2007) Acid sphingomyelinase inhibition suppresses lipopolysaccharide-mediated release of inflammatory cytokines from macrophages and protects against disease pathology in dextran sulphate sodium-induced colitis in mice. Immunology 122:54–64
112. Mizushima T, Ito T, Kishi D, Kai Y, Tamagawa H, Nezu R, Kiyono H, Matsuda H (2004) Therapeutic effects of a new lymphocyte homing reagent FTY720 in interleukin-10 gene-deficient mice with colitis. Inflamm Bowel Dis 10:182–192
113. Deguchi Y, Andoh A, Yagi Y, Bamba S, Inatomi O, Tsujikawa T, Fujiyama Y (2006) The S1P receptor modulator FTY720 prevents the development of experimental colitis in mice. Oncol Rep 16:699–703
114. Maines LW, Fitzpatrick LR, French KJ, Zhuang Y, Xia Z, Keller SN, Upson JJ, Smith CD (2008) Suppression of ulcerative colitis in mice by orally available inhibitors of sphingosine kinase. Dig Dis Sci 53:997–1012
115. Manicassamy S, Reizis B, Ravindran R, Nakaya H, Salazar-Gonzalez RM, Wang YC, Pulendran B (2010) Activation of beta-catenin in dendritic cells regulates immunity versus tolerance in the intestine. Science 329:849–853
116. Keerthivasan S, Aghajani K, Dose M, Molinero L, Khan MW, Venkateswaran V, Weber C, Emmanuel AO, Sun T, Bentrem DJ et al (2014) β-Catenin promotes colitis and colon cancer through imprinting of proinflammatory properties in T cells. Sci Transl Med 6:225
117. Salim T, Sand-Dejmek J, Sjolander A (2014) The inflammatory mediator leukotriene D(4) induces subcellular beta-catenin translocation and migration of colon cancer cells. Exp Cell Res 321:255–266
118. Bäckhed F, Ley R, Sonnenburg J, Peterson D, Gordon J (2005) Host-bacterial mutualism in the human intestine. Science 307:1915–1920
119. Savage D (1977) Microbial ecology of the gastrointestinal tract. Annu Rev Microbiol 31:107–133
120. Xu J, Gordon JI (2003) Inaugural Article: Honor thy symbionts. Proc Natl Acad Sci 100:10452–10459

121. Turnbaugh PJ, Ley RE, Hamady M, Fraser-Liggett CM, Knight R, Gordon JI (2007) The human microbiome project. Nature 449:804–810
122. Larsen N, Vogensen F, van den Berg F, Nielsen D, Andreasen A, Pedersen B, Abu Al-Soud W, Sorensen S, Hansen L, Jakobsen M (2010) Gut Microbiota in Human Adults with Type 2 Diabetes Differs from Non-Diabetic Adults. PLoS One 5, e9085
123. Bergman EN (1990) Energy contributions of volatile fatty-acids from the gastrointestinal-tract in various species. Physiol Rev 70:567–590
124. Duncan S, Hold G, Barcenilla A, Stewart C, Flint H (2002) Roseburia intestinalis sp nov., a novel saccharolytic, butyrate-producing bacterium from human faeces. Int J Syst Evol Microbiol 52:1615–1620
125. Gibson G, Roberfroid M (1995) Dietary modulation of the human colonic microbiota: introducing the concept of prebiotics. J Nutr 125:1401–1412
126. Hylemon P, Harder J (1999) Biotransformation of monoterpenes, bile acids, and other isoprenoids in anaerobic ecosystems. FEMS Microbiol Rev 475–488
127. Chen T, Isomaki P, Rimpilainen M, Toivanen P (1999) Human cytokine responses induced by gram-positive cell walls of normal intestinal microbiota. Clin Exp Immunol 118:261–267
128. Louis P, Young P, Holtrop G, Flint HJ (2010) Diversity of human colonic butyrate-producing bacteria revealed by analysis of the butyryl-CoA: acetate CoA-transferase gene. Environ Microbiol 12:304–314
129. Munoz-Tamayo R, Laroche B, Walter E, Dore J, Duncan SH, Flint HJ, Leclerc M (2011) Kinetic modelling of lactate utilization and butyrate production by key human colonic bacterial species. FEMS Microbiol Ecol 76:615–624
130. Ghosh S, Dai C, Brown K, Rajendiran E, Makarenko S, Baker J, Ma C, Halder S, Montero M, Ionescu VA et al (2011) Colonic microbiota alters host susceptibility to infectious colitis by modulating inflammation, redox status, and ion transporter gene expression. Am J Physiol Gastrointest Liver Physiol 301:G39–G49
131. Lam YY, Ha CWY, Campbell CR, Mitchell AJ, Dinudom A, Oscarsson J, Cook DI, Hun NH, Caterson ID, Holmes AJ, Storlien LH (2012) Increased gut permeability and microbiota change associate with mesenteric fat inflammation and metabolic dysfunction in diet-induced obese mice. PLoS One 7
132. Gambero A, Marostica M, Abdalla Saad MJ, Pedrazzoli J (2007) Mesenteric adipose tissue alterations resulting from experimental reactivated colitis. Inflamm Bowel Dis 13: 1357–1364
133. Holmes E, Li JV, Athanasiou T, Ashrafian H, Nicholson JK (2011) Understanding the role of gut microbiome–host metabolic signal disruption in health and disease. Trends Microbiol 19:349–359
134. Brown K, DeCoffe D, Molcan E, Gibson DL (2012) Diet-induced dysbiosis of the intestinal microbiota and the effects on immunity and disease. Nutrients 4:1095–1119
135. Dicksved J, Halfvarson J, Rosenquist M, Jarnerot G, Tysk C, Apajalahti J, Engstrand L, Jansson JK (2008) Molecular analysis of the gut microbiota of identical twins with Crohn's disease. ISME J 2:716–727
136. Spor A, Koren O, Ley R (2011) Unravelling the effects of the environment and host genotype on the gut microbiome. Nat Rev Microbiol 9:279–290
137. Arthur JC, Perez-Chanona E, Muhlbauer M, Tomkovich S, Uronis JM, Fan TJ, Campbell BJ, Abujamel T, Dogan B, Rogers AB et al (2012) Intestinal inflammation targets cancer-inducing activity of the microbiota. Science 338:120–123
138. Wu S, Rhee K-J, Albesiano E, Rabizadeh S, Wu X, Yen H-R, Huso DL, Brancati FL, Wick E, McAllister F et al (2009) A human colonic commensal promotes colon tumorigenesis via activation of T helper type 17 T cell responses. Nat Med 15:1016–1022
139. Yamamoto ML, Maier I, Dang AT, Berry D, Liu J, Ruegger PM, Yang J, Soto PA, Presley LL, Reliene R et al (2013) Intestinal bacteria modify lymphoma incidence and latency by affecting systemic inflammatory state, oxidative stress, and leukocyte genotoxicity. Cancer Res 73:4222–4232

140. Yoshimoto S, Loo TM, Atarashi K, Kanda H, Sato S, Oyadomari S, Iwakura Y, Oshima K, Morita H, Hattori M et al (2013) Obesity-induced gut microbial metabolite promotes liver cancer through senescence secretome. Nature 499:97–101
141. Dennis KL, Wang Y, Blatner NR, Wang S, Saadalla A, Trudeau E, Roers A, Weaver CT, Lee JJ, Gilbert JA et al (2013) Adenomatous polyps are driven by microbe-instigated focal inflammation and are controlled by IL-10-producing T cells. Cancer Res 73:5905–5913
142. An D, Oh SF, Olszak T, Neves JF, Avci FY, Erturk-Hasdemir D, Lu X, Zeissig S, Blumberg RS, Kasper DL (2014) Sphingolipids from a symbiotic microbe regulate homeostasis of host intestinal natural killer T cells. Cell 156:123–133
143. Rueda R, Sabatel JL, Maldonado J, Molina-Font JA, Gil A (1998) Addition of gangliosides to an adapted milk formula modifies levels of fecal Escherichia coli in preterm newborn infants. J Pediatr 133:90–94
144. de La Serre CB, Ellis CL, Lee J, Hartman AL, Rutledge JC, Raybould HE (2010) Propensity to high-fat diet-induced obesity in rats is associated with changes in the gut microbiota and gut inflammation. Am J Physiol Gastrointest Liver Physiol 299:G440–G448
145. Kim KA, Gu W, Lee IA, Joh EH, Kim DH (2012) High fat diet-induced gut microbiota exacerbates inflammation and obesity in mice via the TLR4 signaling pathway. PLoS One 7, e47713
146. Bardou M, Barkun AN, Martel M (2013) Obesity and colorectal cancer. Gut 62:933–947

Role of Sphingolipids in Liver Cancer

Carmen Garcia-Ruiz, Albert Morales, and José C. Fernández-Checa

Abstract Hepatocellular carcinoma (HCC) is the most common form of liver cancer and one of the leading causes of cancer-related deaths in the world. Unfortunately, current therapy is inefficient and advanced HCC is highly resistant to chemotherapy, making early diagnosis crucial for survival. HCC develops in the course of chronic liver disease and inflammation. Progressive non-alcoholic fatty liver disease (NAFLD) is a major cause of chronic liver disease and HCC and NAFLD incidence is expected to grow due to its association with obesity and type 2 diabetes. Sphingolipids (SLs) biology has evolved from the inceptive view of being considered mere structural components of membrane bilayers to the current status of critical second messengers involved in the regulation of myriads cell functions, including cell death pathways. A crucial mechanism underlying the therapeutic potential of cancer treatment involves ceramide species, which stand as the basis for the mode of action of chemotherapy and radiotherapy. However, many solid tumors, including HCC, develop strategies that counter this increase in ceramides, which blunt therapy efficacy and promotes treatment resistance. Understanding these pathways may provide novel strategies to exploit the potential of ceramide and SLs to multiply the therapeutic effect of chemotherapy in the treatment of liver cancer. The present chapter summarizes these pathways and highlights the potential of combinational therapy based on maneuvers to increase SLs to combat liver cancer.

C. Garcia-Ruiz • A. Morales
Department of Cell Death and Proliferation, Instituto Investigaciones Biomédicas de Barcelona, CSIC, Barcelona, Spain

Liver Unit-Hospital Clinic-IDIBAPS and Centro de Investigación Biomédica en Red (CIBERehd), Barcelona, Spain

J.C. Fernández-Checa (✉)
Department of Cell Death and Proliferation, Instituto Investigaciones Biomédicas de Barcelona, CSIC, Barcelona, Spain

Liver Unit-Hospital Clinic-IDIBAPS and Centro de Investigación Biomédica en Red (CIBERehd), Barcelona, Spain

University of Southern California Research Center for Liver Diseases, Keck School of Medicine, USC, Los Angeles, CA, USA
e-mail: checa229@yahoo.com

Y.A. Hannun et al. (eds.), *Bioactive Sphingolipids in Cancer Biology and Therapy*, DOI 10.1007/978-3-319-20750-6_9

Keywords Ceramides • Ganglioside GD3 • Mitochondria • Hepatocellular carcinoma • Nonalcoholic fatty liver disease

1 Chronic Liver Disease: A Driving Path to Liver Cancer Development

Liver cancer remains a major health problem worldwide and one of the most common cancers in the World. Its predominant form, hepatocellular carcinoma (HCC) is the third leading cause of cancer-related mortality, exhibits resistance to chemotherapy, particularly in advanced stages, and its clinical manageability is dependent on early diagnosis [1, 2]. HCC normally arises in an environment of chronic inflammation and fibrosis, characteristic of advanced chronic liver disease, such as nonalcoholic fatty liver disease (NAFLD). NAFLD comprises a spectrum of liver disorders beginning with liver steatosis that can progress to steatohepatitis and culminate in cirrhosis and HCC [3, 4]. Associated with obesity and insulin resistance, NAFLD is a major health concern worldwide and a leading cause of liver transplantation and HCC. Unfortunately therapies for advanced and unresectable HCC are inefficient and the development of systemic agents has been challenging. The recommended treatment with sorafenib, a mulkinase inhibitor first used in the treatment of renal carcinoma [5], entails a significant survival of 2–3 months in patients with advanced HCC [2]. Moreover, recent findings with brivanib, a dual inhibitor of vascular-endothelial growth factor and fibroblast growth factor receptors that are implicated in HCC pathogenesis in the treatment of advanced HCC in patients have been disappointing [6]. This scenario of increasing incidence of HCC in the future and inadequate treatment demands for the characterization and discovery of more efficient therapeutic combinations.

Sphingolipids (SLs) are a family of lipids, which contain sphingosine base and an amide-linked fatty acyl chain that have an important structural function in cellular membranes. Their characteristic association with cholesterol defines specific domains of membrane bilayers that exhibit unique physical properties where key signaling platforms operate in the regulation of important cell processes, including proliferation, senescence, differentiation and death pathways. Among SLs ceramides are the best characterized and most studied prototype due to their description as a second messengers of cell death in response to stress, apoptotic triggers and chemotherapy and in the regulation of metabolism [7–10]. The function of ceramides as a trigger of cell death pathways underlies the action of many antitumoral therapies, including ionizing radiation and chemotherapy. Therefore, understanding the regulation and metabolism of ceramides may have an important impact in the design of more efficient combinational therapeutic treatments against HCC. In this review, we will summarize the functions of SLs, their role in chronic liver disease preceding HCC and the impact of modulating SLs in the treatment of HCC.

2 Pathways of Ceramide Generation and Metabolism

Cells generate and metabolize ceramides by different pathways [11, 12]. While this particular topic has been covered in greater detail in a separate chapter, we will briefly summarize some aspects of ceramide metabolism that may be of relevance in the context of liver cancer. The predominant pathway of ceramides generation occurs in the endoplasmic reticulum (ER). In this pathway, the amino acid serine is conjugated with the fatty acid palmitoyl CoA by the rate-limiting enzyme serine palmitoyl transferase (SPT), generating sphinganine, which is then acylated by (dihydro)ceramide synthase (CerS) to dihydroceramide. The subsequent dehydrogenation of dihydroceramide catalyzed by dihydroceramide desaturase (DES) generates ceramides. Of relevance to hepatocarcinogenesis, previous studies indicated the stimulation of SPT and hence SLs metabolism in Morris hepatoma cells [13]. Interestingly, six different CerS have been identified [11, 14], which exhibit tissue specific expression and variable substrate selectivity, thereby providing the basis for tissue specific synthesis of ceramides with varying acyl chain lengths. Despite this defined specific profile of ceramides synthesized by the different CerS, there are compensatory mechanisms that offset the absence of specific ceramide species; moreover, an increase in a particular CerS may regulate a specific ceramide pool that may affect the integrity and function of individual cell compartments, such as lysosomes, ER or mitochondria. For instance, mice deleted for CerS2 exhibit a compensatory increase in the levels of C16 and sphinganine in the liver [15]. These changes in ceramides homeostasis translate to increased rates of hepatocyte apoptosis, mitochondrial dysfunction and mitochondrial generation of ROS, as well as proliferation that progress to widespread formation of nodules of regenerative hepatocellular hyperplasia in aged mice. Progressive hepatomegaly and noninvasive HCC are observed in 10-months old CerS2 null mice [15].

The control of ceramides synthesis is regulated by substrate availability and membrane lipid composition [16, 17]. The availability of the substrate palmitoyl-CoA enhances the *de novo* synthesis of ceramides in the ER. Since free palmitate levels rise in obesity, metabolic syndrome and related disorders (i.e., NAFLD), sphingolipid flux through the *de novo* pathway is enhanced [18, 19].

While the generation of ceramides by *de novo* synthesis is slow but sustained, cells can instantaneously generate ceramides by a mechanism involving sphingomyelin (SM) hydrolysis by activation of sphingomyelinases (SMases). In response to many deleterious stimuli causing stress, apoptosis, chemotherapy and cell death, cells activate SMases leading to a rapid and transient release of ceramides in specific sites that engage particular signaling pathways [11, 12, 20, 21]. Several mammalian SMases have been characterized, which are classified according to their optimal pH (alkaline, neutral or acid) with a differential participation in cell death. While neutral sphingomyelinase (NSMase)-induced ceramide generation has been described as a critical lipid mediator in inflammatory diseases and *X. laevis* oocyte maturation [22, 23], acid SMase (ASMase) has been extensively described as a signaling intermediate in cell death pathways and metabolic liver diseases [24–30]. ASMase catalyzes the formation of ceramide from SM primarily within the endo-lysosomal compartment, although it can be secreted extracellularly through Golgi trafficking as a secretory ASMase (S-SMase) [20, 21]. Of relevance for liver diseases, ASMase knockout mice have been shown resistant to Fas or TNF-mediated liver failure [25, 26, 31], ischemia/reperfusion [32], hepatocyte apoptosis due to Cu^{2+} accumulation, characteristic of Wilson disease [27] and diet-induced steatohepatitis [29, 30] indicating that the ASMase/Ceramide pathway is critical in liver pathophysiology.

Ceramides can be transformed in other derivatives and complex SLs by different pathways. The *de novo* synthesized ceramides in the ER can be transported to the Golgi by a non-vesicular mechanism involving the protein CERT, where serves as the precursor for SM synthesis by SM synthases (SMS). There are two SMS, SMS1 and SMS2, which are localized in distinct cellular locations, Golgi and plasma membrane, respectively [33, 34].

In the Golgi, ceramides can also be transformed into complex glycosphingolipids and gangliosides due to the addition of glucose to ceramides to generate glucosylceramide (GlcCer) catalyzed by glucosylceramide synthase (GCS) (Fig. 1). Non-vesicular transport of GlcCer from its site of synthesis (early Golgi) to distal Golgi is carried out by FAPP2 contributing to the synthesis of glycosphingolipids [35]. The subsequent addition of sialic acid moiety to GM3 generates ganglioside GD3, catalyzed by GD3 synthase (Fig. 1). Ganglioside GD3 is a bioactive lipid with a prominent role in cell death pathways by targeting mitochondria, which is of relevance in liver pathologies [12]. Therefore, the regulation of ceramide along with

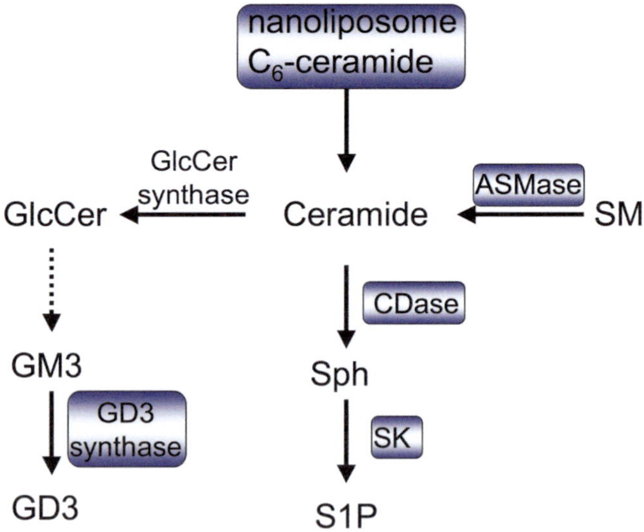

Fig. 1 Ceramide metabolism and role in HCC. Ceramides are metabolized by several pathways including its deacylation by ceramidases to sphingosine (Sph), which can be phosphorylated to S1P by sphingosine kinase (SK). In addition, following transport to Golgi it is converted to glucosylceramide (GlcCer) by GlcCer synthase, which is further processed to glycosphingolipids such as GD3 synthesized from GM3 by GD3 synthase. The generation of ceramides by nanoliposomes or ASMase regulate HCC chemotherapy and in vivo tumor growth

the subsequent GD3 generation in the Golgi may constitute an effective measure to control hepatocellular cancer cell death [36].

The predominant pathway of ceramides catabolism is its deacylation to sphingosine by ceramidases (CDases), followed by the phosphorylation of sphingosine to sphingosine-1-phosphate (S1P) by sphingosine kinases (SK) (Fig. 1). Three CDase isoforms with distinctive pH optimum, acid, alkaline and neutral have been described [37, 38], of which the acidic CDase is the isoform that has been more clearly involved in the regulation of progression of various types of cancers [39, 40]. S1P plays an antiapoptotic role and hence is considered a prosurvival SLs due to a combined mechanism of action including the antagonism of ceramide-mediated cell death pathways from inside cells, as well as by activating proliferative pathways from outside the cell via five specific G protein-coupled S1P receptors (S1PR1-5) [41–43].

S1P can then be irreversibly cleaved by the S1P-lyase to phosphoethanolamine and hexadecenal [44]. Alternatively, S1P can be dephosphorylated by sphingosine phosphatases (SPP) back to sphingosine and thereby recycled for ceramide formation [45]. S1P-lyase, SPP and SK2 are localized at the ER and regulate intracellular levels of S1P [46, 47]. As S1P and its metabolic precursor sphingosine exert opposing functions in the regulation of cell survival, the activities of these enzymes are critical for determination of the cellular fate and they can become potential targets in cancer cell biology.

3 Role and Mechanisms of Sphingolipids in Cell Death

In addition to the cardinal role of mitochondria in energy generation, they also play a strategic function in the regulation of cell death, including apoptosis (caspase-dependent and independent) or necrosis. Apoptosis is a specific form of cell death characterized by biochemical events that ultimately lead to cell fragmentation into compact membrane-enclosed structures, called "apoptotic bodies" that are taken up by neighboring cells and phagocytes, that normally prevent inflammation and tissue damage [48]. Apoptosis is induced via two main routes involving either the mitochondria (the intrinsic pathway) or the activation of death receptors (the extrinsic pathway). While ceramide has been shown to mediate death receptor-mediated apoptosis, the role of ceramide in regulating the intrinsic pathway may be of particular relevance in cancer cell biology. In addition to mitochondrial targeting, ceramide also induces ER stress and regulates autophagy, while inactivation of survival pathways by gangliosides modulates susceptibility to cell death in response to stress (Fig. 2).

3.1 *Mitochondrial Targeting*

Consistent with the recognized role of ceramide as an intermediate of apoptosis, pioneering work established the ability of ceramide to interact with components of the mitochondrial electron transport chain accounting for the stimulation of ROS,

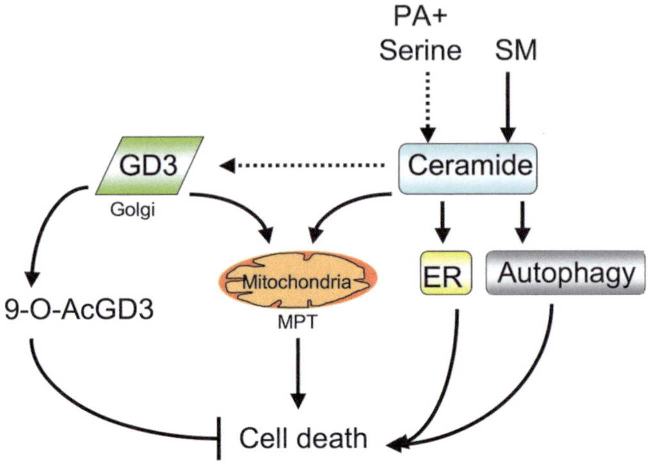

Fig. 2 Cell death pathways triggered by ceramide. Ceramide generation by *de novo* synthesis of SM hydrolysis by ASMase can elicit the activation of cell death pathways by targeting mitochondria, ER stress or autophagy. In addition, ceramide conversion to GD3 targets mitochondria to cause MPT and cell death. The acetylation of GD3 to 9-O-acetyl-GD3 disables the apoptotic potential of GD3

mitochondrial depolarization and mitochondrial dysfunction [49–51]. These effects contributed to the mitochondrial membrane permeabilization and release of cytochrome c into the cytosol leading to the apoptosome assembly and activation. In cell-free assays using purified rat liver mitochondria it was first shown that the addition of ceramide C2 induced ROS generation predominantly from the complex III of the mitochondrial electron carriers, thus contributing to the mitochondrial depolarization and dysfunction observed during cell death [49]. This outcome paralleled the findings of purified mitochondria from cells exposed to stress and apoptosis triggers, such as TNF, Fas, or UV irradiation [49, 52]. These results imply that the stimulated ceramide levels induced by TNF traffic to mitochondria or that ceramide is generated in situ in these organelles. In line with the latter possibility Birbes et al. observed that the enforced mitochondrial targeting of NSMase in MCF7 cells resulted in mitochondrial ceramide increase that caused cytochrome c release and apoptotic cell death [53]. Moreover, it has been shown that ceramide can be generated in these organelles by a reverse ceramidase activity or a ceramide synthase [54, 55], implying that mitochondria have the ability to generate ceramide *de novo.* These observations suggest the existence of an independent and highly regulated SL metabolism in mitochondria that contribute to the accumulation of ceramide in these organelles modulating cell death pathways. Mitochondria are functionally and physically associated with heterotypic membranes, and the mitochondrial apoptosis pathway is regulated by members of the Bcl-2 family proteins, most prominently by Bax and Bak that control mitochondrial outer membrane permeabilization, cytochrome c release and apoptosis. The dissociation of heterotypic membranes from mitochondria inhibited Bak/Bax-dependent cytochrome c release. Recent evidence has shown that SL metabolism plays a key role in mitochondrial apoptosis by regulating Bax/Bak activation [56]. Furthermore, S1P and hexadecenal cooperated specifically with Bak and Bax, respectively, to induce mitochondrial outer membrane permeabilization and apoptosis. These findings suggest that SL metabolism cooperates with Bak and Bax to promote the mitochondrial pathway of apoptosis. Moreover, it has been demonstrated that ceramide generation in the mitochondrial outer membrane of mammalian cells upon irradiation, forms a platform into which Bax inserts, oligomerizes and functionalizes as a pore, causing mitochondrial membrane permeabilization and apoptosis [57].

Besides the regulation of mitochondrial apoptosis, a novel function of ceramides in cell death has described the targeting of mitochondria to autophagosomes leading to massive mitophagy [58]. Treatment of human cancer cells with C(18)-pyridinium ceramide treatment or endogenous C(18)-ceramide generation by CerS1 expression caused autophagic cell death, independent of apoptosis. C(18)-ceramide-induced lethal autophagy by targeting of mitochondria to LC3B-II-containing autophagolysosomes (mitophagy) through direct interaction between ceramide and LC3B-II, resulting in Drp1-dependent mitochondrial fission, leading to inhibition of mitochondrial function and oxygen consumption.

3.2 ER Stress, Autophagy and Lysosomal Membrane Permeabilization

The ER plays a critical homeostatic function in cells, mainly in the control of protein and lipid synthesis and traffic. Disruption in this function leads to a specific stress response called the unfolded protein response (UPR) with the aim to reinstate homeostasis, which has been reviewed elsewhere [59, 60]. ER membranes are rich in gangliosides and *de novo* ceramides synthesis occurs in the ER, implying that SLs regulate UPR and the ER stress. For instance, accumulation of GluCer, which is the main storage product in Gaucher disease, increases Ca^{2+} mobilization from intracellular stores in cultured neurons via ryanodine receptor-mediated Ca^{2+} release from the ER, while microsomes from the brain of a Sandhoff disease mouse model that accumulate GM2 show reduced rates of Ca^{2+} uptake due to reduced Ca^{2+}-ATPase SERCA activity [61, 62]. Of relevance for cancer it has been shown that the *de novo* synthesized ceramides elicit an ER stress-dependent death mechanism in glioma cells in response to tetra-dydrocannabinol that is inhibited upon inhibition of SPT. [63] Ceramides synthesized *de novo* upregulated the levels of p8 (also known as candidate of metastasis 1), leading to the activation of ATF-4 and stress-regulated protein tribble homolog 3 (TRB-3), events prevented by ISP-1, a SPT inhibitor. These effects were reproduced by C2-ceramide and blocked by pharmacological inhibition of the mitochondrial respiratory chain, supporting an ER-mitochondria cross-talk in ceramide-induced cell death. Interestingly, this ceramide-ER stress axis mediated cell death seems to be specific for transformed or malignant cancer cells but not normal cells.

Autophagic cell death is a distinctive form of cell death known as type II programmed cell death, in which autophagy triggers cell death pathways [64]. The characteristics of autophagic cell death include its independence from apoptosis, and prevention by agents that suppress autophagy. Sustained autophagy-mediated cell death is caused by irreversible cellular atrophy and dysfunction due to the loss of cytosolic components and organelles. Ceramides regulate autophagy in a multidisciplinary fashion, including activation the phosphatase PP2A [65], suppression of mTOR activity in a PP1/PP2A dependent manner [66], inhibition of amino acid transporters leading to AMPK-dependent autophagy induction [67], and by enhancing Beclin1 expression [68]. Furthermore, ceramide-induced ER stress also contributes to autophagy induction [69]. The ultimate effect of ceramide-mediated autophagy in cell death is cell type specific and stimuli dependent [70]. For example, in the HCC cell line Hep3B treatment with ceramide C2 induced lethal autophagy by a mechanism involving JNK activation, which upregulated Beclin1 expression [71].

Lysosomal membrane permeabilization (LMP) has been involved in apoptotic and non-apoptotic cell death by the release of lysosomal proteases. Furthermore, LMP is a mechanism involved in lipotoxicity of relevance in NAFLD [72] and lysosomal cholesterol accumulation due to increase in lysosomal SM due to ASMase deficiency has been recently shown to reduce LMP and palmitic acid induced apoptosis in primary hepatocytes [30]. Intriguingly, increasing evidence indicates that tumor cell lysosomes are more fragile than normal lysosomes and are more susceptible to LMP and hence tumor demise. Recent studies described that ASMase inhibition emerges as a novel mechanism to destabilize cancer cell lysosomes to induce cell death [73]. Cationic amphiphilic drugs (CAD) inhibit ASMase by interfering the binding of ASMase to its essential lysosomal cofactor, bis (monoacylglycerol)phosphate, increase lysosomal SM content and induce LMP and tumor cell death and reduced tumor growth *in vivo* and revert multidrug resistance. Quite intriguingly, despite the high affinity of SM for cholesterol [74, 75], the increase in lysosomal SM levels upon ASMase inhibition by CAD was not accompanied by enhanced cholesterol levels, which are known to regulate LMP, as described in fibroblasts treated with U18666A [76]. While further understanding of the mechanisms involved in the fragility of tumor lysosomes and LMP is needed, database search demonstrated a highly significant cancer-associated reduction in ASMase expression in several microarray studies comparing mRNA levels in tumors originating from gastrointestinal tract, liver, head and neck, kidney, pancreas, cervix, lung and brain [73].

4 Sphingolipids in Liver Disease Preceding HCC

The role of SLs in progressive NAFLD and liver cancer remains to be fully characterized. Ceramide has been mainly studied in the context of glucose homeostasis and insulin signalling, which is closely associated with NAFLD [10]. Moreover, deletion of CerS2 causes liver injury and compensatory hepatocellular proliferation progressing to regenerative hepatocellular hyperplasia and HCC [15]. Moreover, the role of NSMase in NAFLD has been limited to the study of FAN (factor associated with NSMase activation) in a murine model of NASH induced by feeding mice a methionine and choline deficient (MCD) diet [77]. These studies indicate that FAN knockout mice fed MCD exhibit signs of NAFLD, demonstrating that FAN is dispensable in this model of NASH. In contrast to the contribution of these pathways of ceramide generation to NAFLD, recent findings have reported a critical role for ASMase in metabolic liver diseases and in the development of NAFLD [29, 30].

4.1 ASMase Promotes NASH and Liver Fibrosis

Nonalcoholic steatohepatitis (NASH) is an advanced stage of NAFLD and is characterized by steatosis, hepatocellular injury and inflammation. While fibrosis (generally in zone 3) is not strictly necessary for NASH diagnosis, its presence reflects disease progression. Recent findings have revealed a key role for ASMase in NASH. ASMase overexpression has been reported in adipose tissue of *ob/ob* mice, mice fed an MCD diet, and in liver samples from patients with NASH [78–80]. ASMase deletion superimposed on the genetic background of LDL receptor deficiency prevented diet-induced hyperglycemia [81]. *ASMase*$^{-/-}$ mice fed HFD for 12 weeks were resistant to HFD-induced activation of lipogenic enzymes, including Srebp-1c, Acc, Dgat-2 and Fas [30]. These findings translated to resistance to HFD-mediated steatosis, with similar findings observed when mice were fed MCD diet. Furthermore, ASMase inhibition with amitriptyline protected wild type mice against HFD-induced steatosis, liver injury, inflammation and fibrosis. ASMase has been recently described to regulate HSC activation and hence liver fibrogenesis. Selective stimulation of ASMase, but not NSMase, occurs during the transdifferentiation/activation of primary mouse HSCs to myofibroblast-like cells, coinciding with the processing of downstream effectors cathepsin B (CtsB) and D (CtsD) [80]. ASMase inhibition or genetic knockdown by small interfering RNA blunted CtsB/D processing and prevented the activation and proliferation of mouse and human HSCs, revealing a key role for ASMase in HSC biology and liver fibrosis. In line with these findings, it has been shown that amitriptyline, an ASMase inhibitor, reduces established hepatic fibrosis induced by CCl_4 in mice [82].

4.2 GD3 Acetylation Regulates Fibrosis

The pattern of gangliosides synthesized in the liver may be cell-dependent, with hepatic stellate cells (HSC) expressing a distinct and selective GSLs synthesis ability during activation [83]. The total ganglioside content in a myofibroblastic GRX cell line has been reported to be approximately half the ganglioside content as compared to lipocytes. This outcome mirrored the activity of the GM2 synthase, indicating that the control of this difference probably operates at the level of synthesis rather than of the turnover of gangliosides.

Using a large scale sequencing of a 3′-directed cDNA library during HSC activation, it has been reported that activated HSC cells express *O*-acetyl disialoganglioside synthase (OAcGD3S) resulting in increased levels of O-acetyl GD3 in active HSC cells as well as in human cirrhotic livers [33]. These observations are of potential significance not only because acetylated gangliosides have been reported as mitogens and stimulate cell proliferation [84] but also because acetylation of GD3 impairs its apoptotic potential [85]. Since GD3 ganglioside promotes apoptosis via a dual mechanism involving mitochondrial targeting and suppression of NF-κB survival program [86–88], the acetylation of GD3 during HSC activation may contribute to liver fibrogenesis and apoptosis resistance. Thus, targeting OAcGD3S may arise as a novel promising approach in the treatment of liver fibrosis by accelerating or sensitizing activated HSC to apoptotic pathways.

5 Role of Sphingolipids in HCC and Metastasis

As mentioned above SLs play a key role in the activation of cell death and many cancer cell therapies such as chemotherapy and ionizing radiation exert their therapeutic effects by increasing ceramides levels in cancer cells. Hence, strategies that modulate SLs, in particular ceramides content may be of potential relevance to combat cancer. However, the efficacy of these approaches aimed to raise intratumoral ceramide levels may be dependent on the singularity of cancer cells to regulate ceramide homeostasis and the type of chemotherapy used. Moreover, the balance of ceramide to S1P generation is known to modulate cell fate in response to cancer therapy and alterations in SLs metabolism may thus contribute to multidrug resistance in cancer cells [11]. Thus, in the following sections we summarize the strategies and pathways that regulate ceramide homeostasis in HCC, which in turn determine the efficacy chemotherapy treatment (Fig. 1).

5.1 Acid Ceramidase

CDases modulate the ceramide/S1P ratio through phosphorylation of sphingosine by SKs, and thus these ceramide metabolizing enzymes promote carcinogenesis and determine the efficacy of cancer therapy. Acid CDase inhibition by the ceramide analogue, B13, has been reported to induce apoptosis in cultured human colon cancer cells and diminish liver metastases *in vivo* [89]. However, the use of B13 as an acid CDase inhibitor may have limited therapeutic potential due to its poor ability to traffic to lysosomes to target the resident acid CDase. Interestingly, cancer cells evade chemotherapy-mediated cell death by specifically inducing acid CDase. For instance, anthracyclines, such as daunorubicin, have been shown to activate acid CDase by a post-transcriptional-dependent mechanism in HCC cells as well as in primary cells from murine liver tumors, and acid CDase silencing or pharmacological inhibition with N-oleoylethanolamine (NOE) enhanced the ceramide to S1P ration, sensitizing HCC cells [12]. The enforced ceramide increase following acid CDase antagonism targeted mitochondria resulting in reduced in vivo tumor cell growth after daunorubicin treatment. Although these findings imply that CDases promote carcinogenesis through the enhanced conversion of the proapoptotic SL ceramide to its antiapoptotic derivative S1P, recent observations have shown that a combination of tumor necrosis factor-α (TNF) and cycloheximide (CHX) induced apoptosis in HTC cells, a rat hepatoma line, by a mechanism dependent on sphingosine generation and LMP activation [90]. Acid CDase silencing prevented TNF/CHX-mediated LMP and HTC apoptosis by blocking sphingosine formation. Thus, targeting CDase may be an attractive strategy to enhance cancer therapy.

5.2 *Inhibition of CerS by Fumonisin B1*

Fumonisin B_1 (FB1) is the most important component of the fungus *Fusarium verticillioides*, which can contaminate corn and other cereals. FB1 is hepatotoxic and nephrotoxic with the earliest histological changes apoptosis followed by regenerative cell proliferation. Dietary exposure to FB1 has been linked to human cancer in certain parts of the world, and treatment with FB1 causes oval cell proliferation and liver tumors in rats. For instance, FB1 treatment in male F344 rats caused OV-6+ oval cell proliferation, correlating with the appearance of hepatic adenomas and cholangiofibromas [91]. Moreover, FB1 has been shown to induce global DNA hypomethylation and histone demethylation in HepG2 cells that causes chromatin instability and may lead to liver tumourigenesis [92]. However, case-control studies nested within two large cohorts in China showed no statistically significant association between FB1 exposure and HCC, thus discarding the participation of FB1 in human liver cancer [93]. These findings, although somewhat preliminary, do not support the association between FB1 and HCC. The mechanism underlying the promotion of HCC by FB1 is poorly understood. As FB1 prevents SLs neosynthesis by inhibiting CerS [94], it can be speculated that the prevention of ceramide generation by FB1 may contribute to its hepatocarcinogenic effect. However, inhibition of ceramide synthesis by FB1 is accompanied by increased sphingosine levels and activation of other SLs metabolizing enzymes, including ASMase as well as the sphingosine kinase 1 [95]. Besides CerS inhibition, FB1 has been described to target other pathways, inhibiting protein serine/threonine phosphatases, particularly PP5 while stimulating the activation of mitogen-activated protein kinase [96, 97]. Hence, further research is needed to characterize the role of FB1 in human liver cancer and the underlying mechanisms.

5.3 *Neutral and Acid Sphingomyelinases*

Integrative genomic methylation analysis of human HCC samples identified several tumor suppressors genes, including NSMase2 [98]. Overexpression of NSMase2 by stable transfection of inducible constructs into an HCC cell line reduced cell proliferation, while knocking down NSMase2 expression with small hairpin RNA promoted cell invasion and migration in vitro and increased the formation of tumors after subcutaneous injection or orthotopic transplantation into mice, confirming the role of NSMase2 as a tumor suppressor gene in HCC. Low levels of NSMase2 were associated with early recurrence of HCC after curative surgery in an independent patient cohort. Moreover, sodium selenite has been shown to induce cell death in different types of cancer cells, including HCC, by a mechanism involving NSMase. The ability of selenite to reduce the viability of the HCC cell line Huh7 was accompanied by increased levels of long chain C14-, C16-, C18- and C18:1- ceramides. Inhibition of NSMase with 3-O-methylsphingosine significantly reduced the cytotoxic effect of selenite [99]. The cytotoxic effect of selenite was potentiated by the SK1 inhibitor SK1-II and was specific for HCC cell lines but not to the non-tumorigenic hepatocyte cell line MIHA. These findings suggest that a combination of selenite and SK1 inhibitors could of potential to kill liver cancer cells at doses that are non-toxic to normal liver cells.

ASMase transcription is regulated by the methylation status in CpG sites [100]. The epigenetic regulation of ASMase by methylation has been shown in lymphoblasts from patients with deficient ASMase expression [101]. Moreover, incubation of lymphoblasts with the demethylating agent 5-aza-2′-deoxycytidine increased the expression of ASMase. A recent study explored the potential of recombinant ASMase in the treatment of HCC in combination with sorafenib, the only approved systemic treatment for unresectable HCC. Recombinant ASMase/sorafenib treatment reduced the viability of Huh7 liver cancer cells more than sorafenib. In vivo, using a subcutaneous Huh7 tumor model, mouse survival was increased and proliferation in the tumors decreased to a similar extent in both sorafenib and recombinant ASMase/sorafenib treatment groups [102]. However, combination of recombinant ASMase and sorafenib treatment significantly lowered tumor volume, increased tumor necrosis, and decreased tumor blood vessel density compared to sorafenib. Quite intriguingly, these results were obtained despite poor delivery of recombinant ASMase to the tumors. The mechanism underlying the low incorporation of recombinant ASMase in the tumors remains unknown, although it could be mediated by low expression of the mannose receptor, which is involved in the endocytosis of ASMase.

Furthermore, recent findings examined the role of ASMase in host defense against liver metastasis. The involvement of ASMase in liver metastasis of colon cancer was analyzed in mice deficient in ASMase inoculated with SL4 colon cancer cells to produce metastatic liver tumors. ASMase$^{-/-}$ mice exhibited enhanced tumor growth and reduced macrophage accumulation in the tumor, accompanied by decreased numbers of hepatic myofibroblasts (hMFs), which express tissue inhibitor of metalloproteinase 1 (TIMP1), around the tumor margin [103]. Tumor growth was increased by macrophage depletion or by Timp1 deficiency, but was decreased by hepatocyte-specific ASMase overexpression, which was associated with increased S1P production. These findings indicate that ASMase in the liver inhibits tumor growth through cytotoxic macrophage accumulation and TIMP1 production by hMFs in response to S1P. Targeting ASMase may represent a new therapeutic strategy for treating liver metastasis of colon cancer.

5.4 Gangliosides

Early finding have suggested a glycolipid-sorting defect in hepatocarcinogenesis where gangliosides, and especially monosialogangliosides, e.g., GM3, are diverted to internal membranes rather than being correctly transported to the cell surface. For instance, while 70–80 % of the lipid bound sialic acid residues were associated with plasma membrane in normal liver tissue, this percentage reduced significantly in liver tumor cells [104]. Moreover, recent observations indicated that GM3-suppressed hepatoma cell motility and migration by a mechanism involving the inhibition of phosphorylation of EGFR and the activity of PI3K/AKT signaling pathway [105]. Using two related sublines derived from murine ascites hepatoma cell lines Hca-F25, selected for their markedly different metastatic potential to lymph nodes, it was found that the two cell lined differed in the amount of GM3. The low metastatic cell line (HcaP) contained high levels of ganglioside GM3, whereas ganglioside GM2 was the preferential species found in the high metastatic cell line (HcaF). Suppression of GM3 enhanced the mobility and migration of the low metastatic HcaP cells in vitro, while increasing the GM3 content in high metastatic HcaF cells by addition of exogenous GM3 inhibited the mobility and migration.

As mentioned above, ganglioside GD3 has been identified as a lipid death effector, with a dual mechanism involving its interaction with mitochondria leading to activation of apoptotic pathways and the suppression of survival programs mediated by NF-κB activation. To explore the potential effect of this dual nature of GD3 in HCC, the role of GD3 synthase overexpression in the susceptibility to hypoxia was examined in HCC cells and in vivo tumor growth [106]. Overexpression of GD3 synthase increases the levels of GD3, which is synthesized from endogenous GM3, rendering Hep3B cells susceptible to hypoxia-induced ROS generation by suppressing the hypoxia-mediated NF-κB activation via tyrosine kinase Src, which results in lower expression of the κB-dependent antioxidant Mn-SOD. Moreover, GD3 synthase overexpression reduces tumor growth in vivo in Hep3B-GD3 xenografts. In line with these findings in HCC, a worst prognosis for breast cancer patients correlates with low tumor expression of GD3 synthase [107]. Thus, these findings identify GD3 as a potential relevant therapeutic agent that exploits hypoxia as a cancer-reducing factor due to the overproduction of ROS.

5.5 Nanoliposomal Ceramide

Given the function of ceramide as an effective cell death trigger recruiting diverse pathways and mechanisms, strategies that result in its accumulation may stand as an interesting approach to enhance the efficacy of current cancer therapy. Recent advances in nanotechnology have illustrated the feasibility of generating nanoliposomes that encapsulate hydrophobic compounds, like ceramide. The FDA has approved several nanoliposomal formulations containing anti-cancer drugs.

One of the features of cancer is a metabolic reprogramming that allows better adaptation to keep an unlimited growth and survival even in suboptimal conditions such as nutrient deprivation. Cancer cells exhibit the ability to avidly take up glucose and convert it to lactate, even in the presence of sufficient oxygen. This altered glycolytic dependency known as Warburg effect favors less efficient generation of ATP compared to the oxidative phosphorylation process, which occurs in normal cells. In this regard, recent observations have shown that C6-ceramide nanoliposomes target the Warburg effect in chronic lymphocytic leukemia [108]. The efficacy of nanoliposomal ceramide formulation has been tested in HCC as well [109]. Nanoliposomal C6-ceramide, but not ghost nanoliposomes (without ceramide), induced apoptotic cell death of SK-HEP-1 cells in vitro, concomitant with an accumulation of cells in the G2 phase of the cell cycle and decreased phosphorylation of AKT. In vivo administration of ceramide nanoliposomes to mice reduced SK-HEP-1 tumor cell growth in xenografts associated with decreased tumor cell vascularization and proliferation. Moreover, ceramide delivery by nanoliposomes has been recently shown to potentiate the anticancer effect of current drugs, such as vinblastine, a tubulin binding alkaloid that disrupts microtubule function, in hepatocarcinoma and colorectal cancer models [110]. Nanoliposomal C6-ceramide and vinblastine synergistically killed HCC and colorectal cancer cells in vitro and reduced tumor growth in vivo. These effects were accompanied by a disruption in autophagy flux and increased p62 levels. Although vinblastine's effect on autophagy is non-specific, likely reflecting disruption of microtubule formation, Beclin-1 silencing suppressed combination treatment-induced apoptosis in vitro, suggesting that the synergistic antitumor activity of the nanoliposomal C6-ceramide and vinblastine combination is potentially mediated by an autophagy mechanism. Moreover, it has been shown that the combination of nanoliposomal ceramide and sorafenib synergistically decrease melanoma and breast cancer cell survival [111], suggesting that a similar approach may be of benefit in the treatment of HCC. However, as cancer cells develop strategies to decrease ceramide levels by metabolizing it into other non-apoptotic derivatives, such as S1P, the combination of sorafenib with inhibitors of SK may be of potential relevance in HCC. In this regard, recent findings have explored the efficacy of a SK2 inhibitor, ABC204649, with sorafenib in hepatocellular carcinoma xenografts [112]. Combining ABC294640 and sorafenib led to a decrease in the levels of phosphorylated ERK in SK-HEP-1 cells and reduced in vivo HepG2 cell xenografts. These findings correlated with a decrease in the levels of S1P in plasma of mice treated daily with 100 mg/kg of ABC294640 for 5 weeks, which was not affected by co-administration of sorafenib.

6 Concluding Remarks

Although a significant progress has been made in the treatment of liver cancer and HCC in particular, with the introduction of sorafenib, the incidence of HCC is expected to grow in the near future as it is driven by the increasing incidence of

NAFLD worldwide. Besides genetic studies linking the pattern of gene expression with prognosis and disease recurrence, a better understanding of the molecular pathways, genetic and epigenetic reprogramming, abnormal signaling pathways and lipid homeostasis contributing to HCC may foster the identification of novel therapeutic approaches. In spite of the well-known ability of ceramides to activate cell death pathways, mediating the mechanism of action of many chemotherapeutics, the exploitation of forced generation of ceramides alone or in combination with FDA approved therapies such sorafenib has been explored in a very limited fashion as potential therapy for HCC. In addition to basic studies aimed to uncover the molecular metabolic alterations of sphingolipid metabolism in HCC models, the characterization of a sphingolipidomic signature in samples from HCC patients in different stages of the disease may be required to maximize the potential of targeting sphingolipids in combination therapy to combat HCC. With the development of new technology including the cluster TOF-SIMS imaging in HCC samples may pave the way for a better characterization of the profile of SLs metabolites to tailor more specific approaches exploiting the ability of ceramide to kill tumor cells.

Acknowledgments The work was supported by grants SAF-2011-23031, SAF-2012-34831 from Plan Nacional de I+D, Spain, the center grant P50-AA-11999 Research Center for Liver and Pancreatic Diseases funded by NIAAA/NIH, a grant from Fundació Marató de TV3, La Mutua Madrileña, PI11/0325 (META) grant from the Instituto Salud Carlos III, and by the support of CIBERehd.

References

1. Motola-Kuba D, Zamora-Valdes D, Uribe M, Mendez-Sanchez N (2006) Hepatocellular carcinoma. An overview. Ann Hepatol 5:16–24
2. Gish RG, Finn RS, Marrero JA (2013) Extending survival with the use of targeted therapy in the treatment of hepatocellular carcinoma. Clin Adv Hematol Oncol 11(Suppl 5):1–22
3. Tilg H, Diehl AM (2000) Cytokines in alcoholic and nonalcoholic steatohepatitis. N Engl J Med 343:1467–1476
4. Ratziu V, Bellentani S, Cortez-Pinto H, Day C, Marchesini G (2010) A position statement on NAFLD/NASH based on the EASL 2009 special conference. J Hepatol 53:372–384
5. Schöffski P, Dumez H, Clement P, Hoeben A, Prenen H, Wolter P, Joniau S, Roskams T, Van Poppel H (2006) Emerging role of tyrosine kinase inhibitors in the treatment of advanced renal cell cancer: a review. Ann Oncol 17:1185–1196
6. Bolos D, Finn RS (2014) Systemic therapy in HCC: lessons from brivanib. J Hepatol 61:947–950
7. Ryland LK, Fox TE, Liu X, Loughran TP, Kester M (2011) Dysregulation of sphingolipid metabolism in cancer. Cancer Biol Ther 11:138–149
8. Ogretmen B, Hannun YA (2004) Biologically active sphingolipids in cancer pathogenesis and treatment. Nat Rev Cancer 4:604–616
9. Ponnusamy S, Meyers-Needham M, Senkal CE, Saddoughi SA, Sentelle D, Selvam SP, Salas A, Ogretmen B (2010) Sphingolipids and cancer: ceramide and sphingosine-1-phosphate in the regulation of cell death and drug resistance. Future Oncol 6:1603–1624
10. Chavez JA, Summers SA (2012) A ceramide-centric view of insulin resistance. Cell Metab 15:585–594

11. Hannun YA, Obeid LM (2008) Principles of bioactive lipid signalling: lessons from sphingolipids. Nat Rev Mol Cell Biol 9:139–150
12. Morales A, Lee H, Goñi F, Kolesnick R, Fernandez-Checa J (2007) Sphingolipids and cell death. Apoptosis 12:923–939
13. Williams RD, Nixon DW, Merrill AH Jr (1984) Comparison of serine palmitoyltransferase in Morris hepatoma 7777 and rat liver. Cancer Res 44(5):1918–1923
14. Mizutani Y, Kihara A, Chiba H, Tojo H, Igarashi Y (2008) 2-Hydroxy-ceramide synthesis by ceramide synthase family: enzymatic basis for the preference of FA chain length. J Lipid Res 49:2356–2364
15. Pewzner-Jung Y, Brenner O, Braun S, Laviad EL, Ben-Dor S, Feldmesser E, Horn-Saban S, Amann-Zalcenstein D, Raanan C, Berkutzki T, Erez-Roman R, Ben-David O, Levy M, Holzman D, Park H, Nyska A, Merrill AH Jr, Futerman AH (2010) A critical role for ceramide synthase 2 in liver homeostasis: II. Insights into molecular changes leading to hepatopathy. J Biol Chem 285:10911–10923
16. Bartke N, Hannun YA (2009) Bioactive sphingolipids: metabolism and function. J Lipid Res 50:S91–S96
17. Breslow DK, Weissman JS (2010) Membranes in balance: mechanisms of sphingolipid homeostasis. Mol Cell 40:267–279
18. Bikman BT, Summers SA (2011) Ceramides as modulators of cellular and whole-body metabolism. J Clin Invest 121:4222–4230
19. Holland WL, Summers SA (2008) Sphingolipids, insulin resistance, and metabolic disease: new insights from in vivo manipulation of sphingolipid metabolism. Endocr Rev 29:381–402
20. Smith EL, Schuchman EH (2008) The unexpected role of acid sphingomyelinase in cell death and the pathophysiology of common diseases. FASEB J 22:3419–3431
21. Canals D, Perry DM, Jenkins RW, Hannun YA (2011) Drug targeting of sphingolipid metabolism: sphingomyelinases and ceramidases. Br J Pharmacol 163:694–712
22. Angulo S, Morales A, Danese S, Llacuna L, Masamunt MC, Pultz N, Cifone MG, De Simone C, Delgado S, Vila J, Panes J, Donskey C, Fernandez-Checa JC, Fiocchi C, Sans M (2011) Probiotic sonicates selectively induce mucosal immune cells apoptosis through ceramide generation via neutral sphingomyelinase. PLoS One 6, e16953
23. Coll O, Morales A, Fernández-Checa JC, Garcia-Ruiz C (2007) Neutral sphingomyelinase-induced ceramide triggers germinal vesicle breakdown and oxidant-dependent apoptosis in Xenopus laevis oocytes. J Lipid Res 48:1924–1935
24. Marí M, Fernández-Checa JC (2007) Sphingolipid signalling and liver diseases. Liver Int 27:440–450
25. Garcia-Ruiz C, Colell A, Mari M, Morales A, Calvo M, Enrich C, Fernandez-Checa JC (2003) Defective TNF-α-mediated hepatocellular apoptosis and liver damage in acidic sphingomyelinase knockout mice. J Clin Invest 111:197–208
26. Lin T, Genestier L, Pinkoski MJ, Castro A, Nicholas S, Mogil R, Paris F, Fuks Z, Schuchman EH, Kolesnick RN, Green DR (2000) Role of acidic sphingomyelinase in Fas/CD95-mediated cell death. J Biol Chem 275:8657–8663
27. Lang PA, Schenck M, Nicolay JP, Becker JU, Kempe DS, Lupescu A, Koka S, Eisele K, Klarl BA, Rubben H, Schmid KW, Mann K, Hildenbrand S, Hefter H, Huber SM, Wieder T, Erhardt A, Haussinger D, Gulbins E, Lang F (2007) Liver cell death and anemia in Wilson disease involve acid sphingomyelinase and ceramide. Nat Med 13:164–170
28. Mari M, Colell A, Morales A, Paneda C, Varela-Nieto I, Garcia-Ruiz C, Fernandez-Checa JC (2004) Acidic sphingomyelinase downregulates the liver-specific methionine adenosyltransferase 1A, contributing to tumor necrosis factor-induced lethal hepatitis. J Clin Invest 113:895–904
29. Fernandez A, Matias N, Fucho R, Ribas V, Von Montfort C, Nuño N, Baulies A, Martinez L, Tarrats N, Mari M, Colell A, Morales A, Dubuquoy L, Mathurin P, Bataller R, Caballeria J, Elena M, Balsinde J, Kaplowitz N, Garcia-Ruiz C, Fernandez-Checa JC (2013) ASMase is required for chronic alcohol induced hepatic endoplasmic reticulum stress and mitochondrial cholesterol loading. J Hepatol 59:805–813

30. Fucho R, Martinez L, Baulies A, Torres S, Tarrats N, Fernandez A, Ribas V, Astudillo AM, Balsinde J, Garcia-Roves P, Elena M, Bergheim I, Lotersztajn S, Trautwein C, Appelqvist H, Paton AW, Paton JC, Czaja MJ, Kaplowitz N, Fernandez-Checa JC, Garcia-Ruiz C (2014) Asmase regulates autophagy and lysosomal membrane permeabilization and its inhibition prevents early stage nonalcoholic steatohepatitis. J Hepatol 61(5):1126–1134
31. Paris F, Grassmé H, Cremesti A, Zager J, Fong Y, Haimovitz-Friedman A, Fuks Z, Gulbins E, Kolesnick R (2001) Natural ceramide reverses Fas resistance of acid sphingomyelinase −/− hepatocytes. J Biol Chem 276:8297–8305
32. Llacuna L, Mari M, Garcia-Ruiz C, Fernandez-Checa JC, Morales A (2006) Critical role of acidic sphingomyelinase in murine hepatic ischemia-reperfusion injury. Hepatology 44:561–572
33. Lee SH, Seo GS, Park P-H, Choi J-Y, Park YN, Kim HK, Chae K-S, Sohn DH (2003) Increased expression of O-acetyl disialoganglioside synthase during rat liver fibrogenesis relates to stellate cell activation. Biochem Biophys Res Commun 303:954–961
34. Huitema K, Van Den Dikkenberg J, Brouwers JFHM, Holthuis JCM (2004) Identification of a family of animal sphingomyelin synthases. EMBO J 23:33–44
35. D'angelo G, Polishchuk E, Tullio GD, Santoro M, Campli AD, Godi A, West G, Bielawski J, Chuang C-C, Van Der Spoel AC, Platt FM, Hannun YA, Polishchuk R, Mattjus P, De Matteis MA (2007) Glycosphingolipid synthesis requires FAPP2 transfer of glucosylceramide. Nature 449:62–67
36. Merritt WD, Morre DJ, Keenan TW (1978) Gangliosides of liver tumors induced by N-2-fluorenylacetamide. II. Alterations in biosynthetic enzymes. J Natl Cancer Inst 60:1329–1337
37. Hassler DF, Bell RM (1993) Ceramidases: enzymology and metabolic roles. Adv Lipid Res 26:49–57
38. Mao C, Obeid LM (2008) Ceramidases: regulators of cellular responses mediated by ceramide, sphingosine and sphingosine-1-phosphate. Biochim Biophys Acta 1781:424–434
39. Morales A, Paris R, Villanueva A, Llacuna L, Garcia-Ruiz C, Fernandez-Checa JC (2006) Pharmacological inhibition or small interfering RNA targeting acid ceramidase sensitizes hepatoma cells to chemotherapy and reduces tumor growth in vivo. Oncogene 26:905–916
40. Ramirez De Molina A, De La Cueva A, Machado-Pinilla R, Rodriguez-Fanjul V, Gomez Del Pulgar T, Cebrian A, Perona R, Lacal JC (2012) Acid ceramidase as a chemotherapeutic target to overcome resistance to the antitumoral effect of choline kinase α inhibition. Curr Cancer Drug Targets 12:617–624
41. Cuvillier O, Pirianov G, Kleuser B, Vanek PG, Coso OA, Gutkind JS, Spiegel S (1996) Suppression of ceramide-mediated programmed cell death by sphingosine-1-phosphate. Nature 381:800–803
42. Perez GI, Knudson CM, Leykin L, Korsmeyer SJ, Tilly JL (1997) Apoptosis-associated signaling pathways are required for chemotherapy-mediated female germ cell destruction. Nat Med 3:1228–1232
43. Pyne NJ, Pyne S (2010) Sphingosine 1-phosphate and cancer. Nat Rev Cancer 10:489–503
44. Van Veldhoven PP (2000) Sphingosine-1-phosphate lyase. Methods Enzymol 311:244–254
45. Le Stunff H, Giussani P, Maceyka M, Lépine S, Milstien S, Spiegel S (2007) Recycling of sphingosine is regulated by the concerted actions of sphingosine-1-phosphate phosphohydrolase 1 and sphingosine kinase 2. J Biol Chem 282:34372–34380
46. Liu H, Sugiura M, Nava VE, Edsall LC, Kono K, Poulton S, Milstien S, Kohama T, Spiegel S (2000) Molecular cloning and functional characterization of a novel mammalian sphingosine kinase type 2 isoform. J Biol Chem 275:19513–19520
47. Spiegel S, Milstien S (2003) Sphingosine-1-phosphate: an enigmatic signalling lipid. Nat Rev Mol Cell Biol 4:397–407
48. Taylor RC, Cullen SP, Martin SJ (2008) Apoptosis: controlled demolition at the cellular level. Nat Rev Mol Cell Biol 9:231–241
49. García-Ruiz C, Colell A, Marí M, Morales A, Fernández-Checa JC (1997) Direct effect of ceramide on the mitochondrial electron transport chain leads to generation of reactive oxygen species: role of mitochondrial glutathione. J Biol Chem 272:11369–11377

50. Gudz TI, Tserng K-Y, Hoppel CL (1997) Direct inhibition of mitochondrial respiratory chain complex III by cell-permeable ceramide. J Biol Chem 272:24154–24158
51. Quillet-Mary A, Jaffrézou J-P, Mansat V, Bordier C, Naval J, Laurent G (1997) Implication of mitochondrial hydrogen peroxide generation in ceramide-induced apoptosis. J Biol Chem 272:21388–21395
52. Dai Q, Liu J, Chen J, Durrant D, Mcintyre TM, Lee RM (2004) Mitochondrial ceramide increases in UV-irradiated HeLa cells and is mainly derived from hydrolysis of sphingomyelin. Oncogene 23:3650–3658
53. Birbes H, Luberto C, Hsu Y-T, El Bawab S, Hannun YA, Obeid LM (2005) A mitochondrial pool of sphingomyelin is involved in TNFalpha-induced Bax translocation to mitochondria. Biochem J 386:445–451
54. El Bawab S, Roddy P, Qian T, Bielawska A, Lemasters JJ, Hannun YA (2000) Molecular cloning and characterization of a human mitochondrial ceramidase. J Biol Chem 275:21508–21513
55. Bionda C, Portoukalian J, Schmitt D, Rodriguez-Lafrasse C, Ardail D (2004) Subcellular compartmentalization of ceramide metabolism: MAM (mitochondria-associated membrane) and/or mitochondria? Biochem J 382:527–533
56. Chipuk JE, Mcstay GP, Bharti A, Kuwana T, Clarke CJ, Siskind LJ, Obeid LM, Green DR (2012) Sphingolipid metabolism cooperates with BAK and BAX to promote the mitochondrial pathway of apoptosis. Cell 148:988–1000
57. Lee H, Rotolo JA, Mesicek J, Penate-Medina T, Rimner A, Liao W-C, Yin X, Ragupathi G, Ehleiter D, Gulbins E, Zhai D, Reed JC, Haimovitz-Friedman A, Fuks Z, Kolesnick R (2011) Mitochondrial ceramide-rich macrodomains functionalize Bax upon irradiation. PLoS One 6, e19783
58. Sentelle RD, Senkal CE, Jiang W, Ponnusamy S, Gencer S, Panneer Selvam S, Ramshesh VK, Peterson YK, Lemasters JJ, Szulc ZM, Bielawski J, Ogretmen B (2012) Ceramide targets autophagosomes to mitochondria and induces lethal mitophagy. Nat Chem Biol 8:831–838
59. Garcia-Ruiz C, Mato JM, Vance D, Kaplowitz N, Fernández-Checa JC (2015) Acid sphingomyelinase-ceramide system in steatohepatitis: a novel target regulating multiple pathways. J Hepatol 62(1):219–233
60. Harding HP, Zhang Y, Zeng H, Novoa I, Lu PD, Calfon M, Sadri N, Yun C, Popko B, Paules R, Stojdl DF, Bell JC, Hettmann T, Leiden JM, Ron D (2003) An integrated stress response regulates amino acid metabolism and resistance to oxidative stress. Mol Cell 11:619–633
61. Lloyd-Evans E, Pelled D, Riebeling C, Bodennec J, De-Morgan A, Waller H, Schiffmann R, Futerman AH (2003) Glucosylceramide and glucosylsphingosine modulate calcium mobilization from brain microsomes via different mechanisms. J Biol Chem 278:23594–23599
62. Pelled D, Lloyd-Evans E, Riebeling C, Jeyakumar M, Platt FM, Futerman AH (2003) Inhibition of calcium uptake via the sarco/endoplasmic reticulum Ca^{2+}-ATPase in a mouse model of Sandhoff disease and prevention by treatment with N-butyldeoxynojirimycin. J Biol Chem 278:29496–29501
63. Carracedo A, Lorente M, Egia A, Blázquez C, García S, Giroux V, Malicet C, Villuendas R, Gironella M, González-Feria L, Piris MÁ, Iovanna JL, Guzmán M, Velasco G (2006) The stress-regulated protein p8 mediates cannabinoid-induced apoptosis of tumor cells. Cancer Cell 9:301–312
64. Gozuacik D, Kimchi A (2007) Autophagy and cell death. Curr Top Dev Biol 78:217–245
65. Schubert KM, Scheid MP, Duronio V (2000) Ceramide inhibits protein kinase B/Akt by promoting dephosphorylation of serine 473. J Biol Chem 275:13330–13335
66. Edinger AL (2009) Starvation in the midst of plenty: making sense of ceramide-induced autophagy by analyzing nutrient transporter expression. Biochem Soc Trans 37:253–258
67. Guenther GG, Peralta ER, Rosales KR, Wong SY, Siskind LJ, Edinger AL (2008) Ceramide starves cells to death by downregulating nutrient transporter proteins. Proc Natl Acad Sci U S A 105:17402–17407

68. Scarlatti F, Bauvy C, Ventruti A, Sala G, Cluzeaud F, Vandewalle A, Ghidoni R, Codogno P (2004) Ceramide-mediated macroautophagy involves inhibition of protein kinase B and up-regulation of Beclin 1. J Biol Chem 279:18384–18391
69. Young MM, Kester M, Wang H-G (2013) Sphingolipids: regulators of crosstalk between apoptosis and autophagy. J Lipid Res 54:5–19
70. Jiang W, Ogretmen B (2014) Autophagy paradox and ceramide. Biochim Biophys Acta 1841:783–792
71. Li X, Xu M, Pitzer A, Xia M, Boini K, Li P-L, Zhang Y (2014) Control of autophagy maturation by acid sphingomyelinase in mouse coronary arterial smooth muscle cells: protective role in atherosclerosis. J Mol Med 92:473–485
72. Li ZZ, Berk M, Mcintyre TM, Gores GJ, Feldstein AE (2008) The lysosomal-mitochondrial axis in free fatty acid-induced hepatic lipotoxicity. Hepatology 47:1495–1503
73. Petersen NHT, Olsen OD, Groth-Pedersen L, Ellegaard A-M, Bilgin M, Redmer S, Ostenfeld MS, Ulanet D, Dovmark TH, Lønborg A, Vindeløv SD, Hanahan D, Arenz C, Ejsing CS, Kirkegaard T, Rohde M, Nylandsted J, Jäättelä M (2013) Transformation-associated changes in sphingolipid metabolism sensitize cells to lysosomal cell death induced by inhibitors of acid sphingomyelinase. Cancer Cell 24:379–393
74. Slotte JP (1999) Sphingomyelin–cholesterol interactions in biological and model membranes. Chem Phys Lipids 102:13–27
75. Ridgway ND (2000) Interactions between metabolism and intracellular distribution of cholesterol and sphingomyelin. Biochim Biophys Acta 1484:129–141
76. Appelqvist H, Nilsson C, Garner B, Brown AJ, Kågedal K, Öllinger K (2011) Attenuation of the lysosomal death pathway by lysosomal cholesterol accumulation. Am J Pathol 178:629–639
77. Mas E, Danjoux M, Garcia V, Carpentier S, Ségui B, Levade T (2013) The pro-inflammatory action of tumour necrosis factor-α in non-alcoholic steatohepatitis is independent of the NSMAF gene product. Dig Liver Dis 45:147–154
78. Samad F, Hester KD, Yang G, Hannun YA, Bielawski J (2006) Altered adipose and plasma sphingolipid metabolism in obesity: a potential mechanism for cardiovascular and metabolic risk. Diabetes 55:2579–2587
79. Caballero F, Fernández A, Matías N, Martínez L, Fucho R, Elena M, Caballeria J, Morales A, Fernández-Checa JC, García-Ruiz C (2010) Specific contribution of methionine and choline in nutritional nonalcoholic steatohepatitis: impact on mitochondrial S-adenosyl-L-methionine and glutathione. J Biol Chem 285:18528–18536
80. Moles A, Tarrats N, Morales A, Domínguez M, Bataller R, Caballería J, García-Ruiz C, Fernández-Checa JC, Marí M (2010) Acidic sphingomyelinase controls hepatic stellate cell activation and in vivo liver fibrogenesis. Am J Pathol 177:1214–1224
81. Deevska GM, Rozenova KA, Giltiay NV, Chambers MA, White J, Boyanovsky BB, Wei J, Daugherty A, Smart EJ, Reid MB, Merrill AH, Nikolova-Karakashian M (2009) Acid sphingomyelinase deficiency prevents diet-induced hepatic triacylglycerol accumulation and hyperglycemia in mice. J Biol Chem 284:8359–8368
82. Quillin RC 3rd, Wilson GC, Nojima H, Freeman CM, Wang J, Schuster RM, Blanchard JA, Edwards MJ, Gandhi CR, Gulbins E, Lentsch AB (2014) Inhibition of acidic sphingomyelinase reduces established hepatic fibrosis in mice. Hepatol Res 45:305
83. Andrade CMB, Trindade VMT, Cardoso CCA, Ziulkoski AL, Trugo LC, Guaragna RM, Borojevic R, Guma FCR (2003) Changes of sphingolipid species in the phenotype conversion from myofibroblasts to lipocytes in hepatic stellate cells. J Cell Biochem 88:533–544
84. Gocht A, Rutter G, Kniep B (1998) Changed expression of 9-O-acetyl GD3 (CDw60) in benign and atypical proliferative lesions and carcinomas of the human breast. Histochem Cell Biol 110:217–229
85. Malisan F, Franchi L, Tomassini B, Ventura N, Condo I, Rippo MR, Rufini A, Liberati L, Nachtigall C, Kniep B, Testi R (2002) Acetylation suppresses the proapoptotic activity of GD3 ganglioside. J Exp Med 196:1535–1541

86. García-Ruiz C, Colell A, París R, Fernández-Checa JC (2000) Direct interaction of GD3 ganglioside with mitochondria generates reactive oxygen species followed by mitochondrial permeability transition, cytochrome c release, and caspase activation. FASEB J 14:847–858
87. García-Ruiz C, Colell A, Morales A, Calvo MA, Enrich C, Fernández-Checa JC (2002) Trafficking of ganglioside GD3 to mitochondria by tumor necrosis factor-α. J Biol Chem 277:36443–36448
88. Rippo MR, Malisan F, Ravagnan L, Tomassini B, Condo I, Costantini P, Susin SA, Rufini A, Todaro M, Kroemer G, Testi R (2000) GD3 ganglioside directly targets mitochondria in a bcl-2-controlled fashion. FASEB J 14:2047–2054
89. Selzner M, Bielawska A, Morse MA, Rüdiger HA, Sindram D, Hannun YA, Clavien P-A (2001) Induction of apoptotic cell death and prevention of tumor growth by ceramide analogues in metastatic human colon cancer. Cancer Res 61:1233–1240
90. Ullio C, Casas J, Brunk UT, Sala G, Fabriàs G, Ghidoni R, Bonelli G, Baccino FM, Autelli R (2012) Sphingosine mediates TNFα-induced lysosomal membrane permeabilization and ensuing programmed cell death in hepatoma cells. J Lipid Res 53:1134–1143
91. Lemmer ER, Vessey CJ, Gelderblom WC, Shephard EG, Van Schalkwyk DJ, Van Wijk RA, Marasas WF, Kirsch RE, Hall PL (2004) Fumonisin B1-induced hepatocellular and cholangiocellular tumors in male Fischer 344 rats: potentiating effects of 2-acetylaminofluorene on oval cell proliferation and neoplastic development in a discontinued feeding study. Carcinogenesis 25(7):1257–1264
92. Chuturgoon A, Phulukdaree A, Moodley D (2014) Fumonisin B1 induces global DNA hypomethylation in HepG2 cells - an alternative mechanism of action. Toxicology 315:65–69
93. Persson EC, Sewram V, Evans AA, London WT, Volkwyn Y, Shen YJ, Van Zyl JA, Chen G, Lin W, Shephard GS, Taylor PR, Fan JH, Dawsey SM, Qiao YL, McGlynn KA, Abnet CC (2012) Fumonisin B1 and risk of hepatocellular carcinoma in two Chinese cohorts. Food Chem Toxicol 50:679–683
94. Wang E, Norred WP, Bacon CW, Riley RT, Merrill AH Jr (1991) Inhibition of sphingolipid biosynthesis by fumonisins. Implications for diseases associated with Fusarium moniliforme. J Biol Chem 266(22):14486–14490
95. He Q, Suzuki H, Sharma N, Sharma RP (2006) Ceramide synthase inhibition by fumonisin B1 treatment activates sphingolipid-metabolizing systems in mouse liver. Toxicol Sci 94(2):388–397
96. Wattenberg EV, Badria FA, Shier WT (1996) Activation of mitogen-activated protein kinase by the carcinogenic mycotoxin fumonisin B1. Biochem Biophys Res Commun 227(2):622–627
97. Fukuda H, Shima H, Vesonder RF, Tokuda H, Nishino H, Katoh S, Tamura S, Sugimjra T, Nagao M (1996) Inhibition of protein serine/threonine phosphatases by fumonisin B1, a mycotoxin. Biochem Biophys Res Commun 220(1):160–165
98. Revill K, Wang T, Lachenmayer A, Kojima K, Harrington A, Li J, Hoshida Y, Llovet JM, Powers S (2013) Genome-wide methylation analysis and epigenetic unmasking identify tumor suppressor genes in hepatocellular carcinoma. Gastroenterology 145:1424–1435, e25
99. Chatzakos V, Rundlöf AK, Ahmed D, De Verdier PJ, Flygare J (2012) Inhibition of sphingosine kinase 1 enhances cytotoxicity, ceramide levels and ROS formation in liver cancer cells treated with selenite. Biochem Pharmacol 84:712–721
100. Jenkins RW, Idkowiak-Baldys J, Simbari F, Canals D, Roddy P, Riner CD, Clarke CJ, Hannun YA (2011) A novel mechanism of lysosomal acid sphingomyelinase maturation: requirement for carboxyl-terminal proteolytic processing. J Biol Chem 286:3777–3788
101. Simonaro CM, Park J-H, Eliyahu E, Shtraizent N, Mcgovern MM, Schuchman EH (2006) Imprinting at the SMPD1 locus: implications for acid sphingomyelinase-deficient Niemann-Pick disease. Am J Hum Genet 78:865–870
102. Savic R, He X, Fiel I, Schuchman EH (2013) Recombinant human acid sphingomyelinase as an adjuvant to sorafenib treatment of experimental liver cancer. PLoS One 8, e65620
103. Osawa Y, Suetsugu A, Matsushima-Nishiwaki R, Yasuda I, Saibara T, Moriwaki H, Seishima M, Kozawa O (2013) Liver acid sphingomyelinase inhibits growth of metastatic colon cancer. J Clin Invest 123:834–843

104. Morré DJ, Wilkinson FE, Keenan TW (1990) Gangliosides depleted in plasma membrane are directed to internal membranes of rat hepatomas: Evidence for a glycolipid sorting defect in hepatocarcinogenesis. Biochem Biophys Res Commun 169:192–197
105. Huang X, Li Y, Zhang J, Xu Y, Tian Y, Ma K (2013) Ganglioside GM3 inhibits hepatoma cell motility via down-regulating activity of EGFR and PI3K/AKT signaling pathway. J Cell Biochem 114:1616–1624
106. Lluis JM, Llacuna L, Von MC, Barcena C, Enrich C, Morales A, Fernandez-Checa JC (2009) GD3 synthase overexpression sensitizes hepatocarcinoma cells to hypoxia and reduces tumor growth by suppressing the cSrc/NF-kappaB survival pathway. PLoS One 4, e8059
107. Ruckhaberle E, Karn T, Rody A, Hanker L, Gatje R, Metzler D, Holtrich U, Kaufmann M (2009) Gene expression of ceramide kinase, galactosyl ceramide synthase and ganglioside GD3 synthase is associated with prognosis in breast cancer. J Cancer Res Clin Oncol 135:1005 –1013
108. Ryland LK, Doshi UA, Shanmugavelandy SS, Fox TE, Aliaga C, Broeg K, Baab KT, Young M, Khan O, Haakenson JK, Jarbadan NR, Liao J, Wang H-G, Feith DJ, Loughran TP, Liu X, Kester M (2013) C6-ceramide nanoliposomes target the Warburg effect in chronic lymphocytic leukemia. PLoS One 8:e84648/1–e84648/15, 15 pp
109. Tagaram HRS, Divittore NA, Barth BM, Kaiser JM, Avella D, Kimchi ET, Jiang Y, Isom HC, Kester M, Staveley-O'Carroll KF (2011) Nanoliposomal ceramide prevents in vivo growth of hepatocellular carcinoma. Gut 60:695–701
110. Adiseshaiah PP, Clogston JD, Mcleland CB, Rodriguez J, Potter TM, Neun BW, Skoczen SL, Shanmugavelandy SS, Kester M, Stern ST, Mcneil SE (2013) Synergistic combination therapy with nanoliposomal C6-ceramide and vinblastine is associated with autophagy dysfunction in hepatocarcinoma and colorectal cancer models. Cancer Lett 337:254–265
111. Tran MA, Smith CD, Kester M, Robertson GP (2008) Combining nanoliposomal ceramide with sorafenib synergistically inhibits melanoma and breast cancer cell survival to decrease tumor development. Clin Cancer Res 14:3571–3581
112. Beljanski V, Lewis CS, Smith CD (2011) Antitumor activity of sphingosine kinase 2 inhibitor ABC294640 and sorafenib in hepatocellular carcinoma xenografts. Cancer Biol Ther 11:524–534

Pancreatic Cancer and Sphingolipids

Samy A.F. Morad and Myles C. Cabot

Abstract Sphingolipid metabolism is an area of cancer science that has recently risen to prominence in the laboratory and in the clinical arena. This is because ceramide, the aliphatic backbone of sphingolipids, can act as a powerful tumor censor; a favorable attribute. However, cancer cells are equipped to efficiently inactivate ceramide via various metabolic routes: glycosylation to form glucosylceramide, and hydrolysis, which generates sphingosine and then, through the action of sphingosine kinase, sphingosine 1-phosphate, a tumor cell mitogen. These actions severely blunt ceramide anticancer properties, thwarting potential utility in treatment, and pancreatic cancer, a drug resistant, severely aggressive malignancy, is no exception. In this chapter we will review pancreatic cancer epidemiology, classical treatment options, and progression, and survey sphingolipid metabolism in pancreatic cancer as it relates to the action and efficacy of anticancer drugs. The work will conclude with ideas on how targeting ceramide metabolism could reveal new vulnerabilities for countering growth of this intractable malignancy.

Keywords Pancreatic cancer • Ceramide • Sphingolipids • Ceramide metabolism • Ceramidase • Glucosylceramide • Glucosylceramide synthase • Sphingomyelin • Sphingomyelinase • C6-ceramide

S.A.F. Morad
Department of Biochemistry and Molecular Biology, Brody School of Medicine, East Carolina Diabetes and Obesity Institute, East Carolina University, 115 Heart Drive, Greenville, NC 27835, USA

Department of Pharmacology, Faculty of Veterinary Medicine, South Valley University, Qena 83523, Egypt
e-mail: morads@ecu.edu

M.C. Cabot (✉)
Department of Biochemistry and Molecular Biology, Brody School of Medicine, East Carolina Diabetes and Obesity Institute, East Carolina University, 115 Heart Drive, Greenville, NC 27835, USA
e-mail: cabotm@ecu.edu

Y.A. Hannun et al. (eds.), *Bioactive Sphingolipids in Cancer Biology and Therapy*, DOI 10.1007/978-3-319-20750-6_10

1 Pancreatic Cancer: Historical Perspective

Cancer of the pancreas was first described in 1835 [1] and again shortly thereafter, in 1858 [2]. The earlier work described carcinoma of the pancreas as being scarcely distinguishable from inflammation of the pancreas, but detailed insidious commencement, greater duration, increase suffering, absence of fever, and steady augmenting emaciation as hallmark features of carcinoma "[*sic*]". For treatment, it was noted that general bleeding was seldom necessary, and instead recommendation was made to apply a dozen leeches to the abdomen immediately over the seat of the pancreas. This procedure could be repeated from time to time if pain or tumors remain. We can assume these clinicians were not aware that leech saliva contains lipase, cholesterol esterase, free fatty acids, and phosphatidic acid [3, 4], which might indicate the presence of phospholipases. Unfortunately, current day literature makes no mention of sphingolipid-metabolizing machinery in leech saliva, which would be apropos in the current context. Also of note was the utility of poultices and fomentations, but caution was given regarding severe purging, which should be avoided. One hundred years passed before surgical intervention for treatment of pancreatic cancer was described [5, 6]. However, with surgery, significant morbidity and mortality were common, and cure rates were dismal.

2 Epidemiology in Brief

Pancreatic cancer is the fourth leading cause of death from cancer in men and women in the United States. For all stages, the 5-year survival rate is very poor, slated at approximately 6 %. This has not changed in the last 50 years. In 2013, approximately 45,000 people were diagnosed, and of those cases only 15 % survived. The life-time risk of developing pancreatic cancer is 1 in 78; risk increases with age with a steep increase after the age of 50. The disease rarely occurs before age 45 and is more common among men. At diagnosis, greater than 90 % of patients present with metastatic disease.

Some identifiable risk factors for pancreatic cancer include obesity, cigarette smoking, diet high in animal fats, and exposure to nickel and chlorinated hydrocarbons. African–Americans have the highest rates of pancreatic cancer; however, studies suggest that this may be a result of prevalence of risk factors in the African–American population [7]. Obesity also elevates risk, in particular abdominal obesity [8]. Interestingly, Vitamin D has been shown to exert a protective role in pancreatic cancer with links to sun exposure [9, 10].

Pancreatitis, first described in 1889 [11], is the topmost risk factor for pancreatic ductal adenocarcinoma [12]. Early on, it was believed that pancreatitis was the result of pancreas auto-digestion [13]; this idea was bolstered by subsequent work showing that pancreatic secretions were involved in the digestion of proteins, carbohydrates, and fats [14]. From this enlightenment, came the notion that pancreatic juices could be given to patients with pancreatitis to ameliorate steatorrhea [15].

These early observations on pancreatitis and pancreatic auto-digestion have been supported by subsequent investigations. Pancreatitis occurs when pancreatic digestive enzymes are activated within the pancreas causing localized tissue damage (see Sect. 4 for discussion of pancreatic digestive enzymes and their activation). Pancreatitis can be either acute, which occurs suddenly and from which most patients recover completely, or chronic, which results from continual damage to the pancreas, usually from alcohol ingestion. A body of work has implicated lipopolysaccharides, from gut-derived bacteria, as a potential co-factor of alcoholic pancreatitis [16]. The molecular pathways between chronic pancreatitis and cancer of the pancreas are complex and follow a proinflammatory route that includes such players as K-ras, COX-2, NF-κB, ROS, and proinflammatory cytokines [17]. A number of studies have demonstrated a strong association between chronic pancreatitis and pancreatic cancer [12, 18–20]. Because clinical evidence has confirmed correlations between inflammation and pancreatic tumorigenesis, the association between alcoholic pancreatitis, a prevalent form of pancreatitis, with increased risk for pancreatic cancer, appears clear.

3 Treatment Options

Discovered and developed by scientists at Eli Lilly and Company, gemcitabine (2′,2′-difluorodeoxycytidine) (GEM) (Gemzar) received FDA approval in 1996 as first-line treatment for patients with locally advanced (nonresectable Stage II or Stage III) or metastatic (Stage IV) adenocarcinoma of the pancreas in patients previously treated with fluorouracil (5-FU). GEM is a nucleoside analog in which the hydrogen atoms on the 2′ carbon of deoxycytidine are replaced by fluorine atoms. GEM primarily blocks DNA synthesis by inhibiting DNA polymerase activity. The drug is not active until taken up by cells where it is phosphorylated to its active metabolites by deoxycytidine kinase to difluorodeoxycytidine diphosphate and difluorodeoxycytidine triphosphate, metabolites that interfere at different steps in the synthesis of DNA. Resistance to GEM, however, is almost universal, posing problems for patients and clinicians alike. Resistance to GEM is thought to be linked to defects in the intrinsic mitochondrial pathway of programmed cell death, thus providing an entrée and rationale for introduction of ceramide, a bioactive sphingolipid whose apoptosis-inducing pathways converge on mitochondria (Fig. 1) [21]. As depicted, ceramide signaling consists of a complex arrangement of players that propel mitochondrial-orchestrated end-point apoptosis either by caspase-independent, endonuclease G (ENDOG)- and apoptosis-inducing factor (AIF)-associated events or by caspase-dependent, inhibitor of apoptosis protein (IAP)-driven pathways. Upstream, ceramide can mediate the activity of BCL-2 family members, downregulate AKT to elicit BAX translocation, activate p38 MAPK, or activate JNK.

A pivotal study [22] matching up GEM and bolus 5-FU, clearly defined a role for GEM as first-line treatment for patients with advanced pancreatic cancer.

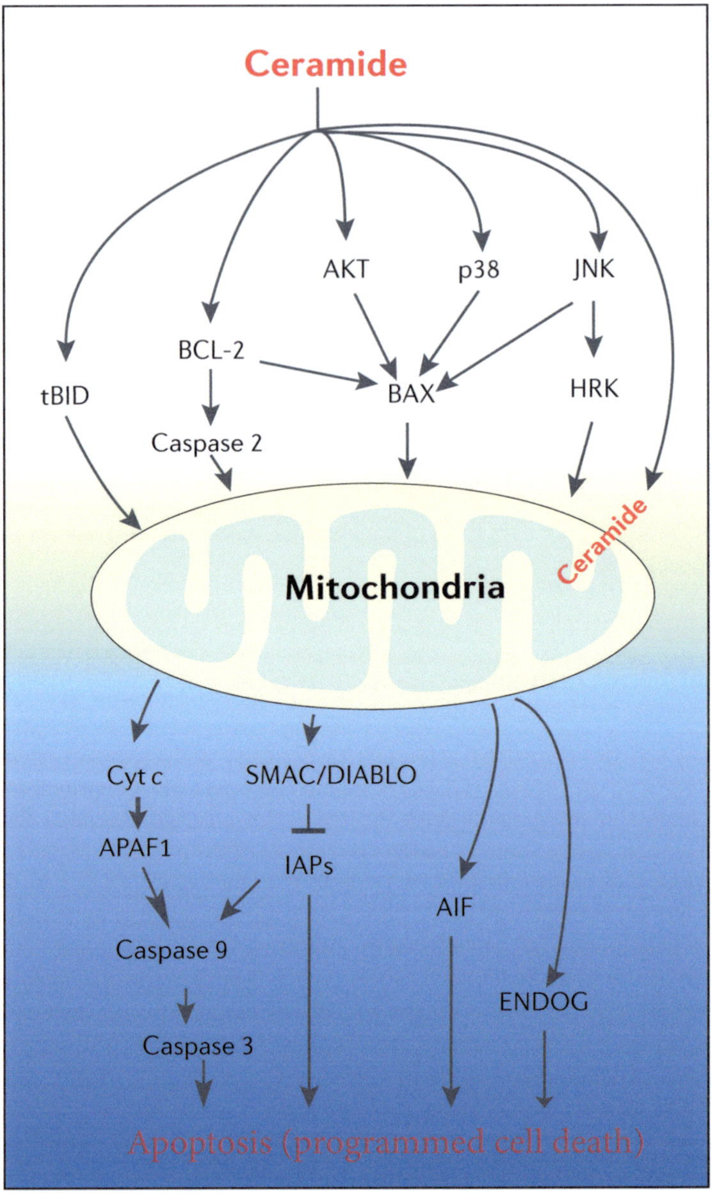

Fig. 1 The intrinsic mitochondrial pathway of ceramide-regulated signaling and the induction of apoptosis. The intrinsic pathway of ceramide signaling converges on mitochondria and has upstream and downstream elements, as depicted. This pathway is mainly regulated by caspases and BCL-2 (anti-apoptotic) protein family members. Loss of mitochondrial outer membrane potential, due to upstream mitochondrial insult, gives rise to release of pro-apoptotic proteins, such as cytochrome c, apoptosis inducing factor, and other players as indicated. This loss of mitochondrial membrane potential is a key element in apoptotic signaling by ceramide. *t-BID* membrane-targeting death ligand, *BCL-2* B-cell lymphoma 2 protein regulator of apoptosis, *AKT* protein kinase B, *p38* mitogen-activated protein kinase, *BAX* member of Bcl-2 family of proteins, *Cyt c* cytochrome c, *AIF* apoptosis-inducing factor, *SMAC* second mitochondria-derived activator of caspase, *DIABLO* direct inhibitor of apoptosis protein (IAP)-binding protein with low PI, *ENDOG* endonuclease G, *APAF1* apoptotic protease-activating factor 1, *HRK* harakiri, *JNK* JUN N-terminal kinase

Subsequent studies partnering GEM with a platinum drug, 5-FU, or irinotecan, demonstrated no survival benefit in Phase III trials. Of note, regarding cytotoxic chemotherapy, is a newer regimen, FOLFIRINOX (Fluorouracil, Leucovorin, Oxaliplatin, Irinotecan), which in Phase II/III trials showed superiority to GEM in all outcome measures [23].

Presently we are seeing a new, emerging standard for treatment of metastatic pancreatic cancer, Nab-paclitaxel (Nano-albumin-bound paclitaxel). This agent, FDA-approved for use in metastatic breast cancer as well as locally advanced or metastatic non-small cell lung cancer, is showing promise as a single agent and in combination with GEM in patients with advanced pancreatic cancer [24]. Poor drug delivery has always been a factor in treatment failure in pancreatic cancer, and it has been suggested that Nab-paclitaxel may diminish delivery roadblock.

On the science horizon, ceramide, a pro-apoptotic sphingolipid (SL) metabolite, has generated great interest in the area of cancer therapy [21, 25–30]. One exciting aspect of using ceramide as chemotherapeutic is the preferential selectivity for inducing apoptosis in cancer cells. For example, the short chain, cell-permeable ceramide analog, C6-ceramide, induces cell growth arrest and apoptosis in breast cancer cells and melanoma, but not in non-transformed mammary gland epithelial cells or melanocytes [31–33]. Use of ceramide alone or in combination, and targeting SL metabolism are at the threshold of utility in treatment of pancreatic cancer [34, 35]. These strategies will be discussed in a subsequent section.

Gangliosides, complex glycosphingolipids that contain oligosaccharides and *N*-acetylneuraminic acid (sialic acid), are concentrated on cell surfaces; the ceramide moiety of gangliosides is imbedded in the plasma membrane with the oligosaccharides located on the extracellular surface. Gangliosides have long been considered attractive targets for cancer immunotherapy, and a large body of work suggests that glycolipids could be relevant antigens for production of vaccines [36–39]. In this regard, work by Chu et al. [40] demonstrated that pancreatic adenocarcinomas express gangliosides (GM2) and sialyl Lewis antigens on the cell surface. Specifically, the study revealed that patients with unresectable tumors had increased serum levels of total gangliosides compared to patients with resectable tumors. It was suggested that increased serum levels of gangliosides, and specifically GM2, reflected shedding and release from the surface of tumor cells, and the production of IgM antibody against GM2 and GD1b indicated that these gangliosides are immunogenic antigens and potential targets for active immunotherapy.

Surgery has also come a long way with improvements in surgical techniques, staging, and diagnosis. However, only about 20 % of patients are eligible for the Whipple procedure, also known as pancreaticoduodenectomy. The overall five-year survival rate is only about 20 % (20–30 % in node-negative patients, 10 % in node-positive patients). Regardless of node status, most patients will receive chemotherapy after surgery, hence the need to zero-in on more efficacious drug treatment options.

4 Pancreas Anatomy and Function and Pancreatic Cancer Progression

The pancreas, located in the abdomen and positioned snugly against the greater curvature of the stomach, is part of the digestive system. The organ is normally 6–10 in. (18–25 cm) in length, and consists of the head, the largest part, located close to the duodenum, the body or mid-section, and a tail portion. The pancreas is both an endocrine and an exocrine gland. The endocrine function produces hormones, insulin and glucagon, that are released into circulation to control glucose metabolism. As an exocrine gland, the pancreas produces and secretes digestive juices into the duodenum, juices that in a healthy individual amount to approximately 1 L per day. Ductal cells and acinar cells of the pancreas contribute to the production of the digestive enzymes, which include trypsinogen, chymotrypsinogen, carboxypeptidase, pancreatic lipase that catalyzes hydrolysis of triacylglycerols into fatty acids and glycerol, cholesterol esterase that degrades cholesteryl esters to fatty acid and cholesterol, phospholipases, mainly of the A_2 type that are selective for the sn-2 position of glycerophospholipids and produce lysophospholipids, and amylase. The enzymes produced are inactive (zymogens) and are activated in the small bowel.

The majority of tumors of the pancreas arise in the head (78 %); the body and tail each account for 11 %. Symptoms may vary widely, based on the location of the tumor. Tumors located in the head of the pancreas often present earlier due to biliary obstruction. Because tumors in the body and tail often remain asymptomatic until late-stage disease, cure is rarely possible.

Early morphological studies provided the first evidence for distinctive precursor lesions to infiltrating adenocarcinoma of the pancreas [41]. Later clinical studies reported and supported the idea that atypical papillary duct lesions in pancreata forecast development of infiltrating ductal pancreatic adenocarcinoma [42, 43]. Molecular analyses provided further proof that ductal lesions are the precursors to pancreatic cancer, as nearly all of the genetic mutations arising in ductal adenocarcinomas are present in duct lesions [44]. In brief, pancreatic lesions exhibit activating point mutations in the K-*ras* oncogene and overexpress the HER-2/*neu* gene product that can activate a number of signaling pathways including MAPK, PI3K/Akt, and STAT [45]. Inactivation/mutation of *p* 16 tumor suppressor gene is also a feature of pancreatic cancer, not appearing in histologically normal pancreatic ductal epithelium; mutations of the *p* 16 gene occur in pancreatic cancer at a rate higher than that in any other tumor type. *p* 16 is one of the inhibitors of cyclin-dependent kinases, enzymes that drive cell division. Occurring late in the development of pancreatic neoplasia is loss of tumor suppressor genes *p* 53, *DCP4* and *BRCA2*.

5 Sphingolipid Biology and Chemistry

Sphingolipids are a class of lipids containing an organic, aliphatic amino alcohol, primarily sphingosine or sphinganine (Fig. 2). Sphingolipids comprise a second major class of lipids found in cell membranes, the other major classes being phospholipids, which are glycerol-containing, and the sterols. Sphingolipids do not contain a glycerol backbone. However, sphinganine does house a 3-carbon unit similar to "tri-hydroxy" glycerol but retains only two alcohols with the middle

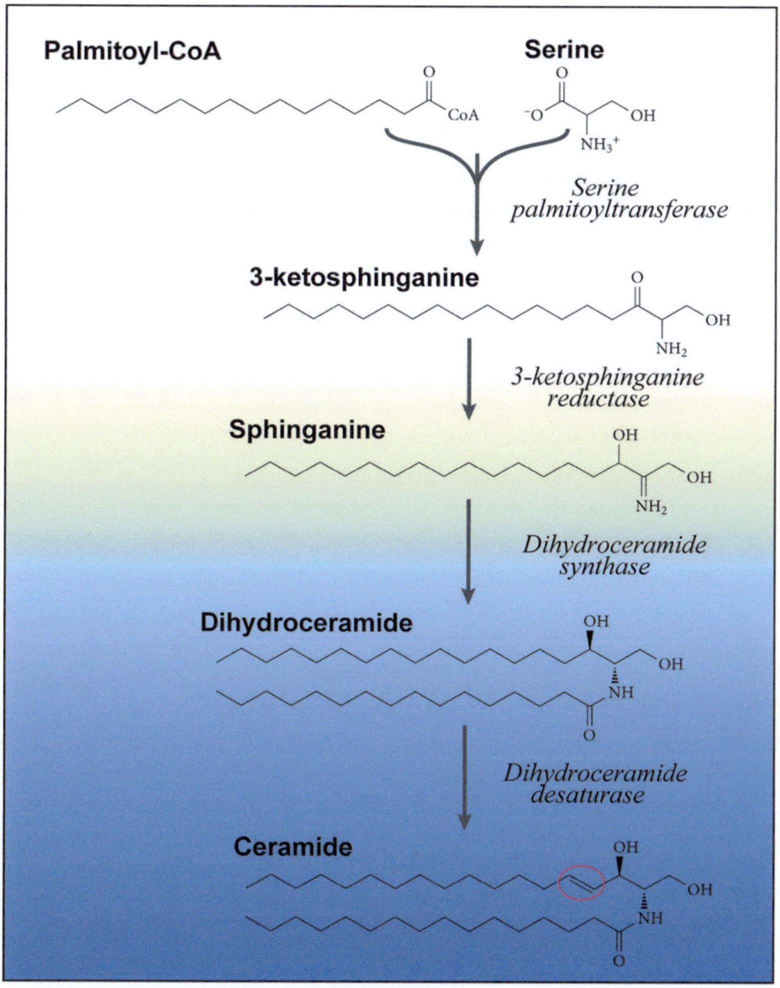

Fig. 2 *De novo* synthesis of ceramide. Ceramide synthesis begins with condensation of a fatty acyl CoA, (palmitoyl-CoA) with serine, catalyzed by serine palmitoyltransferase. The final step is addition of the 4,5-*trans* double bond (denoted by red circle), by the enzyme dihydroceramide desaturase

carbon being attached to an amine. Sphinganine with a fatty acid attached to the amine by an amide bond is known as acyl sphinganine or dihydroceramide (Fig. 2). The fatty acids joined to sphinganine are generally 16–26 carbon atoms in length. Ceramide is synthesized *de novo* in the endoplasmic reticulum. The steps begin with condensation of serine and palmitoyl-CoA, catalyzed by serine palmitoyltransferase (SPT). The product, 3-ketosphinganine contains 18 carbons and is reduced to sphinganine by 3-ketosphinganine reductase. The next enzyme, often referred to as ceramide synthase, does not generate ceramide but produces its saturated precursor, dihydroceramide, a step catalyzed by dihydroceramide synthase. Dihydroceramide synthases, which currently number six (CERS1-6), have preferences for specific acyl-CoAs and direct the assembly of a variety of molecular species of dihydroceramides. Lastly, although dihydroceramide is nearly identical in structure to ceramide, it lacks the 4,5-*trans* double bond (Fig. 2, circled in red) that is inserted by dihydroceramide desaturase to form ceramide. There exists great molecular species diversity among ceramides, and several cellular responses relegated to ceramide are known to be species specific [46, 47].

Ceramide, a potent tumor suppressor and understudy in the present review, is the precursor of complex sphingolipids such as sphingomyelin (SM), catalyzed by sphingomyelin synthase (SMS) and glucosylceramide (GC). GC assembly is catalyzed by glucosylceramide synthase (GCS) (Fig. 3). GCS, together with the multidrug resistance protein, P-glycoprotein [48], is important in regulating tumor cell sensitivity to ceramide and to chemotherapy [49–54]. Further glycosylation and enzymatic modification of GC leads to the generation of gangliosides and globosides, whereas the addition of a galactose to ceramide constitutes step one in the synthesis

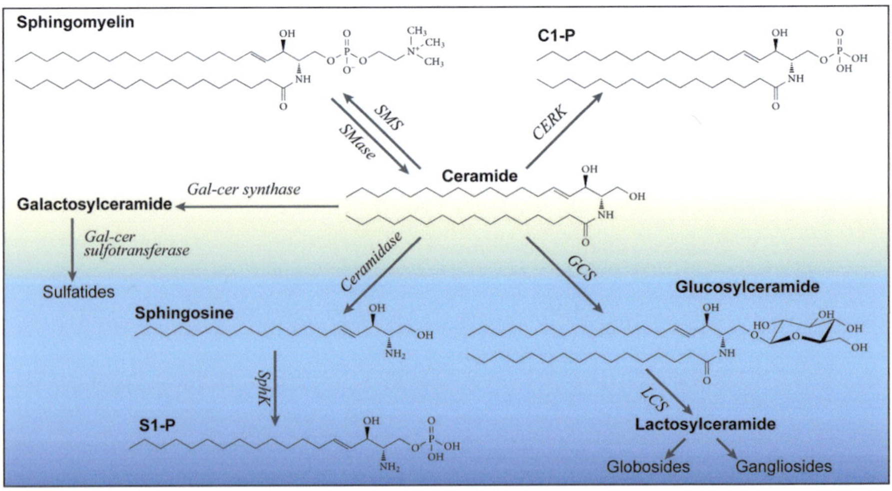

Fig. 3 Ceramide constructive and destructive metabolism. Enzymes of ceramide metabolism are shown in *italics*. *SMS* sphingomyelin synthase, *SMase* sphingomyelinase, *CERK* ceramide kinase, *Gal-cer synthase* galactosylceramide synthase, *GCS* glucosylceramide synthase, *SphK* sphingosine kinase, *LCS* lactosylceramide synthase

of sulfatides (Fig. 3). Ceramide is also phosphorylated by ceramide kinase to ceramide 1-phosphate [55], and ceramide can be hydrolyzed by ceramidases to component sphingosine and fatty acid.

Sphingolipid metabolism is an area of cancer science that has risen to clinical prominence over the last score of years. This is because ceramide, the aliphatic backbone of SL, acts as a powerful tumor suppressor, whereas its glycosylated product, GC formed by GCS, is anti-apoptotic and a biomarker of multi-drug resistance, as identified in the mid-1990s, by Cabot and colleagues [56, 57]. Ceramidase, in particular acid ceramidase (AC), another sentinel enzyme regulator of cancer cell growth, has recently been identified as candidate gene for development of new cancer diagnostics and touted as a therapeutic target in metastatic cancer [58, 59]. Much like GCS, AC dampens the tumor suppressor properties of ceramide via ceramide deacylation, leading to the generation of sphingosine that is phosphorylated to form sphingosine 1-phosphate (S 1-P) (Fig. 3), a mitogen that can contribute to enhanced cancer cell growth [60]. In some instances, sphingosine liberated through ceramide hydrolysis, can be acylated by ceramide synthase to regenerate ceramide, a metabolic avenue known as the salvage pathway of ceramide synthesis [61]. Thus, SL metabolism is a dynamic process with complex orchestration, impact, and clinical applications, presenting druggable targets for exploration and application. These observations lend an enlightening stepping stone to future studies aimed at deciphering the role of SL in development of pancreatic cancer.

Appropriate for the above discussion is a recent study by Jiang et al. [62], on SL metabolism in patients with pancreatic cancer. Using a mass spectrometry-based lipidomic approach, the study revealed for the first time unique correlations between levels of ceramide species and metastatic disease. For example, results of subgroup analysis showed that patients with lymph node-positivity had markedly higher levels of C16:0 and C24:1 ceramide species in their tumor specimens compared to node-negative patients and patients with pancreatitis. Additionally, sphingosine, S 1-P, and glycosylated ceramide were elevated in plasma of patients with nodal disease.

6 Anticancer Drugs and Sphingolipid Metabolism in Pancreatic Cancer

Many anticancer agents themselves are apt ceramide generators, a property that has been shown to contribute to cytotoxicity [63]. Table 1 lists several such agents, some of which promote formation of ceramide in pancreatic cancer cells, the type of agent responsible, and the intracellular target that is pinpointed. This brand of agent-induced ceramide production could be termed multi-hit because multiple enzyme targets serve as starting points for the production of ceramide: the *de novo* avenue through SPT and through dihydroceramide synthase, and also activation of acid sphingomyelinase (SMase) (see Figs. 2 and 3).

Table 1 Agents that generate ceramide in cancer cells

Agents	Type	Target	References
PSC 833	P-glycoprotein antagonist	Serine palmitoyltransferase	[71]
4-HPR	Vitamin A analog	Serine palmitoyltransferase, Dihydroceramide synthase	[58, 59]
Etoposide	Topoisomerase inhibitor	Serine palmitoyltransferase, Sphingomyelinases	[61, 109]
Gemcitabine	Nucleoside analog	Sphingomyelinases	[29, 93]
Cannabinoids	Act on cannabinoid receptors	Serine palmitoyltransferase, Sphingomyelinases	[83, 110, 111]
Bortezomib	Proteasome inhibitor	*De novo* synthesis pathway	[87]

Intracellular ceramide can be generated *de novo* in various types of cancer cells by fenretinide, also known as 4-HPR [64, 65], a vitamin A analog in clinical trials [66], etoposide [67], arsenic trioxide [68], anthracyclines [69], paclitaxel [70], and the P-glycoprotein antagonist/multidrug resistance modulator, SDZ PSC 833 (PSC 833) (Valspodar) [71]. DT_{388}-GM-CSF, a targeted fusion toxin constructed by conjugation of human granulocyte-macrophage colony stimulating factor (GM-CSF) with the catalytic and translocation domain of diphtheria toxin, promotes ceramide generation via SMase [72]. The ceramide-generating properties of GEM in human pancreatic cancer cell lines, BxPC-3 and PANC-1, have been demonstrated in work by Guillermet-Guibert et al. [34], an interesting finding in light of the importance of GEM in treatment of pancreatic cancer. Etoposide, although not commonly used in treatment of pancreatic cancer, has been paired with GEM in a Phase II trial for locally advanced or metastatic disease [73], and single agent etoposide has been shown to induce apoptosis in PANC-1 cells through mitochondrial-, caspase-dependent pathways [74]; however, the extent to which ceramide participates in the biochemistry of response was not ascertained. Compared with other neoplasms, the field is fallow regarding knowledge of anticancer agents that induce ceramide formation in pancreatic cancer.

PSC 833, unlike its chemical cousin, cyclosporin A, has been shown to be a powerful activator of cellular ceramide formation, irrespective of P-glycoprotein expression [75, 76]. In MCF-7 human breast cancer cells, PSC 833 exposure lead to a dose-dependent (1–10 μM range) reduction in cell viability [75] as it did in the multidrug resistant cervical carcinoma cell line, KB-V1 [76]. Ceramide generation by PSC 833 could be halted by fumonisin B1, a ceramide synthase inhibitor, and by the SPT inhibitor, myriocin [77], indicating that ceramide was produced by the *de novo* avenue.

Unlike other cancer cell lines previously tested, PANC-1 cells have been shown to be relatively refractory to PSC 833 exposure, despite multifold increases in dihydroceramide and ceramide levels [77]. In both PANC-1 and MIA-PaCa-2 cells, ceramide generation in response to PSC 833 was accompanied by concomitant formation of (dihydro)glucosylceramides and (dihydro)sphingomyelins, demarking robust conversion of (dihydro)ceramides to higher metabolites. This anabolism or "constructive metabolism" was speculated to mute ceramide-induced cytotoxicity. Noteworthy, however, was that PSC 833 exposure produced decreases in DNA synthesis and in retinoblastoma (RB) phosphorylation in pancreatic cancer cells, consistent with cell cycle arrest, as was demonstrated at G1 [77]. The cytostatic effect of PSC 833 could be converted to a cytotoxic end-point by inhibition of AC, which resulted in increased expression of LC3, an autophagosome marker, indicative of autophagic response, whereas inhibition of GCS was without influence. Overall, this suggests that AC plays an important role in pancreatic cancer cell homeostasis. AC has of late gained wide attention in the cancer field [59, 78, 79]; however, there are few reports documenting AC expression and activity in pancreatic cancer, and thus no information on its possible relationship to the intractability of this neoplasm. Thus, AC could be a formidable and promising target for study in pancreatic cancer.

4-HPR is active as a single agent in Hs 766T human pancreatic cancer cells and has demonstrated synergy in PANC-1 when co-administered with L-THREO-DIHYDROSPHINGOSINE (safingol) [80], a protein kinase C inhibitor and inhibitor of sphingosine kinase. In a comprehensive study in pancreatic cancer cells, Messner and Cabot [65] showed that 4-HPR triggered PANC-1 cell death by apoptosis associated with caspase activation and that sensitivity appeared to be mediated not by ceramide but by generation of reactive oxygen species (ROS), a known mechanism of action of 4-HPR [81]. It was further demonstrated that cell death was also accompanied by increased phosphorylation of JNK and p38, and decreased ERK phosphorylation (see Fig. 4, signaling pathways and targets). Importantly, the study showed that clinically relevant concentrations of 4-HPR elicited responses in PANC-1 and MIA-PaCa-2 cells.

Cannabinoids are the active compounds extracted from cannabis [82]; they have wide pharmacological potential, including antitumor activity, and function through targeting cannabinoid receptors (CB), of which there are two, CB1 and CB2 [83]. These receptors are highly expressed in human pancreatic tumor cell lines and tumor biopsies, compared to normal pancreatic tissue [84]. Cannabinoids are considered a new family of potential antitumor agents in pancreatic cancer [85, 86]. p8 is a stress-induced protein, a downstream target of ceramide [87] and a key player in the regulation of apoptosis in pancreatic tumor cells (Fig. 4) [86, 88]. In MIA-PaCa-2 and PANC-1 cell lines, cannabinoid administration induced apoptosis and increased ceramide levels in association with upregulated mRNA levels of p8. Interestingly, these effects were prevented by either blockade of the CB2 receptor or by pharmacological inhibition of *de novo* ceramide synthesis, indicating for the first time, in pancreatic cancer, that ceramide is implicated in CB receptor–mediated apoptosis [86], a finding in agreement with a study in glioma cells [89]. Supportive is the work of Salazar et al. [90], who found that delta(9)-tetrahydrocannabinol, the active component of marijuana, induced ceramide accumulation and eukaryotic translation initiation factor 2alpha (elF2alpha) phosphorylation that activated an ER stress response that promoted autophagic glioma cell death. The response was driven by tribbles homolog 3-dependent (TRB3-dependent) inhibition

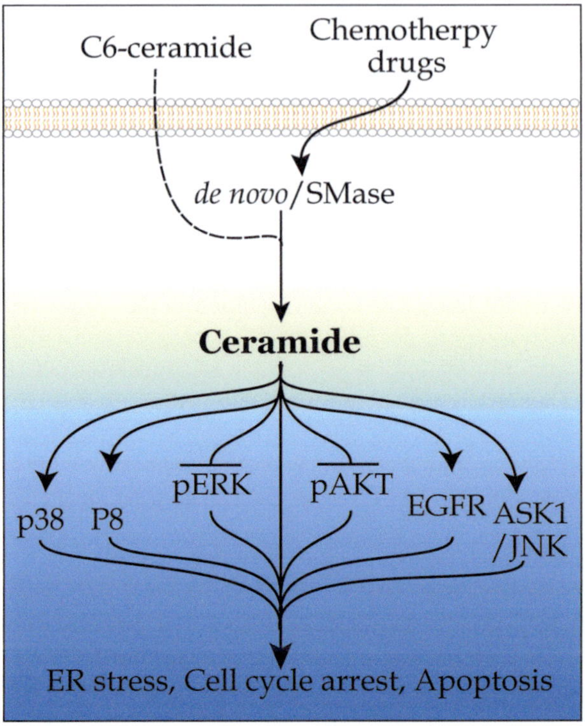

Fig. 4 Utility of ceramide in the targeting prominent signaling pathways in pancreatic cancer cells. Intracellular increases in ceramide levels, either via exogenous addition of C6-ceramide or by chemotherapy-induced activation of *de novo* and sphingomyelinase pathways, can result in blocked phosphorylation of ERK and AKT and in activation of JNK, p21, and P8. These "stop and go" signals converge to induce ER stress, cell cycle arrest, and apoptosis. *SMase* sphingomyelinase, *ASK1* Apoptosis signal-regulating kinase 1, *JNK* JUN N-terminal kinase, *pERK* phospho-ERK; extracellular signal-regulated kinase, *pAKT* phospho-AKT (protein kinase B), *EGFR* epidermal growth factor receptor

of the AKT/mammalian target of rapamycin complex 1 (mTORC1) axis. This group also showed that autophagy was upstream of apoptosis in cannabinoid-induced human cancer cell death and that activation of the above prescribed pathway was necessary for the *in vivo* antitumor activity of cannabinoids. In summation, these studies indicate that the cannabinoid receptor/ceramide/p8/ER stress cascade (Fig. 4) is a promising target in cancer, inclusive of pancreatic cancer, and moreover open doors to further investigations on the use of cannabinoids to treat pancreatic cancer.

Bortezomib (Velcade, PS-341), a protease inhibitor, displays antitumor activities against many types of cancer [91–93]. Recent study results are encouraging for use of bortezomib in treatment of pancreatic cancer, as this agent has shown significant antiproliferative activity in pancreatic cancer cells [94–96]. The antitumor activity was shown to be associated with induction of apoptosis, ER stress, ASK1/JNK pathway activation, and ceramide generation (Fig. 4). It appears that activation of the ASK1/JNK pathway is a key signaling element that regulates bortezomib-induced cell death [94, 97, 98]. As this pathway is downstream of ceramide, the possibility exists that bortezomib induces cell death via generation of ceramide. And, indeed, Gong et al. [87], have shown that bortezomib-induced JNK activation, cytotoxicity, and apoptosis, were duly blocked by introduction of the *de novo* ceramide synthesis inhibitor fumonosin B1, a strong indication that ceramide regulates bortezomib-induced cell fate via an ASK1/JNK signaling avenue (Fig. 4).

In addition to ceramide-generating agents, chemical analogs of ceramide have been investigated as therapeutics in pancreatic cancer. Recently, an interesting study was presented delineating the cytotoxic effects of a cationic analog of ceramide in pancreatic cancer cells [99]. The analog, LCL 124 [(2S,3S,4E)-2-N-[6′-(1″-pyridinium)-hexanoyl-sphingosine bromide)], was shown to accumulate in mitochondria and initiate apoptosis via mitochondrial depolarization with little activity in non-malignant pancreatic ductal epithelial cells. The analog was also effective in inhibiting xenograft growth, *in vivo*, and shown to selectively accumulate in tumors. Notably, GEM-resistant pancreatic cancer cells became more sensitive to sequential exposure to LCL 124, suggesting utility for this ceramide analog in GEM resistance. These studies are apropos in light of what can be termed the ceramide rheostat, wherein a delicate balance between ceramide and S 1-P is necessary to promote the tumor-suppressor role of ceramide and downplay the mitogenic impact of the latter, again and indication that sphingolipid metabolism is a dynamic process with complex orchestration.

As in other cancers, dysregulation of sphingolipid metabolism is a also a property in pancreatic cancer [62]. Further, and as discussed in the section below, excess ceramide generation appears to be a critical facet in the induction of apoptosis in pancreatic cancer cells in response to chemotherapy [100, 101], a response that merits exploitation in the clinical realm, especially in light of the need to develop new therapeutics for cancer of the pancreas.

7 Ceramide and Targeting Ceramide Metabolism as Therapeutic Approach in Pancreatic Cancer

A theme and multiple variations has been composed regarding ceramide and the manipulation of ceramide metabolism as a therapeutic entree in pancreatic cancer. These variations consist of, (1) direct supplementation with the more soluble, membrane-permeable forms of ceramide, C6-ceramide for example [102], (2) co-administration of a short-chain ceramide with a chemotherapy staple such as paclitaxel [103] or the histone deacetylase inhibitor, trichostatin A [104], (3) circuitous and novel avenues such as co-administration of SM (as the ceramide source) and GEM, a regimen that is synergistic for ceramide-mediated apoptosis in pancreatic cancer cells lines [100], (4) emerging combinatorial therapies that utilize nanoliposomal C6-ceramide with either GEM or an inhibitor of GCS to propel

Table 2 Targeting ceramide metabolism—a strategy for overcoming drug resistance in pancreatic cancer

Target	Agent	Modulator	Response	References
Glucosylceramide synthase	Gemcitabine, Bortezomib, Trichostatin A	PDMP, D-MAPP, PPMP	Cell death	[87, 94, 97]
Ceramidase	Gemcitabine, PSC 833	D-MAPP, DM102	Cell death	[71, 94]
Sphingosine kinase	Gemcitabine, Bortezomib	SKI-178, siRNA	Cell death	[29, 87]
Sphingomyelinase	Sphingomyelin	Gemcitabine	Cell death	[93, 94]

PPMP D-threo-1-phenyl-2-palmitoylamino-3-morpholino-1-propanol, d-*MAPP* D-erythro-2-tetradecanoylamino-1-phenyl-1-propanol, *siRNA* small interfering RNA, *SKI-178* sphingosine kinase inhibitor, *DM102* a pivaloylamide of a 2-substituted aminoethanol, acid ceramidase (AC) inhibitor

programmed cell death cascades [35], or (5) the use of carbon nanotubes engineered to carry multiple agents such as C6-ceramide and paclitaxel [105]. Table 2 provides a listing of agents and modulators that can be employed in combination regimens to magnify cellular response to ceramide; the points are discussed subsequently.

Earlier studies in PANC-1 cells demonstrated that C6-ceramide was a viable substrate for SMS, GCS, and ceramidase, emphasizing the difficulties inherent in the use of single agent ceramide analogs in pancreatic cancer [102]. The study showed that PANC-1 cells, in response to low-dose (0.6 μM) C6-ceramide, converted the majority of the supplement to C6-SM, whereas in response to high-dose C6-ceramide (12 μM), cells converted the majority of supplemented lipid to C6-GC. It is significant to acknowledge that when challenged with high-dose C6-ceramide, the preferred direction of cellular metabolism was glycosylation, an enzymatic avenue, which we propose, was taken in an effort to buffer deleterious effects of excess short-chain sphingomyelins that could become membrane-resident and disruptive to membrane integrity. PANC-1 cells as well hydrolyzed C6-ceramide, as evidenced by large increases in production of intracellular long-chain ceramide by way of the salvage pathway that operates via liberation of sphingosine from ceramide by ceramidase and subsequent re-acylation by ceramide synthase [61, 102].

In addition to the negative therapeutic impact imparted by efficient ceramide metabolic machinery, as related above, it has been suggested that ceramide generated in pancreatic cancer cells in response to various stimuli might not have access to effector molecules responsible for the initiation of apoptosis (Fig. 1) [34, 100]. Moreover, the levels of ceramide generated by exposure to chemotherapeutic agents may be insufficient to elicit anti-proliferative response in pancreatic cancer. In a scenario like this, the ceramide clearance pathways likely play a pivotal role in response thresholds, because the newly generated and/or supplemented ceramide can be rapidly cleared. In addition, limited SM pools, if ceramide is generated via the SMase pathway, could dampen responses. Thus, the installation of inhibitors of ceramide metabolism or supplementing to enrich SM pool size, could serve to boost the ceramide response (Table 2). These tactics have been investigated and are discussed below.

Some of the first work to demonstrate a role for ceramide in pancreatic cancer cell death originated from Modrak et al. [100]. These authors showed that addition of SM (egg yolk) to cultures of PANC-1 cells increased GEM sensitivity; this was accompanied by loss of mitochondrial membrane permeability, a hallmark of ceramide cytotoxicity and signaling, and increased Annexin V binding, indicative of apoptosis. The authors were able to show that GEM activated acid SMase in PANC-1 cells (depicted in Fig. 5). It was proposed that addition of exogenous SM enriched endogenous SM pools and created a rich field for endogenous SMase, a manipulation that afforded continuous replenishment with ceramide. The authors

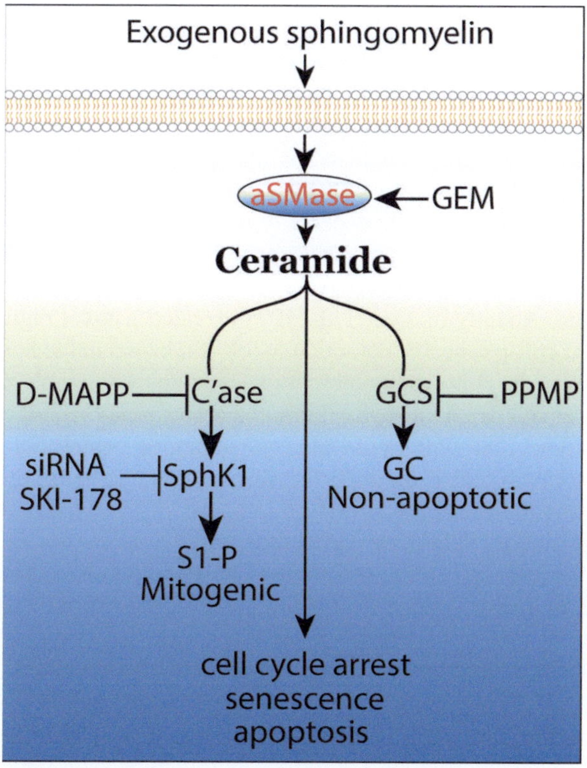

Fig. 5 Manipulation of ceramide genesis and metabolism in pancreatic cancer cells and impact on cell fate. In this depiction, addition of exogeneous SM contributes to ceramide production via GEM-induced activation of acid SMase. Ceramide signaling can be intensified by introduction of enzyme inhibitors at junctures shown. *aSMase* acid sphingomyelinase, *GEM* gemcitabine, d-MAPP D-*erythro*-2-tetradecanoylamino-1-phenyl-1-propanol, *C'ase* ceramidase, *siRNA* small interfering RNA, *SKI-178* sphingosine kinase inhibitor, *SphK1* sphingosine kinase-1, *S1-P* sphingosine 1-phosphate, *GCS* glucosylceramide synthase, *PPMP* D-*threo*-1-phenyl-2-palmitoylamino-3-morpholino-1-propanol, *GC* glucosylceramide

suggested that some tumor cells could be intrinsically resistant to chemotherapy because of limited pools of signaling SM, and that supplementing these pools would provide fodder for SMase-orchestrated ceramide generation and apoptosis. In subsequent studies, it was shown that ceramide elicited GEM-induced senescence *and* apoptosis in human pancreatic cancer cells [101], validating that ceramide levels are critical in cell fate decisions in pancreatic cancer, with cell cycle progression being uninhibited at low ceramide levels, senescence induced at moderate levels, and apoptosis initiated at higher levels [101]. By manipulating ceramide metabolism to improve cytotoxic response [25], it was subsequently shown that blocking ceramide glycosylation by using D-*threo*-1-phenyl-2-palmitoylamino-3-morpholino-1-propanol (PPMP) and inhibiting ceramidase activity by addition of D-*erythro*-2-tetradecanoylamino-1-phenyl-1-propanol (D-MAPP), increased GEM cytotoxicity in PANC-1 cells (Fig. 5) [101].

The employment of albumin-bound nab-paclitaxel for first-line treatment of advanced pancreatic cancer in the United States represents a significant step forward, and in combination with GEM, this regimen delivers a new therapeutic with significant improvement in survival [106]. In the challenging quest to improve cancer treatment, scientists and clinicians alike seek drug combinations that can offer enhanced efficacy. Such can be realized by using agents that hit multiple targets to converge on cell death via separate avenues. A paclitaxel-ceramide combination is a good example of this approach; especially in light of the findings that paclitaxel promotes ceramide formation in cancer cells [70]. Paclitaxel targets microtubules, polymerizes tubulin, and blocks cell cycle at G2/M, thus cells cannot form normal mitotic apparatus. A study by Qui et al. [107] revealed a novel regimen consisting of paclitaxel and cell-permeable ceramide, work conducted in the human pancreatic cancer cell line, L3.6. This drug combination, as opposed to single agents, induced pancreatic cancer cell death through differential activation of EGFR-mediated MAP kinases. We believe that exploration of novel ceramide-containing drug regimens that deliver assorted blows is an important avenue that can be followed to advance therapy for pancreatic cancer.

Sphingosine kinase (SphK), of which there are two isozymes, is a key enzyme that regulates the ratio of cellular ceramide to S1-P. Sphingosine kinase 1 (SphK1) is upregulated in many cancers [60, 108], and studies show that this enzyme occupies a central role in signaling events that dictate cell survival in pancreatic cancer [34, 94]. Appropriate in this regard, it has been shown that the ceramide/S1-P ratio is a critical biosensor for predicting pancreatic cancer cell sensitivity to GEM. For example, pancreatic cancer cell resistance to GEM has been demonstrated to be correlated with SphK1 activity, wherein high SphK1 activity was associated with GEM resistance [34]. Moreover, it was shown that SphK1 activity and protein expression increased in pancreatic cells treated with either GEM [34] or bortezomib [94], important findings revealing a pivotal role for SphK1 and S1-P in survival of pancreatic cancer cells. These studies went on to demonstrate that limiting SphK1 enzymatic activity by using specific inhibitors such as SKI-178 or anti-SphK1-based RNA interference, sensitized pancreatic cancer cells to GEM- and bortezomib-induced apoptotic cell death. Importantly, these studies reveal that resistance to conventional chemotherapeutics in pancreatic cancer is in part related to Sphk1 activity, which we suggest is a crucial weakness or Achilles heel in pancreatic cancer.

8 Nanoliposomal Ceramide as an Emerging Therapeutic in Pancreatic Cancer

One exciting aspect of using ceramide as a chemotherapeutic is the preferred selectivity for inducing programmed cell death in cancer cells [67, 109, 110]. Mechanisms underlying this selectivity are not well understood; however, enhanced pro-mitogenic signaling cascades, such as protein kinase C, ERK, and Akt that are activated or overexpressed in cancers, may play a role. Another viable mechanism for selectivity is ceramide' s targeting of the Warburg effect [111], a metabolism exhibited by most tumors manifest by a preferential switch to glycolysis.

Despite the increased solubility and membrane permeability of short-chain ceramide, therapeutic efficacy can be limited somewhat by poor bioavailability, a tendency to precipitate in biological fluids, and constructive and destructive metabolism. To improve solubility, delivery, and stability, investigators have turned to nano solutions as a solution [33], and have produced and tested formulations of nanoliposomal C6-ceramide alone or partnered with agents that either deter ceramide metabolism or complement activity; many of these regimens operate synergistically. Excellent examples of effective drug combinations in pancreatic cancer have been presented in investigations of nanoliposomal C6-ceramide (Lip-C6) paired with either GEM or the GCS inhibitor, D-*threo*-1-phenyl-2-decanoyl-3-morpholino-1-propanol (PDMP). For example, Jiang et al. [35], using 80 nm diameter liposomal formulations that incorporate 30 mol% C6-ceramide, demonstrated that whereas single agent Lip-C6 was minimally cytotoxic in PANC-1 cells that are known to be GEM-resistant, efficacy was achieved by partnering Lip-C6 with either sub-therapeutic doses of GEM, or with PDMP, which blocks ceramide glycosylation, a chief ceramide-detoxification pathway in pancreatic cancer cells [102]. These same combinations inhibited tumor growth, in vivo, in human pancreatic tumor xenografts in athymic, nude mice. Fittingly, it has been shown that inhibition of Akt/PI3 kinase by small molecules can synergize with GEM to induce apoptosis in human pancreatic cancer cell lines [112–114]. Whereas GEM was shown ineffective in diminishing Akt phosphorylation, as would be expected for a DNA-intercalating agent, inhibition of Akt phosphorylation was efficiently affected by Lip-C6 administration, an action that significantly enhanced sensitivity to GEM, in this "GEM-insensitive" model [35]. Overall, these studies show that rationally designed combinatorial therapies have the potential to achieve synergy in treatment of pancreatic cancer (Fig. 4).

9 Future Directions

Whereas a significant amount of effort has been directed toward studies of bioactive SL in a myriad of cancer types, including breast, colorectal, prostate, neuroblastoma, and hematological malignancies, a review of the literature easily reveals a scarcity of work in the area of pancreatic cancer; whether this is due to the lower numbers that are affected by pancreatic adenocarcinoma compared to other cancers, the intractability and frustration associated with this cancer, less SL machinery of which to take advantage, or funding priorities, is unknown to these authors. However, as discussed here, there have been a number of interesting studies that present inroads for further exploitation of SL metabolism and the roles and function of SL in development and treatment of pancreatic cancer. The schematic in Fig. 4 shows some of the most prominent signaling pathways in pancreatic cancer and illustrates how ceramide might be employed to trigger specific cell death cascades. Surgery, chemotherapy, and radiation are first-line therapeutic avenues for pancreatic cancer; however, it is clear that other strategies must be explored in order to attain more favorable clinical results. We believe that in-depth studies of SL metabolism, the incorporation of SL metabolism inhibitors (Fig. 5), and the design of novel ceramide "bullets" will reveal vulnerabilities in pancreatic cancer.

References

1. Bigsby J (1835) Observations, pathological and therapeutic, on diseases of the pancreas. Edinburgh Med Surg J 44:85–102
2. Da Costa JM (1858) Cancer of the pancreas. North Am Med Chir Rev 2:883
3. Baskova IP, Iusupova GI, Nikonov GI (1984) [Lipase and cholesterol-esterase activity in the salivary gland secretions of the medical leech Hirudo medicinalis]. Biokhimiia 49(4):676–678
4. Rabinowitz JL (1996) Salivary lipid profiles of the leech (Hirudo medicinalis). Lipids 31(8):887–888
5. Whipple AO, Parsons WB, Mullins CR (1935) Treatment of carcinoma of the ampulla of Vater. Ann Surg 102(4):763–779
6. Brunschwig A (1937) Resection of head of pancreas and duodenum for carcinoma—pancreatoduodenectomy. Surg Gynecol Obstet 65:681–684
7. Silverman DT, Hoover RN, Brown LM, Swanson GM, Schiffman M, Greenberg RS et al (2003) Why do Black Americans have a higher risk of pancreatic cancer than White Americans? Epidemiology 14(1):45–54
8. Arslan AA, Helzlsouer KJ, Kooperberg C, Shu XO, Steplowski E, Bueno-de-Mesquita HB et al (2010) Anthropometric measures, body mass index, and pancreatic cancer: a pooled analysis from the Pancreatic Cancer Cohort Consortium (PanScan). Arch Intern Med 170(9):791–802
9. Mohr SB, Garland CF, Gorham ED, Grant WB, Garland FC (2010) Ultraviolet B irradiance and vitamin D status are inversely associated with incidence rates of pancreatic cancer worldwide. Pancreas 39(5):669–674
10. Bao Y, Ng K, Wolpin BM, Michaud DS, Giovannucci E, Fuchs CS (2010) Predicted vitamin D status and pancreatic cancer risk in two prospective cohort studies. Br J Cancer 102(9):1422–1427

11. Fitz RH (1889) Acute pancreatitis: a consideration of pancreatic hemorrhage, hemorrhagic, suppurative, and gangrenous pancreatitis, and of disseminated fat-necrosis. Boston Med Surg J 120(8):181–187, 205–207, 29–35
12. Lowenfels AB, Maisonneuve P, Cavallini G, Ammann RW, Lankisch PG, Andersen JR et al (1993) Pancreatitis and the risk of pancreatic cancer. International Pancreatitis Study Group. N Engl J Med 328(20):1433–1437
13. Chiari H (1896) Über die Selbstverdauung des menschlichen Pankreas. Z Heilk 17:69–96
14. Opie EL (1901) The etiology of acute hemorrhagic pancreatitis. Johns Hopkins Hosp Bull 12:182–188
15. Comfort MW, Gambill EE, Baggenstoss AH (1946) Chronic relapsing pancreatitis; a study of 29 cases without associated disease of the biliary or gastrointestinal tract. Gastroenterology 6:376–408
16. Vonlaufen A, Spahr L, Apte MV, Frossard JL (2014) Alcoholic pancreatitis: a tale of spirits and bacteria. World J Gastrointest Pathophysiol 5(2):82–90
17. Ling S, Feng T, Jia K, Tian Y, Li Y (2014) Inflammation to cancer: the molecular biology in the pancreas (Review). Oncol Lett 7(6):1747–1754
18. Ekbom A, McLaughlin JK, Karlsson BM, Nyren O, Gridley G, Adami HO et al (1994) Pancreatitis and pancreatic cancer: a population-based study. J Natl Cancer Inst 86(8): 625–627
19. Talamini G, Falconi M, Bassi C, Sartori N, Salvia R, Caldiron E et al (1999) Incidence of cancer in the course of chronic pancreatitis. Am J Gastroenterol 94(5):1253–1260
20. Bracci PM, Wang F, Hassan MM, Gupta S, Li D, Holly EA (2009) Pancreatitis and pancreatic cancer in two large pooled case-control studies. Cancer Causes Control 20(9):1723–1731
21. Morad SA, Cabot MC (2013) Ceramide-orchestrated signalling in cancer cells. Nat Rev Cancer 13(1):51–65
22. Burris HA 3rd, Moore MJ, Andersen J, Green MR, Rothenberg ML, Modiano MR et al (1997) Improvements in survival and clinical benefit with gemcitabine as first-line therapy for patients with advanced pancreas cancer: a randomized trial. J Clin Oncol 15(6):2403–2413
23. Conroy T, Desseigne F, Ychou M, Bouche O, Guimbaud R, Becouarn Y et al (2011) FOLFIRINOX versus gemcitabine for metastatic pancreatic cancer. N Engl J Med 364(19): 1817–1825
24. Von Hoff DD, Ramanathan RK, Borad MJ, Laheru DA, Smith LS, Wood TE et al (2011) Gemcitabine plus nab-paclitaxel is an active regimen in patients with advanced pancreatic cancer: a phase I/II trial. J Clin Oncol 29(34):4548–4554
25. Senchenkov A, Litvak DA, Cabot MC (2001) Targeting ceramide metabolism–a strategy for overcoming drug resistance. J Natl Cancer Inst 93(5):347–357
26. Barth BM, Cabot MC, Kester M (2011) Ceramide-based therapeutics for the treatment of cancer. Anticancer Agents Med Chem 11(9):911–919
27. Gouaze-Andersson V, Cabot MC (2011) Sphingolipid metabolism and drug resistance in hematological malignancies. Anticancer Agents Med Chem 11(9):891–903
28. Liu J, Beckman BS, Foroozesh M (2013) A review of ceramide analogs as potential anticancer agents. Future Med Chem 5(12):1405–1421
29. Saddoughi SA, Ogretmen B (2013) Diverse functions of ceramide in cancer cell death and proliferation. Adv Cancer Res 117:37–58
30. Dimanche-Boitrel MT, Rebillard A, Gulbins E (2011) Ceramide in chemotherapy of tumors. Recent Pat Anticancer Drug Discov 6(3):284–293
31. Stover TC, Sharma A, Robertson GP, Kester M (2005) Systemic delivery of liposomal short-chain ceramide limits solid tumor growth in murine models of breast adenocarcinoma. Clin Cancer Res 11(9):3465–3474
32. Tran MA, Smith CD, Kester M, Robertson GP (2008) Combining nanoliposomal ceramide with sorafenib synergistically inhibits melanoma and breast cancer cell survival to decrease tumor development. Clin Cancer Res 14(11):3571–3581
33. Stover T, Kester M (2003) Liposomal delivery enhances short-chain ceramide-induced apoptosis of breast cancer cells. J Pharmacol Exp Ther 307(2):468–475

34. Guillermet-Guibert J, Davenne L, Pchejetski D, Saint-Laurent N, Brizuela L, Guilbeau-Frugier C et al (2009) Targeting the sphingolipid metabolism to defeat pancreatic cancer cell resistance to the chemotherapeutic gemcitabine drug. Mol Cancer Ther 8(4):809–820
35. Jiang Y, DiVittore NA, Kaiser JM, Shanmugavelandy SS, Fritz JL, Heakal Y et al (2011) Combinatorial therapies improve the therapeutic efficacy of nanoliposomal ceramide for pancreatic cancer. Cancer Biol Ther 12(7):574–585
36. Ravindranath MH, Gonzales AM, Nishimoto K, Tam WY, Soh D, Morton DL (2000) Immunology of gangliosides. Indian J Exp Biol 38(4):301–312
37. Fredman P, Hedberg K, Brezicka T (2003) Gangliosides as therapeutic targets for cancer. BioDrugs 17(3):155–167
38. Terando AM, Faries MB, Morton DL (2007) Vaccine therapy for melanoma: current status and future directions. Vaccine 25(Suppl 2):B4–B16
39. Durrant LG, Noble P, Spendlove I (2012) Immunology in the clinic review series; focus on cancer: glycolipids as targets for tumour immunotherapy. Clin Exp Immunol 167(2):206–215
40. Chu KU, Ravindranath MH, Gonzales A, Nishimoto K, Tam WY, Soh D et al (2000) Gangliosides as targets for immunotherapy for pancreatic adenocarcinoma. Cancer 88(8): 1828–1836
41. Sommers SC, Meissner WA (1954) Unusual carcinomas of the pancreas. AMA Arch Pathol 58(2):101–111
42. Brat DJ, Lillemoe KD, Yeo CJ, Warfield PB, Hruban RH (1998) Progression of pancreatic intraductal neoplasias to infiltrating adenocarcinoma of the pancreas. Am J Surg Pathol 22(2):163–169
43. Brockie E, Anand A, Albores-Saavedra J (1998) Progression of atypical ductal hyperplasia/carcinoma in situ of the pancreas to invasive adenocarcinoma. Ann Diagn Pathol 2(5): 286–292
44. Hruban RH, Goggins M, Parsons J, Kern SE (2000) Progression model for pancreatic cancer. Clin Cancer Res 6(8):2969–2972
45. Day JD, Digiuseppe JA, Yeo C, Lai-Goldman M, Anderson SM, Goodman SN et al (1996) Immunohistochemical evaluation of HER-2/neu expression in pancreatic adenocarcinoma and pancreatic intraepithelial neoplasms. Hum Pathol 27(2):119–124
46. Hannun YA, Obeid LM (2011) Many ceramides. J Biol Chem 286(32):27855–27862
47. Pinto SN, Silva LC, Futerman AH, Prieto M (2011) Effect of ceramide structure on membrane biophysical properties: the role of acyl chain length and unsaturation. Biochim Biophys Acta 1808(11):2753–2760
48. Gottesman MM, Ling V (2006) The molecular basis of multidrug resistance in cancer: the early years of P-glycoprotein research. FEBS Lett 580(4):998–1009
49. Liu YY, Han TY, Giuliano AE, Cabot MC (2001) Ceramide glycosylation potentiates cellular multidrug resistance. FASEB J 15(3):719–730
50. Radin NS (1999) Chemotherapy by slowing glucosphingolipid synthesis. Biochem Pharmacol 57(6):589–595
51. Messner MC, Cabot MC (2010) Glucosylceramide in humans. Adv Exp Med Biol 688:156–164
52. Gouaze-Andersson V, Cabot MC (2006) Glycosphingolipids and drug resistance. Biochim Biophys Acta 1758(12):2096–2103
53. Morad SA, Madigan JP, Levin JC, Abdelmageed N, Karimi R, Rosenberg DW et al (2013) Tamoxifen magnifies therapeutic impact of ceramide in human colorectal cancer cells independent of p53. Biochem Pharmacol 85(8):1057–1065
54. Chapman JV, Gouaze-Andersson V, Cabot MC (2010) Expression of P-glycoprotein in HeLa cells confers resistance to ceramide cytotoxicity. Int J Oncol 37(6):1591–1597
55. Wijesinghe DS, Lamour NF, Gomez-Munoz A, Chalfant CE (2007) Ceramide kinase and ceramide-1-phosphate. Methods Enzymol 434:265–292
56. Lavie Y, Cao H, Bursten SL, Giuliano AE, Cabot MC (1996) Accumulation of glucosylceramides in multidrug-resistant cancer cells. J Biol Chem 271(32):19530–19536

57. Lucci A, Cho WI, Han TY, Giuliano AE, Morton DL, Cabot MC (1998) Glucosylceramide: a marker for multiple-drug resistant cancers. Anticancer Res 18(1B):475–480
58. Musumarra G, Barresi V, Condorelli DF, Scire S (2003) A bioinformatic approach to the identification of candidate genes for the development of new cancer diagnostics. Biol Chem 384(2):321–327
59. Zeidan YH, Jenkins RW, Korman JB, Liu X, Obeid LM, Norris JS et al (2008) Molecular targeting of acid ceramidase: implications to cancer therapy. Curr Drug Targets 9(8): 653–661
60. Orr Gandy KA, Obeid LM (2013) Targeting the sphingosine kinase/sphingosine 1-phosphate pathway in disease: review of sphingosine kinase inhibitors. Biochim Biophys Acta 1831(1):157–166
61. Ogretmen B, Pettus BJ, Rossi MJ, Wood R, Usta J, Szulc Z et al (2002) Biochemical mechanisms of the generation of endogenous long chain ceramide in response to exogenous short chain ceramide in the A549 human lung adenocarcinoma cell line. Role for endogenous ceramide in mediating the action of exogenous ceramide. J Biol Chem 277(15):12960–12969
62. Jiang Y, DiVittore NA, Young MM, Jia Z, Xie K, Ritty TM et al (2013) Altered sphingolipid metabolism in patients with metastatic pancreatic cancer. Biomolecules 3(3):435–448
63. Dimanche-Boitrel MT, Rebillard A (2013) Sphingolipids and response to chemotherapy. Handb Exp Pharmacol 216:73–91
64. Wang H, Maurer BJ, Reynolds CP, Cabot MC (2001) N-(4-hydroxyphenyl)retinamide elevates ceramide in neuroblastoma cell lines by coordinate activation of serine palmitoyltransferase and ceramide synthase. Cancer Res 61(13):5102–5105
65. Messner MC, Cabot MC (2011) Cytotoxic responses to N-(4-hydroxyphenyl)retinamide in human pancreatic cancer cells. Cancer Chemother Pharmacol 68(2):477–487
66. Maurer BJ, Kang MH, Villablanca JG, Janeba J, Groshen S, Matthay KK et al (2013) Phase I trial of fenretinide delivered orally in a novel organized lipid complex in patients with relapsed/refractory neuroblastoma: a report from the New Approaches to Neuroblastoma Therapy (NANT) consortium. Pediatr Blood Cancer 60(11):1801–1808
67. Perry DK, Carton J, Shah AK, Meredith F, Uhlinger DJ, Hannun YA (2000) Serine palmitoyltransferase regulates de novo ceramide generation during etoposide-induced apoptosis. J Biol Chem 275(12):9078–9084
68. Zhang S, Zhou J, Zhang C, Wu H, Wang Y, Bian J et al (2012) Arsenic trioxide inhibits HCCLM3 cells invasion through de novo ceramide synthesis and sphingomyelinase-induced ceramide production. Med Oncol 29(3):2251–2260
69. Rath G, Schneider C, Langlois B, Sartelet H, Morjani H, Btaouri HE et al (2009) De novo ceramide synthesis is responsible for the anti-tumor properties of camptothecin and doxorubicin in follicular thyroid carcinoma. Int J Biochem Cell Biol 41(5):1165–1172
70. Charles AG, Han TY, Liu YY, Hansen N, Giuliano AE, Cabot MC (2001) Taxol-induced ceramide generation and apoptosis in human breast cancer cells. Cancer Chemother Pharmacol 47(5):444–450
71. Wang H, Giuliano AE, Cabot MC (2002) Enhanced de novo ceramide generation through activation of serine palmitoyltransferase by the P-glycoprotein antagonist SDZ PSC 833 in breast cancer cells. Mol Cancer Ther 1(9):719–726
72. Senchenkov A, Han TY, Wang H, Frankel AE, Kottke TJ, Kaufmann SH et al (2001) Enhanced ceramide generation and induction of apoptosis in human leukemia cells exposed to DT(388)-granulocyte-macrophage colony-stimulating factor (GM-CSF), a truncated diphtheria toxin fused to human GM-CSF. Blood 98(6):1927–1934
73. Melnik MK, Webb CP, Richardson PJ, Luttenton CR, Campbell AD, Monroe TJ et al (2010) Phase II trial to evaluate gemcitabine and etoposide for locally advanced or metastatic pancreatic cancer. Mol Cancer Ther 9(8):2423–2429
74. Zhang SH, Huang Q (2013) Etoposide induces apoptosis via the mitochondrial- and caspase-dependent pathways and in non-cancer stem cells in Panc-1 pancreatic cancer cells. Oncol Rep 30(6):2765–2770

75. Cabot MC, Han TY, Giuliano AE (1998) The multidrug resistance modulator SDZ PSC 833 is a potent activator of cellular ceramide formation. FEBS Lett 431(2):185–188
76. Cabot MC, Giuliano AE, Han TY, Liu YY (1999) SDZ PSC 833, the cyclosporine A analogue and multidrug resistance modulator, activates ceramide synthesis and increases vinblastine sensitivity in drug-sensitive and drug-resistant cancer cells. Cancer Res 59(4):880–885
77. Morad SA, Messner MC, Levin JC, Abdelmageed N, Park H, Merrill AH Jr et al (2013) Potential role of acid ceramidase in conversion of cytostatic to cytotoxic end-point in pancreatic cancer cells. Cancer Chemother Pharmacol 71(3):635–645
78. Liu X, Cheng JC, Turner LS, Elojeimy S, Beckham TH, Bielawska A et al (2009) Acid ceramidase upregulation in prostate cancer: role in tumor development and implications for therapy. Expert Opin Ther Targets 13(12):1449–1458
79. Camacho L, Meca-Cortes O, Abad JL, Garcia S, Rubio N, Diaz A et al (2013) Acid ceramidase as a therapeutic target in metastatic prostate cancer. J Lipid Res 54(5):1207–1220
80. Maurer BJ, Melton L, Billups C, Cabot MC, Reynolds CP (2000) Synergistic cytotoxicity in solid tumor cell lines between N-(4-hydroxyphenyl)retinamide and modulators of ceramide metabolism. J Natl Cancer Inst 92(23):1897–1909
81. Hail N Jr, Kim HJ, Lotan R (2006) Mechanisms of fenretinide-induced apoptosis. Apoptosis 11(10):1677–1694
82. Walsh D, Nelson KA, Mahmoud FA (2003) Established and potential therapeutic applications of cannabinoids in oncology. Support Care Cancer 11(3):137–143
83. Pertwee RG, Howlett AC, Abood ME, Alexander SP, Di Marzo V, Elphick MR et al (2010) International Union of Basic and Clinical Pharmacology. LXXIX. Cannabinoid receptors and their ligands: beyond CB(1) and CB(2). Pharmacol Rev 62(4):588–631
84. Brandi J, Dando I, Palmieri M, Donadelli M, Cecconi D (2013) Comparative proteomic and phosphoproteomic profiling of pancreatic adenocarcinoma cells treated with CB1 or CB2 agonists. Electrophoresis 34(9-10):1359–1368
85. Dando I, Donadelli M, Costanzo C, Dalla Pozza E, D'Alessandro A, Zolla L et al (2013) Cannabinoids inhibit energetic metabolism and induce AMPK-dependent autophagy in pancreatic cancer cells. Cell Death Dis 4, e664
86. Carracedo A, Gironella M, Lorente M, Garcia S, Guzman M, Velasco G et al (2006) Cannabinoids induce apoptosis of pancreatic tumor cells via endoplasmic reticulum stress-related genes. Cancer Res 66(13):6748–6755
87. Vasseur S, Vidal Mallo G, Fiedler F, Bodeker H, Canepa E, Moreno S et al (1999) Cloning and expression of the human p8, a nuclear protein with mitogenic activity. Eur J Biochem 259(3):670–675
88. Malicet C, Lesavre N, Vasseur S, Iovanna JL (2003) p8 inhibits the growth of human pancreatic cancer cells and its expression is induced through pathways involved in growth inhibition and repressed by factors promoting cell growth. Mol Cancer 2:37
89. Sanchez C, de Ceballos ML, Gomez del Pulgar T, Rueda D, Corbacho C, Velasco G et al (2001) Inhibition of glioma growth in vivo by selective activation of the CB(2) cannabinoid receptor. Cancer Res 61(15):5784–9
90. Salazar M, Carracedo A, Salanueva IJ, Hernandez-Tiedra S, Lorente M, Egia A et al (2009) Cannabinoid action induces autophagy-mediated cell death through stimulation of ER stress in human glioma cells. J Clin Invest 119(5):1359–1372
91. Xing SS, Shen CC, Godard MP, Wang JJ, Yue YY, Yang ST et al (2014) Bortezomib inhibits C2C12 growth by inducing cell cycle arrest and apoptosis. Biochem Biophys Res Commun 445(2):375–380
92. Kretowski R, Borzym-Kluczyk M, Cechowska-Pasko M (2014) Efficient induction of apoptosis by proteasome inhibitor: bortezomib in the human breast cancer cell line MDA-MB-231. Mol Cell Biochem 389(1–2):177–185
93. Hou DR, Huang AC, Shiau CW, Wang CY, Yu HC, Chen KF (2013) Bortezomib congeners induce apoptosis of hepatocellular carcinoma via CIP2A inhibition. Molecules 18(12):15398–15411
94. Gong L, Yang B, Xu M, Cheng B, Tang X, Zheng P et al (2014) Bortezomib-induced apoptosis in cultured pancreatic cancer cells is associated with ceramide production. Cancer Chemother Pharmacol 73(1):69–77

95. Wang H, Cao Q, Dudek AZ (2012) Phase II study of panobinostat and bortezomib in patients with pancreatic cancer progressing on gemcitabine-based therapy. Anticancer Res 32(3): 1027–1031
96. Nawrocki ST, Carew JS, Dunner K Jr, Boise LH, Chiao PJ, Huang P et al (2005) Bortezomib inhibits PKR-like endoplasmic reticulum (ER) kinase and induces apoptosis via ER stress in human pancreatic cancer cells. Cancer Res 65(24):11510–11519
97. Lauricella M, Emanuele S, D'Anneo A, Calvaruso G, Vassallo B, Carlisi D et al (2006) JNK and AP-1 mediate apoptosis induced by bortezomib in HepG2 cells via FasL/caspase-8 and mitochondria-dependent pathways. Apoptosis 11(4):607–625
98. Yu C, Rahmani M, Dent P, Grant S (2004) The hierarchical relationship between MAPK signaling and ROS generation in human leukemia cells undergoing apoptosis in response to the proteasome inhibitor Bortezomib. Exp Cell Res 295(2):555–566
99. Beckham TH, Lu P, Jones EE, Marrison T, Lewis CS, Cheng JC et al (2013) LCL124, a cationic analog of ceramide, selectively induces pancreatic cancer cell death by accumulating in mitochondria. J Pharmacol Exp Ther 344(1):167–178
100. Modrak DE, Cardillo TM, Newsome GA, Goldenberg DM, Gold DV (2004) Synergistic interaction between sphingomyelin and gemcitabine potentiates ceramide-mediated apoptosis in pancreatic cancer. Cancer Res 64(22):8405–8410
101. Modrak DE, Leon E, Goldenberg DM, Gold DV (2009) Ceramide regulates gemcitabine-induced senescence and apoptosis in human pancreatic cancer cell lines. Mol Cancer Res 7(6):890–896
102. Chapman JV, Gouaze-Andersson V, Messner MC, Flowers M, Karimi R, Kester M et al (2010) Metabolism of short-chain ceramide by human cancer cells–implications for therapeutic approaches. Biochem Pharmacol 80(3):308–315
103. Gopal P, Rehman RU, Chadha KS, Qiu M, Colella R (2006) Matrigel influences morphology and cathepsin B distribution of prostate cancer PC3 cells. Oncol Rep 16(2):313–320
104. Zhu QY, Wang Z, Ji C, Cheng L, Yang YL, Ren J et al (2011) C6-ceramide synergistically potentiates the anti-tumor effects of histone deacetylase inhibitors via AKT dephosphorylation and alpha-tubulin hyperacetylation both in vitro and in vivo. Cell Death Dis 2, e117
105. Wu CH, Cao C, Kim JH, Hsu CH, Wanebo HJ, Bowen WD et al (2012) Trojan-horse nanotube on-command intracellular drug delivery. Nano Lett 12(11):5475–5480
106. Cecco S, Aliberti M, Baldo P, Giacomin E, Leone R (2014) Safety and efficacy evaluation of albumin-bound paclitaxel. Expert Opin Drug Saf 13(4):511–520
107. Qiu L, Zhou C, Sun Y, Di W, Scheffler E, Healey S et al (2006) Paclitaxel and ceramide synergistically induce cell death with transient activation of EGFR and ERK pathway in pancreatic cancer cells. Oncol Rep 16(4):907–913
108. Shida D, Takabe K, Kapitonov D, Milstien S, Spiegel S (2008) Targeting SphK1 as a new strategy against cancer. Curr Drug Targets 9(8):662–673
109. Pan S, Brentnall TA, Kelly K, Chen R (2013) Tissue proteomics in pancreatic cancer study: discovery, emerging technologies, and challenges. Proteomics 13(3–4):710–721
110. Bose R, Verheij M, Haimovitz-Friedman A, Scotto K, Fuks Z, Kolesnick R (1995) Ceramide synthase mediates daunorubicin-induced apoptosis: an alternative mechanism for generating death signals. Cell 82(3):405–414
111. Ryland LK, Doshi UA, Shanmugavelandy SS, Fox TE, Aliaga C, Broeg K et al (2013) C6-ceramide nanoliposomes target the Warburg effect in chronic lymphocytic leukemia. PLoS One 8(12), e84648
112. Ito D, Fujimoto K, Mori T, Kami K, Koizumi M, Toyoda E et al (2006) In vivo antitumor effect of the mTOR inhibitor CCI-779 and gemcitabine in xenograft models of human pancreatic cancer. Int J Cancer 118(9):2337–2343
113. Takeda A, Osaki M, Adachi K, Honjo S, Ito H (2004) Role of the phosphatidylinositol 3′-kinase-Akt signal pathway in the proliferation of human pancreatic ductal carcinoma cell lines. Pancreas 28(3):353–358
114. Bondar VM, Sweeney-Gotsch B, Andreeff M, Mills GB, McConkey DJ (2002) Inhibition of the phosphatidylinositol 3′-kinase-AKT pathway induces apoptosis in pancreatic carcinoma cells in vitro and in vivo. Mol Cancer Ther 1(12):989–997

Sphingolipids in the Pathogenesis of Head and Neck and Lung Cancers: Translational Aspects for Therapy and Biomarker Development

Can E. Senkal and Lina M. Obeid

Abstract Sphingolipids have garnered tremendous attention in the last two decades as a class of the bioactive lipids that regulate multiple biological events in addition to being important for membrane structures. With the identification of enzymes involved in metabolism of various bioactive molecules of the sphingolipid pathway, a better and broader understanding has been gained of how cells control signaling events, such as proliferation and apoptosis, through regulating levels of sphingolipids. This chapter will focus on recent findings relating to functions of sphingolipids in the pathogenesis of lung and head and neck cancers. Specifically, recent discoveries on the roles of sphingolipids in the regulation of head and neck squamous cell carcinoma and lung cancers will be discussed. Moreover, recent efforts on the identification of sphingolipid metabolism related biomarker development will be deliberated.

Keywords Head and neck squamous cell carcinoma • Lung cancer • Ceramide • Sphingolipids • Sphingosine-1-phosphate • Ceramide synthase • Sphingosine kinase • Ceramidase • Glucosylceramide synthase • Cancer therapy

C.E. Senkal (✉)
Department of Medicine, Stony Brook University, Health Science Center, Stony Brook University Medical Center, HSC Level 15, Room 15023, Stony Brook, NY 11794-8155, USA
e-mail: Can.Senkal@stonybrookmedicine.edu

L.M. Obeid
Department of Medicine, Stony Brook University, Health Science Center, Stony Brook University Medical Center, HSC Level 15, Room 15023, Stony Brook, NY 11794-8155, USA

Northport Veterans Affairs Medical Center, Northport, NY, USA

Stony Brook University, Health Science Center, Stony Brook University Medical Center, HSC Level 4, Rook 179, Stony Brook, NY 11794-8430, USA
e-mail: Lina.Obeid@stonybrookmedicine.edu

Y.A. Hannun et al. (eds.), *Bioactive Sphingolipids in Cancer Biology and Therapy*, DOI 10.1007/978-3-319-20750-6_11

1 Head and Neck Squamous Cell Carcinoma (HNSCC) and Lung Cancer Epidemiology

Head and neck squamous cell carcinoma (HNSCC), which represents about 85 % of all head and neck cancers, is the sixth most common cancer amongst men and the eighth for women. With nearly 43,000 new patients diagnosed each year, HNSCC embodies the 6 % of newly diagnosed cancers in the USA. Every year approximately 12,000 deaths are estimated to occur due to HNSCC [1, 2]. The important risk factors of developing HNSCC are exposure to heavy smoking, alcohol consumption, and human papilloma virus (HPV) infections. Historically, smoking is strongly linked to the development of HNSCC in the larynx, hypolarynx, oral cavity, and floor of the mouth [3]. On the other hand, HPV infections are linked to cancers of the tonsils and base of the tongue. More than 70 % of the HNSCC of oropharynx tumors have been shown to have HPV infection [4]. In the clinic, HNSCC is usually presents as locally advanced disease with stage III-IV classification. Traditionally, the treatment of HNSCC has been surgical resection of the tumor, radiotherapy, combination chemotherapy, and chemo-radiation therapy [5]. Despite advances in therapeutic modalities, the 5-year survival rate of HSNCC remains about 50 % [3]. Moreover, more that 50 % of newly diagnosed patients with HNSCC are predicted to have a relapse of localized tumor or a metastatic secondary site cancer. In addition, patients with recurrent and/or metastatic HNSCC are known to have poor prognosis with a median survival time of less than a year [6].

As for lung cancer, it is the leading cause of cancer related deaths worldwide [7]. Approximately 224,000 patients with new lung cancer diagnoses and about 160,000 lung cancer related deaths are estimated for 2014 in the USA. Currently, lung cancer is the second most diagnosed cancer amongst men and women [8]. The primary risk factor for developing lung cancer is smoking. Nearly, 85 % of newly diagnosed lung cancer cases are classified under non-small-cell lung cancer (NSCLC), which is comprised of adenocarcinoma, squamous carcinoma, and large-cell carcinoma, according to histologic analysis [9]. Adenocarcinomas and squamous cell carcinoma represent about 50 % and 40 % of NSCLC cases, respectively [10, 11]. In the clinic, NSCLC predominantly presents with unresectable stage III disease [9]. Importantly, the 5-year survival rate for NSCLC is less that 15 % and this rate has not improved significantly in the last few decades [12].

2 Sphingolipid Chemistry and Metabolism

Ceramide lies at the center of sphingolipid metabolism and is composed of a fatty acyl group linked to the sphingosine backbone with an amide linkage (Table 1). Biologically, ceramide has been shown to regulate anti-proliferative cellular responses such as, apoptosis, senescence, growth arrest, and autophagy [13, 14]. Ceramide can be synthesized de novo, or by the hydrolysis of complex sphingolipids such as sphingolmyelin (SM) and glucosylceramide (GlcCer) (Fig. 1).

Table 1 Chemical structures of sphingosine, ceramide, ceramidoids (Pyridinium-ceramides), and cited inhibitors of sphingolipid metabolism

Compound	Structure	Mode of action	References
Sphingosine	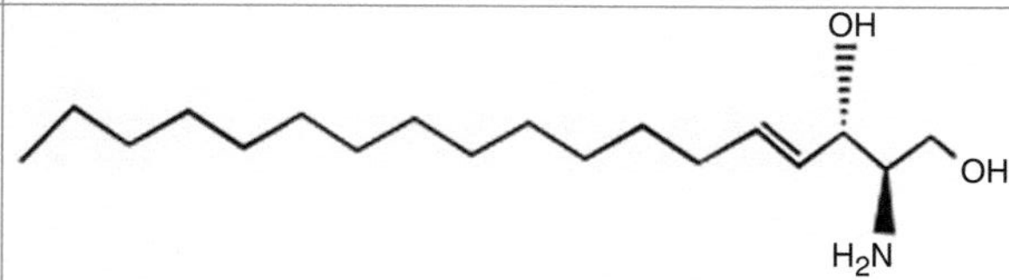	Sphingosine	[43]
C_{16}-ceramide	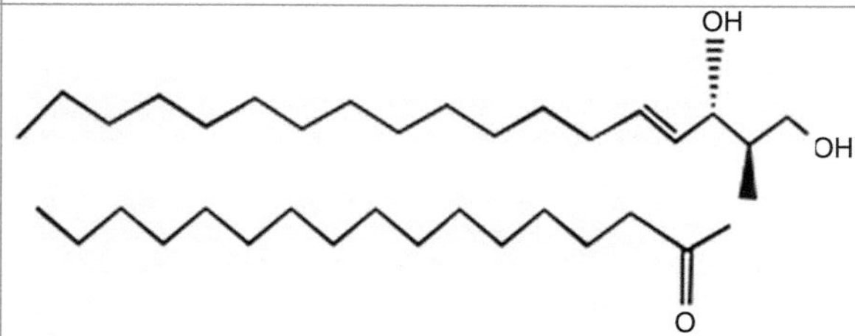	Ceramide	[13, 14]
LCL-124 (L-t-C_6-pyr-cer)	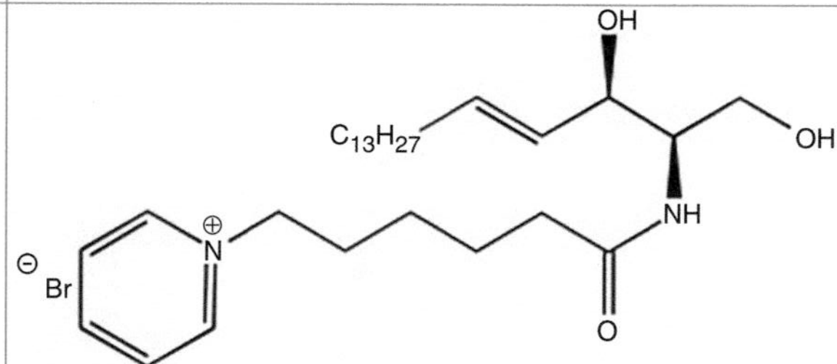	Ceramide analog	[67]

(continued)

Table 1 (continued)

Compound	Structure	Mode of action	References
D-e-C_{16}-pyr-cer	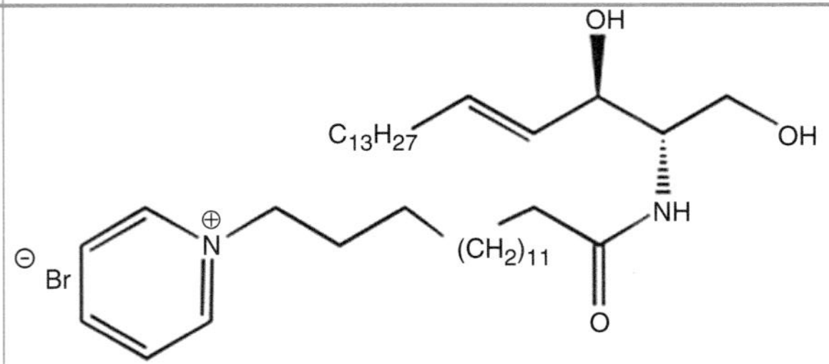	Ceramide analog	[56, 72]
D-e-C_{18}-pyr-cer	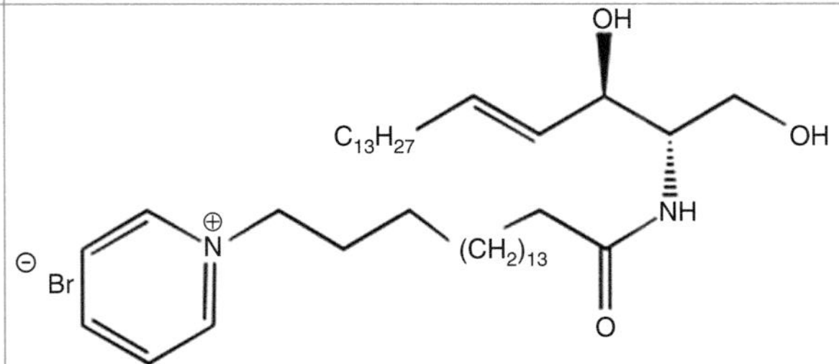	Ceramide analog	[56]
LCL29 (D-e-C_6-pyr-cer)	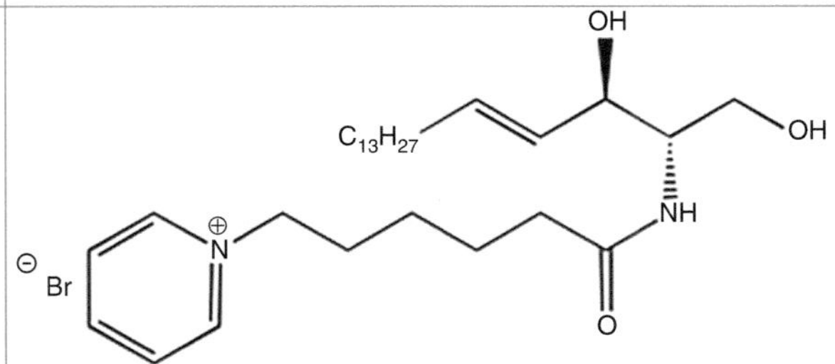	Ceramide analog	[71]

FTY-720	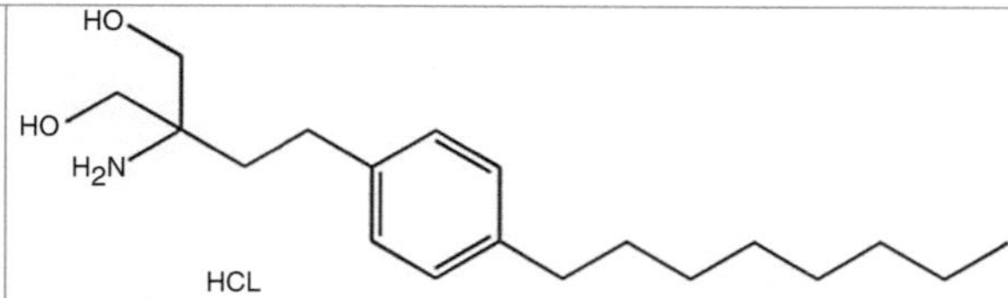	Sphingosine analog	[77, 78]
PDMP	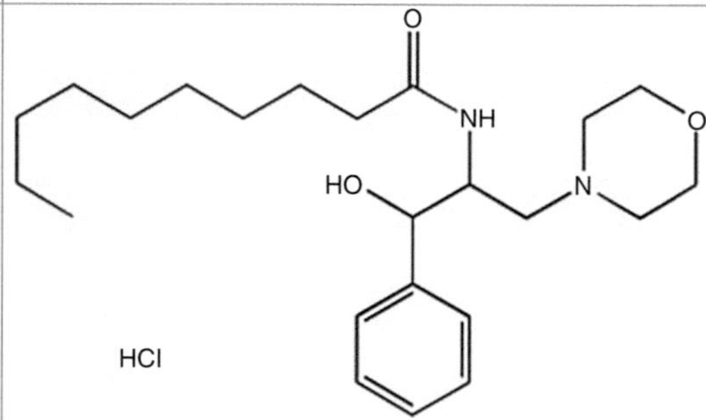	GCS inhibitor	[85, 86]
NVP-231	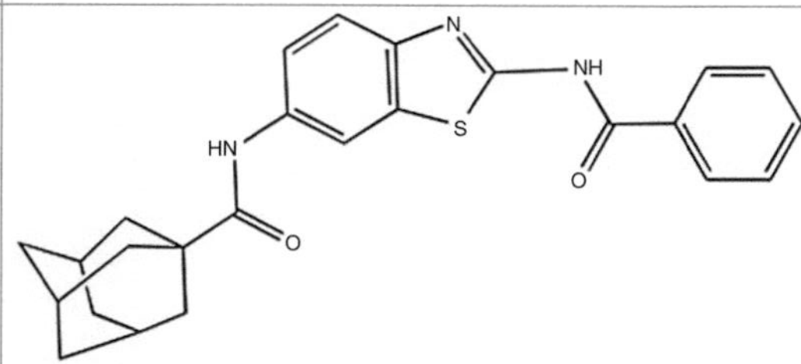	CERK inhibitor	[89]

(continued)

Table 1 (continued)

Compound	Structure	Mode of action	References
B13	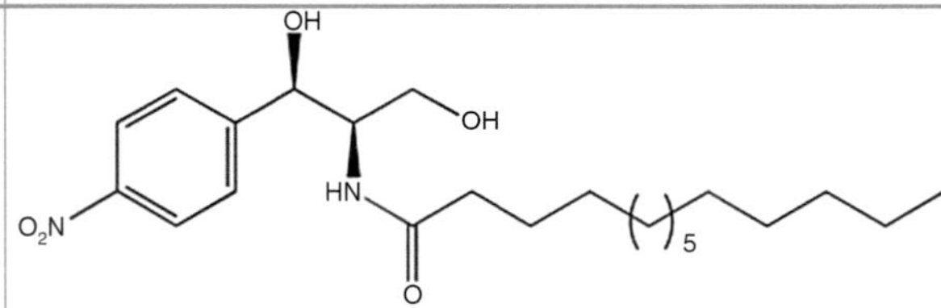	Ceramidase inhibitor	[92]
LCL-204	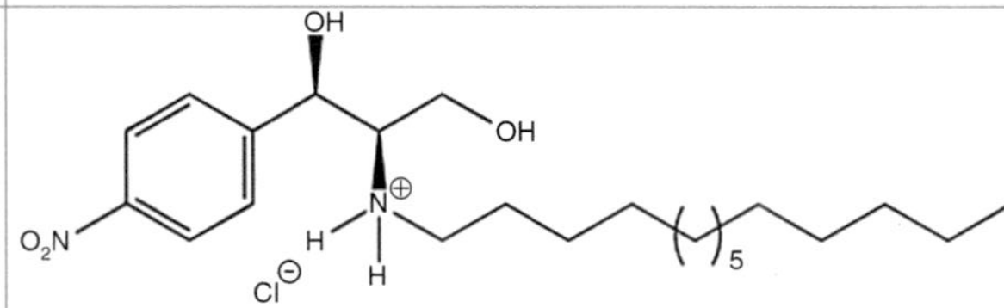	ASAH1 (acid ceramidase) inhibitor	[90, 91]
XM-462	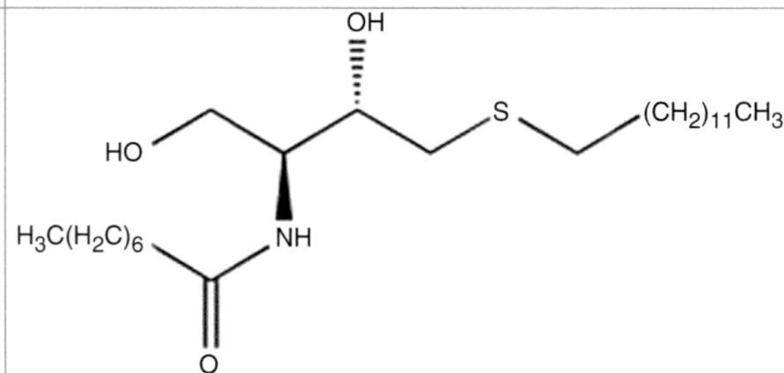	Desaturase inhibitor	[93, 94]

SK1-I	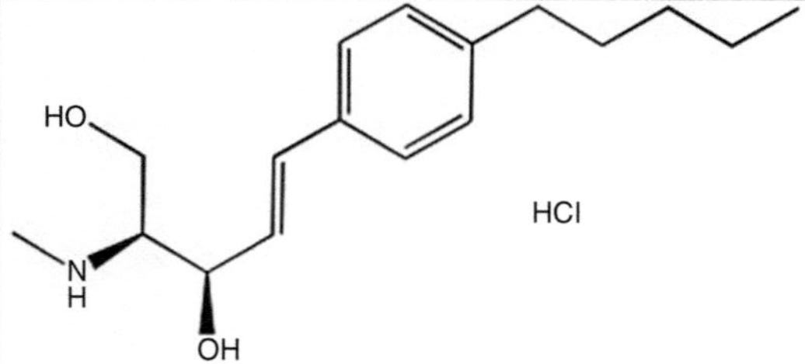	Sphingosine kinase 1 inhibitor	[101]
SKI-II	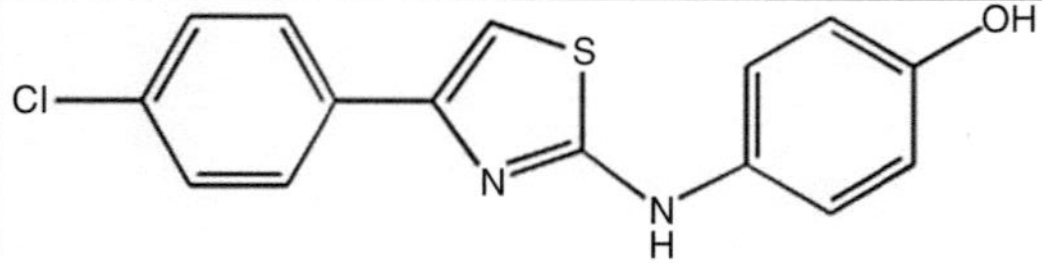	Sphingosine kinase 1 inhibitor	[102]

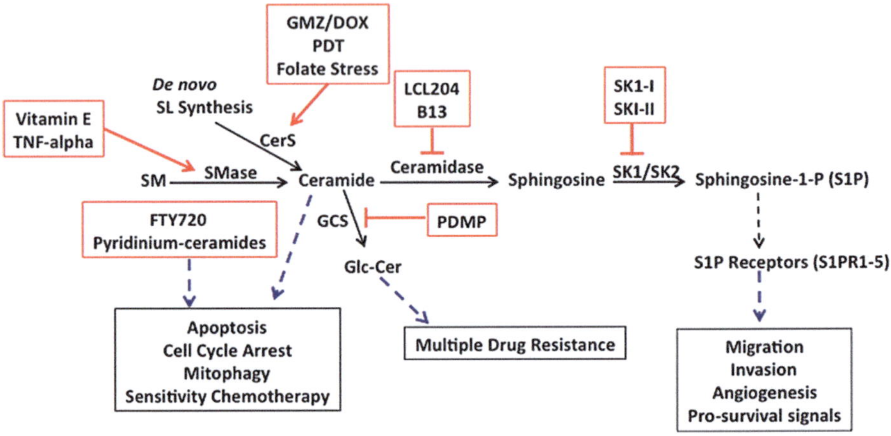

Fig. 1 Schematic representation of bioactive sphingolipids in HNSCC and lung cancer pathogenesis and therapy. The central molecules of the sphingolipid pathway, ceramides, can be synthesized by CerS enzymes in the de novo pathway or by the hydrolysis of SM. The cellular biological responses to elevated ceramides are generally in anti-proliferative nature and include apoptosis and cell cycle arrest. Generation of Glc-Cer and S1P are associated with multiple drug resistance and pro-survival, respectively. While chemotherapy and cellular stressors can elevate ceramides by inducing its production via the de novo pathway or SM hydrolysis, ceramide analogs can also mimic the actions of endogenous ceramides. Moreover, several inhibitors, such as B13 and PDMP have been shown to intervene with the clearance of ceramides. In addition, inhibition of SK1/2 and S1P generation can lead to prevention of metastasis and invasion of cancer cells

De novo generation of ceramide, which takes place in the endoplasmic reticulum (ER), starts with the condensation of serine and palmitoyl-CoA by the action of serine palmitoyltransferase (SPT) resulting in 3-ketosphinganine (3-ketodihydrosphingosine) [15]. SPT has been implicated as one of the rate limiting steps of de novo ceramide production. 3-ketodihydrosphingosine reductase (KDHR) converts 3-ketodihydrosphingosine to dihydrosphingosine (dhSph). 3-Ketodihydrosphingosine is a very short-lived metabolite, which suggests that KDHR has a very high affinity for 3-ketodihydrosphingosine [16].

The acylation of dhSph and generation of dihydroceramide is carried out by (Dihydro) ceramide synthases (CerS). After the identification of *longevity assurance gene-1* (*LAG1*) as the de novo ceramide synthase in *Saccharomyces cerevisiae* [17, 18], mammalian homologues of LAG1 were cloned and shown to regulate fatty acyl CoA dependent dihydroceramide generation. One of the mouse homologues, upstream of growth and differentiation 1 (UOG1), also known as LASS1 was demonstrated to specifically generate dihydroceramide with an 18-carbon fatty acyl chain [19]. Later studies revealed several other LASS homologues (LASS1-6) and they were re-named as CerS1-6 [20, 21]. Notably, CerS enzymes have specific preferences for different chain length-containing fatty acyl-CoAs to generate distinct ceramide species with different fatty acyl chains. For example, while CerS1 is mainly responsible for the generation of C18-ceramide [19], CerS5 and CerS6 catalyze the formation of mainly C16-ceramide [22, 23]. On the other hand, very long chain (C22–24) ceramide generation is carried out by CerS2 [24]. Moreover, CerS3 catalyzes the generation of ultra long chain ceramides (C26 and longer) [25, 26]. In addition to being critical enzymes of the *de novo* pathway, CerS proteins also utilize sphingosine (Sph) that is produced from the salvage (recycling) pathway to generate ceramide [27–29]. The final step of ceramide generation in the de novo pathway is the generation of the double bond between the carbons 4 and 5 of the dihydrosphingosine backbone of dihydroceramide. Desaturation and generation of ceramide from dihydroceramide is carried out by dihydroceramide desaturase 1 (DES1) [30, 31].

Ceramide can also be generated by the hydrolysis of complex sphingolipids such as sphingomyelin and glucosylceramide. Hydrolysis of sphingomyelin (SM) to generate ceramide and phosphocholine is catalyzed by sphingomyelinases (SMase) [32]. On the other hand, cleavage of glucose moiety from glucosylceramide (GlcCer) and production of ceramide is carried out by cerebrosidase.

Ceramide is also a precursor for the synthesis of complex sphingolipids. The synthesis of SM is carried out by SM synthase (SMS) enzymes. SMS enzymes transfer the phosphocholine group form phosphatidylcholine (PC) to C-1 position of ceramide to generate SM and DAG [33]. SMS enzymes can regulate the levels of bioactive lipids DAG and ceramide, with opposing survival and anti-proliferative roles, respectively [34]. Importantly, for SM synthesis to take place at the Golgi, ceramide needs to be transported by CERT from ER to Golgi [35, 36]. Ceramide can also be glycosylated to form GlcCer and galactosylceramide by glucosylceramide synthase (GCS) and ceramide galactosyltransferase (CGT) [37]. Notably, GCS has been implicated in multiple drug resistance of several cancers [38]. In addition to phosphocholine and glycosyl modifications, ceramide can also be phosphorylated to form ceramide-1-phosphate (C1P). The phosphorylation of ceramide is carried out by ceramide kinase (CERK) and C1P has been implicated in arachidonic acid release [39, 40] by directly interacting with cytosolic phospholipase A2 (cPLA2) [41, 42].

Clearance of ceramide from cells is initiated by ceramidase enzymes (CDase) (Fig. 1). CDase hydrolyzes the amide bond between the sphingosine backbone and the fatty acyl, generating free fatty acyl and sphingosine [43]. The resulting sphingosine then can be phosphorylated by the action of sphingosine kinase (SK) enzymes to generate sphingosine-1-phosphate (S1P). There are two sphingosine kinases that are products of two different genes and have different cellular localization, namely SK1 and SK2 [44, 45]. While SK1 is mainly cytosolic and can translocate to the plasma membrane upon activation by ERK [46, 47], SK2 is localized to nucleus [48] and membrane structures around the nucleus (nuclear envelope) [49]. Importantly, S1P can act on one of five G protein-coupled S1P receptors (S1PR1-5) and activate downstream signaling cascades. The signaling via S1PR receptors can control important cellular biologies such as, lymphocyte egress, blood vessel formation, and metastasis [50]. Because of its anti-apoptotic and pro-angiogenic nature, S1P has been suggested to be a good target for treatment of many pathophysiologic conditions including cancer [51, 52]. The removal of S1P from cellular systems is catalyzed by sphingosine-1-phosphate lyase (SPL). The products of S1P degradation by SPL are ethanolaminephosphate and hexadecenal [53]. In addition, complete recycling of hexadecenal to glycerolipid formation had been identified recently [54].

3 Ceramides and Their Analogs in HNSCC and Lung Cancer Therapy

3.1 Role of Endogenous Ceramides in Mediating Cancer Therapy

After the identification of ceramide as a bioactive lipid with anticancer properties, cellular functions of ceramide and the other sphingolipids in the regulation of cancer growth and response to therapeutic agents were studied in multiple cancer types including HNSCC and lung cancers (Fig. 1).

In multiple HNSCC cell lines, overexpression of CerS1, which specifically generates C_{18}-ceramide, induced cell death. In addition, FDA approved chemotherapeutic agents gemcitabine (GMZ) and doxorubicin (DOX) in combination were shown to increase CerS1 mRNA and activity. Moreover, downregulation of CerS1 with short interfering RNA (siRNA) partially prevented caspase 9 and 3 activation, as well as, C_{18}-ceramide increases in response to GMZ/DOX combination treatment [55]. In other models of HNSCC, employing tetracycline inducible stable expression of CerS1, C_{18}-ceramide induced cell death was mechanistically linked to induction of mitophagy, selective degradation of mitochondria by autophagosome/lysosome system. Specifically, CerS1 generated C_{18}-ceramide was shown to be localized to mitochondria and physically associate with microtubule associated light chain protein B (LC3B) to target mitochondria for degradation via mitophagy [56].

Involvement of CerS1 generated C_{18}-ceramide in Pc4 photodynamic therapy (PTD) of HNSCC was also revealed. It was shown that PTD induced accumulation of C_{18}- and C_{20}-ceramides and caspase 3 like DVEDase activity was prevented in cells transfected with CerS1 siRNA [57]. In another study, PTD in combination with tyrosine kinase inhibitor (TKI) desatinib, was shown to induce CerS1 mRNA and C_{18}-ceramide generation in HNSCC cells [58]. In addition to PTD, CerS1 was also implicated in sensitization of cancer cells to cisplatin-induced cell death [59].

Ceramide synthase 6 (CerS6) generated C_{16}-ceramide was reported to be important for the regulation of endoplasmic reticulum calcium homoeostasis in HNSCC cells. Briefly, it was shown that alteration of C_{16}-ceramide levels by down-regulation of CerS6 activates the ATF6 arm of the ER-stress pathway. Mechanistically ATF6 activation was linked to alterations in ER/Golgi membrane network and calcium release from ER [60, 61]. CerS6 was also found to be important for folate stress induced apoptosis in lung cancer cells. Specifically, folate deficiency induced CerS6 mRNA upregulation in a p53 dependent manner and prevention of de novo ceramide generation by fumonisin B1 (FB1) prevented folate stress induced apoptosis [62]. These two studies suggest that CerS6 and C_{16}-ceramide have distinct roles in mediating apoptosis in HNSCC and lung cancer cell lines.

Formation of ceramide through the hydrolysis of sphingomyelin is also linked to the regulation of anti-tumor effects of vitamin E and tumor necrosis factor-alpha (TNFα) in HNSCC and lung cancer cells. Specifically, vitamin E induces acidic and neutral sphingomyelinase activation, ceramide generation and apoptosis in HNSCC cells and in vivo xenografts [63]. On the other hand, in another study employing A549 lung cancer cells, TNF-α was discovered to induce the activation of neutral sphingolmyelinase 2 (nSMase2) in a time and dose dependent manner. Upon activation, nSMase2 generated ceramide was shown to induce ICAM and VCAM expression and modulate cell adhesion [64]. Moreover, nSMase generated ceramide was found to initiate apoptotic cell death in H1299 lung cancer cells upon overexpression of p53 and induce apoptosis [65].

3.2 Anti-Cancer Mechanism of Ceramides

Multiple studies were aimed at identifying the molecular mechanisms of how ceramides exert their anti-cancer effects and molecular targets that are modulated directly by ceramide have been identified. One of the first downstream targets of ceramides to be discovered is protein phosphatase 1 (PP1) and protein phosphatase 2A (PP2A) [66]. Mechanistically, increased endogenous ceramides was shown to activate PP1 and cause dephosphorylation of serine-arginine rich (SR) proteins in A549 lung cancer cells [67]. In a more recent study, direct interaction of ceramide with inhibitor 2 of PP2A (I2PP2A) was found. Interestingly, it was shown that in addition to short-chain C_6-ceramide, endogenous C_{18}-ceramide, but not C_{16}-Ceramide, associated with I2PP2A. Importantly, interaction and inhibition of I2PPA by ceramide triggered PP2A activation and in turn, induced cell death in lung cancer cells [68].

3.3 Sphingolipid Analogs in Cancer Therapy

The use of ceramide analogs for the treatment of HNSCC and lung cancers was also shown to have promise both in vitro and in vivo xenograft studies. One strategy is to conjugate the pyridinium ring to ceramide and construct cationic ceramide analogues (ceramidoids) (Table 1) that are water soluble, therefore more bioavailable, and are targeted to cancer cell mitochondria because of their positive charge [69]. Importantly, one of the ceramidoids, L-threo-C_6-pyridinium-ceramide (LCL124), was found to accumulate in mitochondria of HNSCC cell lines within 6 h and inhibited cell growth [70]. In addition, after the determination of maximum tolerated dose and bio-distribution of LCL124 in mice, researches also showed that LCL124 inhibited HNSCC xenograft growth in SCID mice alone or in combination with GMZ. Importantly, LCL124 was shown to accumulate in the tumor xenografts after systemic administration, suggesting that tumor targeting of ceramides by chemical modification is possible [71]. Moreover, LCL124 was also shown to accumulate in mitochondria and induce cell death in pancreatic cancer and insulinoma cells [72, 73]. In addition to short chain ceramidoids, C_{16}- and C_{18}-pyridinium-ceramides were also shown to induce HNSCC cell death through mitophagy [56].

In different sets of studies, ceramide analogs were revealed to enhance therapeutic cancer cell killing effects of PTD. Mice bearing SCCIV syngeneic oral squamous cell carcinoma were treated with PTD and LCL29, a C_6-ceramide analog (Table 1), in combination and it was found that combining LCL29 with PTD enhanced intracellular calcium release in HNSCC and amplified the apoptotic effects of PTD [74]. In another study, C_{16}-pyridinium-ceramide analog also enhanced the apoptotic effects of Pc4 PTD [75].

3.4 *Role of Ceramide Metabolizing Enzyme Inhibitors in Cancer Therapy*

While initially identified as an immunosuppressant by inhibiting the S1P receptors and lymphocyte egress [76], the sphingosine analog Fingolimod (FTY720) (Table 1), was shown to induce growth retardation of Lewis Lung carcinoma (LLC1) xenografts [77]. More recently, the growth inhibitory actions of FTY720 were found to be linked to activation of PP2A. Mechanistically, FTY720 associated with I2PP2A and blocked the inhibition of PP2A causing oncogenic myc dephosphorylation and degradation [78].

As several lines of evidence form multiple research groups showed that formation of ceramide is one of the critical events taking place during cell death and/or growth inhibition in response to multiple stimuli, prevention of clearance of ceramide by inhibitors, or inhibition of SK1/S1P signaling has also been shown to be a valid strategy to sensitize HNSCC or lung cancer cells to anti-cancer therapy [79, 80] (Fig. 1).

After the initial identification of glucosylceramide (GlcCer) as one the factors involved in the development of multiple drug resistance [81] and metastasis [82], several studies focused on validation of GCS and its downstream metabolites as targets to prevent multiple drug resistance and to enhance therapeutic efficiency of chemotherapeutic agents [83]. For example, ganglioside GM3 production was linked to attained resistance to etoposide and doxorubicin by up-regulating Bcl-2 expression in 3LL Lewis lung carcinoma cells [84]. Furthermore, pharmachological inhibition of GCS with PDMP enhanced radiosensitizing effects of vinorelbine in lung adenocarcinoma cells [85]. In another study, generation of de novo ceramide was responsible for caspase 3 dependent apoptotic cell death in response to doxorubicin and camptothecin treatment in follicular thyroid carcinoma cells. Moreover, inhibition of GCS by PDMP leading to accumulation of ceramide enhanced the effects of chemotherapeutic agents [86].

While discovered to regulate cPLA2 mediated arachidonic acid release [87], C1P was also implicated in the regulation of lung cancer growth such that inhibition of C1P generation by siRNA mediated knockdown of ceramide kinase (CERK) elevated ceramide levels in cells and in turn induced apoptotic cell death, showing that CERK is an anti-cancer target candidate [88]. Along with this notion, recently, a CERK inhibitor (NVP-231) was developed and found to sensitize lung cancer cells to staurosporine-induced cell death [89].

In order to prevent the clearance of ceramide, the ceramidase enzymes were targeted in many studies. For example, inhibition of acid ceramidase (ASAH1) by either siRNA or a specific inhibitor LCL204, sensitized HNSCC cells to FasL induced apoptosis in cell and xenograft models [90]. In lung cancers, ASAH1 was associated with resistance to choline kinase (ChoKα) inhibitors and ASAH1 inhibition synergistically sensitizes lung cancer cells to the antiproliferative effect of ChoKα inhibitors [91]. Moreover, the ceramidase inhibitor B13 and its sulfonamide analogs were also found to inhibit lung cancer cell growth [92].

Another approach to interfere with sphingolipid metabolism to induce cancer cell death was to increase the accumulation of dihydroceramides. This was achieved by the development of dihydroceramide desaturase inhibitor XM462 [93]. Importantly, the anti-tumor activity of XM462 in lung cancer cells was also demonstrated [94].

3.5 *Sphingosine-1-Phosphate and S1P Receptor Signaling*

The role of sphingosine-1-phosphate in HNSCC carcinogenesis was defined using chemical carcinogenesis models. An established carcinogen, 4-nitroquinone-1-oxide (4-NQO) induced HNSCC formation was monitored in wild type and SK1 knockout mice. The data showed that the genetic loss of SK1 significantly reduced HNSCC incidence, multiplicity and volume, compared to wild type control mice [95]. In addition to oncogenesis, SK1/S1P was also shown to be involved in governing HNSCC invasion and metastasis. Briefly, SK1 generated S1P was shown to act through sphingosne-1-phosphate receptor 1 (S1P1) and induce IL-6/STAT3 signaling for initiation of cellular invasion [96].

Targeting the SK1/S1P signaling was also implied to enhance the effects of anticancer therapy and overcome resistance [97]. Notably, inhibition of SK1 by siRNA in HNSCC cell lines was found to enhance the anti-proliferative effects of radiation therapy both in vitro and in vivo xenograft studies [98]. Moreover, the use of SK1 siRNA nanocomplexed to gold nanorod particles in xenograft studies was shown to efficiently downregulate SK1 and radiosensitize HNSCC tumors [99]. In other studies, inhibition of SK1 in H358 lung cancer cells using shRNA decreased cellular proliferation and sensitized cells to taxol [100]. Moreover, the pharmacological inhibitor of SK1, SK1-I (Table 1), enhanced sensitivity of NSCLC cells to docetaxel both in vitro and in vivo by impacting anti-apoptotic functions of the PI3K/Akt/NF-κB signaling pathway [101]. Furthermore, inhibition of SK1 with SKI-II sensitized HNSCC cells to cell death in response to cetuximab, an EGFR inhibitor [102].

The activation of S1P receptors by S1P is important for many cancer pathologies including inhibition of apoptosis, chemoresistance, and migration/invasion. For example, $S1P_1$ activation by S1P was identified to render lung fibroblast cells resistant to apoptosis. Mechanistically, it was shown that, $S1P_1$ suppressed caspase 3 activation by causing anti-apoptotic Mcl1 protein accumulation [103]. Moreover, in a recent study, lung colonization of cancer cells in an in vivo model of metastasis was linked to activation of $S1P_2$ [104]. Furthermore, in another report, activation of $S1P_3$ was linked to increased epidermal growth factor receptor (EGFR) transcription through activation of Rho kinase (ROCK) in lung adenocarcinoma cells [105].

Cellular transport of S1P, in addition to its production by sphingosine kinases and signaling through S1P receptors is also implicated as a target to modulate cancer cell migration/invasion. S1P is transported from cells to the extracellular space by the action of a specific transporter, spns2 [106]. Recently, modulation of spns2 level was found to modulate the migratory phenotype of lung cancer cells. Specifically, knockdown of spns2 induced an elevation of cellular S1P levels and enhanced migration of NSCLC cells [107].

4 Sphingolipids as Biomarkers in HNSCC and Lung Cancer

In an effort to identify the molecular mechanisms of carcinogenesis, improve the criteria for better diagnosis, and predict the outcomes and response to therapy, multiple approaches including genomic and proteomic profiling of tumor samples have been in the forefront of cancer medicine. As the machinery and methodologies became more developed, lipidomic profiling of cancers is emerging to be promising for biomarker determination.

The initial sphingolipidomic profiling of HNSCC was carried out by Koybasi et al. [108] and defects in the ceramide metabolism were identified. HNSCC tumor and normal adjacent tissues (32 pairs) from patients were analyzed for the amounts of long chain ceramides using LC/MS and specifically C_{18}-ceramide levels were shown to be decreased compared to their normal adjacent controls in majority of the pairs [108]. Mechanistically, the decreased C_{18}-ceramide was linked to reduced CerS1 mRNA levels resulting from increased miR-574-5p expression and HDAC1 activity [109]. In another study, possible links between the clinical parameters in HNSCC and tumor ceramide levels were investigated. Interestingly, decreased C_{18}-ceramide content of the tumor tissues, compared to normal adjacent, was significantly correlated with lymphovascular spread and pathologic nodal metastasis [110], indicating that sphingolipidomics can serve as powerful tool to better characterize HNSCC in the clinic.

Additionally, serum sphingolipid levels were measured from HNSCC patients who received gemcitabine (GMZ) and doxorubicin (DOX) combination treatment as a phase II clinical trial, after failing initial platinum based therapy. The study showed that elevated C_{18}-ceramide levels in serum correlated with complete response, partial response, and stable disease conditions after the conclusion of the trial [111]. Moreover, in an epidemiologic study serum from lung cancer patients was analyzed for sphingolipid content. Interestingly, increased plasma S1P and total ceramide content in lung cancer cases compared to age-matched controls were associated with increased future risk for developing lung cancer [112]. Furthermore, in a study published by a different research group, serum lipid metabolites of lung cancer patients and healthy individuals were measured using Fourier transform ion cyclotron resonance coupled to mass spectrometry. Univariate analysis of their results identified that sphingomyelin (SM) levels were associated with lung cancer progression and C_{16}-SM levels could be used as a biomarker to segregate lung cancer patients from healthy controls [113]. Total sphingolipid levels can be identified and measured in tissues or in serum by LC/MS; however the information obtained does not contain any localization component relative to histopathology of the tissue. In order to overcome this handicap, high-resolution MALDI imaging mass spectrometry (MALDI-IMS) has been developed and was shown to effectively localize and quantitate multiple sphingolipid species on tissues including lung cancers [114].

In addition to a lipidomics approach, there are examples of the use of histological analysis of lung cancers for enzymes of the sphingolipid pathway. Johnson, et al. compared the immunological staining of sphingosine kinase 1 (SK1) in tumors versus matched normal tissues from lung cancer patients and reported strongly positive SK1 staining in tumors compared to normal tissues [115]. Moreover, tissue microarrays found that expression levels of SK1 were significantly increased in HNSCC tumor tissues compared to normal controls. Furthermore, increased protein expression was discovered to correlate with increased SK1 mRNA levels as determined by laser capture microdissection coupled to real-time PCR. Notably, elevated SK1 expression was associated with shorter patient survival [116]. In a more recent study, sphingosine kinase 2 (SK2) expression levels in 180 NSCLC patient paraffin embedded tissue samples were detected using immunohistochemistry. Importantly, increased SK2 expression correlated with histology grade, lymph node status, clinical stage, and tumor size [117]. Furthermore, also in NSCLC, glucosylceramide synthase (GCS) expression levels were discovered to be increased in cancerous tissues compared to matched normal adjacent control tissues. Also, the study identified that increased GCS expression was associated with lymph node metastasis and chemotherapy resistance [118].

In summary, profiling of sphingolipids in HNSCC and lung cancer patients using LC/MS and/or MALDI-IMS and identification of the expression levels of enzymes of sphingolipid metabolism in patient samples pose a great potential for identification and development of novel sphingolipid biomarkers that are relevant to the clinical characteristics of the diseases.

5 Conclusions and Future Directions

It is clear that a significant amount of work had been done in order to identify the cellular functions of sphingolipids in the pathogenesis and therapy of HNSCC and lung cancers. Many studies identified the mechanisms by which sphingolipids modulate cellular pathways important for cancer cell growth, invasion, and apoptosis. While some of these studies are still descriptive, others present direct cellular targets that can be selected as possible therapeutic foci for treating HNSCC and lung cancers. In addition, many critical enzymes of the sphingolipid family, such as SK1, have been identified as anti-cancer targets themselves. Furthermore, with recent advances in lipidomic technologies, members of the sphingolipid family can be detected and identified from clinical samples and associations with clinical parameters can be made, implicating sphingolipids as novel biomarkers. However, it is apparent that there is little or no effort from clinicians managing HNSCC or lung cancer patients at the "bed side" to incorporate the knowledge about the sphingolipids to patient care. Therefore, it is critical to identify the defects in the field that are delaying the advancement of "sphingo-therapies" to the use in clinic.

Acknowledgement This work was supported by: NIH research grants GM062887 and P01CA097132, and Veterans Affairs Merit Award to LMO.

References

1. Siegel R, Naishadham D, Jemal A (2013) Cancer statistics, 2013. CA Cancer J Clin 63(1):11–30
2. Gregoire V, Lefebvre JL, Licitra L, Felip E, E.-E.-E.G.W. Group (2010) Squamous cell carcinoma of the head and neck: EHNS-ESMO-ESTRO Clinical Practice Guidelines for diagnosis, treatment and follow-up. Ann Oncol 21(Suppl 5):v184–v186
3. Jemal A, Thun MJ, Ries LA, Howe HL, Weir HK, Center MM, Ward E, Wu XC, Eheman C, Anderson R, Ajani UA, Kohler B, Edwards BK (2008) Annual report to the nation on the status of cancer, 1975-2005, featuring trends in lung cancer, tobacco use, and tobacco control. J Natl Cancer Inst 100(23):1672–1694
4. Syrjanen S, Lodi G, von Bultzingslowen I, Aliko A, Arduino P, Campisi G, Challacombe S, Ficarra G, Flaitz C, Zhou HM, Maeda H, Miller C, Jontell M (2011) Human papillomaviruses in oral carcinoma and oral potentially malignant disorders: a systematic review. Oral Dis 17(Suppl 1):58–72
5. Denaro N, Russi EG, Adamo V, Merlano MC (2014) State-of-the-art and emerging treatment options in the management of head and neck cancer: news from 2013. Oncology 86(4): 212–229
6. Haddad RI, Shin DM (2008) Recent advances in head and neck cancer. N Engl J Med 359(11):1143–1154
7. World Health Organization Cancer (2013) Available from http://www.who.int/mediacentre/factsheets/fs297/en/
8. Siegel R, Ma J, Zou Z, Jemal A (2014) Cancer statistics, 2014. CA Cancer J Clin 64(1): 9–29
9. Ettinger DS, Akerley W, Borghaei H, Chang AC, Cheney RT, Chirieac LR, D'Amico TA, Demmy TL, Govindan R, Grannis FW Jr, Grant SC, Horn L, Jahan TM, Komaki R, Kong FM, Kris MG, Krug LM, Lackner RP, Lennes IT, Loo BW Jr, Martins R, Otterson GA, Patel JD, Pinder-Schenck MC, Pisters KM, Reckamp K, Riely GJ, Rohren E, Shapiro TA, Swanson SJ, Tauer K, Wood DE, Yang SC, Gregory K, Hughes M (2013) National comprehensive cancer, Non-small cell lung cancer, version 2.2013. J Natl Compr Canc Netw 11(6):645–653, quiz 653
10. Davidson MR, Gazdar AF, Clarke BE (2013) The pivotal role of pathology in the management of lung cancer. J Thorac Dis 5(Suppl 5):S463–S478

11. Langer CJ, Besse B, Gualberto A, Brambilla E, Soria JC (2010) The evolving role of histology in the management of advanced non-small-cell lung cancer. J Clin Oncol 28(36):5311–5320
12. Molina JR, Yang P, Cassivi SD, Schild SE, Adjei AA (2008) Non-small cell lung cancer: epidemiology, risk factors, treatment, and survivorship. Mayo Clin Proc 83(5):584–594
13. Hannun YA, Obeid LM (2008) Principles of bioactive lipid signalling: lessons from sphingolipids. Nat Rev Mol Cell Biol 9(2):139–150
14. Mullen TD, Hannun YA, Obeid LM (2012) Ceramide synthases at the centre of sphingolipid metabolism and biology. Biochem J 441(3):789–802
15. Merrill AH Jr (2011) Sphingolipid and glycosphingolipid metabolic pathways in the era of sphingolipidomics. Chem Rev 111(10):6387–6422
16. Kihara A, Igarashi Y (2004) FVT-1 is a mammalian 3-ketodihydrosphingosine reductase with an active site that faces the cytosolic side of the endoplasmic reticulum membrane. J Biol Chem 279(47):49243–49250
17. Guillas I, Kirchman PA, Chuard R, Pfefferli M, Jiang JC, Jazwinski SM, Conzelmann A (2001) C26-CoA-dependent ceramide synthesis of Saccharomyces cerevisiae is operated by Lag1p and Lac1p. EMBO J 20(11):2655–2665
18. Jazwinski SM, Conzelmann A (2002) LAG1 puts the focus on ceramide signaling. Int J Biochem Cell Biol 34(11):1491–1495
19. Venkataraman K, Riebeling C, Bodennec J, Riezman H, Allegood JC, Sullards MC, Merrill AH Jr, Futerman AH (2002) Upstream of growth and differentiation factor 1 (uog1), a mammalian homolog of the yeast longevity assurance gene 1 (LAG1), regulates N-stearoyl-sphinganine (C18-(dihydro)ceramide) synthesis in a fumonisin B1-independent manner in mammalian cells. J Biol Chem 277(38):35642–35649
20. Riebeling C, Allegood JC, Wang E, Merrill AH Jr, Futerman AH (2003) Two mammalian longevity assurance gene (LAG1) family members, trh1 and trh4, regulate dihydroceramide synthesis using different fatty acyl-CoA donors. J Biol Chem 278(44):43452–43459
21. Pewzner-Jung Y, Ben-Dor S, Futerman AH (2006) When do Lasses (longevity assurance genes) become CerS (ceramide synthases)?: insights into the regulation of ceramide synthesis. J Biol Chem 281(35):25001–25005
22. Lahiri S, Futerman AH (2005) LASS5 is a bona fide dihydroceramide synthase that selectively utilizes palmitoyl-CoA as acyl donor. J Biol Chem 280(40):33735–33738
23. Mizutani Y, Kihara A, Igarashi Y (2005) Mammalian Lass6 and its related family members regulate synthesis of specific ceramides. Biochem J 390(Pt 1):263–271
24. Laviad EL, Albee L, Pankova-Kholmyansky I, Epstein S, Park H, Merrill AH Jr, Futerman AH (2008) Characterization of ceramide synthase 2: tissue distribution, substrate specificity, and inhibition by sphingosine 1-phosphate. J Biol Chem 283(9):5677–5684
25. Mizutani Y, Kihara A, Igarashi Y (2006) LASS3 (longevity assurance homologue 3) is a mainly testis-specific (dihydro)ceramide synthase with relatively broad substrate specificity. Biochem J 398(3):531–538
26. Jennemann R, Rabionet M, Gorgas K, Epstein S, Dalpke A, Rothermel U, Bayerle A, van der Hoeven F, Imgrund S, Kirsch J, Nickel W, Willecke K, Riezman H, Grone HJ, Sandhoff R (2012) Loss of ceramide synthase 3 causes lethal skin barrier disruption. Hum Mol Genet 21(3):586–608
27. Ogretmen B, Pettus BJ, Rossi MJ, Wood R, Usta J, Szulc Z, Bielawska A, Obeid LM, Hannun YA (2002) Biochemical mechanisms of the generation of endogenous long chain ceramide in response to exogenous short chain ceramide in the A549 human lung adenocarcinoma cell line. Role for endogenous ceramide in mediating the action of exogenous ceramide. J Biol Chem 277(15):12960–12969
28. Kitatani K, Idkowiak-Baldys J, Hannun YA (2008) The sphingolipid salvage pathway in ceramide metabolism and signaling. Cell Signal 20(6):1010–1018
29. Sultan I, Senkal CE, Ponnusamy S, Bielawski J, Szulc Z, Bielawska A, Hannun YA, Ogretmen B (2006) Regulation of the sphingosine-recycling pathway for ceramide generation by oxidative stress, and its role in controlling c-Myc/Max function. Biochem J 393(Pt 2):513–521

30. Kraveka JM, Li L, Szulc ZM, Bielawski J, Ogretmen B, Hannun YA, Obeid LM, Bielawska A (2007) Involvement of dihydroceramide desaturase in cell cycle progression in human neuroblastoma cells. J Biol Chem 282(23):16718–16728
31. Michel C, van Echten-Deckert G, Rother J, Sandhoff K, Wang E, Merrill AH Jr (1997) Characterization of ceramide synthesis. A dihydroceramide desaturase introduces the 4,5-trans-double bond of sphingosine at the level of dihydroceramide. J Biol Chem 272(36):22432–22437
32. Airola MV, Hannun YA (2013) Sphingolipid metabolism and neutral sphingomyelinases. Handb Exp Pharmacol 215:57–76
33. Holthuis JC, Luberto C (2010) Tales and mysteries of the enigmatic sphingomyelin synthase family. Adv Exp Med Biol 688:72–85
34. Villani M, Subathra M, Im YB, Choi Y, Signorelli P, Del Poeta M, Luberto C (2008) Sphingomyelin synthases regulate production of diacylglycerol at the Golgi. Biochem J 414(1):31–41
35. Hanada K, Kumagai K, Yasuda S, Miura Y, Kawano M, Fukasawa M, Nishijima M (2003) Molecular machinery for non-vesicular trafficking of ceramide. Nature 426(6968):803–809
36. Hanada K, Kumagai K, Tomishige N, Yamaji T (2009) CERT-mediated trafficking of ceramide. Biochim Biophys Acta 1791(7):684–691
37. Vitner EB, Futerman AH (2013) Neuronal forms of Gaucher disease. Handb Exp Pharmacol 216:405–419
38. Liu YY, Hill RA, Li YT (2013) Ceramide glycosylation catalyzed by glucosylceramide synthase and cancer drug resistance. Adv Cancer Res 117:59–89
39. Pettus BJ, Bielawska A, Subramanian P, Wijesinghe DS, Maceyka M, Leslie CC, Evans JH, Freiberg J, Roddy P, Hannun YA, Chalfant CE (2004) Ceramide 1-phosphate is a direct activator of cytosolic phospholipase A2. J Biol Chem 279(12):11320–11326
40. Stahelin RV, Subramanian P, Vora M, Cho W, Chalfant CE (2007) Ceramide-1-phosphate binds group IVA cytosolic phospholipase a2 via a novel site in the C2 domain. J Biol Chem 282(28):20467–20474
41. Wijesinghe DS, Massiello A, Subramanian P, Szulc Z, Bielawska A, Chalfant CE (2005) Substrate specificity of human ceramide kinase. J Lipid Res 46(12):2706–2716
42. Lamour NF, Stahelin RV, Wijesinghe DS, Maceyka M, Wang E, Allegood JC, Merrill AH Jr, Cho W, Chalfant CE (2007) Ceramide kinase uses ceramide provided by ceramide transport protein: localization to organelles of eicosanoid synthesis. J Lipid Res 48(6):1293–1304
43. Mao C, Obeid LM (2008) Ceramidases: regulators of cellular responses mediated by ceramide, sphingosine, and sphingosine-1-phosphate. Biochim Biophys Acta 1781(9): 424–434
44. Heffernan-Stroud LA, Obeid LM (2013) Sphingosine kinase 1 in cancer. Adv Cancer Res 117:201–235
45. Gandy KA, Obeid LM (2013) Regulation of the sphingosine kinase/sphingosine 1-phosphate pathway. Handb Exp Pharmacol 216:275–303
46. Pitson SM, Moretti PA, Zebol JR, Lynn HE, Xia P, Vadas MA, Wattenberg BW (2003) Activation of sphingosine kinase 1 by ERK1/2-mediated phosphorylation. EMBO J 22(20): 5491–5500
47. Stahelin RV, Hwang JH, Kim JH, Park ZY, Johnson KR, Obeid LM, Cho W (2005) The mechanism of membrane targeting of human sphingosine kinase 1. J Biol Chem 280(52): 43030–43038
48. Hait NC, Allegood J, Maceyka M, Strub GM, Harikumar KB, Singh SK, Luo C, Marmorstein R, Kordula T, Milstien S, Spiegel S (2009) Regulation of histone acetylation in the nucleus by sphingosine-1-phosphate. Science 325(5945):1254–1257
49. Maceyka M, Sankala H, Hait NC, Le Stunff H, Liu H, Toman R, Collier C, Zhang M, Satin LS, Merrill AH Jr, Milstien S, Spiegel S (2005) SphK1 and SphK2, sphingosine kinase isoenzymes with opposing functions in sphingolipid metabolism. J Biol Chem 280(44): 37118–37129

50. Maceyka M, Harikumar KB, Milstien S, Spiegel S (2012) Sphingosine-1-phosphate signaling and its role in disease. Trends Cell Biol 22(1):50–60
51. Kunkel GT, Maceyka M, Milstien S, Spiegel S (2013) Targeting the sphingosine-1-phosphate axis in cancer, inflammation and beyond. Nat Rev Drug Discov 12(9):688–702
52. Orr Gandy KA, Obeid LM (2013) Targeting the sphingosine kinase/sphingosine 1-phosphate pathway in disease: review of sphingosine kinase inhibitors. Biochim Biophys Acta 1831(1): 157–166
53. Kumar A, Saba JD (2009) Lyase to live by: sphingosine phosphate lyase as a therapeutic target. Expert Opin Ther Targets 13(8):1013–1025
54. Kihara A (2014) Sphingosine 1-phosphate is a key metabolite linking sphingolipids to glycerophospholipids. Biochim Biophys Acta 1841(5):766–772
55. Senkal CE, Ponnusamy S, Rossi MJ, Bialewski J, Sinha D, Jiang JC, Jazwinski SM, Hannun YA, Ogretmen B (2007) Role of human longevity assurance gene 1 and C18-ceramide in chemotherapy-induced cell death in human head and neck squamous cell carcinomas. Mol Cancer Ther 6(2):712–722
56. Sentelle RD, Senkal CE, Jiang W, Ponnusamy S, Gencer S, Selvam SP, Ramshesh VK, Peterson YK, Lemasters JJ, Szulc ZM, Bielawski J, Ogretmen B (2012) Ceramide targets autophagosomes to mitochondria and induces lethal mitophagy. Nat Chem Biol 8(10): 831–838
57. Separovic D, Breen P, Joseph N, Bielawski J, Pierce JS, Van Buren E, Gudz TI (2012) siRNA-mediated down-regulation of ceramide synthase 1 leads to apoptotic resistance in human head and neck squamous carcinoma cells after photodynamic therapy. Anticancer Res 32(7):2479–2485
58. Separovic D, Breen P, Boppana NB, Van Buren E, Joseph N, Kraveka JM, Rahmaniyan M, Li L, Gudz TI, Bielawska A, Bai A, Bielawski J, Pierce JS, Korbelik M (2013) Increased killing of SCCVII squamous cell carcinoma cells after the combination of Pc 4 photodynamic therapy and dasatinib is associated with enhanced caspase-3 activity and ceramide synthase 1 upregulation. Int J Oncol 43(6):2064–2072
59. Min J, Mesika A, Sivaguru M, Van Veldhoven PP, Alexander H, Futerman AH, Alexander S (2007) (Dihydro)ceramide synthase 1 regulated sensitivity to cisplatin is associated with the activation of p38 mitogen-activated protein kinase and is abrogated by sphingosine kinase 1. Mol Cancer Res 5(8):801–812
60. Senkal CE, Ponnusamy S, Bielawski J, Hannun YA, Ogretmen B (2010) Antiapoptotic roles of ceramide-synthase-6-generated C16-ceramide via selective regulation of the ATF6/CHOP arm of ER-stress-response pathways. FASEB J 24(1):296–308
61. Senkal CE, Ponnusamy S, Manevich Y, Meyers-Needham M, Saddoughi SA, Mukhopadyay A, Dent P, Bielawski J, Ogretmen B (2011) Alteration of ceramide synthase 6/C16-ceramide induces activating transcription factor 6-mediated endoplasmic reticulum (ER) stress and apoptosis via perturbation of cellular Ca2+ and ER/Golgi membrane network. J Biol Chem 286(49):42446–42458
62. Hoeferlin LA, Fekry B, Ogretmen B, Krupenko SA, Krupenko NI (2013) Folate stress induces apoptosis via p53-dependent de novo ceramide synthesis and up-regulation of ceramide synthase 6. J Biol Chem 288(18):12880–12890
63. Gu X, Song X, Dong Y, Cai H, Walters E, Zhang R, Pang X, Xie T, Guo Y, Sridhar R, Califano JA (2008) Vitamin E succinate induces ceramide-mediated apoptosis in head and neck squamous cell carcinoma in vitro and in vivo. Clin Cancer Res 14(6):1840–1848
64. Clarke CJ, Truong TG, Hannun YA (2007) Role for neutral sphingomyelinase-2 in tumor necrosis factor alpha-stimulated expression of vascular cell adhesion molecule-1 (VCAM) and intercellular adhesion molecule-1 (ICAM) in lung epithelial cells: p38 MAPK is an upstream regulator of nSMase2. J Biol Chem 282(2):1384–1396
65. Soans E, Evans SC, Cipolla C, Fernandes E (2014) Characterizing the sphingomyelinase pathway triggered by PRIMA-1 derivatives in lung cancer cells with differing p53 status. Anticancer Res 34(7):3271–3283

66. Kishikawa K, Chalfant CE, Perry DK, Bielawska A, Hannun YA (1999) Phosphatidic acid is a potent and selective inhibitor of protein phosphatase 1 and an inhibitor of ceramide-mediated responses. J Biol Chem 274(30):21335–21341
67. Chalfant CE, Ogretmen B, Galadari S, Kroesen BJ, Pettus BJ, Hannun YA (2001) FAS activation induces dephosphorylation of SR proteins; dependence on the de novo generation of ceramide and activation of protein phosphatase 1. J Biol Chem 276(48):44848–44855
68. Mukhopadhyay A, Saddoughi SA, Song P, Sultan I, Ponnusamy S, Senkal CE, Snook CF, Arnold HK, Sears RC, Hannun YA, Ogretmen B (2009) Direct interaction between the inhibitor 2 and ceramide via sphingolipid-protein binding is involved in the regulation of protein phosphatase 2A activity and signaling. FASEB J 23(3):751–763
69. Szulc ZM, Bielawski J, Gracz H, Gustilo M, Mayroo N, Hannun YA, Obeid LM, Bielawska A (2006) Tailoring structure-function and targeting properties of ceramides by site-specific cationization. Bioorg Med Chem 14(21):7083–7104
70. Rossi MJ, Sundararaj K, Koybasi S, Phillips MS, Szulc ZM, Bielawska A, Day TA, Obeid LM, Hannun YA, Ogretmen B (2005) Inhibition of growth and telomerase activity by novel cationic ceramide analogs with high solubility in human head and neck squamous cell carcinoma cells. Otolaryngol Head Neck Surg 132(1):55–62
71. Senkal CE, Ponnusamy S, Rossi MJ, Sundararaj K, Szulc Z, Bielawski J, Bielawska A, Meyer M, Cobanoglu B, Koybasi S, Sinha D, Day TA, Obeid LM, Hannun YA, Ogretmen B (2006) Potent antitumor activity of a novel cationic pyridinium-ceramide alone or in combination with gemcitabine against human head and neck squamous cell carcinomas in vitro and in vivo. J Pharmacol Exp Ther 317(3):1188–1199
72. Beckham TH, Lu P, Jones EE, Marrison T, Lewis CS, Cheng JC, Ramshesh VK, Beeson G, Beeson CC, Drake RR, Bielawska A, Bielawski J, Szulc ZM, Ogretmen B, Norris JS, Liu X (2013) LCL124, a cationic analog of ceramide, selectively induces pancreatic cancer cell death by accumulating in mitochondria. J Pharmacol Exp Ther 344(1):167–178
73. Syed I, Szulc ZM, Ogretmen B, Kowluru A (2012) L-threo-C6-pyridinium-ceramide bromide, a novel cationic ceramide, induces NADPH oxidase activation, mitochondrial dysfunction and loss in cell viability in INS 832/13 beta-cells. Cell Physiol Biochem 30(4):1051–1058
74. Korbelik M, Zhang W, Separovic D (2011) Amplification of cancer cell apoptosis in photodynamic therapy-treated tumors by adjuvant ceramide analog LCL29. Lasers Surg Med 43(7):614–620
75. Separovic D, Saad ZH, Edwin EA, Bielawski J, Pierce JS, Buren EV, Bielawska A (2011) C16-ceramide analog combined with Pc 4 photodynamic therapy evokes enhanced total ceramide accumulation, promotion of DEVDase activation in the absence of apoptosis, and augmented overall cell killing. J Lipids 2011:713867
76. Brinkmann V, Billich A, Baumruker T, Heining P, Schmouder R, Francis G, Aradhye S, Burtin P (2010) Fingolimod (FTY720): discovery and development of an oral drug to treat multiple sclerosis. Nat Rev Drug Discov 9(11):883–897
77. Schmid G, Guba M, Papyan A, Ischenko I, Bruckel M, Bruns CJ, Jauch KW, Graeb C (2005) FTY720 inhibits tumor growth and angiogenesis. Transplant Proc 37(1):110–111
78. Saddoughi SA, Gencer S, Peterson YK, Ward KE, Mukhopadhyay A, Oaks J, Bielawski J, Szulc ZM, Thomas RJ, Selvam SP, Senkal CE, Garrett-Mayer E, De Palma RM, Fedarovich D, Liu A, Habib AA, Stahelin RV, Perrotti D, Ogretmen B (2013) Sphingosine analogue drug FTY720 targets I2PP2A/SET and mediates lung tumour suppression via activation of PP2A-RIPK1-dependent necroptosis. EMBO Mol Med 5(1):105–121
79. Beckham TH, Elojeimy S, Cheng JC, Turner LS, Hoffman SR, Norris JS, Liu X (2010) Targeting sphingolipid metabolism in head and neck cancer: rational therapeutic potentials. Expert Opin Ther Targets 14(5):529–539
80. Norris JS (2013) The sphingolipid response to cancer therapy. Adv Cancer Res 117: xiii–xvii
81. Ogretmen B, Hannun YA (2001) Updates on functions of ceramide in chemotherapy-induced cell death and in multidrug resistance. Drug Resist Updat 4(6):368–377

82. Inokuchi J, Jimbo M, Momosaki K, Shimeno H, Nagamatsu A, Radin NS (1990) Inhibition of experimental metastasis of murine Lewis lung carcinoma by an inhibitor of glucosylceramide synthase and its possible mechanism of action. Cancer Res 50(20):6731–6737
83. Morad SA, Cabot MC (2013) Ceramide-orchestrated signalling in cancer cells. Nat Rev Cancer 13(1):51–65
84. Noguchi M, Kabayama K, Uemura S, Kang BW, Saito M, Igarashi Y, Inokuchi J (2006) Endogenously produced ganglioside GM3 endows etoposide and doxorubicin resistance by up-regulating Bcl-2 expression in 3LL Lewis lung carcinoma cells. Glycobiology 16(7): 641–650
85. Chiu WH, Chen HH, Chang JY, Luo SJ, Li CL, Chen CL, Su WC, Lin CF (2014) Inhibiting glucosylceramide synthase facilitates the radiosensitizing effects of vinorelbine in lung adenocarcinoma cells. Cancer Lett 349(2):144–151
86. Rath G, Schneider C, Langlois B, Sartelet H, Morjani H, Btaouri HE, Dedieu S, Martiny L (2009) De novo ceramide synthesis is responsible for the anti-tumor properties of camptothecin and doxorubicin in follicular thyroid carcinoma. Int J Biochem Cell Biol 41(5): 1165–1172
87. Lamour NF, Chalfant CE (2008) Ceramide kinase and the ceramide-1-phosphate/cPLA2alpha interaction as a therapeutic target. Curr Drug Targets 9(8):674–682
88. Mitra P, Maceyka M, Payne SG, Lamour N, Milstien S, Chalfant CE, Spiegel S (2007) Ceramide kinase regulates growth and survival of A549 human lung adenocarcinoma cells. FEBS Lett 581(4):735–740
89. Pastukhov O, Schwalm S, Zangemeister-Wittke U, Fabbro D, Bornancin F, Japtok L, Kleuser B, Pfeilschifter J, Huwiler A (2014) The ceramide kinase inhibitor NVP-231 inhibits breast and lung cancer cell proliferation by inducing M phase arrest and subsequent cell death. Br J Pharmacol 171(24):5829
90. Elojeimy S, Liu X, McKillop JC, El-Zawahry AM, Holman DH, Cheng JY, Meacham WD, Mahdy AE, Saad AF, Turner LS, Cheng J, A Day T, Dong JY, Bielawska A, Hannun YA, Norris JS (2007) Role of acid ceramidase in resistance to FasL: therapeutic approaches based on acid ceramidase inhibitors and FasL gene therapy. Mol Ther 15(7):1259–1263
91. Ramirez de Molina A, de la Cueva A, Machado-Pinilla R, Rodriguez-Fanjul V, Gomez del Pulgar T, Cebrian A, Perona R, Lacal JC (2012) Acid ceramidase as a chemotherapeutic target to overcome resistance to the antitumoral effect of choline kinase alpha inhibition. Curr Cancer Drug Targets 12(6):617–624
92. Lee SK, Park SM, Im C (2011) Cytotoxicities and quantitative structure activity relationships of B13 sulfonamides in HT-29 and A549 cells. Korean J Physiol Pharmacol 15(6):423–429
93. Munoz-Olaya JM, Matabosch X, Bedia C, Egido-Gabas M, Casas J, Llebaria A, Delgado A, Fabrias G (2008) Synthesis and biological activity of a novel inhibitor of dihydroceramide desaturase. ChemMedChem 3(6):946–953
94. Camacho L, Simbari F, Garrido M, Abad JL, Casas J, Delgado A, Fabrias G (2012) 3-Deoxy-3,4-dehydro analogs of XM462. Preparation and activity on sphingolipid metabolism and cell fate. Bioorg Med Chem 20(10):3173–3179
95. Shirai K, Kaneshiro T, Wada M, Furuya H, Bielawski J, Hannun YA, Obeid LM, Ogretmen B, Kawamori T (2011) A role of sphingosine kinase 1 in head and neck carcinogenesis. Cancer Prev Res (Phila) 4(3):454–462
96. Tamashiro PM, Furuya H, Shimizu Y, Kawamori T (2014) Sphingosine kinase 1 mediates head & neck squamous cell carcinoma invasion through sphingosine 1-phosphate receptor 1. Cancer Cell Int 14(1):76
97. Tamashiro PM, Furuya H, Shimizu Y, Iino K, Kawamori T (2013) The impact of sphingosine kinase-1 in head and neck cancer. Biomolecules 3(3):481–513
98. Sinha UK, Schorn VJ, Hochstim C, Chinn SB, Zhu S, Masood R (2011) Increased radiation sensitivity of head and neck squamous cell carcinoma with sphingosine kinase 1 inhibition. Head Neck 33(2):178–188

99. Masood R, Roy I, Zu S, Hochstim C, Yong KT, Law WC, Ding H, Sinha UK, Prasad PN (2012) Gold nanorod-sphingosine kinase siRNA nanocomplexes: a novel therapeutic tool for potent radiosensitization of head and neck cancer. Integr Biol (Camb) 4(2):132–141
100. Kotelevets N, Fabbro D, Huwiler A, Zangemeister-Wittke U (2012) Targeting sphingosine kinase 1 in carcinoma cells decreases proliferation and survival by compromising PKC activity and cytokinesis. PLoS One 7(6), e39209
101. Song L, Xiong H, Li J, Liao W, Wang L, Wu J, Li M (2011) Sphingosine kinase-1 enhances resistance to apoptosis through activation of PI3K/Akt/NF-kappaB pathway in human non-small cell lung cancer. Clin Cancer Res 17(7):1839–1849
102. Schiefler C, Piontek G, Doescher J, Schuettler D, Misslbeck M, Rudelius M, Haug A, Reiter R, Brockhoff G, Pickhard A (2014) Inhibition of SphK1 reduces radiation-induced migration and enhances sensitivity to cetuximab treatment by affecting the EGFR / SphK1 crosstalk. Oncotarget 5(20):9877–9888
103. Rutherford C, Childs S, Ohotski J, McGlynn L, Riddick M, MacFarlane S, Tasker D, Pyne S, Pyne NJ, Edwards J, Palmer TM (2013) Regulation of cell survival by sphingosine-1-phosphate receptor S1P1 via reciprocal ERK-dependent suppression of Bim and PI-3-kinase/protein kinase C-mediated upregulation of Mcl-1. Cell Death Dis 4, e927
104. Ponnusamy S, Selvam SP, Mehrotra S, Kawamori T, Snider AJ, Obeid LM, Shao Y, Sabbadini R, Ogretmen B (2012) Communication between host organism and cancer cells is transduced by systemic sphingosine kinase 1/sphingosine 1-phosphate signalling to regulate tumour metastasis. EMBO Mol Med 4(8):761–775
105. Hsu A, Zhang W, Lee JF, An J, Ekambaram P, Liu J, Honn KV, Klinge CM, Lee MJ (2012) Sphingosine-1-phosphate receptor-3 signaling up-regulates epidermal growth factor receptor and enhances epidermal growth factor receptor-mediated carcinogenic activities in cultured lung adenocarcinoma cells. Int J Oncol 40(5):1619–1626
106. Kawahara A, Nishi T, Hisano Y, Fukui H, Yamaguchi A, Mochizuki N (2009) The sphingolipid transporter spns2 functions in migration of zebrafish myocardial precursors. Science 323(5913):524–527
107. Bradley E, Dasgupta S, Jiang X, Zhao X, Zhu G, He Q, Dinkins M, Bieberich E, Wang G (2014) Critical role of spns2, a sphingosine-1-phosphate transporter, in lung cancer cell survival and migration. PLoS One 9(10), e110119
108. Koybasi S, Senkal CE, Sundararaj K, Spassieva S, Bielawski J, Osta W, Day TA, Jiang JC, Jazwinski SM, Hannun YA, Obeid LM, Ogretmen B (2004) Defects in cell growth regulation by C18:0-ceramide and longevity assurance gene 1 in human head and neck squamous cell carcinomas. J Biol Chem 279(43):44311–44319
109. Meyers-Needham M, Ponnusamy S, Gencer S, Jiang W, Thomas RJ, Senkal CE, Ogretmen B (2012) Concerted functions of HDAC1 and microRNA-574-5p repress alternatively spliced ceramide synthase 1 expression in human cancer cells. EMBO Mol Med 4(2):78–92
110. Karahatay S, Thomas K, Koybasi S, Senkal CE, Elojeimy S, Liu X, Bielawski J, Day TA, Gillespie MB, Sinha D, Norris JS, Hannun YA, Ogretmen B (2007) Clinical relevance of ceramide metabolism in the pathogenesis of human head and neck squamous cell carcinoma (HNSCC): attenuation of C(18)-ceramide in HNSCC tumors correlates with lymphovascular invasion and nodal metastasis. Cancer Lett 256(1):101–111
111. Saddoughi SA, Garrett-Mayer E, Chaudhary U, O'Brien PE, Afrin LB, Day TA, Gillespie MB, Sharma AK, Wilhoit CS, Bostick R, Senkal CE, Hannun YA, Bielawski J, Simon GR, Shirai K, Ogretmen B (2011) Results of a phase II trial of gemcitabine plus doxorubicin in patients with recurrent head and neck cancers: serum C(1)(8)-ceramide as a novel biomarker for monitoring response. Clin Cancer Res 17(18):6097–6105
112. Alberg AJ, Armeson K, Pierce JS, Bielawski J, Bielawska A, Visvanathan K, Hill EG, Ogretmen B (2013) Plasma sphingolipids and lung cancer: a population-based, nested case-control study. Cancer Epidemiol Biomarkers Prev 22(8):1374–1382
113. Guo Y, Wang X, Qiu L, Qin X, Liu H, Wang Y, Li F, Wang X, Chen G, Song G, Li F, Guo S, Li Z (2012) Probing gender-specific lipid metabolites and diagnostic biomarkers for lung

cancer using Fourier transform ion cyclotron resonance mass spectrometry. Clin Chim Acta 414:135–141

114. Jones EE, Dworski S, Canals D, Casas J, Fabrias G, Schoenling D, Levade T, Denlinger C, Hannun YA, Medin JA, Drake RR (2014) On-tissue localization of ceramides and other sphingolipids by MALDI mass spectrometry imaging. Anal Chem 86(16):8303–8311
115. Johnson KR, Johnson KY, Crellin HG, Ogretmen B, Boylan AM, Harley RA, Obeid LM (2005) Immunohistochemical distribution of sphingosine kinase 1 in normal and tumor lung tissue. J Histochem Cytochem 53(9):1159–1166
116. Facchinetti MM, Gandini NA, Fermento ME, Sterin-Speziale NB, Ji Y, Patel V, Gutkind JS, Rivadulla MG, Curino AC (2010) The expression of sphingosine kinase-1 in head and neck carcinoma. Cells Tissues Organs 192(5):314–324
117. Wang Q, Li J, Li G, Li Y, Xu C, Li M, Xu G, Fu S (2014) Prognostic significance of sphingosine kinase 2 expression in non-small cell lung cancer. Tumour Biol 35(1):363–368
118. Zhang C, Lin X, Song Y, Zhang X, Li H, Wang Q (2014) Overexpression of glucosylceramide synthase and its significance in the clinical outcome of non-small cell lung cancer. Chin Med J (Engl) 127(17):3071–3076

The Role and Function of Sphingolipids in Glioblastoma Multiforme

Loubna Abdel Hadi*, **Clara Di Vito***, **Giovanni Marfia**, **Stefania Elena Navone**, **Rolando Campanella**, **and Laura Riboni**

Abstract Aberrations in sphingolipid metabolism and thus levels have been implicated in promoting the aggressiveness of glioblastoma multiforme, one of the most lethal cancers in humans. A major player is sphingosine-1-phosphate, that pressures GBM cells to exhibit its hallmarks, leading to increased proliferation, invasiveness, stemness, angiogenesis and death resistance, this indicating a fine balance and interplay between S1P function and this malignancy. To the opposite GBM are organized to maintain low their ceramide and sphingomyelin levels, which in turn lead to a loss of growth control and to a gain of death resistance. While the mechanisms of these alterations are emerging, the sphingolipid signaling pathway has been implicated in controlling GBM action and mass, and in mediating the link of malignancy. Here we describe and discuss the current understanding on how GBM cells arm themselves with the abilities of manipulating sphingolipids, especially sphingosine-1-phosphate and ceramide, and how these alterations, through differential interactions, regulate different signaling

---

* These authors contributed equally to this work.

L. Abdel Hadi • C. Di Vito
Department of Medical Biotechnology and Translational Medicine, LITA-Segrate, University of Milan, Milan, Italy
e-mail: loubna.abdelhadi@unimi.it; clara.divito@unimi.it

G. Marfia • S.E. Navone
Neurosurgery Unit, Laboratory of Experimental Neurosurgery and Cell Therapy, Fondazione IRCCS Cà Granda Ospedale Maggiore Policlinico, University of Milan, Milan, Italy
e-mail: giovanni.marfia@unimi.it; stefania.navone@unimi.it

R. Campanella
Neurosurgical Unit, San Carlo Borromeo Hospital, Milan, Italy
e-mail: rolando.campanella@unimi.it

L. Riboni (✉)
Department of Medical Biotechnology and Translational Medicine, LITA-Segrate, University of Milan, via Fratelli Cervi, 93, 20090 Segrate, Milan, Italy
e-mail: laura.riboni@unimi.it

Y.A. Hannun et al. (eds.), *Bioactive Sphingolipids in Cancer Biology and Therapy*, DOI 10.1007/978-3-319-20750-6_12

pathways, and integrate GBM function and mass, thus providing molecular cues for GBM properties and progression. It is a future challenge unrevealing how the multiforme features of sphingolipid signaling could be effectively manipulated as strategies to optimize the efficacy and selectivity of future therapies for GBM.

Keywords Glioblastoma multiforme • Glioblastoma hallmarks • Sphingolipids • Sphingosine-1-phosphate • Ceramide • Sphingosine kinase • Ceramide synthase • Sphingomyelinase • Glucosylceramide synthase • Sphingomyelin synthase

Abbreviations

2OHOA	2-Hydroxyoleic acid
A-ceramidase	Acid ceramidase
A-SMase	Acid sphingomyelinase
Bcl2L13	B-cell lymphoma 2-like 13
bFGF	Basic fibroblast growth factor
CD95L	CD95 ligand
CerS	Ceramide synthase
ECM	Extracellular matrix
EGF	Epidermal growth factor
ER	Endoplasmic reticulum
GlcCer	Glucosylceramide
GSCs	Glioblastoma stem-like cells
HIF	Hypoxia inducible factor
IL	Interleukin
N-SMase	Neutral sphingomyelinase
PAI-1	Plasminogen activator inhibitor-1
PAS	Plasminogen activator system
PERK	Protein kinase R-like endoplasmic reticulum kinase
PKCδ	Protein kinase C delta
PLD	Phospholipase D
PRKD2	Protein kinase D2
PTEN	Phosphatase and tensin homolog located on chromosome TEN
S1P	Sphingosine-1-phosphate
$S1P_{1-5}$	Sphingosine-1-phosphate receptors 1–5
SPP2	Sphingosine-1-phosphate phosphatase 2
THC	Tetrahydrocannabinol
TMZ	Temozolomide
TNFα	Tumor necrosis factor α
VEGF	Vascular endothelial growth factor

1 Overview of GBM and Sphingolipids

Glioblastoma multiforme (GBM), the highest grade of gliomas (World Health Organization Grade IV), is the most common and aggressive primary central nervous system malignancy in adult humans. In sharp contrast to the rapidly growing knowledge of GBM pathogenesis and relevant progress achieved in terms of outcomes for different cancers, GBM remains a serious threat and is widely incurable [1]. Indeed, despite the introduction of aggressive and multi-modal therapy with neurosurgery, radiation, and chemotherapy, current treatment regimens have modest benefits, and recurrence of GBM is nearly universal. The overall median survival time of adult patients remains as short as 12–14 months [2, 3], only 3–5 % of patients surviving more than 3 years [4]; patients with recurrent GBM fare even worse, with a median survival of only 5–7 months with optimal therapy [5].

GBM may arise from lower grade gliomas over time (secondary GBM), but most of these tumors (>90 %) seem to arise de novo (primary GBM), with no prior symptoms or evidence of lower grade tumor. Increasing evidence suggests that GBM usually follows the cancer stem cell model where a small population of cancer cells, called GBM stem-like cells (GSCs), possess transcriptional and epigenetic programs that endow them with stemness properties [6]. Different biological properties of GSCs render such a population a critical contributor to tumor initiation, recurrence, metastasis, and thus poor patient outcome [7].

As its multiforme name implies, GBM exhibits complex features and remarkable interpatient and intratumor heterogeneity. GBM heterogeneity is evident at different levels: (a) macroscopically, with regions of necrosis and hemorrhage; (b) microscopically, with pleomorphic cells and nuclei, regions of microvascular proliferation, and pseudopalisanding necrosis; (c) cellular, with multiple distinct cell types and subclones existing within the tumor cell population, and a repertoire of recruited normal cells that contribute to the tumor microenvironment and participate in heterotypic interactions with one another; and (d) molecular, with genetic mutations, amplifications and aberrations affecting multiple signal transduction and metabolic pathways [8, 9]. GBM heterogeneity emerges as a direct implication of the GSC model, and indicates that cell populations with different properties exist within the same tumor [10]. Of relevance, notwithstanding their multiple heterogeneity and different underlying genetic mutations and molecular characteristics, different GBMs display the same histopathologic features, and behave in a clinically indistinguishable manner [11, 12].

Although the mechanisms of GBM onset and progression remain unclear, recent advances in the understanding of the aberrations of signaling pathways associated with GBM have opened a new window in the comprehension of GBM's aggressive nature and in the identification of new targets for treatment of resistant GBM cells [13]. Indeed, multiple oncogenic signaling aberrations are associated with GBM, and drive its aggressive behavior. Among them, key signaling pathways involved in the control of cell proliferation, motility and fate, including p53, tumor suppressor retinoblastoma, PI3K (phosphatidylinositol 3-kinase)/Akt/PTEN (phosphatase and tensin homolog, located on chromosome TEN), and the RAS/MAPK signaling pathways are the most frequent [14].

An increasing amount of evidence indicates that also alterations in sphingolipid signaling play important roles in GBM, participating in its progression and malignant properties. Sphingolipids are one of the major lipid families in human cells, with a wide range of functions in the organization of cell membranes, cell-cell recognition, and signal transduction, strictly related to cancer [15].

Of particular relevance in the world of sphingolipids is the persistent battle carried out by two molecular forces that act as bioactive mediators with antagonist, opposite roles. These are ceramide and sphingosine-1-phosphate (S1P) that are still vying for control of both normal and tumor cell fate. Indeed, the ceramide doctrine is 'fight to death', thus to suppress the tumors, whereas the S1P one is 'survive to extend'. This conflict have led to the so-called 'sphingolipid rheostat' [16], that acts as a binary switch between survival and death in cell decisions, and where the relative levels of these lipids are important determinants. Strategic in, and peculiar of, this molecular battle is the fact that these two fighters, ceramide and S1P, are strictly interconnected in their metabolic and signaling pathways, both being either precursor or product of common metabolic pathways. Consequently, the role played by the different enzymes involved in their metabolic production/degradation, and by the two molecules to influence the cell signaling network is crucial for designing the final winner.

This chapter describes the battle scenes observed in one of the deadliest cancer type, the GBM, and in the roles of ceramide/S1P in the spectrum of aberrant molecular mechanisms associated with GBM pathobiology. Even if less investigated, we will also discuss the role of more complex sphingolipids, namely sphingomyelin and glucosylceramide (GlcCer), often disregarded, but of relevance not only for their interconnected metabolism with ceramide and S1P, but also for their specific functional properties.

This chapter concentrates on two broad parts: (1) 'basic aspects', involved in the nuts and bolts of molecular mechanisms and signaling that help sphingolipids and their balance operate and interact with other signaling pathways in GBM, and (2) 'clinical aspects', a part with more clinical components, that describes the findings in GBM specimens from patients, providing possible translational cues, and reports on the diagnostic/prognostic potentials of sphingolipids, as well as emerging promising therapeutic approaches attempting to target sphingolipid pathways.

For readers who are not familiar with the sphingolipid family, propaedeutic to this chapter are concepts on sphingolipid structure, metabolism, signaling and functions, described in the first part of this book.

2 Sphingolipid Aberrations in GBM

To reinforce their malignancy and their capacity to survive and colonize, GBMs adapt diverse 'crafty' strategies and, among them, exhibit aberrations in sphingolipid metabolism and thus altered sphingolipid levels. A key point appears to reside in the down-regulation of ceramide levels. Indeed, ceramide levels are lower in human GBM tissue compared to normal surrounding brain tissue, and the decrease in ceramide in GBMs is directly related to histological grade and patient survival [17]. The decrease in total ceramides with increasing glioma grade was found parallel to an increase in S1P content, ceramide and S1P levels being on average twofold lower and ninefold higher, respectively in GBM tissues than in normal gray matter [18].

It is important to recall that ceramide is heterogeneous in its structure, as fatty acids with different acyl chain lengths, double bonds and hydroxylations can be present in the molecule [19, 20]. Notably, this heterogeneity depends not only on the availability of different fatty acyl precursors, but mainly on the action of specific enzymes controlling the production of specific ceramide species [21]. Differences with regard to the fatty acyl chain distributions of ceramides were detected among several glioma cell lines using liquid chromatography tandem mass spectrometry [22], suggesting heterogeneity in ceramide subfamilies in different GBMs. Of interest, a specific reduction in C_{18} ceramide was recently shown in human gliomas as a function of their malignancy grade, and, notwithstanding their great heterogeneity, low levels of C_{18} ceramide were common to different GBMs [18]. In head and neck squamous cell carcinoma, C_{18} ceramide was found to be the only form of ceramide whose levels were decreased, and its levels inversely correlated with metastasis [23], suggesting that reduced C_{18} ceramide levels may also contribute to GBM malignancy. In addition, human GBM specimens and tumor cells have markedly lower levels of sphingomyelin than non-tumor cells and normal brain, respectively [24, 25], and a recent report showed that, besides low ceramide, decreased sphingomyelin levels are associated with the GBM tumorigenic transformation [26].

Recent studied provided evidence about possible molecular explanations responsible for the aberrant levels of ceramide and S1P in GBMs (Fig. 1). Indeed, acid ceramidase (A-ceramidase), which converts ceramide to sphingosine, was found to be significantly up-regulated in GBM specimens [18]. Moreover, Jensen et al. [27] showed that B-cell lymphoma 2-like 13 (Bcl2L13), an atypical member of the Bcl-2 family overexpressed in GBM, acts as ceramide synthase (CerS) inhibitor.

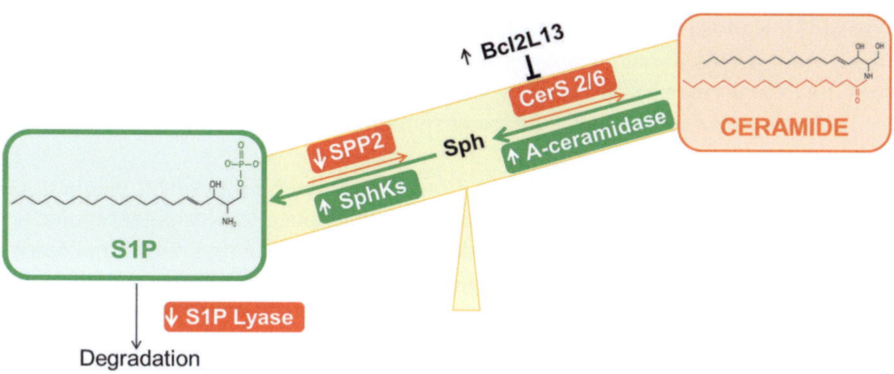

Fig. 1 Multiple dysregulations in sphingolipid metabolism occur in GBM and lead to unbalanced S1P/ceramide rheostat in favor of S1P

Interestingly, through its binding to, and inhibiting CerS2 and CerS6 activity, Bcl2L13 functions as an anti-apoptotic protein by protecting the mitochondrial membrane integrity.

With regard to the increased level of S1P, a likely explanation should reside in the high expression of both SphK1 and SphK2 in GBMs [28–30]. In addition to SphK up-regulation, it was found that the S1P phosphatase 2 (SPP2), a S1P specific phosphohydrolase localized to the endoplasmic reticulum (ER) [31], is significantly down-regulated in GBM as compared to normal gray matter samples [18]. Of note, SPP1 expression has been shown to increase ceramide levels at the ER by the recycling of its product sphingosine [32], suggesting that the loss of SPP2 expression in GBM not only increases S1P but also drives ceramide levels down. Of relevance is also the finding that the chromosomal region containing the gene for S1P lyase is deleted in human GBMs [33], most likely favoring high levels of S1P in this cancer.

Thus it emerged that GBM is equipped to down-regulate ceramide and up-regulate S1P to drive a systematic shift in sphingolipid balance which may favor aggressiveness and protection from death (Fig. 1). In agreement, using a mathematical model to construct how sphingolipid metabolism is altered in GBM cells compared to normal astrocytes, Mora et al. [34] observed that a key difference is the preferential channeling of complex sphingolipids into S1P synthesis in GBM cells, whereas in normal astrocytes, sphingosine formed in the lysosomes is mainly recycled into ceramide.

S1P exerts multiple roles through its five specific receptors ($S1P_{1-5}$) [35] and $S1P_{1-3}$ and $S1P_5$ receptors were found in human GBM specimens [28, 36, 37]. In a recent study, Quint et al. [30] reported that the mRNAs for all these receptors are up-regulated in human GBM specimens compared to normal brain, with increasing order of magnitude from primary, up to recurrent and secondary GBMs.

3 Sphingolipid Molecules in the Hallmark Traits of GBM

As neoplasia, GBM exhibits hallmark capabilities that are crucial to cancer phenotypes [38]. Hallmark features of GBM include uncontrolled proliferation, invasiveness, stemness, intense angiogenesis and death resistance; all these features account for GBM's resistance toward therapy and poor prognosis [39, 40].

The emerging scenario is that different sphingolipids act as crucial players in the hallmarks of GBM (Fig. 2).

Impressively, the simple lysosphingolipid S1P emerges as a critical mediator of the multiple hallmark capabilities of GBM, being supplied as pluripotent molecule to the GBM microenvironment, and displaying a spectrum of GBM hallmark-facilitating activities. Indeed, S1P appears to act as growth factor that sustains proliferative signaling, as a pro-invasion factor that favors GBM recurrence and metastasis, as a stemness-promoting factor that fosters limitless growth and self-renewal, as a proangiogenic factor that promotes GBM vascularization, and as a survival factor that limits GBM cell death and confers therapy resistance. With multiple S1P weapons, GBM fights to gain its aberrant cancer properties, against different

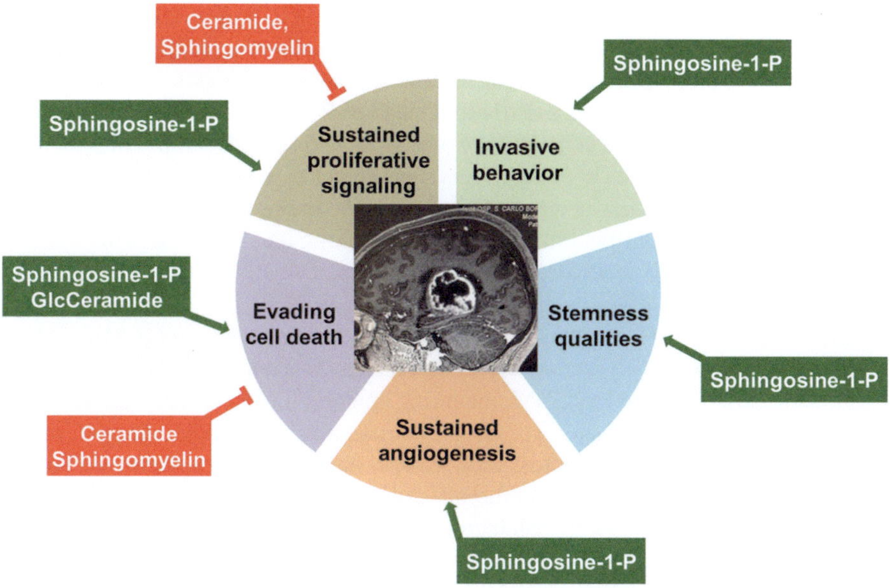

Fig. 2 Different sphingolipids act as crucial players in the hallmarks of GBM

tumor suppressors, among them ceramide. Moreover, emerging evidence indicates a role for complex sphingolipids, such as GlcCer and sphingomyelin in participating to this scenario (Fig. 2).

4 Sphingolipids in GBM Proliferation

GBM is a highly proliferative tumor, and the individual growth potential of a GBM results in regrowth, i.e., relapse, after a variable (case-specific) time, which is inversely correlated to their cell proliferation index. In agreement, the histopathologic assessment in GBM diagnosis includes the identification of a high tumor cell mitotic index [41]. Moreover, the capacity for extensive proliferation and self-renewing is a key determinant of GSCs too [42, 43].

GBM cells exhibit the capability to sustain their proliferative potential in an uncontrolled and unlimited fashion by different mechanisms [38]. These include: (1) autocrine proliferative stimulation, as they spontaneously produce growth factors, to which they respond via the expression of cognate receptors; (2) stimulation of noncancer cells within their microenvironment to export appropriate mitogenic stimuli; (3) displaying of elevated receptor levels at their cell surface, being thus highly responsive to growth factors; and (4) constitutive activation of components of signaling pathways operating downstream of these receptors, being thus independent in their activation from exogenous signals. Of relevance, sphingolipids and their signaling pathways, particularly with the participation of S1P and ceramide as mediators, emerge as crucial for providing regulatory contribution to all these mechanisms.

Changes in cellular metabolism associated with cell proliferation are tightly linked to the synthesis of membrane sphingolipids (and phospholipids), supplying cells with material to build plasma membrane and the various membranes of intracellular organelles in growing and dividing cells [44]. During normal cell division, sphingolipid metabolism is activated not only to provide a basic supply of structural and functional metabolites but also to serve as second messengers [45].

So far, it is S1P that has been shown to act as the major pro-proliferative sphingoid in normal cells, but also in cancer ones. In particular, in the nervous system, S1P has been reported to favor propagation of astrocytes [46], and also of GBM cells [36, 37, 47].

Excessive growth factor signaling is one crucial component in GBM development [48]. Several growth factors use S1P as a signaling mediator [49] and, among them, those strictly related to GBM, including epidermal growth factor (EGF) [50], basic fibroblast growth factor (bFGF), and vascular endothelial growth factor (VEGF) [46, 51, 52]. These factors induce rapid and transient activation of SphK1 and its subsequent translocation to the plasma membrane and S1P export, and can exert also long-term effects, by enhancing the SphK1 expression (Fig. 3).

More than 10 years ago, Van Brocklyn et al. [47] demonstrated that GBM cell lines express mRNA encoding the $S1P_{1-3}$ and $S1P_5$ receptors, and respond mitogenically to nanomolar concentrations of S1P. These receptors were found to be expressed in human GBM specimens [28, 36, 37], suggesting that S1P signaling through its receptors may contribute to GBM growth in vivo. Indeed, it was found that the proliferating effect of S1P involves Gi-coupled receptors, and activation of extracellular signal-regulated (ERK/MAP) kinase contributes to mitogenic S1P signaling, by both PI3K-dependent and independent pathways [28, 53]. Among S1P receptors, $S1P_1$ is a transcriptional target for signal transducer and activator of transcription 3 (Stat3), whose persistent activation is critical for the growth of a range of cancers. In turn, $S1P_1$ signaling is necessary for persistent Stat3 activation, forming a positive feedback loop that fuels the development and growth of different cancers [54]. $S1P_1$ expression was found to be either down-regulated in GBM specimens, and this down-regulation enhances tumor cell proliferation and correlates with shorter survival of GBM patients [36, 37], or to be up-regulated [18, 30]. Whether these opposite findings depend on the heterogeneity of GBM or experimental conditions remains to be clarified.

It was shown that while the overexpression $S1P_{1-3}$ contributes to S1P-stimulated growth, that of $S1P_5$ decreases cell proliferation in GBM cells [55], suggesting that a balance between $S1P_{1-3}$ and $S1P_5$ is involved in the control of GBM proliferation. However, in conflict with the anti-growth action of $S1P_5$ in cultured GBM cells, GBM patients overexpressing $S1P_5$ were reported to have worse prognosis [30], indicating that a more complex situation occurs. In agreement, it was shown that the proliferative role of S1P in GBM cells could also occur by mechanisms independent from its specific receptors, and most probably by its intracellular actions (Fig. 3). In particular, on investigating the role of SphK activity in tumor necrosis factor α (TNFα)-stimulated GBM cell proliferation, it was found that TNFα-stimulated DNA synthesis required SphK1 activity, and this promoted Akt phosphorylation,

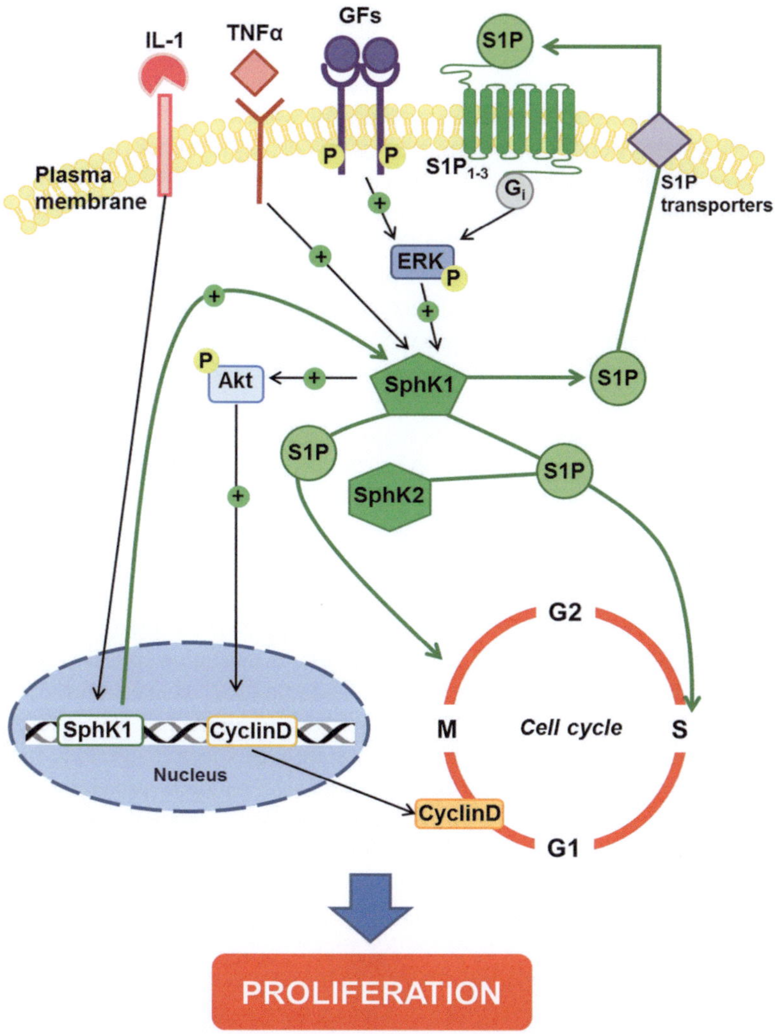

Fig. 3 S1P produced in response to extracellular stimuli promotes GBM cell proliferation in both receptor-dependent and -independently fashion

cyclin D expression and cell growth, without the intervention of the S1P-Gi-coupled receptors [56]. Notably, Akt phosphorylation also promotes the vesicular transport of ceramide from ER to Golgi apparatus in GBM cells, promoting ceramide metabolism to complex sphingolipids [57], and thus favoring low ceramide levels in GBM. Nitric oxide causes inhibition of ER/Golgi traffic of ceramide and, by leading to ceramide accumulation in the ER, it exerts an inhibitory effect on GBM cell growth [58].

Recent studies showed that 2-hydroxyoleic acid (2OHOA) (Minerval®), a potential antitumor compound, causes cell cycle arrest and death in GBM cells [25]. In particular, 2OHOA treatment of GBM cells strongly increased sphingomyelin level, by rapid activating of sphingomyelin synthases (SMS), and SMS inhibition diminished 2OHOA effect on cell cycle. It was then demonstrated that sphingomyelin accumulation at the plasma membrane is crucial in GBM cell growth, as it causes an increase in membrane global order, resulting in modified lipid raft properties and leading to inhibition of cyclin D and cell cycle arrest [26].

Strong evidence demonstrates that SphK1 has an important growth-regulatory role in different cancer cells, and this enzyme is critically involved in the mitogenic and oncogenic action of S1P in human GBM cells. Indeed, SphK1 inhibition results in growth arrest of GBM cells both in vitro and in vivo [28, 29, 56, 59]. SphK1 was found to be involved in the proliferative action of interleukin 1 (IL-1), a pro-inflammatory cytokine secreted by the majority of GBMs [60]. Indeed it was found that IL-1 up-regulates SphK1, but not SphK2, at the transcription level, resulting in a correlation between its level and that of SphK1 in GBM cells [61]. Surprisingly, although SphK2 expression has been shown to be unrelated to survival in GBM patients, SphK2 knockdown inhibits GBM cell proliferation more potently than did SphK1 knockdown [28], suggesting that, besides SphK1, also endogenous SphK2 influences cell growth and aggressive behavior of GBM. In agreement, the participation of both SphK1 and SphK2 isoforms to the proliferation of GBM was documented in a recent report describing SphK effects on GBM cell proliferation under hypoxia [62]. This study showed that both SphK1 and SphK2 are involved in GBM cell proliferation in hypoxic conditions, but act through distinct signaling pathways and mechanisms of action. Indeed, the SphK1-dependent proliferative effect required ERK activation, mediated by S1P receptors, whereas ERK was not involved in the SphK2-dependent one. Moreover, when SphK1 or SphK2 were inhibited, the cell cycle arrested in G2/M phase or in S phase, respectively. Thus, despite SphK1 mediates proliferative signaling in GBM, as it occurs in various cancer cells [63], inhibitory strategies that target both SphKs appear to be required to halt the proliferative potential of GBM cells, particularly in hypoxic regions.

5 Sphingolipids and GBM Invasive Behavior

GBM is an aggressive tumor, that not only proliferates rapidly but also exhibits a high propensity to invade the surrounding parenchyma and to widely disseminate within normal brain [7]. In fact, individual GBM cells can be found many centimeters away from the primary tumor site, often crossing great distances into the contralateral hemisphere, and represent sources for tumor recurrences after resection or radiation [64]. Using in vivo multiphoton laser scanning microscopy, it was shown that a single or a few GBM invasive cells are capable of remodeling and multiplying the host vasculature, thus improving the conditions for GBM to thrive [65]. Moreover, GBM can spread to nearly any tissue in the body via hematogenous and/or lymphatic dissemination [66].

The invasion of GBM cells requires the degradation of the extracellular matrix (ECM), which depends on the activation/inhibition of proteinases and their inhibitors, respectively. This process includes two main proteolytic systems: the plasminogen activator system (PAS), which controls the activation of the proteinase plasmin from inactive plasminogen, and the matrix metalloproteinases and their inhibitors [67]. The attachment at the leading edge of a migrating cell, and the concurrent detachment at its trailing edge are imperative for efficient migration and invasion (Fig. 4). More importantly, the enhanced expression of the PAS components may provide attachment or detachment depending on the ratio of urokinase plasminogen activator (uPA) and its receptor (uPAR), and plasminogen activator inhibitor-1 (PAI-1), at opposite cell edges. The uPA–uPAR binding leads to the degradation of the ECM, and provides attachment via the uPA/uPAR/integrins complex. In contrast, the enhanced expression of PAI-1 on the trailing edge of migrating cells induces the internalization of the uPA/uPAR/integrin complex, and results in cell detachment [68], a crucial step for the cell to proceed with its invasion.

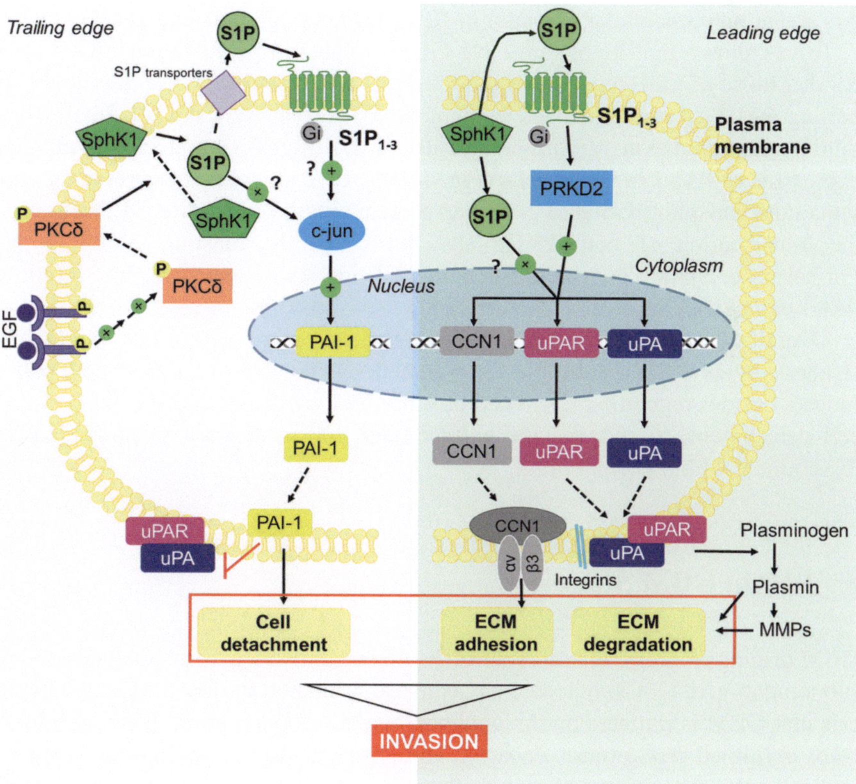

Fig. 4 SphK1/S1P/S1PRs axis contributes to GBM invasivity by regulating the attachment (at the leading edge) (*right*) and the detachment (at the trailing edge) (*left*) of a migrating cell through plasminogen activator system and matrix metalloproteinases

Different studies demonstrated that S1P acts as important signal to regulate both migration and invasiveness of GBM cells, mainly through receptor-dependent mechanisms. The three S1P receptors commonly expressed in GBM, $S1P_{1-3}$, have different effects on GBM cell motility, $S1P_1$ and $S1P_3$ stimulating, and $S1P_2$ decreasing cell migration [53]. The $S1P_2$-mediated inhibition of GBM cell migration is associated with activation of Rho A, and suppression of Rac1 through Rho signaling pathways [69]. Instead, all the three receptors contribute, in a coordinated fashion, to promote GBM invasiveness ([53, 70]; Young and [71]) (Fig. 4, right). This occurs through a SphK1-dependent up-regulation of uPA and its receptor, as well as secretion of the pro-invasive molecule CCN1 [72–74]. Only $S1P_1$ and $S1P_2$ receptors contribute to CCN1 induction, whereas all $S1P_{1-3}$ receptors cooperate to induce expression of members of the uPA system, with $S1P_1$ being the most potent. The relevance of S1P-induced invasiveness of GBM cells is supported by evidence that neutralizing antibodies directed against uPA or CCN1 significantly decrease both basal and S1P-stimulated GBM cell invasiveness, and that SphK inhibition potently blocks uPA activity and GBM invasion [74]. Recently protein kinase D2 (PRKD2), a PKC isoform highly expressed in GBM [75], was identified as a target activated by S1P receptors and as an intracellular mediator of S1P-induced migration and invasion [76].

It was shown that highly invasive GBMs express high levels of PAI-1, which parallel those of EGF receptor, and both levels are related to worse prognosis, supporting PAI-1 as important determinant of GBM invasiveness [77]. Interestingly, SphK1 was found as an important mediator of EGF signaling that leads to enhanced expression of PAI-1 in GBM cells (Fig. 4, left). EGF signaling occurs via the rapid, sequential phosphorylation of c-Src and protein kinase C delta (PKCδ), followed by the translocation of both PKCδ and SphK1 to the plasma membrane [78]. Furthermore, SphK1 is crucial for both EGF-induced c-Jun phosphorylation and PAI-1 expression.

Overall, the SphK/S1P/S1P receptor signaling contributes to favor GBM invasiveness by providing multiple coordinated mechanisms of GBM cell control. Indeed, besides regulating GBM cell motility, S1P efficiently participate to promote cell detachment, to enhance the extracellular matrix degradation, and finally to disseminate within normal brain.

6 S1P in GBM Stemness

GBM origin, progression and recurrence are currently largely attributed to the rare subpopulation of GSCs. Indeed, increasing experimental studies support the hypothesis that GBM is initiated and maintained by this subpopulation of GBM cells that owns unlimited proliferation capacity, ability to self-renew, multipotent properties, invasive behavior, evasion of cell death and intrinsic resistance to the current GBM treatments [79–81]. GSCs are able to generate and propagate new tumors when transplanted, and to differentiate into the different cellular subtypes that form the tumor mass [82].

Growing evidence implicates a complex interplay between GSCs and their microenvironment in determining GSC properties and maintenance ([83], [84]). GSCs are localized in specialized niches within the tumor, namely hypoxic and perivascular microenvironments. These niches are composed by differentiated cell types that secrete several factors that are crucial for GSC homeostasis and regulate the balance between self-renewal and differentiation. Among its multiple roles in GBM phenotype, S1P has emerged as a key mediator in determining GSC stemness and aggressiveness and as crucial component of the GSC niche. Indeed, as described below, different pieces of evidence suggest a chief role for S1P as an autocrine/paracrine signal connecting GSCs to their niche.

We recently demonstrated that GSCs are equipped with efficient molecular machinery which allow them to rapidly synthesize S1P from sphingosine, as well as to export the newly produced S1P in their microenvironment [85, 86]. GSCs appear to share with normal cells of the nervous system, such as neurons and astrocytes, and also human GBM cells the capability of releasing S1P in the extracellular milieu [18, 87, 88]. Of relevance, the amount of extracellular S1P provided by GSCs was found to be markedly higher (about tenfold) than that by astrocytes and non-stem GBM cells [85]. It is worth noting that, among stem cells, the export of S1P seems specific to GSCs, as neural stem cells have been reported to be uncapable of this export [89]. Thus it appears that, different from neural stem cells, GSCs possess the unique ability to enrich their extracellular milieu with S1P, that in turn could act as a autocrine/paracrine signal favoring tumor progression.

The efficiency of S1P export by GSCs is strictly dependent on the availability of its substrate sphingosine [86], that can be released from necrotic cells. The enrichment of GSCs in perinecrotic regions [90], suggests that in these locations S1P biosynthesis and release occurs very rapidly and to a high extent in GSCs, providing these cells a favorable environment. Of interest, the proliferative and stemness properties of GSCs are related to a gain in extracellular S1P, paralleled by a down-regulation of ceramide levels due to its increased flux to complex sphingolipids [86], indicating that the imbalance of the sphingolipid rheostat, with the S1P domination in the GSC niche, provides an advantage to the stemness qualities of GSCs.

Differently from GBM cells, GSCs express all the five S1P receptors, including $S1P_4$, this last being absent in cells of the nervous system and GBM cells. $S1P_1$, $S1P_2$, and $S1P_4$ are highly expressed in GSCs [91]. As described below, emerging literature supports that S1P, after receptor binding, exerts different actions on GSCs which favor and potentiate their malignant features.

A key determinant of GSCs is their capacity for extensive proliferation and self-renewing [42, 43]. When exposed to exogenous S1P, GSCs switch to a faster growth mode implying that S1P can act an autocrine signal for GSCs to grow and proliferate. Indeed, S1P sends a proliferative signal to GSCs that is strong enough to promote cell cycle and progression into G2/M phase of GSCs [86]. A consistent body of studies indicates that EGF is an important factor in the GSC environment, supporting GSC growth and inhibiting their spontaneous differentiation [92, 93]. Noteworthy, EGF receptor III variant level, particularly expressed in GSCs and crucial for their stemness and tumorigenic potential [92, 94], is directly related to

SphK1 activity, that in turn determines a proliferative and survival advantage [95]. Furthermore, EGF and bFGF, which are crucial factors supporting GSC growth and stemness, promote S1P export by GSCs [86]. S1P proliferative action in GSCs seems dependent on its receptor binding, as it is inhibited by FTY720, a sphingosine analogue that, after cellular internalization and phosphorylation, is released extracellularly and down-regulates S1P receptors, acting as functional antagonist of S1P [96].

Interestingly, besides changing the growth pattern of GSCs, S1P induced elevated levels of different stem cell markers in GSCs, thus enhancing their stemness phenotype [86]. Since S1P did not induce stemness marker expression in non-stem primary GBM cells, it is likely that the stemness promoting action of S1P resides in its ability to favor the selective expansion of GSCs.

Moreover, S1P was shown to induce GSCs migration, that was more pronounced than in the bulk of GBM cells [34, 91], and SphK inhibition promoted neurosphere dissociation, and GSC death [34]. In addition, the administration of FTY720 decreased GSC proliferation, invasiveness and viability, and enhanced the cytotoxic effect of temozolomide (TMZ), the mainstay of GBM therapy, promoting survival in a GBM xenograft mouse model [97].

Very recently, we reported that GSCs derived from U87 GBM cells and those isolated from a human GBM specimen can release S1P extracellularly, and that S1P acts as a first messenger to enhance GSC chemoresistance [85].

Overall, our current knowledge implicates the imbalance of the sphingolipid rheostat in favour of S1P as an important cue in determining GSC characteristics, with extracellular S1P emerging as a key modulator of proliferation, migration, stemness, and therapy resistance in these stem cells, thus reinforcing the role of this sphingoid in GBM progression and malignancy.

7 Sphingolipids in GBM Angiogenesis

GBM is one of the most highly vascularized tumor in man, and its blood vessels show enhanced endothelial cell proliferation, which is a key feature of their classification in the WHO grading system. Moreover, a marked increase in angiogenesis, that is in new blood vessel formation from preexisting ones, plays a critical role in the biological behavior of GBM, being crucial for its growth, survival and colonization in the brain [98].

Angiogenesis is a complex process with the involvement of several cells, particularly GSCs and endothelial cells, and different molecular mediators [99].

The main trigger of angiogenesis processes in GBM is hypoxia, mainly localized in the center part of the tumor, and driven primarily by VEGF. GBM cell response and adaptation to low oxygen conditions is primarily controlled by hypoxia inducible factors (HIFs). These are transcription factors which promote adaptation to hypoxic conditions through the regulation of more than 150 genes involved mainly in angiogenesis, metabolism, proliferation and cell migration [100]. VEGF is induced by HIFs and is up-regulated in GBM, where it regulates endothelial cell survival, proliferation, permeability, and migration, with the end result of production of immature, highly permeable blood vessels [101].

Importantly, under hypoxia SphK1 up-regulates HIF-1 in GBM cells, by stabilizing HIF-1α through the Akt pathway and preventing its proteasomal degradation [102]. It was also shown that HIF-2α up-regulation of SphK1 in GBM cells leads to enhancement of neovascularization [103], and its product S1P acts as potent mediator of angiogenesis as VEGF [104], most likely through its specific receptor $S1P_1$ [49]. Moreover, S1P is a downstream target of different pro-angiogenic growth factors such as VEGF, EGF, PDGF, bFGF and pro-angiogenic cytokines such as IL-6 and IL-8 leading to a positive signaling loop [105].

Interestingly, it was found that in GBM cells SphK1 down-regulation, but not that of SphK2, reduced extracellular S1P levels, which in turn decreased both migration and tube formation in co-cultured vascular endothelial cells [88]. In addition, VEGF down-regulation was unable to block the angiogenic switch triggered by GBM-derived S1P.

In GBM cells S1P is able to initiate endothelial cell sprouting, which is representative of a multistep angiogenic process, and SphK1 plays an essential role in regulating paracrine angiogenesis, favoring the early sprouting response of endothelial cells by increasing both number and length of sprouts [18]. Notably this effect was induced by S1P released by GBM cells, and was not affected by VEGF [18], suggesting that S1P signaling is required even when potent angiogenic factors such as VEGF are present.

8 Sphingolipids in GBM Cell Death

Sphingolipid imbalance is deeply implicated in protecting cells from death, favoring GBM survival even in diverse, cytotoxic conditions. Before analyzing the sphingolipid alterations underlying the propensity of GBM cells to stay alive, we describe how sphingolipids, especially ceramide can act as an effective mediator of GBM death.

Ceramide was identified as 'the tumor suppressor lipid' more than 15 years ago [106], and a huge amount of literature reports on the apoptotic role of ceramide in cancer cells, including GBM cells. Indeed it is widely recognized that ceramide acts as a crucial mediator in cancer cell death, acting in different death mechanisms by regulating an 'orchestra' of signal molecules involved in the control of cell fate [107, 108], including that of GBM cells. As described below, ceramide plays an important role in the execution of GBM cell death induced by different agents, such as radiation, chemotherapeutics and proapoptotic TNF family ligands.

8.1 Ceramide in GBM Apoptosis

The apoptotic machinery is hijacked in GBM, and the ability of GBM cells to maintain their ceramide levels low appears crucial in favoring their survival. Several studies on GBM cells revealed that ceramide acts as cellular apoptotic messenger in the two principal pathways of caspase-3 activation, the key event of apoptotic death. Indeed ceramide participates as crucial mediator to the 'extrinsic apoptosis pathway' that is triggered by the engagement of cell surface 'death receptors' by exogenous signals, such as CD95 ligand (CD95L) and TNFα, as well as the 'intrinsic apoptosis pathway' that is provoked by various forms of stress, such as radiations and chemotherapeutic drugs.

8.1.1 Ceramide in the Extrinsic Apoptotic Pathway of GBM Cells

One of the first indicators of ceramide's involvement in GBM apoptosis was prompted by evidence that the TNF receptor CD95 is highly expressed in GBM. These studies demonstrated that radio-resistant GBM cell lines are susceptible to CD95-mediated cell death, and ceramide is required as the mediator of apoptosis [109, 110] (Fig. 5). After inducing apoptosis in GBM cells by an anti-CD95 antibody, ceramide formation by caspase-8-dependent activation of acid sphingomyelinase (A-SMase), but not neutral sphingomyelinase (N-SMase), was observed [111]. CD95-induced ceramide formation and apoptosis was shown to proceed regardless of p53 status or reactive oxygen species (ROS). A mechanism by which ceramide acts as pro-apoptotic molecule and is essential to CD95-mediated apoptosis in GBM cells was provided in a study were ceramide was found to down-regulate cFLIP, the cellular FLICE (caspase 8) inhibitory protein, a protein able to inhibit CD95-mediated apoptosis [112]. Notably, the CD95 receptor is highly expressed in GBMs and is almost exclusively located in perinecrotic areas, where cells undergoing apoptosis accumulate. This suggests that CD95-mediated cell death may play a role in the pathogenesis of necrosis which constitutes a histological hallmark of GBM.

In GBM cells, the extrinsic pathway of apoptosis can be also activated via the TNFRα/TNFα receptor-ligand system, and ceramide is involved in this pathway too [113] (Fig. 5). Two separate signaling cascades, ROS-dependent (p53-induced), and ROS-independent, contribute to the TNFα-induced apoptotic death of human GBM cells. Interestingly, the p53-induced ROS-dependent signaling cascade is triggered by ceramide formation by the activation of N-SMase, whereas the A-SMase activation is independent of functional p53 [114]. Mitochondria participate in TNFα-induced apoptosis of human glioma cells by cytochrome c release through ceramide generated via p53-mediated ROS-dependent and -independent pathways, both of which are initiated by the activation of caspase-8.

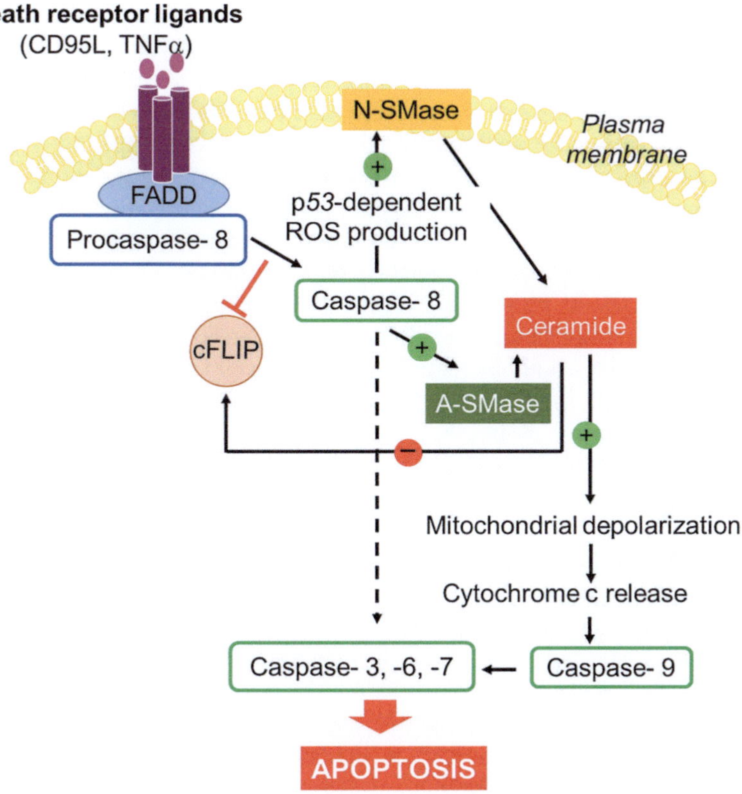

Fig. 5 The activation of death receptors results in caspase 8-dependent activation of SMases, that leads to ceramide formation, which in turn acts as a trigger of GBM cell apoptotic death

8.1.2 Ceramide in the Apoptotic Intrinsic Pathway of GBM Cells

Killing GBM cells by diverse cytotoxic approaches such as γ-irradiation and anticancer drugs was shown to be mediated through induction of apoptosis. The underlying mechanism for initiation of an apoptotic response upon cytotoxic therapy can be different for various stimuli and is only partially understood. However, damage to DNA or to other critical molecules and/or subcellular structures of GBM cells appears to be a common early hit by some inducers which is then propagated by the cellular stress response. Among the multiple stress-inducible molecules, ceramide has a profound impact on apoptotic pathways in GBM and thereby its management. Several studies reported the pioneering role of ceramide in driving apoptosis by affecting the mitochondria, the 'engine of the apoptotic intrinsic pathway', as well as regulating different intracellular effectors that mediate the activation of the apoptotic process.

In GBM cells, γ-radiations and diverse antineoplastic drugs and stress stimuli require ceramide as an effector of their apoptotic effects. Mutations of tumor suppressor p53 are common in GBM, and the sensitivity to γ-radiations of GBM cell lines is influenced by their p53 status, cells with functional p53 being more resistant to γ-radiation than those with mutant p53 [115]. It was reported that γ-irradiation induced apoptosis in human GBM cells with mutant p53 by activating A-SMase, and the formed ceramide triggers caspase activation [116, 117]. on the flip side, cells with wild-type p53 avoid the pro-apoptotic ceramide signal through both blockade of A-SMase and up-regulation of A-ceramidase, a ceramide degrading enzyme. These results support that ceramide functions as mediator of γ-radiation-induced p53-independent apoptosis in human GBM cells, and suggest that aberrant ceramide signaling—both increased A-ceramidase expression and blockage of A-SMase activation—underly the γ-radiation resistance of GBM cells with wild type p53. Of interest, it has been recently demonstrated that cell permeable C6-ceramide can effectively kill human GBM cells by inducing a p53-dependent apoptotic pathway, suggesting that p53 can be downstream of the ceramide-induced apoptosis [118].

In contrast to these findings, a recent study showed that lentivirally induced A-SMase overexpression does not sensitize LNT-229 GBM cells to irradiation, despite leading to increased A-SMase activity and ceramide level [119]. Since the LNT-229 cell line is wild-type p53, it appears likely that up-regulation of A-SMase alone is not sufficient *per se*, and the activity of A-ceramidase might be relevant to avoid γ-radiation-induced and ceramide-mediated apoptosis.

Diverse chemotherapeutic agents, including gemcitabine, doxorubicin, cisplatin, and etoposide, activate sphingolipid signaling to produce ceramide and finally induce apoptosis in cultured GBM cells. In particular, a study implicated A-SMase in apoptotic ceramide generation and in sensitizing GBM cells to gemcitabine and doxorubicin [120]. After increasing the expression and activity of A-SMase via reactive oxygen species (ROS), the Authors demonstrated a significant sensitization of U373MG GBM cells to drug treatment, and suggested that chemotherapeutic drugs activate this enzyme via ROS. In another study investigating the mechanism of gemcitabine-triggered glioma cell death, it was shown that this drug exerts apoptotic effects in GBM cells by activating A-SMase in the lysosomes, where the enzyme normally resides. In this site, the accumulated ceramide directly interacts with the aspartic protease cathepsin D, and mediates its activation. Cathepsin D then translocates to the cytosol and cleaves Bid to truncated Bid, which activates Bax to initiate the intrinsic apoptosis pathway [121, 122]. The finding that cathepsin D is a direct ceramide target to induce cell death in GBM is of relevance because this protease is overexpressed in this cancer, its serum level being predictive of poor prognosis and aggressive tumor behavior [123].

Ceramide has also been closely implicated in GBM cell apoptosis induced by cisplatin, a common drug used in the treatment of malignant brain tumors through activation of caspase-3-like proteases [124]. Inhibition of N-SMase suppressed cisplatin-induced apoptosis, suggesting a pro-apoptotic implication of this SMase during cisplatin treatment. Moreover, treatment of resistant glioma cells with the

A-ceramidase inhibitor N-oleoylethanolamine increased their sensitivity to cisplatin, suggesting that A-ceramidase is involved in cisplatin resistance too [124]. The role of N-SMase in generating pro-apoptotic ceramide was also observed after treatment of C6 glioma cells with etoposide, a topoisomerase II inhibitor acting as potent apoptotic inducer in several cancer types [113]. The Authors reported that etoposide treatment of GBM cells induces ceramide generation through both activation of N-SMase and increase in serine palmitoyl transferase activity. In turn, elevated ceramide acts as mediator of etoposide-induced apoptosis through the increase in the Bax/Bcl-2 ratio, followed by cytochrome c release from mitochondria and caspases-9 and -3 activation [113].

Furthermore, the de novo synthesis of ceramide has been shown as crucial for cannabinoid-triggered ER stress and apoptosis in GBM cells [125], suggesting that not only ceramide generation from sphingomyelin, but also its new synthesis are effective in generating the apoptotic ceramide signal.

8.2 Ceramide in Necrotic and Autophagic Death of GBM Cells

Although apoptotic cell death is a recognized mechanism of ceramide-induced death, ceramide is capable of activating cell death by totally distinct mechanisms from apoptosis in GBM cells, among them necrotic and autophagic death. It appears that ceramide causes either apoptotic or non-apoptotic cell death depending on the cell type and the experimental conditions, suggesting that different GBM cells and environmental conditions may influence the mechanisms by which ceramide exerts its fatal effect.

As pertaining to necrosis, ceramide was shown to induce a caspase-independent cell death with necrotic-like morphology in some human GBM cell lines, by inhibiting the Akt protein kinase [126], or the anti-apoptotic transcriptional factor NF-κB [127].

Autophagy is a Janus-faced process able to regulate the balance between cell survival and death, being able to act either as a protective mechanism, that may prevent cell death in stress conditions, or as a direct cause of cell death, and both may be associated with apoptosis in some model systems [128]. In cancer cells like GBM cells, autophagy occurs in response to anticancer therapies, and can overcome the therapeutically induced stress and cellular damage, or can be the mechanism responsible for cell death; the switch from survival to death likely depends on the cellular stress threshold [129]. Of interest, recent findings demonstrated that GBMs have lower expression of autophagy-related proteins when compared with low-grade gliomas [130] and that progression of astrocytic tumors is associated with a decrease in autophagic capacity [131]. Several studies have also shown that modulation of autophagy sensitizes brain tumor cells to standard chemotherapy and radiotherapy induced death.

In several GBM cell lines, ceramide was shown to trigger autophagic death by enhancing the expression of the mitochondria-associated cell death protein BNIP3 that induces non-apoptotic cell death through decreasing mitochondrial membrane potential [132]. Similarly, treatment of malignant glioma cells with arsenic trioxide up-regulates BNIP3 to induce caspase-independent cell death [133, 134].

Interestingly, enforced expression of IL-24 in GBM cells induced autophagy, and this was shown to overlap with activation of the pro-apoptotic pathway culminating in cell death [135]. Ceramide was found to be important as a mediator of IL-24, promoted the generation of reactive oxygen species (ROS), and resulting in autophagic GBM cell death [136]. In particular, induced expression of IL-24 in GBM cells results in accumulation of IL-24 protein in the ER. A series of subsequent events follows, including activation of the stress kinase PERK (protein kinase R–like ER kinase), cytosolic Ca^{2+} increase, ceramide and ROS generation. PERK inhibition blocked ceramide generation, which was critical for Ca^{2+} elevation and subsequent ROS formation [136]. It was also demonstrated that histone deacetylase inhibitors enhance toxicity of IL-24 in invasive primary human GBM cells by synergizing with IL-24 killing effect [137]. Interestingly, enhanced lethality correlated with increased autophagy that was dependent on the expression of CerS6. In addition, exposure of human GBM cells to TMZ promoted autophagic cell death by the production of ceramide [138].

Similar to IL-24, de novo synthesis of ceramide has been shown as crucial for Δ9-tetrahydrocannabinol (THC)-triggered ER stress and GBM cell death [125]. The mechanism of THC antitumoral action was shown to rely on the ceramide-dependent activation of an ER stress-related signaling route that leads to the upregulation of the transcriptional coactivator p8 and its target the pseudokinase tribbles homologue 3 [139], which promoted autophagy and was indispensable for the pro-apoptotic and antitumoral action of THC [140, 141]. Moreover, the combined administration of THC and TMZ produced strong antitumoral action in glioma xenografts, even in TMZ-resistant GBMs, and activation of autophagy plays a crucial role on the mechanism of action of this drug combination too [142]. All these findings support the notion that the combined administration of TMZ and cannabinoids could be therapeutically exploited for the management of GBM.

Ceramide was also involved in antagonizing the protective role of autophagy. Indeed, in a recent study examining the effect of perifosine in GBM cells, Qin et al. [143] found that this Akt inhibitor induced a potent protective autophagy, and that inhibition of autophagy led to enhanced perifosine-induced cytotoxicity by apoptosis. Of relevance, when perifosine was administered in combination with C6-ceramide, it was able to induce GBM cell apoptosis. Thus a potential role for ceramide is emerging in the inhibition of oncogenic autophagy and the sensitization of glioma cells to apoptosis.

Of interest, increased level of sphingomyelin at the plasma membrane through activation of SMS was described as a further mechanism for inducing GBM cell autophagy. Indeed sphingomyelin accumulation at the plasma membrane was found crucial for inducing a sequence of events in GBM cells such as growth inhibition, differentiation and autophagic death [25]. These effects appeared to be mediated by perturbations of membrane properties consequent to increased sphingomyelin levels, which resulted in modifications of signal transduction pathways leading to decreased survival and increased pro-death signals [26].

9 Sphingolipids in GBM Cell Survival and Death Resistance

As described above, ceramide plays a key role as mediator of GBM cell death, and GBM cells appear to exhibit different strategies to avoid this. Among these mechanisms, ceramide down-regulation, which limits this critical pro-apoptotic mediator, appears crucial in GBM. The overexpression of Bcl2L13, a CerS inhibitor, does contribute to the low levels of pro-apoptotic ceramide species in GBM [27]. Notably, Bcl2L13, through its binding to and inhibiting CerS2 and CerS6 activity, acts as an anti-apoptotic protein by protecting the mitochondrial membrane integrity.

Besides keeping down their ceramide, GBM uses S1P to maintain and promote cell survival. In GBM cells S1P is a chief 'fighter' promoting survival pathways and antagonizing the players of death signaling, including those involving ceramide. In facts, the increase of S1P synthesis and level in both GBM cell lines and human GBM is critical strategy to survive and GBM enhances its death resistance by up-regulating S1P. Moreover, patients with GBM display a high expression levels of SphK1, and the expression of this enzyme is correlated with a significant poor patient survival [28].

Overall it appears that the ceramide/S1P imbalance confers to GBM cells a potent mechanism for protection against death, sustaining their survival propensity (Fig. 6).

Phospholipase D (PLD) is a recognized survival mediator in GBM, and promotes survival signaling through direct regulation of Akt [144, 145] demonstrated that PLD is implicated in the ceramide-induced apoptotic process in C6 glioma cells,

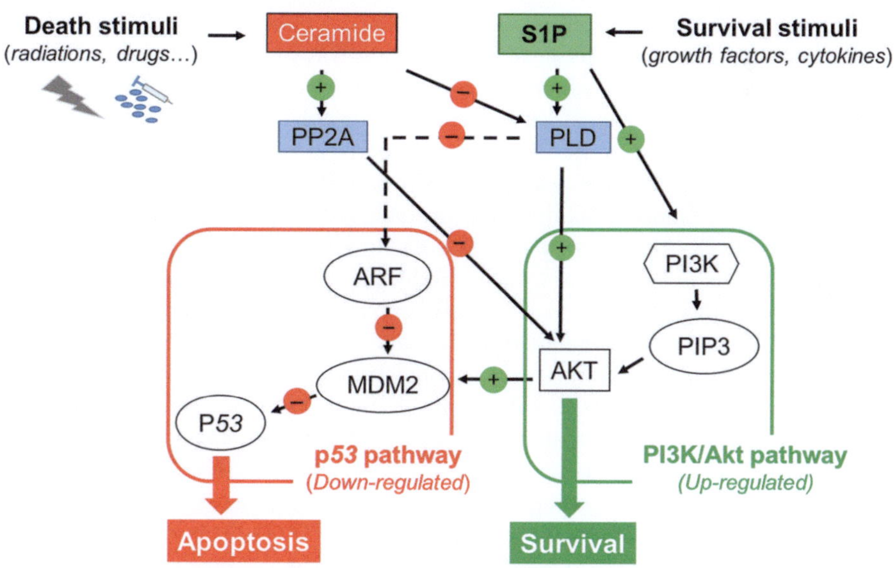

Fig. 6 Antagonism between S1P and ceramide in apoptosis/survival of GBM and their cross-talk with GBM deregulated signaling pathways controlling cell fate

showing that ceramide exerts multiple effects to down-regulate PLD and thus favor cell death. Indeed, ceramide was able to inhibit the GTPγS-dependent activation of PLD, to suppress the ADP-ribosylation factor 1 activation of membrane-associated PLD, as well as to down-regulate PLD expression.

Alterations in signaling are of crucial importance in the biology of GBM, and overall activation of signal transduction pathways differs markedly between GBM and the normal brain. A chief pathway altered in GBM is the PI3K/Akt/PTEN signaling. Mutations of PTEN are found in approximately 70–90 % of GBM [40], and particularly in primary GBM. Due to this loss, the PI3K/Akt pathway is up-regulated, and Akt appears to control crucial functions related to GBM malignant features, including their resistance to cytotoxic treatments [14, 40].

An inverse correlation between ceramide level and Akt activation has been reported in human GBM cells when exposed to ceramide, suggesting that ceramide-induced Akt inactivation is one of the contributing mechanisms by which ceramide causes GBM cell death [146]. In agreement, in human GBM cells, ceramide was shown to inhibit the constitutive phosphorylation and activity of the pro-survival kinase Akt [146] through the activation of protein phosphatase 2A. Of note, Akt activation promotes the ER to Golgi vesicular transport of ceramide in human GBM cells, resulting in a reduction in the ceramide levels at the ER and increased synthesis of complex sphingolipids [57]. Thus it appears that ceramide-induced cell death may be inhibited by Akt activation, suggesting the survival role of PI3K/Akt signaling in GBM may reside, at least in part, on its facilitating effects on ceramide transport and thus metabolic utilization.

It is recognized that S1P can exert its biological effects both intracellularly, by acting as a second messenger, and in the extracellular milieu, where it mainly acts as a ligand for the specific cell surface G-protein coupled receptors S1P receptors [147]. High expression of $S1P_1$ in GBM specimens was found to correlate with poor patient prognosis [36, 37]. Moreover, blocking S1P receptors in GBM cells was revealed as an efficient pro-apoptotic strategy. Indeed, the S1P functional antagonist FTY720 induced apoptosis of GBM cells by promoting FAK dephosphorylation, cutting off the FAK-PI3K pathway, and finally activating caspase-6 [148].

Opposite to ceramide's actions, S1P exerts a protective role in GBM cells by activating the PI3K/Akt pathway. Indeed, SKI, an SphK1 inhibitor induced apoptosis in GBM cells [149], was shown to inhibit Akt phosphorylation, and this was functional to reduce tumor growth rate and enhance mouse survival in orthotopic GBMs [59]. As a mechanism underlying the anti-apoptotic role of S1P, Guan et al. [150] showed that SphK1 action blocks apoptosis in GBM cells by down-regulation of Bim, a pro-apoptotic Bcl-2 family member, and this was mediated through activation of PI3K/Akt/FOXO3a signaling. The final effect was suppression of UV radiation- and adriamycin-induced apoptosis, i.e., promotion of GBM cell survival [150]. Moreover, through binding to $S1P_3$, exogenous S1P was shown to stimulate the phospholipase C/Ca^{2+} system and phospholipase D (PLD) in C6 glioma cells [151], thus opposing the effect of ceramide and favoring cell survival.

Multidrug resistance poses a major problem in the treatment of GBM. Besides its intrinsic resistance to therapeutic stimuli toxicity, GBM acquires resistance to the chemotherapeutic(s) used during treatment. To do so, it adopts new strategies to limit or circumvent death, and finally to exhibit enhanced death resistance.

Ceramide metabolism in malignant cells has gained considerable interest as a key contributor to chemoresistance. The upregulation of GlcCer synthase has been implicated as a major chemoresistance mechanism.

One key mechanism adopted by GBM cells to avoid drug toxicity is the rapid consumption of ceramide to prevent cell death. In particular, ceramide consumption pathways are mediated by A-ceramidase and GlcCer synthase, which lead to formation of sphingosine (and then S1P) and GlcCer, respectively.

Some studies reported that ceramide consumption by GlcCer synthase is responsible for the resistance of GBM cells to some chemotherapeutic drugs, and that inhibition of GlcCer synthase restores sensitivity of resistant GBM cells. Dumitru et al. [122] showed that gemcitabine triggers ceramide accumulation and cell death in the bulk population of GBM cells, but not in cells selected for resistance to the drug. It was found that resistant cells rapidly consume ceramide upon gemcitabine treatment, and the activity of GlcCer synthase mediates ceramide consumption. In addition, selection of drug-resistant GBM cell by gradual exposure to increasing drug concentrations, revealed an up-regulation of GlcCer synthase protects GBM cells against autophagic and apoptotic death induced by TMZ and paclitaxel, and contributes to their chemoresistance [138].

Some studies have shown the correlation between SphK1 and S1P in chemoresistance due to ceramide/S1P imbalance. SphK1, S1P and its receptors play a vital role in induction of cancer chemoresistance. Selection of TMZ-resistant GBM cells has been shown to be associated not only with an increase in GlcCer synthase [138], but also with SphK1 expression [149]. However, in spite of GSC inhibition hindering resistance to TMZ-induced autophagic death [138], SphK1 down-regulation was found to have a similar proapoptotic effect on TMZ-sensitive and -resistant cells [149]. A possible explanation of this difference may reside in the heterogeneity of GBM cells, and particularly in the different intrinsic resistance of the GBM models used in these studies, which is also reflected by the different mechanisms of death, autophagic and apoptotic, exerted by TMZ and SphK1 inhibition.

Overall, these studies on cell drug sensitivity suggest that GBM cells take advantage of ceramide-removing and S1P-enriching pathways in order to survive the stress of chemotherapy.

10 Clinical Aspects

10.1 Sphingolipids and Sphingolipid-Related Enzymes as Possible Diagnostic/Prognostic Markers

As reported above, diverse alterations of sphingolipid signaling have been described in tumor specimens from GBM patients and some of them emerged as potential markers for clinical prognosis (Table 1).

Table 1 Sphingolipid levels and sphingolipid-related proteins correlated with diagnosis, prognosis and survival in GBM patients

Sphingolipids and related proteins	Variation	Clinical correlation	Note	References
Ceramide	Decreased levels[a,b]	Poor patient survival	Inversely related to histological grade	[17]
			Specific reduction in C18-ceramide	[18]
S1P	Increased levels[b]	Poor patient survival	Increase in S1P content with increasing glioma grade	[18]
Acid ceramidases	Increased expression[b]	?	Inversely related to histological grade	[18]
CerS2/6	Decreased activity[a]	?	Decrease is parallel by over-expression of the CerS inhibitor Bcl2L13	[27]
SphK1	Increased expression[a,b]	Poor patient survival	Directly related to histological grade; positive correlation between SphK1 mRNA and S1P level	[18, 28–30, 36, 37, 59]
SphK2	Increased expression in primary GBM; decreased expression in recurrent and secondary GBM[a]	Poor prognosis in patient with primary GBM	SphK2 decrease[b] was observed in GBM by [18]	[30]
SGGP	Decreased expression[b]	Poor patient survival	Inversely related to S1P levels	[18]
S1P$_{1-3}$	Increased expression in primary, through recurrent, to secondary GBM[a,b]	Poor patient survival	In contrast, S1P1 was found down-regulated and inversely correlated to patient survival by Yoshida et al. [36, 37]	[18] [30]
S1P$_{4}$	High and exclusive expression in GSCs	?	Possible indicator of GBM stemness	[88]
S1P$_{5}$	Increased expression in primary, through recurrent, to secondary GBM[a,b]	Poor patient survival	S1P5 was identified as an independent prognostic factor	[18, 30]

[a]vs. healthy brain
[b]vs. normal gray matter

Analysis of ceramide expression in GBM showed an inverse correlation between ceramide level and disease progression as well as patient survival [17], suggesting the measure of ceramide level could be a tool in GBM prognostication. Although Abuhusain et al. [18] reported an increase of A-ceramidase in GBM, so far the role of the enzymes/proteins involved in ceramide metabolism and their potential use in clinical prognosis remains to be investigated.

Clinical correlation supports S1P as a GBM promoter. Accordingly, a marked increase (about tenfold) in S1P level was shown in GBM tissues when compared to normal gray matter [18]. This elevation, together with increased expression of SphK1 strongly correlates with short survival rate in patients with GBM [28, 29, 37, 59]. In addition, a higher expression of $S1P_5$ was directly associated with poor survival in GBM patients, and among different receptors $S1P_5$ was found as an independent prognostic factor in multivariate analysis [30].

10.2 GBM Therapy: The Potential Application of Sphingolipid Targeting Drugs in GBM

GBM remains a major health issue, its localization in the brain, its invasive behavior, and its extreme therapy resistance making it one of the most dreaded forms of cancer. The overall median survival time is as short as 12–14 months [2, 3] despite aggressive therapy, actually including the combined actions of neurosurgery, radiation, and chemotherapeutic drugs. The alkylating cytostatic drug TMZ constitutes, in combination with radiotherapy, the current standard of care for GBM [152, 153]. However, the action of TMZ may be counteracted in the tumors by expression of the DNA repair enzyme MGMT which repairs the TMZ-induced DNA lesion [154], and almost all GBM patients undergo tumor recurrence, 90 % of GBMs recurring at the primary site [7].

A lot of effort has been made in the last decades to identify molecular targets and signaling pathways to which GBM cells are 'addicted', and subsequently to render these cells more susceptible towards therapy-induced death. However, the effectiveness of current therapies is limited resulting in disease recurrence, and interventions are complicated by resistance mechanisms towards therapeutic agents building up a threshold for therapeutic efficacy. The highly infiltrative nature of GBM makes complete resection with clean margins nearly impossible, and appears a major reason for recurrence. In addition, extensive regions of hypoxia in GBMs limit the efficacy of radiotherapy by decreasing the generation of DNA-damaging free radicals [155].

The slow development of an effective therapy is in sharp contrast to the rapidly growing knowledge of the molecular pathogenesis of the disease. In the last years, the key for having an improved and greater control on GBM growth and relapse emerged to be the development of a therapeutic that is able to kill GBM cells efficiently by modulating critical signaling pathways on which GBM cells rely for their survival. However, different attempts at targeting a single molecule or pathway revealed modest efficacy or failure, possibly due to molecular heterogeneity among patients, within a single tumor and the existence of parallel or compensatory pathways [156]. It is now believed that truly effective therapies will result from the use of divergent targeting approach, with complementary combinations of targeted agents, or combination of targeted agents with other treatment modalities, and also with patient-specific treatments [157].

It is in this complex and intricate context that novel therapeutic strategies against GBM have been proposed to target sphingolipids. Based on the knowledge that GBM cells enforce their aggressiveness and develop resistance to therapy by arming themselves with the abilities to enhance elevations of S1P and to avoid of those of ceramide, targeting different players of sphingolipid signaling is emerging to be a promising strategy to improve the response to cancer therapy. Different molecules have been proven to antagonize the disrupted metabolism of ceramide and S1P in cancer; these are different agents that, by down-regulating S1P and/or up-regulating ceramide are able to exert toxic effects in GBM.

As discussed above, ceramide metabolism is tightly regulated in GBM cells, elevated ceramide often resulting from treatment with anticancer drugs and also irradiation, and being important for modulating responses to therapy. Moreover, limiting ceramide generation is one of the mechanisms exploited by GBM cells in order to evade therapeutic treatments. This supports the rationale for the potential benefits of a ceramide-based chemotherapy in patients harboring GBM.

The development of ceramide-based therapies has been somewhat limited by the innate resistance of cancer cells, which is driven by the up-regulated metabolic clearance of Cer or by its conversion to pro-mitogenic and pro-survival S1P, making it difficult to attain therapeutic levels of Cer. In this context, the increase of GlcCer synthase represents an important strategy to survive in GBM. It was shown that drug-resistant GBM cells have higher levels of GlcCer than sensitive ones, and GlcCer synthase inhibition in vitro and in vivo can restore sensitivity of drug-resistant GBM cells to cytotoxic insults of different proapoptotic and proautophagic drugs [138, 158, 159]. The mechanism by which GlcCer synthase might regulate drug sensitivity biochemically remains to be clarified. At present, two mechanisms appear feasible. First, based on the critical death-inducing role of ceramide, it is likely that the inhibition of this enzyme results in an increase in ceramide, which acts as death mediator. However it is also conceivable that GlcCer synthase inhibition results in the failure of the formation of GlcCer, which by itself may act as pro-survival mediator [160].

Overall, as described above, targeting the proteins that regulate the ceramide level, as the trio composed of GlcCer synthase, A-ceramidase and SMS by agents such as *N*-(*n*-nonyl)deoxygalactonojirimycin, cannabinoids and 2OHOA appears promising as a useful tool to manipulate and halt the pro-survival GBM properties and to induce GBM cell death. It is worth noting that a large number of pharmacological and molecular approaches have been investigated to increase ceramide and to modulate sphingolipid signaling in different cancer cells [107, 161, 162], but the potential action of most of them remains to be investigated in GBMs.

As discussed above, S1P signaling is implicated in multiple GBM cell properties, which are associated with their hallmarks. Due to these important roles of S1P, targeting its signaling appears a potential target for cancer therapy. Since S1P is generated from sphingosine by SphKs, decreasing SphK activity and the factors that regulate the balance of the ceramide-S1P rheostat towards ceramide may be candidates for anti-cancer drug development. Among them the SphK1 inhibitor SKI [163] was shown to inhibit colony formation and activate caspase-3 in both

TMZ-sensitive and -resistant cells [149], and to prevent tumor growth and induce apoptosis in GBM xenografts [59]. However in a recent study it was shown that the targeting GBM with novel, more effective inhibitors of SphKs is without relevant effect on GBM cell viability [164], leading to a contrasting scenario, and suggesting uncertainty of non-in vivo approaches.

The S1P specific monoclonal antibody LT1009 has been formulated for phase I clinical trials in cancer, and was shown to reduce tumor progression and eliminate measurable tumors in murine xenograft and allograft models [165]. Even if not yet tested in GBM, this antibody should be taken in consideration as effective for GBM therapy.

One of the most promising treatment options for patients with GBM appears to be the sphingosine analogue FTY720 (Fingolimod®), currently used in clinical trials as an immunosuppressant in multiple sclerosis and organ transplant rejection [166]. FTY720 is phosphorylated by SphK2 in vivo, and down-regulates the expression of S1P receptors [96]. FTY720 has emerged as a potential effective drug for several cancers, causing suppression of tumor growth, as well as induction of apoptosis in multiple tumor cell types [167], including GBM cells and stem cells [86, 95, 97]. In spite the down-regulation of S1P receptors appears of relevance in the anti-cancer effects of FTY720, this drug is also able to exert effects on S1P formation, by promoting the proteolysis of SK1 [168]. It is worth mentioning that one mechanism contributing to the effectiveness of FTY720 in many cancer models is its ability to down-regulate nutrient transporters also via ceramide generation [169]. Besides its capacity to antagonize the survival propensity of GBM, FTY720 exhibits different features, such as low toxicity and the ability to cross the blood–brain barrier and to accumulate at high levels in the brain [170], which appear of relevance for making it an attractive candidate for GBM therapy.

11 Conclusions

Taken together, evidence from numerous studies indicates that SLs play significant roles in determining GBM cell properties and fate. However, there are still several discrepancies as well as unanswered questions. Among them, little is known about the mechanisms underlying ceramide and S1P dysregulation, and if or how GBM cells can compensate for the lack of a single component of their signaling. It is also unclear how the multiplicity of functional actions of S1P is utilized by GBM as signal enhancement for their microenvironment, and the possible molecular circuits that integrate sphingolipids with other cell-regulatory elements of GBM. Last, but not least, it remains largely unknown what the impact and mechanisms of sphingolipid signaling on the intra- and inter-tumor heterogeneity of GBMs is, and if this further strengthens tumor growth and increases chances to escape therapy.

Although hard work is required for unraveling the multiple roles of the sphingolipid machinery malfunctions in GBMs, sphingolipid targeting strategies appear promising not only due to the observed beneficial effects in GBM models, but also due to the multi-functional potential of these approaches. Indeed, the multifaceted metabolism and the multitarget regulatory properties of ceramide and S1P, together with their intricate interconnection appear promising target qualities for helping 'terminate' what once was defined—and until now remains—'the terminator' [11]. It is a future challenge to unravel how the multiple features of sphingolipid signaling could be effectively manipulated as strategies to halt or slow down the multiforme aggressiveness of GBMs, particularly if applied in combination with other drugs.

In conclusion, a dedicated understanding of sphingolipid aberrations in GBM cells and GSCs is mandatory for a better knowledge of the biology of GBM, for identifying novel vulnerabilities, for developing effective therapeutic strategies, and, hopefully, for eradicating, or at least slowing, this lethal cancer. Rigorous scientific methods will be needed to do so, and more definitive solutions will depend on the ability of sphingolipid basic science to point diagnosis and treatment into these new directions. A major challenge remains to be the translation of pre-clinical findings to clinical trials for GBMs. Although the challenges are still substantial, if we build on past advances, there is every reason for optimism that novel breakthrough in the sphingolipid scenario of GBM will occur in the foreseeable future, with an impact.

References

1. Siegel R, Ward E, Brawley O, Jemal A (2011) Cancer statistics, 2011: the impact of eliminating socioeconomic and racial disparities on premature cancer deaths. CA Cancer J Clin 61:212e36
2. Johnson DR, O'Neill BP (2012) Glioblastoma survival in the United States before and during the temozolomide era. J Neurooncol 107:359–364
3. Serwer LP, James CD (2012) Challenges in drug delivery to tumors of the central nervous system: an overview of pharmacological and surgical considerations. Adv Drug Deliv Rev 64:590–597
4. Krex D, Klink B, Hartmann C et al (2007) Long-term survival with glioblastoma multiforme. Brain 130:2596–2606
5. Henriksson R, Asklund T, Poulsen HS (2011) Impact of therapy on quality of life, neurocognitive function and their correlates in glioblastoma multiforme: a review. J Neurooncol 104:639–646
6. Dick JE (2008) Stem cell concepts renew cancer research. Blood 112:4793–4807
7. Wen PY, Kesari S (2008) Malignant gliomas in adults. N Engl J Med 359:492–507
8. Brennan C, Momota H, Hambardzumyan D et al (2009) Glioblastoma subclasses can be defined by activity among signal transduction pathways and associated genomic alterations. PLoS One 4, e7752
9. Huse JT, Holland EC (2010) Targeting brain cancer: advances in the molecular pathology of malignant glioma and medulloblastoma. Nat Rev Cancer 10:319–331
10. Siebzehnrubl FA, Reynolds BA, Vescovi A et al (2011) The origins of glioma: e pluribus unum? Glia 59:1135–1147
11. Holland EC (2000) Glioblastoma multiforme: the terminator. Proc Natl Acad Sci U S A 97:6242–6244
12. Maher EA, Furnari FB, Bachoo RM et al (2001) Malignant glioma: genetics and biology of a grave matter. Genes Dev 15:1311–1333

13. Charles NA, Holland EC, Gilbertson R et al (2011) The brain tumor microenvironment. Glia 59:1169–1180
14. Mao H, LeBrun DG, Yang J et al (2012) Deregulated signaling pathways in glioblastoma multiforme: molecular mechanisms and therapeutic targets. Cancer Invest 30:48–56
15. Hannun YA, Obeid LM (2008) Principles of bioactive lipid signalling. Lessons from sphingolipids. Nat Rev Mol Cell Biol 9:139–150
16. Cuvillier O, Pirianov G, Kleuser B et al (1996) Suppression of ceramide-mediated programmed cell death by sphingosine-1-phosphate. Nature 381:800–803
17. Riboni L, Campanella R, Bassi R et al (2002) Ceramide levels are inversely associated with malignant progression of human glial tumors. Glia 39:105–113
18. Abuhusain HJ, Matin A, Qiao Q et al (2013) A metabolic shift favoring sphingosine 1-phosphate at the expense of ceramide controls glioblastoma angiogenesis. J Biol Chem 288:37355–37364
19. Merrill AH Jr, Stokes TH, Momin A et al (2009) Sphingolipidomics: a valuable tool for understanding the roles of sphingolipids in biology and disease. J Lipid Res 50:S97–S102
20. Hannun YA, Obeid LM (2011) Many ceramides. J Biol Chem 286:27855–27862
21. Park JW, Park WJ, Futerman AH (2013) Ceramide synthases as potential targets for therapeutic intervention in human diseases. Biochim Biophys Acta 1841:671–681
22. Sullards MC, Wang E, Peng Q, Merrill AH Jr (2003) Metabolomic profiling of sphingolipids in human glioma cell lines by liquid chromatography tandem mass spectrometry. Cell Mol Biol (Noisy-le-Grand) 49:789–797
23. Karahatay S, Thomas K, Koybasi S et al (2007) Clinical relevance of ceramide metabolism in the pathogenesis of human head and neck squamous cell carcinoma (HNSCC): attenuation of C(18)-ceramide in HNSCC tumors correlates with lymphovascular invasion and nodal metastasis. Cancer Lett 256:101–111
24. Campanella R (1992) Membrane lipid modifications in human gliomas of different degree of malignancy. J Neurosurg Sci 36:11–25
25. Barceló-Coblijn G, Martin ML, de Almeida RF et al (2011) Sphingomyelin and sphingomyelin synthase (SMS) in the malignant transformation of glioma cells and in 2-hydroxyoleic acid therapy. Proc Natl Acad Sci U S A 108:19569–19574
26. Terés S, Lladó V, Higuera M et al (2012) 2-Hydroxyoleate, a nontoxic membrane binding anticancer drug, induces glioma cell differentiation and autophagy. Proc Natl Acad Sci U S A 109:8489–8494
27. Jensen SA, Calvert AE, Volpert G et al (2014) Bcl2L13 is a ceramide synthase inhibitor in glioblastoma. Proc Natl Acad Sci U S A 111:5682–5687
28. Van Brocklyn JR, Jackson CA, Pearl DK et al (2005) Sphingosine kinase-1 expression correlates with poor survival of patients with glioblastoma multiforme: roles of sphingosine kinase isoforms in growth of glioblastoma cell lines. J Neuropathol Exp Neurol 64:695–705
29. Li J, Guan HY, Gong LY et al (2008) Clinical significance of sphingosine kinase-1 expression in human astrocytomas progression and overall patient survival. Clin Cancer Res 14: 6996–7003
30. Quint K, Stiel N, Neureiter D et al (2014) The role of sphingosine kinase isoforms and receptors S1P1, S1P2, S1P3, and S1P5 in primary, secondary, and recurrent glioblastomas. Tumour Biol 35(9):8979–8989. doi:10.1007/s132770142172
31. Ogawa C, Kihara A, Gokoh M, Igarashi Y (2003) Identification and characterization of a novel human sphingosine-1-phosphate phosphohydrolase, hSPP2. J Biol Chem 278: 1268–1272
32. Le Stunff H, Giussani P, Maceyka M et al (2007) Recycling of sphingosine is regulated by the concerted actions of sphingosine-1-phosphate phosphohydrolase 1 and sphingosine kinase 2. J Biol Chem 282:34372–34380
33. Steck PA, Ligon AH, Cheong P et al (1995) Two tumor suppressive loci on chromosome 10 involved in human glioblastomas. Genes Chromosomes Cancer 12:255–261
34. Mora R, Dokic I, Kees T et al (2010) Sphingolipid rheostat alterations related to transformation can be exploited for specific induction of lysosomal cell death in murine and human glioma. Glia 58:1364–1383

35. Rosen H, Goetzl EJ (2005) Sphingosine-1-phosphate and its receptors: an autocrine and paracrine network. Nat Rev Immunol 5:560–570
36. Yoshida Y, Nakada M, Sugimoto N et al (2010) Sphingosine-1-phosphate receptor type 1 regulates glioma cell proliferation and correlates with patient survival. Int J Cancer 126: 2341–2352
37. Yoshida Y, Nakada M, Harada T et al (2010) The expression level of sphingosine-1-phosphate receptor type 1 is related to MIB-1 labeling index and predicts survival of glioblastoma patients. J Neurooncol 98:41–47
38. Hanahan D, Weinberg RA (2011) Hallmarks of cancer: the next generation. Cell 144: 646–674
39. Furnari FB, Fenton T, Bachoo RM et al (2007) Malignant astrocytic glioma: genetics, biology, and paths to treatment. Genes Dev 21:2683–2710
40. Ohgaki H, Kleihues P (2009) Genetic pathways to primary and secondary glioblastoma. Am J Pathol 170:1445–1453
41. Persano L, Rampazzo E, Basso G, Viola G (2013) Glioblastoma cancer stem cells: role of the microenvironment and therapeutic targeting. Biochem Pharmacol 85:612–622
42. Al-Hajj M, Becker MW, Wicha M et al (2004) Therapeutic implications of cancer stem cells. Curr Opin Genet Dev 14:43–47
43. Oliver TG, Wechsler-Reya RJ (2004) Getting at the root and stem of brain tumors. Neuron 42:885–888
44. Holthuis JC, Pomorski T, Raggers RJ et al (2001) The organizing potential of sphingolipids in intracellular membrane transport. Physiol Rev 81:1689–1723
45. Lebman DA, Spiegel S (2008) Cross-talk at the crossroads of sphingosine-1-phosphate, growth factors, and cytokine signaling. J Lipid Res 49:1388–1394
46. Bassi R, Anelli V, Giussani P (2006) Sphingosine-1-phosphate is released by cerebellar astrocytes in response to bFGF and induces astrocyte proliferation through Gi-protein-coupled receptors. Glia 53:621–630
47. Van Brocklyn JR, Letterle CA, Snyder PJ, Prior TW (2002) Sphingosine-1-phosphate stimulates human glioma cell proliferation through Gi-coupled receptors: role of ERK MAP kinase and phosphatidylinositol 3-kinase β. Cancer Lett 181:195–204
48. Swartling FJ, Hede S-M, Weiss WA (2013) What underlies the diversity of brain tumors? Cancer Metastasis Rev 32:5–24
49. Spiegel S, Milstien S (2003) Sphingosine-1-phosphate: an enigmatic signalling lipid. Nat Rev Mol Cell Biol 4:397–407
50. Sukocheva O, Wadham C, Holmes A et al (2006) Estrogen transactivates EGFR via the sphingosine 1-phosphate receptor Edg-3: the role of sphingosine kinase-1. J Cell Biol 173: 301–310
51. Riboni L, Viani P, Bassi R et al (2000) Biomodulatory role of ceramide in basic fibroblast growth-factor induced proliferation of cerebellar astrocytes in primary culture. Glia 32: 137–145
52. LaMontagne K, Littlewood-Evans A, Schnell C et al (2006) Antagonism of sphingosine-1-phosphate receptors by FTY720 inhibits angiogenesis and tumor vascularization. Cancer Res 66:221–231
53. Van Brocklyn JR, Young N, Roof R (2003) Sphingosine-1-phosphate stimulates motility and invasiveness of human glioblastoma multiforme cells. Cancer Lett 199:53–60
54. Lee H, Deng J, Kujawski M et al (2010) STAT3-induced S1PR1 expression is crucial for persistent STAT3 activation in tumors. Nat Med 16:1421–1428
55. Young N, Van Brocklyn JR (2007) Roles of sphingosine-1-phosphate (S1P) receptors in malignant behavior of glioma cells. Differential effects of S1P2 on cell migration and invasiveness. Exp Cell Res 313:1615–1627
56. Radeff-Huang J, Seasholtz TM, Chang JW et al (2007) Tumor necrosis factor-α-stimulated cell proliferation is mediated through sphingosine kinase-dependent Akt activation and cyclin D expression. J Biol Chem 282:863–870
57. Giussani P, Brioschi L, Bassi R et al (2009) Phosphatidylinositol 3-kinase/AKT pathway regulates the endoplasmic reticulum to Golgi traffic of ceramide in glioma cells: a link

between lipid signaling pathways involved in the control of cell survival. J Biol Chem 284:5088–5096
58. Viani P, Giussani P, Brioschi L et al (2003) Ceramide in nitric oxide inhibition of glioma cell growth. Evidence for the involvement of ceramide traffic. J Biol Chem 278:9592–9601
59. Kapitonov D, Allegood JC, Mitchell C et al (2009) Targeting sphingosine kinase 1 inhibits Akt signaling, induces apoptosis, and suppresses growth of human glioblastoma cells and xenografts. Cancer Res 69:6915–6923
60. Lu T, Tian L, Han Y et al (2007) Dose-dependent cross-talk between the transforming growth factor-β and interleukin-1 signaling pathways. Proc Natl Acad Sci U S A 104:4365–4370
61. Paugh BS, Bryan L, Paugh SW et al (2009) Interleukin-1 regulates the expression of sphingosine kinase 1 in glioblastoma cells. J Biol Chem 284:3408–3417
62. Zhang H, Li W, Sun S et al (2012) Inhibition of sphingosine kinase 1 suppresses proliferation of glioma cells under hypoxia by attenuating activity of extracellular signal-regulated kinase. Cell Prolif 45:167–175
63. Pyne NJ, Tonelli F, Lim KG et al (2012) Sphingosine 1-phosphate signalling in cancer. Biochem Soc Trans 40:94–100
64. Glas M, Rath BH, Simon M et al (2010) Residual tumor cells are unique cellular targets in glioblastoma. Ann Neurol 68:264–269
65. Winkler F, Kienast Y, Fuhrmann M et al (2009) Imaging glioma cell invasion in vivo reveals mechanisms of dissemination and peritumoral angiogenesis. Glia 57:1306–1315
66. Kalokhe G, Grimm SA, Chandler JP et al (2012) Metastatic glioblastoma: case presentations and a review of the literature. J Neurooncol 107:21–27
67. Giese A, Bjerkvig R, Berens ME, Westphal M (2003) Cost of migration: invasion of malignant gliomas and implications for treatment. J Clin Oncol 21:1624–1636
68. Czekay RP, Aertgeerts K, Curriden SA, Loskutoff DJ (2003) Plasminogen activator inhibitor-1 detaches cells from extracellular matrices by inactivating integrins. J Cell Biol 160: 781–791
69. Malchinkhuu E, Sato K, Maehama T et al (2008) S1P2 receptors mediate inhibition of glioma cell migration through Rho signaling pathways independent of PTEN. Biochem Biophys Res Commun 366:963–968
70. Malchinkhuu E, Sato K, Horiuchi Y et al (2005) Role of p38 mitogen-activated kinase and c-Jun terminal kinase in migration response to lysophosphatidic acid and sphingosine-1-phosphate in glioma cells. Oncogene 24:6676–6688
71. Van Brocklyn JR (2007) Sphingolipid signaling pathways as potential therapeutic targets in gliomas. Mini Rev Med Chem 7:984–990
72. Bryan L, Paugh BS, Kapitonov D et al (2008) Sphingosine-1-phosphate and interleukin-1 independently regulate plasminogen activator inhibitor-1 and urokinase-type plasminogen activator receptor expression in glioblastoma cells: implications for invasiveness. Mol Cancer Res 6:1469–1477
73. Walsh CT, Radeff-Huang J, Matteo R et al (2008) Thrombin receptor and RhoA mediate cell proliferation through integrins and cysteine-rich protein 61. FASEB J 22:4011–4021
74. Young N, Pearl DK, Van Brocklyn JR (2009) Sphingosine-1-phosphate regulates glioblastoma cell invasiveness through the urokinase plasminogen activator system and CCN1/Cyr61. Mol Cancer Res 7:23–32
75. Azoitei N, Kleger A, Schoo N et al (2011) Protein kinase D2 is a novel regulator of glioblastoma growth and tumor formation. Neuro Oncol 13:710–724
76. Bernhart E, Damma S, Wintersperger A et al (2013) Protein kinase D2 regulates migration and invasion of U87MG glioblastoma cells in vitro. Exp Cell Res 319:2037–2048
77. Muracciole X, Romain S, Dufour H et al (2002) PAI-1 and EGFR expression in adult glioma tumors: toward a molecular prognostic classification. Int J Radiat Oncol Biol Phys 52: 592–598
78. Paugh BS, Paugh SW, Bryan L et al (2008) EGF regulates plasminogen activator inhibitor-1 by a pathway involving c-Src, PKCδ, and sphingosine kinase 1 in glioblastoma cells. FASEB J 22:455–465

79. Vescovi AL, Galli R, Reynolds BA (2006) Brain tumour stem cells. Nat Rev Cancer 6:425–436
80. Cheng L, Wu Q, Guryanova OA et al (2011) Elevated invasive potential of glioblastoma stem cells. Biochem Biophys Res Commun 406:643–648
81. Lathia JD, Gallagher J, Myers JT et al (2011) Direct in vivo evidence for tumor propagation by glioblastoma cancer stem cells. PLoS One 6, e24807
82. Chen R, Nishimura MC, Bumbaca SM et al (2010) A hierarchy of self-renewing tumor-initiating cell types in glioblastoma. Cancer Cell 17:362–375
83. Charles NA, Holland EC, Gilbertson R, Glass R, Kettenmann H (2012) The brain tumor microenvironment. Glia 60:502–514
84. Filatova A, Acker T, Garvalov BK (2013) The cancer stem cell niche(s): the crosstalk between glioma stem cells and their microenvironment. Biochim Biophys Acta 1830:2496–2508
85. Riccitelli E, Giussani P, Di Vito C et al (2013) Extracellular sphingosine-1-phosphate: a novel actor in human glioblastoma stem cell survival. PLoS One 8, e68229
86. Marfia G, Campanella R, Navone SE et al (2014) Autocrine/paracrine sphingosine-1-phosphate fuels proliferative and stemness qualities of glioblastoma stem cells. Glia. doi:10.1002/glia.22718
87. Anelli V, Bassi R, Tettamanti G et al (2005) Extracellular release of newly synthesized sphingosine-1-phosphate by cerebellar granule cells and astrocytes. J Neurochem 92: 1204–1215
88. Anelli V, Gault CR, Snider AJ, Obeid LM (2010) Role of sphingosine kinase-1 in paracrine/transcellular angiogenesis and lymphangiogenesis in vitro. FASEB J 24:2727–2738
89. Kimura A, Ohmori T, Ohkawa R et al (2007) Essential roles of sphingosine 1-phosphate/S1P1 receptor axis in the migration of neural stem cells toward a site of spinal cord injury. Stem Cells 25:115–124
90. Seidel S, Garvalov BK, Wirta V et al (2010) A hypoxic niche regulates glioblastoma stem cells through hypoxia inducible factor 2 alpha. Brain 133:983–995
91. Annabi B, Lachambre MP, Plouffe K et al (2009) Modulation of invasive properties of CD133+ glioblastoma stem cells: a role for MT1-MMP in bioactive lysophospholipid signaling. Mol Carcinog 48:910–919
92. Li G, Chen Z, Hu YD et al (2009) Autocrine factors sustain glioblastoma stem cell self-renewal. Oncol Rep 21:419–424
93. Soeda A, Inagaki A, Oka N et al (2008) Epidermal growth factor plays a crucial role in mitogenic regulation of human brain tumor stem cells. J Biol Chem 283:10958–10966
94. Stockhausen MT, Kristoffersen K, Stobbe L, Poulsen HS (2014) Differentiation of glioblastoma multiforme stem-like cells leads to downregulation of EGFR and EGFRvIII and decreased tumorigenic and stem-like cell potential. Cancer Biol Ther 15:216–224
95. Estrada-Bernal A, Lawler SE, Nowicki MO et al (2011) The role of sphingosine kinase-1 in EGFRvIII-regulated growth and survival of glioblastoma cells. J Neurooncol 102:353–366
96. Gräler MH, Goetzl EJ (2004) The immunosuppressant FTY720 down-regulates sphingosine 1-phosphate G-protein-coupled receptors. FASEB J 18:551–553
97. Estrada-Bernal A, Palanichamy K, Ray Chaudhury A, Van Brocklyn JR (2012) Induction of brain tumor stem cell apoptosis by FTY720: a potential therapeutic agent for glioblastoma. Neuro Oncol 14:405–415
98. Linkous AG, Yazlovitskaya EM (2011) Angiogenesis in glioblastoma multiforme: navigating the maze. Anticancer Agents Med Chem 11:712–718
99. Bulnes S, Bengoetxea H, Ortuzar N et al (2012) Angiogenic signalling pathways altered in gliomas: selection mechanisms for more aggressive neoplastic subpopulations with invasive phenotype. J Signal Trans 2012, e597915
100. Krock BL, Skuli N, Simon MC (2011) Hypoxia-induced angiogenesis: good and evil. Genes Cancer 2:1117–1133
101. Onishi M, Ichikawa T, Kurozumi K, Date I (2011) Angiogenesis and invasion in glioma. Brain Tumor Pathol 28:13–24
102. Ader I, Brizuela L, Bouquerel P et al (2008) Sphingosine kinase 1: a new modulator of hypoxia inducible factor 1α during hypoxia in human cancer cells. Cancer Res 68:8635–8642

103. Anelli V, Gault CR, Cheng AB, Obeid LM (2008) Sphingosine kinase 1 is up-regulated during hypoxia in U87MG glioma cells. Role of hypoxia-inducible factors 1 and 2. J Biol Chem 283:3365–3375
104. Liu Y, Wada R, Yamashita T et al (2000) Edg-1, the G protein-coupled receptor for sphingosine-1-phosphate, is essential for vascular maturation. J Clin Invest 106:951–961
105. Shu X, Wu W, Mosteller RD, Broek D (2002) Sphingosine kinase mediates vascular endothelial growth factor-induced activation of ras and mitogen-activated protein kinases. Mol Cell Biol 22:7758–7768
106. Hannun YA (1997) Apoptosis and the dilemma of cancer chemotherapy. Blood 89: 1845–1853
107. Morad SAF, Cabot MC (2013) Ceramide-orchestrated signalling in cancer cells. Nat Rev Cancer 13:51–65
108. Giussani P, Tringali C, Riboni L et al (2014) Sphingolipids: key regulators of apoptosis and pivotal players in cancer drug resistance. Int J Mol Sci 15:4356–4392
109. Yount GL, Levine KS, Kuriyama H et al (1999) Fas (APO-1/CD95) signaling pathway is intact in radioresistant human glioma cells. Cancer Res 59:1362–1365
110. Wagenknecht B, Roth W, Gulbins E et al (2001) C2-ceramide signaling in glioma cells: synergistic enhancement of CD95-mediated, caspase-dependent apoptosis. Cell Death Differ 8:595–602
111. Sawada M, Nakashima S, Kiyono T et al (2002) Acid sphingomyelinase activation requires caspase-8 but not P53 nor reactive oxygen species during Fas-induced apoptosis in human glioma cells. Exp Cell Res 273:157–168
112. Yoon G, Kim KO, Lee J et al (2002) Ceramide increases Fas-mediated apoptosis in glioblastoma cells through FLIP down-regulation. J Neurooncol 60:135–141
113. Sawada M, Nakashima S, Banno Y et al (2000) Ordering of ceramide formation, caspase activation, and Bax/Bcl-2 expression during etoposide-induced apoptosis in C6 glioma cells. Cell Death Differ 7:761–772
114. Sawada M, Kiyono T, Nakashima S et al (2004) Molecular mechanisms of TNF-alpha-induced ceramide formation in human glioma cells: P53-mediated oxidant stress-dependent and -independent pathways. Cell Death Differ 11:997–1008
115. England B, Huang T, Karsy M (2013) Current understanding of the role and targeting of tumor suppressor p53 in glioblastoma multiforme. Tumour Biol 34:2063–2074
116. Hara S, Nakashima S, Kiyono T et al (2004) P53-independent ceramide formation in human glioma cells during gamma-radiation-induced apoptosis. Cell Death Differ 11:853–861
117. Hara S, Nakashima S, Kiyono T et al (2004) Ceramide triggers caspase activation during gamma-radiation-induced apoptosis of human glioma cells lacking functional p53. Oncol Rep 12:119–123
118. Banerjee HN, Blackshear M, Williams J et al (2012) C6 Ceramide induces p53 dependent apoptosis in human astrocytoma grade 4 (glioblastoma multiforme) cells. J Cancer Sci Ther 4:12–20
119. Gramatzki D, Herrmann C, Happold C et al (2013) Glioma cell death induced by irradiation or alkylating agent chemotherapy is independent of the intrinsic ceramide pathway. PLoS One 8, e63527
120. Grammatikos G, Teichgraber V, Carpinteiro A et al (2007) Overexpression of acid sphingomyelinase sensitizes glioma cells to chemotherapy. Antioxid Redox Signal 9:1449–1956
121. Heinrich M, Neumeyer J, Jakob M et al (2004) Cathepsin D links TNF-induced acid sphingomyelinase to Bid-mediated caspase-9 and -3 activation. Cell Death Differ 11:550–563
122. Dumitru CA, Sandalcioglu IE, Wagner M et al (2009) Lysosomal ceramide mediates gemcitabine-induced death of glioma cells. J Mol Med 87:1123–1132
123. Fukuda ME, Iwadate Y, Machida T et al (2005) Cathepsin D is a potential serum marker for poor prognosis in glioma patients. Cancer Res 65:5190–5194
124. Noda S, Yoshimura S, Sawada M et al (2001) Role of ceramide during cisplatin-induced apoptosis in C6 glioma cells. J Neurooncol 52:11–21
125. Gomez del Pulgar T, Velasco G, Sanchez C et al (2002) De novo-synthesized ceramide is involved in cannabinoid-induced apoptosis. Biochem J 363:183–188

126. Mochizuki T, Asai A, Saito N et al (2002) Akt protein kinase inhibits non-apoptotic programmed cell death induced by ceramide. J Biol Chem 277:2790–2797
127. Kim WH, Choi CH, Kang SK et al (2005) Ceramide induces non-apoptotic cell death in human glioma cells. Neurochem Res 30:969–979
128. Choi AMK, Ryter SV, Levine B (2013) Autophagy in human health and disease. N Engl J Med 368:651–662
129. Mathew R, Karantza-Wadsworth V, White E (2007) Role of autophagy in cancer. Nat Rev Cancer 7:961–967
130. Pirtoli L, Cevenini G, Tini P et al (2009) The prognostic role of Beclin 1 protein expression in high-grade gliomas. Autophagy 5:930–936
131. Huang X, Bai HM, Chen L et al (2010) Reduced expression of LC3B-II and Beclin 1 in glioblastoma multiforme indicates a down-regulated autophagic capacity that relates to the progression of astrocytic tumors. J Clin Neurosci 17:1515–1519
132. Daido S, Kanzawa T, Yamamoto A et al (2004) Pivotal role of the cell death factor BNIP3 in ceramide-induced autophagic cell death in malignant glioma cells. Cancer Res 64: 4286–4293
133. Kanzawa T, Kondo Y, Ito H et al (2003) Induction of autophagic cell death in malignant glioma cells by arsenic trioxide. Cancer Res 63:2103–2108
134. Kanzawa TL, Zhang LC, Xiao IM et al (2005) Arsenic trioxide induces autophagic cell death in malignant glioma cells by upregulation of mitochondrial cell death protein BNIP3. Oncogene 24:980–991
135. Park MA, Yacoub A, Sarkar D et al (2008) PERK-dependent regulation of MDA-7/IL-24-induced autophagy in primary human glioma cells. Autophagy 4:513–515
136. Yacoub A, Hamed HA, Allegood J et al (2010) PERK-dependent regulation of ceramide synthase 6 and thioredoxin play a key role in mda-7/IL-24-induced killing of primary human glioblastoma multiforme cells. Cancer Res 70:1120–1129
137. Hamed HA, Yacoub A, Park MA et al (2013) Histone deacetylase inhibitors interact with melanoma differentiation associated-7/interleukin-24 to kill primary human glioblastoma cells. Mol Pharmacol 84:171–181
138. Giussani P, Bassi R, Anelli V et al (2012) Glucosylceramide synthase protects glioblastoma cells against autophagic and apoptotic death induced by temozolomide and paclitaxel. Cancer Invest 30:27–37
139. Carracedo A, Lorente M, Egia A et al (2006) The stress-regulated protein p8 mediates cannabinoid-induced apoptosis of tumor cells. Cancer Cell 9:301–312
140. Salazar M, Carracedo A, Salanueva IJ et al (2009) Cannabinoid action induces autophagy-mediated cell death through stimulation of ER stress in human glioma cells. J Clin Invest 119:1359–1372
141. Salazar M, Carracedo A, Salanueva IJ et al (2009) TRB3 links ER stress to autophagy in cannabinoid anti-tumoral action. Autophagy 5:1048–1049
142. Torres S, Lorente M, Rodríguez-Fornés F et al (2011) A combined preclinical therapy of cannabinoids and temozolomide against glioma. Mol Cancer Ther 10:90–103
143. Qin LS, Yu ZQ, Zhang SM et al (2013) The short chain cell-permeable ceramide (C6) restores cell apoptosis and perifosine sensitivity in cultured glioblastoma cells. Mol Biol Rep 40:5645–5655
144. Bruntz RC, Taylor HE, Lindsley CW, Brown HA (2014) Phospholipase D2 mediates survival signalling through direct regulation of Akt in glioblastoma cells. J Biol Chem 289:600–616
145. Yoshimura S, Sakai H, Ohguchi K, Nakashima S et al (1997) Changes in the activity and mRNA levels of phospholipase D during ceramide-induced apoptosis in rat C6 glial cells. J Neurochem 69:713–720
146. Zinda MJ, Vlahos CJ, Lai MT (2001) Ceramide induces the dephosphorylation and inhibition of constitutively activated Akt in PTEN negative U87mg cells. Biochem Biophys Res Commun 280:1107–1115
147. Maceyka M, Harikumar KB, Milstien S, Spiegel S (2012) Sphingosine-1-phosphate signaling and its role in disease. Trends Cell Biol 22:50–60

148. Sonoda Y, Yamamoto D, Sakurai S et al (2001) FTY720, a novel immunosuppressive agent, induces apoptosis in human glioma cells. Biochem Biophys Res Commun 281:282–288
149. Bektas M, Johnson SP, Poe WE et al (2009) A sphingosine kinase inhibitor induces cell death in temozolomide resistant glioblastoma cells. Cancer Chemother Pharmacol 64:1053–1058
150. Guan H, Song L, Cai J et al (2011) Sphingosine kinase 1 regulates the Akt/FOXO3a/Bim pathway and contributes to apoptosis resistance in glioma cells. PLoS One 6, e19946
151. Sato K, Ui M, Okajima F (2000) Differential roles of Edg-1 and Edg-5, sphingosine 1-phosphate receptors, in the signaling pathways in C6 glioma cells. Mol Brain Res 85: 151–160
152. Stupp R, Mason WP, van den Bent MJ et al (2005) Radiotherapy plus concomitant and adjuvant temozolomide for glioblastoma. New Engl J Med 352:987–996
153. Preusser M, de Ribaupierre S, Wohrer A et al (2011) Current concepts and management of glioblastoma. Ann Neurol 70:9–21
154. Hegi ME, Diserens AC, Gorlia T et al (2005) MGMT gene silencing and benefit from temozolomide in glioblastoma. N Engl J Med 352:997e1003
155. Hall E, Giaccia A (2006) Radiobiology for the radiologist, 6th edn. Lippincott, Williams & Wilkins, Philadelphia
156. Chamberlain M (2011) Evolving strategies: future treatment of glioblastoma. Expert Rev Neurother 11:519–532
157. Rekers H, Sminia P, Peters GJ (2011) Towards tailored therapy of glioblastoma multiforme. J Chemother 23:187–199
158. Mégalizzi V, Mathieu V, Mijatovicz T et al (2007) 4-IBP, a Δ1 receptor agonist, decreases the migration of human cancer cells, including glioblastoma cells, in vitro and sensitizes them in vitro and in vivo to cytotoxic insults of proapoptotic and proautophagic drugs. Neoplasia 9:358–369
159. Dumitru CA, Weller M, Gulbins E (2009) Ceramide metabolism determines glioma cell resistance to chemotherapy. J Cell Physiol 221:688–695
160. Reynolds CP, Maurer BJ, Kolesnick RN (2004) Ceramide synthesis and metabolism as a target for cancer therapy. Cancer Lett 206:169–180
161. Adan-Gokbulut A, Kartal-Yandim M, Iskender G, Baran Y (2013) Novel agents targeting bioactive sphingolipids for the treatment of cancer. Curr Med Chem 20:108–122
162. Truman JP, García-Barros M, Obeid LM, Hannun YA (2014) Evolving concepts in cancer therapy through targeting sphingolipid metabolism. Biochim Biophys Acta 1841:1174–1188
163. French KJ, Schrecengost RS, Lee BD et al (2003) Discovery and evaluation of inhibitors of human sphingosine kinase. Cancer Res 63:5962–5969
164. Rex K, Jeffries S, Brown ML et al (2013) Sphingosine kinase activity is not required for tumor cell viability. PLoS One 8, e68328
165. Visentin B, Vekich JA, Sibbald BJ et al (2006) Validation of an anti-sphingosine-1-phosphate antibody as a potential therapeutic in reducing growth, invasion, and angiogenesis in multiple tumor lineages. Cancer Cell 9:225–238
166. Horga A, Montalban X (2008) FTY720 (fingolimod) for relapsing multiple sclerosis. Expert Rev Neurother 8:699–714
167. Huwiler A, Pfeilschifter J (2008) New players on the center stage: sphingosine 1-phosphate and its receptors as drug targets. Biochem Pharmacol 75:1893–1900
168. Adachi K, Chiba K (2008) FTY720 story. Its discovery and the following accelerated development of sphingosine 1-phosphate receptor agonists as immunomodulators based on reverse pharmacology. Perspect Medicin Chem 1:11–23
169. Romero Rosales K, Singh G, Wu K et al (2011) Sphingolipid-based drugs selectively kill cancer cells by down-regulating nutrient transporter proteins. Biochem J 439:299–311
170. Miron VE, Schubart A, Antel JP (2008) Central nervous system-directed effects of FTY720 (fingolimod). J Neurol Sci 274:13–17

Part II
Emerging Technologies for Sphingolipid Detection and Prediction of Modulation of Sphingolipid Metabolism

Systems Biology Approaches for Studying Sphingolipid Signaling

Xinghua Lu, W. Jim Zheng, and Yusuf A. Hannun

Abstract The importance of sphingolipid metabolism and bioactive sphingolipid products in cancer development and disease progression is now well established (see other chapters in this volume). However, much remains unknown in terms of the signaling mechanisms and downstream pathways affected by sphingolipids in cancers, such as which sphingolipid pathway is involved, which sphingolipid species contributes to a specific cancer's biologic response, and which regulatory pathways are modulated by these bioactive sphingolipids. Due to the connectivity of sphingolipid metabolic network, it is difficult to address the above questions using conventional experimental approaches because manipulation of any one enzyme often results in a metabolic "ripple effect" across the network that are not easy to predict. In this chapter, we introduce a systems biology framework for deciphering signaling roles of distinct sphingolipid species and enzymes and for exploring novel mechanisms involving these pathways. These approaches have been successfully developed, applied, and validated in yeast systems. One component of the framework involves systematically perturbing sphingolipid metabolism using physiological and pharmacological approaches while monitoring lipidomic and transcriptomic responses to the perturbations are monitored; then different computational models are designed and applied to reveal relationship between specific sphingolipid species with transcriptomic modules, as a means to reveal signaling roles of distinct sphingolipid species. This component has successfully identified specific signaling pathways regulated by phytosphingosine-1-phosphate and clearly demonstrated that distinct ceramides encode disparate cellular signals. Another component of the

X. Lu (✉)
Department of Biomedical Informatics, University of Pittsburgh,
5607 Baum Blvd, Pittsburgh, PA 15206, USA
e-mail: xinghua@pitt.edu

W.J. Zheng
School of Biomedical Informatics, University of Texas Health Science Center
at Houston, 7000 Fannin St. Suite 600, Houston, TX 77030, USA
e-mail: Wenjin.J.Zheng@uth.tmc.edu

Y.A. Hannun
Stony Brook Cancer Center, Stony Brook University, Stony Brook, NY, USA
e-mail: yusuf.hannun@stonybrookmedicine.edu

Y.A. Hannun et al. (eds.), *Bioactive Sphingolipids in Cancer Biology and Therapy*, DOI 10.1007/978-3-319-20750-6_13

framework concentrates on systematically identifying novel genes that could influence the activity of sphingolipid pathways. More specifically, we mined the literatures of yeast genes, constructed an Ontology Fingerprint of for each gene and developed an Ontology Fingerprint derived gene network, which were used to discover novel genes that modulates the activities of the sphingolipid pathway. This approach expanded our knowledge of the sphingolipid pathway by finding novel genes whose functions have not been associated with the pathway before. In summary, we show that systems biology approaches can effectively complement the experimental research and have a great potential to enhance the research of sphingolipids in cancers.

Keywords Sphingolipids • Systems biology • Modeling • Signal transduction

1 Introduction

The importance of sphingolipid signaling in cancer is now well known (as illustrated by the scope of this volume). However, much remains to be learned about the exact cellular signaling mechanisms through which this family of bioactive lipids contributes to the development of cancers and maintenance of cancer cell homeostasis. In this respect, the fundamental questions include: Whether different sphingolipid species contribute to distinct cancer process? If yes, in which signaling pathways do the distinct sphingolipid species are involved? How are these pathways regulated? What is the impact of perturbed sphingolipid metabolism on the pathways? How sphingolipid-mediated signaling contributes to cancer development and progression?

Since sphingolipid metabolism forms a highly connected network rather than a linear pathway, perturbing any part of sphingolipid network, for example overexpression of an enzyme in a tumor, often leads to broad changes in multiple sphingolipid metabolites. This convoluted lipidomic landscape makes it difficult to pinpoint which sphingolipid species are the "drivers" contributing to a specific cancer process in tumors, which are 'reactive' changes, and which are "passengers", and as such the connectivity of metabolites defies conventional experimental approaches attempting to pinpoint driver lipid species. This would be of high significance as it not only informs us as to what lipid to focus on in elucidating the mechanisms involved, but also in defining 'vulnerability points' for intervention. Such considerations invite system-level approaches that can make use of the increasing availability of 'omic' type data such as sphingolipidomic measurements, transcriptomic responses, systematic screens for gene function, systematic studies on protein–protein interactions, and many others. In accordance, in the past decade we have developed a systems biology framework to address these challenges. The principle of applying systems biology techniques to decipher the signaling roles of distinct sphingolipids has been successfully applied in yeast (*Saccharomyces cerevisiae*) cell systems, and we believe that these principles can be applied in the cancer research setting. In particular, with a large volume of cancer genomic data, such as those from The Cancer Genome Atlas (TCGA) and International Cancer Genome Consortium (ICGC) projects, it is foreseeable that systems biology approaches developed in our studies will play a more significant role in deciphering the role of sphingolipid metabolisms in cancers. This could be particularly useful after we used the Ontology Fingerprint based systems biology approach to identify novel genes that could influence the activity of the sphingolipid pathway, as the cancer genome data could not only help us to identify novel genes linking cancer and the sphingolipid pathway, but also help us to decipher mechanisms and missing links between cancer and the known sphingolipid genes.

2 Strategies of System Biology Study of Sphingolipid Signaling

Systematic perturbation, such as yeast and drosophila deletion libraries, has long been exploited by researchers as a powerful tool to investigate signal transduction pathways in model organisms [1–5]. However, the conventional approach of perturbing a single gene and monitor certain phenotypes as reflections of the functional roles of the gene cannot be directly applied to study the signaling roles of metabolites, because perturbing metabolic enzyme often results in a "ripple" effect across an entire metabolic network; that is, manipulating an enzyme not only affects the concentrations of the substrates and products of the enzyme but also recursively affects the metabolites connected to the substrates and products. Under such a situation, it is difficult to pinpoint which metabolite, among many affected, is responsible for an observed phenotype. To address this challenge in our setting, we hypothesized that, if a sufficient number of physiological and pharmacological perturbations is applied to the sphingolipid metabolic system such that different parts of the metabolic network are decoupled under certain perturbations, computational approaches can be applied to deconvolute the signaling roles of sphingolipid. For example, deletion of yeast sphingosine kinases *LCB4* and *LCB5* disrupts the metabolic connection (thereby the statistical correlation) between sphingoid bases and sphingosine-1-phosphate. We further hypothesized that, since many signal transduction pathways eventually regulate cellular functions through regulating gene expression, transcriptomic data collected under different perturbation conditions are the readouts of the states of cellular signaling pathways. As such, if the concentration of a sphingolipid species is, under various perturbation conditions, consistently associated with the differential expression of a set of genes involved in related function (a molecular phenotype), it is likely that the lipid species is involved in the signaling pathway regulating the differentially expressed genes (DEGs).

Based on the above hypotheses, we set out to study the signaling roles of different sphingolipids in two studies [6, 7]. In the first study, we concentrated on determining the signaling roles of phytosphingosine-1-phosphate (PHS1P); in the second study, we concentrated on determining whether distinct ceramides have disparate signaling pathways. The signaling roles of sphingosine-1-phosphate in mammalian cells and in cancers are well established, but it is unclear whether the two sphingoid base-1-phosphate species, dihydrosphingosine-1-phosphate[1] and phytosphingosine-1-phosphate (PHS1P), play any signaling role in yeast because yeast cells lack the homologue of mammalian G-protein-coupled sphingosine-1-phosphate receptor. On the hand, ceramide biosynthesis in yeast generates more than 30 distinct species that can be identified by contemporary mass spectroscopy-based lipidomic approaches [8]; in mammals, the total number of ceramide species may exceed 200 [9]. While it is well accepted that ceramides, as a family of bioactive lipids, participate in important biological processes of yeast cells, it is unclear whether distinct ceramide species play different signaling roles; in other words, whether the diverse ceramide species underlie the apparently multifaceted functions attributed to ceramides in yeast.

In both studies, we adopted systematic perturbation approaches to disrupt the sphingolipid metabolic network. In the PHS1P study [10], yeast cells were subjected to heat stress and myriocin treatments in wild type (*wt*), and in the *lcb4Δ/lcb5Δ*, and the *dpl1Δ* deletion strains, and microarray data were collected from each combination of conditions. Heat stress induces a large number of DEGs as well as abroad changes in sphingolipids. In fact, de novo synthesis of sphingolipids by the serine palmitoyltransferase (SPT), which can be blocked by myriocin, is a required program for yeast to survive the stress [11–13]. Therefore treatment with myriocin in the presence and absence of heat stress separates sphingolipid-mediated DEGs from those regulated by other pathways during heat stress. Double deletion of sphingoid (dihydro- and phyto-sphingosine) kinases *LCB4* and *LCB5* eliminates the production of PHS1P whereas deletion of the sphingoid base phosphate lyase *DPL1* prevents the degradation of PHS1P. These manipulations created certain conditions in which PHS1P was decoupled from the rest of sphingolipids, allowing us to deconvolute the signaling roles of the species. In the ceramides signaling study [7], we also employed heat stress and myriocin treatments to induce sphingolipid changes; in addition, we developed a novel approach to perturb ceramide metabolism by treating cells with free fatty acid myristate to perturb abundance of ceramides with specific lengths of acyl side chain.

We simultaneously collected lipidomic and transcriptomic data from the above experiments, and we further integrated existing genomic data regarding transcription factor (TF) binding sites to perform integrative analysis. Figure 1 (reproduced from [10]) shows the overall scheme of our approaches. In this figure, we collected lipidomic data (Fig. 1a), transcriptomic data (Fig. 1b) and existing genomic data (Fig. 1c); we then combined the lipidomic and transcriptomic data to perform correlation analysis to identify genes that are significantly correlated with specific sphingolipids, and we paid special attention to the genes correlated to PHS1P, which was the focus of the study (Fig. 1d); we combined the knowledge of TF-binding sites and transcriptomic data to infer the activation states of TFs under each experimental condition using a Bayesian regression model [10] (Fig. 1e); we performed logistic regression analysis to detect if any TFs are likely regulated by PHS1P (Fig. 1f); we identified a specific pathway and performed validation experiments to validate that the target genes were indeed regulated by PHS1P (Fig. 1g).

[1] Due to some technical difficulties and low levels, our mass spectrometry measurement of the dihydrosphingosine-1-phosphate was not accurate; therefore we only studied the signaling role of phytosphingosine-1-phosphate.

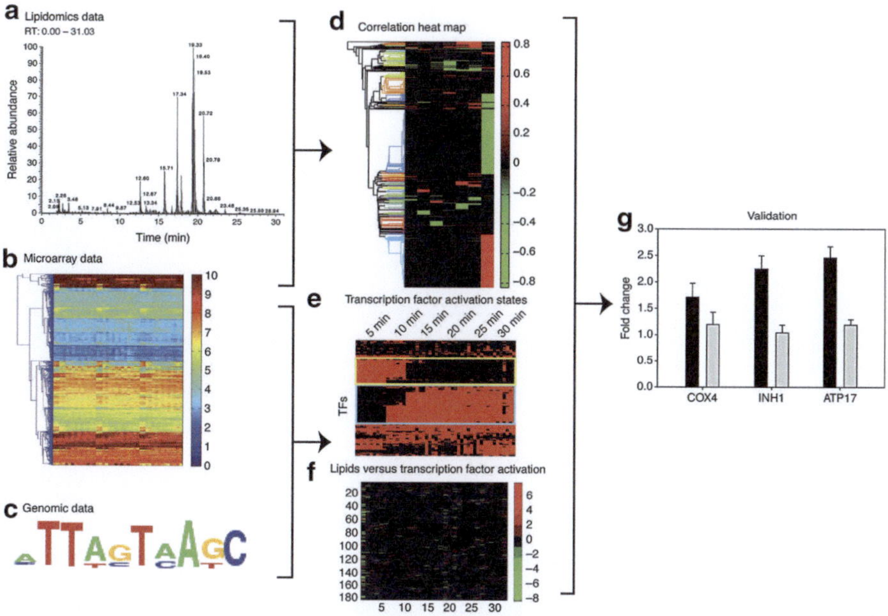

Fig. 1 Overview of the integrative systems approach. (**a–c**) The lipidomics, transcriptomic, and genomic data were collected from experiments and existing databases. (**d**) Integrating the lipidomic and transcriptomic data in a correlation analysis led to a gene-versus-lipid correlation coefficient matrix, shown as a heat map. (**e**) Genomic and transcriptomic data were combined to infer the activation states of TFs under each experiment, shown as a TF-versus-condition heat map representing the activation states in (**e**). The inferred activation states of TFs from (**e**) were combined with lipidomic data (**a**) to model the relationship between lipid mass and activation of TFs, shown as a heat map representing the significant logistic parameters in (**f**). The results from (**e**) and (**f**) resulted in the hypothesis that PHS1P mediated regulation of a subset of genes through activation of the HAP complex, which was tested in a series of genetic and pharmacological experiments (**g**)

3 Systematic Perturbations Creates Sufficient Conditions to Deconvolute Signaling Roles of Sphingolipids

In the above two studies [7, 10], the heat stress or manipulation of individual enzymes, e.g., the deletion of *LCB4* and *LCB5*, led to broad changes in lipidomic profiles beyond the substrates and products of the perturbed enzymes. However, when sufficient numbers of well-designed perturbations were applied to the metabolic system, it was possible to generate lipidomic profiles such that some normally highly correlated lipid species were decoupled. For example, in the ceramides signaling study [7], systematic perturbations were able to induce distinct profiles separating ceramides with different specific modifications, e.g., phytoceramides (with a 4-OH) versus dihydroceramides, as well as to separate ceramides with the same head groups but with different *N*-acyl side chains, e.g., dihydroceramides with C14–C16 side chains versus those with C18–C20 side chains, as shown in Fig. 2 (reproduced from [7]). In particular, using consensus-clustering analysis [14], we showed that ceramide

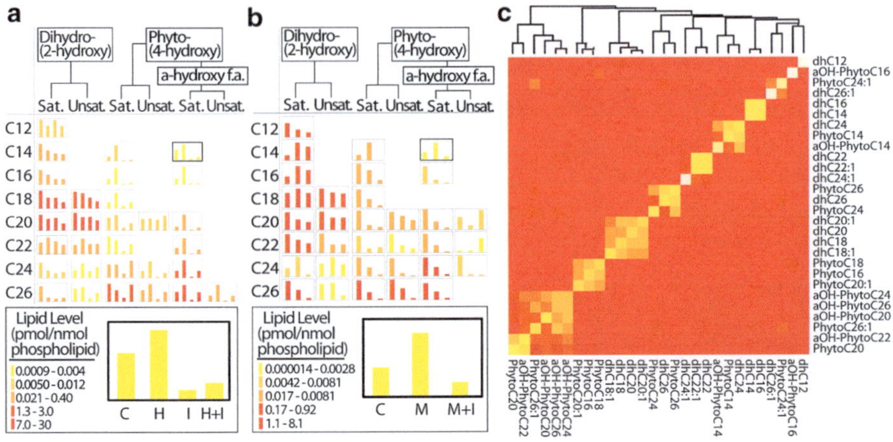

Fig. 2 Lipid groups revealed through systematic perturbation of ceramide metabolism. (**a**, **b**) Ceramide profiles under different experimental conditions in which sphingolipid metabolism is perturbed by the following factors: control (C), heat stress (H), ISP1 treatment (I), heat shock in the presence of ISP1 (H+I); control in low free fatty acid medium (C), myristate supplement (M), and myristate treatment in the presence of ISP1 (M+I). The insets in (**a**) and (**b**) show the enlarged results for C14 saturated a-hydroxyl ceramide under the above seven conditions, where total amount of the species in control is normalized to 1. (**c**) Lipid groups revealed through consensus clustering. The pseudo-colored heat map shows the tendency of different lipids being assigned to a common cluster during repeated re-sampling and clustering analyses of lipid profiles across the above seven experimental conditions. A red cell indicating two lipids is rarely assigned to a common cluster during analysis; a non-red cell reflects the degree of tendency of two lipids belonging to a common cluster (with a spectrum from *orange* to *yellow*, and to *white* reflecting increasing tendency)

profiles (the abundance of a ceramide species across all experimental conditions) were organized into clusters, with each cluster consisting of lipids that are metabolically inseparable. The results indicate that, to a certain degree, systematic perturbation disrupted the correlation (convolution) structure of sphingolipids and rendered them independent of each other under specific conditions, and reduced covariance attenuates the confounding effects of lipids when we set out to investigate the relationships among sphingolipids and transcriptomic responses to such perturbations.

4 Distinct Signaling Roles of Sphingolipids are Revealed by Identifying Specific Transcriptomic Responses

As mentioned before, the central hypothesis of our framework is that, if the covariance patterns of sphingolipid are disrupted by systematic perturbation, one can discern the cellular signals encoded by changing levels of a lipid species by examining if it is significantly associated with a particular cellular or molecular phenotypes. In our studies, we used transcriptomic measurements as the molecular phenotypes reflecting the state of signaling pathways, and we examined correlation between sphingolipid abundance and gene as a means to discover regulatory relationship between lipid and transcribed genes. Note that, when we observe a significant correlation between a sphingolipid species and the expression value of a gene, it is more likely that the sphingolipid regulates the expression of the gene rather than the reverse, because the abundance of lipids were manipulated by multiple experimental perturbations specifically targeted at sphingolipid metabolic systems, i.e., randomized trials. We measure the strength of association between sphingolipids and gene expression using different approaches, including linear correlation analysis as well as non-linear information-theory-based metrics [15]. The results revealed many significant correlations (linear or nonlinear) between specific sphingolipids, including PHS1P and ceramide species or ceramide groups, and gene expression despite the diversity of lipid and gene response to different perturbations, thus supporting the hypotheses that causal relationships exist between these sphingolipids and genes.

5 Functional Analyses Support the Notion That Distinct Sphingolipids Regulate Specific Signaling Pathways

Observing the correlation between a specific sphingolipids and a set of genes indicates that the lipid may participate in a signaling pathway regulating the genes. We then applied different functional analyses to search for the evidence for delineating sphingolipid-mediated signaling pathways. For the PHS1P study [10], we developed and applied a Bayesian regression model to infer which TFs were activated and their relationship to the sphingolipid-sensitive genes under each experiment condition (Fig. 1e), and we then applied logistic regression to further identify the sphingolipids that likely regulate the activities of TFs. This led to the discovery that PHS1P concentration is significantly associated with the activities of the HAP transcription factor complex (including *HAP2* and *HAP4*), which regulates the expression of a set of mitochondrial proteins [16]. We then set out to experimentally validate this hypothesis, and the results strongly support it, see [10] for details.

In the ceramides signaling study [7], we identified genes likely regulated by specific ceramide groups (from Fig. 2) through a regularized regression model known as *elastic net* [17], using expression value of a gene as dependent variable (target) and concentrations of the ceramide groups (each group is represented as the averaged abundance of the species within it) as independent variables (predictors). This model is capable of identifying the ceramide that is most predictive of the expression value of a DEG, thus potentially regulating the expression of the gene. After identify the candidate target genes for each ceramide group, we further performed the Gene Ontology analysis of the genes to divide genes into non-disjoint functional modules, and we searched for TFs that potentially regulate the genes within such a module. We were able to identify many such functional modules and their corresponding TF regulators.

The above analysis revealed that, indeed, distinct ceramide groups are associated with different genes and functional modules (Fig. 3, reproduced from [7]), indicating

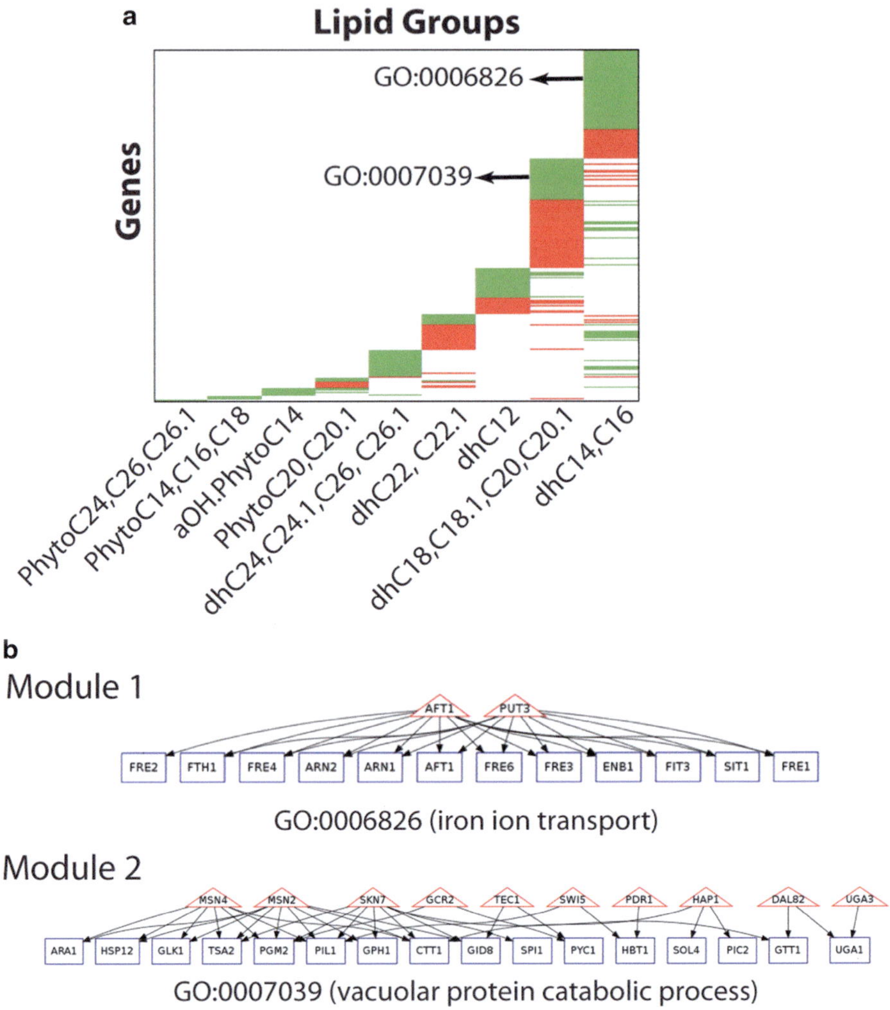

Fig. 3 Modeling relationship between lipidomic and gene expression data. (**a**) Organizing genes demonstrating significant correlation with specific ceramides. Genes (rows) are organized according to their association with the different lipid subgroups. A green block represents a set of genes negatively correlated to a lipid, and a red block represents a set of genes positively correlated to a lipid. Examples of major enriched GO terms within gene blocks are shown. (**b**) Defining pathways of specific biologic modules that respond to specific ceramides, perform related functions, and share transcription factors. Two example modules are shown (see http://www.dbmi.pitt.edu/publications/YeastCeramideSignaling for all modules). Rectangles represent lipid-correlated genes, triangles indicate the transcription factors shared by the genes; an edge from a transcription factor to a gene indicates that the gene has the binding sites for the transcription factor in its promoter. The function performed by the genes in a module is represented with a GO term

that different ceramide exert disparate signaling effects. For example, our study revealed, for the first time, that heat stress suppresses the abundance of dihydroceramides through activation of alkaline dihydraceramidase (encoded by *YDC1* gene), which mediate differential expression of a large numbers genes; more intriguingly, we showed that dihydroceramides with different *N*-acyl side chains, e.g., long-chain dihydroceramides (LC-HDCs) with C14–C16 side chains versus very-long-chain dihydroceramides (VLC-DHCs) with C18–C20 side chains, regulate different functional modules, indicating they participate in different pathways. We focused on the two gene modules that were negatively correlated to LC-DHCs or VLC-DHCs respectively. We predicted that increasing the respective lipids would repress genes in the corresponding modules and would produce phenotypes mimicking those resulting from deletion of module genes. We performed experiments to validate the predictions, and the results support our hypotheses (Fig. 4, reproduced from [7]). These findings revealed previously unappreciated complexity of ceramide-mediated signaling, in that almost all moieties of ceramide molecules may contribute to different signaling function of the ceramides.

6 Identify Novel Players of the Sphingolipid Pathway from the Ontology Fingerprint Derived Gene Networks

While the experimental approach described above to perturb sphingolipid pathway helps us to understand pathway and lipid function, other computational methods may allow us to infer novel genes that could regulate the pathway; thus, providing an additional dimension to expand our scope of understanding pathway regulation. In this section, we describe a literature based systems biology approach to identify novel genes that may have impact on sphingolipid metabolism beyond those genes encoding known sphingolipid enzymes.

The complexity of lipid metabolism far exceeds a simple collection of individual signaling modules such that one agonist may regulate several enzymes or the bioactive product of one enzyme (e.g., ceramide) may serve as a substrate for another enzyme generating a different bioactive molecule (such as diacylglycerol or sphingosine). Instead, the complexity of lipid metabolism serves to provide a highly regulated and coordinated network of bioactive molecules with distinct and overlapping functions. This network then serves to *integrate* and *coordinate* complex responses of cells to various agents and environmental stimuli. Such complexity raises a critical question that could help us fully understand the underlying mechanisms of this complex regulation: are there any novel pathway players beyond those genes we already know that can modulate lipid metabolism?

Identifying novel players for a biological pathway is not a trivial task, and several approaches based on gene networks have been conducted to identify these novel players of known pathways. One theoretical foundation of systems biology is that genes work in concert to execute cellular functions. Therefore, we hypothesize that putting pathway genes in a gene network could help to identify novel players of the pathway based on the network substructure and connectivity.

Networks built upon gene–gene interactions occur in various forms and have been investigated in many different ways. The protein–protein interaction (PPI)

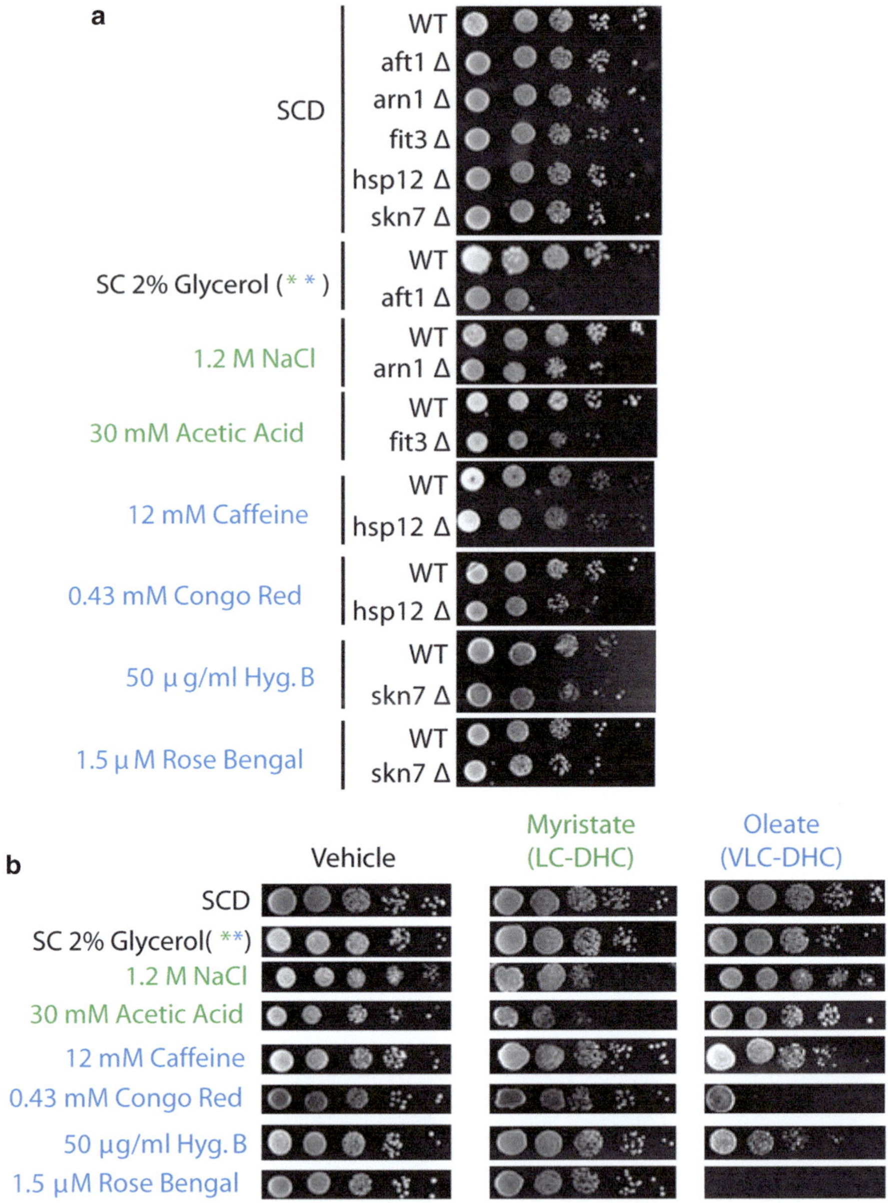

Fig. 4 Experimental validation of lipid-dependent phenotypes. (**a**) Confirmation of published genetic phenotypes as positive controls. Published phenotypes for deletion mutants from the LC-DHC-sensitive gene module (green font) or the VLC-DHC-sensitive gene module (blue font) were used to predict ceramide and fatty acid-specific growth defects. One phenotype was confirmed for each treatment employed. Conditions were selected from the literature based on phenotypes of genes within each module [5, 23–28]. (**b**) Validation of LC-DHC- or VLC-DHC-sensitive phenotypes. Rows: specific phenotypes predicted to manifest in response to C14 or C18:1 dihydroceramides, given the indicated treatment condition. Cells were spotted onto agar containing specified treatment plus vehicle (0.1 % ethanol), or saturating (1 mM) myristate or oleate. SCD is no treatment. Spots represent 1:10 serial dilutions of a single mid-log culture. Green font: phenotypes of the LC-DHC-sensitive module predicted to be induced by myristate treatment. Blue font: phenotypes of the VLC-DHC-sensitive module predicted to be induced by oleate treatment, and 2 % glycerol is associated with both modules

network measures the physical interaction of gene products to generate an interaction network, and approaches based on this have been used in species ranging from bacteria all the way to human. In yeast, another approach to study gene–gene interaction is the genetic interaction among non-essential genes—genes that can be deleted from the genome without impairing a cell's viability. Genetic interaction refers to the deletion of two non-essential genes together resulting in severe consequences, ranging from defects in growth to inviability (synthetic lethality), and such genetic interaction is an indication of functional relevance [18]. Genome wide investigation of gene–gene interactions have been performed [19], and the results from these studies are compiled into the BioGRID database [20]. While these networks captures certain aspects of biology and helped to make new discoveries, there is a lack of comprehensiveness in their content due to the fact that each of these networks focuses on a single aspect of biology such as protein–protein interaction.

The scientific literature as a comprehensive source of information could overcome the networks built upon a single biological character such as protein–protein interaction. Publications about a gene studies the gene's many aspects, and provides an overall picture of how that gene functions, what gene products it interacts with, and how the gene is regulated, to name a few. To overcome this drawback of the current networks focusing on a single aspect of biology, we developed the Ontology Fingerprints, upon which we constructed a novel gene network [21]. Ontology Fingerprint for a gene is a set of Gene Ontology (GO) terms that are overrepresented in the PubMed abstracts linked to the gene together with the terms' enrichment p-value. Because the fingerprint is extracted from all the publications about that gene—it captures comprehensive information. Such comprehensive characterization of a gene has been proven to effectively assess the relevance between genes and diseases [21].

The Ontology Fingerprint derived network is constructed by comparing the Ontology Fingerprints between two genes across the whole genome. For each gene pair in the network, a weight is generated from the similarity of the genes' fingerprints. This weighted network is rich in subnetwork structures, and initial evaluation showed that the subnetwork structures have functional implication as genes that are functionally relevant tend to be in the same subnetwork structure. The network is typical of a biological network as it is scale-free with some interesting hub genes. Table 1 shows the top five enriched GO terms among the hub genes in the network. Furthermore, the relationship between two genes in the network encompasses comprehensive information captured in Gene Ontology and summarized through all the

Table 1 Top 5 enriched GO terms among the hub genes in the yeast gene network

GO ID	GO term	FDR
GO:0030687	Preribosome, large subunit precursor	1.19E-12
GO:0042273	Ribosomal large subunit biogenesis	6.37E-08
GO:0030684	Preribosome	7.19E-08
GO:0005730	Nucleolus	7.42E-08
GO:0006364	rRNA processing	8.93E-08

The enriched GO terms among the hub genes were identified by DAVID [22]

As shown in Table S5, the top 5 enriched GO terms are related to preribosome, nucleolus and rRNA, indicating the biological function of translation

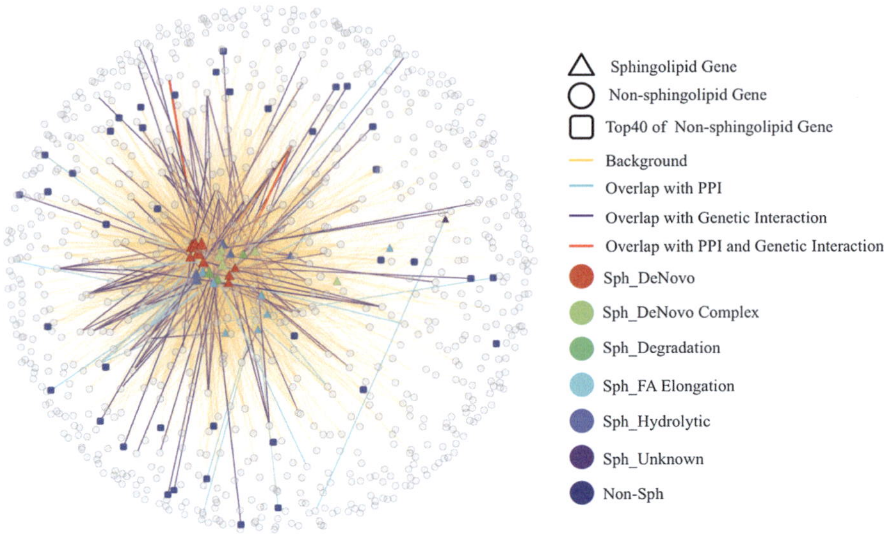

Fig. 5 A subnetwork of the known sphingolipid genes (*triangle*) and the candidate genes (*circle and square*), as well as connections between the two group genes. The functional groups of sphingolipid genes and interaction types between sphingolipid genes and non-sphingolipid genes are highlighted by *different colors*

publications about the two genes. Such comprehensiveness could capture the biological relevance more precisely than a single factor such as protein–protein interaction, thus it has the potential to identify other novel genes relevant to the sphingolipid pathway. This notion is supported by the observation that functionally relevant genes form subnetwork structures in this network—an indication that network structure and function are closely related in our Ontology Fingerprint derived gene network.

In order to find novel players of the sphingolipid metabolic pathway, we first identified 30 genes that are known to belong to the pathway. We then put these sphingolipid genes in the Ontology Fingerprint derived gene network. Given our observation that functionally relevant genes form subnetwork structure in this network, we used these genes as a bait to fish out the associated subnetwork structure (Fig. 5) by calculating a Total Score. This subnetwork was then used to identify the novel players of the sphingolipid metabolic pathway.

In this subnetwork we identified, the 30 yeast sphingolipid genes were classified into six functional categories including de novo biosynthesis, synthesis of complex sphingolipids, fatty acid elongation, sphingolipid degradation, hydrolytic function, and unknown functions according to the annotation by a group of sphingolipid pathway experts. The proportion of connections in the sub-network that overlapped with biological interactions (either PPI or genetic interactions) is remarkably high among top 40 candidate genes predicted by Total Score. This demonstrated that the network does capture the biological associations, especially among highly weighted

Category	Term	RT	Genes
SP_PIR_KEYWORDS	Acyltransferase	RT	
GOTERM_BP_FAT	protein amino acid acylation	RT	
INTERPRO	Acyl-CoA N-acyltransferase	RT	
UP_SEQ_FEATURE	domain:N-acetyltransferase	RT	
INTERPRO	GCN5-related N-acetyltransferase	RT	
GOTERM_MF_FAT	N-acetyltransferase activity	RT	
GOTERM_MF_FAT	N-acyltransferase activity	RT	
GOTERM_BP_FAT	protein amino acid acetylation	RT	
SP_PIR_KEYWORDS	transferase	RT	
GOTERM_MF_FAT	acetyltransferase activity	RT	
GOTERM_BP_FAT	histone acetylation	RT	
GOTERM_BP_FAT	covalent chromatin modification	RT	
GOTERM_BP_FAT	histone modification	RT	
GOTERM_BP_FAT	chromatin organization	RT	
SP_PIR_KEYWORDS	chromatin regulator	RT	
GOTERM_BP_FAT	chromatin modification	RT	

Fig. 6 Functional categories of the experimentally validated genes influencing sphingolipid pathway activity in vivo

edges. Based on the functional classification of the known sphingolipid genes and the functional association between sphingolipid genes and candidate genes in the network, the potential roles of candidate genes in the yeast sphingolipid pathway were inferred. For instance, gene TGL5, the top 1 ranked candidate gene, is functionally associated with 25 out of 30 sphingolipid genes; the GO enrichment test revealed that "membrane lipid metabolic process" is the most enriched functions among the 25 sphingolipid genes, indicating that gene TGL5 is probably related to the yeast sphingolipid pathway via the membrane lipid metabolic process, consistent with the fact that TGL5 is a bifunctional enzyme with triacylglycerol lipase and lysophosphatidic acid acyltransferase activity involved in triacylglycerol mobilization and it is integral to membranes.

From the genes in the subnetwork structure associated with the sphingolipid pathway genes, we removed genes associated with any publication mentioning sphingolipid or ceramide to find novel players of the sphingolipid metabolic pathway. The remaining genes were subjected to experimental analysis, and this resulted in 14 novel genes whose deletion could influence the in vivo activity of the sphingolipid pathway as measured by the changed sensitivity of deletion strain to myriocin.

The functions of these novel genes are also interesting (Fig. 6). For example, one set of these genes relates to chromatin organization and histone modification. Recent experimental evidence shows that sphingolipid function could influences cell cycle checkpoint while no underlying mechanism has been identified for such an observation. The finding that genes related to the chromatin organization and histone modification influence the sphingolipid pathway in vivo could be a hint of the missing link between the sphingolipid pathway and the cell cycle. The finding could provide some new directions in deciphering the roles of sphingolipid in cell cycle check point. Finding like this in human could help to identify novel therapeutic targets surrounding the sphingolipid pathway for cancer treatment.

Another interesting finding is the role of the Pfa4 gene in modulating the sphingolipid pathway. While deleting pfa4 could result in an increased sensitivity of cells to myriocin, adding exogenous sphingolipids actually reversed the effect, indicating that the influence of pfa4 on sphingolipid pathway is lipid related.

Combining literature mining with ontology and systems biology is a powerful tool and could help to identify novel pathway players without performing laborious and time consuming experiments. The method could help to expand our knowledge about sphingolipid pathway regulation and its role in cellular functions. The new discovery could also help to shape new directions of the future research.

7 Discussion and Future Directions

The studies summarized above demonstrate the value and capability of systems biology approaches to decipher signaling roles of distinct sphingolipids in yeast and define novel regulatory components. Combining systematic perturbation and unbiased systems biology approaches enabled us to reveal the complexity of sphingolipid signaling in yeast and provided mechanistic evidence to support the notion that sphingolipids play essential roles during responses to environmental stresses in yeast cells. The very principles applied in these studies are applicable in mammalian cells to study the roles of sphingolipids in cellular signaling under physiological and pathological conditions such as cancers. In a foreseeable future, genome sequencing and transcriptomic profiles from tumors will be routinely acquired from individual tumors in a clinical setting, and there is a need to also examine the lipidomic profiles from a reasonably large number of tumors to investigate the impact of genomic alterations of sphingolipid metabolic enzymes on lipidomic changes and their relationship to transcriptomic profiles. Such data will provide a foundation to investigate the role of sphingolipids in cancer development and maintenance of cancer cell homeostasis.

References

1. Breslow DK et al (2008) A comprehensive strategy enabling high-resolution functional analysis of the yeast genome. Nat Methods 5(8):711–718
2. Ryder E et al (2007) The DrosDel deletion collection: a Drosophila genomewide chromosomal deficiency resource. Genetics 177(1):615–629
3. Hughes TR et al (2000) Functional discovery via a compendium of expression profiles. Cell 102(1):109–126
4. Giaever G et al (2002) Functional profiling of the Saccharomyces cerevisiae genome. Nature 418(6896):387–391
5. Lu S et al (2013) From data towards knowledge: Revealing the architecture of signaling systems by unifying knowledge mining and data mining of systematic perturbation data. PLoS One 8, e61134

6. Cowart LA et al (2010) Revealing a signaling role of phytosphingosine-1-phosphate in yeast. Mol Syst Biol 6:349
7. Montefusco DJ et al (2013) Distinct signaling roles of ceramide species in yeast revealed through systematic perturbation and systems biology analyses. Sci Signal 6(299):rs14
8. Kolter T (2011) A view on sphingolipids and disease. Chem Phys Lipids 164(6):590–606
9. Hannun YA, Obeid LM (2011) Many ceramides. J Biol Chem 286(32):27855–27862
10. Cowart LA et al (2010) Revealing a signaling role of phytosphingosine-1-phosphate in yeast. Mol Syst Biol 6:349. doi:10.1038/msb.2010.3
11. Dickson RC, Sumanasekera C, Lester RL (2006) Functions and metabolism of sphingolipids in Saccharomyces cerevisiae. Prog Lipid Res 45(6):447–465
12. Jenkins GM et al (2002) Acute activation of de novo sphingolipid biosynthesis upon heat shock causes an accumulation of ceramide and subsequent dephosphorylation of SR proteins. J Biol Chem 277(45):42572–42578
13. Jenkins GM, Hannun YA (2001) Role for de novo sphingoid base biosynthesis in the heat-induced transient cell cycle arrest of Saccharomyces cerevisiae. J Biol Chem 276(11): 8574–8581
14. Monti S et al (2003) Consensus clustering: a resampling-based method for class discovery and visualization of gene expression microarray data. Mach Learn 52(1):91–118
15. Reshef DN et al (2011) Detecting novel associations in large data sets. Science 334 (6062):1518–1524
16. Forsburg SL, Guarente L (1989) Identification and characterization of HAP4: a third component of the CCAAT-bound HAP2/HAP3 heteromer. Genes Dev 3(8):1166–1178
17. Friedman J, Hastie T, Tibshirani R (2010) Regularization paths for generalized linear models via coordinate descent. J Stat Softw 33(1):1–22
18. Boone C, Bussey H, Andrews BJ (2007) Exploring genetic interactions and networks with yeast. Nat Rev Genet 8(6):437–449
19. Tong AH et al (2004) Global mapping of the yeast genetic interaction network. Science 303(5659):808–813
20. Breitkreutz BJ et al (2008) The BioGRID interaction database: 2008 update. Nucleic Acids Res 36(Database issue):D637–D640
21. Tsoi LC et al (2009) Evaluation of genome-wide association study results through development of ontology fingerprint. Bioinformatics 25(10):1314–1320
22. da Huang W, Sherman BT, Lempicki RA (2009) Systematic and integrative analysis of large gene lists using DAVID bioinformatics resources. Nat Protoc 4(1):44–57
23. Sambade M et al (2005) A genomic screen for yeast vacuolar membrane ATPase mutants. Genetics 170(4):1539–1551
24. Mira N, Teixeira M, Sa-Correia I (2010) Adaptation and tolerance to weak acid stress in Saccharomyces cerevisiae: a genome-wide view. OMICS 14(5):525–540
25. Motshwene P et al (2004) LEA (late embryonic abundant)-like protein Hsp 12 (heat-shock protein 12) is present in the cell wall and enhances the barotolerance of the yeast Saccharomyces cerevisiae. Biochem J 377(3):769–774
26. Karreman RJ, Lindsey GG (2007) Modulation of Congo-red-induced aberrations in the yeast Saccharomyces cerevisiae by the general stress response protein Hsp12p. Can J Microbiol 53(11):1203–1210
27. Li S et al (2002) The eukaryotic two-component histidine kinase Sln1p regulates OCH1 via the transcription factor, Skn7p. Mol Biol Cell 13(2):412–424
28. Brombacher K et al (2006) The role of Yap1p and Skn7p-mediated oxidative stress response in the defence of Saccharomyces cerevisiae against singlet oxygen. Yeast 23(10):741–750

Qualitative and Quantitative Measurements of Sphingolipids by Mass Spectrometry

Nadia A. Rana, Ashutosh Singh, Maurizio Del Poeta, and Yusuf A. Hannun

Abstract Sphingolipids (SLs) serve the dual roles of acting as structural entities in cellular membranes as well as bioactive signaling molecules that modulate signal transduction. As the already immense database of identified bioactive SL subspecies continues to expand, the need for structure-specific identification and quantification continues to rise. The characterization and analysis of the sphingolipidome by mass spectrometry has advanced steadily over the last 20 years with the aid of improvements in technological advancements in instrumentation, coupled with optimization of lipid extraction methodologies, and an increasing library of available reference standards. Pivotal advances in sphingolipidomics include the adoption of soft ionization techniques, including electrospray ionization (ESI), tandem mass spectrometry (MS/MS), and matrix-assisted laser desorption ionization (MALDI), as well as the use of multiple reaction monitoring (MRM), all of which have aided in improving the quality of analysis of often complex lipid extracts from mammalian, yeast, and even plant cells. In this chapter we explore qualitative and quantitative mass spectrometry methods used for structural elucidation and quantitation of sphingolipids found in cells as well as tissues. Sections included here detail extraction and HPLC methodologies, in vitro labeling techniques, use of internal and calibration lipid standards for quantitation, and data analysis of sphingolipids derived from mammalian and yeast sources.

Keywords Lipidomics • ESI-MS • Sphingolipids • Mass spectrometry • Liquid chromatography • Tandem mass spectrometry

N.A. Rana
Department of Medicine, Stony Brook University, Stony Brook, NY, USA
e-mail: Nadia.Rana@stonybrookmedicine.edu

A. Singh • M. Del Poeta
Department of Molecular Genetics and Microbiology, Stony Brook University, Stony Brook, NY, USA
e-mail: Ashutosh.Singh@stonybrook.edu; Maurizio.DelPoeta@stonybrook.edu

Y.A. Hannun (✉)
Stony Brook Cancer Center, Stony Brook University, Stony Brook, NY, USA
e-mail: yusuf.hannun@stonybrookmedicine.edu

Y.A. Hannun et al. (eds.), *Bioactive Sphingolipids in Cancer Biology and Therapy*, DOI 10.1007/978-3-319-20750-6_14

1 Overview

Over the last two decades, research from several laboratories has established that while the role of sphingolipids (SL) includes serving as important structural entities, it in fact extends far beyond this scaffolding function due to the demonstration that they can also act as dynamic biologically active signaling molecules whose fluxes are key modulators of metabolism, cell stress response, and cell signaling. Therefore, the ability to monitor these fluxes is important in understanding underlying regulatory mechanisms of action. The number of bioactive sphingolipid subspecies within each class of sphingoid bases is significantly great. The highly interconnected nature of this network of lipids has necessitated the development of increasingly structure-focused qualitative and quantitative analyses of multiple components of this network. When a particular species' levels are perturbed within this web-like network, it causes shifts in the levels of many other species or nodes within the network, a metabolic ripple effect. Thus the importance of sphingolipidomics has only increased over the last decade, and will continue to do so as signaling and function of these molecular components continue to be elucidated. Mass spectrometry offers complementary structural and quantitative analyses that are enabling novel approaches to the study of bioactive sphingolipids.

2 Classes of Sphingoid Bases and Ceramides

2.1 Structural Characteristics

The diverse structural nature of lipids necessitates their broad classification. This is classically based upon common biological precursors and/or shared physical and chemical properties of said species, leading to class identity derived from shared head groups, fatty acyl/chain length, and additional moieties. A broad-based systematic classification of bioactive lipids, along with their functions, has been reviewed in recent years [1, 2]. Among these classes are sterols, fatty acids, glycerophospholipids, and sphingolipids. Briefly, sterols are bioactive lipids found in many metabolic processes in mammalian (cholesterol, fat-soluble vitamins, steroid hormones), plant cells (campesterol, sitosterol), as well as fungi (ergosterol). Glycerophospholipids are defined by the hallmark phosphate moiety esterified to a hydroxyl group of its glycerol backbone, coupled with acylation or alkylation within the backbone. Sphingolipids (SLs) are biologically active molecules that are found in all eukaryotes, and to a lesser extent prokaryotes and viruses. With greater than 400 possible head-groups, many of which exist in combination with several carbon backbone chain length possibilities (in some cases chain length permutations exist in the double digits for a given head-group) and additional modifications (such as desaturation, hydroxylation), to say the size of the lipidome is expansive is an understatement. The focus of this section will be primarily sphingolipid classification.

2.2 Grouping by Class and Involvement in Signaling Cascades and Regulatory Roles

Long-chain bases (LCBs) or sphingoid bases, are the initial products of de novo SL biosynthesis, and are the foundational components of all SLs. Sphingosine, also known as D-erythro-sphingosine, is the most abundant of the sphingoid bases found in mammals. The full scientific name is (2S,3R, 4E)-2-aminoaoctadec-4-ene-1,3-diol, but a more commonly used shorthand to the longer sphingoid base nomenclature exists for convenience. This nomenclature involves using a letter designation as a prefix to denote the number of hydroxyl groups present within the molecule (d=di, t=tri), then the number of carbons in the backbone of the compound, followed by unsaturations following a colon (along with their stereochemistry if known). In the case of sphingosine derived from mammalian cells, this would be denoted as d18:1transD4, or the even more abbreviated d18:1, as the stereochemistry and positional unsaturation are well-known. With its flexibility, this shorthand is able to denote additional hydroxylation sites, differing carbon backbone lengths, as well as other structural hallmarks. In mammals, SLs are comprised of a sphingoid base backbone, consisting of either sphingosine (Sph) [(2S, 3R, 4E)-2-amino-1,3-dihydroxyalkenes], or its saturated precursor sphinganine (Sa) [2S, 3R)-2-amino-1,3-dihydroxy-alkanes]. Moreover, the most abundant SL species are those comprised of an 18 carbon backbone [2-amino-1,3-dihydroxyocadecene and 2-amino-1,3-dihydroxyoctadecane, or 18C-Sph and 18C-Sa respectively]. In yeast, the endogenous sphingolipids are comprised of either the sphingoid base Sa, or phytosphingosine (phytoSph) [(2S, 3R, 4R)-2-amino-1,3,4-trihydroxyalkanes], with the more abundant species comprised of 18 carbon phytosphingosine as their core building block [2-amino-1,3,4-trihydroxyoctadecane]. In addition to the core sphingoid base, SLs are comprised of a variable head group moiety attached to the primary hydroxyl group, and an N-acyl group. These sphingoid bases are the core components of all SLs, and include a broad spectrum of alkanes and alkenes ranging in chain length from approximately 14–22 carbons in length, and are known to vary in positional unsaturation, hydroxylation, and methylation positions within this carbon backbone (Fig. 1).

Sphingolipids can be further categorized into bioactive signaling molecules, including ceramides (Cer) and ceramide-1-phosphates, sphingoid bases, and sphingoid base-1-phosphates. Mammalian ceramides are N-acyl-derivatives, and can be sub-classified as *N*-acylsphingosines (Ceramides) *N*-acylsphinganines (dihydroceramides), and *N*-acyl-4D-hydroxysphinganines (phytoceramides). Phytoceramides are not to be confused with alpha-hydroxyceramides, which bear a hydroxyl group on the fatty acyl moiety. The fatty acid components of ceramides exist in aliphatic chain lengths ranging from 14 to 26 carbons (and even longer in some tissues such as skin), typically found as either completely saturated or containing a single unsaturation. Nomenclature designates sphingoid base first, then fatty acid attributes following either semicolon or backslash. The sphingosine-based ceramide *N*-(octadecanoyl)-sphing-4-enine would be designated as

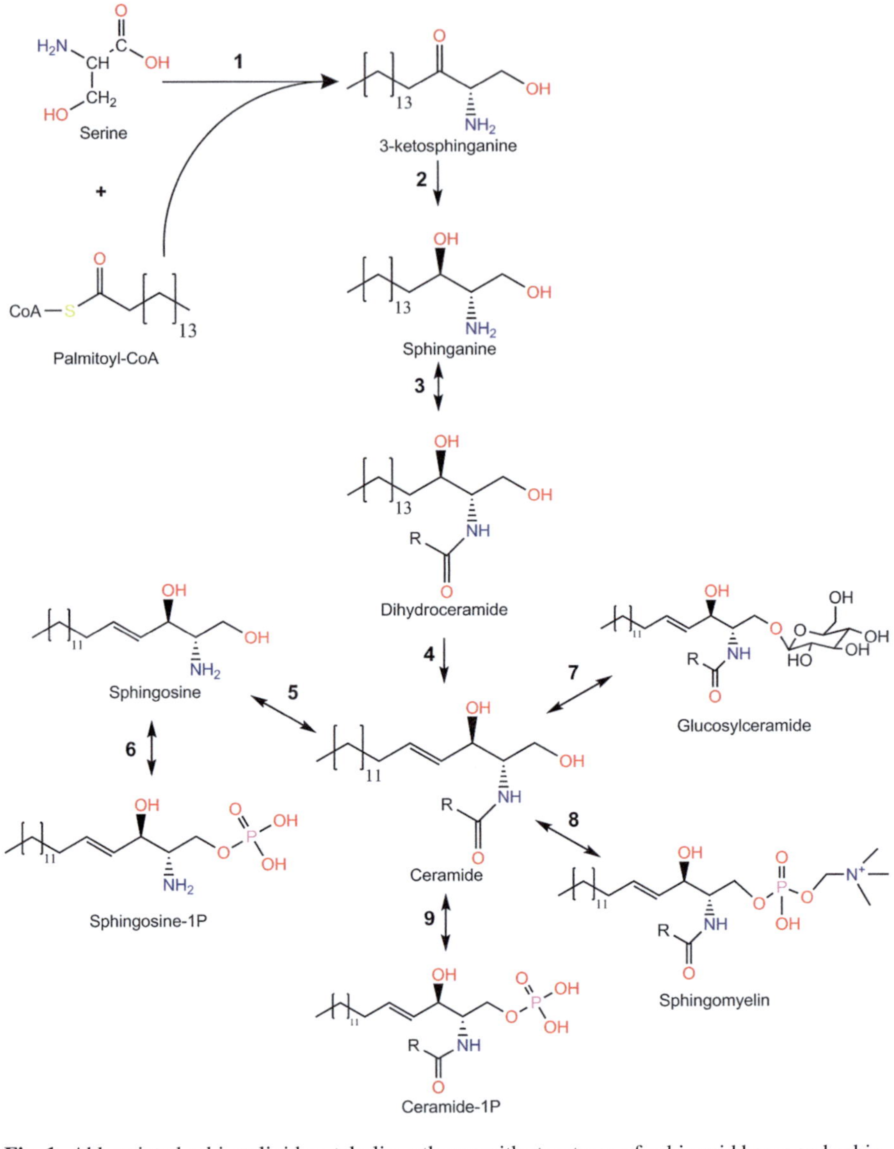

Fig. 1 Abbreviated sphingolipid metabolic pathway with structures of sphingoid bases and sphingolipids. Enzymatic reactions are numbered as follows, with enzymes responsible for reverse reactions indicated in brackets: (*1*) Serine Palmitoyl transferase, (*2*) 3-ketosphinganine reductase, (*3*) dihydroceramide synthases [ceramidases], (*4*) dihydroceramide desaturases, (*5*) ceramidase [ceramide synthases], (*6*) sphingosine kinase [sphingosine-1-phosphate phosphatase], (*7*) glucosylceramide synthase [β-glucocerebrosidase], (*8*) sphingomyelin (SM) synthase [SMases], (*9*) ceramide kinase [ceramide-1-phosphate phosphatase]

Cer(d18:1/18:0), whereas its sphinganine-based dihydroceramide counterpart, *N*-(ocatdecanoyl)-sphinganine, would be abbreviated Cer(d18:0/18:0. Saturated trihydroxies (t18:0) are also known as phytosphingosine, and are found in yeast and plant cells, and in low levels in animal tissues, including in the cells of the intestine, kidney, and skin.

Moreover, "complex" sphingolipids include phosphosphingolipids and glycosphingolipids, and contain a ceramide backbone, along with either a phospho- or glycosyl- headgroup moiety at the 1-hydroxy position. Ceramide phosphates and ceramide phosphocholines, or sphingomyelins (SM) as they are more readily known, make up the major subgroups of phosphosphingolipids. Glucose (glc) and galactose (gal) are the most abundant glycans that comprise glycosphingolipids, defined by their beta-glycosidic bond, and which can be further elongated with additional glycans or other functional groups. Other neutral glycosphingolipids bear other uncharged sugars including *N*-acetylglucosamine (GlcNAc), *N*-acetylgalactosamine (GalNAc), and fucose (fuc). Glycosphingolipids can also be acidic, bearing a charged Sialic acid, phosphate or sulfate functional moieties. Glycosphingolipid nomenclature and corresponding structural information details are coordinated between the glycobiology and lipid fields, and can be found both with the Consortium of Functional Glycomics as well as the LIPID MAPS Consortium websites (www.glycoforum.gr.jp and www.sphingomap.org).

2.3 *Deoxysphingolipids*

Serine Palmitoyl Transferase (SPT) is the enzyme responsible for the committed step of SL biosynthesis, and is essential for the generation and regulation of SLs. The reaction involves the condensation of Serine with Palmitoyl-CoA to generate the intermediate 3-keto-dihydrosphingosine followed by dihydrosphingosine. The ability for SPT to utilize amino acids other than Serine has been observed via examination of a point mutation in the catalytic site of SPT known to cause the human disease HSAN1 [3–5]. These groups demonstrated this mutant SPT has a shifted substrate specificity that allows for the use of the non-canonical amino acids Alanine (Ala) and Glycine (Gly) to generate two novel deoxysphingolipids (dSLs): 1-deoxysphinganine (dSa) and 1-deoxymethylsphignanine (dmSa). SPT substrate promiscuity was first suggested in regards to fatty acid alternatives to Palmitoyl-CoA. Previous work by the Hannun laboratory demonstrated that SPT is able to use Stearoyl-CoA as a substrate in the generation of C20 sphingolipid species, due to the presence of a small subunit, Tsc3p, an 80 amino acid peptide associated with SPT, and was later confirmed in mammals by Han et al. [6, 7]. These data widen the array of known sphingoid bases, and open the possibility of an even larger, as yet undiscovered set of sphingoid bases ranging in chain length and functional groups present.

3 Existing Methodologies for Detection of Bioactive SLs

3.1 Classical Radiolabeling and Thin Layer Chromatography (TLC)

The classical methodology for detection of sphingoid bases has been metabolic radiolabeling coupled with TLC and autoradiography. This has involved introduction of heavy-labeled precursor compounds of sphingolipid biosynthesis to cells (traditionally Serine or palmitate to follow incorporation and synthesis from the committed step of the pathway), and has allowed for analysis and quantitation of lipid contents in both mammalian and yeast cells. While this technique does allow for following of bulk SL class production, it does have several drawbacks. One is the low level of uptake and subsequent incorporation of heavy label into different cell types. Next are the low resolution capabilities and low sensitivity levels of detection of both TLC and autoradiography. In addition, though TLC does allow for resolution of classes of sphingoid bases, identification of individual species can be cumbersome and dependent upon existing standards being available for the species of interest in sufficient quantities for detection. Finally, there are the drawbacks of generation of radioactive waste in the laboratory. Heavier labels tend to give improved resolution, but this comes at the expense of exposure to more harmful heavy isotopes, as well as their disposal. It is for these reasons that mass spectrometry has grown increasingly attractive as an alternative methodology for detection and quantitation of SLs. It offers detection with lower amounts of starting material, higher sensitivity of low abundance species, the option of label-free quantitation, higher resolution of individual subspecies of sphingoid bases within a molecular class, simultaneous qualitative and quantitative analyses of different classes of sphingoid bases within a single analysis, and high-throughput quantitative analyses of samples within an experimental set.

3.2 High Performance Liquid Chromatography (HLPC)

Sphingolipids are capable of being resolved by HPLC using a reverse-phase C8 or C18 column, in conjunction with either a binary gradient of methanol and water for complex mixtures (e.g., cell lysates), or an isocratic one for less complex samples (e.g., those containing fairly purified components from in vitro reactions). Gradients can be fortified with ammonium formate and formic acid when HLPLC is coupled to mass spectrometry to enhance ionization efficiency. HPLC is a useful technique when combined with heavy metabolic radiolabeling, or when lipids are derivatized with a chromophore at their free amine, to aid in their detection by ultraviolet (UV)/visible or fluorescence spectroscopy [8]. Drawbacks of this methodology include the need to use labeling for detection purposes, as well as large quantities of starting material necessary for detection of derivatized materials, and the fact that the compounds being studied are being chemically modified, which add both complexity and additional steps to the sample and data analyses. For additional details regarding derivatization, see Sect. 3.3.2.3.

Evaporative Light Scattering Detection (ELSD) is another alternative that was designed for use with HPLC, and can be used to analyze non-UV absorbing substances, without the need for derivatization [9]. Here, a heated nebulizer atomizes analyte eluate and signal is generated based on light scattering. Though signal is produced in a direct relationship to the number of particles present, shortcomings include a lower sensitivity for lower molecular weight materials, interference from salt-containing solvents, and that this instrument does not produce a linear response. The most significant drawback of this technique is the lack of specificity in detection, as any particle may contribute to signal regardless of structure.

3.3 Mass Spectrometry (MS)

A significant step towards bringing quantitative and comprehensive lipid analyses into the realm of—omics studies has been cross-class specific lipid structural assignment within a single sample, coupled with reproducible quantitative measurements of levels of these lipids. This was a major step forward for two reasons. First, it imparted exactitude to the species being reported (specific levels reported for the subspecies, i.e., C16, C18, C24-Ceramide; in contrast to levels of ceramide as a bulk class), allowing for more focused directions for follow-up studies as well as examination of the biological activity of several subspecies being regulated. Second, it moved away from monitoring a single compound thought to be of significance in a given study, and began examining global changes of an entire network and the implications and repercussions of a single player within that network having ripple effects on those other players in a more dynamic fashion. This opened up the opportunity to look at the interplay of sphingoid bases within the network, and treated the nodes of the pathway less as static entities and more as the dynamic molecules they are. Specifically in the case of sphingolipidomics, this has involved defining the sphingoid base backbone, the N-linked fatty acids, the headgroups, and other potential modifications along the long chain bases. Class-specific lipidomics by HPLC-ESI-MS (HPLC-electrospray ionization mass spectrometry) has been used to identify sphingoid bases, ceramides, hexosylceramides (HexCer), SMs, and glycosphingolipids, and has been evolving over the last two decades [10–16].

An exhaustively nuanced and thorough nomenclature exists, though an important caveat that should be mentioned is that it is not always possible to definitively demonstrate, solely by MS techniques alone, every characteristic of a molecule's structural architecture, including the stereochemistry within the compound as well as positions of unsaturation along the carbon backbone. Using biochemical studies as a 'database' of sorts, we can favor one predicted structure over another based on those previously reported within a species or a cell line. This is a routine and established practice for the other—omics fields, including genomics, glycomics, and especially proteomics. In a case where the stereochemistry is both in question and of potential biological significance, synthesized compounds can be utilized and classical biochemical techniques employed to attempt to elucidate the positional stereochemistry.

MS offers the ability to carry out a combination of structurally specific and quantitative analyses of bioactive subspecies of sphingolipids. This "sphingolipidomics" analysis can be conducted by several different approaches of sample preparation, instrumentation, HPLC conditions and ionization modes, all designed to cater to analysis of particular SL classes. A hallmark of mass spectrometry is that amongst instruments, sources, methodologies, and even compound classes, ionization efficiencies will vary, sometimes significantly. These alternative methods have been developed and refined based on structural characteristics of the different sphingolipid classes to limit this variability depending upon MS instrumentation: sphingoid base, polar head group, and N-acyl chain length and saturation state are the major structural hallmarks that help define the above-stated parameters for a given class of molecules. Later, methods for accounting for remaining ionization efficiency differences based upon chemical compound characteristics through selection of internal standards will also be discussed.

3.3.1 Extraction Methodologies

Bligh and Dyer and Other Modern Lipid Extraction Protocols

The importance of extraction methodology for accurate and reproducible lipid measurements cannot be understated. It is essential to empirically test efficiency of extraction for different sample types for one's system (solid samples from mammalian cells and tissue, as well as yeast cells; versus liquid samples derived from serum, plasma, and other biological fluids). The original extraction solvents of the early 1900s were very harsh and composed of ethanol and diethyl ether (3:1), and involved heating samples to approximately 60 °C over several hours [17]. As drawbacks of the extraction method became apparent, including peroxidation in mammalian lipids, inefficient lipid extraction, and phospholipid fluxes in plant samples attributed to residual phospholipase activity, an alternative method was proposed by Folch and published in 1951. This involved the classic chloroform methanol extraction, where the combination of the nonpolar solvent and polar alcohol allowed for highly efficient extraction of total lipids, including the phospholipids, resulting in a more complete overall extraction when used in Folch's 2:1 ratio at room temperature, in combination with large volume aqueous washes to purify out contaminant cell remnants (i.e., proteins, DNA, carbohydrates). Although chloroform:methanol extractions were highly efficient, the issue of emulsion formation was a significant drawback, and eventually led to the replacement of this extraction method by one that is most widely used and is considered the traditional lipid extraction methodology: Bligh and Dyer. In 1959, Bligh and Dyer published their lipid extraction methodology for use on marine tissue, specifically muscle tissue from fish. The innovation here was accounting for the presence of significant aqueous content of the samples for a two-step extraction, and using that to generate an initial chloroform:methanol:water ratio of 1:2:0.8, and a final ratio of 2:2:1.8. This allowed for initial generation of a single extraction phase, followed by biphasic extraction through the addition of water and chloroform, where lipids are recovered from the lower chloroform phase and non-lipid sample contents remain the upper methanol:water portion. After this water:methanol wash, the lower chloroform phase is dried under nitrogen, and the dried extracts are stored at −20 °C or colder. Currently, the Bligh and Dyer extraction, sometimes accompanied with minor variations of the standard protocol, is still widely in use for SL sample analyses, as this extraction method is compatible with several differential downstream applications including TLC, HPLC, and MS.

Drawbacks include extraction efficiency differences for polar lipid species, including the phosphorylated sphingoid bases sphingosine-1-phosphate and dihydrosphingosine-1-phosphate, as well as lyso-sphingomyelin. In response to the need for more efficient and reproducible extraction and detection of the broadest and most inclusive list of sphingoid bases, base hydrolysis has been included as part of the first chloroform extraction to cleave ester linkages of glycerolipids and other co-extracted compounds that act as major contaminants concomitantly extracted with sphingoid bases. This mild alkaline hydrolysis is highly recommended when monitoring SM levels, as it depletes the interfering compound phosphatidylcholine from the final extract. This is important for MS analysis, where overlapping precursor ion masses and similar fragmentation patterns complicate accurate peak identification and quantitation. In addition, a possible undesired consequence of base hydrolysis is the inadvertent hydrolysis of 1-O-acyl-ceramide species. Even if one is not interested in monitoring fluxes in this class of compounds, the effects of their hydrolysis will inadvertently affect ceramide levels observed by inflating them significantly. This inflation can be taken advantage of if one is interested in levels of this class of compounds, as one can differentially treat control and experimental sample sets with and without base hydrolysis, and account for experimental fluxes in levels of specific 1-O-acyl-ceramides by measuring the difference in ceramide levels due to the base hydrolysis. Additional sensitivity for these polar species can be achieved through adjustments to HPLC gradient coupled with scan ranges and MRM (multiple reaction monitoring) segmentation (as discussed in Sect. 3.4.5).

A number of alternative solvent systems that are suitable for lipid extractions for a multitude of lipid classes have also fallen into routine use. These include the Folch, original Bligh and Dyer, acidified Bligh and Dyer, methanol-tert-butyl methyl ether (TBME) and hexane-isopropanol. Currently Bligh and Dyer remains the extraction solvent of choice for broad range lipidomics studies, though Folch extractions are still popular for some tissue extractions. TBME has gained popularity in recent years due to lack of toxic chlorinated solvents in the extraction solvent, and has been successfully used for highly aqueous samples for sphingolipidomics studies, providing lipid profiles similar to Bligh and Dyer extractions for human plasma. It is difficult to compare efficiencies from different studies conducted by different groups, as extraction solvents, as well as MS mobile phases and instrumentation will all factor in ionization efficiency, separation, fragmentation, and detection capabilities. An in depth analysis of efficacy of differential extraction solvents across lipid classes was conducted recently to provide a baseline comparison of solvent extraction efficiencies [18]. Reis et al. found that although the major lipid classes showed small if not negligible differences in extraction efficiency when sample source, solvent volumes as per each extraction protocol, and ionization efficiency on various MS instrumentation are taken into account, less abundant species did show more significant yields in a solvent-specific manner. Folch and acidified Bligh and Dyer were found to yield the highest extraction efficiency across the broadest range of lipid classes, and these studies make the case for optimization of extraction conditions in a lipid class-dependent manner.

In our laboratory, we currently use a one-phase (azeotropic) solvent extraction which utilizes ethyl acetate: isopropanol: water at a 60:30:10 ratio for cell and tissue extractions, and at 85:15:0 for aqueous/media samples, all by volume (Fig. 2) (For recent examples, see [19, 20]). This system allows for total lipid extraction with minimal loss of the labile O-acyl groups and notoriously difficult to recover S1P and dhS1P species [21–23].

3.3.2 In Vitro Labeling

17C-Labeling Methodology

Taking advantage of synthetically generated sphingosine (2-amino-1,3-dihydroxyheptadecene) and sphinganine (2-amino-1,3-dihydroxyheptadecane) whose sphingoid base consists of an aliphatic chain length of 17 carbons instead of the naturally occurring 18 carbon backbone, we can metabolically label downstream products of some enzymes within the sphingolipid pathway, such as ceramide synthases and sphingosine kinase [24]. Cells are treated with C17-Sphingosine at a concentration dependent upon the cell line utilized, typically in the micromolar range, either directly in media, or concomitantly with other treatments. Harvested cells are washed with PBS, and cell pellets are subjected to standard lipid extraction using ethyl acetate: isopropanol: water as described above (60:30:10 (v/v)) using 50 pmol each of 13C16Cer and 13C22Cer as internal standards at the time of extraction. The calibration curve for this method consists of the unnatural C17/C16, C17/C18, C17/C24, and C17/C24:1 Ceramides (now commercially available from Avanti Polar Lipids), to maximize the number of different chain length ceramides accurately quantifiable. A similar calibration curve that is comprised of the C18 dihydroceramides can also be used to measure these additional species of interest within the same experimental set.

^{13}C and Deuterated SL Precursor Metabolic Labeling

Metabolic labeling using chemical isotopes of lipid precursors, such as deuterated Serine or [U-^{13}C]-Palmitate, allow for monitoring of turnover and kinetics of formation for a broad range of lipid classes, including sphingolipids, phospholipids, and

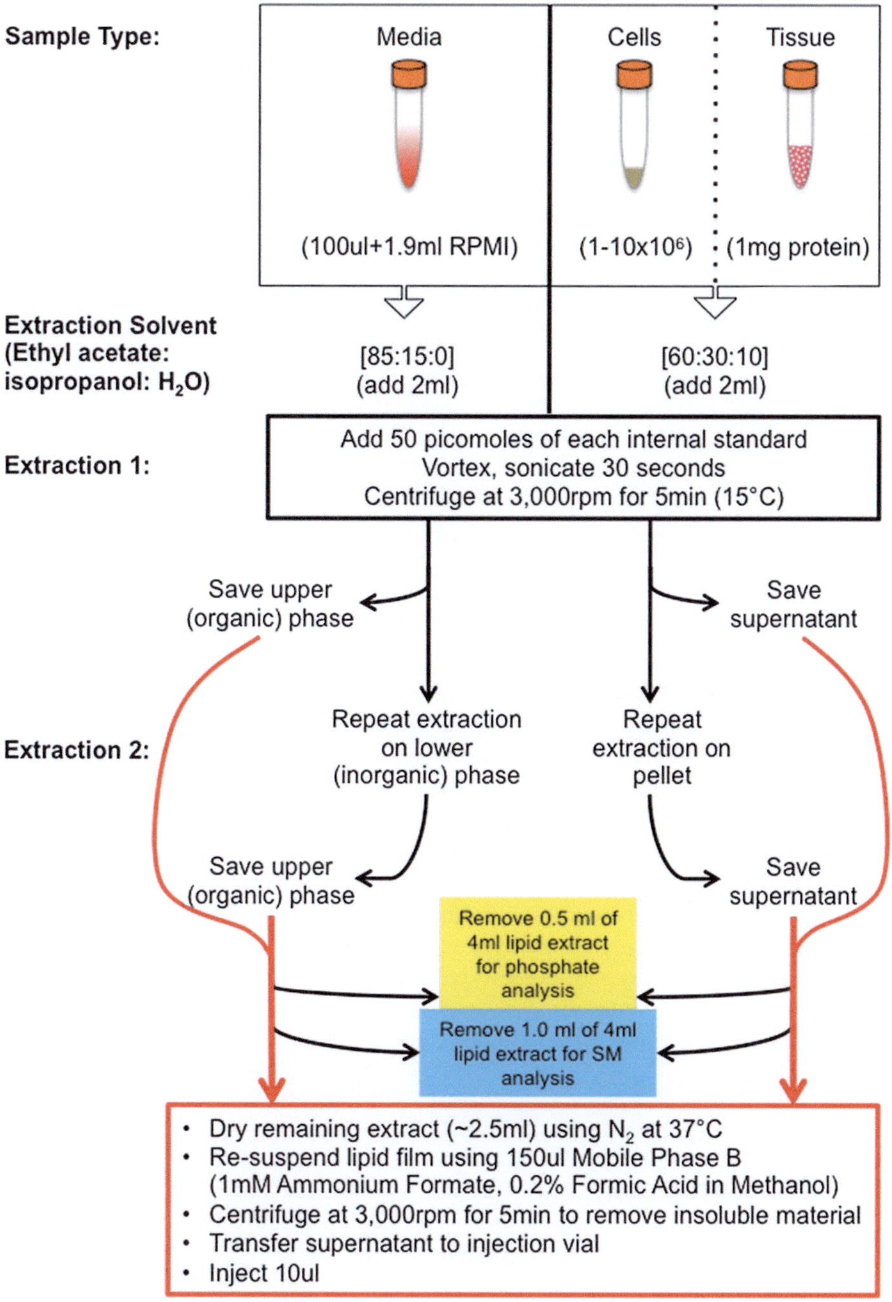

Fig. 2 Lipid extraction procedure. Samples from media, cell, or tissue sources are extracted using ethyl acetate: isopropanol: water extraction solvent in different ratios depending on sample type. All samples are spiked with 50 picomoles of internal standards of interest prior to extraction 1 (exception is 250 picomoles for SM samples). Following a second extraction, supernatants or organic phases (dependent upon whether samples were from media or cell/tissue sources) are pooled together, and aliquots for phosphate and SM analyses are removed prior to drying down lipid extract. Dried lipid film is then resuspended in MS Mobile Phase B prior to injection on a triple quadrupole mass spectrometer

acyl-CoA. This would involve culturing cells in media containing the labeled precursor and BSA for a short pulse or up to 24 h, extracting lipids as described, and then analyzing for mass shifts between standard peaks and the heavier isotopic peak, which should differ by several mass units. In the case of D3-Serine, which bears three positions of deuteration and therefore increases mass of species it is incorporated into by 3 Da, this peak shift should be detectable on a triple quad instrument. For smaller mass shifts, a more sensitive and mass accurate mass spectrometer, such as an Orbitrap, would be more appropriate [25, 26]. For additional details regarding labeling methods, see Sect. 5.

Derivatization

A decade ago, it was necessary to derivatize some lipid classes in order to aid in their ionization for gas chromatography-MS (GC-MS). Due to the inverse relationship between lipid size and vaporizability [27], bulky lipids such as the eicosanoids and polar ones such as sterols required not only derivatization in order to achieve vaporization and transition to the gas phase, but different subspecies required distinct methods of derivatization in order to allow for efficient volatility of the compounds of interest [28–30]. Today, ESI-MS circumvents this issue by allowing for liquid chromatographic (LC) separation and introduction of the analytes in aqueous form at the source interface. For neutral lipids whose ionization remains problematic even when carrying out ESI-MS, these species can be detected in positive ion mode via cationic adducts (ammonium, sodium) [31–34]. Derivatization is still carried out in some instances where samples need to be analyzed initially by HPLC/UV and later by MS analysis. In this case, derivatization may still be necessary to achieve adequate chromatographic separation as well as detection of SLs (see HPLC-ESI-MS). While derivatization does not inherently interfere with mass spectrometry, it can complicate lipidome-type analyses, due to higher amounts of fragmentation generating more complex spectra. For these reasons, it is preferred to keep compounds in their intact (underivatized) form, to avoid these analytical complications.

3.4 Instrumentation

3.4.1 Electron Ionization MS (EI-MS)

Structural determination of ceramides and glycosphingolipids were first carried out using EI-MS in the late 1960s [35–39]. EI-MS allowed for intact characterization of sphingolipids through observation of diagnostic fragmentation patterns, but it also required derivatization using permethyl ethers or trimethylsilyl ethers to increase their volatility by counteracting the polar nature of these bulky molecules that were otherwise difficult to ionize to the gaseous phase [40–43]. Considered a “hard” ionization method due to its utilization of high energy to force fragmentation, a major drawback of this method included overly extensive, inconsistent fragmentation which led to some molecular ions not being observed [44]. Though lowering the fragmentation energy did allow for improved fragmentation for some molecules, this was at the expense of a loss of intensity from decreased ionization efficiency at the lower electron energy level. Another shortcoming was the low resolution and complicated spectra generated from complex mixtures of sphingolipids. Product ions from various precursors (multiple sphingoid bases, N-acyl chains and headgroup species from different lipids present in the sample) produced complex spectra.

3.4.2 Electrospray Ionization MS (ESI-MS)

ESI-MS was first developed in the late 1960s, but its first practical applications arose by way of John B. Fenn in the 1980s, and its advent revolutionized both qualitative and quantitative mass spectrometry across the—omics disciplines, and paved the way for modern lipidomics capabilities [27]. ESI-MS involves sphingolipid analytes in solvent being introduced by infusion to the ion source of the mass spectrometer via a narrow metal capillary or "needle", concomitant with the application of high potential to said needle. This high potential can be either positive or negative in nature, and selection of which ion mode to utilize is dependent upon lipid class(es) being analyzed. As volatile solvent filled with mostly cationic or anionic analytes flows through and exits the tip of the needle, the solvent stream bursts into an aerosol of fine droplets and is exposed to both high potential and vacuum pressure as the droplets enter the orifice of the mass spectrometer. As the charged droplets get further desolvated in the presence of N_2 gas, they form still smaller droplets, continuing to subdivide further still until completely desolvated. The result is the production of a charged solute/analyte that exists as a gas-phase solute ion (Fig. 3). This "softer" ionization technique yields primarily intact molecular ions that have not undergone "in-source" fragmentation (i.e., fragmentation within the ion source that occurs prior to complete desolvation, where it is not desirable for our purposes, as it occurs prior to precursor ion detection in Q1) [10, 14, 45–47].

3.4.3 HPLC-ESI-MS

HPLC is a widely used separation technique that allows for isolation, purification and resolution of compounds using liquid chromatography in tandem with various detectors. Previously, derivatization was necessary to carry out HPLC with lipids in order to allow for UV detection. This was necessitated due to the fact that intact natural SLs absorb mainly in the 200–210 nm range, which overlaps with the absorption range of routinely used lipid-compatible solvent systems, causing a masking effect by the solvent and preventing efficient detection [48, 49].

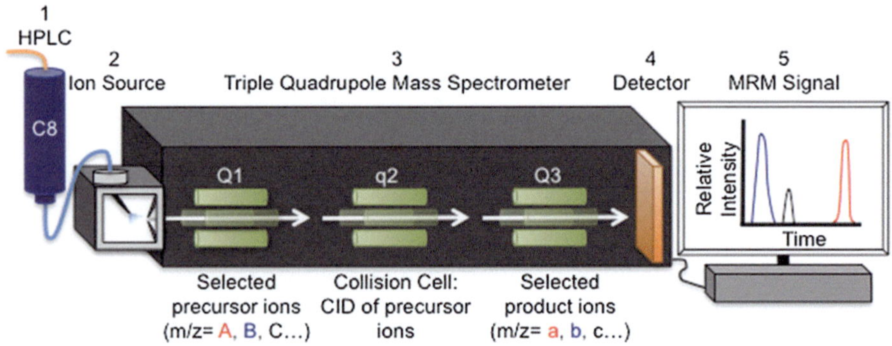

Fig. 3 MS sample analysis on QqQ. MS samples are separated on a C8 column, and eluent is sprayed from the ion source and desolvated before entering the ion transfer tube and subsequently the three series of quadrupoles (Q1, q2, Q3). Q1 is where precursor ions are selected based on m/z values inputted (A, B, C …). Precursor ions continue to the collision cell (q2) where they undergo fragmentation by CID. Resulting fragmentation products that correspond to inputted product ion m/z values (a, b, c…) are separated by Q3 and measured by the detector, and are reported as relative intensity over the course of the chromatographic run

The coupling of HPLC online with ESI-MS has taken advantage of the separation power of the liquid chromatography system prior to infusion of analytes into the mass analyzer. This has allowed for separation of lipid subclasses, as well as individual species separation and desalting of samples (cations can cause ion suppression), the amalgamation of which has resulted in enhanced sensitivity. In addition, this separation promotes a less complex mixture of analytes eluting off the column and entering the mass spectrometer for identification and quantification at any given time over the course of a sample run, reducing compound-to-compound interferences as well as ion suppression.

3.4.4 Solvent Systems

When designing a gradient system, it is important to develop a method compatible with an array of compounds of varying properties (polar to nonpolar, water soluble and insoluble) in order to capture the entire lipidome in a single profile. Normal phase and reverse phase HPLC are both used for ESI-MS of lipids, depending on the nature of the lipid classes being monitored [10, 23, 50–53]. Normal phase is utilized for separation of species by headgroup (i.e., Cer, Cer-P, SM, HexosylCer), whereas it is more standard practice to use reverse phase when separating based on carbon backbone length, as well as the saturation of N-acyl chains (i.e., Sph, dhSph, Sph1P, dhSph1P, Cer-P). Complete chromatographic separation of sphingolipid subclasses can be difficult, but it is not essential for accurate detection and quantitation when using HPLC-ESI-MS, with the understanding that each of the overlapping/co-eluting species has a unique precursor ion m/z, as well as unique product ion m/z. It is the uniqueness of these precursor and product ion transition pairs that allow the mass spectrometer to distinguish similar compounds that may be co-eluting. A notable caveat to complete chromatographic separation not always being essential is that care must be paid to monitor for potential isotopic effects that can skew quantitative results for compounds of similar m/z values. Sphingosine (Sph) and dihydrosphingosine (dhSph) are a classical example of this issue. Sph has a precursor ion mass to charge ratio of 300 while dhSph is 302 m/z. Each sphingoid base also has a unique transition ion; 282 and 284 m/z, respectively. The issue lies in the [M+2] isotope of sphingosine, which has an approximately 2.0 % abundance naturally, and this Sph isotope's precursor and product ion m/z values are identical to those of dhSph [54]. Further compounding the issue is the fact that dhSph is of significantly lower abundance in most biological samples (up to an order of magnitude). In this case, it is ideal to adjust gradient elution of the LC mobile phase system to increase separation and resolve these species chromatographically [55].

3.4.5 Scan Modes

There are different types of mass analyzers that can be coupled to LC systems, including ion traps and quadrupole systems, which range in mass accuracy and application. Here we will discuss the basic parameters of triple quadrupole instruments. Triple quadrupole mass spectrometers are ideal for sphingolipidomics quantitation as they provide reasonable mass accuracy and allow for MRM, where a set of precursor and product ion pairs for sphingolipids of interest are selected and fragmented by the instrument. They contain three quadrupoles (Q1, q2, Q3), where the first and last cells (Q1, Q3) are mass analyzers, and the central quadrupole (q2) is a collision cell that utilizes radiofrequency (RF) only and induces fragmentation by Collision Induced Dissociation (CID) via a collision gas (Fig. 3) [44]. During a sample run, there are several types of scan modes that the tandem mass spectrometer can be utilized for, and they are described here:

Product/Fragment Ion Scan

In product ion scans, Q1 acts as a filter, where a parent or "precursor" ion of selected m/z value is identified and monitored only (MS1). The selected ion undergoes CID within q2, resulting in fragmentation into daughter or "product" ions that act as a fingerprint that helps define the identity of the precursor ion by providing structural information. These product ions are then analyzed within Q3 over a user-defined mass range and measured by the detector (MS2).

Precursor/Parent Ion Scan

In precursor ion scan mode, Q1 scans precursor ions across a user-specified mass range, subjects these ions to fragmentation in q2, and then Q3 acts as the filter, monitoring only a specific diagnostic "product" ion of a selected m/z, and reporting data only for those precursor ions that resulted in formation of the specified product ion m/z.

Neutral Loss Scan

In neutral loss scans, both Q1 and Q3 are scanned together, but with a constant mass offset between them. This is particularly useful in discovery work, where one can monitor a mixture of unknown analytes losing a common fragment, e.g., water molecule (a standard diagnostic used for monitoring sphingolipids), which presents as a neutral loss of 18 mass units, to determine the presence of SLs. Neutral loss scans can also be used to monitor other labile moieties such as glycan modifications which can also result in an observable neutral loss.

MRM Scanning

MRM takes advantage of being able to define specific precursor ion m/z for Q1 and pairs it with a specific product ion m/z for Q3, allowing for quantitative analysis of sphingolipids identified by any of the above mentioned scan modes. The advantage of MRM is that, as the name implies, multiple precursor and product ion pairs (i.e., multiple sphingolipids) can be monitored during the course of a single chromatographic run, allowing for complex samples containing multiple sphingolipid classes and subspecies to be quantitated for these transitions simultaneously.

As product ion abundance is dependent upon efficiency of fragmentation, adjustment of collision energies (CE) can optimize fragmentation, and this CE optimization should be carried out for each class of SLs being analyzed, as well as for subspecies within classes. The general trend observed across laboratories is that ionization parameters for sphingoid bases of 17–20 carbons tend to be similar. Unsaturations, additional hydroxyl modifications, and N-methylation, along with increased carbon chain length tend to require increases in CE in order to achieve adequate CID. Diagnostic fragmentation patterns for sphingoid bases are single and double-dehydration products (Fig. 4). The ratio between these products can be shifted to bias one over the other by adjusting CE values as well, and this can be beneficial depending upon which SL class one is monitoring, and whether there are overlapping precursor/product ion transition m/z values and isotopic effect is a concern.

3.5 Lipid Standards

3.5.1 Internal Standards

The selection of which internal standards to include within a single experiment, an experimental set, and even within project(s) is a significant technical detail that can greatly improve the quality of the single experiment, consistency within an experimental set, and allow an investigator the ability to return back to a previously prepared and analyzed set of samples, this time examining the same treatment conditions for changes in the levels of a different class of lipids, without

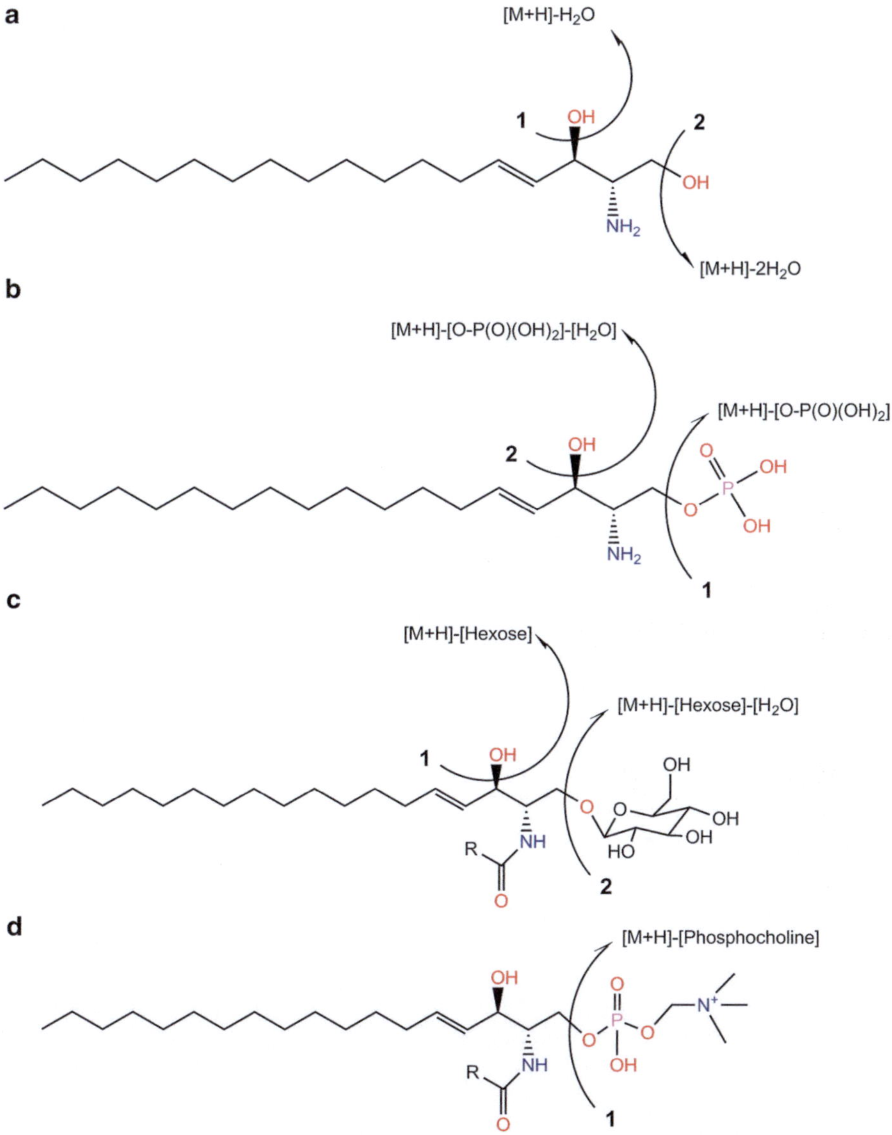

Fig. 4 Standard MS/MS fragmentation of class-representative sphingolipids. Fragmentation of long-chain base sphingosine whose parent ion is 300 m/z (panel **a**) results in a water loss at the 3-postion (*1*) yielding a 282 m/z ion and a second dehydration at the 1-position (*2*) resulting in a major product ion of 264 m/z. Long-chain base phosphate sphingosine-1-phosphate (panel **b**) undergoes loss of phosphate at the 1-position (*1*) followed by dehydration at the 3-position. (**c**) Galactosylceramide (panel **c**) undergoes loss of hexose at the 3-position, and dehydration at the 1-position (**d**) Sphingomyelin loses its choline group at the 1-position, resulting in the diagnostic fragmentation product ion of 184 m/z

having to re-culture and treat cells, harvest, re-extract, normalize, and finally quantify additional species of their new focus/interest. The consensus across research groups including the LIPID MAPS consortium has been that use of a single internal standard as a single point calibrant is ill-advised [22]. The issue here lies in the ionization efficiency and fragmentation of that selected internal standard being able to produce an MS response level equivalent to all molecular species within the same class. This is a poor practice as these responses are known to vary greatly based on several structural characteristics of the compound in question, including number of carbon atoms in the sphingoid base backbone, presence of unsaturations, and any potential branching (such as those found in yeast phytosphingoid bases). LIPID MAPS advises inclusion of one compound per class of lipid being quantitated to mimic all the analyte classes within a biological sample that are to be monitored. Although a compound chemically identical to the sphingolipids being analyzed would be the ideal internal standard (i.e., ^{13}C-labeled SLs), this is not practical based on synthesis costs and number of compounds that would be necessary for so many different classes of SLs. Most researchers rely on SLs of unnatural chain length fatty acyl groups (i.e., 17C-Sphingosine). Our laboratory has made the use of 17-carbon sphingoid base SLs as internal standards a routine practice for lipidomics experiments, as physicochemical properties are shared with the naturally occurring 18-carbon counterparts and reliably consistent (these include diagnostic fragmentation pattern and retention time and elution order between species) [23].

3.5.2 Calibrants

Quantitation capabilities of laboratories carrying out lipidomics experiments have expanded in recent years with the increased availability of authentic and unnatural standards across the SL classes. Previously this was only achievable by research groups that had a strong synthetic chemistry collaboration to custom synthesize standards for the various classes [23]. While for some subspecies this is still the case, there are currently an everexpanding number of options available from both Avanti Polar Lipids Inc. (Alabaster, AL) and Matreya Inc. (Pleasant Gap, PA). In practice, to achieve the most accurate quantitation for all sphingolipid species of interest, a calibration curve should be generated that contains as many representative species as possible. The working linear range of calibrants should be determined empirically on each instrument to account for differences in solvent systems, ionization efficiency, in-source dehydration, instrument sensitivity and instrument calibration differences. Peak identification is aided by consistent retention times, where elution times of subspecies for which standards are available can be used to calculate predicted elution times for theoreticals (lipid species for which an authentic standard does not exist). As with internal standards, ionization efficiency and fragmentation of standards are a concern for authentic standards too, and has lead to our use of an eight-point calibration curve for each analyte [21]. Authentic standards are prepared with internal standard(s) of choice spiked in, and subjected to extraction and analyzed in positive ion mode by MRM. To generate SL species-specific calibration curves, analyte and internal standard peaks at known retention times are integrated using instrument software (LC Quan for the TSQ Quantum triple quadrupole MS), and analyte/internal standard peak area ratios are plotted for each of the eight concentrations. For those sphingolipids that standards are currently unavailable either commercially or by custom synthesis, the nearest related lipid is assigned for quantitation based on chain length, head group, and other structural hallmarks in common (e.g., C22-Phytoceramide does not have an available standard, and C24-Phytoceramide is utilized for quantitation).

4 Data Processing

Quantitative analysis of lipid levels by mass spectrometry is typically reported in picomoles for a given SL species within a lipid extraction sample injection. Results are typically then normalized to a parameter that is expected not to deviate across samples and treatment conditions. The most commonplace are phospholipid content, total protein, and cell count. Of the three, the most stable appears to be the phospholipid content (reported in nanomoles) by quantitation of phosphate content relative to a standard phosphate analysis curve, and can be carried out readily for mammalian and yeast cells having undergone Bligh and Dyer extraction. Normalization to protein levels (reported in milligrams of total protein) or to total number of cells isolated for lipid extraction and also contained within Bligh and Dyer extracts, are also beneficial stable parameters of normalization. Care should be taken to report these values as percentages or fold-changes relative to the control. In addition, data post-normalization via phosphate levels versus protein levels do look different, though trends are the same [21, 22]. Of note, lipid datasets can also be normalized to the dry weight of extracted lipids, as in the case of yeast and plant samples. However, accurate calculation of the dry weights is often not feasible for mammalian samples due to low post extraction lipid amounts. It is highly recommended that, once a researcher has established a normalization parameter compatible with their experiments, they continue using that parameter for normalization of all subsequent experiments within that data series and study, in order to obtain the most quantitative and comparable datasets. Having the foresight to use the same normalization routinely will also give flexibility to the investigator to compare data sets between experimental sets as well as projects carried out over time, and without having to account for this deviation. If for example a set of samples previously used for one project is analyzed for different lipid profile contents at a later time, but were normalized to different parameters, the results will not be quantitatively consistent.

5 Alternative Methodologies

5.1 Stable Isotope Labeling

Over the years radioisotope labeling involving the incorporation of ^{14}C from labeled acetate has been used extensively to follow lipid metabolism within the cell. Another approach to study lipid metabolic flux is the use of tritiated water. However, both these approaches have certain drawbacks and are now less commonly used. For example, while ^{14}C-acetate is incorporated into the acetyl-CoA pool by the enzyme acetyl-CoA synthetase, ^{14}C-acetate label may be readily diluted by the unlabeled acetate pool of the cell. It is quite difficult to determine the dilution factor in such cases. A similar dilution effect is also observed in the case of tritiated water where the ^{3}H pool is affected by unlabeled hydrogen atoms coming from the NADPH pool of the cell. Other concerns include tedious sample preparation and radiation hazards.

In recent years, stable isotope labeling has provided a suitable alternative to radioactive labels for studying lipid metabolic pathways. Molecules with certain number of incorporated isotope labels (^{13}C or deuterium (D) in case of stable isotopes) into a compound are referred to as mass isotopomers [56–58]. The distribution and enrichment of the label within a mass isotopomer can be determined by its mass spectrum, referred to as mass isotopomer distribution analysis (MIDA) [56–58]. Although the mass isotopomers have similar masses, these may be resolved using ultra-high resolution mass spectrometry, or in some cases by increasing chromatographic separation on the LC system prior to MS analysis. Stable isotope labeling coupled with high throughput mass spectrometry based lipidomics is a robust method to understand the synthesis and turnover of lipids, and is a very useful approach to study SL metabolism.

5.1.1 Stable Isotope Labeled Polar Precursors

Polar precursors like D_2O, ^{13}C-acetate and [U-^{13}C]-glucose are often used to study lipid synthesis and turnover [58]. However, labeled head groups present a more selective approach to profile lipids. For example in the case of SLs, a [U-^{13}C,^{15}N]-serine label can be incorporated at the very beginning of the SL biosynthetic pathway [59] (Fig. 5). This approach allows the coverage of all SL classes. Another stable isotope D_9-choline, which labels phosphocholine, is incorporated into SM by sphingomyelin synthase. The D_9-choline label therefore is more specific towards monitoring SM flux [60]. The lipids can be analyzed by LC-ESI-MS/MS. Polar head group labeling is advantageous over the other labeling methods because: (a) these are water-soluble, therefore are easy and safe for cell and in vivo experiments; (b) single step labeling allows scanning of all the species of the biosynthetic pathway under study. However there are certain disadvantages to the polar head group labels. Polar head group labels often result in smaller mass shifts that are difficult to differentiate from the isotopic peaks of unlabeled lipid species. Also poor labeling

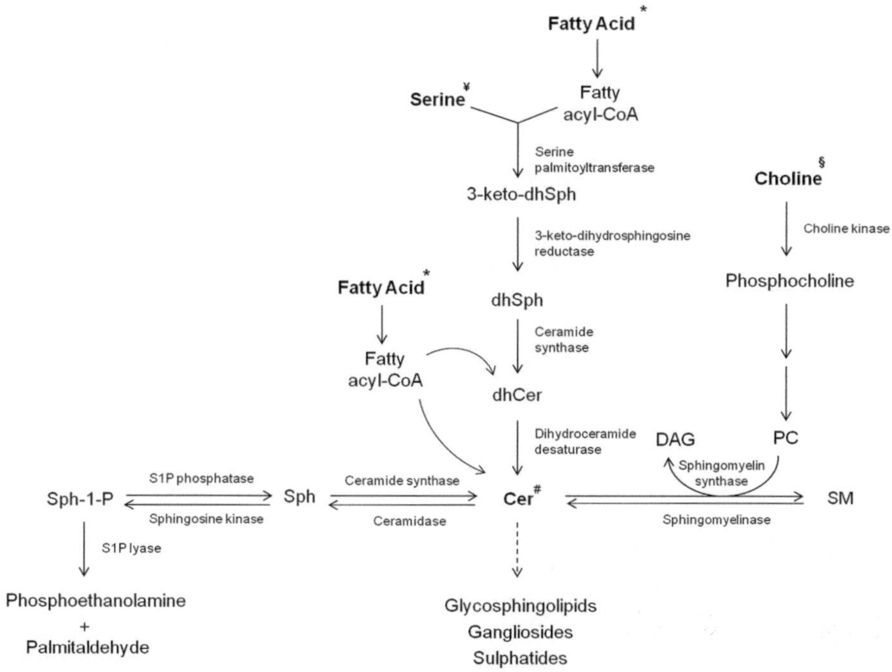

Fig. 5 Stable isotope labeling of sphingolipids. Stable isotope-labeled precursors are shown in *bold* and the examples are represented as: *represents U-^{13}C- or $^{13}C_4$- palmitate; ¥represents U-^{13}C- or ^{15}N-serine; §represents D_9-choline; #represents $^{13}C_{12\text{-}17}$-Cer d18:0/16:0-OH or $^{13}C_4$-Cer d18:1/16:0. *Cer* ceramide, *DihCer* dihydroceramide, *DhSph* dihydrosphingosine, *S-1-P* sphingosine-1-phosphate, *SM* sphingomyelin, *Sph* sphingosine

efficiency may also result in isotopic overlap between labeled and unlabeled species due to low mass shifts. A correction for this overlap is necessary. High-resolution MS separation of labeled and unlabeled species without any fragmentation could be advantageous to this labeling approach. Another disadvantage of serine labeling is that the label might be diluted by the high endogenous serine pool.

5.1.2 Stable Isotope Labeled FA

SL species can also be traced using stable isotope labeled FA. ^{13}C-labeled palmitate is the most commonly used label for studying SL such as Cer, HexCer and SM [61, 62]. In SLs, ^{13}C-labeled palmitate label can be incorporated both in the N-linked acyl chain and the sphingoid base. This results in higher mass shift (16 Da if at one position and 32 Da if at both positions), which is one of the major advantages of this labeling method. The incorporation of label on these positions can be monitored by MS. Similar to serine, the labeled FA pool is diluted by the endogenous palmitate pool, and therefore higher concentrations of the label may be required for efficient labeling. An important factor to consider is choice of position of label on FA, as the FA label might be lost during the FA elongation or desaturation processes, and a labeled FA residue incorporated as an amide may be lost upon CID. Administration of labeled FAs is an issue in vivo experiments and is required to be administered as either lipid emulsions or BSA complex.

5.1.3 Stable Isotope Labeled SL

Stable isotopes labeled SL are not very often used to study SL metabolism. However, labeled SLs such as ^{13}C4-Cer d18:1/16:0 and ^{13}C12-17-Cer and d18:0/16:0-OH have been used to study the uptake and metabolism of these molecules [63, 64]. This labeling approach is limited to analyze specific reactions. Apart from the narrow spectrum of lipid species that can be analyzed using these tracers, administration is another issue and requires vehicles for their delivery such as methyl-β-cyclodextrin, a cyclic oligosaccharide used as a stabilizing agent for hydrophobic molecules. Advanced MS based methodologies are now available to scan various SL groups and the species therein in a high throughput manner [65]. Recently, there has been a movement to develop computational methods that are able to correct for the isotopic overlap between the labeled and unlabeled species, and for the proper integration of these datasets [66]. Overall, the stable isotope labeling combined with high resolution MS presents a robust method to analyze SL metabolism.

5.2 *Mass and Relative Elution Time (MRET) Profiling*

MRET profiling is based on the assumption that lipids when resolved in LC may follow a specific elution pattern based on their chain length under isocratic elution conditions. Thus SLs can be characterized using a relationship between mass and retention time (RT). In a recent study Hejazi et al. used C8 column to separate Cer and dhCer standards of 12–24 carbon N-acyl chain lengths, and could show that elution of these lipids follows a quadratic relationship under isocratic conditions [67]. This equation can be used to calculate the theoretical RT for each analyzed lipid species. The difference between observed RT and theoretical RT, the ‘relative RT’, is plotted against m/z to generate a 2D map. Separation of Cer and dhCer species along x and y axis of this 2D plot is based upon the differences in N-acyl chain lengths, number of acyl chain double bonds and hydroxyl groups. This MRET approach based 2D map allows the separation of lipid species for which synthetic or purified standards are not available. MRET combined with the MS is a powerful tool for the separation and structure determination of complex SLs.

6 Conclusions

HPLC-ESI-MS is a powerful method that has allowed for the quantitative analysis of multiple classes of sphingolipid metabolites present within a single biological system. This methodology has been utilized for analyses across a multitude of mammalian cell lines, assorted tissues, as well as in yeast and even plants. As the number of standards available continues to increase and additional novel MS techniques are developed, the field of lipidomics will continue to grow and progress.

References

1. Fahy E, Subramaniam S, Brown HA, Glass CK, Merrill AH Jr, Murphy RC, Raetz CR, Russell DW, Seyama Y, Shaw W, Shimizu T, Spener F, van Meer G, VanNieuwenhze MS, White SH, Witztum JL, Dennis EA (2005) A comprehensive classification system for lipids. J Lipid Res 46:839–861
2. Bartke N, Hannun YA (2009) Bioactive sphingolipids: metabolism and function. J Lipid Res 50(Suppl):S91–S96
3. Gable K, Gupta SD, Han G, Niranjanakumari S, Harmon JM, Dunn TM (2010) A disease-causing mutation in the active site of serine palmitoyltransferase causes catalytic promiscuity. J Biol Chem 285:22846–22852
4. Eichler FS, Hornemann T, McCampbell A, Kuljis D, Penno A, Vardeh D, Tamrazian E, Garofalo K, Lee HJ, Kini L, Selig M, Frosch M, Gable K, von Eckardstein A, Woolf CJ, Guan G, Harmon JM, Dunn TM, Brown RH Jr (2009) Overexpression of the wild-type SPT1 subunit lowers desoxysphingolipid levels and rescues the phenotype of HSAN1. J Neurosci 29: 14646–14651
5. Zitomer NC, Mitchell T, Voss KA, Bondy GS, Pruett ST, Garnier-Amblard EC, Liebeskind LS, Park H, Wang E, Sullards MC, Merrill AH Jr, Riley RT (2009) Ceramide synthase inhibition by fumonisin B1 causes accumulation of 1-deoxysphinganine: a novel category of bioactive 1-deoxysphingoid bases and 1-deoxydihydroceramides biosynthesized by mammalian cell lines and animals. J Biol Chem 284:4786–4795
6. Cowart LA, Hannun YA (2007) Selective substrate supply in the regulation of yeast de novo sphingolipid synthesis. J Biol Chem 282:12330–12340
7. Han G, Gupta SD, Gable K, Niranjanakumari S, Moitra P, Eichler F, Brown RH Jr, Harmon JM, Dunn TM (2009) Identification of small subunits of mammalian serine palmitoyltransferase that confer distinct acyl-CoA substrate specificities. Proc Natl Acad Sci U S A 106: 8186–8191
8. Caligan TB, Peters K, Ou J, Wang E, Saba J, Merrill AH Jr (2000) A high-performance liquid chromatographic method to measure sphingosine 1-phosphate and related compounds from sphingosine kinase assays and other biological samples. Anal Biochem 281:36–44
9. McNabb TJ, Cremesti AE, Brown PR, Fischl AS (1999) The separation and direct detection of ceramides and sphingoid bases by normal-phase high-performance liquid chromatography and evaporative light-scattering detection. Anal Biochem 276:242–250
10. Sullards MC, Merrill AH Jr (2001) Analysis of sphingosine 1-phosphate, ceramides, and other bioactive sphingolipids by high-performance liquid chromatography-tandem mass spectrometry. Science's STKE: signal transduction knowledge environment 2001, p l1
11. Pettus BJ, Bielawska A, Kroesen BJ, Moeller PD, Szulc ZM, Hannun YA, Busman M (2003) Observation of different ceramide species from crude cellular extracts by normal-phase high-

performance liquid chromatography coupled to atmospheric pressure chemical ionization mass spectrometry. Rapid Commun Mass Spectrom 17:1203–1211
12. Kushi Y, Rokukawa C, Numajir Y, Kato Y, Handa S (1989) Analysis of underivatized glycosphingolipids by high-performance liquid chromatography/atmospheric pressure ionization mass spectrometry. Anal Biochem 182:405–410
13. Suzuki M, Yamakawa T, Suzuki A (1991) A micro method involving micro high-performance liquid chromatography-mass spectrometry for the structural characterization of neutral glycosphingolipids and monosialogangliosides. J Biochem 109:503–506
14. Gu M, Kerwin JL, Watts JD, Aebersold R (1997) Ceramide profiling of complex lipid mixtures by electrospray ionization mass spectrometry. Anal Biochem 244:347–356
15. Mano N, Oda Y, Yamada K, Asakawa N, Katayama K (1997) Simultaneous quantitative determination method for sphingolipid metabolites by liquid chromatography/ionspray ionization tandem mass spectrometry. Anal Biochem 244:291–300
16. Liebisch G, Drobnik W, Reil M, Trumbach B, Arnecke R, Olgemoller B, Roscher A, Schmitz G (1999) Quantitative measurement of different ceramide species from crude cellular extracts by electrospray ionization tandem mass spectrometry (ESI-MS/MS). J Lipid Res 40:1539–1546
17. Bloor WR (1928) The determination of small amounts of lipid in blood plasma. J Biol Chem 77:53–73
18. Reis A, Rudnitskaya A, Blackburn GJ, Mohd Fauzi N, Pitt AR, Spickett CM (2013) A comparison of five lipid extraction solvent systems for lipidomic studies of human LDL. J Lipid Res 54:1812–1824
19. Adada MM, Orr-Gandy KA, Snider AJ, Canals D, Hannun YA, Obeid LM, Clarke CJ (2013) Sphingosine kinase 1 regulates tumor necrosis factor-mediated RANTES induction through p38 mitogen-activated protein kinase but independently of nuclear factor kappaB activation. J Biol Chem 288:27667–27679
20. Matmati N, Metelli A, Tripathi K, Yan S, Mohanty BK, Hannun YA (2013) Identification of C18:1-phytoceramide as the candidate lipid mediator for hydroxyurea resistance in yeast. J Biol Chem 288:17272–17284
21. Bielawski J, Pierce JS, Snider J, Rembiesa B, Szulc ZM, Bielawska A (2009) Comprehensive quantitative analysis of bioactive sphingolipids by high-performance liquid chromatography-tandem mass spectrometry. Methods Mol Biol 579:443–467
22. Bielawski J, Pierce JS, Snider J, Rembiesa B, Szulc ZM, Bielawska A (2010) Sphingolipid analysis by high performance liquid chromatography-tandem mass spectrometry (HPLC-MS/MS). Adv Exp Med Biol 688:46–59
23. Bielawski J, Szulc ZM, Hannun YA, Bielawska A (2006) Simultaneous quantitative analysis of bioactive sphingolipids by high-performance liquid chromatography-tandem mass spectrometry. Methods 39:82–91
24. Spassieva S, Bielawski J, Anelli V, Obeid LM (2007) Combination of C(17) sphingoid base homologues and mass spectrometry analysis as a new approach to study sphingolipid metabolism. Methods Enzymol 434:233–241
25. Li J, Hoene M, Zhao X, Chen S, Wei H, Haring HU, Lin X, Zeng Z, Weigert C, Lehmann R, Xu G (2013) Stable isotope-assisted lipidomics combined with nontargeted isotopomer filtering, a tool to unravel the complex dynamics of lipid metabolism. Anal Chem 85:4651–4657
26. Postle AD, Hunt AN (2009) Dynamic lipidomics with stable isotope labelling. J Chromatogr B Analyt Technol Biomed Life Sci 877:2716–2721
27. Fenn JB, Mann M, Meng CK, Wong SF, Whitehouse CM (1989) Electrospray ionization for mass spectrometry of large biomolecules. Science 246:64–71
28. Baranowski R, Pacha K (2002) Gas chromatographic determination of prostaglandins. Mini Rev Med Chem 2:135–144
29. Abidi SL (2001) Chromatographic analysis of plant sterols in foods and vegetable oils. J Chromatogr A 935:173–201
30. Murphy RC, Barkley RM, Zemski Berry K, Hankin J, Harrison K, Johnson C, Krank J, McAnoy A, Uhlson C, Zarini S (2005) Electrospray ionization and tandem mass spectrometry of eicosanoids. Anal Biochem 346:1–42

31. Duffin KL, Henion JD, Shieh JJ (1991) Electrospray and tandem mass spectrometric characterization of acylglycerol mixtures that are dissolved in nonpolar solvents. Anal Chem 63:1781–1788
32. Han X, Gross RW (2003) Global analyses of cellular lipidomes directly from crude extracts of biological samples by ESI mass spectrometry: a bridge to lipidomics. J Lipid Res 44: 1071–1079
33. Hsu FF, Turk J (1999) Distinction among isomeric unsaturated fatty acids as lithiated adducts by electrospray ionization mass spectrometry using low energy collisionally activated dissociation on a triple stage quadrupole instrument. J Am Soc Mass Spectrom 10:600–612
34. Hsu FF, Turk J (1999) Structural characterization of triacylglycerols as lithiated adducts by electrospray ionization mass spectrometry using low-energy collisionally activated dissociation on a triple stage quadrupole instrument. J Am Soc Mass Spectrom 10:587–599
35. Samuelsson K, Samuelsson B (1969) Gas-liquid chromatography-mass spectrometry of cerebrosides as trimethylsilyl ether derivatives. Biochem Biophys Res Commun 37:15–21
36. Samuelsson B, Samuelsson L (1969) Separation and identification of ceramides derived from human plasma sphingomyelins. J Lipid Res 10:47–55
37. Samuelsson B, Samuelsson K (1969) Gas–liquid chromatography-mass spectrometry of synthetic ceramides. J Lipid Res 10:41–46
38. Samuelsson B, Samuelsson K (1968) Gas-liquid chromatographic separation of ceramides as di-O-trimethylsilyl ether derivatives. Biochim Biophys Acta 164:421–423
39. Sweeley CC, Dawson G (1969) Determination of glycosphingolipid structures by mass spectrometry. Biochem Biophys Res Commun 37:6–14
40. Ariga T, Murata T, Oshima M, Maezawa M, Miyatake T (1980) Characterization of glycosphingolipids by direct inlet chemical ionization mass spectrometry. J Lipid Res 21:879–887
41. Ariga T, Yu RK, Suzuki M, Ando S, Miyatake T (1982) Characterization of GM1 ganglioside by direct inlet chemical ionization mass spectrometry. J Lipid Res 23:437–442
42. Murata T, Ariga T, Oshima M, Miyatake T (1978) Characterization of trimethylsilyl derivatives of cerebrosides by direct inlet-chemical ionization mass spectrometry. J Lipid Res 19:370–374
43. Oshima M, Ariga T, Murata T (1977) Combined gas chromatography-chemical ionization mass spectrometry of sphingolipids. I. Glucosyl sphingosine, ceramides and cerebrosides of the spleen in Gaucher's disease. Chem Phys Lipids 19:289–299
44. De Hoffmann E, Stroobant V (2007) Mass spectrometry: principles and applications. Wiley, Chichester
45. Nguyen S, Fenn JB (2007) Gas-phase ions of solute species from charged droplets of solutions. Proc Natl Acad Sci U S A 104:1111–1117
46. Vieu C, Terce F, Chevy F, Rolland C, Barbaras R, Chap H, Wolf C, Perret B, Collet X (2002) Coupled assay of sphingomyelin and ceramide molecular species by gas liquid chromatography. J Lipid Res 43:510–522
47. Isaac G, Bylund D, Mansson JE, Markides KE, Bergquist J (2003) Analysis of phosphatidylcholine and sphingomyelin molecular species from brain extracts using capillary liquid chromatography electrospray ionization mass spectrometry. J Neurosci Methods 128:111–119
48. McHowat J, Jones JH, Creer MH (1996) Quantitation of individual phospholipid molecular species by UV absorption measurements. J Lipid Res 37:2450–2460
49. Christie WW (1992) Detectors for high-performance liquid chromatography of lipids with special reference to evaporative light-scattering detection. Bridgewater/Oily Press, Ayr
50. Adams JM 2nd, Pratipanawatr T, Berria R, Wang E, DeFronzo RA, Sullards MC, Mandarino LJ (2004) Ceramide content is increased in skeletal muscle from obese insulin-resistant humans. Diabetes 53:25–31
51. Zheng W, Kollmeyer J, Symolon H, Momin A, Munter E, Wang E, Kelly S, Allegood JC, Liu Y, Peng Q, Ramaraju H, Sullards MC, Cabot M, Merrill AH Jr (2006) Ceramides and other bioactive sphingolipid backbones in health and disease: lipidomic analysis, metabolism and roles in membrane structure, dynamics, signaling and autophagy. Biochim Biophys Acta 1758:1864–1884

52. Cole RB (1997) Electrospray ionization mass spectrometry. Wiley, New York
53. Dragusin M, Wehner S, Kelly S, Wang E, Merrill AH Jr, Kalff JC, van Echten-Deckert G (2006) Effects of sphingosine-1-phosphate and ceramide-1-phosphate on rat intestinal smooth muscle cells: implications for postoperative ileus. FASEB 20:1930–1932
54. Haynes CA, Allegood JC, Park H, Sullards MC (2009) Sphingolipidomics: methods for the comprehensive analysis of sphingolipids. J Chromatogr B Analyt Technol Biomed Life Sci 877:2696–2708
55. Merrill AH Jr, Sullards MC, Allegood JC, Kelly S, Wang E (2005) Sphingolipidomics: high-throughput, structure-specific, and quantitative analysis of sphingolipids by liquid chromatography tandem mass spectrometry. Methods 36:207–224
56. Ecker J, Liebisch G (2014) Application of stable isotopes to investigate the metabolism of fatty acids, glycerophospholipid and sphingolipid species. Prog Lipid Res 54:14–31
57. Hellerstein MK, Neese RA (1999) Mass isotopomer distribution analysis at eight years: theoretical, analytic, and experimental considerations. Am J Physiol 276:E1146–E1170
58. Lee WN (1996) Stable isotopes and mass isotopomer study of fatty acid and cholesterol synthesis. A review of the MIDA approach. Adv Exp Med Biol 399:95–114
59. Berdyshev EV, Gorshkova IA, Usatyuk P, Zhao Y, Saatian B, Hubbard W, Natarajan V (2006) De novo biosynthesis of dihydrosphingosine-1-phosphate by sphingosine kinase 1 in mammalian cells. Cell Signal 18:1779–1792
60. Schifferer R, Liebisch G, Bandulik S, Langmann T, Dada A, Schmitz G (2007) ApoA-I induces a preferential efflux of monounsaturated phosphatidylcholine and medium chain sphingomyelin species from a cellular pool distinct from HDL(3) mediated phospholipid efflux. Biochim Biophys Acta 1771:853–863
61. Blachnio-Zabielska AU, Persson XM, Koutsari C, Zabielski P, Jensen MD (2012) A liquid chromatography/tandem mass spectrometry method for measuring the in vivo incorporation of plasma free fatty acids into intramyocellular ceramides in humans. Rapid Commun Mass Spectrom 26:1134–1140
62. Haynes CA, Allegood JC, Wang EW, Kelly SL, Sullards MC, Merrill AH Jr (2011) Factors to consider in using [U-C]palmitate for analysis of sphingolipid biosynthesis by tandem mass spectrometry. J Lipid Res 52:1583–1594
63. Fukami H, Tachimoto H, Kishi M, Kaga T, Waki H, Iwamoto M, Tanaka Y (2010) Preparation of (13)C-labeled ceramide by acetic acid bacteria and its incorporation in mice. J Lipid Res 51:3389–3395
64. Tserng KY, Griffin RL (2004) Ceramide metabolite, not intact ceramide molecule, may be responsible for cellular toxicity. Biochem J 380:715–722
65. Brugger B (2014) Lipidomics: analysis of the lipid composition of cells and subcellular organelles by electrospray ionization mass spectrometry. Annu Rev Biochem 83:79
66. Millard P, Letisse F, Sokol S, Portais JC (2012) IsoCor: correcting MS data in isotope labeling experiments. Bioinformatics 28:1294–1296
67. Hejazi L, Wong JW, Cheng D, Proschogo N, Ebrahimi D, Garner B, Don AS (2011) Mass and relative elution time profiling: two-dimensional analysis of sphingolipids in Alzheimer's disease brains. Biochem J 438:165–175

Detection and Distribution of Sphingolipids in Tissue by FTICR MALDI-Imaging Mass Spectrometry

E. Ellen Jones, Shaalee Dworski, Mustafa Kamani, Daniel Canals, Masayuki Wada, Xiang Liu, Yusuf A. Hannun, James Norris, Jeffrey A. Medin, and Richard R. Drake

Abstract MALDI imaging mass spectrometry is an evolving technology capable of simultaneously profiling multiple analytes of interest across a tissue section and aligning their distribution to tissue histopathology. This chapter summarizes a

E.E. Jones • R.R. Drake (✉)
Department of Cell and Molecular Pharmacology, MUSC Proteomics Center and Hollings Cancer Center, Medical University of South Carolina, 173 Ashley Avenue, Charleston, SC 29425, USA
e-mail: joneli@musc.edu; draker@musc.edu

S. Dworski
Institute of Medical Science, University of Toronto, 67 College Street, Toronto, ON M5G 2M1, Canada
e-mail: s.dworski@mail.utoronto.ca

M. Kamani
University Health Network, 67 College Street, Toronto, ON M5G 2M1, Canada
e-mail: Mustafa.kamani@gmail.com

D. Canals • M. Wada
Departments of Medicine and Biochemistry and the Stony Brook Cancer Center, Stony Brook University Medical Center, 101 Nicholls Road, Stony Brook, NY 11794, USA
e-mail: Daniel.canals@stonybrookmedicine.edu; Masayuki.wada@stonybrookmedicine.edu

X. Liu • J. Norris
Department of Microbiology and Immunology and Hollings Cancer Center, Medical University of South Carolina, 173 Ashley Avenue, Charleston, SC 29425, USA
e-mail: liux@musc.edu; norrisjs@musc.edu

Y.A. Hannun
Stony Brook Cancer Center, Stony Brook University, Stony Brook, NY, USA
e-mail: yusuf.hannun@stonybrookmedicine.edu

J.A. Medin
Institute of Medical Science, University of Toronto, 67 College Street, Toronto, ON M5G 2M1, Canada

Department of Medical Biophysics, University of Toronto, 67 College Street, Toronto, ON M5G 2M1, Canada

University Health Network, 67 College Street, Toronto, ON M5G 2M1, Canada
e-mail: jmedin@uhnresearch.ca

Y.A. Hannun et al. (eds.), *Bioactive Sphingolipids in Cancer Biology and Therapy*, DOI 10.1007/978-3-319-20750-6_15

MALDI imaging workflow for on-tissue identification of sphingolipids and glycosphingolipids using tissues derived from mouse models known to accumulate ceramides (Farber disease) and globotriaosylceramides (Fabry disease). A combination of CID and on-tissue enzyme digestions was utilized for structural confirmation prior to being added to the comprehensive sphingolipid and glycosphingolipid library. Two example case studies related to the modulation of sphingolipid metabolism are provided to illustrate the potential applications of MALDI imaging. In the first case study, distinct ceramides were visualized in relation to Lewis lung carcinoma tumors. In the second case study, tissues were derived from a tumor xenograft model treated with a drug targeting the sphingosine-1-phosphate/ceramide nexus. Representative images of ceramide and hexose ceramides in relation to cancer alone or drug distribution correlating to an increase or decrease in sphingosine-1-phosphate or ceramide species are provided. These MALDI imaging workflows can be readily adapted to assess the distribution of sphingolipids and glycosphingolipids in any tissue system of interest.

Keywords MALDI mass spectrometry imaging • Ceramide • Glycosphingolipid • Lysosomal storage disease • Sphingolipid

1 Introduction

As described throughout the chapters in this volume, sphingolipids, glycosphingolipids and their metabolites play a role in many cellular processes including signaling, recognition, proliferation, differentiation, apoptosis, and autophagy. Altered sphingolipid metabolism has been linked to a variety of disorders, including cancer [1–6]. Identification of the specific sphingolipid and glycosphingolipid species associated with cancer development and progression in affected tissues is imperative, as these tumor specific bioactive lipids could serve as biomarkers of disease or novel therapeutic targets [7–9]. Currently, the standard approach for identification of sphingolipids in a tissue sample is liquid chromatography mass spectrometry (LC-MS), which provides a quantitative analysis of the lipids in a sample of interest. A disadvantage of utilizing LC-MS alone, however, is its inability to link a sphingolipid of interest to the histopathology features of the tissue sample. Laser capture microdissection (LCM) is an approach that can be used to identify individual lipids from specifically captured tissue regions/cells, however, a two-dimensional map of the distribution of all detected lipids across an entire tissue section is not possible. Additionally, there has been evidence that the polymer coating used to isolate the cells may contribute interfering ions for downstream mass spectrometry analysis [10]. Alternatively, MALDI-IMS, or matrix assisted laser desorption ionization imaging mass spectrometry, is an emerging technology with the ability to profile the distribution of multiple sphingolipid ions simultaneously across an entire tissue section, directly linked to the histopathology of the tissue [11, 12]. A limitation is that the data generated are more qualitative relative to other more quantitative LC-MS

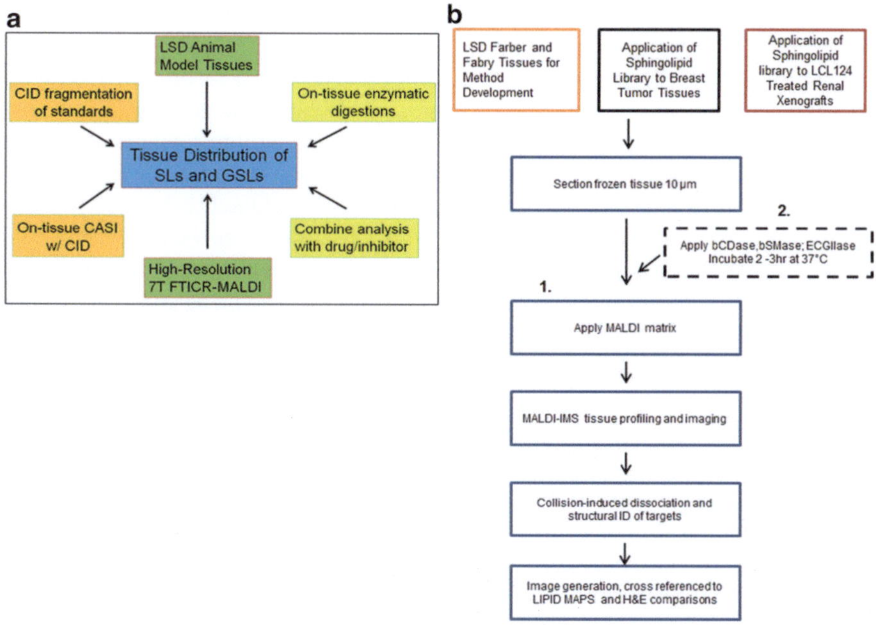

Fig. 1 Overview of the MALDI-IMS approach utilized for construction of a comprehensive sphingolipid tissue library. (**a**) Representative diagram of the systematic approach which was utilized to generate database of sphingolipid and glycosphingolipid species. (**b**) The MALDI-IMS workflow schematic including three different sample tissue sets. For analysis the frozen tissue was sectioned at 10 μm and either (*1*) sprayed with matrix for profiling or (*2*) processed for on-tissue enzyme digestions prior to matrix application and lipid analysis

lipidomic analysis strategies, but both approaches can provide complimentary information. Based on our recent research experience [13], this chapter provides a methodological workflow and summary of approaches for the comprehensive analysis and two-dimensional mapping of sphingolipids and glycosphingolipids in tissues using MALDI-IMS (Fig. 1). Example data generated from brain and kidney tissues derived from lysosomal storage disorder (LSD) models and tumor models are provided.

2 Applying MALDI-IMS to Tissue Profiling of Sphingolipids and Glycosphingolipids

MALDI-IMS workflows were originally developed to profile proteins and peptides within tissue sections, however, new classes of versatile high-resolution (up to 2,500,000) Fourier transform ion cyclotron resonance (FTICR) mass spectrometers have increased the range of biomolecules that can be effectively imaged in tissues. In addition to their high resolving power, FTICR mass spectrometers also have high mass accuracies (<1 ppm) and sensitivities (attomole–femtomole concentrations), characteristics imperative for identification of smaller biomolecules such as sphingolipids [14, 15]. The implementation of these new high-resolution mass spectrometers within imaging workflows has greatly increased the number of bioactive molecules identifiable [16–21] along with the spatial resolution at which they can be imaged. Within imaging workflows, the spatial resolution is one of the limiting factors since this is dependent upon laser size, raster width and matrix type/deposition. Using sublimation, a spatial resolution of 10 μm has been achieved [22], however, this greatly increases the analysis run time. Combined with the size of the target tissue, spatial resolution and analysis time are factors that must be considered in the experimental design. Accordingly, FTICR MALDI-IMS experiments cataloguing libraries of lipids, N-linked glycans, metabolites, and small drug molecules are forthcoming [13, 16, 17, 21, 23].

For example, the tissue distributions of many of the abundant phospholipid classes are being characterized down to variations in fatty acyl chain lengths [19, 24]. The profiling of lower abundance sphingolipid classes such as sphingosine-1-phosphate (S-1-P), the ceramides, and hexose ceramides, however, have proven more challenging due to the low levels at which these sphingolipids exist in most tissue sections. Their low concentration, particularly in the background of more abundant phospholipids, has made structural assignment and on-tissue confirmation difficult.

To address these challenges different MALDI-IMS workflows are being assessed, with success being highly dependent on the starting material and instrumentation [13, 24–31]. In addition to our studies using lysosomal storage disease models [13], MALDI-IMS has been used to identify panels of elevated ceramides in the skin epidermis, including some with very long acyl chains [32]. MALDI-IMS experiments profiling gangliosides have also been achieved in different tissue models of diabetic nephropathy [15], mouse brain [33] and Tay-Sachs disease [20, 27], suggesting that there may be specific populations of gangliosides which are disease or injury specific. An extensive characterization of 544 lipids detected in positive ion mode and negative ion mode using high resolution FTICR-MALDI imaging on a porcine adrenal gland tissue has also recently been reported [19], and included ceramides, sphingomyelins and glycosphingolipids. The success of these studies and data generated has been helpful in highlighting several key factors necessary to expand and improve the MALDI-IMS profiling of sphingolipids in tissues: (1) the ability to initially identify and characterize sphingolipids, and in particular glycosphingolipids, may be largely dependent upon instrumentation and starting material; (2) developing easily adaptable MALDI-IMS workflows to identify the lower abundant sphingolipids is necessary, and (3) a comprehensive sphingolipid and glycosphingolipid reference database of MALDI-IMS data is needed.

To more effectively address these issues, our approach was to synergize these different components into an integrated methods workflow, summarized in Fig. 1a. One primary goal was to generate a comprehensive MALDI ion database of sphingolipid and glycosphingolipid species detected in tissues, including their salt adducts (H+, Na+, K+). Key to these efforts was the use of tissues from mouse models of lysosomal storage disorders (e.g., Farber Disease [34] and Fabry Disease [35]) that are essentially *in situ* biological reservoirs of accumulated sphingolipid and glycosphingolipid species. These tissues further facilitated structural verification of targeted sphingolipids by comparison with chemical standards, on-tissue fragmentation and structural confirmation facilitated by the instrument, and on-tissue biochemical characterization using degradative enzymes (Fig. 1b). Each of these components will be discussed individually or in context with the data examples throughout this chapter. In addition to the Farber and Fabry model tissues, an example application of the workflow to other tissues systems are described for a Lewis lung carcinoma mouse model, and human kidney tumor xenografts treated with a cationic ceramide analogue (LCL124).

3 FT-ICR MALDI-IMS of Farber Disease Mouse Tissues for on-Tissue Identification of Ceramides and Sphingomyelins

Farber disease is an autosomal recessive lysosomal storage disorder, characterized by a deficiency of the enzyme acid ceramidase (ACDase). Farber Disease patients generally suffer from impaired growth and development, shortened life span and intralysosomal accumulation of ceramides in various tissues [34]. The role of ACDase deficiency in the development of the Farber Disease phenotype was recently evaluated using a novel *Asah1* gene 'knock in' Farber disease mouse model that represents the first model of systemic ACDase deficiency [34]. Homozygous *Asah1*$^{P361R/P361R}$ animals exhibit ACDase defects and manifestations including accumulation of tissue ceramides. To aid in the development and establishment of MALDI-IMS workflows for ceramide identification, kidney and brain tissues from homozygous *Asah1*$^{P361R/P361R}$ animals were utilized. The brain was chosen due to its known involvement in disease pathology and histopathologically relevant features, and the kidney was chosen as an organ that is not obviously affected by the disease.

Images of representative sphingolipids in a sagittal tissue slice of a homozygous *Asah1*$^{P361R/P361R}$ Farber mouse brain, including ceramides, sphingomyelin and ceramide-1-phosphate, are shown in Fig. 2. Brains and kidney tissues were sectioned at 10 μm and thaw mounted onto ITO (indium tin oxide) slides. 2,5-Dihydroxybenzoic acid (DHB) matrix was applied at a concentration of 30 mg/mL using an ImagePrep sprayer (Bruker Daltonics). Spectra were acquired across the brain sections using a raster width of 100 μm for the brains and 60 μm for the kidneys. Spectra were loaded into FlexImaging 4.0 (Bruker Daltonics) software where intensities were normalized using root mean squares (RMS). Distribution of a ubiquitous phosphatidylcholine (PC) is included for comparison (Fig. 2c). Glycosphingolipids were also assessed within the Farber model. As MALDI-IMS cannot distinguish between galactose or glucose, the term hexose-ceramide (HexCer) is used to represent either species. In the Farber tissues, glycosphingolipids specific to tissue regions could be identified. For example, in the brain a HexCer 18:1/24:0 was concentrated in the

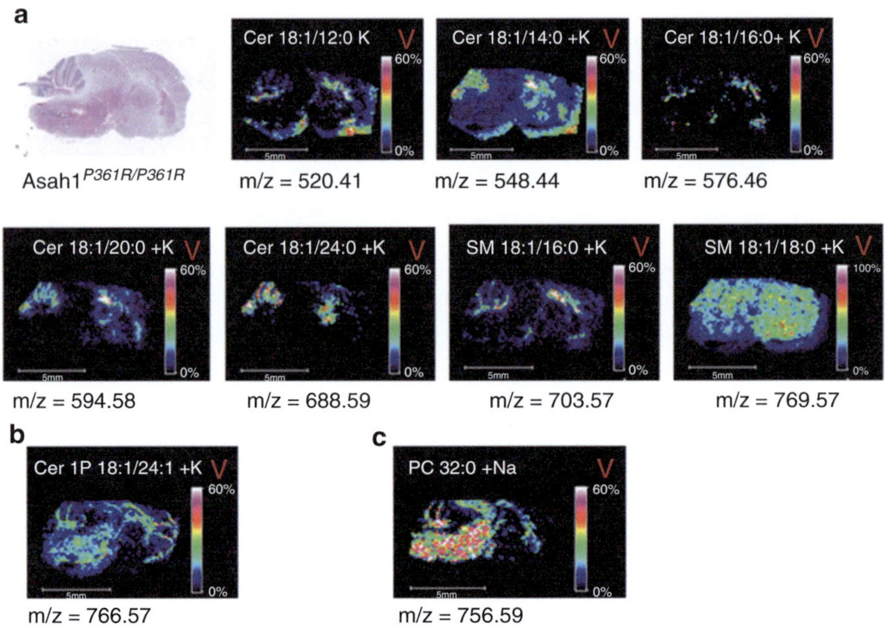

Fig. 2 A panel of sphingolipids identified in the Farber Disease mouse brain with FTICR MALDI-IMS. Brain tissues derived from homozygous *Asah1*$^{P361R/P361R}$ mice were analyzed. (**a**) H&E stain of a serial brain section and representative MALDI-IMS data distribution of a panel of ceramide and sphingomyelin species across the Farber brain. (**b**) Ceramide-1-phosphate distribution in the Farber brain. (**c**) PC 32:0+Na is an abundant brain lipid which is included to demonstrate its homogenous distribution across the Farber brain compared with the ceramide species

cerebellum compared to a HexCer 18:1/22:0, as shown individually and as an overlay of both species (Fig. 3a). This was also true in the Farber kidney, as HexCer 18:1/24:0 is localized in the cortex, while HexCer 18:1/26:0 is in the medulla (Fig. 3b). A Farber model-specific hexose ceramide, HexCer 18:1/24:1, was detected in both brain and kidney as being more abundant than wild-type tissues (Fig. 3c).

4 Approaches to Confirm the Assigned Structures of Tissue Sphingolipids by MALDI

There are multiple approaches to verify the assigned structures of the sphingolipids indicated in Figs. 2 and 3. The general workflow for ceramide and sphingomyelin identification and verification was (1) on-tissue profiling to identify all the sphingolipids present by cross referencing the masses of the precursor ion with those listed in LIPID MAPS database (http://www.lipidmaps.org); (2) on-tissue verification and structural assignment in which the ion of interest was isolated from the tissue and

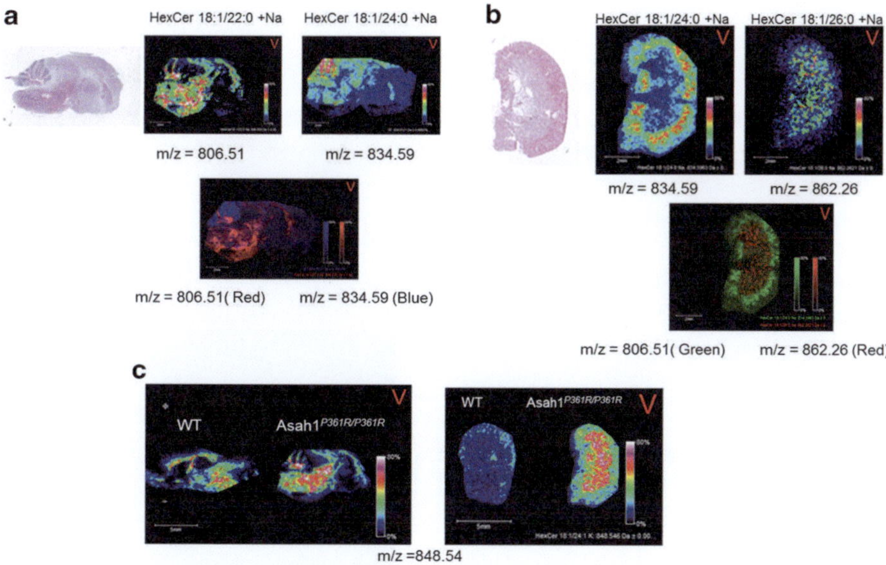

Fig. 3 Glycosphingolipids identified in Farber disease mouse brain and kidney with FTICR MALDI-IMS (**a**) HexCer 18:1/22:0+Na and 18:1/24:0+Na distribution across a Farber brain shows localization of species in different brain regions. This is shown individually and as an overlay of *blue* and *red*. (**b**) The distribution of a HexCer 18:1/24:0+Na and a HexCer 18:1/26:0+Na shows differential distribution in medulla and cortex. This is shown individually and as an overlay. (**c**) HexCer 18:1/24:1+K is elevated in both the Farber brain and kidney in comparison to the wild-type

fragmented using CID, then comparing the ms/ms fragmentation patterns with the species standard; (3) global assessment of the sphingolipid or glycosphingolipid of interest using novel on-tissue enzyme digestions with recombinant bacterial ceramidase (bCDase), recombinant bacterial sphingomyelinase (bSMase), or endoglycoceramidase II (EGCaseII); and (4) comparison of confirmed sphingolipids to tissue histopathology to determine species of interest (Figs. 4 and 5). Inherent in this workflow is the initial use of tissue from a model where the concentration of the target species of interest is known to be elevated, such as the case with the Farber disease model (Fig. 1a).

On-tissue CID fragmentation patterns were compared with those from a commercial standard for the same lipid. This was done using a sphingolipid panel of molecular standards representing ceramide, S-1-P, sphingomyelin, and glycosphingolipid species (data not shown). These standards were spotted on a steel MALDI plate with DHB matrix and analyzed using the Bruker Solarix 7T FTICR mass spectrometer. For comparison, the same standards were measured on a separate MALDI instrument, a Bruker AutoFlexIII MALDI-TOF/TOF in linear and reflectron modes. Use of the FTICR instrument was superior and readily identified the expected parent masses of each sphingolipid standard, while standards on the other

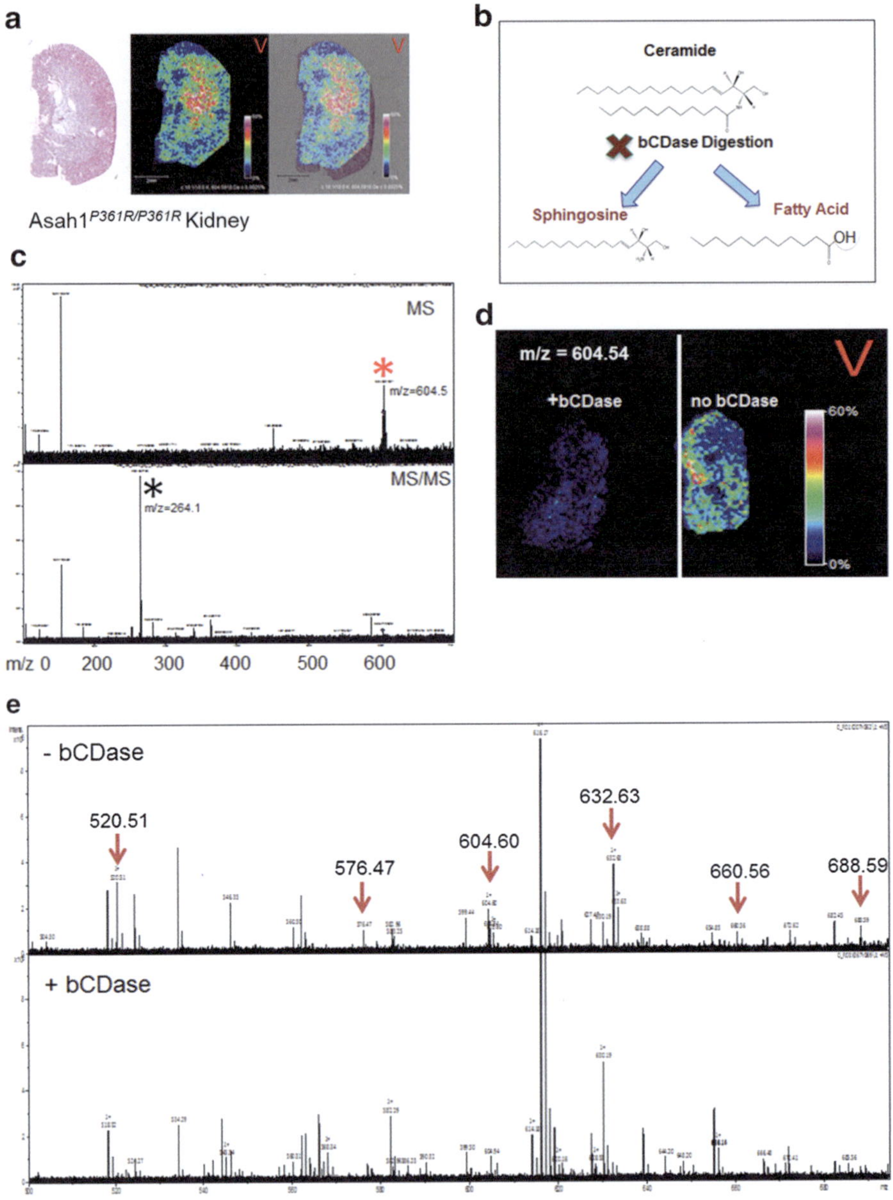

Fig. 4 Representative workflow for ceramide identification (**a**) H&E stain of a serial kidney section. Ceramide confirmation was achieved by (**b**) on-tissue CID and (**c**) on-tissue enzyme digestions with bCDase, which produces a sphingosine and fatty acid. (**d**) Representative picture of the results typically seen following the on-tissue enzyme digestions. Digestions were carried out by mounting mirror image sections on an ITO coated slide. The tissues on the *left* part were sprayed with bCDase while the *right* side was covered prior to enzyme application. (**e**) Representative spectra from a tissue that was treated minus (*top*) or plus (*bottom*) bCDase. In the *top panel* several ceramides are present as indicated with the *red arrows* which are absent or at much lower levels in the tissue digested with bCDase

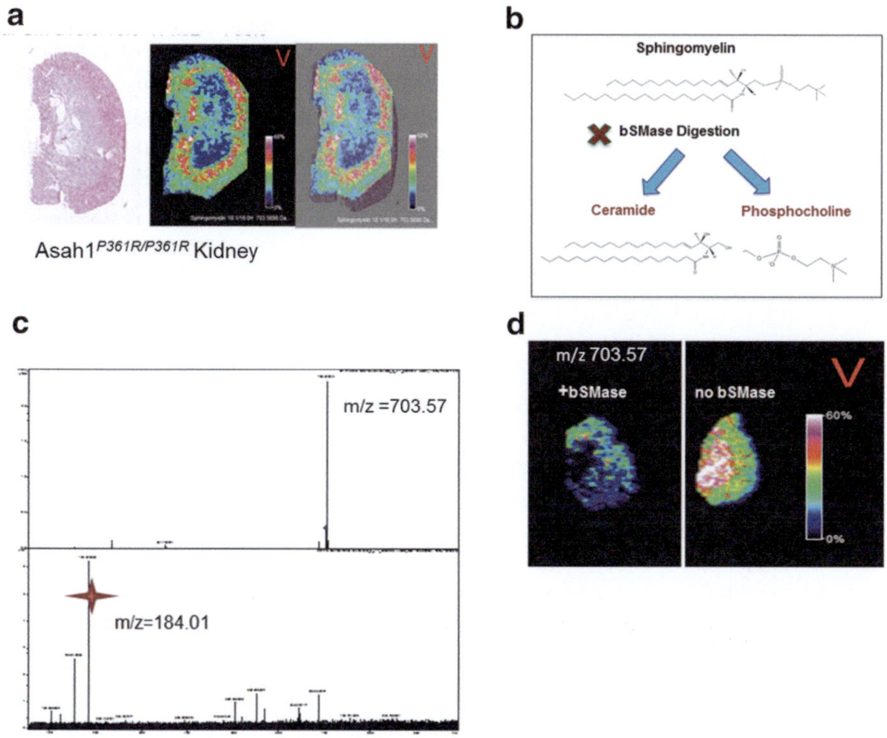

Fig. 5 Representative workflow for sphingomyelin identification. (**a**) H&E stained section of a Farber kidney. (**b**) Sphingomyelins were confirmed using on-tissue CID and enzyme digestions with bSMase. (**c**) bSMase digestion produces a sphingosine and choline-phosphate. (**d**) On-tissue digestions were carried out by thaw mounting mirror image sections on an ITO coated slide. The tissues on the left part were sprayed with bSMase while the right side was covered, prior to enzyme application

platform frequently had loss of water and sometimes greater fragmentation. These effects are related to the different ionization environments of the two instruments. The MALDI-FTICR instrument is configured with a high-pressure collisional cooling source that, following laser ionization, aids in reducing the heat of ionization to keep the molecules intact [36–38]. Standard non-FTICR MALDI instruments are under high vacuum, and the heat of laser ionization is not dissipated, which can lead to degradations like loss of hydroxyls and other fragmentations [36]. This is one aspect of why sphingolipid species, and specifically glycosphingolipids, are less stable in MALDI-TOF instruments without this cooling source. Overall, the configuration of the FTICR instrument facilitates accurate analysis and detection of sphingolipids.

Another function of the MALDI-FTICR instrument that is useful, termed Continuous Accumulation of Selected Ions/Collision-induced dissociation (CASI/CID), is that it allows fragmentation of ions of interest to further confirm the expected structural components. The CASI feature is an ion trap that collects specified ions following laser excitation from a tissue, followed by CID of the captured ions. For all ceramide species tested with an 18:1-sphingoid base, a peak at m/z 264.1 was detected by CID (Fig. 4c), as also reported by other analysis methods [26, 39]. The presence of this m/z=264.1 fragment after CID provides confirmatory evidence of a ceramide species. Fragmentation of sphingomyelins produced CID patterns characterized by the loss of their phosphocholine head group (Fig. 5c), and for S-1-P, loss of phosphate.

Because verification and fragmentation of selected lipids from a tissue can be time consuming, and in the case of ceramides complicated by the presence of neighboring ions from more abundant phospholipids within the same mass range, an on-tissue enzyme digestion with a recombinant bacterial ceramidase (bCDase) [40] was developed and incorporated into the imaging workflow (Fig. 4c). bCDase from *Pseudomonas aeruginosa* cleaves ceramides into a sphingosine and a fatty acid. When sprayed on tissue, the ceramides present will be digested and the loss of their signal can be detected in comparison to untreated tissues (Fig. 4d, e). The experiment was performed by making a mirror image slide in which one piece of tissue was covered, while the other was sprayed with 8 mU bCDase. By comparing the decrease in signal of the ceramides on the treated tissue, the ion can then be verified as a ceramide as opposed to a phospholipid or sphingomyelin. An example is shown in Fig. 4d for a ceramide (d18:1/18:0) present in the Farber mouse kidney tissue. The bCDase digestion provides rapid verification of global ceramides expressed within the tissues, and complements the determinations made by CASI/CID and LIPIDMAPS evaluations.

5 Tissue Imaging of Sphingomyelins

In general, sphingomyelin species are present at much higher concentrations than ceramide species and are easier to detect by MALDI-IMS. They also have odd-number molecular masses, allowing easier identifications relative to the predominantly even-number masses of phospholipids and ceramides. While their detections may be less challenging, a similar workflow as described for ceramides was developed for their identification and verification in tissues (Fig. 5). Fragmentation of sphingomyelins produced CID patterns characterized by the loss of their choline-phosphate head group, of 184 mass units (Fig. 5c). Similar to the bCDase, enzyme digestions with bacterial sphingomyelinase (bSMase) from *Bacillus cereus* [40] were carried out to verify sphingomyelins. bSMase releases choline-phosphate and ceramide as products. As shown in both the image and representative spectra in Fig. 5d, for a sphingomyelin d18:1/16:0 at m/z=703.57, the Farber kidney tissue slice treated with 2 mU bSMase had a significant decrease in the detection of this sphingomyelin ion as compared to the tissue not treated with bSMase. While this approach is not particularly critical for the most common C16 and C18 sphingomyelin species, it is useful for the verification of less abundant, longer chain acyl species.

6 MALDI-IMS of Glycosphingolipids in the Negative and Positive ion Modes

Determining the parent masses of the individual ceramide species in the Farber disease model greatly facilitated subsequent analysis of hexose and lactosylceramide species, the precursor structures for the more complex ganglioside and globotrioacylceramide species. As noted earlier, the hexose and lactosylceramides are readily detected in positive ion mode with DHB matrix (Fig. 3). Addition of 162 Da for hexoses and 324 Da for lactoses to the parent ceramide masses were indicative of these species. These structures can be similarly confirmed by on-tissue CID [13], and characteristically the intact glycan masses can be subtracted from the parent masses in the fragmentation spectra. In addition to the examples shown for the Farber Disease tissues, we have recently reported the detection of hexose and lactosylceramides that accumulate in a mouse model of lupus [41].

For the glycosphingolipids that contain sialic acid residues, including gangliosides, analysis in the negative ion mode is the most effective approach. Some ganglioside species have been previously identified in MALDI imaging experiments [17, 27, 42–45]. The negatively charged gangliosides were also identified in Farber brains in negative ion mode using similar approaches and matrix as utilized with the ceramides and sphingomyelins (Fig. 6). Confirmation of these species was achieved again using CASI/CID to track the loss of the intact glycans as compared to standards. An on-tissue digestion with endoglycoceramidase II (ECGase II) was utilized and aided in distinguishing the gangliosides from other lipids detected in the negative ion mode. ECGase II is an endoglycohydrolase which cleaves the linkage between the oligosaccharide and ceramide of acidic and neutral glycosphingolipids (Fig. 6a). ECGase II enzyme digestions were carried out similar to the bCDase and bSMase experiments, in that adjacent Farber brain tissue slices were mounted on ITO slides, and one slice was digested while the other was covered and protected from digestion. The ganglioside species GM_3, GM_2, GM_1, GD_1 and GT_1 were successfully identified (Fig. 6b), and each species shown was susceptible to on-tissue ECGaseII digestion. A raster width of 80 μm was utilized during analysis. Data was loaded into FlexImaging 4.0 and normalized using RMS as stated previously. There are many uncharacterized ganglioside species that remain to be localized with MALDI-IMS approaches, and we plan similar studies using tissues from animal models of lysosomal storage diseases that accumulate ganglioside species.

Other negative ion mode sphingolipids, such as the sulfatides, can also be identified in these models. Sulfatides are a class of galactosylated ceramides usually with one (or more) 3-O-sulfate groups attached to the sugar, and are highly abundant in brain, retina and kidney [46]. Sulfatide species, especially the ST d18:1/24:1 at m/z = 888.6, are one of the most common negative mode ion sphingolipids reported in MALDI imaging experiments [10, 17, 43, 46–48].

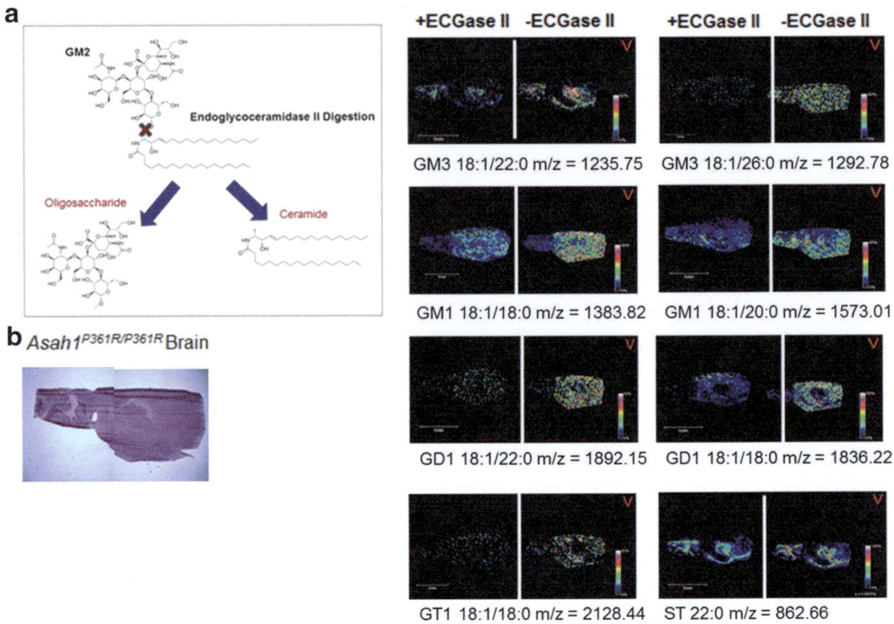

Fig. 6 On-tissue digestion with endoglycoceramidase (ECGase II) on Farber disease brains to identify gangliosides (**a**) ECGase II cleaves gangliosides into an oligosaccharide and a ceramide. (**b**) H&E stained section of Farber brain. (**c**) Adjacent Farber brain sections were mounted on an ITO slide in which the left side of the slide was sprayed with ECGase II while the right side was covered prior to matrix application. GM species were identified by the decrease following digestion. An abundant sulfatide (C22:0) is included to show enzyme specificity and for distribution comparison

7 FTICR MALDI-IMS of Fabry Disease Mouse Tissues to Identify Globotriaosylceramides

Fabry disease is a rare genetic lysosomal storage disorder linked to a variety of symptoms including pain, renal complications and dermatological manifestations. Diagnosis is based on clinical presentation and the presence of a defective alpha-galactosidase enzyme which leads to an accumulation of globotriaosylceramides (Gb_3) in tissues. These species contain one glucose and two galactose residues, biosynthetically derived from galactosylation of lactosylceramides. To characterize the distribution and detection of Gb_3, kidneys from a Fabry disease mouse model were utilized [35]. Kidneys were sectioned to 10 μm, sprayed with 2,5-DHB matrix and analyzed using a raster width of 150 μm. Data was loaded into FlexImaging 4.0 and normalized using RMS. A panel of Fabry specific Gb_3 species of varying chain lengths were identified (Fig. 7b), including the Na ions of Gb_3 C22:0, 24:1 and 24:0. The newly identified globotriaosylceramides were cross-referenced with the literature and correlate with Gb_3 species elevated in the blood and plasma of Fabry disease patients [49].

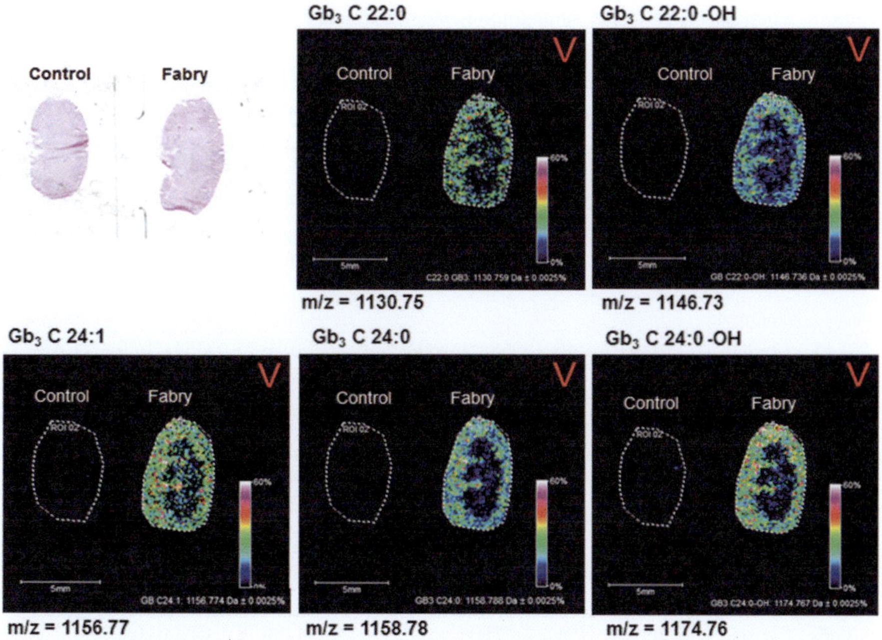

Fig. 7 FTICR MALDI-IMS identifies globotriaosylceramide (Gb_3) species specific to Fabry Disease kidneys. An H&E stain of the serial kidney sections of the control and Fabry disease kidney tissues utilized. The panel of Gb_3 species identified are Fabry specific and include C22:0, C24:1, and C24:0

8 Example Studies Using the MALDI-IMS Sphingolipid Workflows

Summaries of two example case studies are described below as illustrations of how the tissue sphingolipid structural database (Table 1) can be applied to other tissue systems.

Case 1: Identification of Tumor Specific Ceramides in Lewis Lung Carcinoma Model Tissues

A widely used model for testing therapies and mechanisms of metastasis is the Lewis lung carcinoma model in mice [50]. An intravenous tumor cell injection is done resulting in spontaneous, metastasizing tumors forming in the lung. Lungs from two mice with established tumors in the lung were profiled by MALDI-IMS to examine whether ceramide and S-1-P distribution in the tumor could be readily distinguished from normal lung tissue (Fig. 8). Tissues were sectioned and processed according to the previously established methods, using 2,5 DHB as the matrix, and a raster width of 100 μm. It is clear that S-1-P in the tumor tissues could be distinctly identified (Fig. 8b). The distribution of two representative ceramide species (d18:1/18:0 and d18:1/24:1) is also shown. The levels of C24:1 ceramide

Table 1 Cumulative list of the sphingolipids identified using FTICR MALDI-IMS in the LSD tissues

m/z	Sphingosine-1-phosphate	m/z	Hydroxylated ceramides	m/z	Ceramide-1-phosphate
379.25	S-1-P	648.53	Cer(d18:1/20:0)+K OH	656.61	Cer(d18:1/16:0)+K
380.25	S-1-P+H	676.52	Cer(d18:1/22:0)+K OH	740.63	Cer(d18:1/22:0)+K
		704.64	Cer(d18:1/24:0)+K OH	766.57	Cer(d18:1/24:1)+K
m/z	Ceramide			m/z	Globotriaosylceramides
482.44	Cer(d18:1/12:0)+H	m/z	Sphingomyelins	1046.75	Gb3 C16:0+Na
504.43	Cer(d18:1/12:0)+Na	703.57	SM(d18:1/16:0)+H	1074.78	Gb3 C18:0+Na
510.48	Cer(d18:1/14:0)+H	705.58	SM(d18:0/16:0)+H	1102.77	Gb3 C20:0+Na
520.41	Cer(d18:1/12:0)+K	731.61	SM(d18:1/18:0)+H	1128.74	Gb3 C22:1+Na
532.47	Cer(d18:1/14:0)+Na	741.53	SM(d18:1/16:0)+K	1130.75	Gb3 C22:0+Na
538.51	Cer(d18:1/16:0)+H	753.58	SM(18:1/18:0)+Na	1146.73	Gb3 C22:0-+K
548.44	Cer(d18:1/14:0)+K	769.56	SM(d18:1/18:0)+K	1156.77	Gb3 C24:1+Na
560.51	Cer(d18:1/16:0)+Na	797.53	SM(d18:1/20:0)+K	1158.78	Gb3 C24:0+Na
566.55	Cer(d18:1/18:0)+H	m/z	Hexose Ceramides	1174.76	Gb3 C24:0-+K
576.46	Cer(d18:1/16:0)+K	766.52	HexCer (d18:18:0)+K	1186.78	Gb3 C26:0+Na
594.58	Cer(d18:1/20:0)+H	778.54	HexCer (d18:1/20:0)+Na		
604.54	Cer(d18:1/18:0)+K	806.51	HexCer (d18:1/22:0)+Na	m/z	Gangliosides
616.56	Cer(d18:1/20:0)+Na	822.45	HexCer (d18:1/22:1)+K	1152.77	GM3 (18:1/16:0)
622.61	Cer(d18:1/22:0)+H	834.59	HexCer (d18:1/24:0)+Na	1207.78	GM3 (18:1/20:0) -H
632.53	Cer(d18:1/20:0)+K	840.62	HexCer (d18:1/26:0)+H	1235.75	GM3 (18:1/22:0)-H
644.59	Cer(d18:1/22:0)+Na	848.54	HexCer (d18:1/24:1)+K	1292.78	GM3 (18:1/26:0)
648.62	Cer(d18:1/24:1)+H	850.35	HexCer (d18:1/24:0)+K	1355.79	GM2 (18:1/16:0)
650.64	Cer(d18:1/24:0)+H	860.45	HexCer (d18:1/26:1)+Na	1383.82	GM2 (18:1/18:0)
660.56	Cer(d18:1/22:0)+K	862.26	HexCer (d18:1/26:0)+Na	1545.01	GM1 (18:1/18:0)
672.63	Cer(d18:1/24:0)+Na	878.53	HexCer (d18:1/26:1)+K	1573.01	GM1 (18:1/20:0)
676.66	Cer(d18:1/26:1)+H	880.45	HexCer (d18:1/26:0)+K	1892.15	GD1 (18:1/22:0)-H
678.67	Cer(d18:1/26:0)+H			1836.22	GD1 (18:1/18:0)
686.58	Cer(d18:1/24:1)+K			2128.44	GT1 (18:1/18:0)
688.59	Cer(d18:1/24:0)+K				

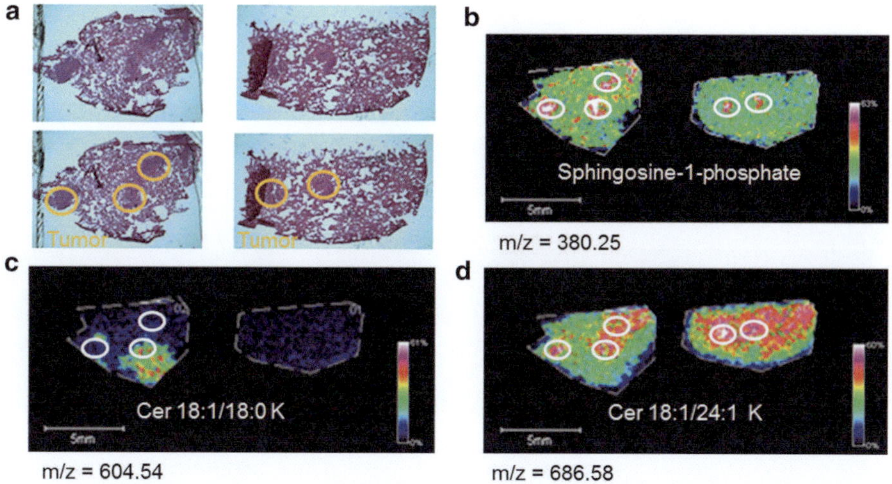

Fig. 8 Lewis lung carcinoma ceramides. (**a**) H&E of two mouse lung tissues with Lewis lung carcinoma tumors. The *circled* regions (*orange*) highlight the tumors. MALDI-IMS distribution of S-1-P (**b**) and two representative ceramides, (**c**) Cer d18:1/18:0 + K and (**d**) Cer d18:1/24:1 + K, are shown

were elevated in the tumor, but also in the adjacent non-tumor regions of the tissues. The C18 ceramide was primarily detected in regions adjacent to the tumor in one of the tissues. While only limited examples are shown, applying MALDI-IMS analysis of tumor tissue models like this one pre- and post-therapy could provide a unique tool to assess the role of ceramides and S-1-P in treatment efficacies.

Case 2: FTICR MALDI-IMS Detection of the Drug LCL124 and its Target Lipids in ACHN Kidney Tumor Xenografts

LCL124 (L-t-C6-CCPS; SPG103) from SphingoGene, Inc. is a cationic ceramide analogue which initiates apoptosis via depolarization of the mitochondria [4]. In a recent study using a pancreatic tumor model with LCL124 treatment, tumors showed drug accumulation and increased apoptosis [4]. LC-MS analysis of the tumors identified a panel of ceramides elevated in response to LCL124. LCL124 distribution in the treated tissues was also visualized with FTICR MALDI-IMS, which could distinguish the small molecule drug at m/z = 475.4. To determine whether changes in the levels of ceramides could be detected after LCL124 treatment, the MALDI-IMS workflows were applied to a human K562 kidney tumor xenograft mouse model, using 2,5 DHB matrix and a raster width of 100 μm. Tumor bearing animals were treated seven times with LCL124 (40 mg/kg, once daily) or PBS. Upon completion of the treatment schedule, tumor xenografts were harvested, and prepared for FTICR MALDI-IMS analysis. LCL124 could be readily detected at m/z = 475.4 in the treated tumor tissue (Fig. 9a). Ceramides of varying chain lengths were identified with higher intensities in the treated tissues, a characteristic which is more prominent in the longer chain Cer 18:1/24:1 K and 18:1/24:0 K

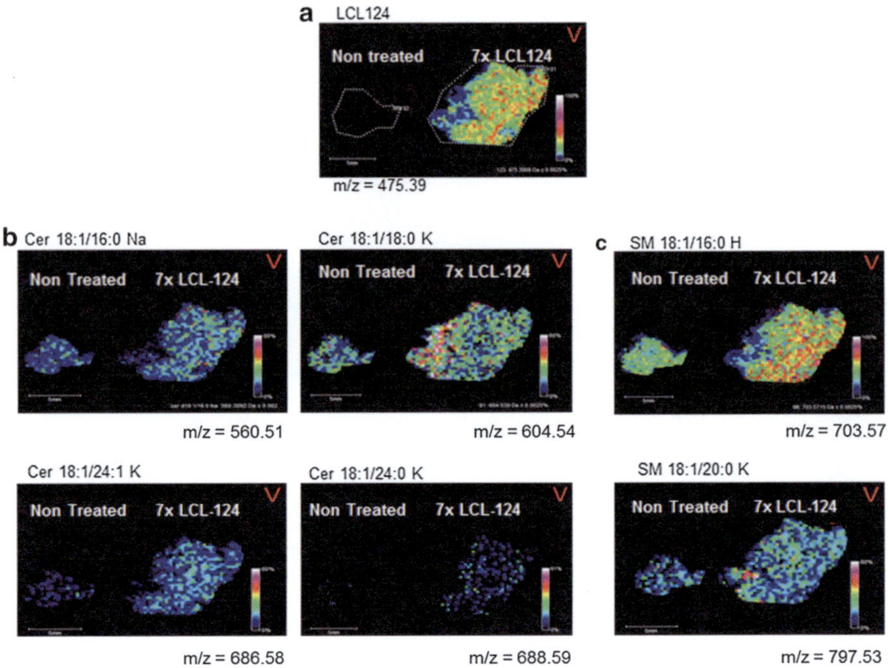

Fig. 9 FTICR MALDI-IMS of ceramides and LCL124 in ACHN mouse xenografts following LCL124 treatment. (**a**) Detection of LCL124 (m/z=475.39) across the non-treated and 7× treated xenografts. (**b**) Panel of ceramides identified in xenografts as compared to distribution of LCL124. (**c**) Sphingomyelins 18:1/16:0+H and 18:1/20:0+K distribution across the xenografts

species (Fig. 9b). Sphingomyelin species were also identified and verified with on-tissue digestions with bSMase (data not shown).

As glucosylceramides are synthesized directly from ceramides, these are additional biomolecules to monitor and detect when assessing modulation of tissue ceramide levels in a treatment regime. Two HexCer species, 18:1/22:0 Na and 18:1/24:1 K, were identified and localized to areas of the tumor where LCL124 is absent (Fig. 10a). Similar to the findings with ceramides, the overlays indicate that LCL124 may be modulating the levels of glucosylceramides (Fig. 10b). Given that the regulation of glucosylceramide synthase and glucosylceramide biosynthesis has been linked to decreased apoptosis and increased chemo-resistance, this was an interesting finding [51]. Studies are ongoing to better define the levels of glucosylceramides with and without drug treatment, as well as defining other glycosphingolipid classes with larger carbohydrate components, such as gangliosides. The cumulative information gathered will be used in conjunction with other mechanistic studies to better define the action of LCL124 in disrupting mitochondrial function in cancer cells.

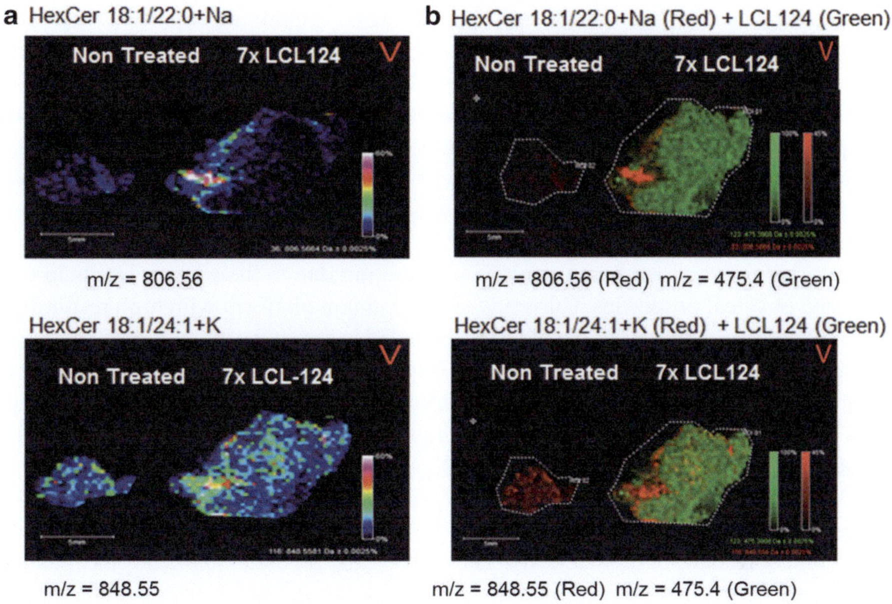

Fig. 10 FTICR MALDI-IMS of hexose ceramides in LCL124 treated ACHN mouse xenografts. (**a**) HexCer 18:1/22:0+Na and HexCer 18:1/24:1+K have higher intensities where LCL124 detection is low as shown individually. (**b**) Co-localization of HexCer 18:1/22:0+Na (*red*) and LCL124 (*green*), and HexCer 18:1/24:1+K (*red*) and LCL124 (*green*)

9 Summary and Extension of the Approach

This chapter focuses on the use of FTICR MALDI-IMS to identify sphingolipids and glycosphingolipids in disease and cancerous tissues. Using tissues derived from two lysosomal storage disorders known to accumulate ceramides and glycosphingolipids, respectively, a comprehensive database was constructed which could then be applied to other tissue studies [13]. The approach used to generate this list of largely uncharged sphingolipids in the Farber and Fabry disease model tissues is currently being applied to other lysosomal storage disease models. There is a particular emphasis on defining ganglioside species in these models.

The extra step of confirming structures using exogenously added enzymes, like bCDase, offers biochemical specificity to the chemical structure information obtained with the mass spectrometer. Given the high mass accuracy and resolving power of the FTICR mass spectrometer whereby analytes can differ by less than 1 amu, this layer of biochemical verification proved beneficial. On-tissue digestions can also be used to quickly sort for candidate analytes of interest in a particular structural class. Quantitative assessment of the analytes detected is limited to ion intensity comparisons in a given tissue, and is affected by tissue type, matrix choice and other factors related to the MALDI process. For low abundance ceramide and glycosphingolipid species, it certainly does not replace the need for quantitative measures provided by LC-MS/MS approaches. Combining the two methods provides synergistic information in relation to ion distribution and relative abundance of individual ions in a given tissue [13].

The LCL124 example indicates the potential utility of MALDI-IMS to detect drug distribution in tissues of interest, along with the molecular effects of the drug linked to its mechanism of action. This will be particularly effective for drugs that modulate lipid pathways, and could include statins and NSAID modulators of prostaglandins. Analyzing bioactive lipid derived components, like prostaglandins and the ceramides described herein, can be challenging due to their lower levels in tissues and physical properties. Currently, an instrument platform with high resolution like that conferred by a MALDI-FTICR is needed to target these types of molecules, especially when linking their distribution with the uptake and localization of drugs. Determining the metabolic correlates of drug distribution and efficacy in target tissues like the sphingolipids and glycosphingolipids described herein will also aid in identifying therapeutic biomarker candidates.

Overall, the functional applications of MALDI-IMS will continue to evolve to better define the tissue distribution of small molecule therapeutic agents and cellular metabolic correlates of their activity. The options and potential applications of using exogenous degradative enzymes added to tissues to quickly confirm structural features of target molecules are only just now being explored. There is great opportunity for expanding these options to meet the continued challenge of using MALDI-IMS for tissue imaging analysis. As instrumentation becomes more sensitive and approaches the capability of single cell level analysis, this type of approach could prove particularly helpful in the data interpretation of molecular changes in response to therapeutic treatments.

References

1. Larsen PJ, Tennagels N (2014) On ceramides, other sphingolipids and impaired glucose homeostasis. Mol Metab 28:252–260
2. Hannon YA, Obeid LM (2011) Many ceramides. J Biol Chem 286:27855–27862
3. Morad SA, Cabot MC (2013) Ceramide-orchestrated signaling in cancer cells. Nat Rev Cancer 13:51–65
4. Beckham TH, Lu P, Jones EE et al (2013) LCL124, a cationic analog of ceramide, selectively induces pancreatic cancer cell death by accumulating in mitochondria. J Pharmacol Exp Ther 344:167–178
5. Saddoughi SA, Ogretman B (2013) Diverse functions of ceramide in cancer cell death and proliferation. Adv Cancer Res 117:37–58
6. Ponnusamy S, Meyers-Needham M, Senkal CE et al (2012) Sphingolipids and cancer: ceramide and sphingosine-1-phosphate in the regulation of cell death and drug resistance. EMBO Mol Med 8:761–765
7. Slotte JP (2013) Biological functions of sphingomyelins. Prog Lipid Res 52:424–437
8. Vouk K, Hevir N, Ribic-Pucelj M et al (2012) Discovery of phosphatidylcholines and sphingomyelins as biomarkers for ovarian endometriosis. Hum Reprod 27:2955–2965
9. Zhu J, Wang Y, Yu Y et al (2013) Aberrant fucosylation of glycosphingolipids in human hepatocellular carcinoma tissues. Liver Int 34:147–160

10. Liu Y, Chen Y, Momin A et al (2010) Elevation of sulfatides in ovarian cancer: an integrated transcriptomic and lipidomic analysis including tissue-imaging mass spectrometry. Mol Cancer 9:186
11. Schwamborn K, Caprioli RM (2010) MALDI Imaging mass spectrometry—painting molecular pictures. Nat Rev Cancer 10:639–646
12. Irie M, Fujimura Y, Yamato M et al (2014) Integrated MALDI-MS imaging and LC-MS techniques for visualizing spatiotemporal metabolomics dynamics in a rat stroke model. Metabolomics 10:473–483
13. Jones EE, Dworski S, Canals D et al (2014) On tissue localization of ceramides and other sphingolipids by MALDI mass spectrometry imaging. Anal Chem 19:8303–8311
14. Stoeckli M, Chaurand P, Hallahan DE et al (2001) Imaging mass spectrometry: a new technology for the analysis of protein expression in mammalian tissues. Nat Med 4:493–496
15. Gessel MM, Norris JL, Caprioli RM (2014) MALDI imaging mass spectrometry: spatial molecular analysis to enable a new age of discovery. J Proteomics 107:71–82
16. Berry KA, Hankin JA, Barkley RM et al (2011) MALDI imaging of lipid biochemistry in tissues by mass spectrometry. Chem Rev 111:6491–6512
17. Angel PM, Spraggins JM, Baldwin HS et al (2012) Enhanced sensitivity for high spatial resolution lipid analysis by negative ion mode matrix assisted laser desorption ionization imaging mass spectrometry. Anal Chem 84:1557–1564
18. Sun N, Ly A, Meding S et al (2014) High-resolution metabolite imaging of light and dark treated retina using MALDI-FTICR mass spectrometry. Proteomics 7–8:913–923
19. Wang X, Han J, Pan J et al (2014) Comprehensive imaging of porcine adrenal gland lipids by MALDI-FTMS using quercetin as a matrix. Anal Chem 86:638–646
20. Chen Y, Liu Y, Sullards MC et al (2010) An introduction to sphingolipid metabolism and analysis by new technologies. Neuromolecular Med 12:306–319
21. Powers TW, Jones EE, Betesh LR et al (2013) A MALDI imaging mass spectrometry workflow for spatial profiling analysis of N-linked glycan expression in tissues. Anal Chem 20:9799–9806
22. Chaurand P, Cornett DS, Angel PM, Caprioli RM (2011) From whole-body sections down to cellular level, multiscale imaging of phospholipids by MALDI mass spectrometry. Mol Cell Proteomics 10:O110.004259
23. Castellino S, Groseclose MR, Sigafoos J et al (2012) Central Nervous system disposition and metabolism of Fosdevirine (GSK2248761), a non-nucleoside reverse transcriptase inhibitor: an LC-MS Matrix-assisted laser desorption/ionization imaging MS investigation into central nervous system toxicity. Chem Res Toxicol 2:241–251
24. Chughtai K, Jiang L, Greenwood TR et al (2013) Mass spectrometry images acylcarnitines, phosphatidylcholines, and sphingomyelin in MDA-MB-231 breast tumor models. J Lipid Res 54:333–344
25. Chaurand P, Cornett DS, Angel PM et al (2011) From whole-body sections down to cellular level, multiscale imaging of phospholipids by MALDI mass spectrometry. Mol Cell Proteomics 2:o110.004259
26. Chen Y, Liu Y, Allegood J et al (2010) Imaging MALDI mass spectrometry of sphingolipids using an oscillating capillary nebulizer matrix application system. Methods Mol Biol 656 :131–146
27. Chen Y, Allegood J, Liu Y et al (2008) Imaging MALDI mass spectrometry using an oscillating capillary nebulizer matrix coating system and its application to analysis of lipids in brain from a mouse model of Tay-Sachs/Sandhoff disease. Anal Chem 80(8):2780–2788
28. Le CH, Han J, Borchers CH et al (2012) Dithranol as a MALDI matrix for tissue imaging of lipids by Fourier transform ion cyclotron resonance mass spectrometry. PLoS One 7(11), e49519
29. Fulop A, Porada MB, Marsching C et al (2013) 4-Phenyl-α-cyanocinnamic acid amide: screening for a negative ion matrix for MALDI-MS imaging of multiple lipid classes. Anal Chem 85:9156–9163

30. Cerruti CD, Touboul D, Guerineau V et al (2011) MALDI imaging mass spectrometry of lipids by adding lithium salts to the matrix solution. Anal Bioanal Chem 401:75–87
31. Goto-Inoue N, Hayasaka T et al (2010) The detection of glycosphingolipids in brain tissue sections by imaging mass spectrometry using gold nanoparticles. J Am Soc Mass Spectrom 11:1940–1943
32. Goto-Inoue N, Hayasaka T, Zaima N et al (2012) Imaging mass spectrometry visualizes ceramides and the pathogenesis of dorfman-chanarin syndrome due to ceramide metabolic abnormality in the skin. PLoS One 7(11), e49519
33. Colsch B, Woods AS (2010) Localization and imaging of sialylated glycosphingolipds in brain tissue sections by MALDI mass spectrometry. Glycobiology 6:661–667
34. Alayoubi AM, Wang JC, Au BC et al (2013) Systemic ceramide accumulation leads to severe and varied pathological consequences. EMBO Mol Med 5:827–842
35. Pacienza N, Yoshitmitsu M, Mizue N et al (2012) Lentivector transduction improves outcomes over transplantation of human HSCs alone in NOD/SCID/Fabry mice. Mol Ther 7:1454–1461
36. Baykut G, Fuchser J, Witt M et al (2002) A combined ion source for fast switching between electrospray and matrix-assisted laser desorption/ionization in Fourier transform ion cyclotron resonance mass spectrometry. Rapid Commun Mass Spectrom 16:1631–1641
37. O'Connor PB, Costello CE (2001) A high pressure matrix-assisted laser desorption/ionization Fourier transform mass spectrometry ion source for thermal stabilization of labile biomolecules. Rapid Commun Mass Spectrom 15:1862–1868
38. O'Connor PB, Mirgorordskaya E, Costello CE (2002) High pressure matrix-assisted laser desorption/ionization Fourier transform mass spectrometry for minimization of ganglioside fragmentation. J Am Soc Mass Spectrom 13:402–407
39. Pettus BJ, Bielawska A, Kroesen BJ et al (2003) Observation of different ceramide species from crude cellular extracts by normal-phase high-performance liquid chromatography coupled to atmospheric pressure chemical ionization mass spectrometry. Rapid Commun Mass Spectrom 17:1203–1211
40. Canals D, Hannun YA (2013) Novel chemotherapeutic drugs in sphingolipid cancer research. Handb Exp Pharmacol 211–38
41. Nowling TK, Mather AR, Thirumagal T et al (2015) Renal glycosphingolipid metabolism is dysfunctional in lupus mice and patients with nephritis. J Am Soc Nephrol 26:1402
42. Whitehead SN, Chan KH, Gangaraju S et al (2011) Imaging mass spectrometry detection of gangliosides species in the mouse brain following transient focal cerebral ischemia and long-term recovery. PLoS One 6, e20808
43. Grove KJ, Voziyan PA, Spraggins JM et al (2014) Diabetic nephropathy induces alterations in the glomerular and tubule lipid profiles. J Lipid Res 55:1375–1385
44. Woods AS, Colsch B, Jackson SN et al (2013) Gangliosides and ceramides change in a mouse model of blast induced traumatic brain injury. ACS Chem Neurosci 4:594–600
45. Richards AL, Lietz CB, Wager-Miller J et al (2012) Localization and imaging of gangliosides in mouse brain tissue sections by laserspray ionization inlet. J Lipid Res 53:1390–1398
46. Honke K, Zhang Y, Cheng X et al (2004) Biological roles of sulfoglycolipids and pathophysiology of their deficiency. Glycoconj J 21:59–62
47. Yuki D, Sugiura Y, Zaima N et al (2011) Hydroxylated and non-hydroxylated sulfatide are distinctly distributed in the human cerebral cortex. Neuroscience 193:44–53
48. Marsching C, Eckhardt M, Grone H et al (2011) Imaging of complex sulfatides SM3 and SB1a in mouse kidney using MALDI-TOF/TOF mass spectrometry. Anal Bioanal Chem 401:53–64
49. Manwaring V, Boutin M, Auray-Blais C (2013) A metabolomic study to identify new globotriaosylceramide-related biomarkers in the plasma of Fabry disease patients. Anal Chem 19:9039–9048
50. Yamaura T, Doki Y, Murakami K, Saiki I (1999) Model for mediastinal lymph node metastasis produced by orthotopic intrapulmonary implantation of lung cancer cells in mice. Hum Cell 12:197–204
51. Giussani P, Tringali C, Riboni L et al (2014) Sphingolipids: key regulators of apoptosis and pivotal players in cancer drug resistance. Int J Mol Sci 3:4356–4392

Sphingoproteomics: Proteomic Strategies to Examine Sphingolipid Biology

Bruce A. Stanley, Tye Deering, and Todd E. Fox

Abstract Interest in sphingolipids has increased in the past couple of decades as the number of biological activities identified has greatly expanded. These include roles in inflammation, proliferation, survival, and metastasis. Sphingolipids can exert these effects through an increasing number of identified interacting cellular targets. To facilitate the understanding of the intrinsic biology of sphingolipids and the development of sphingolipid-based therapeutics, further knowledge is needed. Various analytical protocols assist this endeavor, with mass spectrometry-based techniques seeing increasing usage, especially for measuring steady-state lipid levels. The area of mass spectrometry-based proteomics is also seeing increased usage in the study of lipid biology. This chapter provides an introduction to hypothesis-generating and hypothesis-testing protein-based analytical approaches to investigate sphingolipids and sphingolipid-metabolizing enzymes. These tools can serve to identify how sphingolipids regulate the proteome, to define how post-translational modifications control enzymatic activity, to identify protein–protein and protein–lipid interactions as well as to facilitate inhibitor development, among other concepts. These approaches can help delineate the roles and consequences of perturbations of sphingolipid metabolism in cancer.

Keywords Sphingolipids • Ceramide • Sphingosine kinase • Proteomics • Mass spectrometry • Interactions • Post-translational modifications

B.A. Stanley
Pennsylvania State University College of Medicine, Hershey, PA, USA
e-mail: bstanley@psu.edu

T. Deering • T.E. Fox (✉)
Department of Pharmacology, University of Virginia,
1340 Jefferson Park Avenue, PO Box 800735, Charlottesville, VA, USA
e-mail: tgd4s@virginia.edu; tfox@virginia.edu

Y.A. Hannun et al. (eds.), *Bioactive Sphingolipids in Cancer Biology and Therapy*, DOI 10.1007/978-3-319-20750-6_16

1 Introduction

Mass spectrometry-based lipidomic strategies to investigate sphingolipid metabolism have revealed a diverse array of sphingolipid changes in various cancers. These data largely consist of results from a multitude of cell lines in response to multiple therapeutic strategies and factors believed to regulate cancer progression. The implication that there are specific sphingolipids modulating phenotypic outcomes, such as proliferation, metastasis, and cell death, has greatly increased interest in investigating these molecules. These results also include research into sphingolipid alterations within human specimens, such as breast [1], head and neck [2, 3], lung [4], and pancreatic [5] cancers. These human specimen studies have, in part, challenged conventional dogma, such as ceramides being elevated in cancer, and point to the need for further studies to understand the roles of sphingolipids in cancer biology and cancer therapeutics.

The advantage of analyzing lipid metabolites (lipidomics) themselves is they represent the final outcomes of changes to genotype (e.g. epigenetic, mutation, chromosomal rearrangements), RNA, and protein levels and modifications. However, these steady-state lipid measurements need to be coupled with other analyses such as metabolic flux and protein measurements. This combination will allow a better understanding of the mechanisms of steady-state lipid alterations within *in vitro* and *in vivo* models of cancer as well as clinical specimens. As an example, sphingosine kinases are often implicated for changes in sphingosine-1-phosphate levels; however the contributions of lipid phosphatases and lyases in sphingosine-1-phosphate levels are often overlooked. The objective here is to introduce mass spectrometry-based proteomics strategies as an additional investigative tool to further understand sphingolipid metabolism and mechanisms of action in cancer (Fig. 1). A diverse array of hypothesis-generating and hypothesis-testing techniques exists to aid this endeavor.

Fig. 1 The alphabet soup of mass spectrometry and proteomics

2 Discovery Proteomics

Discovery-based proteomics is typically what people think about in regards to mass spectrometry-based proteomics. Such strategies attempt to take an unbiased approach to examine global differences in protein presence or levels between groups. Discovery proteomics is sometimes referred to as "shotgun" proteomics. The term "shotgun" takes a different meaning here as opposed to shotgun lipidomics. While both are profiling methods, shotgun lipidomics refers to non-chromatographic infusion-based methodologies to introduce the sample into the mass spectrometer [6], whereas shotgun proteomics methods usually employ chromatography.

These approaches have yielded a wealth of information for various diseases and pathways, including cancer. As sphingolipids have been implicated in several facets of cancer progression and as a mediator of therapeutic strategies (i.e. radiation and chemotherapy), these types of approaches have the potential to increase our understanding of how sphingolipid metabolism can regulate the proteome leading to phenotypic changes. These methodologies could be applied to a large variety of sources, including tissues from genetically modified animals, samples arising from gene deletion and knockdown strategies such as siRNA/CRISPR/TALEN-based approaches, and inhibitor treatments to understand the consequences and adaptations of cells to these variations.

There have been few publications that have utilized discovery proteomics in the realm of sphingolipids. Proteomic analysis of C6-ceramide treated HCT-116 cells (a colon carcinoma cell line), was assessed by 2D-gel electrophoresis followed by Matrix-assisted laser desorption/ionization (MALDI)-time of flight (TOF)-MS [7]. This group discovered 43 proteins differentially regulated by C6-ceramide treatment. A similar approach was used with the same cell line in response to C16-ceramide exposure to identify 51 differentially regulated proteins [8]. This included elevated levels of the Bcl-2-associated transcription factor (Btf). Further work showed that Btf regulates p53 and BAX expression to regulate C16-ceramide-induced apoptosis [8]. Proteomic analysis was also undertaken to investigate perturbations of C6 glioma cells in response to 2′-hydroxy C16-ceramide [9]. The 28 proteins changed with short-term treatment suggested further investigations into MAPK and Akt signaling pathways [9].

Proteomics can also be utilized at a subcellular/organelle level. For example, a proteomics screen was used to identify lysosomal proteins elevated in response to the DNA damaging agent, camptothecin [10]. These results demonstrated an increase in prosaposin, a precursor to saposin, which regulates lysosomal glycosphingolipid degradation [10]. Proteomics has also been utilized to identify changes in fibroblast proteins within lipid microdomains upon C8-lactosylceramide treatment [11].

The general workflow for discovery proteomics (Fig. 2) requires the isolation of total protein, which can be cellular, subcellular or extracellular in origin. Most approaches utilize a "bottom-up" strategy, where isolated proteins are subjected to proteolytic digestion, usually with trypsin. A wide variety of published protein isolation and cleanup protocols exist, including organic precipitation of proteins, urea

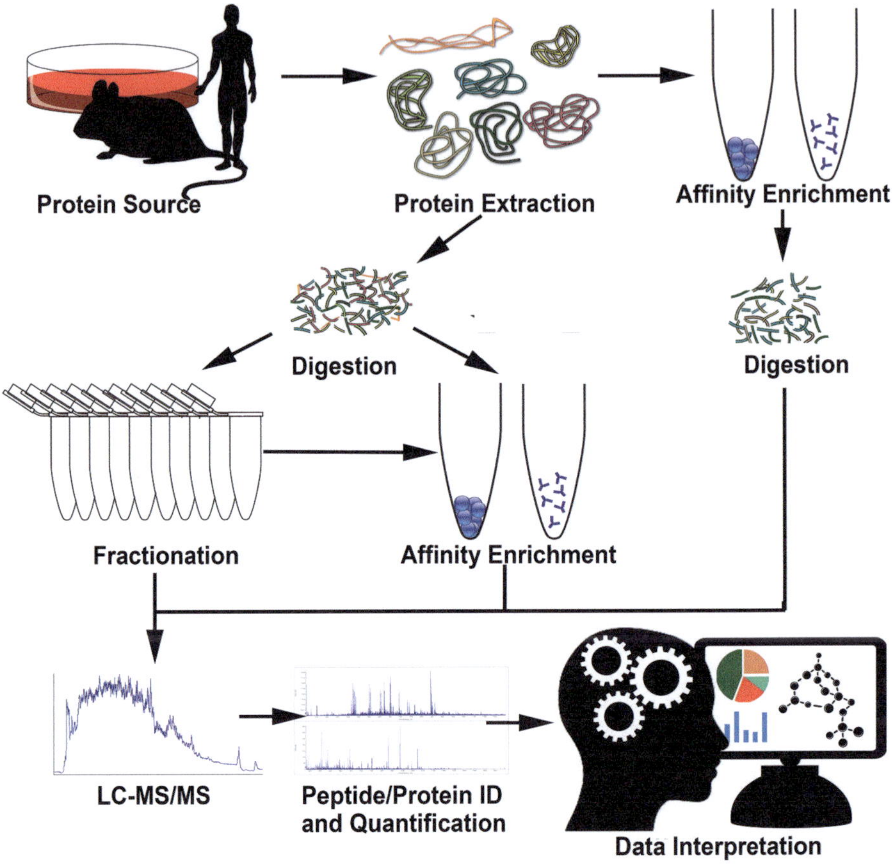

Fig. 2 A general workflow for proteomics strategies. Proteins are derived from a source such as cell culture models, animal models, or clinical specimens. The protein can then be directly subjected to enrichment strategies that include antibodies, lipid, and drug affinity. The enriched samples are then digested and analyzed by mass spectrometry. Alternatively, proteins can be immediately digested and subjected to fractionation techniques (e.g. electrophoresis, chromatography) or enrichments can be done at the peptide level with or without prior fractionation. Samples are then subjected to mass spectrometry analysis. Analysis entails peptide and protein identification, quantification, and data interpretation (e.g. functional enrichment, network analysis)

or other chaotrope denaturation, SDS denaturation, in preparation for proteolytic digestion. The resulting proteolytic peptides can then be used as is (unlabeled) for label-free quantitation strategies, or labeled either pre- or post-digestion for multiplex quantitation using heavy isotopes. Post-digestion labeling strategies may employ options such as isobaric tags (iTRAQ, TMT) or dimethyl labeling, while metabolic ^{15}N-labeling or SILAC (Stable Isotope Labeling by Amino acids in Cell culture) quantitation studies can be performed by growing cells in the presence of heavy-isotope elements (^{15}N) or amino acids (SILAC) rather than adding an isotopically-tagged label post-digestion. In addition, combinations such as SILAC and isobaric tags have been used to obtain up to a 54-plex level of quantification [12]. Stable Isotope LAbeling in Mammals (SILAM) can also be used in animal models via the use of isotopically-labeled lysine or ^{15}N-labeled spirulina as a protein source in the diet [13]. SILAM can theoretically also be coupled with post-digestion isobaric tags for multiplexing. Ultimately, label-free strategies in data-dependent acquisitions are less quantitative, but can be advantageous since label-free approaches have reduced sample preparation complexity and no potential for incomplete labeling with reduced costs. Furthermore, for clinical-specimen-based studies greater statistical power can be more readily obtained. There are isobaric approaches to increase statistical power, such as using the same sample in multiple sets of iTRAQ or TMT-based labeling experiments or to utilize spike-in or super-SILAC [14]. For super-SILAC, a representative cell line that closely matches the source material is labeled via SILAC and then mixed with the sample of interest. The individual protein ratios between samples are then assessed.

After digestion, most approaches utilize a multi-dimension fractionation-based approach to reduce the complexity of peptides that enter the mass spectrometer for analysis. These multiple strategies can employ electrophoresis, strong anionic exchange (SAX), strong cationic exchange (SCX), hydrophilic interaction (HILIC), electrostatic repulsion (ERLIC), or high pH/basic reverse phase chromatography among other techniques in the first dimension. This is usually followed by low pH reverse phase chromatography from which the desalted eluate is more compatible with mass spectrometry for ionization and detection of peptides. The additional work of adding more dimensions of separation can also yield impressive results. A three dimensional chromatography approach (basic pH C18 reverse phase, SAX, acidic pH C18 reverse phase on a 1 m analytical column) identified 11,352 gene products (~70 % of the proteome), which is similar to the coverage provided by RNA-seq strategies [15]. Importantly, this covered a wide dynamic range enabling detection of lower abundance proteins such as several of the sphingolipid metabolizing enzymes, which included peptides from sphingosine kinase 2, alkaline ceramidase, serine palmitoyltransferase, ceramide kinase, several ceramide synthases, among others. Though this approach took 8 days of dedicated instrument time to acquire data, performing the first and/or second dimension fractionation offline could decrease the instrument time. As newer instrumentation become faster, more sensitive, and with increased dynamic range, in depth proteome coverage will become easier.

Currently, two types of mass spectrometers are the predominant workhorses for discovery-based proteomics. These may be Fourier transform ion cyclotron resonance (FT-ICR) or Kingdon/orbitrap-based, which include an orbitrap mass analyzer or hybrid instruments with a Q1 quadrupole in front as well as the newer "Tribrid," containing a quadrupole, orbitrap and linear ion trap. The other major instrument type is the quadrupole-time of flight (qTOF) instrumentation. The use of matrix-assisted laser desorption/ionization (MALDI) has decreased for general proteomics work in recent years, and some major manufacturers have discontinued development. Nonetheless, with proper sample preparation MALDI-based mass spectrometry can currently identify 2000–3000 proteins from a single sample and the different

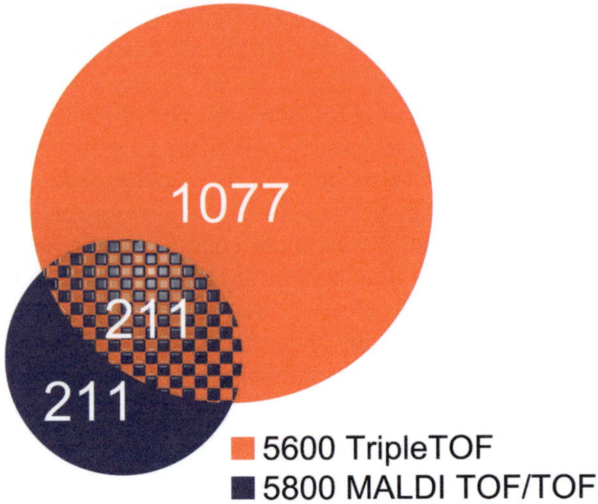

Fig. 3 Complementation of mass spectrometry techniques. Various techniques exist, each with their own advantages and disadvantages. This Venn diagram demonstrates the difference in protein identifications (5 % false discovery rate) of an identical sample run on a nanospray 5600 TripleTOF f (*orange*) or a 5800 MALDI-TOF/TOF (*blue*) instrument. Though the nanospray workflow yielded more protein identifications, there was a subset of proteins that were uniquely identified by the 5800 workflow

ionization mode in MALDI-based instrumentation can offer several advantages. Comparison of a protein digest between a 5600 TripleTOF, a qTOF mass spectrometer and a 5800 MALDI-TOF instrument, both from AB Sciex, emphasizes the synergistic nature of different instrument types (Fig. 3). Though there were close to three times as many total protein identifications from the 5600 proteomics workflow than from the 5800 MALDI workflow, there were however a significant number of strong IDs that came only from the 5800 MALDI workflow. Some existing companies are still working on newer MALDI instrumentation and new MALDI-based instrument companies have also been starting up (SimulTOF Systems). MALDI usage continues to evolve and has seen increased marketing on other uses such as the identification of microorganisms isolated from human specimens and mass spectrometry-based imaging. MALDI also offers several advantages for glycosylated peptides due to the differential fragmentation that can facilitate structural elucidation. In sum, it is a complementary tool and can usually add protein and peptide identification and information not seen from other instruments.

Discovery-based approaches can be further extended to understand perturbations in post-translational modifications (PTMs). There are several strategies to assess phosphorylation, glycosylation, acetylation, nitrosylation, ubiquitination, palmitoylation among other post-translational modifications. These techniques typically employ affinity-based peptide enrichment that can be via antibodies, use of biotin-switch, lectin, and/or metals.

There are several challenges with global post translational modification (PTM) analysis. One of the main challenges is enrichment/isolation from unmodified peptides.

For phosphopeptides, though a large number of proteins are phosphorylated (~30 %), the stoichiometry for a specific site can often be low. Of this, the great bulk of phosphorylated peptide sites are phosphorylated serines. Phosphoamino analysis suggests a ratio of pSer:pThr:pTyr at around 1800:200:1 [16], though up to ~1.5 % pTyr has been reported [17]. The choice of enrichment strategy is dependent on experimental goals and the depth one wants to attempt to uncover. As an example, for phosphoproteomics, first dimensional ERLIC and SCX separation of peptides are complementary with one leading to preferentially acidic and the other basic phosphopeptides. Both strategies, even after further enrichment via immobilized metal ion affinity chromatography (IMAC) and titanium dioxide procedures, will co-purify other modifications such as glycosylation. Since glycosylation is more prevalent than phosphorylation, to maximize the time a mass spectrometer spends on phosphopeptides, incorporation of a glycosidase (e.g. PNGase) can be beneficial. There are also strategies to multiplex different PTM analyses, such as glycosylation and phosphorylation [18] and phosphorylation with acetylation of lysines and ubiquitination [19].

One of the main issues with discovery PTM analysis is that stoichiometric information is needed to put any change into perspective. It needs to be clear if a change in the level of a modified peptide observed between treatments is, in fact, due to a change in the percent of that peptide carrying the PTM (stoichiometry) versus a change in parent protein level. Several strategies exist to overcome this issue. This can be done by assessing the levels of the unmodified peptide if an unenriched (or enriched flow-through) is measured. A challenge with this is that in a discovery mode using data-dependent method (IDA or DDA), MS/MS may not be triggered for sufficient identification and measurement of the unmodified peptide. Thus, further analysis and validation may need to be done through a targeted approach or a data-independent acquisition (DIA) approach such as MS^E on a Waters instrument or SWATH MS (Sequential Window Acquisition of all THeoretical Mass Spectra) [20] or a SWATH-like acquisition on AB Sciex or Thermo accurate mass instruments.

An additional challenge is most PTMs are undefined with regards to function. Therefore, it is not clear if they are activating, priming, inactivating, localization- or stability-conferring, or have some other effect on a given protein. Utilization of bioinformatic-based processing post-acquisition can be used to look for things such as commonalities in phospho-site motifs, or functional enrichment to cluster peptides into families. This information can potentially identify mechanisms of regulation and consequences of these changes. Ultimately, these discovery-based approaches provide a foundation to ask more targeted questions and need to be corroborated with other techniques.

3 Targeted Proteomics

The Merrill laboratory has shown significant parallels between gene expression data (mRNA) and sphingolipid composition [21]. However, the overall correlation between RNA and protein expression is around 0.41 [22, 23]. This low correlation is attributed to factors that include differences in turnover, degradation, and translational control. As there are several sphingolipid proteins that exert the same enzymatic function, such as the ceramidases or the sphingomyelinases, it is important to obtain a larger view of sphingolipid enzymes to correlate with lipid changes. Western blotting has been the traditional method of choice for investigators examining specific protein levels. The challenge with antibody-based approaches for sphingolipid-regulating enzymes is that they are not particularly effective, with a few exceptions, due to a lack of specificity or sensitivity, and in some cases the necessary antibodies have yet to be produced.

The use of targeted proteomics is an option that can allow one to quantify alterations in sphingolipid enzymes even in the absence of specific antibodies. Using selected or multiple reaction monitoring (SRM or MRM), it is possible to multiplex measurements of dozens of proteins within a single analysis. This also offers unparalleled specificity (in addition to quantification) compared to antibody-based approaches. The specificity is obtained by examining multiple peptides (3–4 peptides per protein) and 3–4 transitions (MS/MS fragments) per peptide. This can typically equate to up to 16 MRMs per protein in a typical analysis. Validation can also be obtained by use of a spiked-in stable-isotope labeled peptide, a technique originally termed AQUA (Absolute QUantitation of Proteins) [24]. An extension of this is Stable Isotope Standards and Capture by Anti-Peptide Antibodies (SISCAPA) [25]. Use of the anti-peptide antibodies greatly improves the sensitivity and throughput by enriching for peptides of interest and removing matrix components. This approach allows for the examination of co-elution of the isotopic and endogenous peptide and facilitates normalization and more accurate quantification provided by the combination of the internal standard with an external standard curve. Due to the stable isotope labeling, the mass spectrometer can easily distinguish between the native and isotopically-labeled peptides. As costs of these synthetic peptides can be a roadblock for many when examining a large number of peptides, normalization and relative quantification can also be achieved using an endogenous control akin to what many laboratories do with western blotting.

There are many challenges in MRM applications and there can be a large amount of method development that goes into the process. Not all tryptic peptides are equally detectable, thus the optimal peptides will need to be determined. The mass spectrometer settings will need to be optimized, which includes determination of collisional energies for each peptide fragment and the ideal peptide fragments for quantification. This can be facilitated with software and repositories described further below. On an accurate mass instrument (e.g. qTOFs, orbitraps), the parent peptide can be targeted to readily gather all the product ion information, but the settings will not be ideal for each fragment in a single scan. Ultimately, newer generation triple-quadrupole instrumentation typically offers more sensitivity and faster MS/MS acquisition times for a higher level of targeted multiplexing. On the other hand, high resolution accurate mass instruments offer the capability of data-independent acquisitions. This can allow for the collection of all the ions and fragment ions in a single acquisition allowing for retrospective analyses. In contrast, with a triple quadrupole-based analysis, you can't go back and look for a different peptide of

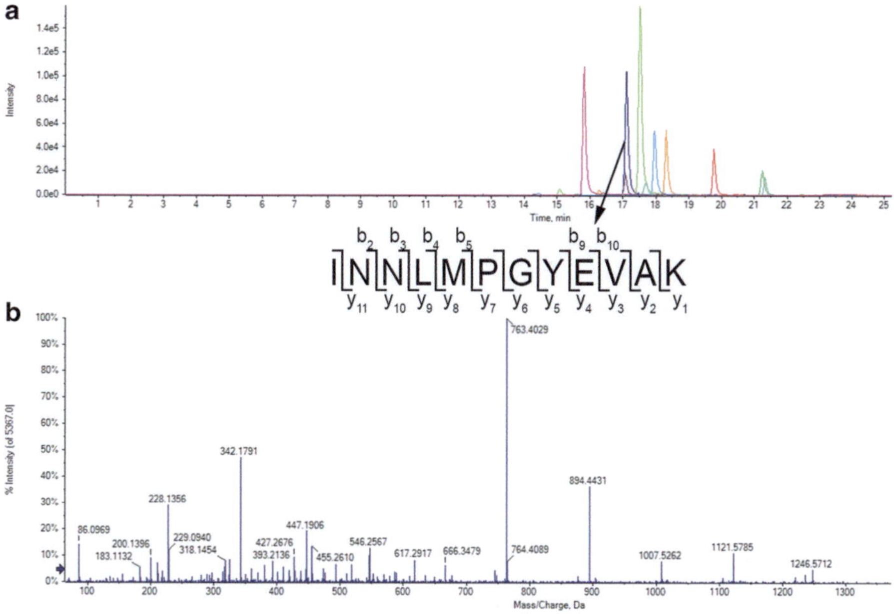

Fig. 4 LC-MS/MS analysis of glucosylceramide synthase. Glucoslyceramide synthase derived from human embryonic kidney cells was digested and analyzed to facilitate development of a targeted proteomics method. (**a**) The chromatograph shows nine readily detectable peptides of glucosylceramide. (**b**) Shows the product ion spectra from one of these peptides and the y and b fragment ions that allow for peptide identification

interest without rerunning a sample. On newer instrumentation, data-independent analyses can provide comparable linear dynamic ranges with a high level of reproducibility. The sensitivity of data-independent-based analyses have been in the past lower than those achievable on triple quadrupole instruments, though this gap has been diminishing. Both of these approaches offer advantages over data-dependent discovery-based approaches. While discovery-focused analyses offer a high number of identifications within a sample, such data-dependent acquisitions show lower levels of reproducibility due to stochastic sampling of peptide ions in different runs, particularly for less abundant peptides. Figure 4 shows some of the process towards developing an assay to assess glucosylceramide synthase.

To facilitate targeted analysis and quantitation of peptides or in data-independent acquisitions when ion libraries are lacking or insufficient, there are several resources that can be utilized to help select good peptides to use. There are several software options, such as Skyline [26] and PeptidePicker [27]. Several repositories exist that can also aid in the process via searching for peptides that other researchers have observed under experimental conditions. These include the Proteomics Database (https://www.proteomicsdb.org) [28], the Human Proteome Map (http://humanproteomemap.org/) [29], and SRMAtlas (http://www.srmatlas.org/) [28]. The complete Human SRM Atlas is currently available as a pre-publication preview. PeptidePicker has the capability of integrating information from several sources, including repositories to expedite the process [27].

Targeted analysis can also be coupled to labeling strategies (iTRAQ, TMT, SILAC) to increase throughput. SILAC can also be utilized with data-independent analysis. Overall, the major roadblock towards comprehensive analysis of several sphingolipid-related proteins is sensitivity. Assuming a good antibody (both specific and high-affinity), an antibody-based technique can often offer more sensitivity, since the mass spectrometry-based detection of sphingolipid enzymes depends both on the ionization efficiency of peptides and the copy number (abundance) of a given protein.

4 Post-Translational Modifications of Sphingolipid-Related Proteins

As with other proteins, sphingolipid-related proteins are modified by post-translational events to regulate aspects such as activity and localization. Various strategies have been put to use over the years to identify such modifications. Bioinformatic strategies examining phosphorylation prediction and species conservation were used to identify putative phosphorylation sites within acid sphingomyelinase [30]. Site-directed mutagenesis helped validate a PKCδ phosphorylation site within this protein [30]. Changes at this site were followed by a phospho-serine PKC substrate antibody on immunopurified acid sphingomyelinase [31]. Metabolic labeling with radiolabeled orthophosphate and site-directed mutagenesis was used to show that the extracellular signal regulated kinases 1 and 2 (ERK1/2) directly phosphorylates SphK1 at Ser225 [32]. Phosphorylation of neutral ceramidase using radiolabeling has also been demonstrated though the specific site(s) were not identified [33]. Another group showed that phosphorylation of Ser354 is involved in neutral ceramidase regulation as determined by site-directed mutagenesis and a serine residue modifier [34]. Modulation of tyrosine kinases through use of inhibitors was shown to regulate ceramide kinases and ceramidase activities by following metabolism of a fluorescent ceramide [35].

Though most work thus far has focused on PTM identification and validation via bioinformatics, labeling, and site-directed mutagenesis, mass spectrometry strategies are now seeing increased usage. Such an approach was utilized to examine ceramide kinase [36]. This work identified serine phosphorylation at positions 340 and 408. Though no function could be identified at position 408, phosphorylation at Ser340 reduced stability of the active enzyme conformation [36].

Assessment of PTMs is not restricted to phosphorylation. Glycosylation of acid ceramidase (aCDase), which is important for enzymatic activity, was demonstrated by enzymatic deglycosylation, and the specific sites determined by site-directed mutagenesis of *N*-glycosylation consensus sequences [37]. A follow-up study further characterized the glycosylation using mass spectrometry to validate the sites of glycosylation with some elucidation of the glycans [38].

A mass spectrometry-based strategy to identify post-translational modifications must take into account several factors. These include the source of protein, proteolytic strategies to obtain wide coverage of the protein or region of interest, accounting for potentially low stoichiometric events, and the strengths and weaknesses of the particular type of mass spectrometric analysis used.

The protein source could be obtained from endogenous or overexpressed sources. Overexpression offers several advantages for obtaining sufficient material for analysis and purification via affinity tags. The major disadvantage of an overexpressed source is that the stoichiometry of a particular PTM may be lower than physiological. This could be attributed to the excess amount of protein relative to the modifying enzyme. Some sources of protein are also not suitable for PTM analysis. One must insure the overexpressing source has the ability to modify the protein appropriately, such as avoiding bacterial expression systems to study phosphorylation of recombinant proteins, since bacterial sources do not contain the requisite kinase activity.

Proteolytic strategies must take into account the peptide size and ionization properties of the peptides produced by a particular proteolytic enzyme. If a specific region is of interest, then *in silico* digests can be performed to determine enzymatic strategies that will produce optimally-sized peptides preferably containing few or no potential PTM-sites other than the PTM site of interest. Looking at multiple proteolytic cleavage patterns is also useful for attempting to examine an entire protein. It is not common for a single enzyme to be suitable for sufficiently covering an entire protein. This can be due to the size of the peptides produced, location of putative cleavage sites, and amino acids near a cleavage site. For example, trypsin predominantly (~95 %) cleaves after arginines and lysines. However, cleavage before a proline is not favored, though such cleavages have been reported [39]. As an example, Ser225 of SphK1 demonstrates the complexity that can arise. This site exists within a 58 amino acid long tryptic peptide that is too big for efficient ionization and fragmentation on most mass spectrometers. However, modification of this site can be observed due to a non-typical trypsin cleavage between a histidine and leucine yielding a much more MS-compatible 16 amino acid long peptide ([40] and unpublished observations). Strategies that utilize different enzymes such as trypsin, chymotrypsin, GluC, AspN, either in separate reactions or in combination are often needed. For example, when phosphorylation of ceramide kinase was examined, 82 % coverage of the total protein was achieved using trypsin, chymotrypsin, and proteinase K digestions [36].

Low stoichiometry can also be a challenge when a modified peptide exists in a low amount. Enrichment strategies, as discussed previously, are suitable here. This can be useful in determination and validation of modified amino acids. Such enrichment strategies are however not suitable for trying to determine whether or how much a particular site's occupancy (modified versus unmodified residue) changes in response to an agonist or other biological condition. An alternative approach is to split a sample in half, remove the modification from one portion (e.g. phosphatase, deglycosylation treatments) leaving the other still attached [41, 42]. The two pools are recombined after isotopic labeling with modifiers such as iTRAQ, TMT, or dimethyl groups. This technique allows for quantitation of low abundance modifications when the modified peptide itself may not be visible on the mass spectrometer (measured as the increase in the unmodified peptide form in the dephosphorylated or deglycosylated aliquot). However, the specific site of modification must already be known, since it cannot be determined in this type of experiment if more than one potential modification site exists.

Lastly, the type of mass spectrometry analysis utilized can have different advantages and disadvantages. While a triple quadrupole-based instrument can typically offer the most sensitivity in targeted analyses, for discovery-based approaches it will lead to more false discovery than an accurate mass and high mass resolution-based instrument. However, a triple quadrupole-based instrument would be suitable for monitoring PTM changes of a known site. This can be coupled with an isotope-labeled peptide for validation and more accurate quantification. Another challenge is that many PTMs are labile. Mass spectrometer-labile PTMs include phosphorylation (more so pSer and pThr than pTyr), glycosylation, sulfonation, and nitrosylation. This means that upon fragmentation and MS/MS spectra accumulation, information on the location of a specific modification can be lost. There are two methods of fragmentation typically used for this analysis, collision-induced dissociation (CID) and electron-based dissociation. If phosphorylation is being examined, multiple serines, threonines and/or tyrosines within a given peptide may preclude site identification if a collision-induced method is utilized. With CID, the collision-based vibrational energy distributes throughout the peptide, and the labile phosphoryl group may fall off before any peptide chain fragments are formed. This results in no peptide fragments still containing the phosphoryl group being visible. Electron-based dissociation, such as electron-collision dissociation (ECD) or electron-transfer dissociation (ETD), fragment the peptide at whatever point they impact it. This prevents distributing the energy through the peptide chain, and can thus preserve the labile PTMs and produce fragment masses still containing the phosphoryl electron transfer dissociationcrapome group whenever the site of electron collision is not at the phosphorylated residue itself. However, the overall efficiency of ECD and ETD is lower than CID. In global analyses collision-based and electron-based can yield different phosphopeptides [43], meaning one technique is not necessarily better than the other, but rather complementary to each other. Furthermore, chromatography can usually separate a peptide that is differentially modified, assisting in validation with isotopically-labeled peptides. For glycosylation, an alternative approach to produce differential fragment ions, such as MALDI, may be needed for elucidation of glycan composition and branch points.

An alternative approach for investigators is to bypass mass spectrometry and make use of existing proteomics mass spectrometry data. Many of these published and unpublished findings are curated at phosphosite.org, which serves as a repository for phosphorylation and other post-translational modifications. Such a strategy is how phosphorylation at Ser115 of the Ceramide transport (CERT) protein was chosen for further investigation. This site was found to be important for interactions with the ER protein VAP and regulation of ceramide trafficking [44]. Though such sources can provide cues for further examination, they are not without their limitations. For example, some modified sites may be the result of false discovery, modifications may not be produced in different model organisms, and most studies utilize trypsin-digested peptides, which will not always yield suitable peptides for analysis of an entire protein.

Whether mass spectrometry-based methods, radiolabeling, bioinformatics or some other approach is used, further validation work will usually be needed. This will involve understanding the function of the modification on the protein and the biological system, sequential and spatial crosstalk amongst other modifications, and the regulatory mechanisms that govern the modification. For phosphorylation, this would include determination of the upstream kinase and phosphatase that regulate the site. In addition to PTMs, other mass spectrometry techniques, such as inductively coupled plasma mass spectrometry (ICP-MS) can be utilized to determine the true metal co-factors involved in regulation the enzymatic activity of sphingolipid enzymes. Such "metalloproteomics" is an emerging field and has expanded what is known about what metals are taken up by the cell and bound to macromolecules within the cell [45–47].

5 Protein–Protein Interactions

Sphingolipid-related proteins, like other proteins, can interact with a wide array of proteins that regulate various aspects such as activity, localization, function and stability. Considering the diverse roles that sphingolipids have in various cellular processes, understanding the composition of protein complexes can shed vital information on cellular function.

Sphingosine Kinase 1 (SphK1) is the most studied sphingolipid-related protein that has been investigated in this regard. Discoveries of SphK1 interacting proteins have used a myriad of techniques. The most prevalent has been the use of a yeast-2-hybrid approach. Using SphK1 as bait, interactions with δ-catenin [48], eukaryotic elongation factor 1A [49], LIM-only factor FHL2 (SLIM3) [50], PECAM-1 [51], SPHK1-interacting protein (SKIP) [52], aminoacylase 1 [53], cytosolic chaperonin CCT (chaperonin containing t-complex polypeptide) [54], RPK118 [55], and Filamin A [56] have been described previously. SphK1 has shown calmodulin-sepharose binding [57], though yeast-2-hybrid studies revealed interactions with Calcium- and Integrin-binding Protein 1 (CIB1, also known as calmyrin and KIP) a protein with 54 % homology to calmodulin [58] and thus may possibly be the true target. The use of the Bovine Viral Diarrhea Virus NS3 protein as bait has also shown to interact with SphK1 [59]. Other approaches have utilized recombinant protein to demonstrate direct phosphorylation, such as with Erk1/2 [32], or dephosphorylating counterpart, PP2A, via a GST-pulldown [60]. Co-immunoprecipitation is another often used tool and has been used to examine SphK1 complexes with PP2A [60], interactions with the tyrosine kinases Lyn [61] and Fyn [62] as well as TRAF2 [63] and co-associating with CD31/PECAM and integrin αvβ3 [64].

These techniques have not been limited to SphK1. Other examples include the detection of heterodimerization of ceramide synthases via co-immunoprecipitation [65], purification of acid ceramidase has implicated binding to saposin D [66], and co-immunoprecipitation has helped reveal FAN (factor associated with neutral sphingomyelinase) as a protein that links neutral sphingomyelinase with the TNF receptor [67].

Use of mass spectrometry has seen wide usage in discovery and dissection of protein-complexes, but the use for examining sphingolipid-related proteins has been limited. One example includes the use of MALDI-TOF to show that heat shock protein 60 co-purifies with neutral sphingomyelinase 2 from brain tissue [68]. The typical mass spectrometry-based approach utilizes affinity purification. This can be an antibody against the endogenous protein, but this requires a high quality antibody. Furthermore, this may require larger-scaling up to purify sufficient protein. A more common approach is the used of overexpressed proteins with affinity tags at one of their termini (e.g. FLAG). Tandem-affinity purification (TAP), where two affinity tags are used, will lead to less non-specific binding than a single affinity tag approach at the potential expense of losing weaker interacting proteins. The use of a tag and the choice of tag location(s) is largely dependent on what, if any, disruption it may have on the protein. Overexpression of protein, compared to examining endogenous protein, will have the potential drawbacks of creating artifacts. These include cellular localization changes, activity changes, perhaps due to a change in stoichiometry with interacting proteins, and non-typical interactions with other proteins.

A general challenge with purification schemes from endogenous or ectopically expressed proteins is non-specific binding. Thus, appropriate controls are necessary to minimize these false positives that result from cell lysis and non-specific interactions with affinity approaches including binding to the resin. Both label-free and labeling techniques are applicable for detection of protein–protein interactions and offer different approaches in minimizing unauthentic interactions. For label-free methods, various statistical approaches exist to examine protein–protein interactions [69]. These can be based on spectral counting (precursor signal intensity), which is defined as the total number of spectra observed for peptides from a specific protein, or on MS1 peak signal quantification. For spectral counting, the number of spectra for an identified protein is compared between a capture and a control immunoprecipitation to determine the background. Extensions of spectral counting include methodologies such as normalized spectral abundance factor (NSAF) [70] and protein abundance factor (PAX, http://pax-db.org/) [71], significance analysis of interactome (SAINT) [72, 73], the Comparative Proteomic Analysis Software Suite (CompPASS) [74], MasterMAP [75], and mass spectrometry interaction statistics (MiST) [76]. These various approaches, and others currently in development, attempt to overcome limitations of spectral counting, account for a wide dynamic range in peptide intensities observed as well as providing confidence scores to detected interactions.

Labeling methods can also be utilized for determination of a protein's interactome. Use of SILAC-based approaches allows for simultaneous sample processing of samples for comparison. This minimizes experimental variations and improves accuracy in relative quantification. Post-immunoaffinity purification labeling through the addition of labels such as iTRAQ, dimethyl, and ICAT can also minimize system (chromatography and mass spectrometer) variability, though the samples are independently prepared and combined prior to sample injection. Techniques to enable distinguishing contaminants from an associated protein also share acronymic-names. These include the isotopic differentiation of interactions as random or targeted (I-DIRT) [77] and the quantitative immunoprecipitation combined with knockdown (QUICK) [78] methods. The QUICK method relies on an RNA interference strategy to reveal the non-specific contaminating proteins. The ratios of the different labels are compared to each other and controls with a user-defined threshold to help define putative interactions.

The CRAPome database (http://www.crapome.org/) serves as an additional resource; a repository of common background contaminants that are typically observed in affinity purification protocols [79]. This resource can be used to assess the presence of a protein within the database and facilitate analysis of user-acquired datasets to help score detected interactions.

An additional challenge associated with identifying protein complex composition is the detection and determination of transient versus stable interactions. Cross-linking-based approaches can be used to elucidate transient or weak interactions that may otherwise be lost during standard affinity purification procedures. Comparing data obtained using SILAC labeled proteins by mixing the heavy and light labeled proteins before (PAM SILAC) or after (MAP SILAC) affinity purification can also be utilized to identify dynamic interactions by examining for exchanging between heavy and light protein complexes [80].

Regardless of the method utilized, validation of interactions is needed. Co-immunoprecipitation of two proteins does not prove a direct protein–protein interaction. Complementary approaches to utilize may include performing a reciprocal immunoprecipitation, use of yeast-2-hybrid, or surface plasmon resonance. These approaches are not limited to discovery of unknown protein complex identification. When some knowledge of binding partners is known, a MRM/SRM or a data-independent analysis such as SWATH can be utilized to quantify differences under various conditions. Perturbations in protein–protein interactions may also lead to clues as to the function of these interactions.

Mass spectrometry can be used to investigate several details of known protein–protein interactions. These include investigating conformational changes, subunit and protein-complex assembly. While the discovery-based approaches provide information on the relative abundance of the complex components, this can be expanded into analyses of undigested protein complexes, which shed light onto the stoichiometry of protein–protein interactions.

Use of ion mobility-MS can be utilized to investigate shape and conformational changes to characterize protein architectures and dynamics. There are three major types of ion mobility; (1) drift tube, (2) traveling wave, and (3) Field Asymmetric Waveform Ion Mobility Spectrometry (FAIMS, also referred to as DMS for differential mobility separation). All three techniques have the ability to separate conformers, though FAIMS may be less able to separate subtle differences than the other two ion mobility types. An advantage of FAIMS/DMS is the ability to separate before Q1, which could allow for selectively transmitting proteins (or other biomolecules) at a particular voltage. For proteomics workflows, this can be used as an orthogonal technique to selectively reduce the complexity of a mixture prior to entry into the mass spectrometer. Use of a drift tube or traveling wave option allows for the calculations of collisional cross-sections (CCSs), which depend strongly on the ion shape. This technique has been used to investigate the assembly of protein complexes as well as the macromolecules themselves [81, 82].

An additional approach to delineate protein–protein and protein subunit associations by mass spectrometry is the use of hydrogen/deuterium exchange [83, 84]. Here, mapping of the interactions is obtained by proteolysis, usually with acidic pH optimum pepsin, after the exchange reaction. Steps such as immediate MS injection by inline pepsin digestion should be taken to minimize back-exchange, which is also discouraged by the acidic medium optimal for pepsin digestion. High resolution can be obtained through the use of fragmentation; however CID can cause the deuterium position to scramble to other locations [85, 86]. On the other hand, fragmentation produced by electron capture dissociation or electron transfer dissociation can produce minimal deuterium movement [87, 88].

Cross-linking strategies can also facilitate elucidation of protein complexes [89–91]. The protein complex is cross-linked to convert non-covalent interactions into covalent bonds. The protein complex is then enzymatically digested. An enrichment step can also be added to improve analyses. Various cross-linking strategies exist towards different functional groups [92]. These typically target primary amines (e.g. *N*-hydroxysuccinimide), though other complementary approaches such as acidic amino acid (aspartic and glutamic acid) cross-linkers [93] also exist and can be coupled to stable isotope-labels to aid in quantification [94]. These types of approaches are suitable for investigating several aspects of sphingolipid enzymes, including subunit assembly of serine palmitoyltransferase and homo/heterodimerization of ceramide synthases.

6 Protein–Lipid Interactions

While sphingolipids have biophysical properties that can influence protein function, they also directly bind to various proteins, act as signaling molecules via first- or second- messenger functions, are metabolized, and are transported. The increasing discovery of lipid–protein binding partners for sphingolipids is of importance to determine mechanisms of actions and the roles of specific sphingolipids in cellular phenotypes.

A wide variety of approaches, including mass spectrometry, have been utilized to assess and identify sphingolipid–protein interactions. We have utilized ceramide-1-phosphate (C1P) containing liposomes to identify C1P-interacting proteins. One of these interacting proteins, annexin-A2, was shown to mediate cellular migration and invasion [95]. These interactions were also validated by surface plasmon resonance, lipid–protein ELISAs, and mass spectrometry detection of C1P from immunoprecipitates. Interactions between TRAF2 and sphingosine-1-phosphate (S1P) were characterized by S1P immobilized on agarose beads, the use of ^{32}P-labelled S1P, and mass spectrometry detection of S1P from immunoprecipitated TRAF2 [96]. Ceramide has been shown to have several binding partners. These include, but are not exclusive to, cathepsin D (identified via ceramide affinity chromatography [97]), c-Raf (via radiolabeled ceramide and modulation of activity [98]), kinase-suppressor of ras (via modulation of activity [99]), PP2A (identified via protein fractionation [100]), and PKCζ (via modulation of recombinant protein activity [101, 102]). Sphingosine interactions with 14-3-3ζ was also detected from 14-3-3ζ immunoprecipitates [103].

Usage of mass spectrometry-based approaches to investigate protein–lipid interactions is very similar to investigation of protein–protein interactions. For discovery approaches, a means of enrichment of target proteins is often needed, and this can be achieved via a positive selection approach. Tactics that utilize liposomes, or nanodiscs [104] can better mimic a membrane environment. Alternatively, a single lipid can be used in an affinity approach using either an affinity tag for purification or linked to a stationary support. As examples, a biotin-C6-ceramide has been used to identify 97 interacting proteins in D6P2T Schwannoma cells [105]. A ceramide linked to a stationary support identified interactions between cathepsin D and ceramide [97]. A biotin-sphingosine was utilized to identify interaction with acidic leucine-rich nuclear phosphoprotein-32A (ANP32A) [106]. While the use of artificial membranes can better recapitulate physiological interactions between proteins and lipids compared to a single pure lipid, this can lead to higher non-specific binding and thus background contaminants can become a factor. Use of a pure lipid that has been modified to facilitate attachment to a stationary support or an affinity tag can potentially alter interactions to a protein. Regardless of the enrichment technique utilized, they are amenable to label-free and labeling techniques to allow for minimizing the impact of background proteins to more confidently identify authentic interactions.

An emerging topic in sphingolipid biology is the roles of specific fatty acyl composition. Ceramide synthases show various specificities towards different fatty-acyl CoAs in the *de novo* pathway to generate distinct sphingolipid species (for review, see [107]). This raises the possibility of protein specificity differences for sphingolipids with different fatty acyl chain length. As examples, glucosylceramide synthase shows a preference for ceramides with shorter fatty acyl lengths [108], whereas ceramide kinase may [109] or may not [110] show a preference for longer (>C12) chain ceramides. The role of the sphingoid backbone (d18:0, d18:1, d18:2, d20:1, etc.) is even less defined. These are important factors as increasing evidence suggests that all ceramides are not equal. Examples include data showing a role for elevated CerS6/C16-ceramide in tumor growth, whereas siRNA-mediated knockdown of CerS6 induced ATF-6 activation and apoptosis [111]. In contrast, an apoptotic role of C16-ceramide and a proliferative role of C24-ceramides have been described [1]. Thus, identification of proteins that show preferential interactions may shed light on different properties of different sphingolipids.

While the above focuses on identifying proteins, mass spectrometry can also be used to identify lipids bound to purified proteins. Some examples include annexin A2-C1P [95] and TRAF2–S1P interactions [96]. These were identified by a targeted MS analysis, but use of various lipid analysis techniques, such as precursor ion scanning, neutral loss scanning, data-dependent, and/or data-independent analyses can be used to identify lipids interacting with proteins. Such approaches can be invaluable since in addition to interactions with sphingolipids, sphingolipid enzymes have been demonstrated to interact with other lipids. For example, the anionic lipids phosphatidic acid was shown to bind SphK1 using lipids-conjugated beads and liposomes [112] and phosphatidylserine via surface plasmon resonance [40].

Use of ion-mobility or hydrogen/deuterium exchange to investigate specific lipid–protein interactions is also possible, but has seen minimal usage. A large number of studies of protein–drug/ligand interactions have been performed with hydrogen/deuterium exchange, though most studies involving lipids have focused on lipid vesicles to investigate membrane binding. For example, both ion mobility (traveling wave) and hydrogen/deuterium exchange were used to examine interactions and secondary structure changes of α-synuclein upon binding to model membranes [113]. Evidence from studies with P-glycoprotein demonstrated that specificity may be possible [114]. These studies using ion mobility-MS revealed a preference for anionic lipids over zwitterionic lipids (phosphatidylcholine and phosphatidylethanolamine). As P-glycoprotein may also serve as a flippase for glucosylceramides [115], these interactions could be probed in a similar manner. Though interactions would need to be preserved in a gas phase environment for ion mobility to be useful, such an environment may be more physiologically relevant due to the dielectric constant of a vacuum having similarity to a membrane environment.

As with protein–protein interactions, these approaches serve as initial discovery tools. Additional validation approaches will still be needed to prove an authentic interaction, such as employing recombinant protein, surface plasmon resonance, activity/functional assays, structural biology studies (e.g. x-ray crystallography, NMR, small-angle scattering), and/or site-directed mutagenesis to produce complementary data.

7 Chemical Proteomics

From initial discovery, drug development can take an average of 10 years with costs between $868 million and $1.24 billion to obtain approval [116]. This can be a huge cost for a drug that doesn't fail until late in the pipeline, or when in clinical use (e.g. Vioxx). A large number of inhibitors have been generated against a variety of sphingolipid enzymes. Though many of these compounds have shown efficacy within *in vitro* and *in vivo* models, the pharmacokinetics, liberation, absorption, distribution, metabolism, and excretion and off-target effects of these compounds are largely unknown. Likewise, the intracellular concentration and metabolism of a compound can lead to changes in selectivity for the target of interest compared to "off-target" molecules. More extensive screening of off-target effects through proteomic analysis could improve the efficiency of weeding out "bad" candidates earlier.

Pointing to the importance of these off-target effects, FTY720 (aka Fingolimod or Gilenya) can modulate S1PRs upon phosphorylation by sphingosine kinase 2 [117]. However, other identified targets include S1P lyase [118], SphK1 [119], ceramide synthases [120] acid sphingomyelinase [103], cPLA2 [121], CB1 cannabinoid receptor [122], and 14-3-3 proteins [103]. A sphingosine kinase inhibitor, SKI-II, has recently been shown to also inhibit dihydroceramide desaturase [123]. Ultimately, the extent of knowledge about the specificity of a given compound is determined by the number of protein targets it is screened against.

There are several proteomic strategies that can be utilized to identify a drug's protein-binding partners. One category of approach is termed "compound-centric chemical proteomics [124, 125]. These approaches are very similar to several of the protein–protein and protein–lipid interaction strategies described above. These methods rely on immobilization of the drug to facilitate affinity purification. Compound immobilization can be via the attachment of specific chemical groups (e.g. carboxyl, hydroxyl, amino) or through chemical modification of the compound to facilitate attachment, such as use of a linker. When a compound needs to be modified for immobilization, it is important to assess this modification with an appropriate biochemical or cell-based assay to ensure that activity is retained. This approach usually requires some knowledge of how the inhibitor interacts with the protein of interest (structure-activity relationship [126]), so that any effects on binding with target proteins upon drug modification are minimized. This approach can potentially eliminate some "off-target" binding to the now-modified region.

Drug immobilization can also be coupled with surface plasmon resonance. After protein capture, proteins are eluted for mass spectrometry-based identification [127]. An advantage of surface plasmon resonance is that other parameters, such as equilibrium constants and kinetic rates can be obtained. Furthermore, the interaction strength can be characterized with increasing salt or other washes. This may also be useful to minimize identification of non-specific proteins.

With a properly immobilized drug-matrix, comparisons between drug–protein with an inactive-drug analog, control matrix, and/or unrelated compounds can be performed to help distinguish a background contaminant from an authentic interaction. As with affinity approaches for protein–protein and protein–lipid interactions, these approaches can employ label-free, or protein/peptide labeling strategies. Likewise, these methods do not prove a direct protein-drug interaction as some detected proteins can be the result of interacting with the already associated true drug-interacting protein.

An alternative approach uses energetics-based target identification by measuring the change in the conformational stabilities of proteins upon ligand binding to identify target proteins. This differs from the affinity-based capture approaches mentioned above, as these methods do not require modification or immobilization of the test ligand. Use of a method termed Stability of Proteins from Rates of Oxidation

(SPROX) depends on the conformation shifting or stabilizing interaction of the drug–protein interaction to keep some methionines protected from oxidation with hydrogen peroxide under titrated conditions where the same methionine-containing peptide is not protected without the compound. These conformation shifts are then measured by mass spectrometry analysis of the peptides and ratio of modification/no modification at particular points in the titration curve [128–130]. So far this approach has only been described in yeast and not in human/mammalian samples, which have at least 3–4 times the number of proteins. This method should be amendable to larger and more complex proteomes, but will require strategies to acquire more in depth proteomics data as discussed above under discovery proteomics.

As "targeted" drugs are often found to be promiscuous, delineation of drug–protein interactions can help enable the design of more effective next-generation drugs by improving the efficacy and minimizing side effects through reducing undesirable off-target interactions. It also offers the potential of identifying the true mechanism of action and alternative therapeutic targets.

8 Conclusions

This chapter outlines the diversity of mass spectrometry-based proteomics approaches to discover and assess perturbations in protein levels, protein–protein interactions, and protein–lipid interactions as well as provide a means of investigating protein-drug interactions and protein function. There are multiple approaches for each experimental question that can be utilized, each with their own advantages and disadvantages, though the main determinants of experimental success is judicial experimental design and computational biology/bioinformatics expertise to facilitate data interpretation. Regardless of the method utilized, mass spectrometry represents just a single tool and other approaches are needed to validate and complement studies as well as to determine the importance of such findings. These strategies will however help address fundamental questions in both basic science and clinical research to better understand the roles of sphingolipids, their metabolizing enzymes, and interacting partners to delineate the roles in cancer.

Acknowledgments This work was supported by the American Cancer Society and P01 CA171983. The authors would like to thank Kevin Fox for the creation of Fig. 1.

References

1. Hartmann D et al (2012) Long chain ceramides and very long chain ceramides have opposite effects on human breast and colon cancer cell growth. Int J Biochem Cell Biol 44(4): 620–628
2. Karahatay S et al (2007) Clinical relevance of ceramide metabolism in the pathogenesis of human head and neck squamous cell carcinoma (HNSCC): attenuation of C(18)-ceramide in

HNSCC tumors correlates with lymphovascular invasion and nodal metastasis. Cancer Lett 256(1):101–111
3. Saddoughi SA et al (2011) Results of a phase II trial of gemcitabine plus doxorubicin in patients with recurrent head and neck cancers: serum C(1)(8)-ceramide as a novel biomarker for monitoring response. Clin Cancer Res 17(18):6097–6105
4. Alberg AJ et al (2013) Plasma sphingolipids and lung cancer: a population-based, nested case-control study. Cancer Epidemiol Biomarkers Prev 22(8):1374–1382
5. Jiang Y et al (2013) Altered sphingolipid metabolism in patients with metastatic pancreatic cancer. Biomolecules 3(3):435–448
6. Han X, Yang K, Gross RW (2012) Multi-dimensional mass spectrometry-based shotgun lipidomics and novel strategies for lipidomic analyses. Mass Spectrom Rev 31(1):134–178
7. Fillet M et al (2005) Differential expression of proteins in response to ceramide-mediated stress signal in colon cancer cells by 2-D gel electrophoresis and MALDI-TOF-MS. J Proteome Res 4(3):870–880
8. Renert AF et al (2009) The proapoptotic C16-ceramide-dependent pathway requires the death-promoting factor Btf in colon adenocarcinoma cells. J Proteome Res 8(10):4810–4822
9. Kota V et al (2013) 2'-hydroxy C16-ceramide induces apoptosis-associated proteomic changes in C6 glioma cells. J Proteome Res 12(10):4366–4367
10. Parent N et al (2009) Proteomic analysis of enriched lysosomes at early phase of camptothecin-induced apoptosis in human U-937 cells. J Proteomics 72(6):960–973
11. Kim SY et al (2009) Proteomic identification of proteins translocated to membrane microdomains upon treatment of fibroblasts with the glycosphingolipid, C8-beta-D-lactosylceramide. Proteomics 9(18):4321–4328
12. Everley RA et al (2013) Increasing throughput in targeted proteomics assays: 54-plex quantitation in a single mass spectrometry run. Anal Chem 85(11):5340–5346
13. McClatchy DB, Yates JR 3rd (2008) Stable isotope labeling in mammals (SILAM). Methods Mol Biol 1156:133–146
14. Pozniak Y, Geiger T (2014) Design and application of super-SILAC for proteome quantification. Methods Mol Biol 1188:281–291
15. Zhou F et al (2013) Genome-scale proteome quantification by DEEP SEQ mass spectrometry. Nat Commun 4:2171
16. Mann M et al (2002) Analysis of protein phosphorylation using mass spectrometry: deciphering the phosphoproteome. Trends Biotechnol 20(6):261–268
17. Lundby A et al (2012) Quantitative maps of protein phosphorylation sites across 14 different rat organs and tissues. Nat Commun 3:876
18. Hao P, Guo T, Sze SK (2011) Simultaneous analysis of proteome, phospho- and glycoproteome of rat kidney tissue with electrostatic repulsion hydrophilic interaction chromatography. PLoS One 6(2), e16884
19. Mertins P et al (2013) Integrated proteomic analysis of post-translational modifications by serial enrichment. Nat Methods 10(7):634–637
20. Gillet LC et al (2012) Targeted data extraction of the MS/MS spectra generated by data-independent acquisition: a new concept for consistent and accurate proteome analysis. Mol Cell Proteomics 11(6):O111.016717
21. Momin AA et al (2011) A method for visualization of "omic" datasets for sphingolipid metabolism to predict potentially interesting differences. J Lipid Res 52(6):1073–1083
22. Lundberg E et al (2010) Defining the transcriptome and proteome in three functionally different human cell lines. Mol Syst Biol 6:450
23. Schwanhausser B et al (2011) Global quantification of mammalian gene expression control. Nature 473(7347):337–342
24. Gerber SA et al (2003) Absolute quantification of proteins and phosphoproteins from cell lysates by tandem MS. Proc Natl Acad Sci U S A 100(12):6940–6945
25. Anderson NL et al (2004) Mass spectrometric quantitation of peptides and proteins using stable isotope standards and capture by anti-peptide antibodies (SISCAPA). J Proteome Res 3(2):235–244

26. Bereman MS et al (2012) The development of selected reaction monitoring methods for targeted proteomics via empirical refinement. Proteomics 12(8):1134–1141
27. Mohammed Y et al (2014) PeptidePicker: a scientific workflow with web interface for selecting appropriate peptides for targeted proteomics experiments. J Proteomics 106C:51–161
28. Wilhelm M et al (2014) Mass-spectrometry-based draft of the human proteome. Nature 509(7502):582–587
29. Kim MS et al (2014) A draft map of the human proteome. Nature 509(7502):575–581
30. Zeidan YH, Hannun YA (2007) Activation of acid sphingomyelinase by protein kinase Cdelta-mediated phosphorylation. J Biol Chem 282(15):11549–11561
31. Parent N et al (2011) Protein kinase C-delta isoform mediates lysosome labilization in DNA damage-induced apoptosis. Int J Oncol 38(2):313–324
32. Pitson SM et al (2003) Activation of sphingosine kinase 1 by ERK1/2-mediated phosphorylation. EMBO J 22(20):5491–5500
33. Franzen R et al (2002) Nitric oxide induces degradation of the neutral ceramidase in rat renal mesangial cells and is counterregulated by protein kinase C. J Biol Chem 277(48): 46184–46190
34. Galadari S et al (2006) Identification of a novel amidase motif in neutral ceramidase. Biochem J 393(Pt 3):687–695
35. Tada E et al (2010) Activation of ceramidase and ceramide kinase by vanadate via a tyrosine kinase-mediated pathway. J Pharmacol Sci 114(4):420–432
36. Chen WQ et al (2010) Ceramide kinase profiling by mass spectrometry reveals a conserved phosphorylation pattern downstream of the catalytic site. J Proteome Res 9(1):420–429
37. Ferlinz K et al (2001) Human acid ceramidase: processing, glycosylation, and lysosomal targeting. J Biol Chem 276(38):35352–35360
38. Schulze H, Schepers U, Sandhoff K (2007) Overexpression and mass spectrometry analysis of mature human acid ceramidase. Biol Chem 388(12):1333–1343
39. Rodriguez J et al (2008) Does trypsin cut before proline? J Proteome Res 7(1):300–305
40. Stahelin RV et al (2005) The mechanism of membrane targeting of human sphingosine kinase 1. J Biol Chem 280(52):43030–43038
41. Pflieger D et al (2008) Quantitative proteomic analysis of protein complexes: concurrent identification of interactors and their state of phosphorylation. Mol Cell Proteomics 7(2): 326–346
42. Glibert P et al (2015) Phospho-iTRAQ: assessing isobaric labels for the large-scale study of phosphopeptide stoichiometry. J Proteome Res 14(2):839–849
43. Frese CK et al (2011) Improved peptide identification by targeted fragmentation using CID, HCD and ETD on an LTQ-Orbitrap Velos. J Proteome Res 10(5):2377–2388
44. Kumagai K, Kawano-Kawada M, Hanada K (2014) Phosphoregulation of the ceramide transport protein CERT at serine 315 in the interaction with VAMP-associated protein (VAP) for inter-organelle trafficking of ceramide in mammalian cells. J Biol Chem 289(15): 10748–10760
45. Cvetkovic A et al (2010) Microbial metalloproteomes are largely uncharacterized. Nature 466(7307):779–782
46. Lothian A et al (2013) Metalloproteomics: principles, challenges and applications to neurodegeneration. Front Aging Neurosci 5:35
47. Yannone SM et al (2012) Metals in biology: defining metalloproteomes. Curr Opin Biotechnol 23(1):89–95
48. Fujita T et al (2004) Delta-catenin/NPRAP (neural plakophilin-related armadillo repeat protein) interacts with and activates sphingosine kinase 1. Biochem J 382(Pt 2):717–723
49. Leclercq TM et al (2008) Eukaryotic elongation factor 1A interacts with sphingosine kinase and directly enhances its catalytic activity. J Biol Chem 283(15):9606–9614
50. Sun J et al (2006) FHL2/SLIM3 decreases cardiomyocyte survival by inhibitory interaction with sphingosine kinase-1. Circ Res 99(5):468–476
51. Fukuda Y et al (2004) Identification of PECAM-1 association with sphingosine kinase 1 and its regulation by agonist-induced phosphorylation. Biochim Biophys Acta 1636(1):12–21

52. Lacana E et al (2002) Cloning and characterization of a protein kinase A anchoring protein (AKAP)-related protein that interacts with and regulates sphingosine kinase 1 activity. J Biol Chem 277(36):32947–32953
53. Maceyka M et al (2004) Aminoacylase 1 is a sphingosine kinase 1-interacting protein. FEBS Lett 568(1–3):30–34
54. Zebol JR et al (2009) The CCT/TRiC chaperonin is required for maturation of sphingosine kinase 1. Int J Biochem Cell Biol 41(4):822–827
55. Hayashi S et al (2002) Identification and characterization of RPK118, a novel sphingosine kinase-1-binding protein. J Biol Chem 277(36):33319–33324
56. Maceyka M et al (2008) Filamin A links sphingosine kinase 1 and sphingosine-1-phosphate receptor 1 at lamellipodia to orchestrate cell migration. Mol Cell Biol 28(18):5687–5697
57. Olivera A et al (1998) Purification and characterization of rat kidney sphingosine kinase. J Biol Chem 273(20):12576–12583
58. Jarman KE et al (2010) Translocation of sphingosine kinase 1 to the plasma membrane is mediated by calcium- and integrin-binding protein 1. J Biol Chem 285(1):483–492
59. Yamane D et al (2009) Inhibition of sphingosine kinase by bovine viral diarrhea virus NS3 is crucial for efficient viral replication and cytopathogenesis. J Biol Chem 284(20): 13648–13659
60. Barr RK et al (2008) Deactivation of sphingosine kinase 1 by protein phosphatase 2A. J Biol Chem 283(50):34994–35002
61. Urtz N et al (2004) Early activation of sphingosine kinase in mast cells and recruitment to FcepsilonRI are mediated by its interaction with Lyn kinase. Mol Cell Biol 24(19):8765–8777
62. Olivera A et al (2006) IgE-dependent activation of sphingosine kinases 1 and 2 and secretion of sphingosine 1-phosphate requires Fyn kinase and contributes to mast cell responses. J Biol Chem 281(5):2515–2525
63. Xia P et al (2002) Sphingosine kinase interacts with TRAF2 and dissects tumor necrosis factor-alpha signaling. J Biol Chem 277(10):7996–8003
64. Gamble JR et al (2009) Sphingosine kinase-1 associates with integrin {alpha}V{beta}3 to mediate endothelial cell survival. Am J Pathol 175(5):2217–2225
65. Laviad EL et al (2012) Modulation of ceramide synthase activity via dimerization. J Biol Chem 287(25):21025–21033
66. Azuma N et al (1994) Stimulation of acid ceramidase activity by saposin D. Arch Biochem Biophys 311(2):354–357
67. Adam-Klages S et al (1996) FAN, a novel WD-repeat protein, couples the p55 TNF-receptor to neutral sphingomyelinase. Cell 86(6):937–947
68. Ahn KH et al (2013) Identification of heat shock protein 60 as a regulator of Neutral Sphingomyelinase 2 and its role in Dopamine uptake. PLoS One 8(6), e67216
69. Miteva YV, Budayeva HG, Cristea IM (2013) Proteomics-based methods for discovery, quantification, and validation of protein-protein interactions. Anal Chem 85(2):749–768
70. Paoletti AC et al (2006) Quantitative proteomic analysis of distinct mammalian Mediator complexes using normalized spectral abundance factors. Proc Natl Acad Sci U S A 103(50):18928–18933
71. Wang M et al (2012) PaxDb, a database of protein abundance averages across all three domains of life. Mol Cell Proteomics 11(8):492–500
72. Teo G et al (2014) SAINTexpress: improvements and additional features in significance analysis of INTeractome software. J Proteomics 100:37–43
73. Choi H et al (2011) SAINT: probabilistic scoring of affinity purification-mass spectrometry data. Nat Methods 8(1):70–73
74. Sowa ME et al (2009) Defining the human deubiquitinating enzyme interaction landscape. Cell 138(2):389–403
75. Rinner O et al (2007) An integrated mass spectrometric and computational framework for the analysis of protein interaction networks. Nat Biotechnol 25(3):345–352

76. Jager S et al (2011) Global landscape of HIV-human protein complexes. Nature 481(7381):365–370
77. Tackett AJ et al (2005) I-DIRT, a general method for distinguishing between specific and nonspecific protein interactions. J Proteome Res 4(5):1752–1756
78. Selbach M, Mann M (2006) Protein interaction screening by quantitative immunoprecipitation combined with knockdown (QUICK). Nat Methods 3(12):981–983
79. Mellacheruvu D et al (2013) The CRAPome: a contaminant repository for affinity purification-mass spectrometry data. Nat Methods 10(8):730–736
80. Wang X, Huang L (2008) Identifying dynamic interactors of protein complexes by quantitative mass spectrometry. Mol Cell Proteomics 7(1):46–57
81. Jurneczko E, Barran PE (2011) How useful is ion mobility mass spectrometry for structural biology? The relationship between protein crystal structures and their collision cross sections in the gas phase. Analyst 136(1):20–28
82. Lanucara F et al (2014) The power of ion mobility-mass spectrometry for structural characterization and the study of conformational dynamics. Nat Chem 6(4):281–294
83. Huang RY, Chen G (2014) Higher order structure characterization of protein therapeutics by hydrogen/deuterium exchange mass spectrometry. Anal Bioanal Chem 406(26):6541–6558
84. Jaswal SS (1834) Biological insights from hydrogen exchange mass spectrometry. Biochim Biophys Acta 6:1188–1201
85. Jorgensen TJ et al (2005) Intramolecular migration of amide hydrogens in protonated peptides upon collisional activation. J Am Chem Soc 127(8):2785–2793
86. Jorgensen TJ et al (2005) Collisional activation by MALDI tandem time-of-flight mass spectrometry induces intramolecular migration of amide hydrogens in protonated peptides. Mol Cell Proteomics 4(12):1910–1919
87. Zehl M et al (2008) Electron transfer dissociation facilitates the measurement of deuterium incorporation into selectively labeled peptides with single residue resolution. J Am Chem Soc 130(51):17453–17459
88. Rand KD et al (2008) Electron capture dissociation proceeds with a low degree of intramolecular migration of peptide amide hydrogens. J Am Chem Soc 130(4):1341–1349
89. Rappsilber J (2011) The beginning of a beautiful friendship: cross-linking/mass spectrometry and modelling of proteins and multi-protein complexes. J Struct Biol 173(3):530–540
90. Stengel F, Aebersold R, Robinson CV (2012) Joining forces: integrating proteomics and cross-linking with the mass spectrometry of intact complexes. Mol Cell Proteomics 11(3):R111.014027<>
91. Bruce JE (2012) In vivo protein complex topologies: sights through a cross-linking lens. Proteomics 12(10):1565–1575
92. Paramelle D et al (2013) Chemical cross-linkers for protein structure studies by mass spectrometry. Proteomics 13(3–4):438–456
93. Leitner A et al (2014) Chemical cross-linking/mass spectrometry targeting acidic residues in proteins and protein complexes. Proc Natl Acad Sci U S A 111(26):9455–9460
94. Fischer L, Chen ZA, Rappsilber J (2013) Quantitative cross-linking/mass spectrometry using isotope-labelled cross-linkers. J Proteomics 88:120–128
95. Hankins JL et al (2013) Ceramide 1-phosphate mediates endothelial cell invasion via the annexin a2-p11 heterotetrameric protein complex. J Biol Chem 288(27):19726–19738
96. Alvarez SE et al (2010) Sphingosine-1-phosphate is a missing cofactor for the E3 ubiquitin ligase TRAF2. Nature 465(7301):1084–1088
97. Heinrich M et al (1999) Cathepsin D targeted by acid sphingomyelinase-derived ceramide. EMBO J 18(19):5252–5263
98. Huwiler A et al (1996) Ceramide-binding and activation defines protein kinase c-Raf as a ceramide-activated protein kinase. Proc Natl Acad Sci U S A 93(14):6959–6963
99. Zhang Y et al (1997) Kinase suppressor of Ras is ceramide-activated protein kinase. Cell 89(1):63–72
100. Galadari S et al (1998) Purification and characterization of ceramide-activated protein phosphatases. Biochemistry 37(32):11232–11238

101. Lozano J et al (1994) Protein kinase C zeta isoform is critical for kappa B-dependent promoter activation by sphingomyelinase. J Biol Chem 269(30):19200–19202
102. Bourbon NA, Yun J, Kester M (2000) Ceramide directly activates protein kinase C zeta to regulate a stress-activated protein kinase signaling complex. J Biol Chem 275(45): 35617–35623
103. Woodcock JM et al (2010) Sphingosine and FTY720 directly bind pro-survival 14-3-3 proteins to regulate their function. Cell Signal 22(9):1291–1299
104. Borch J, Roepstorff P, Moller-Jensen J (2011) Nanodisc-based co-immunoprecipitation for mass spectrometric identification of membrane-interacting proteins. Mol Cell Proteomics 10(7):O110.006775
105. Kota V, Szulc ZM, Hama H (2012) Identification of C(6)-ceramide-interacting proteins in D6P2T Schwannoma cells. Proteomics 12(13):2179–2184
106. Habrukowich C et al (2010) Sphingosine interaction with acidic leucine-rich nuclear phosphoprotein-32A (ANP32A) regulates PP2A activity and cyclooxygenase (COX)-2 expression in human endothelial cells. J Biol Chem 285(35):26825–26831
107. Stiban J, Tidhar R, Futerman AH (2010) Ceramide synthases: roles in cell physiology and signaling. Adv Exp Med Biol 688:60–71
108. Vunnam RR, Radin NS (1979) Short chain ceramides as substrates for glucocerebroside synthetase. Differences between liver and brain enzymes. Biochim Biophys Acta 573(1):73–82
109. Wijesinghe DS et al (2005) Substrate specificity of human ceramide kinase. J Lipid Res 46(12):2706–2716
110. Van Overloop H, Gijsbers S, Van Veldhoven PP (2006) Further characterization of mammalian ceramide kinase: substrate delivery and (stereo)specificity, tissue distribution, and subcellular localization studies. J Lipid Res 47(2):268–283
111. Senkal CE et al (2011) Alteration of ceramide synthase 6/C16-ceramide induces activating transcription factor 6-mediated endoplasmic reticulum (ER) stress and apoptosis via perturbation of cellular Ca2+ and ER/Golgi membrane network. J Biol Chem 286(49): 42446–42458
112. Delon C et al (2004) Sphingosine kinase 1 is an intracellular effector of phosphatidic acid. J Biol Chem 279(43):44763–44774
113. Lee SJ et al (2014) Probing conformational change of intrinsically disordered alpha-synuclein to helical structures by distinctive regional interactions with lipid membranes. Anal Chem 86(3):1909–1916
114. Marcoux J et al (2013) Mass spectrometry reveals synergistic effects of nucleotides, lipids, and drugs binding to a multidrug resistance efflux pump. Proc Natl Acad Sci U S A 110(24):9704–9709
115. Eckford PD, Sharom FJ (2005) The reconstituted P-glycoprotein multidrug transporter is a flippase for glucosylceramide and other simple glycosphingolipids. Biochem J 389(Pt 2):517–526
116. Ciociola AA, Cohen LB, Kulkarni P (2014) How drugs are developed and approved by the FDA: current process and future directions. Am J Gastroenterol 109(5):620–623
117. Allende ML et al (2004) Mice deficient in sphingosine kinase 1 are rendered lymphopenic by FTY720. J Biol Chem 279(50):52487–52492
118. Bandhuvula P et al (2005) The immune modulator FTY720 inhibits sphingosine-1-phosphate lyase activity. J Biol Chem 280(40):33697–33700
119. Tonelli F et al (2010) FTY720 and (S)-FTY720 vinylphosphonate inhibit sphingosine kinase 1 and promote its proteasomal degradation in human pulmonary artery smooth muscle, breast cancer and androgen-independent prostate cancer cells. Cell Signal 22(10):1536–1542
120. Lahiri S et al (2009) Ceramide synthesis is modulated by the sphingosine analog FTY720 via a mixture of uncompetitive and noncompetitive inhibition in an Acyl-CoA chain length-dependent manner. J Biol Chem 284(24):16090–16098
121. Payne SG et al (2007) The immunosuppressant drug FTY720 inhibits cytosolic phospholipase A2 independently of sphingosine-1-phosphate receptors. Blood 109(3):1077–1085

122. Paugh SW et al (2006) Sphingosine and its analog, the immunosuppressant 2-amino-2-(2-[4-octylphenyl]ethyl)-1,3-propanediol, interact with the CB1 cannabinoid receptor. Mol Pharmacol 70(1):41–50
123. Cingolani F et al (2014) Inhibition of dihydroceramide desaturase activity by the sphingosine kinase inhibitor SKI II. J Lipid Res 55(8):1711–1720
124. Wang K et al (2012) Chemistry-based functional proteomics for drug target deconvolution. Expert Rev Proteomics 9(3):293–310
125. Liu Y, Guo M (2014) Chemical proteomic strategies for the discovery and development of anticancer drugs. Proteomics 14(4–5):399–411
126. Lukman S, Verma CS, Fuentes G (2014) Exploiting protein intrinsic flexibility in drug design. Adv Exp Med Biol 805:245–269
127. Huber W, Mueller F (2006) Biomolecular interaction analysis in drug discovery using surface plasmon resonance technology. Curr Pharm Des 12(31):3999–4021
128. Strickland EC et al (2013) Thermodynamic analysis of protein-ligand binding interactions in complex biological mixtures using the stability of proteins from rates of oxidation. Nat Protoc 8(1):148–161
129. Tran DT, Adhikari J, Fitzgerald MC (2014) SILAC-based strategy for proteome-wide thermodynamic analysis of protein-ligand binding interactions. Mol Cell Proteomics 13(7):1800–1813
130. West GM et al (2010) Quantitative proteomics approach for identifying protein-drug interactions in complex mixtures using protein stability measurements. Proc Natl Acad Sci U S A 107(20):9078–9082

Utilization of Caged Ceramide and Ceramide 1-Phosphate Analogs for Monitoring Cellular Events after Photoactivation

Robert Bittman and Antonio Gomez-Muñoz

Abstract Caged compounds possess a photolabile covalent bond. This chapter reviews studies of caged derivatives of two important sphingolipid signaling molecules, ceramide and ceramide 1-phosphate. Biophysical studies were carried out after a 6-bromo-7-hydroxycoumarinyl-ceramide conjugate was inserted into model bilayer membranes. Uncaging with long-wavelength UV light liberated *N*-palmitoylceramide, and reorganization of lipid domains in the bilayer was monitored. Two derivatives of *N*-palmitoyl-ceramide 1-phosphate in which the phosphate group was esterified to a caging group were investigated in macrophages; in one derivative the cage is 7-(*N*,*N*-diethylamino)coumarin (DECM-C1P)while in the other it is a 4-bromo-5-hydroxy-2-nitrobenzhydryl moiety (BHNB-C1P). The caged derivatives were delivered to macrophages in aqueous solution. The photolytic uncaging process then released ceramide 1-phosphate in the cytosol of macrophages, which was accompanied by stimulation of macrophage proliferation, reactive oxygen species production, and other intracellular signaling events. A distinction can thus be made in some cells between extracellular events evoked by ceramide 1-phosphate, as for example by its interaction with a putative cell-surface receptor, from its intracellular bioactivities. These studies show that elevation of ceramide or ceramide 1-phosphate levels by uncaging of their inactive caged forms enable investigations of a wide variety of biophysical and biochemical processes.

Keywords Caged lipids • Coumarin • Light activation • Photolysis • Signaling • Uncaging

R. Bittman (✉)
Department of Chemistry and Biochemistry, Queens College of The City University of New York, Flushing, NY 11367, USA
e-mail: robert.bittman@qc.cuny.edu

A. Gomez-Muñoz
Department of Biochemistry and Molecular Biology, Faculty of Science and Technology, University of the Basque Country, P.O. Box 644, Bilbao 48080, Spain
e-mail: antonio.gomez@ehu.es

Y.A. Hannun et al. (eds.), *Bioactive Sphingolipids in Cancer Biology and Therapy*, DOI 10.1007/978-3-319-20750-6_17

1 Introduction to Caged Biomolecules

This chapter will focus on investigations of ceramide (Cer) and ceramide 1-phosphate (C1P) in cells and model membranes using "caged" derivatives of these sphingolipids. The cage is a photoremovable group that is covalently attached to a biomolecule of interest (denoted as the "ligand" in Fig. 1a). The conjugated derivative is biologically inactive; if the bond between the ligand and the cage masks a charged group in the ligand, the caged compound may become cell permeable. On irradiation with light, the ligand is released from the caged precursor molecule, and its bioactivity can then be assessed.

To extend our interest in studying the biochemical and biophysical behavior of Cer and C1P, we have explored the synthesis and use of caged analogs of Cer and C1P because analogs labeled with a photoactivatable moiety offer the possibility of inserting these compounds into the cytosol of cells or into lipid bilayers. Thus, in this review, the ligands attached to various cages are Cer and dihydro-Cer (Fig. 1b), C1P (Fig. 1c). When the caged compound is activated by absorption of light, the bond between the ligand and the cage is cleaved irreversibly, and the concentration of the free ligand is elevated. The caged Cer and C1P derivatives were synthesized by coupling the cage to a site in the molecule that abolishes its bioactivity. In caged Cer, a carbonate linkage was used to couple the C-1 hydroxyl group of Cer to the cage, since this linkage is known to be susceptible to cleavage by light [1]. In caged C1P, the phosphate headgroup (which, as an anionic group, would block the passive transport of C1P across cell membranes) is esterified to the cage [2].

The criteria for the design of cages include the following features. (a) The ease of chemical modification of the ligand needs to be considered so that, ideally, established methodologies in organic synthesis can be applied to couple the biomolecule with the caging group. (b) The caged compound should have appreciable aqueous solubility and must not self-associate in water at the concentrations employed. After a solution of the caged compound is prepared in subdued light, the caged compound is incubated with cells in the dark to permit passive transport across the plasma membrane, and then uncaging is initiated by subjecting the cell suspension to photo-illumination. (c) The caged molecule must be resistant to endogenous proteases and lipases, and must have a high stability against spontaneous hydrolysis in aqueous buffer. Alternatively, it is possible to employ cages that are released by an endogenous enzyme such as an esterase that removes the cage to release the bioactive ligand by hydrolysis rather than by photolysis. This approach is not discussed in this chapter. (d) The by-product(s) of the photo-irradiation reaction, which is either the cage itself (e.g., a coumarin derivative) or a derivative of the cage molecule, should be shown in a control experiment to be devoid of bioactivity if the caged molecule is to be compatible with biological applications.

The photophysical properties of the caged compound must meet a number of criteria in order for the photolytic reaction to proceed efficiently, releasing the free ligand. The wavelength of the light for excitation of the caged compound should not damage cellular components. Long-wavelength UV light (>350 nm) or visible light is essential in order to avoid cell damage. A brief flash of light of >350 nm or a

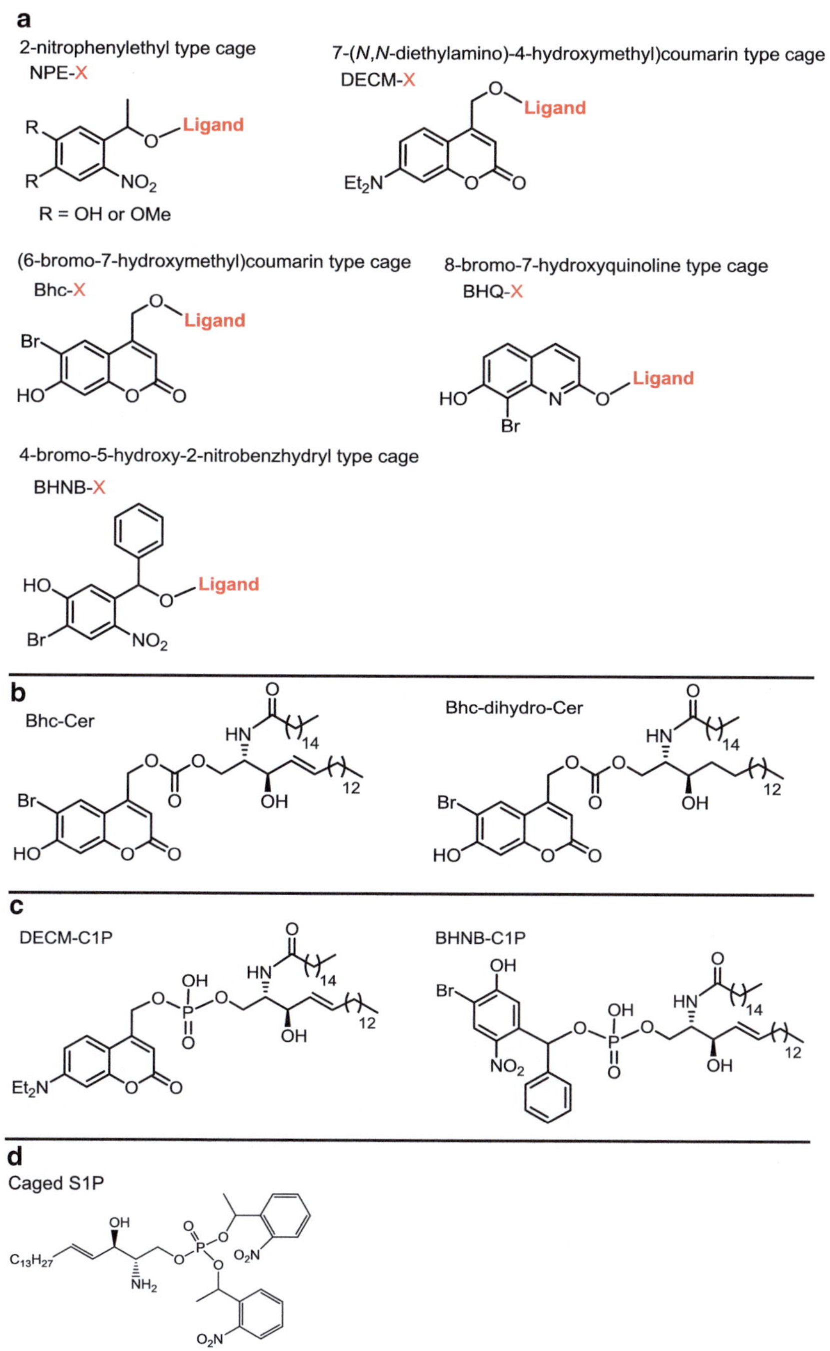

Fig. 1 (**a**) Structures of 2-nitrophenylethyl (NPE), 7-(*N*,*N*-diethylamino)coumarin (DECM), 7-bromo-6-hydroxycoumarin (Bhc), 8-bromo-7-hydroxyquinoline (BHQ), and 4-bromo-5-hydroxy-2-nitrobenzhydryl (BHNB) cages. (**b**) Structures of Bhc-C16-Cer and Bhc-C16-dihydro-Cer. (**c**) C1P analogs bearing a 7-(*N*,*N*-diethylamino)-coumarin (DECM) or 4-bromo-5-hydroxy-2-nitrobenzhydryl (BHNB) group in a photolabile ester bond. The cage masks the ionizable phosphate head group and is released on photolysis. (**d**) Structure of a caged S1P bearing two NPE groups esterified to the phosphate polar head group

prolonged period of irradiation with an apparatus of low-light intensity may be used, depending on the photochemical equipment available to the researcher. Examples of the excitation source include a hand-held UV light set to the long-UV light mode, a non-UV transilluminator, or a Rayonet photoreactor containing numerous lamps and filters providing a wavelength of ~350 nm. An attractive alternative is to use a cage that can be excited with an intense flash of a 2-photon near-infrared laser beam (e.g., ~720–900 nm); then, highly localized release of the bioactive molecule can be achieved. Thus a very short exposure of the caged molecule to a high-intensity light source permits the study of physiological functions in a temporally and spatially confined fashion. Two-photon cages coupled to neurotransmitters have been used to achieve uncaging of neurotransmitters in a localized region (several μm) of tissues or neurons.

Various photocleavable moieties have been used as cages of bioactive molecules for irradiation at >350 nm. The first-generation cages were substituted nitroaromatic compounds such as 2-nitrobenzyl derivatives that are photo-released with a relatively low quantum efficiency. The 1-(2-nitrophenyl)ethyl (NPE) group (Fig. 1a) is an example of such a classical caging group. Cages based on the 6-bromo-7-hydroxycoumarin-4-ylmethyl (Bhc) group [3] and the 8-bromo-7-hydroxyquinolinyl (BHQ) group [4] (Fig. 1b) have improved photolytic efficiency and have been used to prepare many caged biomolecules including lipids, neurotransmitters, nucleotides, nucleic acids, peptides, and proteins. For a recent review of caged lipids and citations of previous reviews of caged compounds, see [5].

The Bhc and BHQ cages have been shown to release various functionalities, such as alcohols, aldehydes, amines, carboxylic acids, diols, ketones, phosphates, and thiols. In addition to having a higher photochemical cleavage efficiency than 2-nitrobenzyl derivatives at wavelengths >350 nm, they can also be released using two-photon excitation conditions. Near-infrared 2-photon photolysis is desirable in studies with tissue preparations because tissue destruction is minimized and deeper penetration into the tissue is possible. The phenolic hydroxyl group of Bhc is partially in the phenolate ion form at pH 7.4, as estimated from the two bands (λ_{max} 338 and ~380 nm) in the UV absorption spectrum of Bhc-dihydro-C16:0-Cer [1]; the longer wavelength band arises from the anionic form. This enhances the hydrophilicity of the caged compound. In medicinal chemistry involving putative drug candidates that are membrane permeable, the lipophilicity of small molecules is frequently estimated by molecular hydrophobicity parameters for 1-octanol/water partition coefficients. The calculated log P (CLogP) value based on chemical structure can be estimated using computational approaches (such as the widely used ChemBioDraw Ultra Suite); programs that use other algorithms for calculating lipophilicity, such as ALOGPS and logD, are also available (http://www.vcclab.org/lab/alogps). The bromine atom lowers the pK_a of the adjacent phenolic hydroxyl group of Bhc and BHQ and thus enhances the aqueous solubility; it also increases the extinction coefficient for light absorption of the chromophore and may contribute to intersystem crossing to the triplet state, which is the photochemically reactive state [6].

Although sphingosine derivatives are not discussed in this chapter, it should be noted that a caged sphingosine 1-phosphate compound featuring two NPE cages esterified to the phosphate group (Fig. 1d) was reported in 1998 [7]. Photolysis of the caged S1P mobilized Ca^{2+} from thapsigarin-sensitive stores in human neuroblastoma (SK-N-MC), human liver carcinoma cell line (HepG2), and human embryonic kidney (HEK-293) cell lines, indicating that S1P acts on intracellular targets [8]. Similarly, photolysis of the caged S1P compound in human endothelial pulmonary cells tightened the endothelial barrier via mobilization of intracellular Ca^{2+} and modulation of various signal transduction pathways; these results are indicative of S1P activities on intracellular targets independent of activation of the five cell-surface S1P receptors [9].

2 Caged Cer

A great deal of attention has been paid to ceramides (*N*-acylsphingosines) because they participate in a multitude of signaling pathways and cellular functions. Studies suggest that some of the functions of ceramides are chain-length dependent [10]. However, the high lipophilicity and low cell permeability of naturally occurring ceramides has spurred investigators seeking to incorporate exogenous Cer into cells to use synthetic Cer analogs with a short (*N*-acetyl to *N*-octanoyl) fatty amide chain. Short-chain ceramides have a higher solubility in water, making them amenable to cell uptake, a requirement for most biological studies. Unfortunately, short-chain ceramides differ from natural ceramides in many functions [11]; for example, they do not form gel-phase domains readily, are unable to stabilize rafts, and fail to displace cholesterol efficiently from liquid-ordered domains. Alternatively, formulations of ceramides with a natural, long *N*-acyl chain have employed mixtures of the hydrocarbon dodecane with ethanol as a means to deliver natural ceramides to cells. This solvent delivery approach may be accompanied by unwanted cellular defects, depending on the dodecane concentration used; concentrations that have been reported range from <0.1 to 2 %. Therefore, we explored the use of caged ceramides with a long *N*-acyl chain as new tools for analyzing Cer-mediated cell signaling events. Our synthetic route [1] allows for the synthesis of Bhc-ceramides with different *N*-acyl chains to be prepared. Future studies may then explore the effects of the various natural ceramides generated by ceramide synthases in various intracellular signaling pathways.

We review here biophysical and biological studies with Bhc-C16:0-Cer (Bhc-Cer) and Bhc-C16:0-dihydro-Cer performed with cells, small unilamellar vesicles (SUVs), and supported bilayers in aqueous solution at neutral pH. Figure 1b shows the structure of Bhc-C16:0-Cer in which the cage is linked via a carbonate tether to the primary hydroxyl group of Cer [1]. Bhc-Cer is stable in aqueous buffer in the dark at room temperature; its half-life is ~30 h, as monitored by HPLC. On photolysis of SUVs containing Bhc-Cer, the C-O bond between the C4-methylene group and the carbonate oxygen of Bhc-Cer is cleaved, yielding Cer; 50 % conversion of

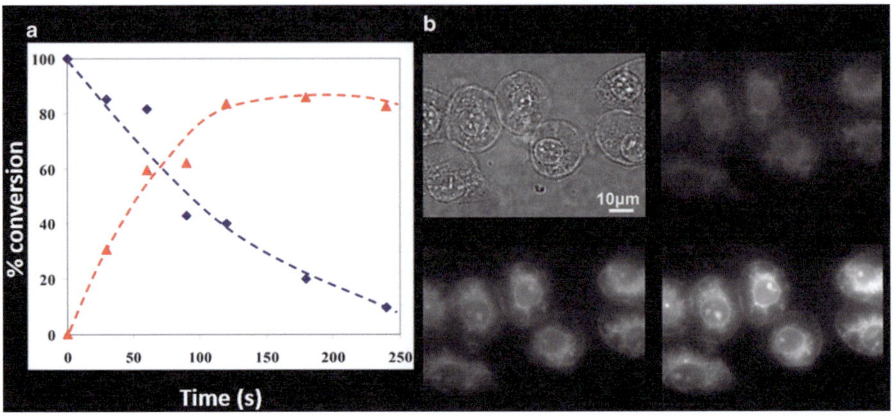

Fig. 2 (**a**) Photolysis of Bhc-C16:0-Cer in SUVs as a function of time. The decrease in the concentration of the caged BHC-Cer (*blue*) and the increase in the concentration of free Bhc (*red*) were quantified by HPLC analysis. The photolysis was carried out using a Rayonet photolysis apparatus at 350 nm. The SUVs were prepared from DOPC, egg SM, cholesterol, and Bhc-C16:0-Cer in a molar ratio of 8:7:4:1. (**b**) Fluorescence microscopic images of J774 macrophages after uptake of Bhc-16:0-Cer (20 μM in KMops, 5 % EtOH) for 2 h in the dark followed by UV irradiation for the following time periods: 0 min (*top*), 3 min (*left-hand side*, *bottom*), and 5 min (*right-hand side*, *bottom*). Initially, the fluorescence intensity in the cytosol is very low because Bhc-Cer has a low quantum yield in water (0.02). However, after a 5-min period of photolysis in the Rayonet apparatus at 350 nm (*right-hand side*, *bottom*), there is a large fluorescence enhancement because free Bhc has a very high quantum yield in water (0.47). (Taken from [1], with permission)

Bhc-Cer to Cer and free Bhc was observed at ~4 min under the photolysis conditions employed (Fig. 2a). The release of C16:0-dihydroceramide from Bhc-C16:0-dihydro-Cer on photolysis at 350 nm for 5 min in KMops buffer (pH 7.4) containing 50 % EtOH was monitored by HPLC [1]. The chromatograms [see reference [1], supporting material] showed that the uncaging of Bhc-C16:0-Cer afforded C16:0-dihydroceramide; the retention time of C16:0-dihydroceramide generated by photolysis was 7.5 min, and the retention time of a commercial sample of C16:0-ceramide was 7.3 min.

The by-product of the photolytic reaction is free Bhc, which is readily detected by HPLC equipped with a fluorescence detector. Figure 2b shows the uptake of Bhc-Cer into the cytosol of macrophages. After a 5-min exposure to 350-nm light, the weakly fluorescent Bhc-Cer (Fig. 2b, top) is replaced by free Bhc (Fig. 2b, bottom), which is highly fluorescent in water.

The photolysis of Bhc-Cer in supported bilayers was also examined (Fig. 3) [12]. Irradiation of a spatially confined area of the supported bilayer resulted in formation of transient Cer-enriched gel-phase domains. Therefore, this method for insertion of Cer into bilayers is an attractive alternative to the use of exogenous Cer or addition of sphingomyelinase to generate Cer in domains of sphingomyelin-containing bilayers [13]. Photocleavage of caged Cer permits one to control the concentration of Cer within localized regions of the membrane, which cannot be achieved with other methods used to elevate the Cer level. The chemical generation of Cer by

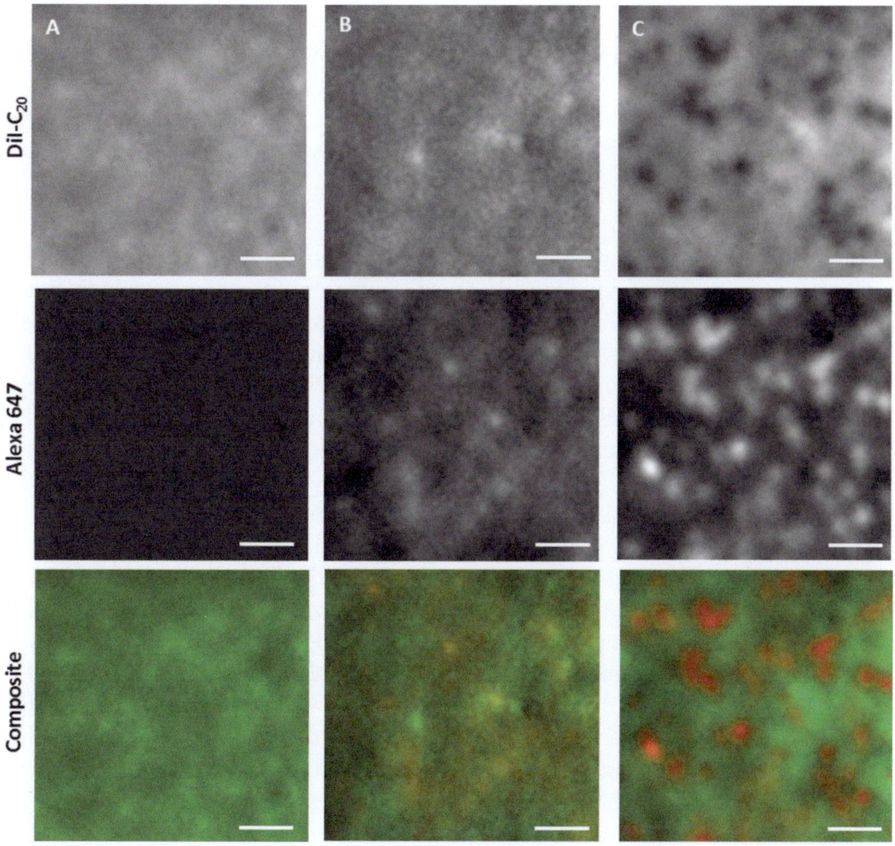

Fig. 3 GM1 is localized in ordered Cer-rich domains after photolysis of Bhc-Cer in planar supported bilayers on mica. The bilayers contained 1-palmitoyl-2-oleoyl-*sn*-glycero-3-phosphocholine (POPC), Bhc-Cer, and GM1 in a molar ratio of 90:10:1. 1,1′-Dieicosanyl-3,3,3′,3′-tetramethylindocarbocyanine perchlorate (DiI-C20) (0.5 mol%) (*green*) was included as a marker of the liquid-disordered phase in the bilayers. As cholera toxin exhibits a high affinity for liquid-ordered domains, recombinant Alexa Fluor 647/Ctx-B conjugate was used as a label for raft domains (*red*). The ganglioside-protein (GM1-Ctx-B) complex is distributed throughout the fluid POPC supported bilayer prior to photolysis. After photolysis (60 min) to release Cer from Bhc-Cer, the Cer-rich domains with GM1 were visualized with Alexa Fluor 647/Ctx-B. Similar results were obtained using Alexa-488 labeled protein. *Scale bar* = 2 μm (Taken from [12], with permission)

reaction of sphingosine with fatty acids [14] or by other reactions [15] is also difficult to control. The role of compartmentalization of Cer in bilayers can also be studied with caged ceramides. As shown in Fig. 3, fluorescence imaging revealed that a conjugate of the ganglioside GM1 with a protein (cholera toxin subunit B, Ctx-B) partitioned into the newly generated Cer-enriched domains after photolysis, as monitored with the Alexa 667-labeled cholera toxin B subunit [12]. Furthermore, in bilayers comprised of both liquid-ordered and liquid-disordered domains, photolysis of Bhc-Cer led to a reorganization of the liquid-ordered domains [16]. Thu this caged Cer is a new probe for biophysical examination of lipid domains in model membranes.

3 Caged C1P

The phosphorylated sphingolipid metabolites sphingosine 1-phosphate (S1P) and C1P regulate a multitude of physiologic and pathologic functions, including cell growth and survival, inflammation, chemotaxis, and modulation of immune responses. Most of the activities of S1P are mediated by its action as an extracellular signaling molecule following binding to a family of five G protein coupled receptors, $S1P_1$–$S1P_5$, at the cell surface; however, S1P signaling also takes place by non-receptor mechanisms when intracellularly generated S1P binds to as yet unidentified targets in the cytosol [17–19].

As with S1P, C1P has also been postulated to act both intracellularly and through interaction with an as yet unidentified plasma membrane receptor. In particular, it has been shown that C1P regulates numerous cellular functions, including arachidonic acid release, mast cell degranulation, calcium ion mobilization, translocation of lipid-metabolizing enzymes, cell proliferation, apoptosis, and cell migration [20–29].

C1P was shown to stimulate glucose uptake and ATP production in RAW264.7 macrophages, suggesting that C1P is a regulator of carbohydrate metabolism [30]. The mechanism by which C1P stimulated glucose incorporation into the macrophages involves translocation of the GLUT3 glucose transporter from the cytosol to the plasma membrane, enhancement of the affinity of the glucose transporter for its substrate, and activation of the phosphatidylinositol 3-kinase/protein kinase B (PKB, also known as Akt) pathway. However, although it is clear that C1P binds to a specific site at the plasma membrane of cells to stimulate cell migration or glucose uptake, a receptor for C1P still remains to be isolated or cloned. Nonetheless, the binding site of C1P is coupled to pertussis toxin-sensitive Gi proteins and was shown not to bind S1P [28, 29].

Most of the cellular functions that are regulated by C1P are associated with tumorigenesis and tumor metastasis. In fact, C1P was first isolated from human leukemia HL-60 cells [31], and Cer kinase, the enzyme responsible for C1P biosynthesis, has been demonstrated to participate in stimulation of human neuroblastoma cell proliferation [32].

4 Caged C1P Derivatives

A caged-C1P analog (DECM-C1P or BHNB-C1P; see structures in Fig. 1c is a powerful molecular tool for three reasons [2]. First, it is soluble in buffer without the need to first dissolve it in harsh organic solvents such as dodecane that might introduce defects in cell membranes. Second, it is cell permeable. Third, when incubated with cells in the dark, it crosses the plasma membrane in its biologically

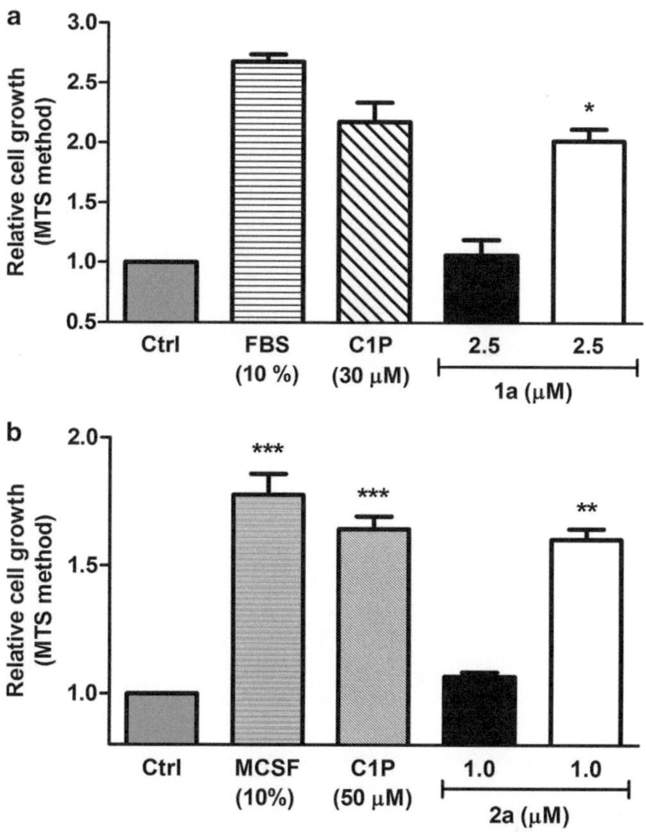

Fig. 4 Delivery of DECM-C1P (**1a**) and BHNB-C1P (**2a**) into RAW 264.7 macrophages (panel **A**) and primary bone marrow derived macrophages (panel **B**) stimulates cell growth. *Open bars*: After the cells were incubated with compound **1a** or **2a** in the dark for 30 min, the cells were exposed to 400–500 nm light for 60 min in a Dark Reader non-UV transilluminator (DR45M from Clare Chemical Research, Denver, CO, USA) equipped with a 9 W lamp. *Filled bars*: The cells were incubated in the dark with the compounds. (A) The cells were incubated for 48 h in the absence of FBS or (B) for 24 h with 1.5 % macrophage colony stimulating factor (MCSF) prior to the addition of the compounds. Mean ± S.D. of three independent experiments (Taken from [2], with permission)

inactive form and thus can be used to distinguish between the intra- and extracellular effects of C1P, especially in cells that cannot be efficiently transfected to overexpress any specific protein of interest. For instance, this is the case of primary bone marrow derived macrophages [33] and possibly many other primary cell types.

Figure 4 displays evidence that BHNB-C1P delivers C1P intracellularly in primary macrophages on photoactivation. Relatively low concentrations of DECM-C1P (compound **1a**, 2.5 μM) or BHNB-C1P (compound **2a**, 1.0 or 2.5 μM) stimulated proliferation of bone marrow derived macrophages or RAW264.7 cells potently upon light irradiation, whereas no effect of either DECM-C1P or BHNB-C1P was observed when the cells were incubated with these compounds in the dark [2].

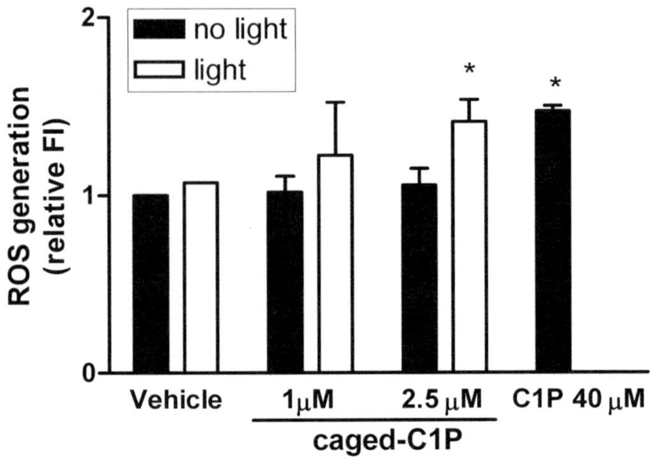

Fig. 5 ROS are generated by C1P acting intracellularly. Macrophages were incubated with 1.0 or 2.5 μM BHNB-C1P or with a concentrated aqueous dispersion of C1P (40 μM). The cells were then exposed to 400–500 nm light for 1 h using a transilluminator (*open bars*) or were allowed to stay in the dark (*filled bars*). Intracellular ROS generation was measured after 8 h of incubation. The results are expressed as mean ± SEM of three independent experiments performed in duplicate, except for the point at 1.0 μM BHNB-C1P, which is the mean ± range of two experiments (Taken from [34], with permission)

C1P-stimulated cell proliferation in primary bone marrow derived macrophages was mediated in part by the production of relatively low concentrations of reactive oxygen species (ROS) [34]. By using the photolabile BHNB-C1P analog, we demonstrated that ROS production was caused by intracellular C1P, since increased ROS generation in the macrophages was observed with light-irradiated BHNB-C1P-loaded cells but not when cells were incubated with BHNB-C1P in the dark (Fig. 5).

Recently, we found that irradiation of BHNB-C1P with 400–500 nm light to release C1P intracellularly caused phosphorylation of ERK1/2 and Akt, two well-known kinases involved in the regulation of cell growth and viability (Fig. 6). This is consistent with the stimulation of cell proliferation by light-irradiated BHNB-C1P-treated cells. However, contrary to exogenous C1P, uncaging of BHNB-C1P in the cytosol failed to stimulate cell migration (unpublished work) or glucose uptake by macrophages, two actions that were demonstrated to be dependent upon interaction of C1P with a specific plasma membrane site [29, 30]. The fact that light-irradiated BHNB-C1P fails to stimulate migration but causes ERK phosphorylation validates the use of this technique to discriminate between the extra- and intracellular effects of C1P. It should be noted that caged compounds bypass cell-surface receptors, and therefore the effects elicited by these compounds can only be observed when they are released into the cytosol after light irradiation. These observations suggest that intracellular accumulation of C1P is absolutely necessary for eliciting cell proliferation. Ongoing experiments in our laboratories indicate that intracellular C1P, but not C1P that was added exogenously as an aqueous suspension, is also necessary for promoting cell survival.

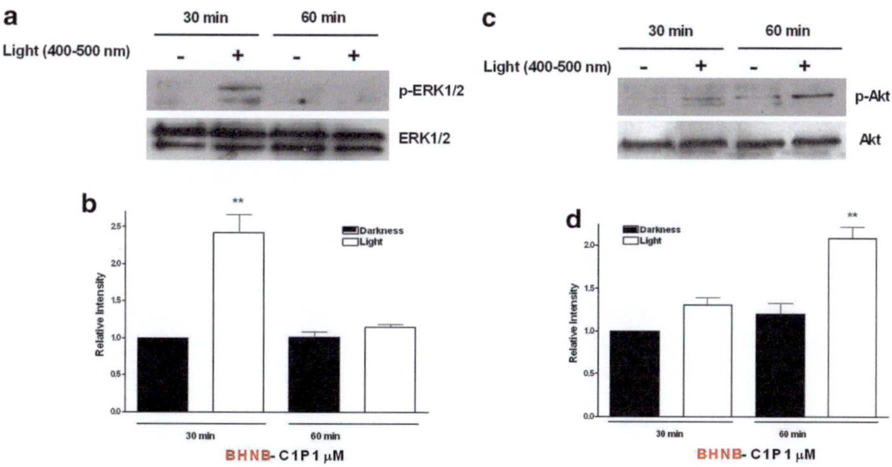

Fig. 6 Stimulation of ERK1/2 and Akt phosphorylation by BHNB-C1P in macrophages. Bone marrow derived macrophages were seeded at 1×10^6 cells/60 mm dish in 1 mL of medium. The cells were incubated in RPMI 1640 without MCSF (Ctrl) or BNHB-C1P (1.0 μM), as indicated. Then, the macrophages were either incubated in the dark (*black bars*) or exposed to 400–500 nm light at the indicated times, using a transilluminator equipped with a 9 W lamp for 60 min at a distance of 1.5 cm at 37 °C, so as to release the C1P into the cytosol (*white bars*). The cells were lysed by sonication. (**a**) ERK1/2 phosphorylation was analyzed by western blotting using a specific antibody to phospho-ERK1/2 (p-ERK1/2). Equal loading of protein was monitored using a specific antibody to total ERK1/2. (**b**) Results of scanning densitometry of exposed film. Similar results were obtained in each of three replicate experiments. Data are expressed as arbitrary units of intensity relative to the time 30 min without the light value and are the mean ± SEM of three replicate experiments (**$p<0.01$). (**c**) Akt phosphorylation was analyzed by western blotting using a specific antibody to phospho-Akt (p-Akt). Equal loading of protein was monitored using a specific antibody to total Akt. (**d**) Results of scanning densitometry of exposed film. Similar results were obtained in each of three replicate experiments. Data are expressed as arbitrary units of intensity relative to the time 30 min without the light value and are the mean ± SEM of three replicate experiments (**$p<0.01$)

5 Conclusions and Future Developments

Non-damaging light irradiation with one- or two-photon excitation triggers the uncaging of the caged Cer and C1P precursors, making possible investigations of the important roles they play in the organization of membranes and in the regulation of intracellular biological processes. The in vitro studies with cell cultures presented in this review show how the utilization of caged C1P derivatives has enhanced our understanding of the functions of C1P. For example, it is now clear that intracellular C1P regulates cell proliferation, whereas cell migration and glucose uptake require the interaction of C1P with a putative plasma membrane receptor. Our results presented in Fig. 6 show that C1P stimulates intracellular cell signaling pathways. Moreover, the ability to prepare caged derivatives of Cer and C1P bearing a variety of long-chain fatty amide chains [1, 2] enables the role of the *N*-acyl group of Cer and C1P in intracellular biological events to be explored. Thus these caged sphingolipids may have utility in a plethora of biochemical applications.

Acknowledegements We are grateful for financial support of the work with caged Cer and caged C1P from the National Institutes of Health (Grant HL-083187 to R.B.) and Grants IT-705-13 from the Departamento de Educación, Universidades e Investigación del Gobierno Vasco (Gazteiz-Vitoria, Basque Country), and S-PE13UN017 from Departamento de Industria, Comercio y Turismo del Gobierno Vasco (Gazteiz-Vitoria, Basque Country) (to A. G. M.).

We are indebted to Dr. Patricia Gangoiti for providing Fig. 6. We also thank our many collaborators, especially Dr. Linda J. Johnston, Dr. Daniel M. Carter Ramirez, Dr. Young Ah Kim, and Dr. Ravi S. Lankalapalli, who worked with us on the caged ceramide and caged ceramide 1-phosphate projects.

References

1. Kim YA, Carter Ramirez DM, Costain WJ, Johnston LJ, Bittman R (2011) A new tool to assess ceramide bioactivity: 6-bromo-7-hydroxycoumarinyl-caged ceramide. Chem Commun 47:9236–9238
2. Lankalapalli RS, Ouro A, Arana L, Gomez-Munoz A, Bittman R (2009) Caged ceramide 1-phosphate analogues: synthesis and properties. J Org Chem 74:8844–8847
3. Furata T, Wang SS-H, Dantzker JL, Dore TM, Bybee WJ, Callaway EM, Denk W, Tsien RY (1999) Brominated 7-hydroxycoumarin-4-ylmethyls: photolabile protecting groups with biologically useful cross-sections for two photon photolysis. Proc Natl Acad Sci U S A 96: 1193–1200
4. Fedoryak OD, Dore TM (2002) Brominated hydroxyquinoline as a photolabile protecting group with sensitivity to multiphoton excitation. Org Lett 4:3419–3422
5. Höglinger D, Nadler A, Schultz C (2014) Caged lipids as tools for investigating cellular signaling. Biochim Biophys Acta 1841:1085–1096
6. Ma J, Rea AC, An H, Ma C et al (2012) Unraveling the mechanism of the photodeprotection reaction of 8-bromo- and 8-chloro-7-hydroxyquinoline caged acetates. Chemistry 18: 6854–6865
7. Qiao L, Kozikowski AP, Olivera A, Spiegel S (1998) Synthesis and evaluation of a photolyzable derivative of sphingosine. Bioorg Med Chem Lett 8:711–714
8. Meyer zu Heringdorf D, Liliom K, Schaefer M, Danneberg K, Jaggar JH, Tigyi G, Jakobs KH (2003) Photolysis of intracellular caged sphingosine-1-phosphate causes Ca^{2+} mobilization independently of G-protein-coupled receptors. FEBS Lett 554:443–449
9. Usatyuk PV, He D, Bindokas V, Gorshkova IA, Berdyshev EV, Garcia JG, Natarajan V (2011) Photolysis of caged sphingosine-1-phosphate induces barrier enhancement and intracellular activation of lung endothelial cell signaling pathways. Am J Physiol Lung Cell Mol Physiol 300:L840–850
10. Grösch S, Schiffmann S, Geisslinger G (2012) Chain length-specific properties of ceramides. Prog Lipid Res 51:50–62
11. Sot J, Goñi FM, Alonso A (2005) Molecular associations and surface-active properties of short- and Long-N-acyl chain ceramides. Biochim Biophys Acta 1711:12–19
12. Carter Ramirez DM, Kim YA, Bittman R, Johnston LJ (2013) Lipid phase separation and protein-ganglioside clustering in supported bilayers are induced by photorelease of ceramide. Soft Matter 9:4890–4899

13. Ira, Zou S, Vanderlip S et al (2009) Enzymatic generation of ceramide induces membrane restructuring: correlated AFM and fluorescence imaging of supported bilayers. J Struct Biol 168:78–89
14. Kunishima M, Tokaji M, Matsuoka K et al (2006) Spontaneous membrane fusion induced by chemical formation of ceramides in a lipid bilayer. J Am Chem Soc 128:14452–14453
15. Shigenaga A, Hirakawa H, Yamamoto J, Ogura K et al (2011) Design and synthesis of caged ceramide: UV-responsive ceramide releasing system based on UV-induced bond cleavage followed by O-N acyl transfer. Tetrahedron 67:3984–3990
16. Carter Ramirez DM, Pitre S, Kim YA, Bittman R, Johnston LJ (2013) Photo-uncaging of ceramides promotes reorganization of liquid-ordered domains in supported lipid bilayers. Langmuir 29:338–347
17. Hannun YA, Obeid LM (2008) Principles of bioactive lipid signalling: lessons from sphingolipids. Nat Rev Mol Cell Biol 9:139–150
18. Fyrst H, Saba JD (2010) An update on sphingosine-1-phosphate and other sphingolipid mediators. Nat Chem Biol 6:489–497
19. Gangoiti P, Camacho L, Arana L, Ouro A, Granado MH, Brizuela L, Casas J, Fabriás G, Abad JL, Delgado A, Gómez-Muñoz A (2010) Control of metabolism and signaling of simple bioactive sphingolipids. Implications in disease. Prog Lipid Res 49:316–334
20. Gomez-Muñoz A (2006) Ceramide 1-phosphate/ceramide, a switch between life and death. Biochim Biophys Acta 1758:2049–2056
21. Gomez-Muñoz A, Gangoiti P, Arana L, Ouro A, Rivera I-G, Ordoñez M, Trueba M (2013) New insights on the role of ceramide 1-phosphate in inflammation. Biochim Biophys Acta 1831:1060–1066
22. Chalfant CE, Spiegel S (2005) Sphingosine 1-phosphate and ceramide 1-phosphate: expanding roles in cell signaling. J Cell Sci 118:4605–4612
23. Lamour NF, Chalfant CE (2005) Ceramide-1-phosphate: the "missing" link in eicosanoid biosynthesis and inflammation. Mol Interv 5:358–367
24. Wijesinghe DS, Lamour NF, Gomez-Munoz A, Chalfant CE (2007) Ceramide kinase and ceramide-1-phosphate. Methods Enzymol 434:265–292
25. Ratajczak MZ, Kim C, Ratajczak J, Janowska-Wieczorek A (2013) Innate immunity as orchestrator of bone marrow homing for hematopoietic stem/progenitor cells. Adv Exp Med Biol 734:219–232
26. Hinkovska-Galcheva V, Boxer LA, Kindzelskii A, Hiraoka M, Abe A, Goparju S, Spiegel S, Petty HR, Shayman JA (2005) Ceramide 1-phosphate, a mediator of phagocytosis. J Biol Chem 280:26612–26621
27. Mitsutake S, Kim TJ, Inagaki Y, Kato M, Yamashita T, Igarashi Y (2004) Ceramide kinase is a mediator of calcium-dependent degranulation in mast cells. J Biol Chem 279:17570–17577
28. Granado MH, Gangoiti P, Ouro A, Arana L, Gonzalez M, Trueba M, Gomez-Munoz A (2009) Ceramide 1-phosphate (C1P) promotes cell migration: involvement of a specific C1P receptor. Cell Signal 21:405–412
29. Arana L, Ordoñez M, Ouro A, Rivera I-G, Gangoiti P, Trueba M, Gomez-Muñoz A (2013) Ceramide 1-phosphate induces macrophage chemoattractant protein-1 release: involvement in ceramide 1-phosphate-stimulated cell migration. Am J Physiol Endocrinol Metab 304:E1213–E1226
30. Ouro A, Arana L, Gangoiti P, Rivera I-G, Ordoñez M, Trueba M, Lankalapalli RS, Bittman R, Gomez-Muñoz A (2013) Ceramide 1-phosphate stimulates glucose uptake in macrophages. Cell Signal 25:786–795
31. Dressler KA, Kolesnick RN (1990) Ceramide-1-phosphate, a novel phospholipid in human leukemia (HL-60) cells. J Biol Chem 265:14917–14921
32. Bini F, Frati A, Garcia-Gil M, Battistini C, Granado M, Martinesi M, Mainardi M, Vannini E, Luzzati F, Caleo M, Peretto P, Gomez-Munoz A, Meacci E (2012) New signalling pathway involved in the anti-proliferative action of vitamin D_3 and its analogues in human neuroblastoma cells. A role for ceramide kinase. Neuropharmacology 63:524–537

33. Comalada M, Xaus J, Sanchez E, Valledor AF, Celada A (2004) Macrophage colony-stimulating factor-, granulocyte-macrophage colony-stimulating factor-, or IL-3-dependent survival of macrophages, but not proliferation, requires the expression of p21 (Waf1) through the phosphatidylinositol 3-kinase/Akt pathway. Eur J Immunol 34:2257–2267
34. Arana L, Gangoiti P, Ouro A, Rivera I-G, Ordoñez M, Trueba M, Lankalapalli RS, Bittman R, Gomez-Muñoz A (2012) Generation of reactive oxygen species (ROS) is a key factor for stimulation of macrophage proliferation by ceramide 1-phosphate. Exp Cell Res 318: 350–360

Part III
Sphingolipid Biology Applied to Therapeutics

Chemotherapy and Sphingolipid Metabolism

Ushma A. Doshi, Jeremy K. Haakenson, Samuel S. Linton, Kimberly Kelly, and Mark Kester

Abstract It is well established that sphingolipids are crucial signal transducers regulating several cellular processes like growth, proliferation, response to stress, differentiation, senescence, autophagy and apoptosis. The critical role of sphingolipids in cancer initiation, progression, pathogenesis and metastasis has led to an expanding interest in developing treatment strategies that perturb native sphingolipid balance towards anti-proliferative and apoptotic signaling by elevating levels of intracellular ceramide. Research has revealed numerous intervention points in the sphingolipid pathway, primarily targeting enzymes of this pathway, that can be exploited alone or in conjunction with chemotherapeutic agents to enhance the effectiveness of current treatment modalities. A multitude of pre-clinical investigations have yielded promising results for treatment with sphingolipid analogs or modulators of sphingolipid metabolism alone or in combination with current cancer therapeutics. This chapter serves to review existing literature on chemotherapy and sphingolipid metabolism with the aim to discuss sphingolipid pathway in the context of cancer therapeutics. We focus on discussing sphingolipid-modulating effects of chemotherapeutic drugs and exploiting potential therapeutic targets in the sphingolipid pathway for effective cancer therapy.

Keywords Sphingolipids • Chemotherapy • Ceramide metabolism • Nanoscale therapeutics

---

U.A. Doshi • J.K. Haakenson • S.S. Linton
Department of Pharmacology, Penn State College of Medicine, Hershey, PA, USA
e-mail: udoshi@hmc.psu.edu; jhaakenson@hmc.psu.edu; sam.linton87@gmail.com

K. Kelly
Department of Biomedical Engineering, University of Virginia, Charlottesville, VA, USA
e-mail: kak3x@eservices.virginia.edu

M. Kester (✉)
Department of Pharmacology, University of Virginia, Charlottesville, VA, USA

Cancer Center, Room 6171-H, Charlottesville, VA 22908, USA
e-mail: mkester@virginia.edu

Y.A. Hannun et al. (eds.), *Bioactive Sphingolipids in Cancer Biology and Therapy*, DOI 10.1007/978-3-319-20750-6_18

1 Introduction

The first reports highlighting the role of sphingolipids in signal transduction were published in the late 1980s. Since then, interest in studying sphingolipid biology, sphingolipid pathways, signal transduction mechanisms and the relevance of sphingolipids in diseases has grown exponentially. The role of sphingolipids in cancer pathogenesis and treatment has been of particular interest to the sphingolipid research community. First reports describing the involvement of sphingolipids in mediating apoptosis in cancer cells came in early 1990s. It was reported that synthetic short chain ceramide analogs like C2-ceramide induced cell death in HL60 leukemic cells and caused internucleosomal DNA fragmentation [1]. Tumor necrosis factor α (TNFα) and ionizing radiation induced apoptotic cell death in cancer cells, which was mediated by sphingomyelin hydrolysis and subsequent ceramide generation [2, 3]. Apoptotic cell death through CD95 crosslinking in U937 cells also utilized the sphingomyelin pathway and depended on ceramide production [4]. The first body of work demonstrating the role of sphingolipids in mediating chemotherapy-induced apoptosis in cancer cells was published in 1995, wherein the authors showed that daunorubicin, a chemotherapeutic drug, induced apoptosis in P388 and U937 leukemia cells by elevating intracellular ceramide levels. This increase in the intracellular ceramide pool was not due to the action of sphingomyelinase, but rather via activation of ceramide synthases, which increased de novo ceramide synthesis within cells [5]. This revelation of a potential role of sphingolipid metabolism in mediating chemotherapy-induced cytotoxicity generated immense interest in the scientific community to decipher the connection between chemotherapy and sphingolipid metabolism. Since then, a large body of research has been conducted to establish an in-depth understanding of how sphingolipid metabolism mediates, enhances, or impedes chemotherapy-induced cytotoxicity, with the goal of identifying critical therapeutic targets and better therapeutic regimens for management and cure of cancer.

Among the major sphingolipids that play a role in regulating cancer cell fate, ceramide is termed as a "tumor suppressor lipid" because of its ability to potentiate signaling cascades that lead to cell death. By contrast, sphingosine-1-phosphate (S1P) is considered as a pro-survival lipid. Thus, in the context of sphingolipids, the ceramide-S1P rheostat dictates cancer cell fate. The role of ceramide in mediating cancer cell death has been extensively studied over the last two decades. Both stimulus-induced intracellular ceramide generation and exogenous cell-permeable short-chain ceramides induce death in cancer cells by apoptosis, necrosis or autophagy. It is well known that ceramide can induce apoptosis in cancer cells through the extrinsic or intrinsic pathway. The extrinsic pathway of apoptosis occurs by the activation of pro-apoptotic receptors at the plasma membrane like CD95, TNFR1, TRAILR1 and TRAILR2. Receptor activation leads to ceramide synthesis at ceramide-enriched membrane platforms by the de novo pathway or activation of sphingomyelinases. The ceramide-enriched membrane platforms act as scaffolds for localization of pro-apoptotic proteins that initiate intracellular signaling for cell death, some of which possess ceramide-binding domains. Similarly, it is known that ceramide mediates apoptosis through the intrinsic pathway by perturbing mitochondrial integrity and function. Ceramide accumulation in the mitochondrial outer membrane leads to formation of ceramide channels that, in conjunction with BAX, induce mitochondrial outer membrane permeabilization (MOMP) [6]. Subsequent apoptotic signaling involves release of pro-apoptotic proteins like cytochrome c from the mitochondria and activation of the caspase cascade. Despite the fact that ceramide is a promoter of autophagy, its role in mediating autophagic cell death is confounded by the role of autophagy in cancer cells, i.e. lethal versus survival autophagy. Increased levels of long-chain ceramides (C14:0–C20:0 ceramides) and especially dihydroceramides have been associated with both lethal and survival autophagy in different cancer cell types [7, 8]. It is believed that the fate of the autophagolysosome dictates the function of autophagy as lethal versus survival. Intracellular generation of sphingosine and S1P in the autophagolysosomes by the hydrolysis of dihydroceramides and ceramides promotes pro-survival autophagy. In contrast, accumulation of dihydroceramides in the autophagolysosomes can enhance autophagolysosomal membrane permeability and cause the release of cathepsins, thereby causing apoptotic cell death [9]. In this case, ceramides promote lethal autophagy in cancer cells. In addition to induction of cell death, ceramide has also been shown to regulate cell cycle and induce cell cycle arrest in cancer cells. Ceramide causes G1 phase arrest by inhibition of cyclin-dependent kinase 2 (CDK2) and by increased association of p21 with CDK2. Additionally, ceramide-mediated suppression of survivin expression induces cell cycle arrest at the G2 phase [7].

Thus, as described above, altering the sphingolipid balance in cancer cells to potentiate cytotoxicity of chemotherapeutic drugs would aim at either elevating pro-apoptotic sphingolipids, especially ceramide, or down-regulating pro-survival sphingolipids such as sphingosine-1-phosphate. This can be achieved by: (i) chemotherapy-induced synthesis of pro-apoptotic ceramides or breakdown of pro-survival sphingolipids; (ii) disruption of ceramide metabolism to enhance ceramide accumulation; and (iii) delivery of exogenous ceramides to induce apoptotic signaling (Fig. 1). This chapter reviews existing literature on chemotherapy and sphingolipid metabolism with the aim to discuss sphingolipid pathway in the context of cancer therapeutics. We will mostly focus on sphingolipid-modulating effects of chemotherapeutic drugs and exploiting potential therapeutic targets in the sphingolipid pathway for effective cancer therapy.

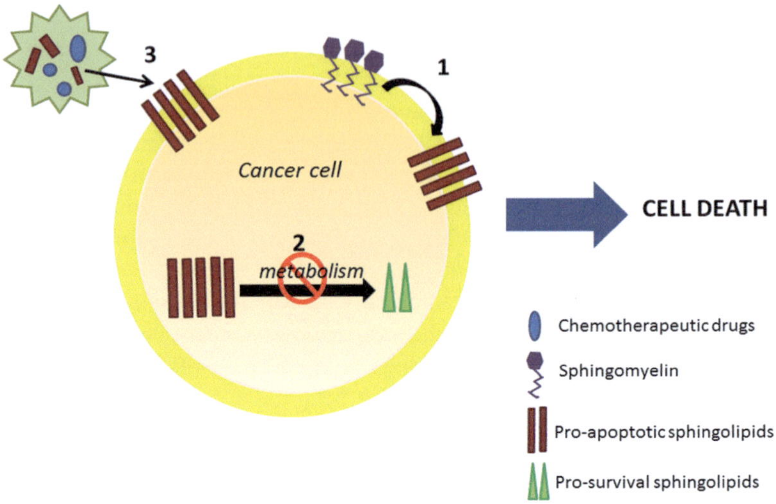

Fig. 1 Strategies to alter the sphingolipid balance in cancer cells to potentiate cytotoxicity of chemotherapeutic drugs. (*1*) Chemotherapy-induced synthesis of pro-apoptotic sphingolipids or breakdown of pro-survival sphingolipids; (*2*) Disruption of metabolism of pro-apoptotic sphingolipids to enhance accumulation; and (*3*) Nanoscale-based delivery of exogenous pro-apoptotic sphingolipids in combination with standard chemotherapeutic drugs to induce apoptotic signaling

2 Chemotherapy-Induced Ceramide Generation

2.1 Ceramide Generation by the De Novo *Pathway*

Many chemotherapeutics increase ceramide levels by upregulating the de novo pathway of ceramide synthesis (Fig. 2). This is accomplished by increasing the activity of serine palmitoyl transferase (SPT), ceramide synthase (CerS), or both. SPT catalyzes the first step in the de novo pathway, adding palmitate to serine to produce 3-ketosphinganine [10]. Later in the pathway, CerS converts sphinganine to dihydroceramide, which can then be converted to ceramide [10]. There are six CerS isoforms (CerS1–6), which are also known as longevity assurance (LASS) genes, with each isoform corresponding to specific resulting carbon chain lengths of ceramide [11]. CerS1 produces C18 ceramide, CerS2 C20–C26, CerS3 C22–C26, CerS4 C18 and C20, CerS5 C16, and CerS6 C14 and C16 ceramide [11]. The reader should also be aware that several inhibitors are routinely used to study the de novo pathway, including fumonisin B1 (FB1), which inhibits CerS; and L-cycloserine and myriocin, which inhibit SPT.

The taxanes docetaxel and paclitaxel are two chemotherapeutic agents that increase ceramide levels by the de novo pathway. Both of these drugs act by preventing microtubule disassembly, leading to cell cycle arrest [12]. Docetaxel, which is used to treat metastatic ovarian, breast, lung, head and neck, and prostate cancer,

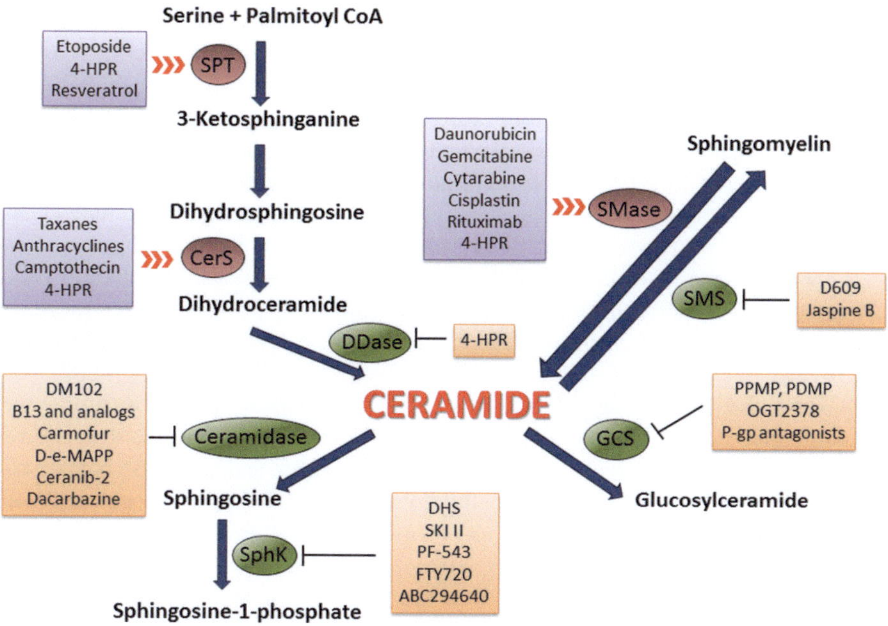

Fig. 2 Points of intervention in the sphingolipid pathway. Enzymes catalyzing ceramide synthesis or ceramide metabolism can be activated or inhibited respectively to cause ceramide accumulation and induce death in cancer cells. *SPT* serine palmitoyl transferase, *CerS* ceramide synthases, *DDase* dihydroceramide desaturase, *SMase* sphingomyelinase, *SMS* sphingomyelin synthase, *SphK* sphingosine kinase, *GCS* glucosyl ceramide synthase, *PPMP* 1-phenyl-2-palmitoylamino-3-morpholino-1-propanol, *PDMP* 1-phenyl-2-decanoylamino-3-morpholino-propanol, *4-HPR* *N*-(4-hydroxyphenyl) retinamide, *DHS* L-*threo*-dihydrosphingosine, *SKI II* 2-(p-hydroxyanilino)-4-(p-chlorophenyl) thiazole

increases CerS1 and CerS2 levels, while decreasing sphingosine kinase 1 (SK-1) levels in prostate cancer cells [13]. It has also been demonstrated that ceramide is a critical regulator of taxane-induced cell death since overexpression of the ceramide transfer protein, CERT, reduces the sensitivity of cells to taxanes [14]. Similarly, paclitaxel-induced apoptosis in breast cancer cells is dependent on ceramide that has been produced by the de novo pathway [15].

In addition to the taxanes, the anthracyclines doxorubicin and daunorubicin also affect ceramide metabolism in tumors. Doxorubicin, which intercalates into DNA, is used to treat malignant lymphoma, breast cancer, small cell lung cancer, a variety of sarcomas, head and neck cancer, and neuroblastoma [12]. In addition to its effects on DNA, doxorubicin also causes a CerS-dependent increase in ceramide levels in neuroepithelioma and neuroblastoma cells [16]. However, inhibition of de novo ceramide generation using FB1 surprisingly has no effect on doxorubicin-induced apoptosis in these cells [16]. On the other hand, in head and neck cancer cells, doxorubicin alone or in combination with gemcitabine causes apoptosis that is dependent on CerS1 and the generation of C18:0 ceramide [17]. This was confirmed in an animal model of head and neck cancer in which doxorubicin in combination with gemcitabine caused a decrease in C16:0 ceramide in the tumor, but an increase in intratumoral CerS1 and C18:0 ceramide [17]. Similar to doxorubicin, camptothecin also exerts antitumor properties in follicular thyroid carcinoma through ceramide accumulation via de novo synthesis. Both of these drugs cause activation of ceramide synthesis without any effects on sphingomyelinases. Apoptosis is mediated by ceramide elevation and can be enhanced by the use of inhibitors of ceramide clearance [18]. Similar to doxorubicin, daunorubicin also acts by intercalating into DNA [12] and induces apoptosis in leukemia cells by de novo ceramide generation and ceramide generation by the action of sphingomyelinase [5]. It is used to treat acute myeloid leukemia (AML) and AIDS-related Kaposi's sarcoma (liposomal formulation) [12].

Besides the taxanes and the anthracyclines, vorinostat in combination with sorafenib also modulates ceramide metabolism. Vorinostat is a histone deacetylase inhibitor that is used to treat cutaneous T-cell lymphoma, and sorafenib is a kinase inhibitor that is used to treat hepatocellular carcinoma (HCC), advanced renal cell cancer, and differentiated thyroid cancer. Vorinostat in combination with sorafenib causes de novo pathway-dependent reactive oxygen species (ROS) production and cell death in HCC cells [19]. This is accompanied by an increase in C16 ceramide, as well as C16, C18, C22, C24:0, and C24:1 dihydroceramide [19].

Another chemotherapeutic agent that increases ceramide levels is fludarabine, an inhibitor of DNA synthesis that is used to treat chronic lymphocytic leukemia (CLL) [12]. Fludarabine causes a de novo pathway-dependent 2.5- to 3-fold elevation in ceramide levels in CLL cells 6 h after treatment. Pretreatment with fumonisin B1 significantly rescues fludarabine-induced ceramide generation and apoptosis [20]. In addition, CLL cells treated with non-physiological, short chain C6 ceramide undergo apoptosis and necrosis [20, 21], suggesting that fludarabine may kill CLL cells via upregulation of ceramide.

Certain chemotherapeutic drugs generate ceramide by acting on SPT. Etoposide-induced apoptosis in human leukemia cells is mediated by ceramide synthesized by the activation of SPT. Ceramide formed by this pathway has distinct functions in this model system as compared to that formed by the sphingomyelinase pathway. Ceramide synthesized by the de novo pathway in this model is not involved in caspase-induced poly(ADP-ribose) polymerase (PARP) cleavage but instead perturbs membrane integrity [22].

In addition to current chemotherapies, there are drugs approved by the United States Food and Drug Administration (FDA) for other uses that are currently being investigated as potential anti-cancer agents. One of these is the COX-2 inhibitor celecoxib, which is currently used to treat pain, inflammation, and arthritis, and to prevent polyposis coli [12]. Recent work indicates that celecoxib decreases cell viability in colorectal carcinoma cells in a CerS6-dependent manner [23]. This is accompanied by an increase in sphingosine, dihydrosphingosine, C14:0 ceramide, C16:0 ceramide and C18:0 ceramide, and a decrease in C24:0 ceramide [23]. Furthermore, C16:0 ceramide is found at elevated levels in tumors treated with celecoxib in an in vivo model of colorectal carcinoma [23]. A number of investigational chemotherapeutics also upregulate ceramide via the de novo pathway. Investigational drugs such as Valspodar, inostamycin, and spisulosine induce apoptosis in a variety of cancer cell types via de novo ceramide generation [24–28]. More recently, the endocannabinoid analog R (+)-methanandamide (RMA) has been shown to increase C16 ceramide levels in neuroglioma cells via the de novo pathway [29]. RMA, which increases CerS3 and CerS6 in mantle cell lymphoma (MCL) cells, also increases C16, C24, and C24:1 ceramide levels via the de novo pathway in MCL cells [30]. This is functionally significant, as RMA-induced cell death in these cells is also dependent on the de novo pathway [30].

In addition to RMA, another investigational drug that increases ceramide levels is fenretinide (4-HPR). It is a synthetic retinoid *N*-(4-hydroxyphenyl) that induces cell death in various cancers like neuroblastoma [31] and in cell lines from cervical carcinoma [32] and acute myeloid leukemia (AML) [33]. Studies have shown that the drug elevates intracellular ceramide levels via de novo synthesis by increasing the activity of both SPT and ceramide synthases [34], and induces p53- and caspase-independent apoptosis in neoplastic cells [35]. Another study reported that 4-HPR functions as a dose-dependent inhibitor of ceramide desaturase [36], indicating that the cytotoxic sphingolipid species may be dihydroceramide or dihydrosphingosine rather than ceramide [37]. Moreover, 4-HPR-induced cytotoxicity is synergistic with inhibitors of ceramide metabolism like sphingosine kinase inhibitors and glucosylceramide synthase inhibitors [38]. Other effects induced by 4-HPR include generation of reactive oxygen species and enhanced expression of LC3B (form II). 4-HPR is currently in clinical trials for ovarian cancer, neuroblastoma, lymphoma and leukemia [39].

Finally, the investigational AMPK inhibitor Compound C has been shown to cause an increase in CerS5, C16:0 ceramide, and C18:0 ceramide, and a decrease in sphingosine in breast cancer cells [40]. Clearly, many current and investigational chemotherapeutics cause an increase in ceramide levels by targeting the de novo pathway. This increase in ceramide, in turn, induces apoptosis in cancer cells.

2.2 Ceramide Generation by Sphingomyelinase Pathway

Independent of the de novo pathway, ceramide can also be generated when sphingomyelinase (SMase) hydrolyzes sphingomyelin to produce ceramide (Fig. 2). There are several sphingomyelinase isoforms, namely: acid sphingomyelinase (aSMase), which is found in the lysosome and can be secreted out of the cell [41]; neutral sphingomyelinase (nSMase1–3), which is found in the cytosol and the plasma membrane [41]; and alkaline sphingomyelinase, which hydrolyzes dietary sphingomyelin in the gut [42]. SMase activity is decreased in colorectal carcinoma [43]. Because it increases ceramide levels and its activity is reduced in cancer, activation of SMase presents a logical target for cancer therapeutics, especially for those intended for colorectal carcinoma.

A number of chemotherapeutics increase ceramide levels in cancer cells by increasing the activity of SMase. In fact, some drugs upregulate ceramide by modulating both the de novo pathway and SMase activity. Daunorubicin is one such drug. Besides upregulating the de novo pathway, it also increases SMase activity in leukemia cells [44]. In addition, in breast cancer cells, daunorubicin causes the transcription factor Sp1 to bind to the nSMase2 gene, leading to increased nSMase2 levels and a nSMase-dependent decrease in cell viability [45]. It should be noted that daunorubicin is currently not approved for the treatment of breast cancer.

Gemcitabine is another drug that targets both the de novo pathway and SMase. Gemcitabine, which blocks DNA synthesis, leading to apoptosis, is used to treat metastatic pancreatic cancer, non-small cell lung cancer, ovarian, bladder, esophageal, and head and neck cancer [12]. It increases aSMase activity in pancreatic cancer cells [46]. In addition, gemcitabine increases aSMase activity in glioma cells, causing an increase in the levels of C16 and C24 ceramide and an aSMase-dependent decrease in cell survival [47]. It is interesting to note that gemcitabine has not been approved for the treatment of glioma.

One chemotherapeutic that elevates ceramide levels via SMase independent of the de novo pathway is cytarabine (also known as ara-C). Cytarabine is a cytidine analog used to treat various forms of leukemia and meningitis [12]. It increases nSMase activity and ceramide levels in acute myelogenous leukemia (AML) cells [48].

Cisplatin is another drug that affects SMase activity. Cisplatin, which is a platinum coordination complex that causes DNA cross-links leading to cell death, is used to treat testicular, ovarian, bladder, head and neck, cervical, endometrial, lung, anal, rectal, esophageal, and central nervous system (CNS) cancer, as well as neoplasms of childhood [12]. It increases ceramide levels and causes apoptosis in glioma cells in a nSMase-dependent manner, wherein it transiently increases aSMase activity and causes it to be redistributed to the plasma membrane [49]. Additionally, a cisplatin-induced increase in SMase activity and the subsequent accumulation of ceramide levels are essential for cytoskeletal remodeling following treatment with cisplatin, such as loss of lamellipodia/filopodia and dephosphorylation and redistribution of the actin-binding protein ezrin [50].

Another chemotherapeutic that is currently used in the clinic and that affects SMase is rituximab, which is a monoclonal antibody to CD20 that is used to treat lymphoma and chronic lymphocytic leukemia [12]. It causes an increase in aSMase activity in lipid rafts, leading to an increase in ceramide levels in B-lymphoma cells, and it inhibits cell growth in these cells [51]. Furthermore, exogenous treatment with C16 ceramide decreases cell viability in this system [51]. Taken together, these findings indicate that rituximab may inhibit B-lymphoma cell proliferation by activating aSMase, thus increasing ceramide levels.

Recently, the investigational drug stichoposide C, a marine triterpene glycoside, has been shown to cause apoptosis in leukemia and colorectal cancer cells in an aSMase- and nSMase-dependent manner [52]. It also inhibits tumor growth in mouse models of leukemia and colorectal cancer [52]. In addition, stichoposide C-treated tumors contained elevated levels of ceramide [52]. Betuletol 3-methyl ether, a natural phenylbenzo-γ-pyrone, is another investigational drug that increases ceramide levels in cancer cells [53]. It causes apoptosis in leukemia cells and increases aSMase activity and ceramide levels [53]. Finally, the investigational drug withanolide D acts by increasing nSMase activity, leading to increased ceramide levels and apoptosis in leukemia cells [54]. It decreases cell viability in leukemia cells but not normal lymphocytes [54]. Furthermore, it induces apoptosis in primary cells from both myeloid and lymphoid leukemia patients [54]. Withanolide D is a good example of an investigational therapy that modulates SMase. While several current chemotherapies target SMase, even more such drugs could be added to the oncology arsenal in the future.

3 Chemotherapy and Ceramide Metabolism

The cytotoxic effects of elevated intracellular ceramide formation can be blunted by the upregulated metabolic clearance of ceramide in neoplastic cells (Fig. 2). Cancer cells overexpress certain ceramide metabolizing enzymes which prevent accumulation of ceramide required for inducing cell death. Thus, another strategy to elevate intracellular ceramide levels in cancer cells is by targeting enzymes catalyzing ceramide clearance like sphingomyelin synthase (SMS), glucosylceramide synthase (GCS), ceramidases, dihydroceramide desaturase and sphingosine kinases [10, 55]. These enzyme inhibitors magnify the effects of ceramide-generating chemotherapies and exogenous ceramides. Mounting evidence points to the use of chemotherapeutic agents, alone or in conjunction with these enzyme inhibitors, as potential treatment modalities for cancer.

3.1 Agents Targeting Sphingomyelin Synthase

Not only can sphingomyelin be converted to ceramide, but the reverse reaction can occur, in which sphingomyelin synthase (SMS) converts ceramide into sphingomyelin [10] (Fig. 2). Increased SMS activity has been found in hepatomas when compared to normal livers [56] and in leukemic cells from chemoresistant patients when compared to chemosensitive patients [57]. Because it metabolizes ceramide to sphingomyelin and is up-regulated in drug-resistant cancer, SMS should be a good target for cancer chemotherapeutics.

D609 is a selective tumor cytotoxic drug that inhibits SMS activity in several tumor cells like human monocytic leukemia [58]. It modulates the activity of SMS to cause elevated ceramides and diacyl glycerol, which mediate D609-induced cytotoxicity [58]. Another natural investigational drug that modulates ceramide levels is Jaspine B, a marine anhydrophytosphingosine [59]. It decreases SMS activity in melanoma cells, leading to increased ceramide levels and apoptosis [59].

Unlike Jaspine B, the investigational chemotherapeutic 2-hydroxyoleic acid (2OHOA) acts by upregulating SMS [60]. It increases SMS activity in glioma cells, leading to increased sphingomyelin levels [60]. It has been speculated that accumulation of sphingomyelin at the plasma membrane after treatment with 2OHOA results in localization of proteins involved in cell apoptosis and differentiation. In addition, 2OHOA inhibits the growth of glioma tumors in vivo and causes glioma cells to differentiate in vitro and in vivo [61].

3.2 Agents Targeting Glycosphingolipid Synthesizing Enzymes

Of the numerous metabolism pathways in sphingolipid biology, formation of glycosphingolipids (GSLs) has been of particular interest to the research community. GSLs are formed by addition of simple or complex saccharides to ceramide species [62]. The simplest members in this class of sphingolipids are cerebrosides, wherein glucose or galactose is attached to the 1-hydroxyl of ceramide via a β-glycosidic bond. Glucosylceramide synthase (GCS) catalyzes the synthesis of GlcCer by transfer of glucose to ceramide. Glucosylceramide (GlcCer) and galactosylceramide act as intermediates in the synthesis of a wide variety of GSLs, such as gangliosides and globosides [63].

GSLs have been implicated in cancer cell processes through regulation of genes associated with proliferation, apoptosis, metastasis, drug resistance, autophagy, and senescence [64]. Hence, most tumors express altered GSL patterns on the cell surface and have elevated GSL biosynthesis as one of the primary mechanisms of ceramide inactivation [65]. For instance, several cancers like ovarian [66, 67] and breast have elevated galactosylceramide synthesis which inhibits apoptosis and plays a role in the survival of metastatic cells in target organs [68]. Additionally, research suggests that ceramide galactosyltransferase levels can be used as an indicator for evaluating tumor aggressiveness and prognosis of lung metastases in breast cancer [69]. Sulfatides, i.e. 3-*O*-sulfogalactosylceramides, are elevated in several cancers including ovarian cancer [66], hepatocellular carcinoma [70] and renal cell carcinoma [71] and might play a possible role in lymph node metastasis of colorectal adenocarcinoma [72]. However, the exact role of sulfatides in cancer initiation and progression still remains unknown [73]. Lactosylceramide has been shown to modulate integrin clustering and internalization via caveolae, and it has been speculated that high levels of lactosylceramide in cancer cells might influence cell attachment events [74, 75].

A growing body of evidence suggests that altered expression of gangliosides might be involved in tumor development and may be potential markers of disease progression. Gangliosides like GD2, GD3 and GM2 are overexpressed in melanoma, neuroblastoma, lymphoma and ovarian cancers [76]. A recent report shows that differential GSL profiles exist in human breast cancer stem cells in comparison to cancer non-stem cells [77]. Breast cancer stem cells express higher levels of GD2, GD3, GM2 and GD1a, lower levels of Gb3Cer and increased mRNA levels of various glycosyltransferases [77]. Data indicate a potential functional role of GD2 and GD3 in cancer stem cell maintenance and tumor progression [77, 78].

Moreover, gangliosides like GD2, GD3, fucosyl-GM1 and *N*-acetyl-GM3 have been highlighted as potential important cancer antigens [79].

GlcCer and GCS levels in cancer cells have been directly correlated to multidrug resistant phenotype. First reports that linked levels of GlcCer, GCS activity and drug resistance demonstrated that overexpression of GCS rendered MCF-7 breast cancer cells resistant to adriamycin [80]. Sensitivity of drug-resistant cancer cells to doxorubicin, paclitaxel and etoposide was restored by transfection of GCS antisense oligonucleotides or use of GCS inhibitors [81, 82]. This correlation between GCS activity and multidrug resistant phenotype has been extended to multiple cancer types like melanoma [83], breast [84], ovarian [63], cervical [63], colon cancers [85] and leukemia [86, 87]. Growing evidence points towards a correlation between GCS and P-glycoprotein (P-gp) in inducing multidrug resistant phenotype. It has been shown that high GCS activity in multidrug-resistant cells coincides with overexpression of P-gp in several cancer types like leukemia, melanoma, colon and breast cancer [62, 65, 88]. Drugs like tamoxifen, verapamil, cyclosporine A and SDZ PSC 833, which target the drug-pumping action of P-gp have been shown to alter levels of GlcCer [10, 82, 89]. Conversely, targeting GCS downregulated P-gp expression and sensitized drug-resistant cancer cells [84, 90]. The role of GCS in modulating drug resistance in cancer may be attributed to the fact that GSLs and GCS upregulate P-gp expression through c-Src and β-catenin signaling pathway [91]. Studies have demonstrated that GlcCer is a substrate of P-gp-mediated translocation across the Golgi membrane and the translocated GlcCer is further used for synthesis of neutral, but not acidic GSLs [92]. Thus, breast, ovarian and epidermoid cancer cells with high P-gp levels also have elevated levels of GlcCer [10]. Despite strong evidence linking GCS and multidrug-resistant phenotype, some reports indicate that this enzyme might not be significant in developing drug resistance in cancer. Certain investigations suggest that alterations in ceramide glycosylation patterns do not sensitize melanoma cells to anticancer drugs [93] and GCS inhibition using selective inhibitors like C9DGJ and C4DGJ do not reverse multidrug-resistant phenotype [94]. It has also been speculated that the role of GCS in regulating drug resistance might be tumor-specific or sub-type specific [10, 93].

Based on the numerous reports addressing the role of GlcCer and GCS in cancer biology and multidrug resistance, several studies have explored GCS inhibition as a potential chemo-sensitization strategy in cancer therapeutics. Two classes of GCS inhibitors have been extensively studied—analogs of D-threo-1-phenyl-2-decanoylamino-3-morpholino-propanol (PDMP) and a group of imino sugars.

PDMP and related compounds including 1-phenyl-2-palmitoylamino-3-morpholino-1-propanol (PPMP) and 1-phenyl-2-hexadecanoylamino-3-pyrrolidino-1-propanol (PPPP) are structurally similar to GCS substrate and are potent competitive inhibitors of GCS. It has been reported that treatment of doxorubicin-resistant breast cancer cells with PPMP enhances sensitivity to vinblastine and paclitaxel by decreasing ganglioside levels, increasing ceramide levels, enhancing uptake of chemotherapy drugs and diminishing expression of P-gp [84]. GCS inhibition-mediated sensitization to chemotherapeutic drugs has also been observed in other cancers like neuroblastoma [95], follicular thyroid carcinoma [18], chronic myeloid leukemia [96, 97], colorectal cancer [98], acute myeloid leukemia [86] and prostate cancer through accumulation of ceramide [13]. Similar sensitizing effects have also been obtained by using agents that downregulate GCS, like oligonucleotides, as in phosphorothionate DNA, or a 2′-*O*-methyl RNA with phosphorothionate DNA [81, 88, 99]. Furthermore, another interesting study reported that GCS suppression caused ceramide accumulation and sensitized ovarian cancer cells to doxorubicin. In addition, the authors observed that restoring active ceramide to cells caused reactivation of wild-type p53 function in p53-mutant cells and induction of p53-mediated apoptosis by activation of genes like Puma, p21Waf1 and Bax [100]. PDMP-induced GCS suppression radiosensitizes lung adenocarcinoma cells to vinorelbine-based chemoradiotherapy [101]. GCS suppression by PDMP or antisense oligos also synergize with other antineoplastic agents like *N*-(4-hydroxyphenyl)-retinamide, safingol [102], fenretinide [103], retinoid [104], lactoferricin [105] and resveratrol [106].

Imino sugars are the other class of GCS inhibitors that are extensively studied. *N*-butyl-deoxynojirimycin (OGT918), OGT2378, C9DGJ and C4DGJ are selective inhibitors of GCS. Studies have demonstrated that OGT2378 treatment causes a delay in melanoma tumor development [107] and a tenfold reduction in tumor growth in a syngeneic, orthotopic murine model [83]. C9DGJ and C4DGJ also sensitize CLL cells to the cytotoxic drug 2-chlorodeoxyadenosine (CdA) and cytostatic drugs like chlorambucil and fludarabine without affecting P-gp functional activity [108]. Genz-123346 and CCG-203586 are new GCS inhibitors currently under investigation [109, 110]. However, it has been reported that the chemosensitizing effects of Genz-123346 are unrelated to lowering glycosphingolipid levels, but instead primarily through modulation of P-gp function [110].

3.3 Agents Targeting Ceramidases

Deacylation of ceramide to sphingosine and a free fatty acid by the action of ceramidases is another pathway of ceramide clearance. There are five ceramidases that are products of different genes: neutral ceramidase, acid ceramidase and three forms of alkaline ceramidase [111]. These enzymes lie at a crucial juncture in the sphingolipid pathway since, in conjunction with sphingosine kinases, they balance the ceramide/sphingosine 1-phosphate (S1P) rheostat in cells (Fig. 2). Neutral and acid ceramidase stimulate cell survival and proliferation by increasing S1P levels [112]. Alkaline ceramidase 1, mainly expressed in the skin, has anti-proliferative and pro-differentiating roles [112]. The role of alkaline ceramidase 2 is cell type-specific and stimulus-dependent. It may promote cell proliferation by augmenting levels of S1P, while it may also induce cell growth arrest and apoptosis by generating sphingosine [112]. Alkaline ceramidase 3 is the only ceramidase that catalyzes the hydrolysis of both unsaturated long-chain dihydroceramides and phytoceramides, and has not been widely studied [112].

Of the five ceramidases, acid ceramidase (AC) has been extensively studied in the context of cancer biology and therapeutics. Acid ceramidase is a lysosomal ceramidase that is aberrantly overexpressed in several cancers such as prostate, head and neck squamous cell carcinoma and leukemia [76, 111]. AC has been shown to play a role in tumor progression, metastasis, and resistance to chemotherapy and radiation therapy [113]. It has been speculated that AC might play an important role in the progression of prostate cancer to androgen-independent stage [113]. Several reports have demonstrated that AC overexpression confers resistance to anti-neoplastic drugs like etoposide, cisplatin, doxorubicin and gemcitabine due to depletion of proapoptotic C16-ceramide, which is induced following treatment with drugs. Moreover, targeting AC with siRNA restores sensitivity to chemotherapy [114]. Lipidomic analysis of AC overexpressing prostate cancer cells revealed lower concentrations of proapoptotic long chain ceramides and elevated levels of very long chain ceramides that might promote cell growth, proliferation and migration [114]. Cells with elevated AC levels have also been shown to possess increased tumorigenic potential in vivo [114]. AC is also overexpressed in leukemic LGLs and inhibition of AC induced apoptosis in these cells [115]. It has been documented that AC inhibition may play a role in converting the cytostatic effect of PSC 833, a P-gp inhibitor to cytotoxic end-point in pancreatic cancer cells [116]. Additionally, AC overexpression that is induced after radiotherapy has been implicated in conferring radio-resistance in irradiated cancer cells and subsequent tumor relapse [117]. Thus, AC has been identified as a potential druggable target to augment the efficacy of chemotherapies and radiation therapy.

Several AC inhibitors have been developed and are currently being investigated for potential to be used in clinic for cancer therapy. B13 and D-MAPP are lipid based inhibitors that are among the earliest AC inhibitors to be studied. Ceramidase inhibition by B13 induces an apoptotic signaling cascade in human colon cancer cells, but has no effect on normal liver cells [118]. Additionally, B13 significantly inhibits growth of colonic tumors in mice and prevents liver metastases from human colon cancer in vivo [118]. B13 also induces apoptosis in human melanoma cells, prostate cancer cells and inhibits growth of xenografted androgen-insensitive prostate tumors by sensitizing them to radiation [119]. Novel small molecule analogs of B13 like LCL85, LCL120, LCL385, LCL284, and LCL204 are potential lead compounds for therapeutic development [120]. For instance, LCL385 sensitizes prostate cancer cells to radiation and significantly decreases tumor xenograft growth [121]. Similarly another small molecule inhibitor, LCL204 induces lysosomal destabilization, cathepsin-dependent AC degradation and subsequent cell death in prostate cancer cells [122]. LCL204 treatment also sensitizes head and neck squamous carcinoma cells to Fas-induced apoptosis in vitro and in vivo, thus suggesting that the combination of FasL gene therapy and LCL204 may become a new treatment option for advanced-stage head and neck cancer [123]. Furthermore, it has been reported that effective inhibition of neutral ceramidase and AC by novel analogs of B13 and LCL464 enhances cell death in breast cancer cell lines [124]. Other AC inhibitors like *N*-oleoylethanolamine (NOE) [125] and D-erythro-2-(*N*-myristoylamino)-1-phenyl-1-propranol (D-MAPP) [126] display potentiating effects to daunorubicin and radiotherapy respectively in both in vitro and in vivo models.

Despite several investigational AC inhibitors, lack of potency and druggability among the current molecules has motivated more research in this field. A recent study reported that the antineoplastic drug carmofur, which is used in the clinic to treat colorectal cancers, is a potent AC inhibitor. The authors synthesized carmofur derivatives by replacing the fluorine atom with substituent groups. Two derivatives, ARN080 and ARN398, demonstrated synergy with 5-FU and taxol to reduce viability of cancer cells in vitro [127]. These studies have led to the identification of substituted 2,4-dioxopyrimidine-1-carboxamides which is a novel class of potent inhibitors of AC [127]. Another novel ceramidase inhibitor under investigation is Ceranib-2. Studies showed that this small molecule inhibitor alone exerted cytostatic effects on ovarian carcinoma cells and induced cell cycle arrest and apoptosis when used in combination with paclitaxel [128]. In vivo, it delayed tumor growth in a syngeneic tumor model [128]. Certain chemotherapeutic drugs like dacarbazine exert cytotoxic effects by degradation of AC. Dacarbazine, commonly used for treatment of metastatic melanoma, causes dose- and time-dependent degradation of AC as a consequence of reactive oxygen species-dependent activation of cathepsin B. This leads to autophagic cell death in melanoma cells which can be partially rescued by inducible overexpression of AC but not of neutral ceramidase [129]. Thus, downregulation of AC alone or in combination with dacarbazine may represent a useful tool in the treatment of metastatic melanoma. Recent studies have shown that commonly used antiestrogens like tamoxifen and toremifene exert cathepsin B-dependent inhibition of AC in cancer cells at single digit micromolar range, thus documenting novel off-target effects of these drugs [130].

In addition to potentiating classical radiotherapy and chemotherapy, AC inhibitors LCL204 and DM102 also induce synergistic cytotoxicity with other agents like viral protein apoptin [131] and C6-ceramide [132] respectively in cancer models.

3.4 Agents Targeting Dihydroceramide Desaturase

Dihydroceramide desaturase (DDase) catalyzes the conversion of dihydroceramide into ceramide by inserting a 4,5-*trans*-double bond to the sphingolipid backbone of dihydroceramide (Fig. 2). Until recently, dihydroceramides were considered to be biologically inactive lipids. However, research in the last decade has demonstrated that long chain dihydroceramides are involved in cell cycle arrest, autophagy, apoptosis and oxidative stress [39]. These studies have revealed the role of DDase in cancer biology and pathogenesis. Short and long chain dihydroceramides have been shown to mitigate ceramide channel formation in the mitochondria [133]. Thus, the ratio of dihydroceramides to ceramides plays a role in inducing apoptosis [133]. Conversely, transfecting neuroblastoma cells with DDase siRNA resulted in elevation of dihydroceramides and inhibited cell growth, induced cell cycle arrest at G0/G1 phase and reduced phosphorylation of retinoblastoma protein [37]. Moreover, certain chemotherapies have been shown to reduce DDase activity, which in turn causes accumulation of endogenous dihydroceramides and their sphingolipid metabolic products [39]. Confluence-induced growth arrest in neuroblastoma cells also involves reduction in DDase activity and subsequent elevation in dihydroceramide levels [134]. However, in contrast, certain studies report that DDase is also suppressed in hypoxic conditions, thus increasing dihydroceramides, and it is speculated that dihydroceramides might in fact favor tumor cell survival in hypoxic conditions via autophagy [39, 135]. Despite the contrasting reports, researchers believe that DDase plays an important role in cell cycle progression and is a potential target for cancer therapy.

4-HPR is the first chemotherapeutic drug found to inhibit DDase. It is a synthetic retinoid *N*-(4-hydroxyphenyl) that induces cell death in various cancers like neuroblastoma [31] and in cell lines from cervical carcinoma [32], oral squamous cell carcinoma [136] and acute myeloid leukemia (AML) [33]. It has been shown that DDase is a direct in vitro target of this drug [36]. 4-HPR causes a post-translational dose-dependent inhibition in DDase activity and a subsequent increase in levels of dihydroceramides in neuroblastoma cells [37]. Furthermore, several reports suggest that DDase inhibition-induced dihydroceramides are further metabolized to dihydrosphingosines and the latter might actually be the sphingolipid species mediating the cytotoxic effects of the drug in some cell lines [136, 137].

Apart from 4-HPR, several other anticancer investigational drugs like resveratrol [138], curcumin [39], celecoxib [39], Δ^9-tetrahydrocannabinol [39] and gamma-tocopherol [39] have been found to inhibit DDase as one of their mechanisms of action.

3.5 *Agents Targeting Sphingosine Kinases*

Sphingosine kinases 1 and 2 (SphK1 and SphK2) are the two sphingosine kinase isozymes that catalyze the formation of sphingosine-1-phosphate (S1P) from sphingosine (Fig. 2). S1P is a bioactive lipid that has been shown to stimulate cell survival [139], motility, and proliferation through activation of five plasma membrane-spanning G protein-coupled receptors ($S1PR_{1-5}$), as well as through unknown intracellular targets. SphK1 is of particular importance in cancer, as there are elevated *SPHK1* mRNA transcript and SphK1 protein levels in cancers of the stomach [140, 141], lung [141, 142], brain [143, 144], colon [141, 145, 146], kidney [141], and breast [141, 147], as well as in non-Hodgkin's lymphomas [148]. Indeed, the importance of SphK1 in cancer is underscored by experiments in which *SPHK1* in cancer cells is genetically knocked down or pharmacologically inhibited. In most cancer cell lines tested, genetic knockdown or chemical inhibition of Sphk1 induces apoptosis. These in vitro data are validated by *SphK1*$^{-/-}$mice, which are highly resistant to colorectal cancer compared with *SphK1*$^{+/+}$ littermate controls [146]. Upregulated SphK/S1P axis activity contributes significantly to enhanced neovascularization of solid tumors [149], as well as increased resistance to radiation [150] and chemotherapy [151]. Therefore, the SphK/S1P axis is critical in cancer progression, and shows promise as a target for the next generation of chemotherapeutic drugs.

SphK1 and SphK2, despite their high sequence homology, have unique differences that underlie their different functions in vivo. Whereas SphK1 is known to promote tumor growth, SphK2 is thought to exert largely pro-apoptotic effects in tumor cells. SphK2 has a proline-rich BH3 motif, through which it binds BCL-X_L, a pro-survival protein, thereby abrogating the pro-survival activity of BCL-X_L [152]. SphK2-mediated release of cytochrome *c* and subsequent caspase-3 activation were shown to occur independently of $S1P_{1-5}$ [152]. However, the effects of SphK2 on tumor growth are equivocal; genetic knockdown of *SPHK2* in vitro sensitizes MCF-7 breast cancer cells to doxorubicin-induced apoptosis [153]. SphK2 also has a nuclear localization signal, whereas SphK1 does not. SphK2 is found in the nucleus, whereas SphK1 is found in the cytosol and translocates to intracellular membranes or the nucleus upon activation. SphK1 contains two nuclear export signals, which shuttle the enzyme from the nucleus [154]. Because of their unique locations in the cell, SphK1 and Sphk2 are thought to generate distinct intracellular S1P pools. Due to the important differences in intracellular localization and activity between SphK1 and SphK2, many drug development efforts in this field have focused on generating drugs that are selective for one SphK isozyme over the other. Indeed, newer SphK inhibitors show better isozyme selectivity and potency, and have contributed to a better understanding of the role of each SphK isozyme in health and disease.

Genetic mouse models indicate that SphK1 appears to be a more promising target for the development of future chemotherapeutic drugs compared to SphK2. One group knocked out both *SphK1* alleles to generate a mouse that showed a 50 % reduction of plasma S1P [155]. Another group knocked out both *SphK2* alleles to create a mouse that showed only a 25 % reduction of plasma S1P [156]. Taken together, these two genetically-modified mouse lines reveal apparently stronger overcompensation by SphK1 when SphK2 is genetically deleted from the germline. Pharmacological inhibition of SphK2 might therefore result in enhanced SphK1 activity, which could support primary tumor growth as well as metastasis. However, a recent study indicated that the increased levels of S1P in *SPHK2*$^{-/-}$cells could not completely rescue the enhanced proliferative and invasive phenotype [157]. Emerging evidence has started to reveal the various physiological functions of SphK2. Of particular importance is the role that SphK2 apparently plays in regulating gene transcription. SphK2-generated S1P inhibits specific histone deacetylases (HDACs) in the nucleus of MCF-7 human breast cancer cells, thereby elevating *p21* and *CFOS* gene transcription, which shows that SphK2 carries out functions in the nucleus [158] that could impact cancer progression. The discovery of additional SphK2-mediated functions will either validate or invalidate SphK2 as a chemotherapeutic target.

The first SphK inhibitors described in the literature are L-threo-dihydrosphingosine (DHS), and *N,N,N*-trimethylsphingosine [159], both of which are sphingosine analogs. Both inhibitors exhibit low potency, which precludes their use as chemotherapeutic drugs. These inhibitors also show significant off-target effects. For example, DHS potently inhibits protein kinase C α (PKC α) [160–162]. Finally, these two compounds exhibit very little selectivity for one SphK isozyme over the other. Many newer SphK inhibitors show greater potency and selectivity.

Research carried out in the early 2000s led to the discovery and development of SphK inhibitors that were more potent than the first generation of SphK inhibitors, as well as more selective for SphK over other kinases such as PKCα and PI3K [141]. Of the inhibitors described in this study, 2-(p-hydroxyanilino)-4-(p-chlorophenyl) thiazole (SKI II) was the most selective toward SphK1. In vitro activity assays carried out with human SphK1 indicated that SKI II had an IC_{50} of 500 nM ± 300 nM [141]. In the human breast cancer line MDA-MB-231, treatment with 66 μM SKI II resulted in a 75 % reduction in cellular S1P levels [141]. SKI II induces apoptosis in a range of different cancer cell types in vitro, and shows low toxicity in mice treated with doses of SKI II up to 75 mg/kg [141]. Although efficacious at inducing apoptosis in a range of cancer cell lines in vitro and selective for SphK over PKCα and PI3K, SKI II is not highly selective for one SphK isozyme over the other. The use of SKI II as a chemotherapeutic drug is limited by its moderate potency. Despite these drawbacks, studies done with SKI II have revealed that SKI II has a unique mechanism of action in addition to directly inhibiting SphK. In 2010, a group reported that treatment of cells of mesenchymal, endothelial, and epithelial origin with SKI II in vitro triggered lysosomal degradation of SphK1 [163]. Other SphK inhibitors, including FTY720 and (*S*)-FTY720 vinylphosphonate [164], have also recently been found to trigger degradation of SphK1 in cells treated in vitro. It is noteworthy that some of the most efficacious SphK inhibitors trigger degradation of SphK1. Future drug development efforts in the field might therefore focus on compounds that show additional mechanisms of action beyond direct inhibition of SphK activity.

Newer sphingoid analogs, such as FTY720 (fingolimod™), have novel mechanisms of action. FTY720 is an FDA-approved immunomodulatory drug for the treatment of relapsing-remitting multiple sclerosis. Upon entering the cell, FTY720 is phosphorylated by SphK2 to generate FTY720-P, which is an agonist at $S1P_{1,3,4,5}$ [165]. FTY720-P is considered a functional antagonist, as it induces internalization and degradation of the $S1P_1$ receptor [166]. FTY720 has been shown to inhibit SphK1 activity [164], as well trigger proteasomal degradation of SphK1 [167]. FTY720 has also been shown to sensitize prostate cancer cells to radiotherapy through inhibition of SphK1 [168]. A recently described derivative of FTY720, known as (S)-FTY720 vinylphosphonate, also inhibits SphK1 catalytic activity and triggers proteasomal degradation of SphK1 [164]. Another drug described recently, a close derivative of FTY720 called (*R*)-FTY720 methyl ether, was designed to selectively target SphK2 [169]. One of the hydroxy groups on FTY720 was substituted with a methoxy group to prevent SphK2-induced phosphorylation of the compound. (*R*)-FTY720 methyl ether was shown to inhibit DNA synthesis in vitro in MCF-7 human breast cancer cells [169]. These newer compounds have the potential to provide a more comprehensive picture of SphK1 and SphK2 in health and disease.

The most potent SphK1-selective inhibitor described to date is PF-543 [170]. PF-543 competes with sphingosine, and was found to inhibit SphK1 with a K_i of 3.6 nM [170]. PF-543 shows 100-fold selectivity for SphK1 over SphK2 [170].

In the human head and neck carcinoma cell line 1483, PF-543 treatment in vitro dramatically reduced S1P levels while raising sphingosine levels. However, reduction of S1P levels with PF-543 in these cells had no significant impact on cell proliferation and survival [170]. The authors of this study reported an increase in ceramides C18:0 and C24:0 16 h after treatment with 3 and 10 μM PF-543. A more comprehensive analysis of ceramide levels could explain why 1483 human head and neck carcinoma cells did not undergo apoptosis upon treatment with PF-543, as C24:0 ceramide has been shown to promote survival [171]. Future studies should look at changes in the levels of pro- and anti-apoptotic ceramides to determine whether changes in pro-apoptotic ceramides in relation to changes in anti-apoptotic ceramides play a significant role in determining cell fate. Taken in the context of other studies that show the important role S1P plays in tumor growth and survival, it stands to reason that SphK1 could have additional, unknown activities besides catalyzing the formation of S1P that contribute to cell proliferation and survival. Finally, PF-543 is not known to induce degradation of SphK1, and this could be a major reason why cells treated with PF-543 do not undergo apoptosis, despite a significant reduction in intracellular S1P levels.

Recent research has highlighted the anti-tumor activity of the SphK2-selective inhibitor, 3-(4-chlorophenyl)-adamantane-1-carboxylic acid (pyridine-4-ylmethyl) amide, also known as ABC294640. This compound is the first SphK2-inhibitor to be described in the literature [172]. Tested in vitro in A-498 human kidney carcinoma cells, PC-3 prostate adenocarcinoma cells, and MDA-MB-231 breast adenocarcinoma cells, this compound promotes autophagy and subsequent nonapoptotic cell death.

An additional strategy for reducing systemic S1P includes administration of monoclonal antibodies that bind and neutralize S1P, thereby decreasing extracellular S1P [173].

4 Sphingolipid Analogs and Combinatorial Therapy for Cancer

Delivering exogenous ceramides to cancer cells is another strategy to perturb sphingolipid levels for inducing apoptosis. However, this strategy is greatly limited by the hydrophobicity and insolubility of ceramide molecules, which in turn restricts delivery in cell culture systems and in vivo. Soluble ceramide analogs have been developed to circumvent this problem. Short chain ceramides like C2-, C6- and C8-ceramides are cytotoxic in multiple cancer models [10]. Chemical modifications of short-chain ceramides improve their solubility, permeability and pharmacokinetics. Certain chemical modifications that have been tested in cancer models include uracil-linked ceramides [174], serinamides [175], serinols [176, 177] and 4,6-diene-ceramides [178]. In addition, cationic water-soluble pyridinium-ceramides have been developed which preferentially accumulate in mitochondria and induce cell death by mitochondrial permeabilization and Bax translocation [179].

These are effective in inducing cell death in head and neck squamous cell cancer [180] and breast cancer cell lines [181]. These analogs also synergize with gemcitabine to cause cell cycle arrest in G0/G1 phase, retard growth and inhibit telomerase activity in human head and neck squamous cell carcinomas in vitro and in vivo [180, 182]. Other structural analogs of ceramide that have shown selective cytotoxicity in drug-resistant human breast cancer cells compared to normal breast epithelial cells include 5R-OH-3E-C_8 ceramide, benzene-C_4-ceramide and adamantyl-ceramide [183]. Additional novel ceramide analogs including AD2646, AD2672, AD2665, AD2646 and AD2687 have cytotoxic effects on leukemic cells [184].

Ceramide-based therapies face challenges like high insolubility and difficulties to design formulations. Nanoscale-based formulations have thus been developed and investigated to deliver these therapies. Nanoemulsions, nanoliposomes, calcium phosphosilicate nanoparticles and biodegradable linear-dendritic nanoparticles are used for delivering ceramide-based therapeutics [185]. Ceramide delivered via nanoscale formulations has been shown to induce cell death selectively in cancer cells while sparing normal cells [185].

Novel oil-in-water nanoemulsions have been evaluated as delivery vehicles for potential combination therapies in vitro. EGFR-targeted nanoemulsions containing myrisplatin and C6-ceramide show synergistic in vitro cytotoxicity in ovarian cancer cells and also possess potential diagnostic capabilities [186]. Similarly, coadministration of paclitaxel and ceramide in nanoemulsion formulations induces enhanced cytotoxicity and apoptotic activity in human glioblastoma cells in vitro [187]. Sustained release of C6-ceramide from thermoresponsive and biodegradable linear-dendritic nanoparticles induces apoptosis in breast adenocarcinoma cells in vitro with hyperthermia, thus presenting a promising formulation to deliver bioactive sphingolipids for treatment of solid tumors in conjunction with hyperthermia [188]. Biodegradable polymeric nanoparticles have also been evaluated for modulation of drug resistance in cancer cells in vitro and in vivo. It has been shown that paclitaxel and tamoxifen administered in biodegradable poly(ethylene oxide)—modified poly(epsilon-caprolactone) (PEO-PCL) nanoparticles possess significant antitumor efficacy in ovarian carcinoma in vitro and in vivo. Tamoxifen in the formulation reverses drug resistance by inhibiting P-gp and GCS, thus elevating intracellular ceramide levels in cancer cells [189]. Paclitaxel and C6-ceramide PEO-PCL nanoparticles have also been used to chemosensitize resistant human ovarian cancer cell lines to induce apoptotic cell death [190] and suppress growth in xenograft tumor models [191]. Additionally, we have demonstrated the cytotoxic effects of C10-ceramide-loaded calcium phosphate nanocomposite particles in drug-sensitive and drug-resistant breast cancer and melanoma cells in vitro [192].

Encapsulation of chemotherapeutic drugs into nanoliposomes has been a largely successful delivery formulation in cancer models in vitro and in vivo. Our lab has extensively studied nanoliposomes as a lucrative delivery formulation for ceramides in several cancer models. We have shown that nanoliposomes loaded with short-chain ceramides suppress tumor growth in models of breast cancer [193, 194], J774 sarcoma [195], melanoma [196], hepatocellular carcinoma [197], large granular lymphocytic (LGL) leukemia [198], chronic lymphocytic leukemia (CLL) [21], natural killer cell leukemia [199] and pancreatic cancer [200]. Mechanistically, the targets of nanoliposomal C6-ceramide include survivin in LGL leukemia [198], GAPDH in CLL [21] and AKT/PKB and Erk in breast cancer [194], hepatocellular carcinoma [197], pancreatic cancer [200] and melanoma models [196]. Our studies and extensive in vivo toxicology studies by the Nanotechnology Characterization Laboratory (http://ncl.cancer.gov/working_technical_reports.asp) have confirmed that C6-ceramide nanoliposomes have minimum adverse toxicity and elicit apoptosis selectively in cancer cells [201]. We have established that selectivity of C6-ceramide nanoliposomes for cancer cells can be attributed to the inhibitory action of ceramide on the Warburg effect prevalent in cancer cells [21].

Our lab has also evaluated C6-ceramide nanoliposomes as a platform for combinatorial therapy with other neoplastic agents. We have shown that C6-ceramide nanoliposomes synergize with encapsulated sorafenib to reduce tumor development in melanoma and breast cancer cells [196], synergize with gemcitabine or liposomal PDMP to exhibit antitumor effects on pancreatic tumor xenografts [200] and synergize with PPMP to induce apoptosis in natural killer cell leukemia [199]. Recently, in collaboration with the Cabot group we showed that nanoliposomes loaded with C6-ceramide and tamoxifen served as a promising regimen for refractory breast cancers like the triple-negative breast cancer [202]. Tamoxifen amplifies C6-ceramide-induced cytotoxicity in the triple-negative breast cancer cells by multiple effects like cell cycle arrest, lysosomal membrane permeability and inhibition of acid ceramidase [202]. C6-ceramide nanoliposomes also exhibit synergy with the autophagy inhibitor vinblastine to induce apoptotic cell death in vitro and in vivo in hepatocarcinoma and colorectal cancer models, potentially mediated by an autophagy mechanism [203]. In addition to chemotherapeutic drugs, multidrug resistance modulators are also favorable adjuvants for C6-ceramide nanoliposomes. This strategy is justified by studies reporting that resistance to C6-ceramide cytotoxicity is a result of expression of P-gp in some cancer cells [204]. P-gp inhibitors also alter sphingolipid levels in cancer cells. For instance, the multidrug resistance modulator SDZ PSC 833 elevates intracellular ceramide levels by inducing the de novo pathway [205]. P-gp antagonists like tamoxifen, verapamil, and cyclosporine A can also be used in conjunction with cytotoxic drugs like doxorubicin to decrease GlcCer, subsequently increasing ceramide levels and sensitizing cells to cytotoxic drugs [25, 206]. In collaboration with the Cabot group, we have shown that C6-ceramide nanoliposomes-mediated cytotoxicity in cancer cell lines can be augmented by P-gp antagonists like tamoxifen, verapamil and VX-710 [207, 208]. C6-ceramide and tamoxifen induced apoptotic cell death in colorectal cancer cells characterized by PARP cleavage, mitochondrial membrane permeabilization, caspase-dependent apoptosis and G1/G2 cell cycle arrest. Moreover, the combinatorial treatment exhibited synergy and induced upregulation of tumor suppressor p53 [207].

Co-administration of paclitaxel and C6-ceramide exhibit synergy to induce cytotoxicity in pancreatic cancer cells via transient activation of EGFR and ERK pathway [209] and ovarian cancer cells [210]. An interesting study revealed that in the absence of paclitaxel, exogenous C6-ceramide enters the cell through a predetermined initiation site of mitosis, or diffuses into cells through water channels and caveolae-mediated endocytosis [210]. However, the combination induces synergistic cytotoxicity in cancer cells as paclitaxel disrupts cytoskeletal proteins, thus enabling an even distribution of C6-ceramide in the cytoplasm of the cells [210]. Other reports have demonstrated that C6-ceramide also synergizes with histone deacetylase inhibitors like trichostatin A to display anticancer effects in in vivo mice xenograft pancreatic and ovarian cancer models [211]. The authors have delineated the mechanism of this synergy and have demonstrated PP1-mediated inactivation of Akt/mTOR and increased α-tubulin acetylation as events causing cancer cell death. Furthermore, the combination resulted in a very pronounced elevation in intracellular ceramide levels and induction of cell death in cancer cells [211]. C6-ceramide also synergizes in inducing apoptotic cell death in leukemia cells with other neoplastic agents like the cationic peptide, bovine lactoferricin [212].

In addition to the pro-apoptotic effects of short chain ceramides, other sphingolipids have been investigated for their cytotoxicity to cancer cells. Safingol, a synthetic L-*threo* isomer of dihydrosphingosine, is the first SphK inhibitor to enter clinical trials as an anticancer agent [213]. Safingol prominently potentiates the antitumor actions of chemotherapeutic agents such as doxorubicin, cisplatin, and mitomycin C in vivo [214]. Phase I clinical trials for refractory adrenocortical cancer evaluating a combination of safingol and cisplatin delivered as an emulsion showed promising results [213]. Furthermore, a novel liposomal formulation of safingol demonstrated high in vivo antitumor efficacy in the treatment of acute myeloid leukemia while limiting adverse effects like hemolysis [215]. An interesting study reported the anticancer activity of another novel sphingolipid analogue, (2S, 3S, 5S)-2-amino-3,5-dihydroxyoctadecane (Enigmol) in colon and prostate cancer [216]. Enigmol undergoes limited metabolism since it is not phosphorylated by sphingosine kinases and is not rapidly N-acylated [216]. It induces cell death in colon and prostate cancer cells and suppresses tumor growth in mouse models of colon and prostate cancer [216]. It has been shown to normalize the atypical accumulation of β-catenin in the nucleus and cytoplasm of colon cancer cells [216]. Thus, it represents a potential orally bioavailable sphingoid base analogue effective against multiple types of cancer.

Another approach to increase intracellular ceramide levels is by increasing sphingomyelin which can then be converted to ceramide by the action of activated SMases. Delivering exogenous sphingomyelin ensures an expansion in the pool of sphingomyelin that is converted to ceramide for signaling apoptosis [217]. Administering sphingomyelin micelles to colonic, pancreatic, melanoma and lymphoma cancer cell lines enhances their chemosensitivity and also potentiates 5-fluorouracil chemotherapy in human colonic xenograft-bearing nude mice [218]. Furthermore, treatment with sphingomyelin micelles synergizes with gemcitabine to induce ceramide-mediated apoptosis in pancreatic cancer [46]. Gemcitabine alone activates aSMase but does not induce apoptosis. However, when augmented with sphingomyelin, the treatment induces ceramide production, mitochondrial depolarization and apoptosis [46]. Sphingomyelin-mediated increase in chemotherapeutic efficacy has also been demonstrated in human colonic tumor xenografts, thus providing additional validation for this approach of therapeutic intervention [219].

In addition to its cytotoxicity, sphingomyelin also enhances the uptake of various lipophilic and amphiphilic drugs in cancer cells. For instance, *N*-hexanoyl sphingomyelin increases the chemotherapeutic efficacy of doxorubicin in cancer cells in vitro by modulating plasma membrane permeability leading to increased drug accumulation in cells [220]. Furthermore, vincristine encapsulated in sphingomyelin liposomes or "sphingosomes" demonstrates significantly greater antitumor efficacy in vitro and in vivo as compared to vincristine alone [221]. Sphingosomal vincristine was used in a Phase II clinical trial for recurrent or refractory adult acute lymphocytic leukemia [221]. Similarly, short-chain GlcCer encapsulated with doxorubicin in lipid nanovesicles enhances the antitumor efficacy of doxorubicin in orthotopic melanoma mouse models by increasing membrane permeability for the drug.

5 Conclusions

The last two decades of sphingolipid research in cancer has revealed the importance of being able to modulate sphingolipid levels in cancer patients to sensitize tumors to current cancer treatments, including chemotherapy and radiation therapy. Extensive research in this field has identified multiple potential druggable targets in the sphingolipid pathway to enhance the cytotoxicity of current chemotherapeutics and to overcome the prime challenge of drug resistance. The exponential growth in the number of studies focused on understanding the intrinsic biology of sphingolipids and exploring sphingolipid-based therapeutics has motivated scientists to develop new analytical techniques to measure steady-state lipid levels. Mass spectrometry-based lipidomics and proteomics have gained wide-spread attention to answer the various complexities of sphingolipid biology like sphingolipid-mediated regulation of the proteome, lipid-lipid interactions, lipid-protein interactions and post-translational modifications of sphingolipid-modifying enzymes. Discovery proteomics has been used to determine global changes in the proteome after treatment with sphingolipid-based therapeutics to study sphingolipid-mediated effects on the proteome. In addition, proteomic analysis has been employed at the level of specific organelles like lysosomes to study changes in the protein profile. In addition to the usage of discovery proteomics for hypothesis-generating projects, scientists are also using targeted proteomics in addition to antibody-based techniques like Western blotting to study aspects of sphingolipid-modifying enzymes like protein level changes, post-translational modifications and changes in subcellular localization. Affinity-based peptide enrichment using antibodies, lectin or biotin-switch is commonly used to enrich sphingolipid-modifying enzymes to study post-translational modifications like phosphorylation, glycosylation, acetylation, ubiquitination and palmitoylation. These studies will give a deeper insight into enzyme activity, turnover and localization. Mass-spectrometric analysis is also used to study protein-protein and protein-lipid interactions in the sphingolipid system to delineate their cellular functions. Additionally, chemical proteomics is used to study the protein-binding partners of chemical inhibitors against sphingolipid enzymes and thus evaluate their specificity and selectivity to the targeted enzyme and to off-target molecules. These studies will certainly help to improve the efficacy and minimizing side effects of these enzyme inhibitors.

Sphingolipid-based therapeutics aim at tipping the sphingolipid balance in cancer cells towards anti-proliferative and apoptotic signaling. As discussed in the chapter, this can be attained by treatment with sphingolipid analogs or modulators of sphingolipid metabolism alone or in combination with current therapeutics. Successful formulation of novel drug-delivery methodologies, coupled with the identification, synthesis and validation of sphingolipid analogs and targets in the sphingolipid metabolism pathway presents a promising therapeutic strategy for cancer treatment.

Acknowledgement This work was supported by National Institutes of Health Grants CA171983 and CA167535 and the State of Pennsylvania CURE grant program. We thank Dr. Andrei V. Khokhlatchev and Dr. Todd Fox at University of Virginia for their discussions and constructive comments.

References

1. Obeid LM, Linardic CM, Karolak LA, Hannun YA (1993) Programmed cell death induced by ceramide. Science 259(5102):1769–1771
2. Jarvis WD, Kolesnick RN, Fornari FA, Traylor RS, Gewirtz DA, Grant S (1994) Induction of apoptotic DNA damage and cell death by activation of the sphingomyelin pathway. Proc Natl Acad Sci U S A 91(1):73–77
3. Haimovitz-Friedman A, Kan CC, Ehleiter D, Persaud RS, McLoughlin M, Fuks Z, Kolesnick RN (1994) Ionizing radiation acts on cellular membranes to generate ceramide and initiate apoptosis. J Exp Med 180(2):525–535
4. Cifone MG, De Maria R, Roncaioli P, Rippo MR, Azuma M, Lanier LL, Santoni A, Testi R (1994) Apoptotic signaling through CD95 (Fas/Apo-1) activates an acidic sphingomyelinase. J Exp Med 180(4):1547–1552
5. Bose R, Verheij M, Haimovitz-Friedman A, Scotto K, Fuks Z, Kolesnick R (1995) Ceramide synthase mediates daunorubicin-induced apoptosis: an alternative mechanism for generating death signals. Cell 82(3):405–414
6. Ganesan V, Perera MN, Colombini D, Datskovskiy D, Chadha K, Colombini M (2010) Ceramide and activated Bax act synergistically to permeabilize the mitochondrial outer membrane. Apoptosis 15(5):553–562. doi:10.1007/s10495-009-0449-0
7. Morad SA, Cabot MC (2013) Ceramide-orchestrated signalling in cancer cells. Nat Rev Cancer 13(1):51–65. doi:10.1038/nrc3398
8. Grosch S, Schiffmann S, Geisslinger G (2012) Chain length-specific properties of ceramides. Prog Lipid Res 51(1):50–62. doi:10.1016/j.plipres.2011.11.001
9. Zheng W, Kollmeyer J, Symolon H, Momin A, Munter E, Wang E, Kelly S, Allegood JC, Liu Y, Peng Q, Ramaraju H, Sullards MC, Cabot M, Merrill AH Jr (2006) Ceramides and other bioactive sphingolipid backbones in health and disease: lipidomic analysis, metabolism and roles in membrane structure, dynamics, signaling and autophagy. Biochim Biophys Acta 1758(12):1864–1884. doi:10.1016/j.bbamem.2006.08.009
10. Ogretmen B, Hannun YA (2004) Biologically active sphingolipids in cancer pathogenesis and treatment. Nat Rev Cancer 4(8):604–616. doi:10.1038/nrc1411

11. Levy M, Futerman AH (2010) Mammalian ceramide synthases. IUBMB Life 62(5):347–356. doi:10.1002/iub.319
12. Brunton LL, Lazo JS, Parker KL (eds) (2006) Goodman and Gilman's the pharmacological basis of therapeutics, 11th edn. McGraw-Hill, New York
13. Bassoy EY, Baran Y (2012) Bioactive sphingolipids in docetaxel-induced apoptosis in human prostate cancer cells. Biomed Pharmacother 66(2):103–110. doi:10.1016/j.biopha.2011.10.003
14. Kolesnick R, Altieri D, Fuks Z (2007) A CERTain role for ceramide in taxane-induced cell death. Cancer Cell 11(6):473–475. doi:10.1016/j.ccr.2007.05.003
15. Charles AG, Han TY, Liu YY, Hansen N, Giuliano AE, Cabot MC (2001) Taxol-induced ceramide generation and apoptosis in human breast cancer cells. Cancer Chemother Pharmacol 47(5):444–450
16. Di Bartolomeo S, Di Sano F, Piacentini M, Spinedi A (2000) Apoptosis induced by doxorubicin in neurotumor cells is divorced from drug effects on ceramide accumulation and may involve cell cycle-dependent caspase activation. J Neurochem 75(2):532–539
17. Senkal CE, Ponnusamy S, Rossi MJ, Bialewski J, Sinha D, Jiang JC, Jazwinski SM, Hannun YA, Ogretmen B (2007) Role of human longevity assurance gene 1 and C18-ceramide in chemotherapy-induced cell death in human head and neck squamous cell carcinomas. Mol Cancer Ther 6(2):712–722. doi:10.1158/1535-7163.MCT-06-0558
18. Rath G, Schneider C, Langlois B, Sartelet H, Morjani H, Btaouri HE, Dedieu S, Martiny L (2009) De novo ceramide synthesis is responsible for the anti-tumor properties of camptothecin and doxorubicin in follicular thyroid carcinoma. Int J Biochem Cell Biol 41(5):1165–1172. doi:10.1016/j.biocel.2008.10.021
19. Park MA, Mitchell C, Zhang G, Yacoub A, Allegood J, Haussinger D, Reinehr R, Larner A, Spiegel S, Fisher PB, Voelkel-Johnson C, Ogretmen B, Grant S, Dent P (2010) Vorinostat and sorafenib increase CD95 activation in gastrointestinal tumor cells through a Ca(2+)-de novo ceramide-PP2A-reactive oxygen species-dependent signaling pathway. Cancer Res 70(15):6313–6324. doi:10.1158/0008-5472.CAN-10-0999
20. Biswal SS, Datta K, Acquaah-Mensah GK, Kehrer JP (2000) Changes in ceramide and sphingomyelin following fludarabine treatment of human chronic B-cell leukemia cells. Toxicology 154(1–3):45–53
21. Ryland LK, Doshi UA, Shanmugavelandy SS, Fox TE, Aliaga C, Broeg K, Baab KT, Young M, Khan O, Haakenson JK, Jarbadan NR, Liao J, Wang HG, Feith DJ, Loughran TP Jr, Liu X, Kester M (2013) C6-ceramide nanoliposomes target the Warburg effect in chronic lymphocytic leukemia. PLoS One 8(12), e84648. doi:10.1371/journal.pone.0084648
22. Perry DK, Carton J, Shah AK, Meredith F, Uhlinger DJ, Hannun YA (2000) Serine palmitoyltransferase regulates de novo ceramide generation during etoposide-induced apoptosis. J Biol Chem 275(12):9078–9084
23. Schiffmann S, Ziebell S, Sandner J, Birod K, Deckmann K, Hartmann D, Rode S, Schmidt H, Angioni C, Geisslinger G, Grosch S (2010) Activation of ceramide synthase 6 by celecoxib leads to a selective induction of C16:0-ceramide. Biochem Pharmacol 80(11):1632–1640. doi:10.1016/j.bcp.2010.08.012
24. Cabot MC, Giuliano AE, Han TY, Liu YY (1999) SDZ PSC 833, the cyclosporine A analogue and multidrug resistance modulator, activates ceramide synthesis and increases vinblastine sensitivity in drug-sensitive and drug-resistant cancer cells. Cancer Res 59(4):880–885
25. Lucci A, Han TY, Liu YY, Giuliano AE, Cabot MC (1999) Multidrug resistance modulators and doxorubicin synergize to elevate ceramide levels and elicit apoptosis in drug-resistant cancer cells. Cancer 86(2):300–311
26. Wang H, Giuliano AE, Cabot MC (2002) Enhanced de novo ceramide generation through activation of serine palmitoyltransferase by the P-glycoprotein antagonist SDZ PSC 833 in breast cancer cells. Mol Cancer Ther 1(9):719–726
27. Kawatani M, Simizu S, Osada H, Takada M, Arber N, Imoto M (2000) Involvement of protein kinase C-regulated ceramide generation in inostamycin-induced apoptosis. Exp Cell Res 259(2):389–397. doi:10.1006/excr.2000.4986

28. Sanchez AM, Malagarie-Cazenave S, Olea N, Vara D, Cuevas C, Diaz-Laviada I (2008) Spisulosine (ES-285) induces prostate tumor PC-3 and LNCaP cell death by de novo synthesis of ceramide and PKCzeta activation. Eur J Pharmacol 584(2–3):237–245. doi:10.1016/j.ejphar.2008.02.011
29. Ramer R, Weinzierl U, Schwind B, Brune K, Hinz B (2003) Ceramide is involved in r(+)-methanandamide-induced cyclooxygenase-2 expression in human neuroglioma cells. Mol Pharmacol 64(5):1189–1198. doi:10.1124/mol.64.5.1189
30. Gustafsson K, Sander B, Bielawski J, Hannun YA, Flygare J (2009) Potentiation of cannabinoid-induced cytotoxicity in mantle cell lymphoma through modulation of ceramide metabolism. Mol Cancer Res 7(7):1086–1098. doi:10.1158/1541-7786.MCR-08-0361
31. Ponzoni M, Bocca P, Chiesa V, Decensi A, Pistoia V, Raffaghello L, Rozzo C, Montaldo PG (1995) Differential effects of N-(4-hydroxyphenyl)retinamide and retinoic acid on neuroblastoma cells: apoptosis versus differentiation. Cancer Res 55(4):853–861
32. Reynolds CP, Maurer BJ, Kolesnick RN (2004) Ceramide synthesis and metabolism as a target for cancer therapy. Cancer Lett 206(2):169–180. doi:10.1016/j.canlet.2003.08.034
33. Jiang L, Pan X, Chen Y, Wang K, Du Y, Zhang J (2011) Preferential involvement of both ROS and ceramide in fenretinide-induced apoptosis of HL60 rather than NB4 and U937 cells. Biochem Biophys Res Commun 405(2):314–318. doi:10.1016/j.bbrc.2011.01.036
34. Wang H, Maurer BJ, Reynolds CP, Cabot MC (2001) N-(4-Hydroxyphenyl)retinamide elevates ceramide in neuroblastoma cell lines by coordinate activation of serine palmitoyltransferase and ceramide synthase. Cancer Res 61(13):5102–5105
35. Maurer BJ, Melton L, Billups C, Cabot MC, Reynolds CP (2000) Synergistic cytotoxicity in solid tumor cell lines between N-(4-hydroxyphenyl)retinamide and modulators of ceramide metabolism. J Natl Cancer Inst 92(23):1897–1909
36. Rahmaniyan M, Curley RW Jr, Obeid LM, Hannun YA, Kraveka JM (2011) Identification of dihydroceramide desaturase as a direct in vitro target for fenretinide. J Biol Chem 286(28):24754–24764. doi:10.1074/jbc.M111.250779
37. Kraveka JM, Li L, Szulc ZM, Bielawski J, Ogretmen B, Hannun YA, Obeid LM, Bielawska A (2007) Involvement of dihydroceramide desaturase in cell cycle progression in human neuroblastoma cells. J Biol Chem 282(23):16718–16728. doi:10.1074/jbc.M700647200
38. Apraiz A, Idkowiak-Baldys JK, Boyano MD, Perez-Yarza G, Hannun YA, Asumendi A (2011) Evaluation of bioactive sphingolipids in 4-HPR-resistant leukemia cells. BMC Cancer 11:477. doi:10.1186/1471-2407-11-477
39. Fabrias G, Munoz-Olaya J, Cingolani F, Signorelli P, Casas J, Gagliostro V, Ghidoni R (2012) Dihydroceramide desaturase and dihydrosphingolipids: debutant players in the sphingolipid arena. Prog Lipid Res 51(2):82–94. doi:10.1016/j.plipres.2011.12.002
40. Jin J, Mullen TD, Hou Q, Bielawski J, Bielawska A, Zhang X, Obeid LM, Hannun YA, Hsu YT (2009) AMPK inhibitor Compound C stimulates ceramide production and promotes Bax redistribution and apoptosis in MCF7 breast carcinoma cells. J Lipid Res 50(12):2389–2397. doi:10.1194/jlr.M900119-JLR200
41. Canals D, Hannun YA (2013) Novel chemotherapeutic drugs in sphingolipid cancer research. Handb Exp Pharmacol 215:211–238. doi:10.1007/978-3-7091-1368-4_12
42. Zhang Y, Cheng Y, Hansen GH, Niels-Christiansen LL, Koentgen F, Ohlsson L, Nilsson A, Duan RD (2011) Crucial role of alkaline sphingomyelinase in sphingomyelin digestion: a study on enzyme knockout mice. J Lipid Res 52(4):771–781. doi:10.1194/jlr.M012880
43. Hertervig E, Nilsson A, Nyberg L, Duan RD (1997) Alkaline sphingomyelinase activity is decreased in human colorectal carcinoma. Cancer 79(3):448–453
44. Jaffrezou JP, Levade T, Bettaieb A, Andrieu N, Bezombes C, Maestre N, Vermeersch S, Rousse A, Laurent G (1996) Daunorubicin-induced apoptosis: triggering of ceramide generation through sphingomyelin hydrolysis. EMBO J 15(10):2417–2424
45. Ito H, Murakami M, Furuhata A, Gao S, Yoshida K, Sobue S, Hagiwara K, Takagi A, Kojima T, Suzuki M, Banno Y, Tanaka K, Tamiya-Koizumi K, Kyogashima M, Nozawa Y, Murate T (2009) Transcriptional regulation of neutral sphingomyelinase 2 gene expression of a human

breast cancer cell line, MCF-7, induced by the anti-cancer drug, daunorubicin. Biochim Biophys Acta 1789(11–12):681–690. doi:10.1016/j.bbagrm.2009.08.006
46. Modrak DE, Cardillo TM, Newsome GA, Goldenberg DM, Gold DV (2004) Synergistic interaction between sphingomyelin and gemcitabine potentiates ceramide-mediated apoptosis in pancreatic cancer. Cancer Res 64(22):8405–8410. doi:10.1158/0008-5472.CAN-04-2988
47. Dumitru CA, Sandalcioglu IE, Wagner M, Weller M, Gulbins E (2009) Lysosomal ceramide mediates gemcitabine-induced death of glioma cells. J Mol Med (Berl) 87(11):1123–1132. doi:10.1007/s00109-009-0514-8
48. Strum JC, Small GW, Pauig SB, Daniel LW (1994) 1-beta-D-Arabinofuranosylcytosine stimulates ceramide and diglyceride formation in HL-60 cells. J Biol Chem 269(22): 15493–15497
49. Noda S, Yoshimura S, Sawada M, Naganawa T, Iwama T, Nakashima S, Sakai N (2001) Role of ceramide during cisplatin-induced apoptosis in C6 glioma cells. J Neurooncol 52(1): 11–21
50. Zeidan YH, Jenkins RW, Hannun YA (2008) Remodeling of cellular cytoskeleton by the acid sphingomyelinase/ceramide pathway. J Cell Biol 181(2):335–350. doi:10.1083/jcb.200705060
51. Bezombes C, Grazide S, Garret C, Fabre C, Quillet-Mary A, Muller S, Jaffrezou JP, Laurent G (2004) Rituximab antiproliferative effect in B-lymphoma cells is associated with acid-sphingomyelinase activation in raft microdomains. Blood 104(4):1166–1173. doi:10.1182/blood-2004-01-0277
52. Yun SH, Park ES, Shin SW, Na YW, Han JY, Jeong JS, Shastina VV, Stonik VA, Park JI, Kwak JY (2012) Stichoposide C induces apoptosis through the generation of ceramide in leukemia and colorectal cancer cells and shows in vivo antitumor activity. Clin Cancer Res 18(21):5934–5948. doi:10.1158/1078-0432.CCR-12-0655
53. Rubio S, Quintana J, Eiroa JL, Triana J, Estevez F (2010) Betuletol 3-methyl ether induces G(2)-M phase arrest and activates the sphingomyelin and MAPK pathways in human leukemia cells. Mol Carcinog 49(1):32–43. doi:10.1002/mc.20574
54. Mondal S, Mandal C, Sangwan R, Chandra S (2010) Withanolide D induces apoptosis in leukemia by targeting the activation of neutral sphingomyelinase-ceramide cascade mediated by synergistic activation of c-Jun N-terminal kinase and p38 mitogen-activated protein kinase. Mol Cancer 9:239. doi:10.1186/1476-4598-9-239
55. Fox TE, Finnegan CM, Blumenthal R, Kester M (2006) The clinical potential of sphingolipid-based therapeutics. Cell Mol Life Sci 63(9):1017–1023. doi:10.1007/s00018-005-5543-z
56. Luberto C, Hannun YA (1998) Sphingomyelin synthase, a potential regulator of intracellular levels of ceramide and diacylglycerol during SV40 transformation. Does sphingomyelin synthase account for the putative phosphatidylcholine-specific phospholipase C? J Biol Chem 273(23):14550–14559
57. Itoh M, Kitano T, Watanabe M, Kondo T, Yabu T, Taguchi Y, Iwai K, Tashima M, Uchiyama T, Okazaki T (2003) Possible role of ceramide as an indicator of chemoresistance: decrease of the ceramide content via activation of glucosylceramide synthase and sphingomyelin synthase in chemoresistant leukemia. Clin Cancer Res 9(1):415–423
58. Meng A, Luberto C, Meier P, Bai A, Yang X, Hannun YA, Zhou D (2004) Sphingomyelin synthase as a potential target for D609-induced apoptosis in U937 human monocytic leukemia cells. Exp Cell Res 292(2):385–392
59. Salma Y, Lafont E, Therville N, Carpentier S, Bonnafe MJ, Levade T, Genisson Y, Andrieu-Abadie N (2009) The natural marine anhydrophytosphingosine, Jaspine B, induces apoptosis in melanoma cells by interfering with ceramide metabolism. Biochem Pharmacol 78(5):477–485. doi:10.1016/j.bcp.2009.05.002
60. Barcelo-Coblijn G, Martin ML, de Almeida RF, Noguera-Salva MA, Marcilla-Etxenike A, Guardiola-Serrano F, Luth A, Kleuser B, Halver JE, Escriba PV (2011) Sphingomyelin and sphingomyelin synthase (SMS) in the malignant transformation of glioma cells and in

2-hydroxyoleic acid therapy. Proc Natl Acad Sci U S A 108(49):19569–19574. doi:10.1073/pnas.1115484108
61. Teres S, Llado V, Higuera M, Barcelo-Coblijn G, Martin ML, Noguera-Salva MA, Marcilla-Etxenike A, Garcia-Verdugo JM, Soriano-Navarro M, Saus C, Gomez-Pinedo U, Busquets X, Escriba PV (2012) 2-Hydroxyoleate, a nontoxic membrane binding anticancer drug, induces glioma cell differentiation and autophagy. Proc Natl Acad Sci U S A 109(22):8489–8494. doi:10.1073/pnas.1118349109
62. Kartal Yandim M, Apohan E, Baran Y (2013) Therapeutic potential of targeting ceramide/glucosylceramide pathway in cancer. Cancer Chemother Pharmacol 71(1):13–20. doi:10.1007/s00280-012-1984-x
63. Liu YY, Hill RA, Li YT (2013) Ceramide glycosylation catalyzed by glucosylceramide synthase and cancer drug resistance. Adv Cancer Res 117:59–89. doi:10.1016/B978-0-12-394274-6.00003-0
64. Patwardhan GA, Liu YY (2011) Sphingolipids and expression regulation of genes in cancer. Prog Lipid Res 50(1):104–114. doi:10.1016/j.plipres.2010.10.003
65. Wennekes T, van den Berg RJ, Boot RG, van der Marel GA, Overkleeft HS, Aerts JM (2009) Glycosphingolipids—nature, function, and pharmacological modulation. Angew Chem Int Ed Engl 48(47):8848–8869. doi:10.1002/anie.200902620
66. Feng D, Zhao WL, Ye YY, Bai XC, Liu RQ, Chang LF, Zhou Q, Sui SF (2010) Cellular internalization of exosomes occurs through phagocytosis. Traffic 11(5):675–687. doi:10.1111/j.1600-0854.2010.01041.x
67. Makhlouf AM, Fathalla MM, Zakhary MA, Makarem MH (2004) Sulfatides in ovarian tumors: clinicopathological correlates. Int J Gynecol Cancer 14(1):89–93
68. Owczarek TB, Suchanski J, Pula B, Kmiecik AM, Chadalski M, Jethon A, Dziegiel P, Ugorski M (2013) Galactosylceramide affects tumorigenic and metastatic properties of breast cancer cells as an anti-apoptotic molecule. PLoS One 8(12), e84191. doi:10.1371/journal.pone.0084191
69. Dziegiel P, Owczarek T, Plazuk E, Gomulkiewicz A, Majchrzak M, Podhorska-Okolow M, Driouch K, Lidereau R, Ugorski M (2010) Ceramide galactosyltransferase (UGT8) is a molecular marker of breast cancer malignancy and lung metastases. Br J Cancer 103(4):524–531. doi:10.1038/sj.bjc.6605750
70. Hiraiwa N, Fukuda Y, Imura H, Tadano-Aritomi K, Nagai K, Ishizuka I, Kannagi R (1990) Accumulation of highly acidic sulfated glycosphingolipids in human hepatocellular carcinoma defined by a series of monoclonal antibodies. Cancer Res 50(10):2917–2928
71. Honke K, Tsuda M, Hirahara Y, Miyao N, Tsukamoto T, Satoh M, Wada Y (1998) Cancer-associated expression of glycolipid sulfotransferase gene in human renal cell carcinoma cells. Cancer Res 58(17):3800–3805
72. Morichika H, Hamanaka Y, Tai T, Ishizuka I (1996) Sulfatides as a predictive factor of lymph node metastasis in patients with colorectal adenocarcinoma. Cancer 78(1):43–47
73. Takahashi T, Suzuki T (2012) Role of sulfatide in normal and pathological cells and tissues. J Lipid Res 53(8):1437–1450. doi:10.1194/jlr.R026682
74. Sharma DK, Brown JC, Cheng Z, Holicky EL, Marks DL, Pagano RE (2005) The glycosphingolipid, lactosylceramide, regulates beta1-integrin clustering and endocytosis. Cancer Res 65(18):8233–8241. doi:10.1158/0008-5472.CAN-05-0803
75. Gouaze-Andersson V, Cabot MC (2006) Glycosphingolipids and drug resistance. Biochim Biophys Acta 1758(12):2096–2103. doi:10.1016/j.bbamem.2006.08.012
76. Ryland LK, Fox TE, Liu X, Loughran TP, Kester M (2011) Dysregulation of sphingolipid metabolism in cancer. Cancer Biol Ther 11(2):138–149
77. Liang YJ, Ding Y, Levery SB, Lobaton M, Handa K, Hakomori SI (2013) Differential expression profiles of glycosphingolipids in human breast cancer stem cells vs. cancer non-stem cells. Proc Natl Acad Sci U S A 110(13):4968–4973. doi:10.1073/pnas.1302825110
78. Battula VL, Shi Y, Evans KW, Wang RY, Spaeth EL, Jacamo RO, Guerra R, Sahin AA, Marini FC, Hortobagyi G, Mani SA, Andreeff M (2012) Ganglioside GD2 identifies breast

cancer stem cells and promotes tumorigenesis. J Clin Invest 122(6):2066–2078. doi:10.1172/JCI59735
79. Cheever MA, Allison JP, Ferris AS, Finn OJ, Hastings BM, Hecht TT, Mellman I, Prindiville SA, Viner JL, Weiner LM, Matrisian LM (2009) The prioritization of cancer antigens: a national cancer institute pilot project for the acceleration of translational research. Clin Cancer Res 15(17):5323–5337. doi:10.1158/1078-0432.CCR-09-0737
80. Liu YY, Han TY, Giuliano AE, Cabot MC (1999) Expression of glucosylceramide synthase, converting ceramide to glucosylceramide, confers adriamycin resistance in human breast cancer cells. J Biol Chem 274(2):1140–1146
81. Liu YY, Han TY, Yu JY, Bitterman A, Le A, Giuliano AE, Cabot MC (2004) Oligonucleotides blocking glucosylceramide synthase expression selectively reverse drug resistance in cancer cells. J Lipid Res 45(5):933–940. doi:10.1194/jlr.M300486-JLR200
82. Liu YY, Han TY, Giuliano AE, Cabot MC (2001) Ceramide glycosylation potentiates cellular multidrug resistance. FASEB J 15(3):719–730. doi:10.1096/fj.00-0223com
83. Weiss M, Hettmer S, Smith P, Ladisch S (2003) Inhibition of melanoma tumor growth by a novel inhibitor of glucosylceramide synthase. Cancer Res 63(13):3654–3658
84. Gouaze V, Liu YY, Prickett CS, Yu JY, Giuliano AE, Cabot MC (2005) Glucosylceramide synthase blockade down-regulates P-glycoprotein and resensitizes multidrug-resistant breast cancer cells to anticancer drugs. Cancer Res 65(9):3861–3867. doi:10.1158/0008-5472.CAN-04-2329
85. Song M, Zang W, Zhang B, Cao J, Yang G (2012) GCS overexpression is associated with multidrug resistance of human HCT-8 colon cancer cells. J Exp Clin Cancer Res 31:23. doi:10.1186/1756-9966-31-23
86. Xie P, Shen YF, Shi YP, Ge SM, Gu ZH, Wang J, Mu HJ, Zhang B, Qiao WZ, Xie KM (2008) Overexpression of glucosylceramide synthase in associated with multidrug resistance of leukemia cells. Leuk Res 32(3):475–480. doi:10.1016/j.leukres.2007.07.006
87. Baran Y, Bielawski J, Gunduz U, Ogretmen B (2011) Targeting glucosylceramide synthase sensitizes imatinib-resistant chronic myeloid leukemia cells via endogenous ceramide accumulation. J Cancer Res Clin Oncol 137(10):1535–1544. doi:10.1007/s00432-011-1016-y
88. Gouaze V, Yu JY, Bleicher RJ, Han TY, Liu YY, Wang H, Gottesman MM, Bitterman A, Giuliano AE, Cabot MC (2004) Overexpression of glucosylceramide synthase and P-glycoprotein in cancer cells selected for resistance to natural product chemotherapy. Mol Cancer Ther 3(5):633–639
89. Goulding CW, Giuliano AE, Cabot MC (2000) SDZ PSC 833 the drug resistance modulator activates cellular ceramide formation by a pathway independent of P-glycoprotein. Cancer Lett 149(1–2):143–151
90. Sun Y, Zhang T, Gao P, Meng B, Gao Y, Wang X, Zhang J, Wang H, Wu X, Zheng W, Zhou G (2010) Targeting glucosylceramide synthase downregulates expression of the multidrug resistance gene MDR1 and sensitizes breast carcinoma cells to anticancer drugs. Breast Cancer Res Treat 121(3):591–599. doi:10.1007/s10549-009-0513-z
91. Liu YY, Gupta V, Patwardhan GA, Bhinge K, Zhao Y, Bao J, Mehendale H, Cabot MC, Li YT, Jazwinski SM (2010) Glucosylceramide synthase upregulates MDR1 expression in the regulation of cancer drug resistance through cSrc and beta-catenin signaling. Mol Cancer 9:145. doi:10.1186/1476-4598-9-145
92. De Rosa MF, Sillence D, Ackerley C, Lingwood C (2004) Role of multiple drug resistance protein 1 in neutral but not acidic glycosphingolipid biosynthesis. J Biol Chem 279(9):7867–7876. doi:10.1074/jbc.M305645200
93. Veldman RJ, Mita A, Cuvillier O, Garcia V, Klappe K, Medin JA, Campbell JD, Carpentier S, Kok JW, Levade T (2003) The absence of functional glucosylceramide synthase does not sensitize melanoma cells for anticancer drugs. FASEB J 17(9):1144–1146. doi:10.1096/fj.02-1053fje
94. Norris-Cervetto E, Callaghan R, Platt FM, Dwek RA, Butters TD (2004) Inhibition of glucosylceramide synthase does not reverse drug resistance in cancer cells. J Biol Chem 279(39):40412–40418. doi:10.1074/jbc.M404466200

95. Dijkhuis AJ, Klappe K, Jacobs S, Kroesen BJ, Kamps W, Sietsma H, Kok JW (2006) PDMP sensitizes neuroblastoma to paclitaxel by inducing aberrant cell cycle progression leading to hyperploidy. Mol Cancer Ther 5(3):593–601. doi:10.1158/1535-7163.MCT-05-0457
96. Camgoz A, Gencer EB, Ural AU, Avcu F, Baran Y (2011) Roles of ceramide synthase and ceramide clearence genes in nilotinib-induced cell death in chronic myeloid leukemia cells. Leuk Lymphoma 52(8):1574–1584. doi:10.3109/10428194.2011.568653
97. Huang WC, Tsai CC, Chen CL, Chen TY, Chen YP, Lin YS, Lu PJ, Lin CM, Wang SH, Tsao CW, Wang CY, Cheng YL, Hsieh CY, Tseng PC, Lin CF (2011) Glucosylceramide synthase inhibitor PDMP sensitizes chronic myeloid leukemia T315I mutant to Bcr-Abl inhibitor and cooperatively induces glycogen synthase kinase-3-regulated apoptosis. FASEB J 25(10): 3661–3673. doi:10.1096/fj.10-180190 fj
98. Litvak DA, Bilchik AJ, Cabot MC (2003) Modulators of ceramide metabolism sensitize colorectal cancer cells to chemotherapy: a novel treatment strategy. J Gastrointest Surg 7(1):140–148, discussion 148
99. Patwardhan GA, Zhang QJ, Yin D, Gupta V, Bao J, Senkal CE, Ogretmen B, Cabot MC, Shah GV, Sylvester PW, Jazwinski SM, Liu YY (2009) A new mixed-backbone oligonucleotide against glucosylceramide synthase sensitizes multidrug-resistant tumors to apoptosis. PLoS One 4(9), e6938. doi:10.1371/journal.pone.0006938
100. Liu YY, Patwardhan GA, Bhinge K, Gupta V, Gu X, Jazwinski SM (2011) Suppression of glucosylceramide synthase restores p53-dependent apoptosis in mutant p53 cancer cells. Cancer Res 71(6):2276–2285. doi:10.1158/0008-5472.CAN-10-3107
101. Chiu WH, Chen HH, Chang JY, Luo SJ, Li CL, Chen CL, Su WC, Lin CF (2014) Inhibiting glucosylceramide synthase facilitates the radiosensitizing effects of vinorelbine in lung adenocarcinoma cells. Cancer Lett 349(2):144–151. doi:10.1016/j.canlet.2014.04.005
102. Maurer BJ, Metelitsa LS, Seeger RC, Cabot MC, Reynolds CP (1999) Increase of ceramide and induction of mixed apoptosis/necrosis by N-(4-hydroxyphenyl)-retinamide in neuroblastoma cell lines. J Natl Cancer Inst 91(13):1138–1146
103. Lovat PE, Di Sano F, Corazzari M, Fazi B, Donnorso RP, Pearson AD, Hall AG, Redfern CP, Piacentini M (2004) Gangliosides link the acidic sphingomyelinase-mediated induction of ceramide to 12-lipoxygenase-dependent apoptosis of neuroblastoma in response to fenretinide. J Natl Cancer Inst 96(17):1288–1299. doi:10.1093/jnci/djh254
104. Di Sano F, Fazi B, Citro G, Lovat PE, Cesareni G, Piacentini M (2003) Glucosylceramide synthase and its functional interaction with RTN-1C regulate chemotherapeutic-induced apoptosis in neuroepithelioma cells. Cancer Res 63(14):3860–3865
105. Furlong SJ, Ridgway ND, Hoskin DW (2008) Modulation of ceramide metabolism in T-leukemia cell lines potentiates apoptosis induced by the cationic antimicrobial peptide bovine lactoferricin. Int J Oncol 32(3):537–544
106. Cakir Z, Saydam G, Sahin F, Baran Y (2011) The roles of bioactive sphingolipids in resveratrol-induced apoptosis in HL60: acute myeloid leukemia cells. J Cancer Res Clin Oncol 137(2):279–286. doi:10.1007/s00432-010-0884-x
107. Guerrera M, Ladisch S (2003) N-Butyldeoxynojirimycin inhibits murine melanoma cell ganglioside metabolism and delays tumor onset. Cancer Lett 201(1):31–40
108. Gerrard G, Butters TD, Ganeshaguru K, Mehta AB (2009) Glucosylceramide synthase inhibitors sensitise CLL cells to cytotoxic agents without reversing P-gp functional activity. Eur J Pharmacol 609(1–3):34–39. doi:10.1016/j.ejphar.2009.03.018
109. Larsen SD, Wilson MW, Abe A, Shu L, George CH, Kirchhoff P, Showalter HD, Xiang J, Keep RF, Shayman JA (2012) Property-based design of a glucosylceramide synthase inhibitor that reduces glucosylceramide in the brain. J Lipid Res 53(2):282–291. doi:10.1194/jlr.M021261
110. Chai L, McLaren RP, Byrne A, Chuang WL, Huang Y, Dufault MR, Pacheco J, Madhiwalla S, Zhang X, Zhang M, Teicher BA, Carter K, Cheng SH, Leonard JP, Xiang Y, Vasconcelles M, Goldberg MA, Copeland DP, Klinger KW, Lillie J, Madden SL, Jiang YA (2011) The chemosensitizing activity of inhibitors of glucosylceramide synthase is mediated primarily through modulation of P-gp function. Int J Oncol 38(3):701–711. doi:10.3892/ijo.2010.888

111. Beckham TH, Cheng JC, Marrison ST, Norris JS, Liu X (2013) Interdiction of sphingolipid metabolism to improve standard cancer therapies. Adv Cancer Res 117:1–36. doi:10.1016/B978-0-12-394274-6.00001-7
112. Mao C, Obeid LM (2008) Ceramidases: regulators of cellular responses mediated by ceramide, sphingosine, and sphingosine-1-phosphate. Biochim Biophys Acta 1781(9):424–434. doi:10.1016/j.bbalip.2008.06.002
113. Liu X, Cheng JC, Turner LS, Elojeimy S, Beckham TH, Bielawska A, Keane TE, Hannun YA, Norris JS (2009) Acid ceramidase upregulation in prostate cancer: role in tumor development and implications for therapy. Expert Opin Ther Targets 13(12):1449–1458. doi:10.1517/14728220903357512
114. Saad AF, Meacham WD, Bai A, Anelli V, Elojeimy S, Mahdy AE, Turner LS, Cheng J, Bielawska A, Bielawski J, Keane TE, Obeid LM, Hannun YA, Norris JS, Liu X (2007) The functional effects of acid ceramidase overexpression in prostate cancer progression and resistance to chemotherapy. Cancer Biol Ther 6(9):1455–1460
115. Shah MV, Zhang R, Irby R, Kothapalli R, Liu X, Arrington T, Frank B, Lee NH, Loughran TP Jr (2008) Molecular profiling of LGL leukemia reveals role of sphingolipid signaling in survival of cytotoxic lymphocytes. Blood 112(3):770–781. doi:10.1182/blood-2007-11-121871
116. Morad SA, Messner MC, Levin JC, Abdelmageed N, Park H, Merrill AH Jr, Cabot MC (2013) Potential role of acid ceramidase in conversion of cytostatic to cytotoxic end-point in pancreatic cancer cells. Cancer Chemother Pharmacol 71(3):635–645. doi:10.1007/s00280-012-2050-4
117. Cheng JC, Bai A, Beckham TH, Marrison ST, Yount CL, Young K, Lu P, Bartlett AM, Wu BX, Keane BJ, Armeson KE, Marshall DT, Keane TE, Smith MT, Jones EE, Drake RR Jr, Bielawska A, Norris JS, Liu X (2013) Radiation-induced acid ceramidase confers prostate cancer resistance and tumor relapse. J Clin Invest 123(10):4344–4358. doi:10.1172/JCI64791
118. Selzner M, Bielawska A, Morse MA, Rudiger HA, Sindram D, Hannun YA, Clavien PA (2001) Induction of apoptotic cell death and prevention of tumor growth by ceramide analogues in metastatic human colon cancer. Cancer Res 61(3):1233–1240
119. Samsel L, Zaidel G, Drumgoole HM, Jelovac D, Drachenberg C, Rhee JG, Brodie AM, Bielawska A, Smyth MJ (2004) The ceramide analog, B13, induces apoptosis in prostate cancer cell lines and inhibits tumor growth in prostate cancer xenografts. Prostate 58(4):382–393. doi:10.1002/pros.10350
120. Szulc ZM, Mayroo N, Bai A, Bielawski J, Liu X, Norris JS, Hannun YA, Bielawska A (2008) Novel analogs of D-e-MAPP and B13. Part 1: synthesis and evaluation as potential anticancer agents. Bioorg Med Chem 16(2):1015–1031. doi:10.1016/j.bmc.2007.08.033
121. Mahdy AE, Cheng JC, Li J, Elojeimy S, Meacham WD, Turner LS, Bai A, Gault CR, McPherson AS, Garcia N, Beckham TH, Saad A, Bielawska A, Bielawski J, Hannun YA, Keane TE, Taha MI, Hammouda HM, Norris JS, Liu X (2009) Acid ceramidase upregulation in prostate cancer cells confers resistance to radiation: AC inhibition, a potential radiosensitizer. Mol Ther 17(3):430–438. doi:10.1038/mt.2008.281
122. Holman DH, Turner LS, El-Zawahry A, Elojeimy S, Liu X, Bielawski J, Szulc ZM, Norris K, Zeidan YH, Hannun YA, Bielawska A, Norris JS (2008) Lysosomotropic acid ceramidase inhibitor induces apoptosis in prostate cancer cells. Cancer Chemother Pharmacol 61(2):231–242. doi:10.1007/s00280-007-0465-0
123. Elojeimy S, Liu X, McKillop JC, El-Zawahry AM, Holman DH, Cheng JY, Meacham WD, Mahdy AE, Saad AF, Turner LS, Cheng J, Day TA, Dong JY, Bielawska A, Hannun YA, Norris JS (2007) Role of acid ceramidase in resistance to FasL: therapeutic approaches based on acid ceramidase inhibitors and FasL gene therapy. Mol Ther 15(7):1259–1263. doi:10.1038/sj.mt.6300167
124. Bhabak KP, Kleuser B, Huwiler A, Arenz C (2013) Effective inhibition of acid and neutral ceramidases by novel B-13 and LCL-464 analogues. Bioorg Med Chem 21(4):874–882. doi:10.1016/j.bmc.2012.12.014
125. Morales A, Paris R, Villanueva A, Llacuna L, Garcia-Ruiz C, Fernandez-Checa JC (2007) Pharmacological inhibition or small interfering RNA targeting acid ceramidase sensitizes

hepatoma cells to chemotherapy and reduces tumor growth in vivo. Oncogene 26(6):905–916. doi:10.1038/sj.onc.1209834

126. Alphonse G, Bionda C, Aloy MT, Ardail D, Rousson R, Rodriguez-Lafrasse C (2004) Overcoming resistance to gamma-rays in squamous carcinoma cells by poly-drug elevation of ceramide levels. Oncogene 23(15):2703–2715. doi:10.1038/sj.onc.1207357
127. Realini N, Solorzano C, Pagliuca C, Pizzirani D, Armirotti A, Luciani R, Costi MP, Bandiera T, Piomelli D (2013) Discovery of highly potent acid ceramidase inhibitors with in vitro tumor chemosensitizing activity. Sci Rep 3:1035. doi:10.1038/srep01035
128. Draper JM, Xia Z, Smith RA, Zhuang Y, Wang W, Smith CD (2011) Discovery and evaluation of inhibitors of human ceramidase. Mol Cancer Ther 10(11):2052–2061. doi:10.1158/1535-7163.MCT-11-0365
129. Bedia C, Casas J, Andrieu-Abadie N, Fabrias G, Levade T (2011) Acid ceramidase expression modulates the sensitivity of A375 melanoma cells to dacarbazine. J Biol Chem 286(32):28200–28209. doi:10.1074/jbc.M110.216382
130. Morad SA, Levin JC, Tan SF, Fox TE, Feith DJ, Cabot MC (2013) Novel off-target effect of tamoxifen—inhibition of acid ceramidase activity in cancer cells. Biochim Biophys Acta 1831(12):1657–1664. doi:10.1016/j.bbalip.2013.07.016
131. Liu X, Elojeimy S, El-Zawahry AM, Holman DH, Bielawska A, Bielawski J, Rubinchik S, Guo GW, Dong JY, Keane T, Hannun YA, Tavassoli M, Norris JS (2006) Modulation of ceramide metabolism enhances viral protein apoptin's cytotoxicity in prostate cancer. Mol Ther 14(5):637–646. doi:10.1016/j.ymthe.2006.06.005
132. Flowers M, Fabrias G, Delgado A, Casas J, Abad JL, Cabot MC (2012) C6-ceramide and targeted inhibition of acid ceramidase induce synergistic decreases in breast cancer cell growth. Breast Cancer Res Treat 133(2):447–458. doi:10.1007/s10549-011-1768-8
133. Stiban J, Fistere D, Colombini M (2006) Dihydroceramide hinders ceramide channel formation: implications on apoptosis. Apoptosis 11(5):773–780. doi:10.1007/s10495-006-5882-8
134. Spassieva SD, Rahmaniyan M, Bielawski J, Clarke CJ, Kraveka JM, Obeid LM (2012) Cell density-dependent reduction of dihydroceramide desaturase activity in neuroblastoma cells. J Lipid Res 53(5):918–928. doi:10.1194/jlr.M019075
135. Devlin CM, Lahm T, Hubbard WC, Van Demark M, Wang KC, Wu X, Bielawska A, Obeid LM, Ivan M, Petrache I (2011) Dihydroceramide-based response to hypoxia. J Biol Chem 286(44):38069–38078. doi:10.1074/jbc.M111.297994
136. Mao Z, Sun W, Xu R, Novgorodov S, Szulc ZM, Bielawski J, Obeid LM, Mao C (2010) Alkaline ceramidase 2 (ACER2) and its product dihydrosphingosine mediate the cytotoxicity of N-(4-hydroxyphenyl)retinamide in tumor cells. J Biol Chem 285(38):29078–29090. doi:10.1074/jbc.M110.105296
137. Wang H, Maurer BJ, Liu YY, Wang E, Allegood JC, Kelly S, Symolon H, Liu Y, Merrill AH Jr, Gouaze-Andersson V, Yu JY, Giuliano AE, Cabot MC (2008) N-(4-Hydroxyphenyl)retinamide increases dihydroceramide and synergizes with dimethylsphingosine to enhance cancer cell killing. Mol Cancer Ther 7(9):2967–2976. doi:10.1158/1535-7163.MCT-08-0549
138. Signorelli P, Munoz-Olaya JM, Gagliostro V, Casas J, Ghidoni R, Fabrias G (2009) Dihydroceramide intracellular increase in response to resveratrol treatment mediates autophagy in gastric cancer cells. Cancer Lett 282(2):238–243. doi:10.1016/j.canlet.2009.03.020
139. Olivera A, Kohama T, Edsall L, Nava V, Cuvillier O, Poulton S, Spiegel S (1999) Sphingosine kinase expression increases intracellular sphingosine-1-phosphate and promotes cell growth and survival. J Cell Biol 147(3):545–558
140. Li W, Yu CP, Xia JT, Zhang L, Weng GX, Zheng HQ, Kong QL, Hu LJ, Zeng MS, Zeng YX, Li M, Li J, Song LB (2009) Sphingosine kinase 1 is associated with gastric cancer progression and poor survival of patients. Clin Cancer Res 15(4):1393–1399. doi:10.1158/1078-0432.CCR-08-1158
141. French KJ, Schrecengost RS, Lee BD, Zhuang Y, Smith SN, Eberly JL, Yun JK, Smith CD (2003) Discovery and evaluation of inhibitors of human sphingosine kinase. Cancer Res 63(18):5962–5969

142. Johnson KR, Johnson KY, Crellin HG, Ogretmen B, Boylan AM, Harley RA, Obeid LM (2005) Immunohistochemical distribution of sphingosine kinase 1 in normal and tumor lung tissue. J Histochem Cytochem 53(9):1159–1166. doi:10.1369/jhc.4A6606.2005
143. Li J, Guan HY, Gong LY, Song LB, Zhang N, Wu J, Yuan J, Zheng YJ, Huang ZS, Li M (2008) Clinical significance of sphingosine kinase-1 expression in human astrocytomas progression and overall patient survival. Clin Cancer Res 14(21):6996–7003. doi:10.1158/1078-0432.CCR-08-0754
144. Van Brocklyn JR, Jackson CA, Pearl DK, Kotur MS, Snyder PJ, Prior TW (2005) Sphingosine kinase-1 expression correlates with poor survival of patients with glioblastoma multiforme: roles of sphingosine kinase isoforms in growth of glioblastoma cell lines. J Neuropathol Exp Neurol 64(8):695–705
145. Kohno M, Momoi M, Oo ML, Paik JH, Lee YM, Venkataraman K, Ai Y, Ristimaki AP, Fyrst H, Sano H, Rosenberg D, Saba JD, Proia RL, Hla T (2006) Intracellular role for sphingosine kinase 1 in intestinal adenoma cell proliferation. Mol Cell Biol 26(19):7211–7223. doi:10.1128/MCB.02341-05
146. Kawamori T, Kaneshiro T, Okumura M, Maalouf S, Uflacker A, Bielawski J, Hannun YA, Obeid LM (2009) Role for sphingosine kinase 1 in colon carcinogenesis. FASEB J 23(2):405–414. doi:10.1096/fj.08-117572
147. Ruckhaberle E, Rody A, Engels K, Gaetje R, von Minckwitz G, Schiffmann S, Grosch S, Geisslinger G, Holtrich U, Karn T, Kaufmann M (2008) Microarray analysis of altered sphingolipid metabolism reveals prognostic significance of sphingosine kinase 1 in breast cancer. Breast Cancer Res Treat 112(1):41–52. doi:10.1007/s10549-007-9836-9
148. Bayerl MG, Bruggeman RD, Conroy EJ, Hengst JA, King TS, Jimenez M, Claxton DF, Yun JK (2008) Sphingosine kinase 1 protein and mRNA are overexpressed in non-Hodgkin lymphomas and are attractive targets for novel pharmacological interventions. Leuk Lymphoma 49(5):948–954. doi:10.1080/10428190801911654
149. Nagahashi M, Ramachandran S, Kim EY, Allegood JC, Rashid OM, Yamada A, Zhao R, Milstien S, Zhou H, Spiegel S, Takabe K (2012) Sphingosine-1-phosphate produced by sphingosine kinase 1 promotes breast cancer progression by stimulating angiogenesis and lymphangiogenesis. Cancer Res 72(3):726–735. doi:10.1158/0008-5472.CAN-11-2167
150. Nava VE, Cuvillier O, Edsall LC, Kimura K, Milstien S, Gelmann EP, Spiegel S (2000) Sphingosine enhances apoptosis of radiation-resistant prostate cancer cells. Cancer Res 60(16):4468–4474
151. Pchejetski D, Doumerc N, Golzio M, Naymark M, Teissie J, Kohama T, Waxman J, Malavaud B, Cuvillier O (2008) Chemosensitizing effects of sphingosine kinase-1 inhibition in prostate cancer cell and animal models. Mol Cancer Ther 7(7):1836–1845. doi:10.1158/1535-7163.MCT-07-2322
152. Liu H, Toman RE, Goparaju SK, Maceyka M, Nava VE, Sankala H, Payne SG, Bektas M, Ishii I, Chun J, Milstien S, Spiegel S (2003) Sphingosine kinase type 2 is a putative BH3-only protein that induces apoptosis. J Biol Chem 278(41):40330–40336. doi:10.1074/jbc.M304455200
153. Sankala HM, Hait NC, Paugh SW, Shida D, Lepine S, Elmore LW, Dent P, Milstien S, Spiegel S (2007) Involvement of sphingosine kinase 2 in p53-independent induction of p21 by the chemotherapeutic drug doxorubicin. Cancer Res 67(21):10466–10474. doi:10.1158/0008-5472.CAN-07-2090
154. Inagaki Y, Li PY, Wada A, Mitsutake S, Igarashi Y (2003) Identification of functional nuclear export sequences in human sphingosine kinase 1. Biochem Biophys Res Commun 311(1): 168–173
155. Allende ML, Sasaki T, Kawai H, Olivera A, Mi Y, van Echten-Deckert G, Hajdu R, Rosenbach M, Keohane CA, Mandala S, Spiegel S, Proia RL (2004) Mice deficient in sphingosine kinase 1 are rendered lymphopenic by FTY720. J Biol Chem 279(50):52487–52492. doi:10.1074/jbc.M406512200
156. Kharel Y, Lee S, Snyder AH, Sheasley-O'Neill SL, Morris MA, Setiady Y, Zhu R, Zigler MA, Burcin TL, Ley K, Tung KS, Engelhard VH, Macdonald TL, Pearson-White S, Lynch KR

(2005) Sphingosine kinase 2 is required for modulation of lymphocyte traffic by FTY720. J Biol Chem 280(44):36865–36872. doi:10.1074/jbc.M506293200
157. Gao P, Smith CD (2011) Ablation of sphingosine kinase-2 inhibits tumor cell proliferation and migration. Mol Cancer Res 9(11):1509–1519. doi:10.1158/1541-7786.MCR-11-0336
158. Hait NC, Allegood J, Maceyka M, Strub GM, Harikumar KB, Singh SK, Luo C, Marmorstein R, Kordula T, Milstien S, Spiegel S (2009) Regulation of histone acetylation in the nucleus by sphingosine-1-phosphate. Science 325(5945):1254–1257. doi:10.1126/science.1176709
159. Truman JP, Garcia-Barros M, Obeid LM, Hannun YA (2013) Evolving concepts in cancer therapy through targeting sphingolipid metabolism. Biochim Biophys Acta 1841(8):1174–1188. doi:10.1016/j.bbalip.2013.12.013
160. Buehrer BM, Bell RM (1992) Inhibition of sphingosine kinase in vitro and in platelets. Implications for signal transduction pathways. J Biol Chem 267(5):3154–3159
161. Hannun YA, Loomis CR, Merrill AH Jr, Bell RM (1986) Sphingosine inhibition of protein kinase C activity and of phorbol dibutyrate binding in vitro and in human platelets. J Biol Chem 261(27):12604–12609
162. Sachs CW, Safa AR, Harrison SD, Fine RL (1995) Partial inhibition of multidrug resistance by safingol is independent of modulation of P-glycoprotein substrate activities and correlated with inhibition of protein kinase C. J Biol Chem 270(44):26639–26648
163. Ren S, Xin C, Pfeilschifter J, Huwiler A (2010) A novel mode of action of the putative sphingosine kinase inhibitor 2-(p-hydroxyanilino)-4-(p-chlorophenyl) thiazole (SKI II): induction of lysosomal sphingosine kinase 1 degradation. Cell Physiol Biochem 26(1):97–104. doi:10.1159/000315110
164. Tonelli F, Lim KG, Loveridge C, Long J, Pitson SM, Tigyi G, Bittman R, Pyne S, Pyne NJ (2010) FTY720 and (S)-FTY720 vinylphosphonate inhibit sphingosine kinase 1 and promote its proteasomal degradation in human pulmonary artery smooth muscle, breast cancer and androgen-independent prostate cancer cells. Cell Signal 22(10):1536–1542. doi:10.1016/j.cellsig.2010.05.022
165. Brinkmann V, Davis MD, Heise CE, Albert R, Cottens S, Hof R, Bruns C, Prieschl E, Baumruker T, Hiestand P, Foster CA, Zollinger M, Lynch KR (2002) The immune modulator FTY720 targets sphingosine 1-phosphate receptors. J Biol Chem 277(24):21453–21457. doi:10.1074/jbc.C200176200
166. Brinkmann V (2009) FTY720 (fingolimod) in multiple sclerosis: therapeutic effects in the immune and the central nervous system. Br J Pharmacol 158(5):1173–1182. doi:10.1111/j.1476-5381.2009.00451.x
167. Adachi K, Chiba K (2008) FTY720 story. Its discovery and the following accelerated development of sphingosine 1-phosphate receptor agonists as immunomodulators based on reverse pharmacology. Perspect Medicin Chem 1:11–23
168. Pchejetski D, Bohler T, Brizuela L, Sauer L, Doumerc N, Golzio M, Salunkhe V, Teissie J, Malavaud B, Waxman J, Cuvillier O (2010) FTY720 (fingolimod) sensitizes prostate cancer cells to radiotherapy by inhibition of sphingosine kinase-1. Cancer Res 70(21):8651–8661. doi:10.1158/0008-5472.CAN-10-1388
169. Lim KG, Sun C, Bittman R, Pyne NJ, Pyne S (2011) (R)-FTY720 methyl ether is a specific sphingosine kinase 2 inhibitor: effect on sphingosine kinase 2 expression in HEK 293 cells and actin rearrangement and survival of MCF-7 breast cancer cells. Cell Signal 23(10):1590–1595. doi:10.1016/j.cellsig.2011.05.010
170. Schnute ME, McReynolds MD, Kasten T, Yates M, Jerome G, Rains JW, Hall T, Chrencik J, Kraus M, Cronin CN, Saabye M, Highkin MK, Broadus R, Ogawa S, Cukyne K, Zawadzke LE, Peterkin V, Iyanar K, Scholten JA, Wendling J, Fujiwara H, Nemirovskiy O, Wittwer AJ, Nagiec MM (2012) Modulation of cellular S1P levels with a novel, potent and specific inhibitor of sphingosine kinase-1. Biochem J 444(1):79–88. doi:10.1042/BJ20111929
171. Sassa T, Suto S, Okayasu Y, Kihara A (2012) A shift in sphingolipid composition from C24 to C16 increases susceptibility to apoptosis in HeLa cells. Biochim Biophys Acta 1821(7):1031–1037. doi:10.1016/j.bbalip.2012.04.008

172. Beljanski V, Knaak C, Smith CD (2010) A novel sphingosine kinase inhibitor induces autophagy in tumor cells. J Pharmacol Exp Ther 333(2):454–464. doi:10.1124/jpet.109.163337
173. Sabbadini RA (2011) Sphingosine-1-phosphate antibodies as potential agents in the treatment of cancer and age-related macular degeneration. Br J Pharmacol 162(6):1225–1238. doi:10.1111/j.1476-5381.2010.01118.x
174. Macchia M, Barontini S, Bertini S, Di Bussolo V, Fogli S, Giovannetti E, Grossi E, Minutolo F, Danesi R (2001) Design, synthesis, and characterization of the antitumor activity of novel ceramide analogues. J Med Chem 44(23):3994–4000
175. Liu J, Antoon JW, Ponnapakkam A, Beckman BS, Foroozesh M (2010) Novel anti-viability ceramide analogs: design, synthesis, and structure-activity relationship studies of substituted (S)-2-(benzylideneamino)-3-hydroxy-N-tetradecylpropanamides. Bioorg Med Chem 18(14): 5316–5322. doi:10.1016/j.bmc.2010.05.044
176. Bieberich E, Kawaguchi T, Yu RK (2000) N-Acylated serinol is a novel ceramide mimic inducing apoptosis in neuroblastoma cells. J Biol Chem 275(1):177–181
177. Bieberich E, Silva J, Wang G, Krishnamurthy K, Condie BG (2004) Selective apoptosis of pluripotent mouse and human stem cells by novel ceramide analogues prevents teratoma formation and enriches for neural precursors in ES cell-derived neural transplants. J Cell Biol 167(4):723–734. doi:10.1083/jcb.200405144
178. Struckhoff AP, Bittman R, Burow ME, Clejan S, Elliott S, Hammond T, Tang Y, Beckman BS (2004) Novel ceramide analogs as potential chemotherapeutic agents in breast cancer. J Pharmacol Exp Ther 309(2):523–532. doi:10.1124/jpet.103.062760
179. Novgorodov SA, Szulc ZM, Luberto C, Jones JA, Bielawski J, Bielawska A, Hannun YA, Obeid LM (2005) Positively charged ceramide is a potent inducer of mitochondrial permeabilization. J Biol Chem 280(16):16096–16105. doi:10.1074/jbc.M411707200
180. Senkal CE, Ponnusamy S, Rossi MJ, Sundararaj K, Szulc Z, Bielawski J, Bielawska A, Meyer M, Cobanoglu B, Koybasi S, Sinha D, Day TA, Obeid LM, Hannun YA, Ogretmen B (2006) Potent antitumor activity of a novel cationic pyridinium-ceramide alone or in combination with gemcitabine against human head and neck squamous cell carcinomas in vitro and in vivo. J Pharmacol Exp Ther 317(3):1188–1199. doi:10.1124/jpet.106.101949
181. Hou Q, Jin J, Zhou H, Novgorodov SA, Bielawska A, Szulc ZM, Hannun YA, Obeid LM, Hsu YT (2011) Mitochondrially targeted ceramides preferentially promote autophagy, retard cell growth, and induce apoptosis. J Lipid Res 52(2):278–288. doi:10.1194/jlr.M012161
182. Rossi MJ, Sundararaj K, Koybasi S, Phillips MS, Szulc ZM, Bielawska A, Day TA, Obeid LM, Hannun YA, Ogretmen B (2005) Inhibition of growth and telomerase activity by novel cationic ceramide analogs with high solubility in human head and neck squamous cell carcinoma cells. Otolaryngol Head Neck Surg 132(1):55–62. doi:10.1016/j.otohns.2004.08.015
183. Crawford KW, Bittman R, Chun J, Byun HS, Bowen WD (2003) Novel ceramide analogues display selective cytotoxicity in drug-resistant breast tumor cell lines compared to normal breast epithelial cells. Cell Mol Biol (Noisy-le-Grand) 49(7):1017–1023
184. Adan-Gokbulut A, Kartal-Yandim M, Iskender G, Baran Y (2013) Novel agents targeting bioactive sphingolipids for the treatment of cancer. Curr Med Chem 20(1):108–122
185. Barth BM, Cabot MC, Kester M (2011) Ceramide-based therapeutics for the treatment of cancer. Anticancer Agents Med Chem 11(9):911–919
186. Ganta S, Singh A, Patel NR, Cacaccio J, Rawal YH, Davis BJ, Amiji MM, Coleman TP (2014) Development of EGFR-targeted nanoemulsion for imaging and novel platinum therapy of ovarian cancer. Pharm Res 31(9):2490–2502. doi:10.1007/s11095-014-1345-z
187. Desai A, Vyas T, Amiji M (2008) Cytotoxicity and apoptosis enhancement in brain tumor cells upon coadministration of paclitaxel and ceramide in nanoemulsion formulations. J Pharm Sci 97(7):2745–2756. doi:10.1002/jps.21182
188. Stover TC, Kim YS, Lowe TL, Kester M (2008) Thermoresponsive and biodegradable linear-dendritic nanoparticles for targeted and sustained release of a pro-apoptotic drug. Biomaterials 29(3):359–369. doi:10.1016/j.biomaterials.2007.09.037

189. Devalapally H, Duan Z, Seiden MV, Amiji MM (2008) Modulation of drug resistance in ovarian adenocarcinoma by enhancing intracellular ceramide using tamoxifen-loaded biodegradable polymeric nanoparticles. Clin Cancer Res 14(10):3193–3203. doi:10.1158/1078-0432.CCR-07-4973
190. van Vlerken LE, Duan Z, Seiden MV, Amiji MM (2007) Modulation of intracellular ceramide using polymeric nanoparticles to overcome multidrug resistance in cancer. Cancer Res 67(10):4843–4850. doi:10.1158/0008-5472.CAN-06-1648
191. Devalapally H, Duan Z, Seiden MV, Amiji MM (2007) Paclitaxel and ceramide co-administration in biodegradable polymeric nanoparticulate delivery system to overcome drug resistance in ovarian cancer. Int J Cancer 121(8):1830–1838. doi:10.1002/ijc.22886
192. Kester M, Heakal Y, Fox T, Sharma A, Robertson GP, Morgan TT, Altinoglu EI, Tabakovic A, Parette MR, Rouse SM, Ruiz-Velasco V, Adair JH (2008) Calcium phosphate nanocomposite particles for in vitro imaging and encapsulated chemotherapeutic drug delivery to cancer cells. Nano Lett 8(12):4116–4121. doi:10.1021/nl802098g
193. Stover T, Kester M (2003) Liposomal delivery enhances short-chain ceramide-induced apoptosis of breast cancer cells. J Pharmacol Exp Ther 307(2):468–475. doi:10.1124/jpet.103.054056
194. Stover TC, Sharma A, Robertson GP, Kester M (2005) Systemic delivery of liposomal short-chain ceramide limits solid tumor growth in murine models of breast adenocarcinoma. Clin Cancer Res 11(9):3465–3474. doi:10.1158/1078-0432.CCR-04-1770
195. Shabbits JA, Mayer LD (2003) High ceramide content liposomes with in vivo antitumor activity. Anticancer Res 23(5A):3663–3669
196. Tran MA, Smith CD, Kester M, Robertson GP (2008) Combining nanoliposomal ceramide with sorafenib synergistically inhibits melanoma and breast cancer cell survival to decrease tumor development. Clin Cancer Res 14(11):3571–3581. doi:10.1158/1078-0432.CCR-07-4881
197. Tagaram HR, Divittore NA, Barth BM, Kaiser JM, Avella D, Kimchi ET, Jiang Y, Isom HC, Kester M, Staveley-O'Carroll KF (2011) Nanoliposomal ceramide prevents in vivo growth of hepatocellular carcinoma. Gut 60(5):695–701. doi:10.1136/gut.2010.216671
198. Liu X, Ryland L, Yang J, Liao A, Aliaga C, Watts R, Tan SF, Kaiser J, Shanmugavelandy SS, Rogers A, Loughran K, Petersen B, Yuen J, Meng F, Baab KT, Jarbadan NR, Broeg K, Zhang R, Liao J, Sayers TJ, Kester M, Loughran TP Jr (2010) Targeting of survivin by nanoliposomal ceramide induces complete remission in a rat model of NK-LGL leukemia. Blood 116(20):4192–4201. doi:10.1182/blood-2010-02-271080
199. Watters RJ, Fox TE, Tan SF, Shanmugavelandy S, Choby JE, Broeg K, Liao J, Kester M, Cabot MC, Loughran TP, Liu X (2013) Targeting glucosylceramide synthase synergizes with C6-ceramide nanoliposomes to induce apoptosis in natural killer cell leukemia. Leuk Lymphoma 54(6):1288–1296. doi:10.3109/10428194.2012.752485
200. Jiang Y, DiVittore NA, Kaiser JM, Shanmugavelandy SS, Fritz JL, Heakal Y, Tagaram HR, Cheng H, Cabot MC, Staveley-O'Carroll KF, Tran MA, Fox TE, Barth BM, Kester M (2011) Combinatorial therapies improve the therapeutic efficacy of nanoliposomal ceramide for pancreatic cancer. Cancer Biol Ther 12(7):574–585
201. Hankins JL, Doshi UA, Haakenson JK, Young MM, Barth BM, Kester M (2013) The therapeutic potential of nanoscale sphingolipid technologies. Handb Exp Pharmacol 215:197–210. doi:10.1007/978-3-7091-1368-4_11
202. Morad SA, Levin JC, Shanmugavelandy SS, Kester M, Fabrias G, Bedia C, Cabot MC (2012) Ceramide—antiestrogen nanoliposomal combinations—novel impact of hormonal therapy in hormone-insensitive breast cancer. Mol Cancer Ther 11(11):2352–2361. doi:10.1158/1535-7163.MCT-12-0594
203. Adiseshaiah PP, Clogston JD, McLeland CB, Rodriguez J, Potter TM, Neun BW, Skoczen SL, Shanmugavelandy SS, Kester M, Stern ST, McNeil SE (2013) Synergistic combination therapy with nanoliposomal C6-ceramide and vinblastine is associated with autophagy dysfunction in hepatocarcinoma and colorectal cancer models. Cancer Lett 337(2):254–265. doi:10.1016/j.canlet.2013.04.034
204. Chapman JV, Gouaze-Andersson V, Cabot MC (2010) Expression of P-glycoprotein in HeLa cells confers resistance to ceramide cytotoxicity. Int J Oncol 37(6):1591–1597

205. Cabot MC, Han TY, Giuliano AE (1998) The multidrug resistance modulator SDZ PSC 833 is a potent activator of cellular ceramide formation. FEBS Lett 431(2):185–188
206. Lavie Y, Cao H, Volner A, Lucci A, Han TY, Geffen V, Giuliano AE, Cabot MC (1997) Agents that reverse multidrug resistance, tamoxifen, verapamil, and cyclosporin a, block glycosphingolipid metabolism by inhibiting ceramide glycosylation in human cancer cells. J Biol Chem 272(3):1682–1687
207. Morad SA, Madigan JP, Levin JC, Abdelmageed N, Karimi R, Rosenberg DW, Kester M, Shanmugavelandy SS, Cabot MC (2013) Tamoxifen magnifies therapeutic impact of ceramide in human colorectal cancer cells independent of p53. Biochem Pharmacol 85(8):1057–1065. doi:10.1016/j.bcp.2013.01.015
208. Chapman JV, Gouaze-Andersson V, Karimi R, Messner MC, Cabot MC (2011) P-glycoprotein antagonists confer synergistic sensitivity to short-chain ceramide in human multidrug-resistant cancer cells. Exp Cell Res 317(12):1736–1745. doi:10.1016/j.yexcr.2011.03.004
209. Qiu L, Zhou C, Sun Y, Di W, Scheffler E, Healey S, Wanebo H, Kouttab N, Chu W, Wan Y (2006) Paclitaxel and ceramide synergistically induce cell death with transient activation of EGFR and ERK pathway in pancreatic cancer cells. Oncol Rep 16(4):907–913
210. Best C, Calianese D, Szulak K, Cammarata G, Brum G, Carbone T, Still E, Higgins K, Ji F, Di W, Wanebo H, Wan Y (2013) Paclitaxel disrupts polarized entry of membrane-permeable C6 ceramide into ovarian cancer cells resulting in synchronous induction of cell death. Oncol Lett 5(6):1854–1858. doi:10.3892/ol.2013.1305
211. Zhu QY, Wang Z, Ji C, Cheng L, Yang YL, Ren J, Jin YH, Wang QJ, Gu XJ, Bi ZG, Hu G, Yang Y (2011) C6-ceramide synergistically potentiates the anti-tumor effects of histone deacetylase inhibitors via AKT dephosphorylation and alpha-tubulin hyperacetylation both in vitro and in vivo. Cell Death Dis 2, e117. doi:10.1038/cddis.2010.96
212. Furlong SJ, Mader JS, Hoskin DW (2006) Lactoferricin-induced apoptosis in estrogen-nonresponsive MDA-MB-435 breast cancer cells is enhanced by C6 ceramide or tamoxifen. Oncol Rep 15(5):1385–1390
213. Dickson MA, Carvajal RD, Merrill AH Jr, Gonen M, Cane LM, Schwartz GK (2011) A phase I clinical trial of safingol in combination with cisplatin in advanced solid tumors. Clin Cancer Res 17(8):2484–2492. doi:10.1158/1078-0432.CCR-10-2323
214. Kedderis LB, Bozigian HP, Kleeman JM, Hall RL, Palmer TE, Harrison SD Jr, Susick RL Jr (1995) Toxicity of the protein kinase C inhibitor safingol administered alone and in combination with chemotherapeutic agents. Fundam Appl Toxicol 25(2):201–217
215. Tan KB, Ling LU, Bunte RM, Chng WJ, Chiu GN (2012) In vivo efficacy of a novel liposomal formulation of safingol in the treatment of acute myeloid leukemia. J Control Release 160(2):290–298. doi:10.1016/j.jconrel.2011.11.002
216. Symolon H, Bushnev A, Peng Q, Ramaraju H, Mays SG, Allegood JC, Pruett ST, Sullards MC, Dillehay DL, Liotta DC, Merrill AH Jr (2011) Enigmol: a novel sphingolipid analogue with anticancer activity against cancer cell lines and in vivo models for intestinal and prostate cancer. Mol Cancer Ther 10(4):648–657. doi:10.1158/1535-7163.MCT-10-0754
217. Modrak DE, Gold DV, Goldenberg DM (2006) Sphingolipid targets in cancer therapy. Mol Cancer Ther 5(2):200–208. doi:10.1158/1535-7163.MCT-05-0420
218. Modrak DE, Lew W, Goldenberg DM, Blumenthal R (2000) Sphingomyelin potentiates chemotherapy of human cancer xenografts. Biochem Biophys Res Commun 268(2):603–606. doi:10.1006/bbrc.2000.2178
219. Modrak DE, Rodriguez MD, Goldenberg DM, Lew W, Blumenthal RD (2002) Sphingomyelin enhances chemotherapy efficacy and increases apoptosis in human colonic tumor xenografts. Int J Oncol 20(2):379–384
220. Veldman RJ, Zerp S, van Blitterswijk WJ, Verheij M (2004) N-Hexanoyl-sphingomyelin potentiates in vitro doxorubicin cytotoxicity by enhancing its cellular influx. Br J Cancer 90(4):917–925. doi:10.1038/sj.bjc.6601581
221. Thomas DA, Sarris AH, Cortes J, Faderl S, O'Brien S, Giles FJ, Garcia-Manero G, Rodriguez MA, Cabanillas F, Kantarjian H (2006) Phase II study of sphingosomal vincristine in patients with recurrent or refractory adult acute lymphocytic leukemia. Cancer 106(1):120–127. doi:10.1002/cncr.21595

Chemical Probes of Sphingolipid Metabolizing Enzymes

Ingrid Nieves, Pol Sanllehí, José Luis Abad, Gemma Fabriàs, Josefina Casas, and Antonio Delgado

Abstract Sphingolipids (SLs) serve the dual roles of acting as structural entities in cellular membranes as well as bioactive signaling molecules that modulate signal transduction. As the already immense database of identified bioactive SL subspecies continues to expand, the need for structure-specific identification and quantification continues to rise. The characterization and analysis of the sphingolipidome by mass spectrometry has advanced steadily over the last 20 years with the aid of improvements in technological advancements in instrumentation, coupled with optimization of lipid extraction methodologies, and an increasing library of available reference standards. Pivotal advances in sphingolipidomics include the adoption of soft ionization techniques, including electrospray ionization (ESI), tandem mass spectrometry (MS/MS), and matrix-assisted laser desorption ionization (MALDI), as well as the use of multiple reaction monitoring (MRM), all of which have aided in improving the quality of analysis of often complex lipid extracts from mammalian, yeast, and even plant cells. In this chapter we explore qualitative and quantitative mass spectrometry methods used for structural elucidation and quantitation of sphingolipids found in cells as well as tissues. Sections included here detail extraction and HPLC methodologies, in vitro labeling techniques, use of internal and calibration lipid standards for quantitation, and data analysis of sphingolipids derived from mammalian and yeast sources.

Keywords Enzyme activity • Chemical probe • Chemical reporter • Fluorescence • Fluorophore • Sphingolipid • Metabolism

---

I. Nieves • J.L. Abad • G. Fabriàs • J. Casas
Consejo Superior de Investigaciones Científicas (CSIC), Institut de Química Avançada de Catalunya (IQAC-CSIC), Departament de Química Biomèdica, Research Unit on Bioactive Molecules (RUBAM), Carrer Jordi Girona 18-26, Barcelona 08034, Spain

P. Sanllehí • A. Delgado (✉)
Consejo Superior de Investigaciones Científicas (CSIC), Institut de Química Avançada de Catalunya (IQAC-CSIC), Departament de Química Biomèdica, Research Unit on Bioactive Molecules (RUBAM), Carrer Jordi Girona 18-26, Barcelona 08034, Spain

Universidad de Barcelona (UB), Facultad de Farmacia, Unidad de Química Farmacéutica (Unidad Asociada al CSIC), Avda. Joan XXIII s/n, Barcelona 08028, Spain
e-mail: delgado@rubam.net

Y.A. Hannun et al. (eds.), *Bioactive Sphingolipids in Cancer Biology and Therapy*, DOI 10.1007/978-3-319-20750-6_19

Abbreviations

Bodipy	4,4-Difluoro-4-bora-3a,4a-diaza-s-indacene
CDases	Ceramidases
Cer	Ceramide
Cer1P	Ceramide 1-phosphate
CerK	Ceramide kinase
CerS	Ceramide synthase
Des1	Dihydroceramide desaturase
dhCer	Dihydroceramide
DNBz	3,5-Dinitrobenzoyl
Flu	Fluorescein
GBA	Glucocerebrosidase
GC	Glucosylceramide
GCS	Glucosylceramide synthase
GSL	Glycosphingolipids
(HMU)PC	6-Hexadecanoylamino-4-methylumbelliferylphosphorylcholine
(HNP)PC	2-Hexadecanoylamino-4-nitrophenylphosphorylcholine
HTS	High throughput screening
iTRAQ	Isobaric tags for relative and absolute quantification
LR	Lissamine-rhodamine
NBD	4-Amino-7-nitro-2,1,3-benzoxadiazole
NDA	Naphthalene-2,3-dicarbaldehyde
NR	Nile Red
OPA	*ortho*-Phthalaldehyde
PC	Phosphocholine
Sa	Sphinganine
SK	Sphingosine kinases
SLs	Sphingolipids
SM	Sphingomyelin
SMase	Sphingomyelinase
SMS	Sphingomyelin synthases
So	Sphingosine
So1P	Sphingosine 1-phosphate
SPL	Sphingosine 1-phosphate lyase

1 Introduction

Sphingolipids (SLs) are ubiquitous natural products that play essential roles in structural biology, cell regulation and signaling [1]. The biosynthesis of SLs in mammals is a complex and well compartmentalized process in which ceramide (Cer) plays a pivotal role [2]. In fact, Cer is not a single entity but a family of closely related molecules that differ in the nature of the acyl chains and in that of

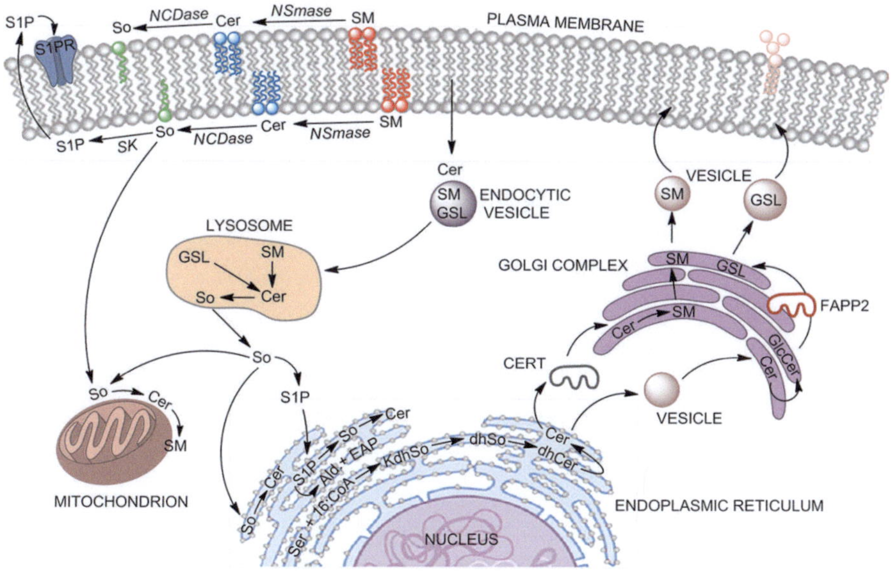

Fig. 1 General overview of sphingolipid metabolism and compartmentalization

the sphingoid base backbone. Thus, in addition to the most abundant C18:1(Δ^4E) system present in mammals, saturated (dihydroceramides), polyunsaturated (such as dienes and trienes), dihydroxylated (phytoceramides), and shortened (C16 and C16:1(Δ^4E)) sphingoid bases [3], as well as the epidermis-unique omega-hydroxyceramides [4] have also been identified in nature. Due to its implications in SL metabolism, Cer should be considered as a "metabolic hub", acting both as precursor of more structurally complex SLs, such as ceramide-1-phosphate (Cer1P), sphingomyelin (SM) or glycosphingolipids (GSL), and also as precursor of simpler SLs arising from the degradative pathway, such as sphingosine (So) and sphingosine-1-phosphate (So1P) (Fig. 1).

The current perception of SL metabolism is that of a highly intricate, interrelated system of enzymes, whose relative activities determine the intracellular concentration of SLs, which can ultimately be responsible for cell fate. In addition, since many pathologies are related to an improper balance of SLs [5], the development of methods for their quantification and the use of specific probes to monitor their intracellular localization and trafficking is gaining importance in contemporary chemical biology and drug design approaches. As extensively discussed in many reviews [6, 7], cancer is one of the pathologies where SLs play a capital role. Guided by these interests, in this chapter (Sect. 2) we present an overview of the most representative chemical probes that have been developed as substrates to monitor the activity of SL metabolizing enzymes. Section 3 is devoted to the description of new approaches to the cellular quantification, localization and trafficking of sphingolipids, some of them wisely adapted from acquired experience in the field of proteomics.

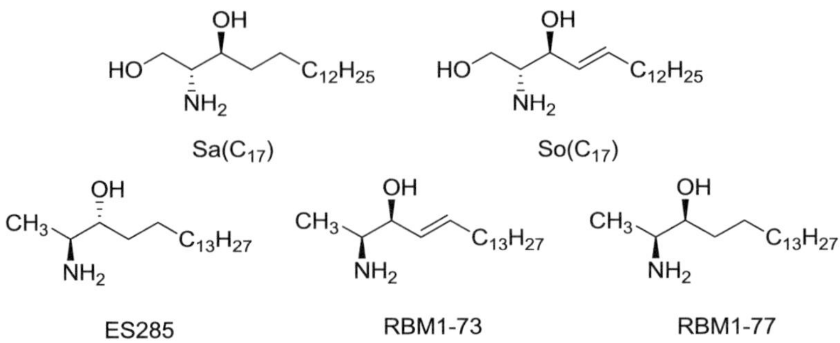

Fig. 2 Minimally modified substrates for the determination of SK and CerS

2 Chemical Probes to Monitor the Activity of SL Metabolizing Enzymes

Until assays based on the use of non-natural substrates were implemented, most enzyme determinations were performed with radioactive substrates. Examples include radiolabeled Cer for ceramide kinase (CerK) [8], ceramidases (CDases) [9, 10], and sphingomyelin synthases (SMS) [11], radiolabeled dihydroceramide (dhCer) for dihydroceramide desaturase (Des1) [12], radiolabeled SM and GlcCer for sphingomyelinases (SMase) [13–19] and glucocerebrosidases (GBA) [20][21], respectively, and [4,5-^{3}H] Sa1P for sphingosine-1-phosphate lyase (SPL) [22, 23]. Assays using radiolabeled cofactors or donors have also been reported, including [γ-32P]-ATP for kinases (CerK [24] and sphingosine kinase, SK [25–27]) and glucose for glucosylceramide synthase (GCS) [28, 29].

The disadvantages of working with radioactive materials have prompted the development of non-natural substrates to monitor SL metabolizing enzymes. The sphingoid base chain shortening is the simplest natural substrate modification so far reported. Spassieva et al. [30] have described the utilization of non-naturally occurring C(17) sphingoid bases in combination with mass spectrometry technology in the assays of ceramide synthase (CerS) and SK (Fig. 2). In a recent contribution [31], a series of stereochemically defined 1-deoxySa and 1-deoxySo were evaluated as probes to unravel CerS activity in intact cells by UPLC-TOF methods. Among the different analogues tested, compounds ES285, RBM1-77 and RBM1-73 (Fig. 2) turned out to be suitable probes for CerS profiling. These probes are metabolically stable at both C1 and at the amide linkage, after CerS acylation, and thus the resulting amide composition reflects the overall CerS activities under a given set of conditions. In particular, compound ES285 (spisulosine) led to the highest acylation rates, thus being the compound of choice as chemical probe to evaluate CerS activity and the distribution of *N*-acylated metabolites under given biological conditions. In another reported assay for CerS, a ω-alkyne-sphinganine is used as substrate for click-based production of fluorescent ceramide products, which are separated from the substrate by TLC for further quantitation [32]. These methods provide high sensitivity and extreme accuracy even when crude extracts are used as enzyme sources. Nevertheless, as discussed in the following sections, fluorescent substrates (Table 1) are amongst the non-natural substrates more extensively used.

Table 1 Probes used to determine enzyme activity in SL metabolism

Probes	Figure	Target	References
(Bodipy)C_{12}Cer	5	CDase	[61]
(Bodipy)C_{12}GC	9	GBAs	[102, 106, 108]
(Bodipy)C_5Cer	5	SMS	[89]
(DMAS)So(Glc)	9	GBAs	[102, 103]
(HMU)PC	8	SMase	[81]
(HNP)OGlc	9	GBAs	[101]
(HNP)PC	8	SMase	[79, 80]
(LR)C_{12}Cer	5	CDase	[61]
(LR)C_{12}GC	9	GBAs	[101, 108]
(LR)O-PAP	9	GBAs	[108]
(NBD)C_{12}Cer	5	CDase	[58, 59, 163]
(NBD)C_{12}CerC_7(NR)	6	CDase	[63]
(NBD)C_{12}GC	9	GBAs	[104, 105]
(NBD)C_{12}SM	5	SMase	[82]
(NBD)C_6Cer	5	CerK	[38, 39]
(NBD)C_6Cer	5	SMS	[87–89]
(NBD)C_6Cer	5	GCS	[94]
(NBD)C_6Cer	5	CDase	[58, 59, 163]
(NBD)C_6dhCer	5	Des1	[72, 73]
(NBD)C_6GC	9	GBAs	[106]
(NR)C_{12}Cer	5	CDase	[62]
(NR)C_{12}CerC_7(NBD)	6	CDase	[63]
(Pyrene)C_{10}CerC_{17}(DNBz)	6	CDase	[64]
(Reso)Glc	9	GBAs	[107]
4-MUG	9	GBAs	[99, 107, 108]
C_{16}CerC_{14}(NR)	5	CDase	[61]
ES285	2	CerS	[31]
Flu(OGlc)$_2$	9	GBAs	[102, 103, 109]
RBM13	7	SPL	[70]
RBM14	7	CDase	[65, 66]
RBM1-73	2	CerS	[31]
RBM1-77	2	CerS	[31]
Sa(C_{17})	2	CerS /SK	[30]
Sa(NBD)	4	CerS	[36]
Sa1P(C_{17})	11	SPL	[115]
So(Biotin)	4	SK	[45]
So(Flu)	4	SK	[53]
So(NBD)	4	SK	[48]
So1[^{33}P](Biotin)	4	SPL	[71]
So1P(NBD)	4	SPL	[68]
So1PC_{14}(Bodipy)	4	SPL	[69]
SoC_{15}(NBD)	4	SK	[46, 47]

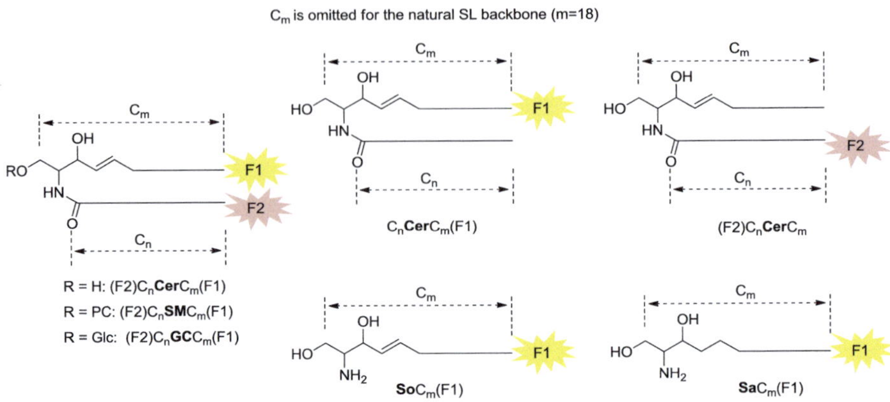

Fig. 3 Notation used in this chapter for the unambiguous description of SL probes

From a structural standpoint, the probes described in the next sections are non-natural SL analogues that incorporate a reporter moiety as part of the structure (see Table 1). For an unambiguous description of these probes, the notations shown in Fig. 3 will be used throughout the chapter.

2.1 *Ceramide Synthases (CerS)*

Ceramide synthases (CerS) catalyze the acylation of sphinganine (Sa) to dhCer in the de novo ceramide synthesis, and that of Sa and So to dhCer and Cer, respectively, in the salvage pathway. In mammals, six different CerS isoforms (CerS1-CerS6) have been identified. Each isoform produces (dh)Cer with characteristic acyl-chain distributions, which may be associated to specific enzyme compartmentalization, acylCoA availability and cell state [33–35]. A fluorescent assay based on the use of commercially available NBD-labeled sphinganine (Sa(NBD), Fig. 4) as CerS substrate has been described [36]. According to the authors, the assay is suitable for the detection of endogenous CerS activity, both in cells or tissue homogenate protein and is more sensitive than the previously reported radioactive assay. Interestingly, Sa(NBD) behaves similarly as the natural substrate in terms of enzyme affinity. The detection and quantification of the resulting dhCer(NBD) is carried out directly on the TLC plate, and the reported detection limit has been estimated in 0.5 pmol.

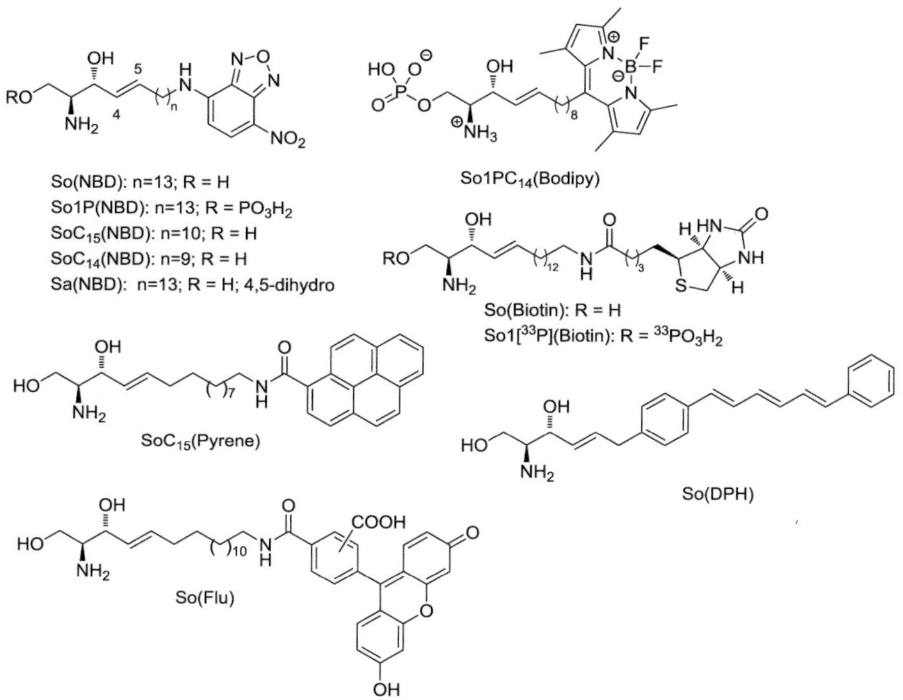

Fig. 4 Fluorescent sphingosine-derived probes

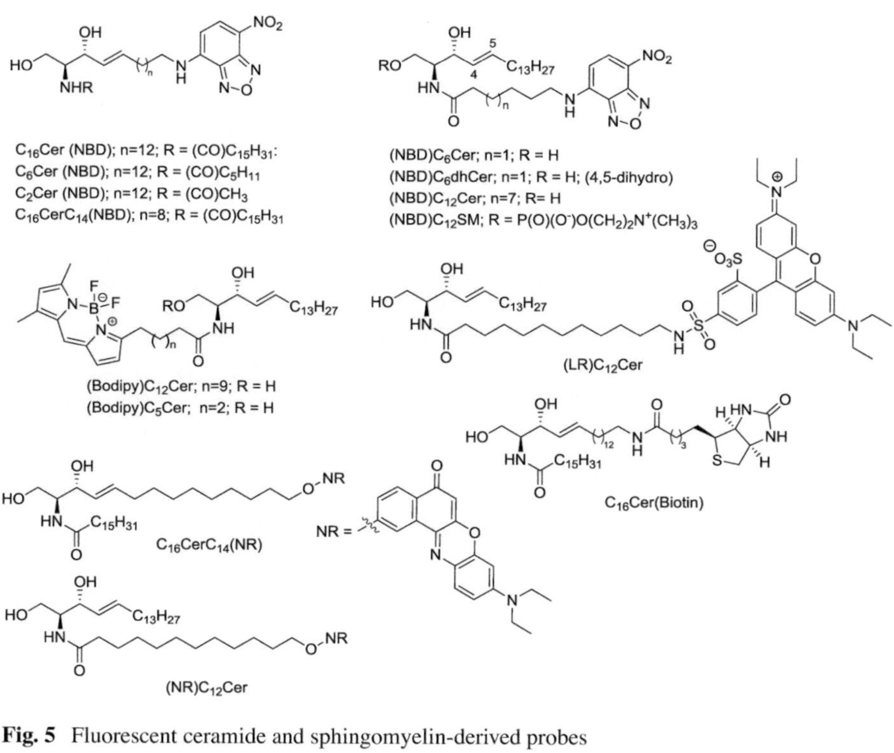

Fig. 5 Fluorescent ceramide and sphingomyelin-derived probes

2.2 Ceramide Kinase (CerK)

Ceramide kinase (CerK) and its reaction product, Cer1P, are implicated in a variety of cellular processes such as proliferation, survival, activation of cytosolic PLA2, inflammation and apoptosis, *inter alia* [37]. Two groups developed independently a method based on the use of (NBD)C_6Cer as substrate (Fig. 5) [38, 39]. Interestingly, despite the steric bulk of the NBD moiety, the fluorescent substrate presents a K_m of 4 μM, similar to that of C_6Cer [38]. The resulting (NBD)C_6CerP can be extracted in the aqueous phase from a buffer/organic system and directly used for on-plate fluorescence quantification [39]. Alternatively, (NBD)C_6CerP can be isolated and separated from the excess substrate by solid phase extraction and further analyzed by TLC [38].

2.3 Sphingosine Kinases (SK)

Sphingosine kinases (SK) phosphorylate So to yield S1P. In mammalian cells, two isoforms of SK have been found, namely SK1 and SK2, which differ in their tissue distribution, cellular expression, and substrate specificities. Despite both enzymes are also able to phosphorylate dihydrosphingosine, SK2 has a broader substrate specificity, catalyzing the phosphorylation of both phytosphingosine and the sphingoid base analog FTY720. However, the functions of SK1 and SK2 are partly redundant, as evidenced by the fact that mice lacking any of the isoforms still maintain a nearly normal phenotype [40]. SK1 has been extensively studied and there is a large body of evidence that proves its role in promoting cell survival, proliferation and neoplastic transformation [41–43]. SK1 is also elevated in many human cancers, which appears to contribute to carcinogenesis, chemotherapeutic resistance and poor patient outcome. SK2, however, has not been as well characterized, and there are contradictions in the key physiological functions that have been proposed for this isoform. However, many emerging studies implicate SK2 in key roles of a variety of diseases, including the development of a range of solid tumours [44]. As a fast and sensitive protocol, the biotinylated So substrate So(Biotin) (Fig. 4) was designed in combination with the capture of the phosphorylated product on a streptavidin-coated membrane [45]. The detection limits of this method are in the low-femtomole range for both recombinant human SK1 and SK2 present in crude cell extracts.

A fluorimetric assay to quantify SK activity based on the use of a NBD-labeled C15 So substrate (SoC_{15}(NBD)) (Fig. 4) has been described [46, 47]. The method is suitable to measure the activity of SK, both from purified preparations and from lysate extracts of mammalian cells [46]. Despite this method represents an improvement with respect to the classical approaches using radiolabeled substrates, it still requires an extraction process of the phosphorylated reaction product for its direct fluorimetric assay in a 96 well plate. However, the use of the longer analogue So(NBD) (Fig. 4) allows the direct measure of SK activity by the increase of fluorescence emission of the resulting phosphorylated probe. This method has been implemented in a 384 well-plate format in a HTS assay for SK, and it is suitable for the screening of SK inhibitors [48].

In addition to NBD [49], dansyl, pyrene and BODIPY derivatives have also been tested as SK substrates. Interestingly, pyrene and dansyl derivatives having comparable chain lengths spacers were phosphorylated by both SK1 and SK2 with an efficiency similar to that of the natural substrate So [50]. A more recent approach to measure SK activity uses the fluorescein-labeled sphingosine So(Flu) as SK substrate (Fig. 4). The phosphate reaction product is separated from the substrate by capillary electrophoresis and quantified by laser induced fluorescence detection [51–53]. Using purified SK2 as enzymatic source, the So(Flu) substrate shows a K_m of 2.8 μM, somewhat higher than that reported for the natural substrate (around 1.0 μM), which is attributed to the steric bulk imposed by the presence of the fluorophore in the probe.

2.4 *Ceramidases (CDases)*

Ceramidases (CDases) are lipolytic amidohydrolases that catalyze the cleavage of Cer into So and fatty acids. According to their optima pH, ceramidases fall into three groups, acidic, neutral and alkaline, with different tissue distribution, subcellular localization and substrate specificity [54]. The implication of ceramidases, mainly the acidic form, in cancer progression has prompted the publications of several reviews [55–57]. The use of the fluorescent (NBD)C_6Cer as ceramidase substrate (Fig. 5) was already reported by Merrill *and col.* in assays carried out in vitro and in intact hepatocytes [58]. In that work, NBD lipids were analyzed by HPLC coupled to a fluorescence detector. The length of the acyl chain connecting the fluorescent NBD moiety with the sphingosine backbone was studied by Tani et al., who found that (NBD)C_{12}Cer (Fig. 5) was hydrolyzed faster in B16 melanoma cell lysates at both pH 8.5 and 4.0 [59] and by the alkaline CDase from *Pseudomonas aeruginosa* [60] than (NBD)C_6Cer. Interestingly, the alkaline bacterial enzyme had also a preferential activity for (NBD)C_{12}Cer over [^{14}C] oleoylsphingosine [60]. Other fluorescent probes reported to measure CDase activity include (Bodipy) C_{12}Cer and the lissamine-rhodamine substituted Cer, (LR)C_{12}Cer (Fig. 5) [61]. In particular, acidic CDase was measured with similar sensitivities with both (Bodipy) C_{12}Cer and the radioactive substrate using mouse kidney extracts as enzyme source. However, (LR)C_{12}Cer was a poorer substrate of acidic CDase, which was attributed to its higher hydrophilic nature in comparison with both the Bodipy and the natural substrates. Very recently, a fluorescent Cer incorporating Nile Red (NR) as an acyl chain fluorophore, (NR)C_{12}Cer, (Fig. 5) has been reported and compared with (NBD)C_{12}Cer as substrate for acid and neutral CDase [62]. Under standard assay conditions, a lower cleavage rate for (NR)C_{12}Cer, as compared to (NBD)C_{12}Cer, was observed. In the same work, new fluorescent substrates incorporating the above fluorescent moieties as part of the sphingoid chain, namely C_{16}CerC_{14}(NR) and

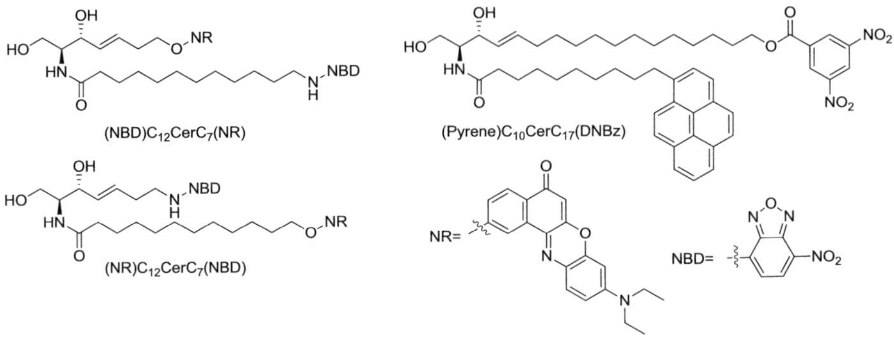

Fig. 6 FRET-based probes for real time determination of CDase activity

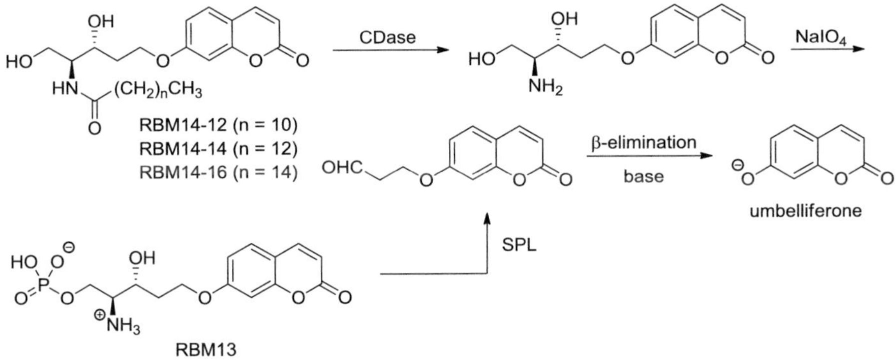

Fig. 7 Fluorogenic coumarin-containing fluorogenic probes to determine CDase and SPL activity

C_{16}CerC_{14}(NBD) (see Fig. 5), were also synthesized and compared for their efficiency towards CDases. In this case, the enzymatic cleavage of both substrates towards acid CDase were comparable, while C_{16}CerC_{14}(NR) was somewhat less efficient as substrate for neutral CDase than C_{16}CerC_{14}(NBD).

The development of fluorogenic substrates (substrates that give rise to a fluorescent readout subsequent to a particular enzymatic reaction) represents a breakthrough in the design of probes suitable for determining enzyme activities. As a continuation of the above work, the same authors reported on two FRET probes, (NBD)C_{12}CerC_7(NR) and (NR)C_{12}CerC_7(NBD) (Fig. 6), for the real time determination of ceramidase activity [63]. The probes were designed by combination of NBD and NR as donor and acceptor FRET pairs, respectively, located as part of the acyl chain and/or the sphingoid base of the Cer substrate. Probe (NBD)C_{12}CerC_7(NR) turned out to be a better substrate than (NR)C_{12}CerC_7(NBD) for ceramidases, with K_m values of 142 μM and 182 μM for recombinant neutral and acid CDase, respectively. In a previous work, another group designed the quenched fluorophore probe (Pyrene)C_{10}CerC_{17}(DNBz) (Fig. 6), whose fluorescence was released after cleavage by *P. aeruginosa* recombinant CDase [64]. After the hydrolytic cleavage, the fluorescence of the released ω-pyrenyldecanoic acid allowed the quantification of the enzyme activity. However, this probe had limited application due to its high hydrophobicity and relatively low quantum yield.

A series of fluorogenic coumarin-containing CDase substrates, amenable for HTS assays, has also been reported [65, 66]. These substrates (compounds RBM14, Fig. 7) are conceptually similar to those described by Reymond and co-workers for other hydrolytic enzymes [67]. After hydrolysis of RBM14 by CDases, oxidation of the resulting vicinal aminodiol renders an intermediate aldehyde, whose subsequent β-elimination under basic conditions liberates the fluorescent umbelliferone reporter (Fig. 7).

Although compounds RBM14 have been reported as substrates of the acidic CDase, recent studies have shown that they are also hydrolyzed by the neutral CDase and at least one of the alkaline CDases (unpublished). The specificity of the substrates towards CDases can be modulated by the length of the acyl chain. Thus, in cells overexpressing acid CDase, the highest rate of hydrolysis was observed for the probe with a dodecanoyl group (RBM14-12), while recombinant human neutral CDase prefers the hexadecanoyl derivative (RBM14-16), and the tetradecanoylamide (RBM14-14) is preferentially hydrolyzed by lysates of neutral CDase-null mouse embryonic fibroblasts at pH 8.5 in the presence of Ca^{2+}. This fluorogenic method is currently in use in the diagnosis of Farber disease, a rare disease characterized by the deficiency of acid CDase [66].

2.5 *Sphingosine-1-phosphate Lyase (SPL)*

Sphingosine-1-phosphate lyase (SPL) is the last enzyme in the catabolism of sphingolipids. It catalyzes the retroaldol cleavage of long chain base phosphates into phosphoethanolamine and a fatty aldehyde (Fig. 1). Similarly as discussed in the preceding sections of this chapter, fluorescent substrates, incorporating a fluorescent reporter as part of the sphingoid base chain, have been developed for this enzyme. This is the case of So1P(NBD) [68] and So1PC_{14}(Bodipy) (Fig. 4). The latter is more extensively used due to its improved photochemical properties [69]. In both cases, the resulting aldehyde is extracted from the incubation mixture and quantified after separation by HPLC coupled to a fluorescence detector. The fluorogenic coumarin-containing substrate RBM13 has also been reported (Fig. 7) [70]. In this case, the released aldehyde undergoes a spontaneous β-elimination to give the fluorescent umbelliferone in a process conceptually similar to that described in Sect. 2.4 for ceramidases. According to the reported protocol, no separation of products is necessary and the assay can be performed in microtiter wells, which is an important improvement in high throughput screening (HTS) assays of putative inhibitors. In a recent report, a high throughput scintillation proximity assay for SPL has been reported. The assay requires the use of So1[^{33}P](Biotin) (Fig. 4) as substrate and employs recombinant human full-length SPL in insect cell membrane preparations to catalyze the conversion of So1[^{33}P](Biotin) to (ω-biotin)-*trans*-2-hexadecenal and [^{33}P] ethanolamine phosphate [71].

An inherent limitation of the above fluorescent or fluorogenic substrates relies on the use of a non-natural SPL substrate and, hence, the presence of a bulky reporter as part of the sphingoid chain. This fact can explain the lower affinity of some of these substrates towards the target enzyme, as evidenced from their higher Km values [68–70] in comparison to that of the natural substrate [23].

2.6 Dihydroceramide Desaturase (Des1)

Desaturation of dihydroceramide at Δ^4 position to introduce a *E*-double bond is the last step in the de novo pathway synthesis of Cer (see Fig. 1). Although originally designed for intracellular transport and metabolism studies of dhCer (see Sect. 3.3), the fluorescent probe (NBD)C_6dhCer (Fig. 5) has also been used to monitor Des1 activity as it has been reported to be a substrate of the target enzyme [72].

The kinetic parameters of (NBD)C_6dhCer desaturation have been determined in rat liver microsomes after incubation with different substrate concentrations and lipid analysis by HPLC coupled to a fluorescence detector. Under these conditions, a K_m of 7.7 μM for the fluorescent substrate was determined [73].

2.7 Enzymes of the Sphingomyelin Cycle

The sphingomyelin cycle comprises the enzymes implicated in the biosynthesis and catabolism of SM from and to Cer.

2.7.1 Sphingomyelinases (SMases)

Sphingomyelinases (SMases) are a group of hydrolases that cleave SM, the most abundant SL in plasma membranes, to form Cer and phosphocholine. According to their pH dependence, different SMases have been identified in mammals. Acid SMase localizes in lysosomes and its deficiency is related to Niemann-Pick disease A and B, a type of sphingolipidosis characterized by an accumulation of SM [74]. In addition, a Zn^{2+}-dependent secretory form of this enzyme has also been found. This form is differentially glycosylated and processed, and is believed to function in inflammatory processes and atherogenesis [75]. On the other hand, neutral SMase is a Mg^{2+}-dependent enzyme located in plasma membranes [76]. In addition, a Mg^{2+}-independent neutral SMase is also found in cytosol [77], and an alkaline

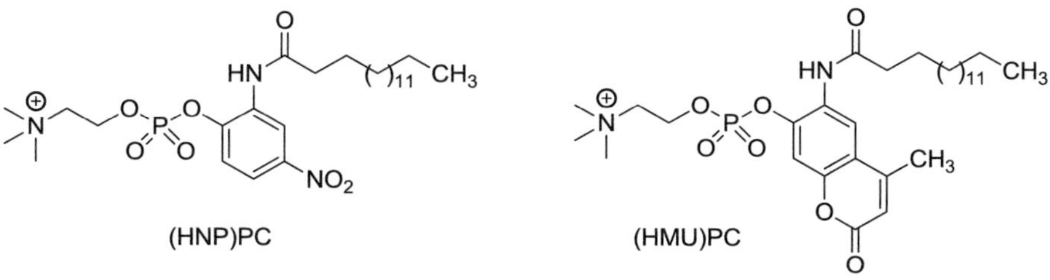

Fig. 8 Fluorogenic substrates to target SMase activity

SMase has been identified in the gastrointestinal tract [78]. The reduction of Cer levels, as a result of SMases inhibition, has emerged as an interesting therapeutic approach for several pathologies where high levels of Cer are involved [55, 76]. In this context, the need for the discovery of new drugs targeting SMases from large libraries of compounds has boosted the development of new methods to monitor the activity of these enzymes. The SMase fluorogenic substrates were preceded by (HNP)PC (Fig. 8), a chromogenic *p*-nitrophenyl derivative. This substrate was reported as a reliable chromogenic reagent for the diagnosis of patients with Niemann-Pick disease and the detection of heterozygous carriers of the disorder [79, 80]. The related fluorogenic substrate (HMU)PC, (Fig. 8) has also been published [81]. Interestingly, this substrate allows the diagnosis of Niemann-Pick patients with a particular mutation by carrying out the fluorogenic assay in the presence of the natural substrate lysosphingomyelin. Other fluorogenic substrates, structurally more similar to the natural one, are NBD acyl chain substituted sphingomyelins (NBD)C_nSM (Fig. 5). In these cases, the enzymatic activity is quantified by measuring the fluorescence of the corresponding hydrolysis product, (NBD)C_nCer, after TLC [82] or HPLC [83] separation.

2.7.2 Sphingomyelin Synthases (SMS)

Sphingomyelin synthases (SMS) are integral membrane enzymes that transfer a phosphorylcholine moiety from phosphatidylcholine to the primary hydroxyl group of Cer, with the concomitant formation of diacylglycerol [84]. Two isoforms of mammalian SMS, with identical enzymatic activities but different localization have been identified. Thus, SMS1 is localized in the Golgi system, whereas SMS2 is localized on plasma membrane [85].

Classical assays for SMS activity are based on the use of (NBD)C_6Cer (Fig. 5) and quantification after TLC [86] or HPLC separation [87–89]. In one of these works [88], a $K_m = 7.50$ μM for this substrate was determined. In addition, (Bodipy)C_5Cer (Fig. 5) has also been reported as substrate to measure SMS activity by quantitative analysis of the corresponding reaction product, (Bodipy)C_5SM, by HPLC [89].

2.8 Enzymes Implicated in Glucosylceramide Metabolism

Glycosylation of Cer at the primary hydroxyl position provides GSL, a large structural family of cell membrane components with important biochemical [90] and biophysical [91] roles. The biosynthesis of most GSL in mammals starts at the cytosolic face of the Golgi complex with glucosylation of Cer to give glucosyl ceramide (GC) in a process catalyzed by GCS. The initially formed GC is the starting building block for the synthesis of complex GSL at the luminal face of the Golgi apparatus from where they are transported to the cell membrane by exocytosis mechanisms. Enzymes implicated in GC metabolism are important targets for the development of therapies for Gaucher disease, a sphingolipidosis resulting from the accumulation of GC in the lysosomes [92], and to reverse multidrug resistance in cancer [93]. The use of fluorescent substrates to measure GCS activity has also been reported. In particular, (NBD)C_6Cer (see Fig. 5) allows the direct visualization and quantification of the corresponding reaction product (NBD)C_6GC after TLC separation and scraping of the spot prior to fluorimetric analysis [94]. The catabolism of GC can take place at different levels. The primary pathway is catalyzed by the lysosomal β-glucocerebrosidase (GBA1), which is defective in Gaucher disease patients [95]. A non lysosomal β-glucosylceramidase (GBA2) is found in the cell membrane,

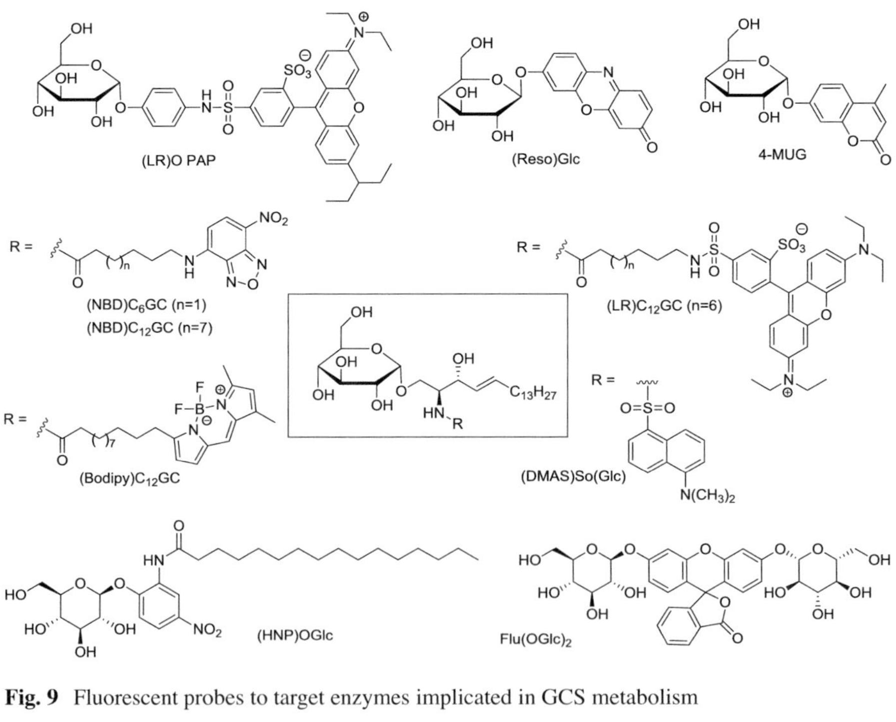

Fig. 9 Fluorescent probes to target enzymes implicated in GCS metabolism

close to the surface, whose activity is linked to SM generation [96]. Its optimum pH is close to neutral, as opposed to GBA1, which exhibits an optimum pH in the acidic region. In addition, a non-specific cytosolic β-glucosidase (GBA3) can also hydrolyze GC albeit with lower specificity due to its character of general hydrolase of xenobiotic glycosides [97].

Several methods have been developed to monitor GBA1. They are driven by the need for diagnostic tools to evaluate the effects of this enzyme in the outcome of Gaucher disease, as well as in discovery processes of pharmacological chaperones for this enzyme [98]. Fluorescent non-natural substrates have been extensively used as probes for GBA1 assays. The application of 4-methylumbelliferyl-β-D-glucopyranoside (4-MUG, Fig. 9) was reported as an efficient alternative to the radiolabeled natural substrate when used alone [99] or in the presence of conduritol-β-epoxide (CBE) to selectively measure GBA1 activity in certain tissues [100]. In related works, the fluorescent analogs 2-hexadecanoylamino-4-nitrophenyl-β-D-glucopyranoside [(HNP)OGlc] [101], 1-*O*-glucosyl-2-*N*-(1-dimethylaminonaphthalene-5-sulfonyl) sphingosine [(DMAS)So(Glc)] [102, 103], (NBD)C_{12}GC [104, 105], (NBD) C_6GC,[106] and (Bodipy)C_{12}GC [106] (Fig. 9) were used as substrate for the diagnosis of Gaucher disease. On incubation with the enzyme, the substrates yield the free fluorescent Cer analogue for its quantification by fluorescence detection HPLC [102, 106]. For diagnostic purposes, (NBD)C_6GC and (Bodipy)C_{12}GC were much more efficient, as judged by their V_{max}/K_m ratio, than the classically used 4-MUG (Fig. 9) [106]. However, this substrate, together with the red fluorogenic substrate resorufin-β-D-glucopyranoside, (Reso)Glc, (Fig. 9) have been adapted to a miniaturized 1536 well-plate format for HTS of GBA1 inhibitors. Interestingly, both assays gave comparable results and helped to avoid artefacts due to autofluorescence or fluorescence quenching by the screened compounds when used in a paired screening [107]. Finally, different fluorescent substrates ((LR)C_{12}GC, (Bodipy)C_{12}GC, (LR)O-PAP, and 4-MUG, Fig. 9) were used to judge the effect of several Gaucher mutations on GBA1 activity. The substrate (LR)C_{12}GC was useful to distinguish between different mutant GBA1 variants overexpressed in a heterologous system [108]. In a different approach, a quantitative fluorescence-activated cell sorter (FACS) assay has been reported for diagnosis of Gaucher disease using fluorescein bis(β-glucopyranoside) (Flu(OGlc)$_2$) (Fig. 9) as substrate, with excellent results [109].

2.9 Natural Substrates Coupled to Chemical or Enzymatic Reactions

A common problem found with the use of fluorogenic enzyme assays based on the use of non-natural substrates in screening tests for inhibitors is the occurrence of false positives, since the affinity of the enzyme towards the artificial substrate may not be the same than that for the natural one. This is a general problem when the screening of potential modulators of a particular enzyme is considered. The activity of an enzyme inhibitor or activator depends on its binding site on the enzyme, which, in turn, modulates the affinity of the substrate and the catalytic ability of the enzyme. For this reason, an apparently potent enzyme inhibitor against an artificial substrate could be less potent or inactive if a natural substrate was used, or vice-versa.

As an illustrative example, recombinant GBA3 was found to hydrolyze artificial substrates such as 4-MUG and (NBD)C_6GC (Fig. 9), but hydrolysis of natural GC was hardly detected [97]. Ideally, the natural substrate itself, or a minimally modified analogue thereof, would be desirable. For this reason, enzymatic activity assays based on the use of the natural substrates is gaining relevance. In general, this strategy requires the use of derivatization reactions, prior to instrumental analysis or enzyme reactions specially designed to quantify the particular reaction products.

2.9.1 Derivatization Reactions

Fluorescence HPLC, after pre-column derivatization of sphingoid bases with naphthalene-2,3-dicarboxaldehyde (NDA) and cyanide ions, has been reported for assays for SK [110] and CDases [111]. According to the authors, the procedure is fast, sensitive (limit of detection of 12.6 nM, based on the NDA-adduct 1, Fig. 10) and reproducible and offers several advantages over other existing methods. Both SK [112] and CDase [113] activities have also been analyzed by derivatization of the resulting sphingoid base with NDA/cyanide or OPA/2-mercaptoethanol and quantification of the formed adducts (Fig. 10) by fluorescence HPLC [112] in a slight modification of a method previously used for amino acid analysis.

OPA derivatization of sphinganine was also applied to measure SPT activity in an HPLC-based assay [114]. In this assay, 3-ketoSa, the direct product of the SPT reaction, was reduced with sodium borohydride to Sa, which was derivatized with OPA and quantified by HPLC.

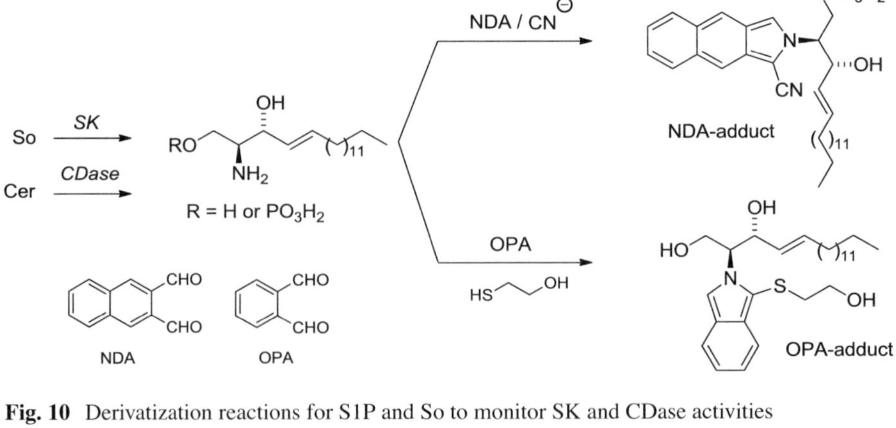

Fig. 10 Derivatization reactions for S1P and So to monitor SK and CDase activities

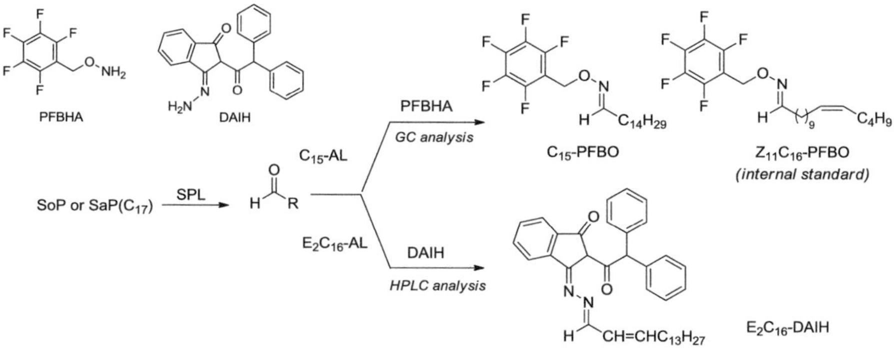

Fig. 11 Use of the natural substrate or a minimally modified probe to monitor SPL activity. *DAIH* 2-diphenylacetyl-1,3-indandione-1-hydrazide, *PFBHA* pentafluorobenzyl hydroxylamine, *PFBO* pentafluorobenzyl oxime, *C_{15}-AL* pentadecanal, *E_2C_{16}-AL* (*E*)-2-hexadecenal, *C_{15}-PFBO* pentadecanal pentafluorobenzyl oxime, *$Z_{11}C_{16}$-PFBO* (*Z*)-11-hexadecenal pentafluorobenzyl oxime, *C16-DAIH* (*E*)-2-hexadecenal 2-diphenylacetyl-1,3-indandione-1-hydrazone

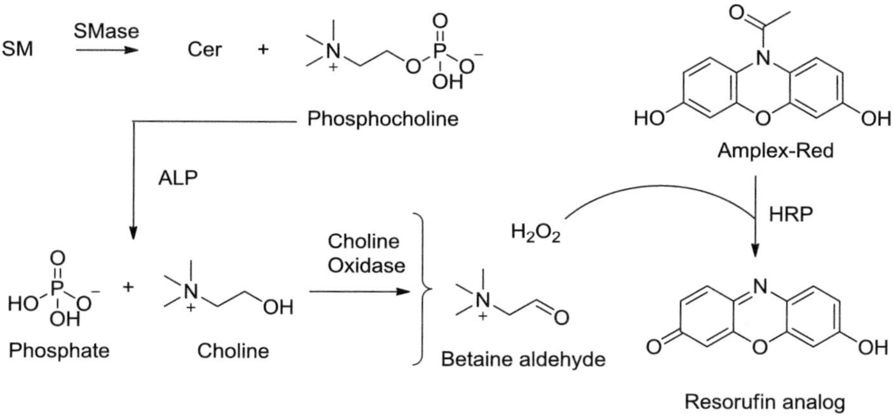

Fig. 12 Enzyme-coupled reaction to monitor SMase activity

A minimally modified C17-dihydrosphingosine-1-phosphate substrate ($Sa1P(C_{17})$, Fig. 11) was tested as SPL activity probe [115]. In this case, quantification of SPL activity is carried out by GC-MS analysis of the aldehyde product (pentadecanal) in the presence of (*Z*)-11 hexadecenal as an internal standard and derivatization to provide the corresponding pentafluorobenzyl oximes (PFBO) (Fig. 11). As expected, the affinity of this minimally modified substrate rises to levels comparable to those of the natural substrate [115]. Similar methods have been reported involving derivatization of (*E*)-2-hexadecenal with semicarbazide and analysis of the resulting semicarbazone by LC-MS/MS [116], 2-diphenylacetyl-1,3-indandione-1-hydrazone for fluorescence HPLC or LC-MS/MS analyses [117] and isonicotinylhydrazide for LC-MS/MS of the resulting isonicotinylhydrazone [118] (Fig. 11). Although not used to measure SPL activity, derivatization of (*E*)-2-

hexadecenal by Hantzsch reaction with cyclohexane-1,3-dione followed by fluorescence detection HPLC of the decahydroacridine derivative has been reported [119].

However, due to the more tedious and time consuming analytical protocols, these substrates are less amenable for HTS analysis than the previously reported method based on the fluorogenic coumarin-containing substrate RBM13 (Fig. 7).

2.9.2 Enzyme Coupled Reactions

An assay for human acid SMase using the natural SM as substrate has been disclosed. The assay is based on the detection of the generated phosphocholine (PC) by the enzyme coupled reporting system depicted in Fig. 12. This method has been

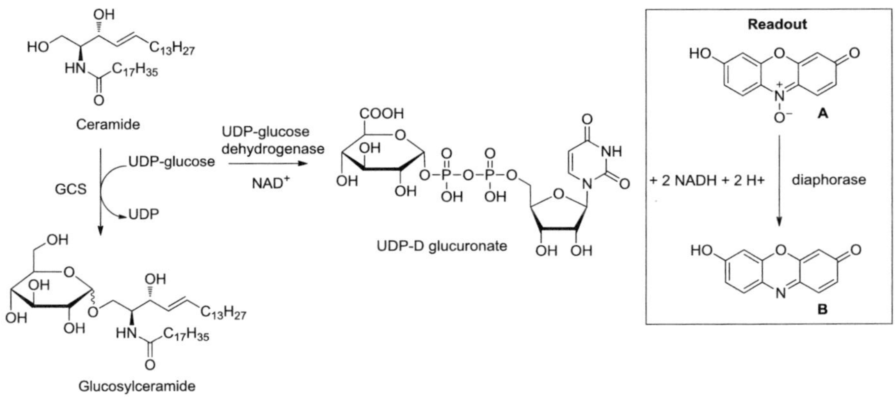

Fig. 13 Enzyme-coupled method used to screen for GCS inhibitors in a HTS assay

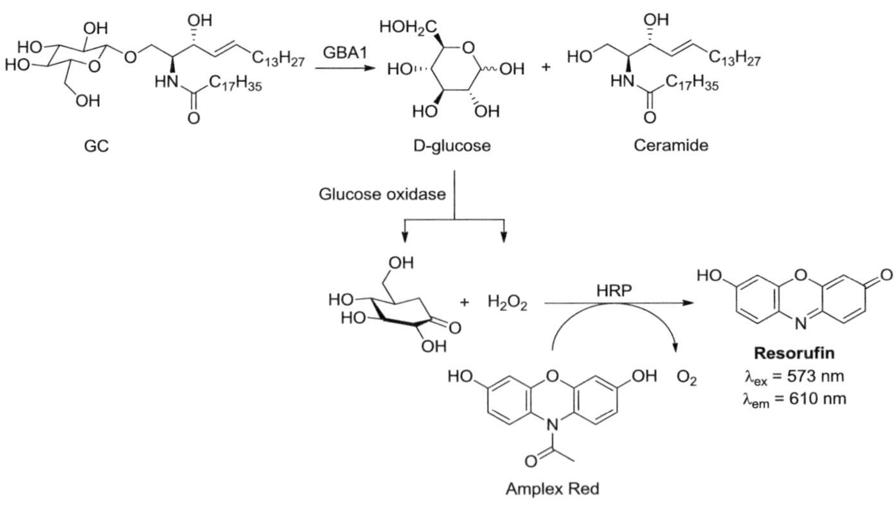

Fig. 14 Glucose oxidase reporting system used in a HTS assay for GBA1 inhibitors

used in the screening of a large library of compounds that led to the identification of some SMase inhibitors in the μM range [120]. Using this assay, the inhibitors showed different activities in comparison those obtained with the fluorogenic assay based on the use of (HMU)PC as substrate (Fig. 8).

Along the same lines, a HTS method has been setup for the screening of large libraries of compounds as GCS inhibitors. In this assay, GCS activity has been measured as the amount of UDP-glucose consumed during the GCS catalyzed reaction. After the reaction quenching, the excess UDP-glucose is transformed by a specific dehydrogenase that produces NADH, which, in turn, reduces resazurin (A) to the fluorescent resorufin (B) in the presence of the enzyme diaphorase (Fig. 13) [121]. A GBA1 assay based on the natural substrate has also been reported. The assay uses an Amplex-red glucose oxidase reporting system in a high throughput assay format (Fig. 14) and it is conceptually similar to that described above by the same authors for SMase [122].

3 Probes for the Quantitation and Trafficking of Sphingolipids

3.1 Amine Reactive Tags (iTRAQ Reagents) for Multiplexed Quantification of Ceramides

SLs are minor lipid cellular components, whose quantification requires the development of accurate and sensitive methods. Despite classical techniques, such as GC [123], HPLC [124–126], and MS [127–129] being widely used, the recent advances in lipidomics have boosted the development of more sophisticated techniques amenable to SL analysis [130–135]. The simultaneous analysis of SLs of different cell populations is of interest in terms of analysis efficiency and equipment optimization. In this context, the use of isobaric tags for relative and absolute quantification (iTRAQ), as developed in the proteomics field to compare the levels of protein expression in multiple samples [136], has been successfully applied to label phosphatidylethanolamine and phosphatidylserine [137] and in the multiplex analysis of Cer and GC, after enzymatic hydrolysis of their *N*-acyl chains with a SL *N*-deacylase [138]. In a recent example, two lactosylCer analogues carrying an azido-BODIPY dye, amenable to further functionalization with an iTRAQ reagent by click chemistry, has been described as a proof of concept for the differential analysis of multiplexed GSL by MS/MS [139].

The iTRAQ reagents consist of a "reporter group", a "balance group" and a "reactive group" in the iTRAQ tag (Fig. 15). The reporter group contains one or more stable ^{15}N or ^{13}C isotopes to form a family of four to eight differently tagged reagents differing by 1–8 mass units. The reactive groups is an *N*-hydroxysuccinimidyl ester, which reacts with the primary amino group of the analyte. Both the reactive and the reporter groups are linked with a balance group that contains the proper type and number of isotopes (^{15}N, ^{13}C and ^{18}O) to account for an identical total mass. In

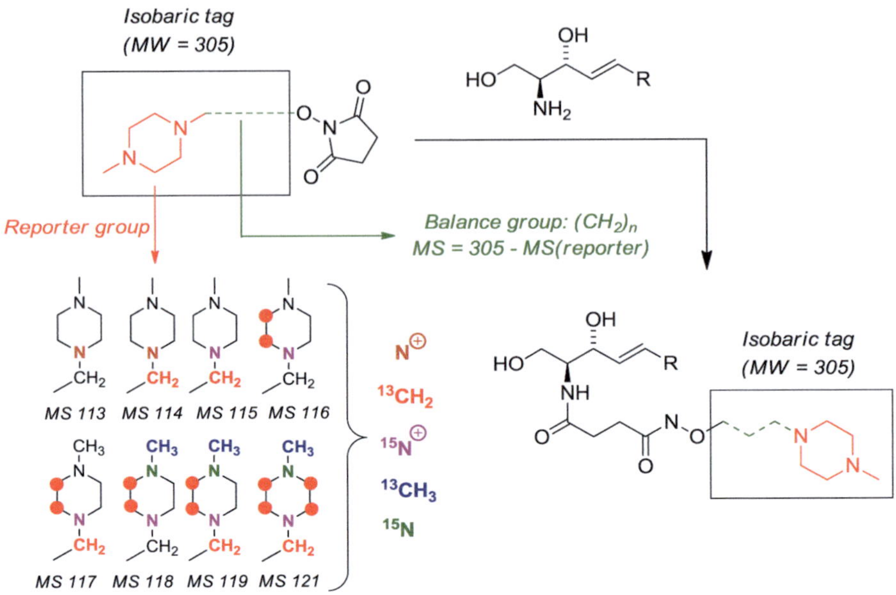

Fig. 15 iTRAQ reagents and application to sphingolipid tagging

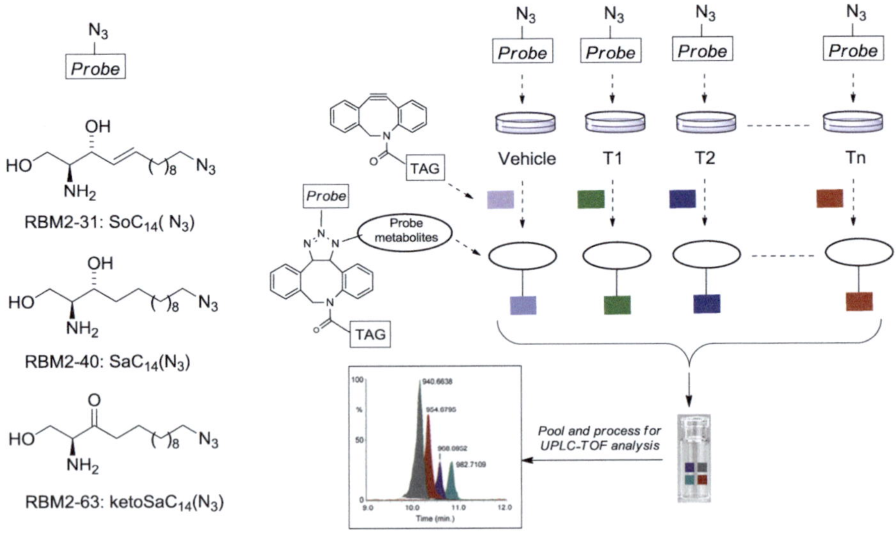

Fig. 16 Azide-tagged SL for the quantitative SL analysis of pooled samples

this way, since each reagent has an identical overall mass, the total intensity of the peaks represents the contribution from multiple samples, thus allowing an enhanced sensitivity. Under appropriate tandem MS-MS analysis, the iTRAQ tags fragment to release tag-specific reporter ions, which can be distinguished by their isotopic composition and are indicative of each original sample.

3.2 Azide-Tagged Sphingolipids for the Quantitative Sphingolipidomics of Pooled Samples

As indicated in Sect. 3.1, there is a need for fast and reliable methods for the simultaneous quantitative sphingolipidomics from different cell populations that are also amenable to HTS formats. In this context, the development of tagged ω-azidoSL probes (RBM2-31, RBM2-40, RBM2-63, Fig. 16) is an interesting alternative [164]. The probes do not affect cell viability and basal SL metabolism at appropriate concentrations. In addition, the reactivity and bioorthogonality of the terminal azido group can be exploited by strain-promoted click chemistry with suitable azodibenzocyclooctyne MS tags to show the usefulness of the approach for the simultaneous sphingolipidome analysis by UPLC-MS of different cell populations after independent cell treatments (Fig. 16). This method is reminiscent of the well-established "stable isotope labeling by amino acids in cell culture" (SILAC) approach for quantitative proteomics [140] and offers new opportunities in sphingolipids analysis.

3.3 Fluorescent Probes for Sphingolipid Localization and Intracellular Trafficking Studies

Investigation of the biological functions, localization and trafficking of SLs requires their detection in their natural compartments, either the cell membranes or the inner organelle. To this end, the high sensitivity of the currently available fluorescence methods [141] makes use of synthetic fluorescent or fluorogenic reporters an attractive option [142].

3.3.1 Fluorophore-Containing Sphingolipids

The use of the diphenyl-1,3,5-hexatrienyl group (DPH) as fluorophore in So (So(DPH), Fig. 4) was first reported in 1992 for distribution and metabolism studies in cultured human skin fibroblasts [143]. Since then, several So derivatives incorporating different fluorescent reporters at the ω-position of the long-chain base, separated by linkers of variable lengths, have been used to study the function and subcellular localization of this kind of SLs (Fig. 4) [47, 50]. In particular, the pyrene labeled sphingosine (SoC_{15}(Pyrene), Fig. 4) was shown to be rapidly incorporated in human endothelial cells and showed predominant distribution to the endoplasmic reticulum and the Golgi apparatus, wherein it is phosphorylated by SK. These data were in line with previously reported rapid uptake and metabolism of SoC_{14}(NBD) (Fig. 4) by cultured cells [144]. Fluorescent SM [145] and Cer analogues [146] have also been reported for trafficking studies. In a recent work, a series of Cer fluorescent probes, derived from an ω-labeled So backbone bearing acyl chains of different length (compounds C_2Cer (NBD), C_6Cer (NBD), C_{16}Cer (NBD), Fig. 5) have been designed and tested as tools to evaluate the binding of potential CERT ligands, being C_{16}Cer (NBD) the most efficient one as CERT substrate [147]. In the same work, the authors also described C_{16}Cer(Biotin) (Fig. 5) to illustrate the versatility of the synthetic approach used and also as a potential probe to further develop a CERT binding assay relying on labelled streptavidin. Fluorescent SLs have also been used as tools for cell membrane labeling and visualization to study membrane trafficking processes, as well as in biophysical studies to characterize the physical properties of lipid bilayers, such as membrane polarity, fluidity, lipid asymmetry, or their diffusion dynamics [148]. For this purpose, fluorescent Cer, SM and GC, generally bearing a Bodipy moiety at the acyl chain terminus have been used (Fig. 5) [149]. The examples shown above are just a sample of the large number of SL analogues described so far that have been modified by incorporation of a fluorescent moiety.

3.3.2 Polyunsaturated Sphingolipids

The use of the generally bulky, aromatic reporters raises some concerns about the potential alteration of the physico-chemical properties of the modified probes in relation to those of the natural SLs [150]. Although this limitation is common to most of the probes reported to date, some fluorescent probes based on the use of linear polyunsaturated systems are revealing as more efficient alternatives to the

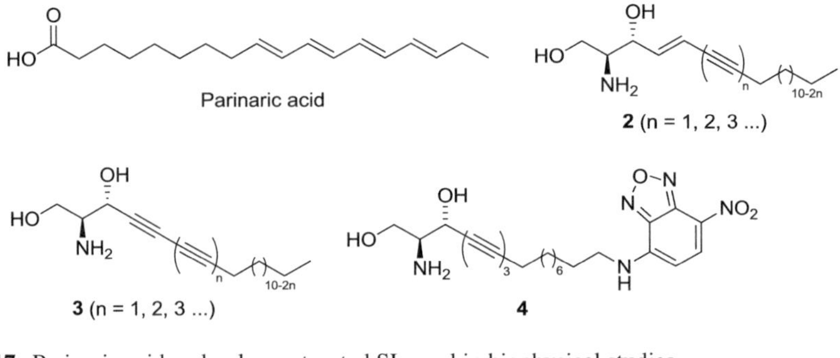

Fig. 17 Parinaric acid and polyunsaturated SL used in biophysical studies

classical aromatic probes. Since some excellent reviews on polyene lipids have been recently published [151, 152], we just want to focus in this section on the application of this strategy to the field of SLs. In general, polyene lipids are intrinsically fluorescent and their photophysical properties depend on the number of conjugated double bonds. Parinaric acid (Fig. 17) is a natural fatty acid with four all(*E*) conjugated double bonds, whose fluorescent properties can be improved by increasing the number of conjugated double bonds. In addition to the number of conjugated double bonds, their configuration (*Z* or *E*) is also a determinant of their overall shape and cellular fate. This is nicely illustrated when comparing natural SM (C_{16}SM) with the analogs (NBD)C_6SM, (Pyrene)C_{12}SM, and (all(*E*)-Pentaene) C_{16}SM [150]. Thus, NBD and BODIPY, two of the most popular and useful fluorophore tags, exert profound changes on the physical properties of the fatty acid to which they are attached. Despite the pyrene derivatives are better mimics in terms of shape, their fluorescent behaviour is complex. For this reason, all(*E*)-pentaene derivatives offer an excellent balance between structural shape and fluorescent properties, including also the possibility to determine protein-lipid interactions due to the overlapping between the pentaene UV excitation and tryptophan fluorescence emission [153]. The alteration of the physical and chemical properties imposed by the presence of unsaturations as part of the sphingoid chain backbone has been exploited in the search of conformationally restrained sphingolipids with modified biochemical properties. For instance, cells treated with the fluorescently labelled NBD-triyne probe 4 (Fig. 17) showed a significantly increased fluorescence signal compared with those treated with So(NBD) (Fig. 4), which has been attributed to the higher ability of the rigid polyyne to cross the cell membrane [154]. These differences in biophysical properties are translated into different pharmacological activity. Thus, the antiproliferative properties of probe 4 (Fig. 17) were superior to those of So(NBD) in most of the cell lines used in this study [154]. In another example, a series of polyyne-containing sphingoid bases (2, 3, Fig. 17) were synthesized via an iterative acetylene homologation sequence [155]. With the exception of polyyne 2a (n = 1), the remaining members of the series showed an antiproliferative effect higher than that of So in human colon cancer HCT116 and human lung cancer A549 cell lines.

3.3.3 Functionalized Sphingolipids Susceptible to Click Reactions

In order to minimally modify the physical properties of the natural sphingolipid, the fluorescent reporter can be introduced, at a later stage of the experiment, by reaction with a suitable probe in a fast and bioorthogonal process. Click chemistry reactions fulfil these requirements and are ideal platforms for this type of experiments, as described in Fig. 18.

An example of this strategy is illustrated by the use of the probes $C_{16}CerC_{14}(\omega N_3)$ and $C_{16}Cer(C_1N_3)$ (1 and 2, Fig. 18) for the *in situ* "click-on" labelling of ceramides in artificial membranes by Cu-catalyzed reaction with the fluorogenic alkyne F [156]. Even if Cu^{2+} at the concentrations used in this work may be toxic to cells,

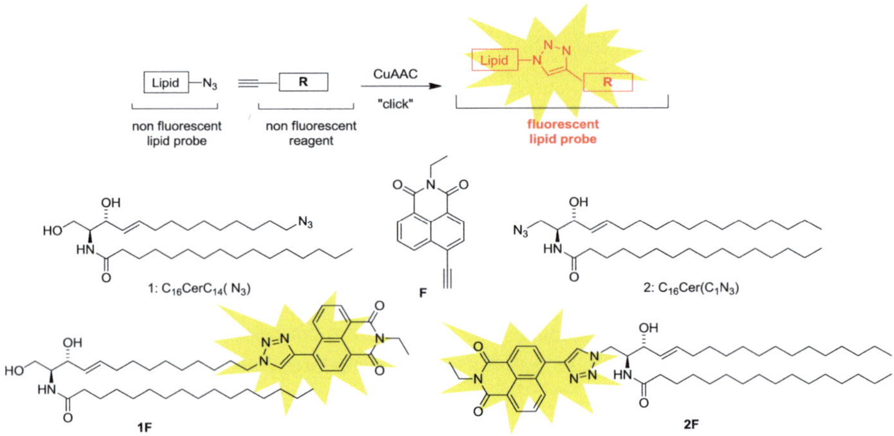

Fig. 18 Introduction of a fluorescent reporter in azido-SL by *in situ* labeling of (N_3)Cer in artificial membranes

localization of ceramide-rich domains in cell membranes could be performed on fixed cell preparations.

The distinctive signature of a fluorogenic reagent is its ability to give rise to a change in fluorescent quantum yield or a change in emissive wave length by reaction with a suitable probe. This last property is fulfilled by alkyne F, which behaves as a "latent fluorophore", whose fluorescence is greatly increased by conjugation with the triazole system resulting from the click reaction. This property is very useful to reduce the expected background fluorescence that would be observed if a conventional fluorescent reporter were used instead. Several examples of the use of fluorogenic reagents have been reported in the literature and excellent reviews and papers on this topic have been published elsewhere [157–162].

4 Summary and Outlook

The development of chemical probes to monitor the activity of SL metabolizing enzymes has classically relied on radioactive substrates. Apart from the safety and regulatory issues associated to the use of radiochemicals, most of these radioassays require the implementation of tedious extraction protocols prior to the scintillation counting stage. Although some interesting improvements to avoid the extraction steps have been developed (see, for example, the use of scintillation proximity counting to monitor SK in ref [26]) techniques for the direct analysis of SL metabolites are widely used. In this context, the simplicity and the sensitivity of fluorimetric analytical techniques have boosted the development of substrates that incorporate a generally bulky, fluorescent reporter as part of the sphingoid structure. By using these probes, several methods to measure the activity of most of the enzymes implicated in SL metabolism have been reported. Interestingly, some of them have been implemented into HTS protocols to screen for potential enzyme inhibitors. Nevertheless, such deeply modified substrates can show significant differences of affinity towards the target enzymes, which can lead to misleading results, especially when the screening of potential enzyme inhibitors is considered. This is one of the reasons that justify the use of the natural or minimally modified substrates as efficient alternatives. Unless the structural modification leads to reaction products easily traceable by MS techniques, a derivatization reaction of the resulting metabolites is required. In this context, copper-free click chemistry is becoming a standard tool in chemical biology for the incorporation of a large variety of chemical tags under biocompatible conditions. The use of controlled chemical reactions in live cells opens extraordinary opportunities for a better understanding of biochemical processes and paves the way for the development of even more sophisticated probes in the future to come.

References

1. Yang J, Yu Y, Sun S, Duerksen-Hughes PJ (2004) Ceramide and other sphingolipids in cellular responses. Cell Biochem Biophys 40:323–350
2. Gault CR, Obeid LM, Hannun YA (2010) An overview of sphingolipid metabolism: from synthesis to breakdown. Adv Exp Med Biol 688:1–23
3. Hannun YA, Obeid LM (2011) Many ceramides. J Biol Chem 286:27855–27862
4. Madison KC (2003) Barrier function of the skin: "la raison d'etre" of the epidermis. J Invest Dermatol 121:231–241
5. Pralhada Rao R, Vaidyanathan N, Rengasamy M, Mammen Oommen A, Somaiya N, Jagannath MR (2013) Sphingolipid metabolic pathway: an overview of major roles played in human diseases. J Lipids Article ID 178910
6. Giussani P, Tringali C, Riboni L, Viani P, Venerando B (2014) Sphingolipids: key regulators of apoptosis and pivotal players in cancer drug resistance. Int J Mol Sci 15:4356–4392
7. Furuya H, Shimizu Y, Kawamori T (2011) Sphingolipids in cancer. Cancer Metastasis Rev 30:567–576
8. Sugiura M, Kono K, Liu H, Shimizugawa T, Minekura H, Spiegel S, Kohama T (2002) Ceramide kinase, a novel lipid kinase. Molecular cloning and functional characterization. J Biol Chem 277:23294–23300
9. Gatt S (1966) Enzymic hydrolysis of sphingolipids. I. Hydrolysis and synthesis of ceramides by an enzyme from rat brain. J Biol Chem 241:3724–3730
10. Sugita M, Williams M, Dulaney JT, Moser HW (1975) Ceramidase and ceramide synthesis in human kidney and cerebellum: description of a new alkaline ceramidase. Biochim Biophys Acta 398:125–131
11. Nikolova-Karakashian M (2000) Assays for the biosynthesis of sphingomyelin and ceramide phosphoethanolamine. In: Merrill A, Hannun Y (eds) Sphingolipid metabolism and cell signaling part A. Methods in enzymology, vol 311. Academic, New York, pp 31–42
12. Schulze H, Michel C, van Echten-Deckert G (2000) Dihydroceramide desaturase. In: Merrill A, Hannun Y (eds) Sphingolipid metabolism and cell signaling part A. Methods in enzymology, vol 311. Academic Press, New York, pp 22–30
13. Duan R-D, Nilsson A (2000) Sphingolipid hydrolyzing enzymes in the gastrointestinal tract. Methods Enzymol 311:276–286

14. Levade T, Leruth M, Graber D, Moisand A, Vermeersch S, Salvayre R, Courtoy PJ (1996) In situ assay of acid sphingomyelinase and ceramidase based on LDL-mediated lysosomal targeting of ceramide-labeled sphingomyelin. J Lipid Res 37:2525–2538
15. Liu B, Hannun YA (2000) Sphingomyelinase assay using radiolabeled substrate. In: Merrill AH, Hannun YA (eds) Sphingolipid metabolism and cell signaling, part A. Methods in enzymology, vol 311. Academic Press, New York, pp 164–167
16. Gulbins E, Kolesnick R (2000) Measurement of sphingomyelinase activity. In: John C, Reed B (eds) Apoptosis, vol 322. Academic Press, New York, pp 382–388
17. Mintzer RJ, Appell KC, Cole A, Johns A, Pagila R, Polokoff MA, Tabas I, Snider RM, Meurer-Ogden JA (2005) A novel high-throughput screening format to identify inhibitors of secreted acid sphingomyelinase. J Biomol Screen 10:225–234
18. Hassler DF, Laethem RM, Smith GK (2000) A high throughput sphingomyelinase assay. Methods Enzymol 311:176–184
19. Barbone AG, Jackson AC, Ritchie DM, Argentieri DC (2000) Robotic assay of sphingomyelinase activity for high throughput screening. In: Merrill AH, Hannun YA (eds) Sphingolipid metabolism and cell signaling, part A. Methods in enzymology, vol 311. Academic Press, New York, pp 168–176
20. Vaccaro AM, Kobayashi T, Suzuki K (1982) Comparison of synthetic and natural glucosylceramides as substrate for glucosylceramidase assay. Clin Chim Acta 118:1–7
21. Strasberg PM, Lowden JA (1982) The assay of glucocerebrosidase activity using the natural substrata. Clin Chim Acta 118:9–20
22. Van Veldhoven PP (2000) Sphingosine-1-phosphate lyase. Methods Enzymol 311:244–254
23. Van Veldhoven PP, Mannaerts GP (1991) Subcellular localization and membrane topology of sphingosine-1-phosphate lyase in rat liver. J Biol Chem 266:12502–12507
24. Bajjalieh S, Batchelor RJ (2000) Ceramide kinase. Methods Enzymol 311:207–215
25. Olivera A, Barlow KD, Spiegel S (2000) Assaying sphingosine kinase activity. In: Merrill AH, Hannun YA, Ávalos M (eds) Sphingolipid metabolism and cell signaling; part A. Methods in enzymology, vol 311. Academic Press, New York, pp 215–223
26. Kharel Y, Mathews TP, Kennedy AJ, Houck JD, Macdonald L, Lynch KR (2011) A rapid assay for assessment of sphingosine kinase inhibitors and substrates. Anal Biochem 411: 230–235
27. Vessey DA, Kelley M, Karliner JS (2005) A rapid radioassay for sphingosine kinase. Anal Biochem 337:136–142
28. Shayman JA, Lee L, Abe A, Shu L (2000) Inhibitors of glucosylceramide synthase. Methods Enzymol 311:373–387
29. Matsuo N, Nomura T, Imokawa G (1992) A rapid and simple assay method for UDP-glucose: ceramide glucosyltransferase. Biochim Biophys Acta 1116:97–103
30. Spassieva S, Bielawski J, Anelli V, Obeid LM (2007) Combination of C(17) sphingoid base homologues and mass spectrometry analysis as a new approach to study sphingolipid metabolism. In: Bown A (ed) Methods in enzymology, vol 434. Academic Press, New York, pp 233–241
31. Abad JL, Nieves I, Rayo P, Casas J, Fabriàs G, Delgado A (2013) Straightforward access to spisulosine and 4,5-dehydrospisulosine stereoisomers: probes for profiling ceramide synthase activities in intact cells. J Org Chem 78:5858–5866
32. Gaebler A, Milan R, Straub L, Hoelper D, Kuerschner L, Thiele C (2013) Alkyne lipids as substrates for click chemistry-based in vitro enzymatic assays. J Lipid Res 54:2282–2290
33. Mullen TD, Hannun YA, Obeid LM (2012) Ceramide synthases at the centre of sphingolipid metabolism and biology. Biochem J 441:789–802
34. Ponnusamy S, Meyers-Needham M, Senkal CE, Saddoughi SA, Sentelle D, Selvam SP, Salas A, Ogretmen B (2010) Sphingolipids and cancer: ceramide and sphingosine-1-phosphate in the regulation of cell death and drug resistance. Future Oncol 6:1603–1624
35. Stiban J, Tidhar R, Futerman AH (2010) Ceramide synthases: roles in cell physiology and signaling. Adv Exp Med Biol 688:60–71

36. Kim HJ, Qiao Q, Toop HD, Morris JC, Don AS (2012) A fluorescent assay for ceramide synthase activity. J Lipid Res 53:1701–1707
37. Wijesinghe DS, Lamour NF, Gomez-Munoz A, Chalfant CE (2007) Ceramide kinase and ceramide-1-phosphate. In: Methods in enzymology, vol 434. Academic Press, New York, pp 265–292
38. Van Overloop H, Van der Hoeven G, Van Veldhoven PP (2012) A nonradioactive fluorimetric SPE-based ceramide kinase assay using NBD-C(6)-ceramide. J Lipids Article ID 404513
39. Don AS, Rosen H (2008) A fluorescent plate reader assay for ceramide kinase. Anal Biochem 375:265–271
40. Fyrst H, Saba JD (2010) An update on sphingosine-1-phosphate and other sphingolipid mediators. Nat Chem Biol 6:489–497
41. Heffernan-Stroud LA, Obeid LM (2013) Sphingosine kinase 1 in cancer. Adv Cancer Res 117:201–235
42. Alshaker H, Sauer L, Monteil D, Ottaviani S, Srivats S, Bohler T, Pchejetski D (2013) Therapeutic potential of targeting SK1 in human cancers. Adv Cancer Res 117:143–200
43. Pyne S, Bittman R, Pyne NJ (2011) Sphingosine kinase inhibitors and cancer: seeking the golden sword of Hercules. Cancer Res 71:6576–6582
44. Neubauer HA, Pitson SM (2013) Roles, regulation and inhibitors of sphingosine kinase 2. FEBS J 280:5317–5336
45. Roberts JL, Moretti PB, Darrow AL, Derian CK, Vadas MA, Pitson SM (2004) An assay for sphingosine kinase activity using biotinylated sphingosine and streptavidin-coated membranes. Anal Biochem 331:122–129
46. Billich A, Ettmayer P (2004) Fluorescence-based assay of sphingosine kinases. Anal Biochem 326:114–119
47. Peters C, Billich A, Ghobrial M, Hogenauer K, Ullrich T, Nussbaumer P (2007) Synthesis of borondipyrromethene (BODIPY)-labeled sphingosine derivatives by cross-metathesis reaction. J Org Chem 72:1842–1845
48. Lima S, Milstien S, Spiegel S (2014) A real-time high-throughput fluorescence assay for sphingosine kinases. J Lipid Res 55:1525–1530
49. French KJ, Zhuang Y, Maines LW, Gao P, Wang W, Beljanski V, Upson JJ, Green CL, Keller SN, Smith CD (2010) Pharmacology and antitumor activity of ABC294640, a selective inhibitor of sphingosine kinase-2. J Pharmacol Exp Ther 333:129–139
50. Ettmayer P, Billich A, Baumruker T, Mechtcheriakova D, Schmid H, Nussbaumer P (2004) Fluorescence-labeled sphingosines as substrates of sphingosine kinases 1 and 2. Bioorg Med Chem Lett 14:1555–1558
51. Yangyuoru PM, Hammonds-Odie L, Mwongela SM (2013) Fluorescent lipids as probes for sphingosine kinase activity by capillary electrophoresis. Methods Mol Biol 984:329–340
52. Yangyuoru PM, Otieno AC, Mwongela SM (2011) Determination of sphingosine kinase 2 activity using fluorescent sphingosine by capillary electrophoresis. Electrophoresis 32: 1742–1749
53. Schnute ME, McReynolds MD, Kasten T, Yates M, Jerome G, Rains JW, Hall T, Chrencik J, Kraus M, Cronin CN, Saabye M, Highkin MK, Broadus R, Ogawa S, Cukyne K, Zawadzke LE, Peterkin V, Iyanar K, Scholten JA, Wendling J, Fujiwara H, Nemirovskiy O, Wittwer AJ, Nagiec MM (2012) Modulation of cellular S1P levels with a novel, potent and specific inhibitor of sphingosine kinase-1. Biochem J 444:79–88
54. Mao C, Obeid LM (2008) Ceramidases: regulators of cellular responses mediated by ceramide, sphingosine, and sphingosine-1-phosphate. Biochim Biophys Acta 1781: 424–434
55. Canals D, Perry DM, Jenkins RW, Hannun YA (2011) Drug targeting of sphingolipid metabolism: sphingomyelinases and ceramidases. Br J Pharmacol 163:694–712
56. Zeidan YH, Jenkins RW, Korman JB, Liu X, Obeid LM, Norris JS, Hannun YA (2008) Molecular targeting of acid ceramidase: implications to cancer therapy. Curr Drug Targets 9:653–661

57. Liu X, Elojeimy S, Turner LS, Mahdy AEM, Zeidan YH, Bielawska A, Bielawski J, Dong J-Y, El-Zawahry AM, Guo G, Hannun YA, Holman DH, Rubinchik S, Szulc Z, Keane TE, Tavassoli M, Norris JS (2008) Acid ceramidase inhibition: a novel target for cancer therapy. Front Biosci 13:2293–2298
58. Nikolova-Karakashian M, Morgan ET, Alexander C, Liotta DC, Merrill AH (1997) Bimodal regulation of ceramidase by interleukin-1 β. J Biol Chem 272:18718–18724
59. Tani M, Okino N, Mitsutake S, Ito M (1999) Specific and sensitive assay for alkaline and neutral ceramidases. J Biochem Biophys Methods 749:746–749
60. Okino N, Tani M, Imayama S, Ito M (1998) Purification and characterization of a novel ceramidase from Pseudomonas aeruginosa. J Biol Chem 273:14368–14373
61. He X, Li CM, Park JH, Dagan A, Gatt S, Schuchman EH (1999) A fluorescence-based high-performance liquid chromatographic assay to determine acid ceramidase activity. Anal Biochem 274:264–269
62. Bhabak KP, Proksch D, Redmer S, Arenz C (2012) Novel fluorescent ceramide derivatives for probing ceramidase substrate specificity. Bioorg Med Chem 20:6154–6161
63. Bhabak KP, Hauser A, Redmer S, Banhart S, Heuer D, Arenz C (2013) Development of a novel FRET probe for the real-time determination of ceramidase activity. Chembiochem 14:1049–1052
64. Nieuwenhuizen WF, van Leeuwen S, Gotz F, Egmond MR (2002) Synthesis of a novel fluorescent ceramide analogue and its use in the characterization of recombinant ceramidase from Pseudomonas aeruginosa PA01. Chem Phys Lipids 114:181–191
65. Bedia C, Casas J, Garcia V, Levade T, Fabrias G (2007) Synthesis of a novel ceramide analogue and its use in a high-throughput fluorogenic assay for ceramidases. Chembiochem 8:642–648
66. Bedia C, Camacho L, Abad JL, Fabrias G, Levade T (2010) A simple fluorogenic method for determination of acid ceramidase activity and diagnosis of Farber disease. J Lipid Res 51:3542–3547
67. Wahler D, Badalassi F, Crotti P, Reymond JL (2001) Enzyme fingerprints by fluorogenic and chromogenic substrate arrays. Angew Chem Int Ed Engl 40:4457–4460
68. Bandhuvula P, Fyrst H, Saba JD (2007) A rapid fluorescence assay for sphingosine-1-phosphate lyase enzyme activity. J Lipid Res 48:2769–2778
69. Bandhuvula P, Li Z, Bittman R, Saba JD (2009) Sphingosine 1-phosphate lyase enzyme assay using a BODIPY-labeled substrate. Biochem Biophys Res Commun 380: 366–370
70. Bedia C, Camacho L, Casas J, Abad JL, Delgado A, Van Veldhoven PR, Fabrias G (2009) Synthesis of a fluorogenic analogue of sphingosine-1-phosphate and its use to determine sphingosine-1-phosphate lyase activity. Chembiochem 10:820–822
71. Kashem MA, Wa C, Wolak JP, Grafos NS, Ryan KR, Sanville-Ross ML, Fogarty KE, Rybina IV, Shoultz A, Molinaro T, Desai SN, Rajan A, Huber JD, Nelson RM (2014) A high-throughput scintillation proximity assay for sphingosine-1-phosphate lyase. Assay Drug Dev Technol 12:1–10
72. Kok JW, Nikolova-karakashian M, Klappe K, Alexander C, Merrill AH (1997) Dihydroceramide biology: structure-specific metabolism and intracellular localization. J Biol Chem 272:21128–21136
73. Munoz-Olaya JM, Matabosch X, Bedia C, Egido-Gabas M, Casas J, Llebaria A, Delgado A, Fabrias G (2008) Synthesis and biological activity of a novel inhibitor of dihydroceramide desaturase. ChemMedChem 3:946–953
74. Schuchman EH (2007) The pathogenesis and treatment of acid sphingomyelinase-deficient Niemann-Pick disease. J Inherit Metab Dis 30:654–663
75. Schissel SL, Jiang X, Tweedie-Hardman J, Jeong T, Camejo EH, Najib J, Rapp JH, Williams KJ, Tabas I (1998) Secretory sphingomyelinase, a product of the acid sphingomyelinase gene, can hydrolyze atherogenic lipoproteins at neutral pH. Implications for atherosclerotic lesion development. J Biol Chem 273:2738–2746
76. Wascholowski V, Giannis A (2001) Neutral sphingomyelinase as a target for drug design. Drug News Perspect 14:581–590

77. Okazaki T, Bielawska A, Domae N, Bell RM, Hannun YA (1994) Characteristics and partial purification of a novel cytosolic, magnesium-independent, neutral sphingomyelinase activated in the early signal transduction of 1 alpha,25-dihydroxyvitamin D3-induced HL-60 cell differentiation. J Biol Chem 269:4070–4077
78. Nyberg L, Duan RD, Axelson J, Nilsson A (1996) Identification of an alkaline sphingomyelinase activity in human bile. Biochim Biophys Acta 1300:42–48
79. Gal AE, Brady RO, Hibbert SR, Pentchev PG (1975) A practical chromogenic procedure for the detection of homozygotes and heterozygous carriers of Niemann-Pick disease. N Engl J Med 293:632–636
80. Gal AE, Fash FJ (1976) Synthesis of 2-*N*-(hexadecanoyl)-amino-4-nitrophenyl phosphorylcholine-hydroxide, a chromogenic substrate for assaying sphingomyelinase activity. Chem Phys Lipids 16:71–79
81. Van Diggelen OP, Voznyi YV, Keulemans JL, Schoonderwoerd K, Ledvinova J, Mengel E, Zschiesche M, Santer R, Harzer K (2005) A new fluorimetric enzyme assay for the diagnosis of Niemann-Pick A/B, with specificity of natural sphingomyelinase substrate. J Inherit Metab Dis 28:733–741
82. Loidl A, Claus R, Deigner HP, Hermetter A (2002) High-precision fluorescence assay for sphingomyelinase activity of isolated enzymes and cell lysates. J Lipid Res 43:815–823
83. López DJ, Egido-Gabas M, López-Montero I, Busto JVV, Casas J, Garnier M, Monroy F, Larijani B, Goñi FMM, Alonso A (2012) Accumulated bending energy elicits neutral sphingomyelinase activity in human red blood cells. Biophys J 102:2077–2085
84. Taniguchi M, Okazaki T (2014) The role of sphingomyelin and sphingomyelin synthases in cell death, proliferation and migration—from cell and animal models to human disorders. Biochim Biophys Acta 1841:692–703
85. Huitema K, Van Den Dikkenberg J, Brouwers JF, Holthuis JC (2004) Identification of a family of animal sphingomyelin synthases. EMBO J 23:33–44
86. Venkataraman K, Futerman AH (2001) Comparison of the metabolism of L-erythro- and L-threo-sphinganines and ceramides in cultured cells and in subcellular fractions. Biochim Biophys Acta 1530:219–226
87. García-Álvarez I, Egido-Gabás M, Romero-Ramírez L, Doncel-Pérez E, Nieto-Sampedro M, Casas J, Fernández-Mayoralas A, Garcia-Alvarez I, Egido-Gabas M, Romero-Ramirez L, Doncel-Perez E, Fernandez-Mayoralas A (2011) Lipid and ganglioside alterations in tumor cells treated with antimitotic oleyl glycoside. Mol Biosyst 7:129–138
88. Deng X, Sun H, Gao X, Gong H, Lu W, Chu Y, Zhou L, Ye D (2012) Development, validation, and application of a novel method for mammalian sphingomyelin synthase activity measurement. Anal Lett 45:1581–1589
89. Zama K, Mitsutake S, Watanabe K, Okazaki T, Igarashi Y (2012) A sensitive cell-based method to screen for selective inhibitors of SMS1 or SMS2 using HPLC and a fluorescent substrate. Chem Phys Lipids 165:760–768
90. Lahiri S, Futerman AH (2007) The metabolism and function of sphingolipids and glycosphingolipids. Cell Mol Life Sci 64:2270–2284
91. Gupta G, Surolia A (2010) Glycosphingolipids in microdomain formation and their spatial organization. FEBS Lett 584:1634–1641
92. Cox TM (2011) Gaucher disease: clinical profile and therapeutic developments. Biologics 4:299–313
93. Gouaze-Andersson V, Cabot MC (2006) Glycosphingolipids and drug resistance. Biochim Biophys Acta 1758:2096–2103
94. Marks DL, Kamisaka PP, Pagano RE (2000) Methods for studying glucosylceramide synthase. Methods Enzymol 311:50–59
95. Edmunds T (2010) Gaucher disease. In: Ramirez-Alvarado M, Kelly JW, Dobson CM (eds) Protein misfolding diseases: current and emerging principles and therapies. Wiley, Hoboken, pp 469–485
96. Boot RG, Verhoek M, Donker-Koopman W, Strijland A, van Marle J, Overkleeft HS, Wennekes T, Aerts JM (2007) Identification of the non-lysosomal glucosylceramidase as beta-glucosidase 2. J Biol Chem 282:1305–1312

97. Dekker N, Voorn-Brouwer T, Verhoek M, Wennekes T, Narayan RS, Speijer D, Hollak CE, Overkleeft HS, Boot RG, Aerts JM (2011) The cytosolic beta-glucosidase GBA3 does not influence type 1 Gaucher disease manifestation. Blood Cells Mol Dis 46:19–26
98. Trapero A, Llebaria A (2013) Glucocerebrosidase inhibitors for the treatment of Gaucher disease. Future Med Chem 5:573–590
99. Peters SP, Coyle P, Glew RH (1976) Differentiation of β-glucocerebrosidase from β-glucosidase in human tissues using sodium taurocholate. Arch Biochem Biophys 175: 569–582
100. Daniels LB, Glew RH, Radin NS, Vunnam RR, Radin S (1980) A revised fluorometric assay for Gaucher's disease using conduritol-β-epoxide with liver as the source of β-glucosidase. Clin Chim Acta 106:155–163
101. Gal AE, Pentchev PG, Fash FJ (1976) A novel chromogenic substrate for assaying glucocerebrosidase activity. Proc Soc Exp Biol Med 153:363–366
102. Naoi M, Kiuchi K, Yagi K (1981) A sensitive assay for glucocerebroside β-glucosidase by high-performance liquid chromatography using 1-O-glucosyl-2-N-(1-dimethylaminonaphthalene-5-sulfonyl)sphingosine as substrate. J Appl Biochem 3:544–551
103. Midorikawa M, Okada S, Yutaka T, Yabuuchi H, Naoi M, Kiuchi K, Yagi K (1985) Assay of glucocerebrosidase using a fluorescent analog of glucocerebroside for the diagnosis of Gaucher disease. Biochem Int 11:327–332
104. Grabowski GA, Dinur T, Gatt S, Desnick RJ (1982) Gaucher type 1 (Ashkenazi) disease: a new method for heterozygote detection using a novel fluorescent natural substrate. Clin Chim Acta 124:123–135
105. Abe A, Shayman JA, Radin NS (1992) Fluorescence assay of glucosylceramide glucosidase using NBD-cerebroside. Lipids 27:1052–1054
106. Hayashi Y, Zama K, Abe E, Okino N, Inoue T, Ohno K, Ito M (2008) A sensitive and reproducible fluorescent-based HPLC assay to measure the activity of acid as well as neutral beta-glucocerebrosidases. Anal Biochem 383:122–129
107. Urban DJ, Zheng W, Goker-Alpan O, Jadhav A, Lamarca ME, Inglese J, Sidransky E, Austin CP (2008) Optimization and validation of two miniaturized glucocerebrosidase enzyme assays for high throughput screening. Comb Chem High Throughput Screen 11:817–824
108. Ron I, Dagan A, Gatt S, Pasmanik-Chor M, Horowitz M (2005) Use of fluorescent substrates for characterization of Gaucher disease mutations. Blood Cells Mol Dis 35:57–65
109. Rudensky B, Paz E, Altarescu G, Raveh D, Elstein D, Zimran A (2003) Fluorescent flow cytometric assay: a new diagnostic tool for measuring β-glucocerebrosidase activity in Gaucher disease. Blood Cells Mol Dis 30:97–99
110. He X, Huang C-L, Schuchman EH (2009) Quantitative analysis of sphingosine-1-phosphate by HPLC after napthalene-2,3-dicarboxaldehyde (NDA) derivatization. J Chromatogr B Analyt Technol Biomed Life Sci 877:983–990
111. He X, Dagan A, Gatt S, Schuchman EH (2005) Simultaneous quantitative analysis of ceramide and sphingosine in mouse blood by naphthalene-2,3-dicarboxyaldehyde derivatization after hydrolysis with ceramidase. Anal Biochem 340:113–122
112. Min J-K, Yoo H-S, Lee E-Y, Lee W-J, Lee Y-M (2002) Simultaneous quantitative analysis of sphingoid base 1-phosphates in biological samples by o-phthalaldehyde precolumn derivatization after dephosphorylation with alkaline phosphatase. Anal Biochem 303:167–175
113. Bernardo K, Hurwitz R, Zenk T, Desnick RJ, Ferlinz K, Schuchman EH, Sandhoff K (1995) Purification, characterization, and biosynthesis of human acid ceramidase. J Biol Chem 270:11098–11102
114. Rütti MF, Richard S, Penno A, von Eckardstein A, Hornemann T (2009) An improved method to determine serine palmitoyltransferase activity. J Lipid Res 50:1237–1244
115. Reina E, Camacho L, Casas J, Van Veldhoven PP, Fabrias G (2012) Determination of sphingosine-1-phosphate lyase activity by gas chromatography coupled to electron impact mass spectrometry. Chem Phys Lipids 165:225–231
116. Berdyshev EV, Goya J, Gorshkova I, Prestwich GD, Byun HS, Bittman R, Natarajan V (2011) Characterization of sphingosine-1-phosphate lyase activity by electrospray ionization-

liquid chromatography/tandem mass spectrometry quantitation of (2E)-hexadecenal. Anal Biochem 408(1):12–18
117. Lüth A, Neuber C, Kleuser B (2012) Novel methods for the quantification of (2E)-hexadecenal by liquid chromatography with detection by either ESI QTOF tandem mass spectrometry or fluorescence measurement. Anal Chim Acta 722:70–79
118. Weiler S, Braendlin N, Beerli C, Bergsdorf C, Schubart A, Srinivas H, Oberhauser B, Billich A (2014) Orally active 7-substituted (4-benzylphthalazin-1-yl)-2-methylpiperazin-1-yl]nicotinonitriles as active-site inhibitors of sphingosine 1-phosphate lyase for the treatment of multiple sclerosis. J Med Chem 57:5074–5084
119. Mezzar S, Schryver E, Van Veldhoven PP, de Schryver E, Van Veldhoven PP (2014) RP-HPLC-fluorescence analysis of aliphatic aldehydes: application to aldehyde-generating enzymes HACL1 and SGPL1. J Lipid Res 55:573–582
120. Xu M, Liu K, Southall N, Marugan JJ, Remaley AT, Zheng W (2012) A high-throughput sphingomyelinase assay using natural substrate. Anal Bioanal Chem 404:407–414
121. Koltun E, Richards S, Chan V, Nachtigall J, Du H, Noson K, Galan A, Aay N, Hanel A, Harrison A, Zhang J, Won K-A, Tam D, Qian F, Wang T, Finn P, Ogilvie K, Rosen J, Mohan R, Larson C, Lamb P, Nuss J, Kearney P (2011) Discovery of a new class of glucosylceramide synthase inhibitors. Bioorg Med Chem Lett 21:6773–6777
122. Motabar O, Goldin E, Leister W, Liu K, Southall N, Huang W, Marugan JJ, Sidransky E, Zheng W (2012) A high throughput glucocerebrosidase assay using the natural substrate glucosylceramide. Anal Bioanal Chem 402:731–739
123. Vieu C, Terce F, Chevy F, Rolland C, Barbaras R, Chap H, Wolf C, Perret B, Collet X (2002) Coupled assay of sphingomyelin and ceramide molecular species by gas liquid chromatography. J Lipid Res 43:510–522
124. Snada S, Uchida Y, Anraku Y, Izawa A, Iwamori M, Nagai Y (1987) Analysis of ceramide and monohexaosyl glycolipid derivatives by high-performance liquid chromatography and its application to the determination of the molecular species in tissues. J Chromatogr 400: 223–231
125. Previati M, Bertolaso L, Tramarin M, Bertagnolo V, Capitani S (1996) Low nanogram range quantitation of diglycerides and ceramide by high-performance liquid chromatography. Anal Biochem 233:108–114
126. Yano M, Kishida E, Muneyuki Y, Masuzawa Y (1998) Quantitative analysis of ceramide molecular species by high performance liquid chromatography. J Lipid Res 39: 2091–2098
127. Kasumov T, Huang H, Chung Y-M, Zhang R, McCullough AJ, Kirwan JP (2010) Quantification of ceramide species in biological samples by liquid chromatography electrospray ionization tandem mass spectrometry. Anal Biochem 401:154–161
128. Scherer M, Leuthäuser-Jaschinski K, Ecker J, Schmitz G, Liebisch G (2010) A rapid and quantitative LC-MS/MS method to profile sphingolipids. J Lipid Res 51:2001–2011
129. Shaner RL, Allegood JC, Park H, Wang E, Kelly S, Haynes CA, Sullards MC, Merrill AH (2009) Quantitative analysis of sphingolipids for lipidomics using triple quadrupole and quadrupole linear ion trap mass spectrometers. J Lipid Res 50:1692–1707
130. Lam SM, Shui G (2013) Lipidomics as a principal tool for advancing biomedical research. J Genet Genomics 40:375–390
131. Wenk MR (2005) The emerging field of lipidomics. Nat Rev Drug Discov 4:594–610
132. van Meer G, Leeflang BR, Liebisch G, Schmitz G, Goni FM (2007) The European lipidomics initiative: enabling technologies. Methods Enzymol 432:213–232
133. Sandra K, Sandra P (2013) Lipidomics from an analytical perspective. Curr Opin Chem Biol 17:847–853
134. Han X, Yang K, Gross RW (2011) Multi-dimensional mass spectrometry-based shotgun lipidomics and novel strategies for lipidomic analyses. Mass Spectrom Rev 31:134–178
135. Schwudke D, Schuhmann K, Herzog R, Bornstein SR, Shevchenko A (2011) Shotgun lipidomics on high resolution mass spectrometers. Cold Spring Harb Perspect Biol 3:a004614. doi:10.1101/cshperspect.a004614

136. Evans C, Noirel J, Ow SY, Salim M, Pereira-Medrano AG, Couto N, Pandhal J, Smith D, Pham TK, Karunakaran E, Zou X, Biggs CA, Wright PC (2012) An insight into iTRAQ: where do we stand now? Anal Bioanal Chem 404:1011–1027
137. Berry KA, Murphy RC (2005) Analysis of cell membrane aminophospholipids as isotope-tagged derivatives. J Lipid Res 46:1038–1046
138. Nabetani T, Makino A, Hullin-Matsuda F, Hirakawa TA, Takeoka S, Okino N, Ito M, Kobayashi T, Hirabayashi Y (2011) Multiplex analysis of sphingolipids using amine-reactive tags (iTRAQ). J Lipid Res 52:1294–1302
139. Son S-H, Daikoku S, Ohtake A, Suzuki K, Kabayama K, Ito Y, Kanie O (2014) Syntheses of lactosyl ceramide analogues carrying novel bifunctional BODIPY dyes directed towards the differential analysis of multiplexed glycosphingolipids by MS/MS using iTRAQ. Chem Commun (Camb) 2:3010–3013
140. Mann M (2006) Functional and quantitative proteomics using SILAC. Nat Rev Mol Cell Biol 7:952–958
141. Van den Wildenberg SM, Prevo B, Peterman EJ (2011) A brief introduction to single-molecule fluorescence methods. Methods Mol Biol 783:81–99
142. Rasmussen JA, Hermetter A (2008) Chemical synthesis of fluorescent glycero- and sphingolipids. Prog Lipid Res 47:436–460
143. Antes P, Schwarzmann G, Sandhoff K (1992) Distribution and metabolism of fluorescent sphingosines and corresponding ceramides bearing the diphenylhexatrienyl (DPH) fluorophore in cultured human fibroblasts. Eur J Cell Biol 59:27–36
144. Hakogi T, Shigenari T, Katsumura S, Sano T, Kohno T, Igarashi Y (2003) Synthesis of fluorescence-labeled sphingosine and sphingosine 1-phosphate; effective tools for sphingosine and sphingosine 1-phosphate behavior. Bioorg Med Chem Lett 13:661–664
145. Gatt S, Dagan A (2012) Cancer and sphingolipid storage disease therapy using novel synthetic analogs of sphingolipids. Chem Phys Lipids 165:462–474
146. Schwarzmann G, Arenz C, Sandhoff K (1841) Labeled chemical biology tools for investigating sphingolipid metabolism, trafficking and interaction with lipids and proteins. Biochim Biophys Acta 2014:1161–1173
147. Combemale S, Santos C, Rodriguez F, Garcia V, Galaup C, Frongia C, Lobjois V, Levade T, Baudoin-Dehoux C, Ballereau S, Genisson Y (2013) A biologically relevant ceramide fluorescent probe to assess the binding of potential ligands to the CERT transfer protein. RSC Adv 3:18970–18984
148. Kleusch C, Hersch N, Hoffmann B, Merkel R, Csiszar A (2012) Fluorescent lipids: functional parts of fusogenic liposomes and tools for cell membrane labeling and visualization. Molecules 17:1055–1073
149. Marks DL, Bittman R, Pagano RE (2008) Use of Bodipy-labeled sphingolipid and cholesterol analogs to examine membrane microdomains in cells. Histochem Cell Biol 130: 819–832
150. Van Meer G, Liskamp RM (2005) Brilliant lipids. Nat Methods 2:14–15
151. Kuerschner L, Thiele C (1841) Multiple bonds for the lipid interest. Biochim Biophys Acta 2014:1031–1037
152. Kuerschner L, Ejsing CS, Ekroos K, Shevchenko A, Anderson KI, Thiele C (2005) Polyene-lipids: a new tool to image lipids. Nat Methods 2:39–45
153. Mateo CR, Souto AA, Amat-Guerri F, Acuña AU (1996) New fluorescent octadecapentaenoic acids as probes of lipid membranes and protein-lipid interactions. Biophys J 71: 2177–2191
154. Lee YM, Lim C, Lee HS, Shin YK, Shin K-O, Lee Y-M, Kim S (2013) Synthesis and biological evaluation of a polyyne-containing sphingoid base probe as a chemical tool. Bioconjug Chem 24:1324–1331
155. Kim S, Lee YM, Kang HR, Cho J, Lee T, Kim D (2007) Synthesis of novel polyyne analogues of sphingoid base via an iterative acetylene homologation sequence. Org Lett 9:2127–2130
156. Garrido M, Abad JL, Alonso A, Goni FM, Delgado A, Montes L-R (2012) In situ synthesis of fluorescent membrane lipids (ceramides) using click chemistry. J Chem Biol 5:119–123

157. Le Droumaguet C, Wang C, Wang Q (2010) Fluorogenic click reaction. Chem Soc Rev 39:1233–1239
158. Qi J, Han MS, Chang YC, Tung CH (2011) Developing visible fluorogenic "click-on" dyes for cellular imaging. Bioconjug Chem 22:1758–1762
159. Li X, Gao X, Shi W, Ma H (2014) Design strategies for water-soluble small molecular chromogenic and fluorogenic probes. Chem Rev 114:590–659
160. Hori Y, Norinobu T, Sato M, Arita K, Shirakawa M, Kikuchi K (2013) Development of fluorogenic probes for quick no-wash live-cell imaging of intracellular proteins. J Am Chem Soc 135:12360–12365
161. Nadler A, Schultz C (2013) The power of fluorogenic probes. Angew Chem Int Ed 52: 2408–2410
162. Yang J, Šečkutė J, Cole CM, Devaraj NK (2012) Live-cell imaging of cyclopropene tags with fluorogenic tetrazine cycloadditions. Angew Chem Int Ed 51:7476–7479
163. Tani M, Kita K, Komori H, Nakagawa T, Ito M (1998) Enzymatic synthesis of omega-amino-ceramide: preparation of a sensitive fluorescent substrate for ceramidase. Anal Biochem 263:183–188
164. Garrido M, Abad JL, Fabrias G, Casas J, Delgado A (2015) Azide-tagged sphingolipids: new tools for metabolic flux analysis. ChemBioChem 16:641–650

Use of Fenretinide to Increase Cytotoxic Dihydroceramides as a Novel Cancer Chemotherapeutic Approach

C. Patrick Reynolds and Barry J. Maurer

Abstract Sphingolipids have been extensively studied as signal transducers regulating stress response, proliferation, and cell death, yet agents that target sphingolipid signaling have been slow to advance to clinical studies in oncology. The role of sphingolipids as critical components of membrane synthesis, combined with the cytotoxicity achieved by ceramides and dihydroceramides, indicate that targeting sphingolipids has significant potential for novel cancer therapeutics. The discovery that high concentrations of the semi-synthetic retinoid, *N*-(4-hydroxyphenyl)retinamide (fenretinide, 4-HPR), can stimulate production of cytotoxic dihydroceramides led to preclinical studies focused on the use of high-dose 4-HPR as a dihydroceramide inducer for cancer therapeutics both as a single agent and in various ceramide-modulating drug combinations. Subsequent studies showed that 4-HPR increased dihydroceramides by simultaneously stimulating the *de novo* dihydroceramide synthesis pathway while inhibiting dihydroceramide desaturase 1 (DES1), preventing forward metabolism of dihydroceramides to ceramides. The high clinical therapeutic index of 4-HPR is due in part to the ability of 4-HPR to stimulate production of dihydroceramides in susceptible cancer cells but not in normal cells. The minimal cytotoxicity of 4-HPR for normal cells combined with improved formulations has enabled 4-HPR plasma concentrations of 10–60 μM to be achieved in clinical trials with minimal systemic toxicity. Durable complete clinical responses to 4-HPR have been observed in recurrent neuroblastoma, recurrent peripheral T cell lymphoma, and recurrent cutaneous T cell lymphoma, with clinical signals of activity also observed in ovarian cancer and gastrointestinal cancers. Future clinical trials will not only employ 4-HPR as a single agent, but will also test 4-HPR in combination with traditional cytotoxic agents, such as vincristine, and with other agents that target sphingolipid metabolism, such as safingol.

C.P. Reynolds, M.D., Ph.D. (✉) • B.J. Maurer, M.D., Ph.D.
Cancer Center, Departments of Cell Biology and Biochemistry, Pediatrics, and Medicine, School of Medicine, Texas Tech University Health Sciences Center, Lubbock, TX USA
e-mail: patrick.reynolds@ttuhsc.edu

Y.A. Hannun et al. (eds.), *Bioactive Sphingolipids in Cancer Biology and Therapy*, DOI 10.1007/978-3-319-20750-6_20

Keywords Ceramides • Dihydroceramides • Fenretinide • T-cell lymphoma • Neuroblastoma • Ovarian cancer safingol

1 Introduction

As reviewed throughout this book, sphingolipids have been extensively studied as signal transducers regulating stress response, proliferation, and cell death. In spite of years of attention to the role of sphingolipids in processes that are important to cancer progression and cancer cell survival, the development of agents that specifically target sphingolipid signaling have been slow to advance to clinical testing in oncology. The key role of sphingolipids as components of membrane synthesis, combined with the cellular cytotoxicity achieved by modulating ceramide(s) and dihydroceramides(s) levels, suggest that targeting sphingolipid pathways may lead to novel cancer therapeutics whose mechanisms of action may enhance or complement currently existing therapies. The discovery that high concentrations of the semi-synthetic retinoid, *N*-(4-hydroxyphenyl)retinamide (fenretinide, 4-HPR), stimulated production and increased levels of various dihydroceramide species in susceptible cancer cell lines suggested that 4-HPR might be used to clinically target sphingolipids for cancer treatment [1]. Efforts quickly focused on formulation development to enable high and extended 4-HPR exposures and on the identification of additional drugs capable of synergizing 4-HPR cytotoxicity through rationale modulation of ceramide-related cellular pathways. This chapter will review preclinical studies of high-dose 4-HPR that have translated to successful early-phase clinical trials. We will also review preclinical studies aimed at supporting early-phase clinical trials testing drug combinations specifically designed to further increase the levels of the cytotoxic dihydroceramides increased by 4-HPR (ex: D-threo-PPMP) or synergize dihydroceramide-based cytotoxicity (e.g., safingol; supplementation of specific fatty acids).

2 Fenretinide

A synthetic retinoid made in the late 1960s, *N*-(4-hydroxyphenyl)retinamide or fenretinide (4-HPR), has been reported to inhibit the growth of head and neck cancer [2], non-small cell lung cancer [3], small cell lung cancer [4, 5], breast cancer [6], ovarian cancer [7–11], prostate cancer [12–14], neuroblastoma [1, 15–18], Ewing's family tumor [19], leukemia [20–27], multiple myeloma [28], pancreatic cancer [29] cell lines *in vitro* at 4-HPR concentrations of 1–10 μM in a dose-dependent manner. 4-HPR has shown activity against leukemia [21] and neuroblastoma [18] cell lines known to be resistant to all-trans retinoic acid (ATRA) or 13-cis-retinoic acid (13-cis-RA).

A variety of intracellular signaling molecules are known to trigger or inhibit cell death via apoptosis [30, 31], including the Bcl-2 family of pro-and anti-apoptotic proteins [32]. However, not all cell death occurs via apoptosis and the malignancy-specific cytotoxicity induced by 4-HPR can involve complementary apoptotic and non-apoptotic cell death mechanisms depending on cancer cell type and the intensity and length of drug exposure [1]. Significantly, 4-HPR cytotoxicity is substantially independent of wild-type p53 function [1, 3, 22]. Initial studies of the mechanism of 4-HPR cytotoxicity focused on the induction of reactive oxygen species (ROS) [1, 21, 33, 34], however, antioxidants only partially abrogate 4-HPR-mediated cytotoxicity in neuroblastoma [1], indicating that other biochemical mechanisms were involved. The demonstration that 4-HPR could significantly increase *de novo* synthesis of ceramide species and cytotoxicity in susceptible neuroblastoma cell lines *in vitro* (but not in non-malignant cells) in both time- and dose-dependent manners suggested ceramide increase as another potential mechanism of action of 4-HPR [1]. This observation was rapidly exploited in preclinical studies that combined 4-HPR with other modulators of ceramide synthesis and metabolism [35]. The ability of 4-HPR to increase levels of certain ceramide species via stimulation of *de novo* ceramide synthesis was demonstrated in cell lines of a variety of cancer cell types [1, 19, 27, 35, 36]. As the methods of sphingolipid analysis progressed from thin-layer chromatography to tandem mass spectrometry, it became clear that the ceramide species increased by 4-HPR were dihydroceramides (i.e., desaturated between Carbons 4 and 5) [37, 38]. In distinct contrast to artificial, short-acyl chain dihydroceramides, the data supported a pro-death anticancer activity related to excessive levels of specific, native, long-acyl chain dihydroceramides [38, 39]. The former observation is consistent with the demonstration that dihydroceramide desaturase (DESI) is inhibited by 4-HPR [40]. Thus, the increase of cytotoxic dihydroceramides in susceptible cancer cells treated with fenretinide results from an increase of *de novo* ceramide synthase activity and a simultaneous decrease of dihydroceramide desaturation to ceramides.

3 Clinical Trials of Fenretinide as a Single Agent

In contrast to 13-cis-retinoic acid (13-cis-RA) and all-trans retinoic acid (ATRA), 4-HPR does not induce differentiation, but causes apoptosis and non-apoptotic cytotoxicity [41]. However, like other retinoids, 4-HPR toxicity in animal studies has been modest [41]. The systemic toxicity of 4-HPR in clinical trials has been minimal and hematologic toxicity infrequent [42–47]. The major clinical toxicity directly attributable to 4-HPR in adults has been transiently impaired night vision due to temporary decrease of plasma retinol levels [48]. In pediatric phase I trials, idiosyncratic cases of pseudotumor cerebri and liver transaminase elevations have also been reported [46]. With a novel, intravenous soybean oil emulsion formulation of 4-HPR that achieved high 4-HPR exposures, vehicle-related hypertriglyceridemia responsive to dose reduction was the major systemic toxicity observed [44].

A single case of fatal hepatic toxicity was seen with intravenous fenretinide emulsion in a pediatric patient that was attributed to a multi-drug interaction involving ceftriaxone and acetaminophen [49].

Initially, the only clinical formulation of 4-HPR was a large, poorly bioavailable capsule. This capsule formulation was developed primarily for chemoprevention studies. A large study employing capsular 4-HPR for chemoprevention of breast cancer demonstrated some signals of activity but was not sufficient to generate follow-on studies [50, 51]. Interestingly, the data from the breast cancer chemoprevention study suggested 4-HPR decreased the incidence of ovarian cancer, at least during the treatment portion of the trial [52]. However, follow-on chemoprevention trials combining oral capsule 4-HPR with tamoxifen were either negative or inconclusive for breast cancer prevention or control [50, 51, 53–55]. Some activity was observed with 4-HPR capsules or topical application in patients with leukoplakia but no definitive studies were carried out [56–59]. Chemoprevention trials of 4-HPR in cervical squamous lesions [60], resected bladder cancer [61], bronchial epithelium metaplasia [62], or prostate cancer [63] were either negative or inconclusive due to low accrual.

Phase II studies of low-dose oral capsule 4-HPR in breast cancer and melanoma [64], or renal cell cancer [65] were negative. Some activity was seen in a study of advanced renal carcinoma combining 4-HPR with paclitaxel and cisplatin, but it was unclear if the responses seen were enhanced by the 4-HPR [66]. A phase II trial of oral capsule 4-HPR in recurrent small cell lung cancer did not achieve objective responses but 24 % of patients showed stable disease [67]. A phase II study of high-sose intermittent capsule r-HPR in recurrent ovarian cancer also did not show objective responses but patients with 4-HPR plasma levels ≥9 μM had a significantly higher survival than patients with lower plasma levels [68]. A high-dose intermittent oral capsule 4-HPR study in prostate cancer patients with rising PSA after completion of anti-androgen therapy failed to show activity [69] while a similar trial in patients with asymptomatic rising PSA prior to anti-androgen therapy did not achieve response endpoints but did demonstrate stable disease in 30 % of patients [70]. It is important to note that the mean peak plasma 4-HPR exposures obtained in all the adult studies employing the oral capsules were <10 μM (often 2–3 μM) apparently due to poor bioavailability of the capsule formulation combined with interpatient variation in drug absorption or metabolism and that low concentrations of 4-HPR do not effectively stimulate production of dihydroceramides in cancer cell lines *in vitro* [1].

4 4-HPR and Neuroblastoma

A major focus of early-phase clinical trials of 4-HPR has been in neuroblastoma due to preclinical data that showed high sensitivity to 4-HPR in neuroblastoma cell lines established at time of progressive disease after multi-agent chemotherapy [1, 18].

Neuroblastoma is a malignant sympathetic nervous system tumor that accounts for 8 % of childhood cancers [71]. High-risk neuroblastoma, defined by age, stage, and *MYCN* oncogene amplification, poses a major therapeutic challenge [72]. For patients with high-risk neuroblastoma, multi-agent therapy, myeloablative consolidation, followed by high-dose, pulse 13-cis-retinoic acid (13-cis-RA) to treat minimal residual disease (MRD), improved event-free survival if utilized before progressive disease [73]. A further improvement in survival was seen with addition of the anti-GD2 antibody ch14.18+cytokines to treat MRD [74] but >40 % of patients still die from progressive disease.

Two different pediatric phase I studies of oral capsule 4-HPR achieved substantial dose escalation [46, 75], approaching the ~10 μM 4-HPR concentrations that demonstrated high activity (and stimulation of *de novo* dihydroceramide synthesis) in neuroblastoma cell lines *in vitro* [1, 36]. Both studies documented a number of patients with stable disease [46, 75], a complete response in a neuroblastoma patient was observed in the study employing two or three doses per day [46], and long-term disease-free survival was seen in the latter study with a recurrent Ewing's sarcoma patient who achieved stable disease on 4-HPR. A phase II study of 4-HPR capsules given at the "maximal practical dose" determined in the multiple-daily dosing phase I study achieved one partial response and seven patients with stable disease, of which one achieved a durable (>5 years) complete response when continued on 4-HPR by compassionate access post-study [47]. Aside from demonstrating signals of activity, these early-phase pediatric studies of capsule 4-HPR demonstrated that, while 4-HPR was well-tolerated, neither study truly achieved a maximal tolerated dose, indicating that formulations with improved bioavailability might be tolerable and produce higher 4-HPR drug exposures and increased anti-tumor activity.

An oral formulation (4-HPR/LXS) was developed as a powder to facilitate administration [76]. The 4-HPR/LXS formulation employs a lipid matrix (Lym-X-Sorb™) to increase bioavailability via gut absorption to the lymphatic system similar to that of a chylomicron [76]. To improve palatability, the powder incorporates vanilla flavoring. The 4-HPR/LXS formulation can be readily given to mice as a slurry in water (by gavage) and it demonstrated significant activity against multiple neuroblastoma mouse xenograft models [76].

A phase I dose-escalation study was undertaken with 4-HPR/LXS in children with neuroblastoma, initially employing a uniform liquid vehicle for administration (Slimfast®) on a BID-TID schedule of 7 days on, 2 weeks off [77]. Overall, the 4-HPR plasma levels obtained using 4-HPR/LXS were significantly higher than with the capsule formulation; four complete responses were observed in the dose escalation phase of the neuroblastoma phase I study [77]. Interpatient differences in obtained peak 4-HPR plasma levels were still observed suggesting the presence of interpatient variation in the rate of 4-HPR metabolism (i.e., 'high' and 'low' metabolizers). Interestingly, preclinical metabolism studies with both human and murine systems identified the potential to further increase 4-HPR plasma levels and tumor exposures by inhibiting P450-mediated 4-HPR metabolism with concurrent low doses of the oral antifungal agent, ketoconazole [78]. The ability of ketoconazole to increase 4-HPR exposures in mice with ketoconazole was demonstrated [78] and the increased 4-HPR exposures achieved using 4-HPR/LXS + ketoconazole was associated with both enhanced 4-HPR activity against neuroblastoma murine xenografts and enhanced accumulation of dihydroceramides in the xenografted tumors [79, 80]. Based on these data, the phase I study of 4-HPR/LXS was amended to test 4-HPR/LXS given with a (mostly) unrestricted choice of vehicles for administration and the addition of concurrent oral ketoconazole at ordinary fungal-treatment doses. Concurrent ketoconazole + 4-HPR/LXS was well-tolerated, increased mean 4-HPR peak plasma levels ~50 %, and demonstrated clinical activity, including two complete responses and a highly encouraging progression-free survival [81]. In this setting, concurrent ketoconazole is postulated to decrease the rate of fenretinide metabolism in patients with 'high metabolizing' P450 enzyme complements. Table 1 summarizes the peak 4-HPR plasma levels and the clinical activity observed in neuroblastoma patients across the series of early-phase studies with the two different formulations of oral 4-HPR.

How much of the preclinical and clinical anti-neuroblastoma activity seen with 4-HPR (used as a single agent) is attributable to an increase of dihydroceramides remains uncertain. Preclinical studies demonstrated that higher 4-HPR concentrations overcome anti-oxidant antagonism of 4-HPR cytotoxicity (but not the ability of the anti-oxidants to diminish reactive oxygen species) suggesting that at higher 4-HPR concentrations the increase of dihydroceramides plays a key role in 4-HPR cytotoxicity [1]. Interestingly, in acute lymphoblastic leukemia (ALL) cell line studies, 4-HPR cytotoxicity correlated only with the increase of a specific subset of long acyl chained dihydroceramides, and supplementation of 4-HPR exposures with specific long chain fatty acids increased both levels of the corresponding dihydroceramides and cytotoxicity [39]. Similar results were observed in neuroblastoma cell lines in association with a slightly different complement of long acyl chain dihydroceramides (Abedi and Maurer, unpublished). Together, these results provide support for a causal link between the increase of specific long acyl chain dihydroceramides and 4-HPR-induced cytotoxicity, possibly in a cancer-type specific manner. Similarly, inhibition of cellular ceramide 'shunting' by inhibition of glucosylceramide synthase (GCS) with such agents as D-*threo*-1-phenyl-2-palmitoylamino-3-morpholino-1-propanol (D-*threo*-PPMP) enhanced the accumulation of ceramide species, including dihydroceramides, and 4-HPR cytoxicity in neuroblastoma [35] and acute lymphoblastic leukemia [27] cell lines, providing additional supportive data for the importance of increased dihydroceramide levels to 4-HPR cytotoxicity.

Table 1 Clinical studies of oral fenretinide in recurrent/refractory high-risk neuroblastoma

Study	Phase	Dosage form	Daily dose mg/ m^2/day (divided)	Mean peak plasma level (max, μM)	Evaluable patients	Objective responses	Stable disease	Reference
CCG 09709	I	Capsule	2475 max (TID)	10±5	30	1 CR	13	46
COG ANBL-0321	II	Capsule	2475/1800 fixed (TID/BID)	8±3	58	1 PR 1 CR[a]	13	47
NANT 2004-04 Escalation	I	LXS powder (specified vehicle)	2210 max (BID)	x	30[a]	4 CR[b]	6[a]	77
NANT 2004-04 Expansion 1	I	LXS powder (liberalized vehicle)	1500 fixed (TID)	12±4	15	1 CR	8	81
NANT 2004-04 Expansion 2	I	LXS powder + ketoconazole (liberalized vehicle)	1500 fixed (TID)	18±4	18	2 CR 1 PR 2 MR	5	81

CR complete response, *PR* partial response, *MR* mixed response

[a]Stable disease while on study, CR observed on continued 4-HPR (compassionate access) post-study

[b]All responses in escalation phase of NANT 2004-04 at ≥774 mg/m^2/day (N=18)

5 Future 4-HPR Studies in Neuroblastoma

Maintenance therapy after completion of myeloablative therapy with 13-cis-retinoic acid [73] interspersed with the ch14.18 antibody to the disialoganglioside GD2 + cytokines to stimulate antibody-dependent cellular cytotoxicity (ADCC) [74] has clearly improved outcome in clinical trials of high-risk neuroblastoma. As mentioned previously, preclinical studies demonstrated that neuroblastoma cell lines selected for resistance to retinoic acid are hypersensitive to 4-HPR [18] and neuroblastoma patients with recurrent disease after 13-cis-RA therapy respond to 4-HPR [77, 81]. Therefore, given the modest observed systemic toxicity of 4-HPR/LXS and clinical activity in recurrent neuroblastoma, the addition of 4-HPR/LXS to maintenance therapy for high-risk neuroblastoma has the potential to further improve outcomes. Furthermore, other laboratory studies suggest that the generation of dihydroceramides by 4-HPR may have benefits beyond the direct cytotoxic effect of 4-HPR for neuroblastoma: 4-HPR has been shown to enhance ADCC of ch14.18 antibody for neuroblastoma cell lines by multiple mechanisms, including the stimulation of increased GD2 expression on the neuroblastoma cell lines (likely due to the enhanced production of dihydrogangliosides resulting from 4-HPR-stimulated increase of dihydroceramides) [82].

Our current preclinical and clinical studies of 4-HPR/LXS + ketoconazole for use in neuroblastoma after disease progression are focused on combining 4-HPR/LXS with novel targeted agents, such as Bcl-2 inhibitors [83], or with microtubule inhibitors [79, 80, 84]. Interestingly, the enhancement of cytotoxicity achieved by combining 4-HPR with microtubule inhibitors appears to be primarily dependent on the ability of 4-HPR to increase reactive oxygen species levels rather than on increased dihydroceramide levels [79, 84]. As modulators of ceramide metabolism, such as PPMP, [27, 35] are developed for clinical trials, multidrug combinations based on 4-HPR that achieve synergistic anti-cancer activity via either or both of the dihydroceramides or reactive oxygen species increasing properties of 4-HPR are likely to emerge.

6 Intravenous Fenretinide

While oral administration of 4-HPR has the advantage of outpatient convenience and lower cost, an intravenous emulsion formulation of 4-HPR was developed in order to insure optimal 4-HPR exposures and to facilitate combination with other agents that require intravenous administration [85]. It was hypothesized that 4-HPR plasma levels >40 μM could be maintained for 5 days, would be well-tolerated, and would provide a high degree of anti-cancer activity. Due to the pharmacokinetic properties of 4-HPR, it was deemed that obtaining such high exposures would require continuous intravenous infusion. A phase I study in recurrent adult hematological malignancies of a novel, soybean oil-based, intravenous emulsion 4-HPR

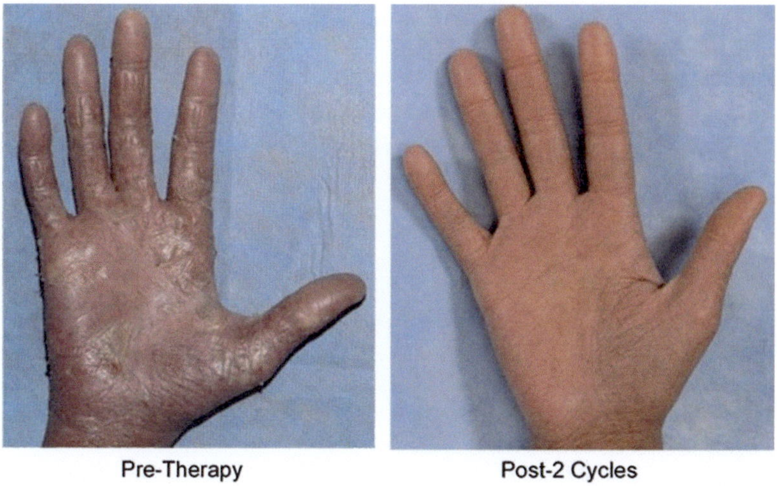

Fig. 1 Response of patient with cutaneous T cell lymphoma to intravenous fenretinide [86]

formulation evidenced minimal systemic toxicity and documented several complete and partial responses in recurrent cutaneous T cell lymphoma and recurrent peripheral T-cell lymphoma [44]. An example of a dramatic response in cutaneous T cell lymphoma to intravenous 4-HPR is shown in Fig. 1 [86]. Plasma levels of ~40–60 μM were achieved. Although, to date, given the limitations of Phase I studies, patient tumors have not been biopsied for the effects on 4-HPR treatment on intratumor ceramide levels, 4-HPR did stimulate high levels of dihydroceramides in T-cell related malignant (but not in non-malignant) lymphoid cell lines *in vitro* [27, 39]. Thus, increased dihydroceramides may be a mechanism by which 4-HPR is clinically active against T cell lymphomas that can be directly tested in future trials. Table 2 summarizes the clinical activity of 4-HPR observed in T cell lymphoma patients in this phase I trial [44]. A companion phase I study in adult solid tumors was also recently completed with clear signals of activity in adenocarcinomas of the gastrointestinal tract. This latter observation was consistent with the prolonged stable disease observed in a patient with a rapidly progressing colon cancer in a phase I adult phase I study of the oral 4-HPR/LXS formulation [87].

6.1 Combinations of 4-HPR with Other Sphingolipid-Modulating Agents: Safingol

Preclinical studies in both pediatric and adult cancers support developing clinical trials testing the combination of 4-HPR with modulators of ceramide synthesis and metabolism. One such agent is safingol, the L-*threo* stereo-chemical variant of the

Table 2 Clinical activity of intravenous emulsion fenretinide in PTCL and CTCL

Diagnosis	Dose (mg/m²/day)	Courses	Response
PTCL (AILT)	1810	6	PRu-PFS 6 months
CTCL-Sezary	1280	26	CR-PFS 36+ months
CTCL	1280 → 905	7+	SD Much improved skin
PTCL (AILT)	905	4	CRu PFS 14+ months
CTCL	905	6	PR PFS 5 months
CTCL	905 → 640	3	SD PFS 3+ months
CTCL	905 → 450	<1,<1	SD 2 months
CTCL	640	12	SD PFS 8 months
PTCL	640	3	SD
PTCL (AILT)	640	2	PD

PTCL peripheral T-cell lymphoma, *CTCL* cutaneous T-cell lymphoma, *AILT* angioimmunoblastic T-cell lymphoma, *CR* complete response, *PR* partial response, *SD* stable disease, *PD* progressive disease, *PFS* progression-free survival, *CRu* complete response, unconfirmed, *PRu* partial response, unconfirmed

normal cellular precursor of dihydroceramides, D-*erythro*-sphinganine [88]. Safingol has been variously reported to be a sphingosine kinase or protein kinase C (PKC) inhibitor at high concentrations but can also function as the sphingoid backbone in place of normal sphinganine and be acylated to various fatty acids to form L-*threo*-dihydroceramides [88]. Safingol (like 4-HPR) has been reported to increase autophagy in cancer cell lines *in vitro*, although the mechanism may not be via modulation of sphingolipids in all cases [89]. A pilot clinical trial of an intravenous safingol formulation combined with doxorubicin (with safingol delivered as a bolus with the intent of achieving plasma levels sufficient to achieve a Protein Kinase C inhibitory effect) demonstrated only modest clinical activity [90]. A phase I clinical trial of intravenous safingol combined with cisplatin and delivered as a bolus intended for sphingosine kinase inhibition demonstrated modest activity [91] and that safingol was well-tolerated at plasma levels higher than those shown to synergize 4-HPR activity pre-clinically [35]. This latter trial also demonstrated a decrease in plasma sphingosine-1-phosphate levels in patients treated with safingol, providing the first *in vivo* evidence for safingol (and/or its metabolites) to potentially act as a clinical inhibitor of sphingosine kinase [91]. Both studies demonstrated that safingol had a fairly short plasma half-life when delivered as an intravenous bolus which may possibly have limited the enhancement of the activity of the administered co-agent through mechanisms depending on PKC or sphingosine kinase inhibition. Based on these clinical studies, and on preclinical studies showing that safingol significantly enhanced 4-HPR cytotoxicity against a broad range of cancer types [35], a phase I trial in adult solid tumors and lymphomas (NCT01553071) of intravenous emulsion 4-HPR combined with intravenous emulsion safingol is intended. This trial will employ an extended low-dose intravenous infusion of safingol to facilitate safingol conversion to L-*threo*-dihydroceramides for enhancement of 4-HPR activity based on the co-presence of both D-erythro- and L-threo-dihydroceramides. Results with this latter study will inform development of future trials combining 4-HPR with other modulators of ceramide synthesis and metabolism.

7 Summary

Preclinical and clinical studies have demonstrated that drugs which manipulate intracellular sphingolipids, with fenretinide being the most widely studied, can have significant anti-cancer activity. A number of other approaches to targeting sphingolipids are underway in many laboratories and future clinical trials will be needed to define their potential as cancer therapeutics and in treating other disease states, such as cystic fibrosis [92] and viral infections [93, 94].

As most anticancer drugs have multiple mechanisms of action that may be dose, time, and tumor-type dependent, it is unclear the extent to which sphingolipids are actually involved in the anti-cancer activity of putative sphingolipid pathway-targeted agents currently in, or approaching, clinical trials. To be definitive, such demonstrations will likely need serial biopsy of patient tumors during treatment under controlled conditions. While obtaining such biopsies is a difficult logistical and ethical undertaking, such studies should be encouraged whenever feasible in clinical trials, especially in lymphoid malignancies where access to cancer cells may be had via peripheral lymph node biopsy. However, with respect to fenretinide and fenretinide-based drug combinations, the observed effects on dihydroceramides, the broad spectrum of anticancer activity observed in preclinical models, and especially the minimal reported effects of 4-HPR on non-cancer cells, suggest the presence of a basic dysregulation of the *de novo* sphingolipid synthetic pathway in many malignant cells. The latter is, perhaps, inherently related to the dysregulation of membrane synthesis necessary for cancer cell proliferation and may serve as the basis for novel sphingolipid-based cancer therapies.

Acknowledgments Supported in part by Cancer Prevention and Research Institute Grants RP10072+CP130023 and by National Cancer Institute grants CA82830+CA161889.

Conflict of Interest The Children's Hospital Los Angeles (CHLA) holds patents and/or patent applications on fenretinide intravenous emulsion, safingol intravenous emulsion, and fenretinide/LYM-XSORB™ (LXS) oral powder. CHLA and co-inventors of fenretinide and safingol formulations, including Drs. Barry J. Maurer and C. Patrick Reynolds, Texas Tech University Health Sciences Center, Lubbock, TX, may potentially benefit financially from the development and future use of those drugs.

References

1. Maurer BJ, Metelitsa LS, Seeger RC, Cabot MC, Reynolds CP (1999) N-(4-hydroxyphenyl) retinamide increases ceramide and reactive oxygen species and induces mixed apoptosis/necrosis in neuroblastoma cell lines. J Natl Cancer Inst 91(13):1138–1146
2. Oridate N, Lotan D, Xu XC, Hong WK, Lotan R (1996) Differential induction of apoptosis by all-trans-retinoic acid and N-(4-hydroxyphenyl)retinamide in human head and neck squamous cell carcinoma cell lines. Clin Cancer Res 2(5):855–863
3. Zou CP, Kurie JM, Lotan D, Zou CC, Hong WK, Lotan R (1998) Higher potency of N-(4-hydroxyphenyl)retinamide than all-trans-retinoic acid in induction of apoptosis in non-small cell lung cancer cell lines. Clin Cancer Res 4(5):1345–1355
4. Kalemkerian GP, Slusher R, Ramalingam S, Gadgeel S, Mabry M (1995) Growth inhibition and induction of apoptosis by fenretinide in small-cell lung cancer cell lines. J Natl Cancer Inst 87:1674–1680
5. Kalemkerian GP, Ou X (1999) Activity of fenretinide plus chemotherapeutic agents in small-cell lung cancer cell lines. Cancer Chemother Pharmacol 43(2):145–150
6. Sheikh MS, Shao ZM, Li XS et al (1995) N-(4-hydroxyphenyl)retinamide (4-HPR)-mediated biological actions involve retinoid receptor-independent pathways in human breast carcinoma. Carcinogenesis 16:2477–2486
7. Illuzzi G, Bernacchioni C, Aureli M et al (2010) Sphingosine kinase mediates resistance to the synthetic retinoid N-(4-hydroxyphenyl)retinamide in human ovarian cancer cells. J Biol Chem 285(24):18594–18602
8. Prinetti A, Basso L, Appierto V et al (2003) Altered sphingolipid metabolism in N-(4-hydroxyphenyl)-retinamide-resistant A2780 human ovarian carcinoma cells. J Biol Chem 278(8):5574–5583
9. Supino R, Crosti M, Clerici M et al (1996) Induction of apoptosis by fenretinide (4HPR) in human ovarian carcinoma cells and its association with retinoic acid receptor expression. Int J Cancer 65:491–497
10. Villani MG, Appierto V, Cavadini E et al (2004) Identification of the fenretinide metabolite 4-oxo-fenretinide present in human plasma and formed in human ovarian carcinoma cells through induction of cytochrome P450 26A1. Clin Cancer Res 10(18 Pt 1):6265–6275
11. Wang H, Zhang Y, Du Y (2013) Ovarian and breast cancer spheres are similar in transcriptomic features and sensitive to fenretinide. Biomed Res Int 2013:510905
12. Pienta KJ, Nguyen NM, Lehr JE (1993) Treatment of prostate cancer in the rat with the synthetic retinoid fenretinide. Cancer Res 53:224–226
13. Wang H, Charles AG, Frankel AJ, Cabot MC (2003) Increasing intracellular ceramide: an approach that enhances the cytotoxic response in prostate cancer cells. Urology 61(5): 1047–1052
14. Gouaze-Andersson V, Flowers M, Karimi R et al (2011) Inhibition of acid ceramidase by a 2-substituted aminoethanol amide synergistically sensitizes prostate cancer cells to N-(4-hydroxyphenyl) retinamide. Prostate 71(10):1064–1073
15. Di Vinci A, Geido E, Infusini E, Giaretti W (1994) Neuroblastoma cell apoptosis induced by the synthetic retinoid N-(4-hydroxyphenyl)retinamide. Int J Cancer 59:422–426
16. Mariotti A, Marcora E, Bunone G et al (1994) N-(4-hydroxyphenyl)retinamide: a potent inducer of apoptosis in human neuroblastoma cells. J Natl Cancer Inst 86:1245–1247

17. Ponzoni M, Bocca P, Chiesa V et al (1995) Differential effects of N-(4-hydroxyphenyl) retinamide and retinoic acid on neuroblastoma cells: apoptosis versus differentiation. Cancer Res 55:853–861
18. Reynolds CP, Wang Y, Melton LJ, Einhorn PA, Slamon DJ, Maurer BJ (2000) Retinoic-acid-resistant neuroblastoma cell lines show altered MYC regulation and high sensitivity to fenretinide. Med Pediatr Oncol 35(6):597–602
19. Batra S, Reynolds CP, Maurer BJ (2004) Fenretinide cytotoxicity in Ewing's sarcoma (ES) and primitive neuroectodermal (PNET) cell lines is blunted by hypoxia but synergistically enhanced by safingol. Cancer Res 64:5415–5424
20. Apraiz A, Idkowiak-Baldys J, Nieto-Rementeria N, Boyano MD, Hannun YA, Asumendi A (2012) Dihydroceramide accumulation and reactive oxygen species are distinct and nonessential events in 4-HPR-mediated leukemia cell death. Biochem Cell Biol 90(2):209–223
21. Delia D, Aiello A, Lombardi L et al (1993) N-(4-hydroxyphenyl)retinamide induces apoptosis of malignant hemopoietic cell lines including those unresponsive to retinoic acid. Cancer Res 53:6036–6041
22. Delia D, Aiello A, Formelli F et al (1995) Regulation of apoptosis induced by the retinoid N-(4-hydroxyphenyl) retinamide and effect of deregulated bcl-2. Blood 85:359–367
23. Faderl S, Lotan R, Kantarjian HM, Harris D, Van Q, Estrov Z (2003) N-(4-hydroxylphenyl) retinamide (fenretinide, 4-HPR), a retinoid compound with antileukemic and proapoptotic activity in acute lymphoblastic leukemia (ALL). Leuk Res 27(3):259–266
24. Jiang L, Pan X, Chen Y, Wang K, Du Y, Zhang J (2011) Preferential involvement of both ROS and ceramide in fenretinide-induced apoptosis of HL60 rather than NB4 and U937 cells. Biochem Biophys Res Commun 405(2):314–318
25. Kang MH, Wan Z, Kang YH, Sposto R, Reynolds CP (2008) Mechanism of synergy of N-(4-hydroxyphenyl)retinamide and ABT-737 in acute lymphoblastic leukemia cell lines: Mcl-1 inactivation. J Natl Cancer Inst 100(8):580–595
26. Morales MC, Perez-Yarza G, Rementeria NN et al (2007) 4-HPR-mediated leukemia cell cytotoxicity is triggered by ceramide-induced mitochondrial oxidative stress and is regulated downstream by Bcl-2. Free Radic Res 41(5):591–601
27. O'Donnell PH, Guo WX, Reynolds CP, Maurer BJ (2002) N-(4-hydroxyphenyl)retinamide increases ceramide and is cytotoxic to acute lymphoblastic leukemia cell lines, but not to non-malignant lymphocytes. Leukemia 16(5):902–910
28. Li X, Ling W, Pennisi A, Khan S, Yaccoby S (2009) Fenretinide inhibits myeloma cell growth, osteoclastogenesis and osteoclast viability. Cancer Lett 284(2):175–181
29. Messner MC, Cabot MC (2011) Cytotoxic responses to N-(4-hydroxyphenyl)retinamide in human pancreatic cancer cells. Cancer Chemother Pharmacol 68(2):477–487
30. Rowan S, Fisher DE (1997) Mechanisms of cell death. Leukemia 11:457–465
31. Cereghetti GM, Scorrano L (2006) The many shapes of mitochondrial death. Oncogene 25(34):4717–4724
32. Kang MH, Reynolds CP (2009) Bcl-2 inhibitors: targeting mitochondrial apoptotic pathways in cancer therapy. Clin Cancer Res 15(4):1126–1132
33. Delia D, Aiello A, Meroni L, Nicolini M, Reed JC, Pierotti MA (1997) Role of antioxidants and intracellular free radicals in retinamide-induced cell death. Carcinogenesis 18(5): 943–948
34. Oridate N, Suzuki S, Higuchi M, Mitchell MF, Hong WK, Lotan R (1997) Involvement of reactive oxygen species in *N*-(4-hydroxyphenyl)retinamide-induced apoptosis in cervical carcinoma cells. J Natl Cancer Inst 89(16):1191–1198
35. Maurer BJ, Melton L, Billups C, Cabot MC, Reynolds CP (2000) Synergistic cytotoxicity in solid tumor cell lines between N-(4-hydroxyphenyl)retinamide and modulators of ceramide metabolism. J Natl Cancer Inst 92(23):1897–1909
36. Wang H, Maurer BJ, Reynolds CP, Cabot MC (2001) N-(4-hydroxyphenyl)retinamide elevates ceramide in neuroblastoma cell lines by coordinate activation of serine palmitoyltransferase and ceramide synthase. Cancer Res 61(13):5102–5105

37. Kraveka JM, Li L, Szulc ZM et al (2007) Involvement of dihydroceramide desaturase in cell cycle progression in human neuroblastoma cells. J Biol Chem 282(23):16718–16728
38. Wang H, Maurer BJ, Liu YY et al (2008) N-(4-hydroxyphenyl)retinamide increases dihydroceramide and synergizes with dimethylsphingosine to enhance cancer cell killing. Mol Cancer Ther 7(9):2967–2976
39. Holliday MW Jr, Cox SB, Kang MH, Maurer BJ (2013) C22:0- and C24:0-dihydroceramides confer mixed cytotoxicity in T-cell acute lymphoblastic leukemia cell lines. PLoS One 8(9), e74768
40. Rahmaniyan M, Curley RW Jr, Obeid LM, Hannun YA, Kraveka JM (2011) Identification of dihydroceramide desaturase as a direct in vitro target for fenretinide. J Biol Chem 286(28):24754–24764
41. Reynolds CP, Lemmons RS (2001) Retinoid therapy of childhood cancer. Hematol Oncol Clin North Am 15:867–910
42. Cobleigh MA, Dowlatshahi K, Deutsch TA et al (1993) Phase I/II trial of tamoxifen with or without fenretinide, an analog of vitamin A, in women with metastatic breast cancer. J Clin Oncol 11:474–477
43. Formelli F, Clerici M, Campa T et al (1993) Five-year administration of fenretinide: pharmacokinetics and effects on plasma retinol concentrations. J Clin Oncol 11(10):2036–2042
44. Mohrbacher A, Kang MH, Yang AS et al (2012) Phase I trial of fenretinide (4-HPR) intravenous emulsion in hematologic malignancies: a California cancer consortium study (PhI-42). J Clin Oncol 30, suppl abstr 8073
45. Torrisi R, Pensa F, Orengo MA et al (1993) The synthetic retinoid fenretinide lowers plasma insulin-like growth factor I levels in breast cancer patients. Cancer Res 53:4769–4771
46. Villablanca JG, Krailo MD, Ames MM, Reid JM, Reaman GH, Reynolds CP (2006) Phase I trial of oral fenretinide in children with high-risk solid tumors: a report from the Children's Oncology Group (CCG 09709). J Clin Oncol 24(21):3423–3430
47. Villablanca JG, London WB, Naranjo A et al (2011) Phase II study of oral capsular 4-hydroxyphenylretinamide (4-HPR/fenretinide) in pediatric patients with refractory or recurrent neuroblastoma: a report from the Children's Oncology Group. Clin Cancer Res 17(21):6858–6866
48. Torrisi R, Parodi S, Fontana V et al (1994) Factors affecting plasma retinol decline during long-term administration of the synthetic retinoid fenretinide in breast cancer patients. Cancer Epidemiol Biomarkers Prev 3:507–510
49. Kang MH, Villablanca JG, Glade Bender JL et al (2014) Probable fatal drug interaction between intravenous fenretinide, ceftriaxone, and acetaminophen: a case report from a New Approaches to Neuroblastoma (NANT) Phase I study. BMC Res Notes 7(1):256
50. Camerini T, Mariani L, De PG et al (2001) Safety of the synthetic retinoid fenretinide: long-term results from a controlled clinical trial for the prevention of contralateral breast cancer. J Clin Oncol 19(6):1664–1670
51. Veronesi U, de Palo G, Marubini E et al (1999) Randomized trial of fenretinide to prevent second breast malignancy in women with early breast cancer. J Natl Cancer Inst 91: 1847–1856
52. De Palo G, Mariani L, Camerini T et al (2002) Effect of fenretinide on ovarian carcinoma occurrence. Gynecol Oncol 86(1):24–27
53. Decensi A, Robertson C, Guerrieri-Gonzaga A et al (2009) Randomized double-blind 2 x 2 trial of low-dose tamoxifen and fenretinide for breast cancer prevention in high-risk premenopausal women. J Clin Oncol 27(23):3749–3756
54. Rao RD, Cobleigh MA, Gray R et al (2011) Phase III double-blind, placebo-controlled, prospective randomized trial of adjuvant tamoxifen vs. tamoxifen and fenretinide in postmenopausal women with positive receptors (EB193): an intergroup trial coordinated by the Eastern Cooperative Oncology Group. Med Oncol 28(Suppl 1):S39–S47
55. Singletary SE, Atkinson EN, Hoque A et al (2002) Phase II clinical trial of N-(4-hydroxyphenyl) retinamide and tamoxifen administration before definitive surgery for breast neoplasia. Clin Cancer Res 8(9):2835–2842

56. Chiesa F, Tradati N, Grigolato R et al (2005) Randomized trial of fenretinide (4-HPR) to prevent recurrences, new localizations and carcinomas in patients operated on for oral leukoplakia: long-term results. Int J Cancer 115(4):625–629
57. Lippman SM, Lee JJ, Martin JW et al (2006) Fenretinide activity in retinoid-resistant oral leukoplakia. Clin Cancer Res 12(10):3109–3114
58. Tradati N, Chiesa F, Rossi N et al (1994) Successful topical treatment of oral lichen planus and leukoplakias with fenretinide (4-HPR). Cancer Lett 76:109–111
59. William WN Jr, Lee JJ, Lippman SM et al (2009) High-dose fenretinide in oral leukoplakia. Cancer Prev Res (Phila) 2(1):22–26
60. Follen M, Atkinson EN, Schottenfeld D et al (2001) A randomized clinical trial of 4-hydroxyphenylretinamide for high-grade squamous intraepithelial lesions of the cervix. Clin Cancer Res 7(11):3356–3365
61. Sabichi AL, Lerner SP, Atkinson EN et al (2008) Phase III prevention trial of fenretinide in patients with resected non-muscle-invasive bladder cancer. Clin Cancer Res 14(1):224–229
62. Kurie JM, Lee JS, Khuri FR et al (2000) N-(4-hydroxyphenyl)retinamide in the chemoprevention of squamous metaplasia and dysplasia of the bronchial epithelium. Clin Cancer Res 6(8):2973–2979
63. Pienta KJ, Esper PS, Zwas F, Krzeminski R, Flaherty LE (1997) Phase II chemoprevention trial of oral fenretinide in patients at risk for adenocarcinoma of the prostate. Am J Clin Oncol 20(1):36–39
64. Modiano MR, Dalton WS, Lippman SM, Joffe L, Booth AR, Meyskens FL Jr (1990) Phase II study of fenretinide (N-[4-hydroxyphenyl]retinamide) in advanced breast cancer and melanoma. Invest New Drugs 8(3):317–319
65. Vaishampayan U, Heilbrun LK, Parchment RE et al (2005) Phase II trial of fenretinide in advanced renal carcinoma. Invest New Drugs 23(2):179–185
66. Otterson GA, Lavelle J, Villalona-Calero MA et al (2005) A phase I clinical and pharmacokinetic study of fenretinide combined with paclitaxel and cisplatin for refractory solid tumors. Invest New Drugs 23(6):555–562
67. Schneider N, Lanz S, Ramer R, Schaefer D, Goppelt-Struebe M (2001) Up-regulation of cyclooxygenase-1 in neuroblastoma cell lines by retinoic acid and corticosteroids. J Neurochem 77(2):416–424
68. Reynolds CP, Frgala T, Tsao-Wei D et al (2007) High plasma levels of fenretinide (4-HPR) were associated with improved outcome in a phase II study of recurrent ovarian cancer: a study by the California cancer consortium. Proc Am Soc Clin Oncol 25(28S):5555
69. Moore MM, Stockler M, Lim R, Mok TS, Millward M, Boyer MJ (2010) A phase II study of fenretinide in patients with hormone refractory prostate cancer: a trial of the Cancer Therapeutics Research Group. Cancer Chemother Pharmacol 66(5):845–850
70. Cheung E, Pinski J, Dorff T et al (2009) Oral fenretinide in biochemically recurrent prostate cancer: a California cancer consortium phase II trial. Clin Genitourin Cancer 7(1):43–50
71. Reynolds CP, Seeger RC (2000) Neuroblastoma. In: Haskell CM (ed) Cancer treatment, 5th edn. W.B. Saunders, Philadelphia, pp 1214–1236
72. Park JR, Bagatell R, London WB et al (2013) Children's Oncology Group's 2013 blueprint for research: neuroblastoma. Pediatr Blood Cancer 60(6):985–993
73. Matthay K, Villablanca JG, Seeger RC et al (1999) Treatment of high risk neuroblastoma with intensive chemotherapy, radiotherapy, autologous bone marrow transplantation, and 13-cis-retinoic acid. N Engl J Med 341(16):1165–1173
74. Yu AL, Gilman AL, Ozkaynak MF et al (2010) Anti-GD2 antibody with GM-CSF, interleukin-2, and isotretinoin for neuroblastoma. N Engl J Med 363(14):1324–1334
75. Garaventa A, Luksch R, Lo Piccolo MS et al (2003) Phase I trial and pharmacokinetics of fenretinide in children with neuroblastoma. Clin Cancer Res 9(6):2032–2039
76. Maurer BJ, Kalous O, Yesair DW et al (2007) Improved oral delivery of N-(4-hydroxyphenyl) retinamide with a novel LYM-X-SORB organized lipid complex. Clin Cancer Res 13(10): 3079–3086

77. Maurer BJ, Kang MH, Villablanca JG et al (2013) Phase I trial of fenretinide delivered orally in a novel organized lipid complex in patients with relapsed/refractory neuroblastoma: a report from the New Approaches to Neuroblastoma Therapy (NANT) consortium. Pediatr Blood Cancer 60(11):1801–1808
78. Cooper JP, Hwang K, Singh H et al (2011) Fenretinide metabolism in humans and mice: utilizing pharmacological modulation of its metabolic pathway to increase systemic exposure. Br J Pharmacol 163(6):1263–1275
79. Lopez-Barcons L, Kang MH, Maurer BJ, Reynolds CP (2012) Abstract 2493: Enhanced activity of fenretinide/-LYM-X-SORB (4-HPR/LXS) oral powder in combination with ketoconazole and vincristine against recurrent neuroblastoma xenografts. In: Proceedings of the 103rd Annual Meeting of the American Association for Cancer Research, Chicago
80. Lopez-Barcons L, Singh H, Maurer BJ, Kang MH, Reynolds CP (2012) Potent antitumor activity of fenretinide/LYM-X-SORB (4-HPR/LXS) oral powder in combination with ketoconazole and vincristine against recurrent neuroblastoma xenografts. Advances in Neuroblastoma Research Abstract POTO 74
81. Marachelian A, Glade-Bender J, Kang MH et al (2014) Fenretinide (4-HPR)/Lym-X-Sorb(LXS) oral powder+ketoconazole in patients with high-risk recurrent or resistant neuroblastoma. A New Approaches to Neuroblastoma Therapy (NANT) consortium trial. J Clin Oncol 32:5s, suppl abstr 10071
82. Shibina A, Seidel D, Somanchi SS et al (2013) Fenretinide sensitizes multidrug-resistant human neuroblastoma cells to antibody-independent and ch14.18-mediated NK cell cytotoxicity. J Mol Med (Berl) 91(4):459–472
83. Fang H, Harned TM, Kalous O, Maldonado V, DeClerck YA, Reynolds CP (2011) Synergistic activity of fenretinide and the Bcl-2 family protein inhibitor ABT-737 against human neuroblastoma. Clin Cancer Res 17(22):7093–7104
84. Reynolds CP, Tran N, Maldonado NV, Hankaldyyan V, Himada H, Aurer BJ (2008) Activity of fenretinide/Lym-X-Sorb oral powder combined with the oral microtubule inhibitor ABT-751 against multidrug-resistant neuroblastoma xenografts. J Clin Oncol 26, abstr 10031
85. Gupta S, Maurer BJ, Reynolds CP, Vishnuvajjala R (2007) Pharmaceutical composition of fenretinide having increased bioavailability and methods of use for same. US Patent 7,169,819
86. Crandon S, Yancey MA (2009) Sezary syndrome: a case study of cutaneous T-cell lymphoma. Clin J Oncol Nurs 13(2):157–159
87. Kummar S, Gutierrez ME, Maurer BJ et al (2011) Phase I trial of fenretinide lym-x-sorb oral powder in adults with solid tumors and lymphomas. Anticancer Res 31(3):961–966
88. Reynolds CP, Maurer BJ, Kolsenick R (2004) Ceramide synthesis and metabolism as a target of cancer chemotherapy. Cancer Lett 206:169–180
89. Coward J, Ambrosini G, Musi E et al (2009) Safingol (L-threo-sphinganine) induces autophagy in solid tumor cells through inhibition of PKC and the PI3-kinase pathway. Autophagy 5(2):184–193
90. Schwartz GK, Ward D, Saltz L et al (1997) A pilot clinical/pharmacological study of the protein kinase C-specific inhibitor safingol alone and in combination with doxorubicin. Clin Cancer Res 3(4):537–543
91. Dickson MA, Carvajal RD, Merrill AH Jr, Gonen M, Cane LM, Schwartz GK (2011) A phase I clinical trial of safingol in combination with cisplatin in advanced solid tumors. Clin Cancer Res 17(8):2484–2492
92. Guilbault C, Wojewodka G, Saeed Z et al (2009) Cystic fibrosis fatty acid imbalance is linked to ceramide deficiency and corrected by fenretinide. Am J Respir Cell Mol Biol 41(1): 100–106
93. Finnegan CM, Rawat SS, Puri A, Wang JM, Ruscetti FW, Blumenthal R (2004) Ceramide, a target for antiretroviral therapy. Proc Natl Acad Sci U S A 101(43):15452–15457
94. Finnegan CM, Blumenthal R (2006) Fenretinide inhibits HIV infection by promoting viral endocytosis. Antiviral Res 69(2):116–123

ERRATUM TO

Basics of Sphingolipid Metabolism and Signalling

Céline Colacios, Frédérique Sabourdy, Nathalie Andrieu-Abadie, Bruno Ségui, and Thierry Levade

Y.A. Hannun et al. (eds.), *Bioactive Sphingolipids in Cancer Biology and Therapy*, DOI 10.1007/978-3-319-20750-6_1

DOI 10.1007/978-3-319-20750-6_21

In Chapter 1 titled "**Basics of Sphingolipid Metabolism and Signalling**", in Figure 5 the text boxes and labels are recurring at the back of the figure. Please find the correct figure below:

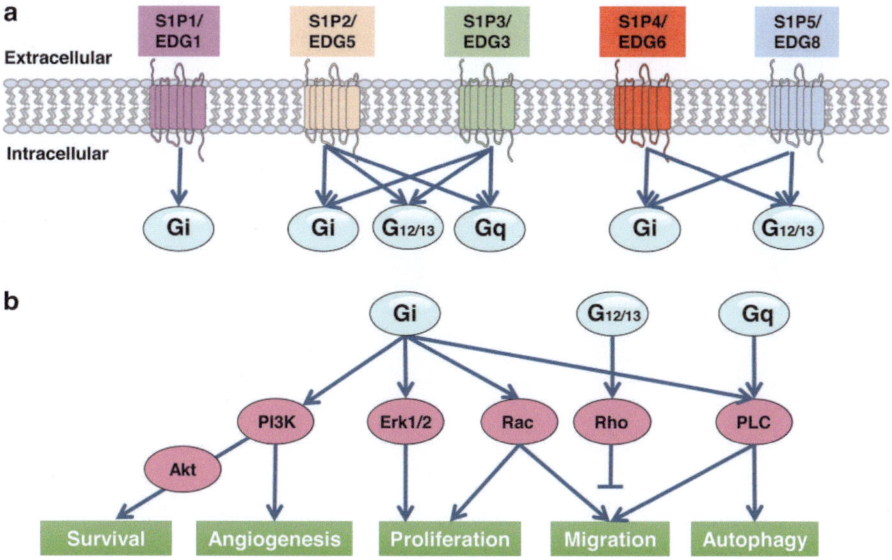

The online version of the original book can be found at http://dx.doi.org/10.1007/978-3-319-20750-6

Y.A. Hannun et al. (eds.), *Bioactive Sphingolipids in Cancer Biology and Therapy*, DOI 10.1007/978-3-319-20750-6_21

ERRATUM TO

Dysregulation of Sphingolipid Metabolism in Melanoma: Roles in Pigmentation, Cell Survival and Tumor Progression

David Garandeau, Marguerite Mrad, Thierry Levade, Cristiana Perrotta, Nathalie Andrieu-Abadie, and Mona Diab-Assaf

Y.A. Hannun et al. (eds.), *Bioactive Sphingolipids in Cancer Biology and Therapy*, DOI 10.1007/978-3-319-20750-6_6

DOI 10.1007/978-3-319-20750-6_21

In Chapter 6 titled **Dysregulation of Sphingolipid Metabolism in Melanoma: Roles in Pigmentation, Cell Survival and Tumor Progression** the order for list of authors should be David Garandeau, Marguerite Mrad, Thierry Levade, Mona Diab-Assaf, Cristiana Perrotta, and Nathalie Andrieu-Abadie

Index

Y.A. Hannun et al. (eds.), *Bioactive Sphingolipids in Cancer Biology and Therapy*, DOI 10.1007/978-3-319-20750-6

MIX
Papier aus verantwortungsvollen Quellen
Paper from responsible sources
FSC® C105338

If you have any concerns about our products,
you can contact us on
ProductSafety@springernature.com

In case Publisher is established outside the EU,
the EU authorized representative is:
Springer Nature Customer Service Center GmbH
Europaplatz 3, 69115 Heidelberg, Germany

Printed by Libri Plureos GmbH
in Hamburg, Germany